Signals and Communication Technology

More information about this series at http://www.springer.com/series/4748

Giovanni Cancellieri

Polynomial Theory of Error Correcting Codes

Springer

Giovanni Cancellieri
Information Engineering
Polytechnic University of Marche
Ancona
Italy

ISSN 1860-4862 ISSN 1860-4870 (electronic)
ISBN 978-3-319-01726-6 ISBN 978-3-319-01727-3 (eBook)
DOI 10.1007/978-3-319-01727-3

Library of Congress Control Number: 2014947130

Springer Cham Heidelberg New York Dordrecht London

Printed on acid-free paper

Springer is part of Springer Science+Business Media (www.springer.com)

To Daniele, my son and pride

Preface

Error correcting codes represent a widely applied technique for assuring reliable electronic communications and data recording. Although often some codes are introduced without the need for a polynomial approach, the attempt to provide a description where polynomial theory is however present could be of some interest. The intent of this book is to develop such theory, in a unitary way, trying to find several conceptual bridges between different classes of codes (block, convolutional, concatenated,...). This goal requires the introduction of some rather unusual mathematical tools (like interleaved polynomial multiplication or interleaved polynomial division), able to support it. Also some structural transformations, like modified code lengthening or modified H-extension, are needed in order to construct a coherent model, able to support noticeable new interpretations.

The introduction of quasi-cyclic codes as a true generalization of the concept of cyclic codes is only an example of the fruitful use of such mathematical tools and geometrical transformations. They are also important for giving an intuitive justification for the encoder circuits to be adopted. State diagrams, constructed on these encoder circuits, contribute to give a better understanding of the properties characterizing the codes under study, besides the measure of the decoding computational complexity.

The distinction between well-designed and not well-designed convolutional codes represents another innovative concept. The latter family of codes is equivalent to catastrophic convolutional codes, but since they are systematic, the catastrophic behavior is no longer a problem. On the other hand, a more than linear increase in the number of low-weight code frames with the number of frame periods remains a drawback for not well-designed convolutional codes, together with some difficulties in their parity check matrix determination and tail-biting arrangement organization. Direct product codes between a pair of block codes are demonstrated to be a subset of not well-designed convolutional codes. Some further conceptual (and rather surprising) bridges between block codes and convolutional codes are constructed.

The treatment is organized maintaining distinct the approaches based on the generator matrix and on the parity check matrix. The reader is gradually guided to the interpretation of such viewpoints. Some concepts, deriving from dual properties, will appear in all their strong efficacy only at this moment. Modified lengthening of cyclic codes and modified H-extension of cyclic codes are dual opportunities leading to a new comprehension of the intrinsic nature of convolutional codes. The former is obtained by acting on the generator matrix, and the latter on the parity check matrix. The same can be made for quasi-cyclic codes. Proper transformation of control symbols into information symbols and vice versa supports the above modifications.

Modern coding (mainly regarding turbo codes and LDPC codes) is a topic, which is faced after a wide application of the above innovative concepts, so allowing a comprehensive understanding of the structures characterizing such codes. For instance, modified H-extension of quasi-cyclic codes offers the possibility of setting up a sort of doubly convolutional LDPC code, in turn interpreted in relation to proper turbo product schemes. The introduction of BPG codes (Binomial Product Generator codes) allows to treat array LDPC codes and some forms of concatenated LDPC codes, together with their convolutional versions, not only by means of the parity check matrix.

The continuous search for ultimate extremely good performance, present in the recent literature, has the consequence of reducing attention to the above theoretic aspects. The dominant use of computer simulations and optimization procedures often entails a nonexhaustive investigation of the true geometrical nature of the code under study. Many code families, apparently very different, on more careful analysis, would appear instead strictly related. All these considerations may produce interesting future theoretical developments. To this purpose, the present comprehensive approach would give a possible contribution.

About 250 Definitions, 300 propositions (Theorems, Corollaries, Lemmas), nearly 500 examples, highly interconnected, can give an idea of the amount of contents treated. In four Appendices some useful concepts, not strictly related to the topics under development, are collected. They are devoted to nonskilled readers, who need auxiliary assistance for framing such theory in a proper context.

Acknowledgments

I wish to thank Marco Baldi for fruitful discussions and continuous help in obtaining most part of the numerical computations, Franco Chiaraluce for his hints and theoretical support, Andrea Carassai for some intuitive views, my students at the Polytechnic University of Marche for their contribution in the organization of some material, and finally Laura Fonti for typewriting help and figure assembling. A particular acknowledgment is devoted to Marco Chiani for having reviewed part of the content, when it was under development. I am also indebted to Rossano Marchesani, Roberto Garello, Torleiv Klove, for our collaboration during the long research activity at the basis of the present work.

Contents

Part I Generator Matrix

Acronyms

ACE	Approximate Cyclic Extrinsic
APP	A Posteriori Probability
ARA	Accumulate Repeat Accumulate
AWGN	Additive White Gaussian Noise
BER	Bit Error Ratio
BPG	Binomial Product Generator
DC-GPC	Direct Concatenation of Generalized Parity Check
DP-GPC	Direct Parallel concatenation of Generalized Parity Check
ExIT	Extrinsic Information Transfer
FEC	Forward Error Correction
FIFO	First-In First-Out
FIR	Finite Impulse Response
GPC	Generalized Parity Check
GR	Generalized Repetition
HEC	H-Extended Cyclic
HEQC	H-Extended Quasi-Cyclic
IIR	Infinite Impulse Response
IRA	Irregular Repeat Accumulate
LC	Lengthened Cyclic
LCM	Least Common Multiple
LDGM	Low Density Generator Matrix
LDPC	Low Density Parity Check
LLR	Logarithm of Likelihood Ratio
LQC	Lengthened Quasi-Cyclic
MDS	Maximum Distance Separable
MHEC	Modified H-Extended Cyclic
MHEQC	Modified H-Extended Quasi-Cyclic
MLC	Modified Lengthened Cyclic
MLQC	Modified Lengthened Quasi-Cyclic
PDC	Progressive Difference Convolutional

PEG	Progressive Edge Growth
P-GPC	Parallelepiped concatenation of Generalized Parity Check
QC	Quasi-Cyclic
RA	Repeat Accumulate
RS	Reed Solomon
RSC	Recursive Systematic Convolutional
SC	Shortened Cyclic
SISO	Soft-In Soft-Output
SNR	Signal-to-Noise Ratio
SOVA	Soft-Output Viterbi Algorithm
SQC	Shortened Quasi-Cyclic

Part I
Generator Matrix

Chapter 1
Generator Matrix Approach to Linear Block Codes

1.1 Additive n × n Linear Transformation of a Binary Sequence

Definition 1.1 An input binary sequence expressed by the row vector $\underline{u}$, having n elements, is subjected to an *additive* $n \times n$ *linear transformation*, described by the $n \times n$ square matrix $\underline{G}$, when the following output is obtained

$$\underline{t} = \underline{u} \cdot \underline{G}, \tag{1.1}$$

where the row vector $\underline{t}$ has n elements too, each one representing a proper ex-OR sum of the elements in $\underline{u}$.

Lemma 1.1 *If the jth column of* $\underline{G}$ *has no 1-symbols, the jth element in* $\underline{t}$ *is a 0-symbol independently of vector* $\underline{u}$.

Proof In matrix product row-by-column (see Appendix A), the jth element of the row vector $\underline{t}$ collects the ex-OR sum of the elements in $\underline{u}$ multiplied by those in the jth column of $\underline{G}$. Since this column contains all null elements, the result is always null. □

Lemma 1.2 *If the ith row of* $\underline{G}$ *has no 1-symbols, the ith element in* $\underline{u}$ *will be always ignored.*

Proof The ith element in $\underline{u}$ participates to the ex-OR sums according to a selection imposed by the columns in $\underline{G}$. If all these columns have their ith element null, the ith element of $\underline{u}$ does not contribute to the output at all. □

The total number of all the possible input vectors $\underline{u}$ is 2^n. The same number is that characterizing all the possible output vectors $\underline{t}$. Nevertheless only sometimes there is a one-to-one correspondence between these two vectors. For instance, in the two cases considered in Lemma 1.1 and Lemma 1.2 this one-to-one correspondence is not verified. The one-to-one correspondence is assured if $\underline{G}$ is non-singular.

G. Cancellieri, *Polynomial Theory of Error Correcting Codes*,
Signals and Communication Technology, DOI 10.1007/978-3-319-01727-3_1

In this situation, it is possible to invert (1.1) and obtain the inverse transformation (see Appendix A for details), expressed by

$$\underline{u} = \underline{t} \cdot \underline{G}^{-1}.$$

Lemma 1.3 *If* $\underline{u}$ *has just one 1-symbol, as its ith element,* $\underline{t}$ *will coincide with the ith row of* $\underline{G}$*.*

Proof This result comes out directly from (1.1). □

Example 1.1 Let us consider an input vector $\underline{u} = [\, a \quad b \quad c \quad d \,]$ and the following 4×4 $\underline{G}$ matrix, where 0-symbols are not represented,

$$\underline{G} = \begin{bmatrix} & 1 & 1 & \\ 1 & & & 1 \\ & & & 1 \\ & 1 & & \end{bmatrix}.$$

The output vector will be $\underline{t} = [\, b \quad (a+d) \quad a \quad (b+c) \,]$.

Definition 1.2 A square $\underline{G}$ matrix is called *permutation matrix* when it exhibits one 1-symbol in each row and in each column. In this case just the order of the elements of $\underline{u}$ is changed in $\underline{t}$, and no ex-OR sums are made. A permutation matrix is non-singular.

Among the possible permutation matrices, one is very particular, and is expressed by the $n \times n$ identity matrix $\underline{I}_n$. In this case we have $\underline{t} = \underline{u}$, which means that this linear transformation does not change the input vector $\underline{u}$.

Let us remove, for a moment, the assumption that $\underline{G}$ is a square matrix, and consider the case in which $\underline{G}$ exhibits more columns than rows. In this situation, some vectors $\underline{t}$ can exist without correspondent vectors $\underline{u}$. The only vectors $\underline{t}$ with correspondent vectors $\underline{u}$ form an ensemble, which can be assumed as a linear block code, and its treatment will be faced starting from the next section.

A particular form for the $n \times n$ $\underline{G}$ matrix is that depicted in Fig. 1.1 up, for $n = 4$, where $a, b, \ldots f$ represent symbols which can be either 0 or 1, whereas under the diagonal made by 1-symbols there are only 0-symbols. The output $\underline{t}$ will result in

$$\underline{t} = [\, u_1 \quad (au_1 + u_2) \quad (bu_1 + cu_2 + u_3) \quad (du_1 + eu_2 + fu_3 + u_4) \,].$$

We can observe how the input elements u_j, $j = 1, 2, \ldots n$, are taken progressively in the computation one at a time. This makes the overall computation clearly easier.

A situation even more particular is that characterized by a unique pattern of 1-symbols which shifts along any row. It is shown in Fig. 1.1 down, assuming, as an example, the pattern 11001 in a 7×7 $\underline{G}$ matrix. The pattern is progressively truncated on the right hand side, up to the condition in which just one 1-symbol remains in the last row. The transformation associated to the $\underline{G}$ matrix in Fig. 1.1 down can be provided by a shift register, where the wires entering an ex-OR sum

Fig. 1.1 Two particular forms for a square $\underline{G}$ matrix

$$\underline{G} = \begin{array}{|cccc|} 1 & a & b & d \\ & 1 & c & e \\ & & 1 & f \\ & & & 1 \end{array}$$

$$\underline{G} = \begin{array}{|ccccccc|} 1 & 1 & 0 & 0 & 1 & & \\ & 1 & 1 & 0 & 0 & 1 & \\ & & 1 & 1 & 0 & 0 & 1 \\ & & & 1 & 1 & 0 & 0 \\ & & & & 1 & 1 & 0 \\ & & & & & 1 & 1 \\ & & & & & & 1 \end{array}$$

are imposed by the 1-symbols of the pattern shifting along the subsequent rows. We can easily imagine the structure of the rectangular $\underline{G}$ matrix formed by all complete patterns in any row, without truncation.

An inversion of (1.1), like that based on multiplication by $\underline{G}^{-1}$, of the type described before, is restricted to the only case for which the two vectors $\underline{u}$ and $\underline{t}$ have the same number of elements n, that is for an $n \times n$ square matrix. It is interesting to observe that matrices like those in Fig. 1.1 are always non-singular and then admit an inverse $\underline{G}^{-1}$. Examples of singular $\underline{G}$ matrix, besides the ones with null rows or columns, are given by matrices with a row or a column linearly dependent on the other rows or columns. The case of a rectangular matrix will be treated in the next section.

1.2 Generator Matrix $\underline{G}$ of a Linear Block Code

Since the maximum number of linearly independent rows in a square $n \times n$ matrix is n (see Lemma A.5 in Appendix A), it is of no interest a case in which $\underline{G}$ exhibits more rows than columns. Let us consider then a $\underline{G}$ matrix in (1.1), rectangular and characterized by k rows, with $k < n$. The input vector $\underline{u}$ exhibits k elements, whereas the output vector $\underline{t}$ is still formed by n elements.

Lemma 1.4 *If the rows of $\underline{G}$ are all linearly independent, there are only k independent elements in $\underline{t}$, and the remaining $(n - k)$ elements depend on the previous ones.*

Proof It is possible to subdivide the elements t_i of vector $\underline{t}$ into two sets: the first one with k elements, and the second one with the remaining $(n - k)$ elements. It is not a restrictive assumption to denote the first set with $i = 1, 2, \ldots k$, and the second set with $i = k + 1, k + 2, \ldots n$. Since the u_j's, $j = 1, 2, \ldots k$, are known constants, we are left with a system of k linearly independent equations in the n unknowns t_i. We are in a situation like that described by Theorem A.4 and subsequent Corollary a in

Appendix A, being $k = \text{rank}\ (\underline{G})$. So it is possible to interpret the $(n-k)$ elements of the second set as depending on the k elements of the first set (and hence on the input u_j's themselves). □

In order to define an inverse transformation, able to give back vector $\underline{u}$ from the knowledge of vector $\underline{t}$, it is necessary to employ, instead of the inverse $\underline{G}^{-1}$, a proper $n \times k$ right-inverse matrix $\underline{G}_{right}^{-1}$ (see Definition A.23 in Appendix A). So we obtain

$$\underline{u} = \underline{t} \cdot \underline{G}_{right}^{-1}. \tag{1.2}$$

Definition 1.3 A *linear block code*, in a binary alphabet, is an ensemble of 2^k permitted sequences (called *code words*) of n binary symbols, properly selected within the total amount of 2^n possible n-tuples.

For further properties of linear block codes, like their synthetic representation by means of the tern $[n,k,d]$, the code rate k/n, minimum distance d, word weight distribution $A(w)$, and the properties deriving directly from linearity, like invariance in the point of view, or code words as ex-OR sum of other code words, the reader is referred to Appendix D.

Definition 1.4 The *generator matrix* $\underline{G}$ of a linear block code is a $k \times n$ matrix, made by all linearly independent rows, each representing a code word, and able to generate the 2^k code words by means of Eq. (1.1), assuming all the possible binary k-tuples as input vectors $\underline{u}$.

Example 1.2 Let us consider the following 3×6 generator matrix

$$\underline{G} = \begin{bmatrix} 1 & & 1 & & 1 & \\ & 1 & & 1 & 1 & \\ 1 & & & 1 & & 1 \end{bmatrix}.$$

The linear block code obtained from it exhibits the following 8 code words: 000000, 001111, 010110, 011001, 100101, 101010, 110011, 111100. Therefore its minimum distance is $d = 3$, and the weight distribution of its code words is $A(0) = 1$, $A(3) = 4$, $A(4) = 3$.

Definition 1.5 A linear block code is called *systematic in strict sense* (*s.s. systematic*) if, in the first k positions of its code words, the input vector $\underline{u}$ is reproduced. It is called *systematic in wide sense* (*w.s. systematic*) if there are k positions, in any code word, where the elements of $\underline{u}$ are reproduced.

Lemma 1.5 *In a linear block code, systematic in strict sense, the* $\underline{G}$ *matrix exhibits a* $k \times k$ *identity sub-matrix* $\underline{I}_k$ *in its left hand side part.*

Proof This result derives from Lemma 1.3, considering the sub-matrix $\underline{I}_k$ as that describing an additive square linear transformation. □

In a linear code, systematic only in wide sense, the $\underline{G}$ matrix exhibits k columns where just one 1-symbol is present. Such 1-symbols belong to different rows and are placed in the columns where information symbols are positioned.

Example 1.3 The following 2×5 generator matrix

$$\underline{G} = \begin{bmatrix} 1 & & 1 & 1 & \\ & 1 & 1 & 1 & 1 \end{bmatrix}$$

describes a linear code, systematic in strict sense. In fact the association between input and output vectors, in its four code words, is as follows

$$\begin{array}{lllll} \underline{u}: & 00, & 01, & 10, & 11, \\ \underline{t}: & 00000, & 01111, & 10110, & 11001. \end{array}$$

Example 1.4 The code described by the generator matrix considered in Example 1.2 is systematic in wide sense. The positions where the elements u_1, u_2, u_3 of the input vector $\underline{u}$ are reproduced in the output vector $\underline{t}$ turn out to be respectively: the third, the second, the sixth.

Example 1.5 A code not systematic in any way is, for instance, that characterized by this 3×4 generator matrix

$$\underline{G} = \begin{bmatrix} 1 & 1 & & \\ & 1 & & 1 \\ & & 1 & 1 \end{bmatrix},$$

in which there are only two columns with just one 1-symbol and the second element of $\underline{u}$, u_2, is never reproduced alone in the output vector $\underline{t}$. Its eight code words are: 0000, 0011, 0101, 0110, 1001, 1010, 1100, 1111. They represent all the possible 2-tuples or 4-tuples, over a total sequence having four positions, besides the all 0-symbol word. The code word weight distribution is then $A(0) = 1$, $A(2) = 6$, $A(4) = 1$.

The properties of a code, expressed by the weight distribution of its code words $A(w)$, are independent of the fact that it is systematic or not. The characteristic to be systematic regards only the possibility of directly recognizing the input binary (information) symbols in precise positions of $\underline{t}$.

About the inverse transformation (1.2), we can observe that it is trivial, if we are with a systematic code ($\underline{G}_{right}^{-1}$ exhibits just one 1-symbols in the rows corresponding to the positions where the elements of $\underline{u}$ are placed and all 0-symbols in any other row). In non-systematic situations the determination of $\underline{G}_{right}^{-1}$ is less straightforward and could present a certain computational complexity, especially when n is high.

Definition 1.6 Two codes, characterized by different $\underline{G}$ matrices, although both $k \times n$, are called *equivalent* if they exhibit the same code word weight distribution.

Given a certain code, it is possible to obtain many codes equivalent to it, by properly processing its $\underline{G}$ matrix.

Lemma 1.6 *The following operations on the* $\underline{G}$ *matrix give a code equivalent to the original one:*

(i) change in the order of the columns,
(ii) change in the order of the rows,
(iii) ex-OR sum of rows.

Proof Operation (*i*) implies a change in the order of the elements of $\underline{t}$. Operation (*ii*) entails a change in the order of the elements of $\underline{u}$. Operation (*iii*) is permitted by linearity and corresponds to a different choice of the code words taken as the rows of $\underline{G}$. Nevertheless, in order to maintain the property that all the rows are linearly independent, it is necessary to follow this procedure: taken two rows in the old generator matrix, their ex-OR sum must be substituted to one of the two. □

Example 1.6 The code represented by the generator matrix introduced in Example 1.5 has a systematic (in strict sense) equivalent, characterized by this generator matrix

$$\underline{G} = \begin{bmatrix} 1 & & & 1 \\ & 1 & & 1 \\ & & 1 & 1 \end{bmatrix}.$$

It has been obtained from the previous one by means of operation (*iii*) described in Lemma 1.6.

Definition 1.7 A generator matrix of the type depicted in Fig. 1.2 up, that is with all 0-symbols under the diagonal of its $k \times k$ left sub-matrix (expressed by all 1-symbols), is called *upper triangular*. The example, where $a, b, \ldots$ can be either 0-symbols or 1-symbols, refers to the case of $k = 4$, and whichever n.

$$\underline{G} = \begin{array}{|ccccccc|} 1 & a & b & d & g & & \\ & 1 & c & e & h & & \\ & & 1 & f & i & & \ldots\ldots \\ & & & 1 & j & & \end{array}$$

$$\underline{G} = \begin{array}{|ccccccccccc|} 1 & 1 & 0 & 0 & 1 & & & & & & \\ & 1 & 1 & 0 & 0 & 1 & & & & & \\ & & 1 & 1 & 0 & 0 & 1 & & & & \\ & & & 1 & 1 & 0 & 0 & 1 & & & \\ & & & & 1 & 1 & 0 & 0 & 1 & & \\ & & & & & 1 & 1 & 0 & 0 & 1 & \\ & & & & & & 1 & 1 & 0 & 0 & 1 \end{array}$$

Fig. 1.2 Upper triangular form for a $\underline{G}$ matrix and example of a 7×11 $\underline{G}$ matrix with a unique 1-symbol pattern shifting along the rows

Remark that the systematic (in strict sense) form of $\underline{G}$ is a particular case of an upper triangular $\underline{G}$. The advantage of choosing an upper triangular form for $\underline{G}$ lies in the fact that the input elements u_j, $j = 1, 2, \ldots k$, are taken sequentially in the computation one at a time. This makes the encoding procedure easier. An upper triangular form for $\underline{G}$ allows also a right-inverse calculation always possible and not too involved. However, it is important to remember that $\underline{G}^{-1}_{right}$, in general, is not unique (see Appendix A).

Example 1.7 The generator matrix introduced in Example 1.5 has an upper triangular form. It is non-systematic, but remains upper triangular. One possible right-inverse matrix associated to it results in

$$\underline{G}^{-1}_{right} = \begin{bmatrix} 1 & & \\ & & \\ & 1 & 1 \\ & 1 & \end{bmatrix}.$$

It has been constructed by imposing satisfaction of constraint $\underline{G} \cdot \underline{G}^{-1}_{right} = \underline{I}_3$ (see Definition A.23 in Appendix A), by few attempts.

A particular situation is that characterized by a unique pattern of 1-symbols which shifts along any row of $\underline{G}$. In this case the patterns will be all complete and invariant. A generator matrix of this type, clearly upper triangular, is shown in Fig. 1.2 down, assuming the same 1-symbol pattern already adopted in Fig. 1.1 down, that is 11001, here in a 7×11 overall $\underline{G}$ matrix. Starting from the first row, and adding one row at a time, it is possible to verify that, independently of the particular pattern, such rows are all linearly independent, because each of them interests a new position on the right. A generator matrix of this type can be implemented by means of a shift register, taking advantage from the polynomial approach to the theory of linear block codes, which will be introduced starting from the next section.

Example 1.8 Let us consider the code described by the generator matrix introduced in Example 1.5 (and treated also in Example 1.6). We can obtain a further equivalent version for this code, by reversing the order of the last two columns. The new form of $\underline{G}$ exhibits a pattern 11 shifting along its rows. The new $\underline{G}$ matrix and its right-inverse respectively are

$$\underline{G} = \begin{bmatrix} 1 & 1 & & \\ & 1 & 1 & \\ & & 1 & 1 \end{bmatrix}, \quad \underline{G}^{-1}_{right} = \begin{bmatrix} 1 & & \\ & & \\ & 1 & \\ & 1 & 1 \end{bmatrix}.$$

It is possible to recognize, in the above three codes, as many equivalent versions of the [4,3,2] binary parity check code (see Appendix D for details).

If we have a non-systematic $\underline{G}$ matrix, it is always possible to partition it into two parts: on the left a $k \times k$ square sub-matrix $\underline{T}$, and on the right a $k \times (n-k)$ rectangular sub-matrix $\underline{P}$. In other terms, we can write $\underline{G} = [\underline{T} \mid \underline{P}]$. The following proposition gives an easy method for obtaining the systematic generator matrix $\underline{G}_{sys}$ of a code equivalent to the original one.

Lemma 1.7 *In order to obtain* $\underline{G}_{sys}$ *for a partitioned generator matrix, provided that sub-matrix* $\underline{T}$ *is non-singular, we can multiply* $\underline{G}$ *on the left by the* $k \times k$ *square matrix* $\underline{T}^{-1}$. *In this situation, a systematic equivalent of the above matrix can be obtained as*

$$\underline{G}_{sys} = \underline{T}^{-1} \cdot \underline{G} = [\underline{I}_k \mid \underline{T}^{-1} \cdot \underline{P}].$$

Proof Let us consider the property shown in Fig. A.3 of Appendix A, assuming in it $\underline{B} = \underline{0}$, $\underline{C} = \underline{T}$, $\underline{D} = \underline{P}$, $\underline{A} = \underline{T}^{-1}$. So, considering also (A.5), we obtain $[\underline{T}^{-1} \cdot \underline{T} \mid \underline{T}^{-1} \cdot \underline{P}] = [\underline{I}_k \mid \underline{T}^{-1} \cdot \underline{P}]$. □

Determination of $\underline{G}_{sys}$ and $\underline{G}_{right}^{-1}$ are clearly related. Sometimes a so simple procedure, like that described in Lemma 1.7, cannot be applied. This occurs, for instance, when $\underline{G}$ is function of a parameter (symbolic generator matrix). Here, after having tested that the right inverse can exist, it is possible to write $\underline{G}$ in Smith form (see Appendix A for details).

1.3 Polynomial Description of the Generator Matrix in a Linear Block Code

Definition 1.8 In a generator matrix $\underline{G}$, characterized by an invariant 1-symbol pattern shifting along its rows, we can express this pattern by means of a polynomial $g(x)$, called the *generator polynomial* of the correspondent block code. This code is named *strict-sense* (briefly *s.s.*) *time-invariant in its* $\underline{G}$ *matrix.*

Theorem 1.1 *In the situation described by Definition 1.8, the matrix product* (1.1), *giving the vector representation of a code word, can be substituted by the polynomial product*

$$t(x) = u(x)g(x), \tag{1.3}$$

where polynomial $u(x)$, *whose degree is at most* $(k-1)$, *collects the elements of vector* $\underline{u}$, *and polynomial* $t(x)$, *whose degree is at most* $(n-1)$, *collects the elements of vector* $\underline{t}$.

Proof Letting $r = \deg[g(x)]$, and labelling g_j, $j = 0, 1, \ldots r$, the coefficients of such polynomial, the constraint $g_0 = g_r = 1$ is to be satisfied in order to delimit the 1-symbol pattern. The following relationship then holds among the code parameters n, k and r,

$$n = k + r. \tag{1.4}$$

In order to simplify polynomial notation, it is convenient to label the elements t_i and u_i, respectively in vector $\underline{t}$ and in vector $\underline{u}$, with subscript i ranging between 0 and the vector length minus one, in reverse order, so obtaining $\underline{t} = [t_{n-1}, \ \ldots, \ t_1, \ t_0]$, $\underline{u} = [u_{k-1}, \ \ldots, \ u_1, \ u_0]$, and correspondently

$$t(x) = t_{n-1}x^{n-1} + \cdots + t_1x + t_0, \quad u(x) = u_{k-1}x^{k-1} + \cdots + u_1x + u_0.$$

After what reported in Appendix B about the polynomial representation of binary sequences and a multiplication performed on them, the following equalities are collected in (1.3)

$$t_{n-1} = u_{k-1}g_r, \qquad t_{n-2} = u_{k-2}g_r + u_{k-1}g_{r-1},$$
$$\cdots\cdots\cdots\cdots$$
$$t_1 = u_1g_0 + u_0g_1, \qquad t_0 = u_0g_0.$$

Exactly the same equalities are originated by the product (1.1) with a $\underline{G}$ matrix in the form described by Definition 1.8. □

Corollary a *It is possible to encode a block code, s.s. time-invariant in its $\underline{G}$ matrix, and characterized by the generator polynomial g(x), by means of a shift register in controller arrangement, whose scheme is depicted in* Fig. 1.3.

Proof From Lemma C.1 of Appendix C, it is possible to calculate a polynomial product, by means of a shift register. This circuit exhibits a feed-forward ex-OR sum, as shown by Fig. C.2. In order to assure (1.3), we have to simply substitute the generator polynomial $g(x)$ to the generic $p(x)$. Polynomial $u(x)$ represents the

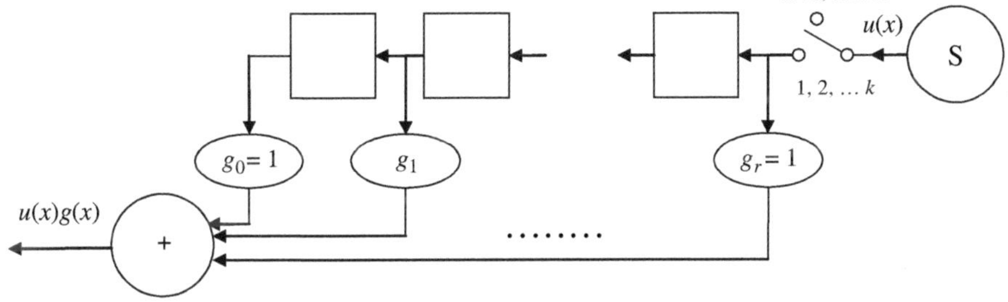

Fig. 1.3 Encoder circuit for a s.s. time invariant block code characterized by generator polynomial $g(x)$

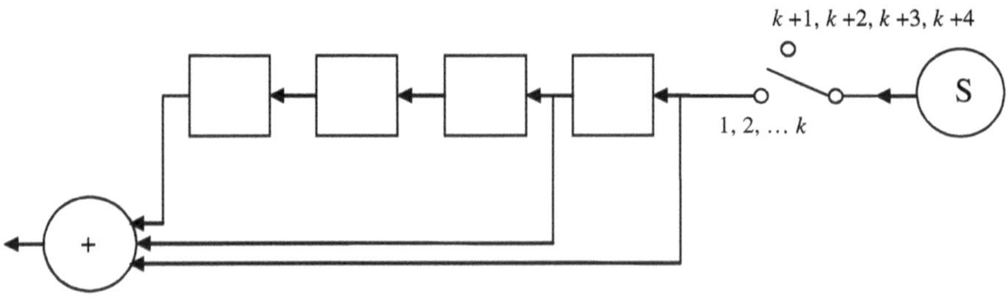

Fig. 1.4 Encoder circuit for the code characterized by $g(x)$ associated to the sequence 11001

symbol sequence outgoing from the information source S. The switch is maintained for k subsequent symbol times connected S to the shift register, and for the remaining r symbol times it produces a cut in this connection, allowing that r subsequent 0-symbols fill the memory elements of the shift register. Since their number is r, after r symbol times the register will be empty, so concluding the encoding procedure. □

Example 1.9 The code characterized by the $\underline{G}$ matrix reported in Fig. 1.2 down, in which the generator polynomial is $g(x) = x^4 + x^3 + 1$, that is the sequence 11001, can be encoded by means of the shift register depicted in Fig. 1.4. Remark that $r = 4$ is a constant, whereas parameter k can be taken as a degree of freedom, controlling the overall cycle of the switch (and hence the length of the code word). In the case of matrix $\underline{G}$ in Fig. 1.2 down, k was assumed to be 7. Nevertheless the encoder circuit does not change, in the presence of a smaller or higher value for k. Simply different block codes, all associated to the same generator polynomial, can be encoded by means of the same circuit.

First of all, we can observe that the encoder circuit based on such a shift register does not allow the construction of a systematic code, in fact the symbols of polynomial $u(x)$ do not appear directly in the output polynomial $t(x)$. To this problem it is possible to give solution, as will be discussed in Sect. 1.5, by a proper pre-processing of the input sequence.

Another important aspect regards the evolution of the code minimum distance d with increasing k, once chosen a certain generator polynomial $g(x)$. We can consider a family of codes, all obtained with the same generator polynomial, and hence with the same encoder circuit, but in which k is different. Remark that, with increasing k, also n increases, the code rate k/n increases too, which are all good features. Nevertheless there is a condition in k, after which the code performance will become very poor, especially if the minimum distance d of our code, for smaller values of k, was high.

Corollary b *Given an s.s. time invariant block code, characterized by the generator polynomial $g(x)$, let us imagine to increase parameter k, and* *correspondently parameter n. Let its minimum distance be d > 2 for small values of n. There will be a maximum length n = N of the code word, able to assure that the minimum distance remains d > 2. It is fixed by the exponent in the binomial* $(x^N + 1)$ *of which g(x) is factor.*

Proof Theorem B.4 in Appendix B states that an exponent N always exists, for which a polynomial, whose degree is finite, is factor of the binomial $(x^N + 1)$. In (1.3), assuming $g(x)$ and its co-factor $\bar{g}(x)$ as $u(x)$ (see Definition B.11 for details), the code word $(x^N + 1)$, whose weight is 2, will appear. Therefore, in order to avoid this abrupt performance deterioration, the maximum value for deg[$u(x)$] has to be $(N - r - 1)$. This limit condition means deg[$t(x)$] = $N - 1$, and hence a code word length $n = N$. □

Example 1.10 Let us consider a family of codes, characterized by $g(x) = x^8 + x^4 + x^2 + x + 1$, that is the sequence 100010111, and a variable k. In Table 1.1 the evolution of the word weight distribution is reported, with n varying from 8 to 18. In any row, the sum of all coefficients $A(w)$ obviously gives 2^k. The generator polynomial assumed here is factor of $(x^{15} + 1)$. For this reason, with $n \geq 16$, there are code words characterized by $w = 2$. At $n = 17$, also a code word with $w = 4$ appears. We can observe that the distribution $A(w)$, with $n = N = 15$, evidenced in the Table, is symmetric. It is important to stress that N is the minimum integer for which $g(x)$ is factor of a binomial of the type $(x^N + 1)$.

Remark that, with increasing k (and hence n), once a certain code word (intended as a permitted pattern of 1-symbols) has appeared, it certainly remains. Moreover it contributes to the weight distribution with an increasing number of replicas, owing to as many shifts along the code word length. This explains the fact that a certain coefficient $A(w)$, except $A(0)$, in Table 1.1, appears as an increasing function of variable n. As a consequence of this property, the minimum distance d can only remain constant or decrease with increasing n. For $n \geq N + 1$, there is an abrupt decrease in d, owing to the appearance of the first 2-weight code word.

Definition 1.9 The s.s. time invariant block code, characterized by a certain generator polynomial $g(x)$, and code word length $n = N$, in the sense described by Corollary b, is called *cyclic code*.

The meaning of such definition will be clarified in the following sections. However it can be considered as a sort of parent code, with respect to the other members of a code family, with the same $g(x)$, but characterized by lower or higher values for n. In this sense, the family of codes described in Example 1.10 contains block codes all derived from a cyclic [15,7,5] primitive BCH code (see Definition D.19 in Appendix D).

Table 1.1 Evolution of the word weight distribution assuming 100010111 as $g(x)$

n	$A(0)$	…	$A(2)$	…	$A(4)$	$A(5)$	$A(6)$	$A(7)$	$A(8)$	$A(9)$	$A(10)$	$A(11)$	$A(12)$	$A(13)$	$A(14)$	$A(15)$	$A(16)$
9	1					1											
10	1					2	1										
11	1					3	3	1									
12	1					5	5	2	1	1	1						
13	1					8	10	4	3	4	2						
14	1					12	18	8	7	12	6						
15	*1*					*18*	*30*	*15*	*15*	*30*	*18*					*1*	
16	1		1			24	42	34	41	96	37	12	6			2	
17	1		2		1	32	58	64	87	108	82	44	23	4	2	4	
18	1		3		3	43	80	109	164	206	179	122	71	23	10	9	1

1.4 Properties of a Linear Block Code Derived from the Structural Characteristics of g(x)

Lemma 1.8 *In a family of codes, s.s. time invariant in* $\underline{G}$ *and characterized by a generator polynomial g(x), the minimum distance d can vary between the following two bounds*

$$2 \leq d \leq \mathrm{w}[g(x)]. \tag{1.5}$$

This property is true for whichever n.

Proof For the lower bound, let us consider Lemma B.3 in Appendix B. For the upper bound, the situation $u(x) = 1$ shows a case in which $d = \mathrm{w}[g(x)]$ is verified. Since the pattern of 1-symbols represented by $g(x)$ is always a code word (from $k = 1$ to whichever k), it is impossible to have $d > \mathrm{w}[g(x)]$. □

Definition 1.10 We can introduce a *minimum distance* d_c *up to the cyclic version* characterizing a family of codes, described by a certain generator polynomial. d_c is the minimum distance reached before the abrupt reduction to $d = 2$, occurring in correspondence with $n \geq N + 1$.

In many situations we have $d_c = \mathrm{w}[g(x)]$. Nevertheless, when $\mathrm{w}[g(x)]$ is particularly high with respect to its degree r, we may have $d_c < \mathrm{w}[g(x)]$.

Example 1.11 Let us consider the generator polynomial $g(x) = x^5 + x^4 + x^3 + x + 1$, that is the sequence 111011, which is factor of the binomial $(x^{31} + 1)$. The evolution of the code word weight distribution, with n varying from 6 to 13, is shown in Table 1.2. Although $\mathrm{w}[g(x)] = 5$, the value $d_c = 3$ is rapidly reached, and then it will remain as the minimum distance up to $n = N = 31$. This family of codes derives from the [31,26,3] Hamming code (see Definition D.18).

Table 1.2 Evolution of the word weight distribution assuming 111011 as $g(x)$

n	A(0)	…	A(3)	A(4)	A(5)	A(6)	A(7)	A(8)	A(9)	A(10)	A(11)
6	1				1						
7	1			1	2						
8	1		1	2	3	1					
9	1		2	3	6	4					
10	1		3	6	11	8	1	1	1		
11	1		4	10	20	16	4	5	4		
12	1		6	16	30	30	18	15	10	2	
13	1		8	24	46	54	46	39	26	10	2

Table 1.3 Evolution of the word weight distribution assuming 11101 as *g*(*x*)

n	*A*(0)	*A*(2)	*A*(4)	*A*(6)	*A*(8)	*A*(10)	*A*(12)	*A*(14)
5	1		1					
6	1		3					
7	*1*		*7*					
8	1	1	11	3				
9	1	2	18	10	1			
10	1	3	29	25	6			
11	1	4	44	58	19	2		
12	1	5	66	114	61	9		
13	1	6	95	212	159	38	1	
14	*1*	*7*	*133*	*371*	*371*	*133*	*7*	*1*

Lemma 1.9 *If the weight w[g(x)] is even, the code word weight distribution A(w) exhibits only even weights w. This property is true for whichever n.*

Proof Let us consider Lemma B.2 in Appendix B, taking $p(x) = u(x)$ and $q(x) = g(x)$. Although w[$u(x)$] can be either even or odd, the product w[$u(x)$]w[$g(x)$] is even, and also the term $2C$ is certainly even. Therefore w[$t(x)$] is even too. □

Example 1.12 Let us take the generator polynomial $g(x) = x^4 + x^3 + x^2 + 1$, that is the sequence 11101, which is factor of the binomial $(x^7 + 1)$. Table 1.3 shows the evolution of the word weight distribution for this family of codes, with *n* ranging from 5 to 14. Up to the cyclic code (whose weight distribution has been evidenced for $n = N = 7$), we have $d_c = 4$ and the only non-null coefficient, besides *A*(0), is *A*(4). This property represents a characteristic of such code, which is a [7,3,4] simplex code (see Example D.24 and Definition D.25 in Appendix D). Then the 2-weight code word appears. In this case, further 4-weight code words, of the type discovered in Example 1.10, are included in coefficient *A*(4), which grows very rapidly. At $n = 2N = 14$ (distribution evidenced too), we have a symmetric weight distribution, although the cyclic code, at $n = N = 7$, does not show such a property. This rather surprising result will be explained in the following.

At this point one could suppose that d_c even is a sufficient condition for *g*(*x*) to have even weight, and hence for all the code words to exhibit even weight as well. But this is not, as the following situation shows.

Example 1.13 A particular family of codes is that originated by the generator polynomial $g(x) = x^6 + x^4 + x^3 + x^2 + 1$, that is the sequence 1011101. It is factor of the binomial $(x^{15} + 1)$. The symmetric structure of *g*(*x*), in spite of the fact that w[$g(x)$] = 5, that is an odd number, allows a reduction of d_c to 4, but not to 3. This is a rare example in which d_c is even, although there are word weights either even or odd. Table 1.4 reports the word weight distributions here obtained, with *n* varying

Table 1.4 Evolution of the word weight distribution taking 1011101 as $g(x)$

n	$A(0)$	…	$A(4)$	$A(5)$	$A(6)$	$A(7)$	$A(8)$	$A(9)$	$A(10)$	$A(11)$	…	$A(15)$
7	1			1								
8	1			2	1							
9	1		1	4	2							
10	1		2	8	4		1					
11	1		5	12	7	4	2		1			
12	1		9	18	13	12	6	2	3			
13	1		15	27	21	27	20	9	7	1		
14	1		22	40	36	56	49	24	20	8		
15	*1*		*30*	*60*	*60*	*105*	*105*	*60*	*60*	*30*		*1*

from 7 to 15. We can observe that the word weight distribution of the cyclic code is here symmetric.

As a general comment, after the analysis of the evolutions exhibited by the above word weight distributions, it is possible to note that the differential increase in coefficients $A(w)$, with w rather small, is greater and greater as n approaches N. This property is typical in block codes, and is due the constant number r of control symbols in the code word. Other types of codes, for instance convolutional codes, where control symbols have a number proportional to the code word length (in general periodically distributed along its length), are characterized by a constant differential increase in coefficients $A(w)$. Appendix D can give further details on this phenomenon, and Chap. 3 contains its explanation.

Lemma 1.10 *If the weight $w[g(x)]$ is even, $g(x)$ contains factor $(x + 1)$.*

Proof This result is due to Lemma B.15 in Appendix B. □

Example 1.14 Polynomial $g(x)$ assumed in Example 1.12 can be factorized as follows $g(x) = x^4 + x^3 + x^2 + 1 = (x + 1)(x^3 + x + 1)$.

Lemma 1.11 *In order to have a symmetric word weight distribution, the code word with all 1-symbols must be present among the words of the considered code.*

Proof Let us label $\underline{0}$ the vector representing the code word with all 0-symbols, and $\underline{1}$ the vector representing the code word with all 1-symbols. Vector $\underline{0}$ and vector $\underline{1}$ form a symmetric pair in the $A(w)$ distribution. Then, if there is the code word represented by $\underline{t}$, owing to linearity, there will be also the code word represented by $(\underline{1} + \underline{t})$. Such two code words form a symmetric pair in the resulting $A(w)$ distribution. □

There are some important consequences of Lemma 1.11, regarding the situation characterized by a symmetric $A(w)$ distribution, and the possibility that it can appear for $n < N$, $n = N$, or $n > N$. Expressing Lemma 1.11 in other terms, $g(x)$ must be factor of a certain unitary polynomial over the length considered (see Definition B.13 in Appendix B). If N is odd, and is the product of more than one odd prime integers (see Lemma B.21 in Appendix B), as many unitary polynomials are present in the factorization of $(x^N + 1)$. Nevertheless, $g(x)$ cannot be a factor in one of them, because N is, by assumption, the smallest exponent of a binomial dividable by $g(x)$. So $g(x)$ will be factor of the overall $u_N(x)$. The same situation occurs if N is prime, where $u_N(x)$ is the unique unitary polynomial in the factorization of $(x^N + 1)$. Thus, if $g(x)$ has odd weight, the symmetric distribution $A(w)$ is reached, but just at the cyclic code condition $n = N$. This property then repeats at any value of n integer multiple of N. On the other hand, if $g(x)$ has even weight, and N is still odd, $u_N(x)$ cannot be present among the code words of the cyclic code with length N, owing to Lemma 1.9. Then this cyclic code cannot exhibit a symmetric distribution $A(w)$. Such symmetric distribution will appear only at the condition $n = 2N$, so confirming what shown in Table 1.3, for the simplex code.

A different situation occurs if N is even. Here at least two factors $(x + 1)$ are present in $(x^N + 1)$. Certainly all of them is included in $g(x)$, otherwise the exponent N of our binomial could not be the smallest one in the situation assumed. In such condition, the weight of $g(x)$ is certainly even, but this is not enough for assuring always the existence of $u_N(x)$ among the words of our code, as the following situation shows.

Example 1.15 Let us consider the generator polynomial $g(x) = (x + 1)(x^3 + 1) = x^4 + x^3 + x + 1$, which is factor of $(x^6 + 1)$. The four code words of its cyclic version are 000000, 011011, 101101, 110110. In order to admit a unitary polynomial as a code word, and hence a symmetric distribution of word weight coefficients $A(w)$, it is necessary to reach the condition $n = 2N$, like in the case of the simplex code, treated in Example 1.12. See also Example D.16, in Appendix D, where a similar property is discussed for the parity check code.

Before concluding this section, it is useful to pay attention to the fact that there are two particular families of codes, for which N can be any integer number. Since the decomposition $x^N + 1 = (x + 1)u_N(x)$ is always possible (Lemma B.14 in Appendix B), such two factors can represent the generator polynomials of two families of codes always present. Assuming $g(x) = x + 1$, we have the so called binary parity check codes, assuming $g(x) = u_n(x)$, we have the so called binary repetition codes (see Appendix D for details on their definitions and properties). The first ones exhibit $d = 2$ over any length n. The second ones are characterized by the only two code words formed by all 0-symbols and all 1-symbols. The latter code word has weight $w = n$. Beyond this condition, which expresses the cyclic version ($n = N$), that is for $n > N$, also in the binary repetition code a 2-weight word appears, as usual.

1.5 Systematic Encoder Circuit

In many applications (not only from the point of view of the decoder, but also, for instance, in order to implement composite encoding procedures, like parallel or serial code concatenation), it is useful to find the information symbols directly in the encoded binary sequence $t(x)$, precisely in its first k positions. Looking at Lemma 1.7, we need a pre-processing of the input sequence, able to realize the effect produced by sub-matrix $\underline{T}^{-1}$. We have the advantage of treating an s.s. time-invariant code in its $\underline{G}$ matrix. Remark that the $k \times k$ sub-matrix $\underline{T}$, in this case characterized by an upper triangular form, can be interpreted as an additive linear transformation (see Definition 1.1), easily invertible. We will perform this matrix inversion by means of polynomial theory, considering polynomial division as the inverse operation with respect to polynomial multiplication.

From another point of view, an additive linear transformation, like that described by a square matrix $\underline{G}$ of the type shown in Fig. 1.1 up, can be interpreted as the $\underline{G}$ matrix of a degenerate code having $r = 0$, that is with no control symbols at all. This additive linear transformation operates simply a different mapping of the information binary symbol. The weight distribution of the n-length sequences after this transformation is obviously coincident with that of the n-length sequences before it, and trivially results in $A(w) = \binom{n}{w}$, $w = 0, 1, 2, \ldots n$. The only difference is in the meaning that we assign to the elements of the sequence. The same relationship holds between a certain block code and its systematic version.

Lemma 1.12 *For an s.s. time-invariant block code in its $\underline{G}$ matrix, another form for writing* (1.3), *putting in evidence the information sequence represented by polynomial $i(x)$, in the first k positions of the code word, is*

$$t(x) = x^r i(x) + c(x), \tag{1.6}$$

where polynomial $c(x)$ expresses the sequence of r control symbols. The following relationships between polynomial $i(x)$ and polynomial $u(x)$ hold

$$u(x) = Q\left[\frac{x^r i(x)}{g(x)}\right], \quad c(x) = R\left[\frac{x^r i(x)}{g(x)}\right]. \tag{1.7}$$

Proof From Lemma B.6 in Appendix B, letting $p(x) = x^r i(x)$ and $z(x) = g(x)$, we obtain

$$x^r i(x) + R\left[\frac{x^r i(x)}{g(x)}\right] = Q\left[\frac{x^r i(x)}{g(x)}\right] g(x),$$

which joins (1.3), (1.6) and (1.7). □

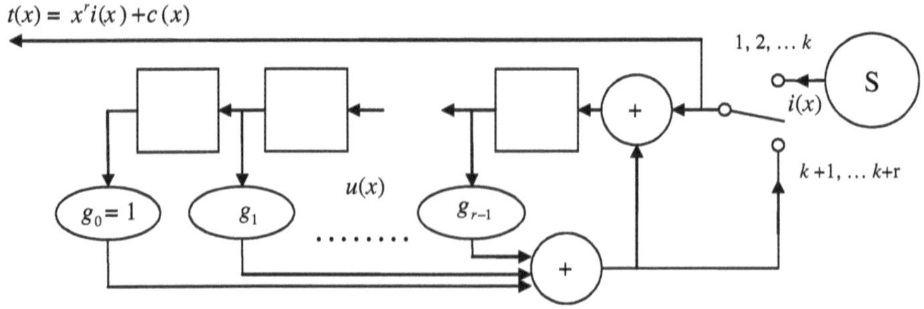

Fig. 1.5 Systematic encoder circuit based on a shift register able to calculate the quotient

As already suggested, it is possible to pre-process $i(x)$, which is the sequence outgoing from the information source, according to (1.6), and obtain, at a middle stage, its quotient $u(x)$. After this, the product $u(x)g(x)$ can be calculated. Equivalently, it is possible to calculate the remainder expressed by the second of (1.7) and attach it at the end of the sequence $i(x)$ itself. On the basis of these two principles, as many circuital schemes can be arranged, in order to obtain a systematic encoder circuit for a block code described by a generator polynomial $g(x)$.

Lemma 1.13 *An electronic circuit, able to perform systematic encoding, according to* (1.6), *is shown in* Fig. 1.5 (Cancellieri and Marchesani 2000). *It is based on the calculation of the polynomial division quotient, and exploits a shift register in controller arrangement. The switch is maintained for k symbol times connected to the information source, and then, for r symbol times, to the feed-back wire.*

Proof Comparison with Fig. C.5 in Appendix C clearly shows that the feed-back ex-OR sum, which begins immediately, gives the quotient $u(x)$. It shifts along the memory elements. At the same time, $i(x)$ is sent to the output. After k symbol times, owing to switch turning off, the remainder $c(x)$ follows $i(x)$ towards the output, and progressively the memory elements becomes empty. □

Remark that the wire correspondent to $g_r = 1$ is absent, as imposed by the convention on the electronic computation of a polynomial division (see Appendix C).

Lemma 1.14 *Another electronic circuit, able to perform systematic encoding, according to* (1.6), *is shown in* Fig. 1.6 (MacWilliams and Sloane 1977). *It is based on the calculation of the polynomial division remainder, and exploits a shift register in observer arrangement. There are two switches. The one immediately before the channel is maintained for k symbol times connected to the information source, and for the remaining r symbol times to the output of the shift register. The other switch enables feed-back, and then, during the last r symbols times, it will make the shift register progressively empty.*

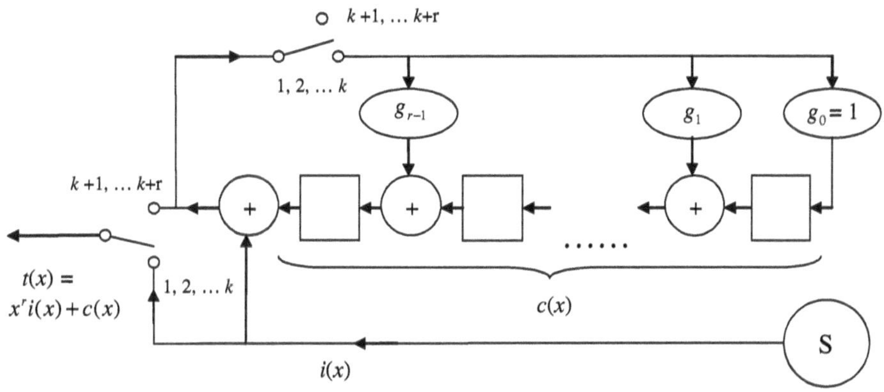

Fig. 1.6 Systematic encoder circuit based on a shift register able to calculate the remainder

Proof Comparison with Fig. C.6 in Appendix C clearly shows that the feed-back ex-OR sum, during the last r symbol times, gives the remainder $c(x)$. It is attached at the end of $i(x)$, which has been already sent to the output. In this case, the quotient $u(x)$ would be formed at the switch contact closer to the channel, before that it is connected (during the first k symbol times). □

Also here the wire correspondent to $g_r = 1$ is absent. In both these circuits, there is not a true pre-processing of $i(x)$, but division and multiplication are made at the same time.

In the following, we will use only the circuit realized in controller arrangement, that is the one depicted in Fig. 1.5, because the explicit calculation of the quotient $u(x)$ will allow very interesting generalizations. It is however important to stress that an encoder circuit so simple, that is based on a unique shift register, is possible only because the block code considered is s.s. time-invariant in its $\underline{G}$ matrix (Definition 1.8), that is it exhibits a $\underline{G}$ matrix where a unique polynomial $g(x)$ shifts. More general codes, in which some different generator polynomials are periodically assumed, as rows of $\underline{G}$, can be named wide-sense time-invariant (or briefly w.s. time-invariant) in $\underline{G}$. For them a more complicated encoder circuit is needed, and will be treated in Chap. 2. In the next two sections, instead, we will consider some types of time-varying block codes in their $\underline{G}$ matrix.

1.6 Code Concatenation: Effects on $\underline{G}$ Matrix

Definition 1.11 When at least two codes are employed on the same sequence of information symbols, we have an operation called *composite coding*. Such codes are named *component codes*.

Clearly a composite code, even when the component codes are s.s. time-invariant in their $\underline{G}$ matrices (see Definition 1.8), is no longer s.s. time-invariant in its overall $\underline{G}$ matrix. The resulting code rate decreases, but the minimum distance typically increases. In the framework of composite coding, code concatenation plays an important role. It is necessary to distinguish several types of code concatenation, according to the following main choices:

(i) parallel concatenation,
(ii) serial concatenation,
(iii) use of a single code word or of an ensemble of code words,
(iv) direct concatenation or concatenation with an interleaver.

This section deals only with single code word use. The employment of ensembles of code words, involved contemporarily in the concatenation, will be discussed in Sect. 1.15 and subsequent sections.

Definition 1.12 A *direct parallel concatenation of single code words*, taken from two s.s. time-invariant codes, described respectively by the generator polynomials $g_a(x)$ and $g_b(x)$, leads to a resulting code word of the type

$$t(x) = x^{r_a+r_b}i(x) + x^{r_b}R\left[\frac{x^{r_a}i(x)}{g_a(x)}\right] + R\left[\frac{x^{r_b}i(x)}{g_b(x)}\right], \tag{1.8}$$

where $i(x)$ is the information sequence, $r_a = deg[g_a(x)]$ and $r_b = deg[g_b(x)]$.

The choice performed in the code word organization expressed by (1.8), which has a systematic form, is made in such a way as to report the information symbols (just one time) in the first k positions, then the r_a control symbols due to code a, finally the r_b control symbols due to code b. An equivalent code is obtained inverting the order of the two sequences of control symbols. This means that the two component codes are interchangeable.

Lemma 1.15 *Code rate k/n and minimum distance d of the composite code, described by* (1.8), *are so constrained*

$$\frac{k}{n} = \frac{k}{k + r_a + r_b}, \qquad d \geq \max\{d_a, d_b\}, \tag{1.9}$$

where d_a, d_b are the minimum distances of the two component codes, in correspondence of the same number k of information symbols.

Proof The first of (1.9) derives from the trivial observation that k remains constant, whereas the total number of control symbols becomes $r = r_a + r_b$. The second of (1.9) can be demonstrated considering that the most critical situation is that in which $i(x)$ coincides with a code word of one of the two component codes. Nevertheless, owing to the addition of the control symbols due to the other component code (obviously different) the largest one between the two minimum distances is reached. □

Clearly a direct parallel concatenation is efficient when the two component codes are remarkably different, that is their code words exhibit different patterns of control symbols, to be attached to a same information sequence $i(x)$, at least for low values of w[$i(x)$]. In this case d can remarkably increase with respect to the lower bound in (1.9).

Example 1.16 Taking $g_a(x) = x^3 + x + 1, g_b(x) = x^3 + x^2 + 1$, and $k = 2$, the two codes, taken separately, are characterized by the following sets of four code words respectively

$$\begin{array}{llll} 00.000, & 01.011, & 10.110, & 11.101, \\ 00.000, & 01.101, & 10.111, & 11.010, \end{array}$$

where a dot separates information and control symbols for allowing an immediate interpretation. Their parallel concatenation yields the following set of four resulting code words

$$00.000.000, \quad 01.011.101, \quad 10.110.111, \quad 11.101.010.$$

The minimum distance d then has been increased from 3 to 5. This benefit has been gained at the expense of the code rate, which decreases from 2/5 to 1/4.

Example 1.17 Let us consider again the situation described in Example 1.16. There are several $\underline{G}$ matrices which can be associated to the code resulting from this direct parallel concatenation. First of all, we consider the $\underline{G}$ matrix of the two component codes, in their form where the respective generator polynomials shift along the rows. They are

$$\underline{G}_a = \begin{bmatrix} 1 & 0 & 1 & 1 & \\ & 1 & 0 & 1 & 1 \end{bmatrix}, \quad \underline{G}_b = \begin{bmatrix} 1 & 1 & 0 & 1 & \\ & 1 & 1 & 0 & 1 \end{bmatrix}.$$

We can construct the resulting generator matrix, letting $\underline{G}_a$ on the left and inserting below the remainder of the division by $g_b(x)$ of the polynomial occupying the only first k positions. Following this approach, we have

$$R\left[\frac{x^4}{x^3 + x^2 + 1}\right] = x^2 + x + 1, \quad R\left[\frac{x^3}{x^3 + x^2 + 1}\right] = x^2 + 1,$$

and hence

$$\underline{G} = \left[\begin{array}{cc|ccc|ccc} 1 & 0 & 1 & 1 & 0 & 1 & 1 & 1 \\ & 1 & 0 & 1 & 1 & 1 & 0 & 1 \end{array}\right].$$

The procedure here introduced is quite general, and holds also for any set of component codes, leading to an overall time-varying generator matrix.

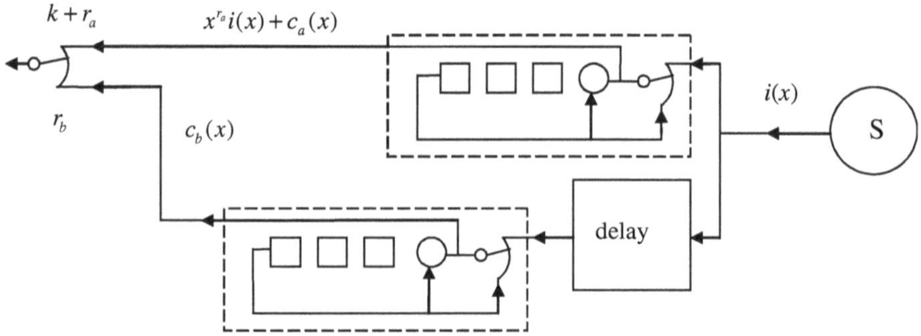

Fig. 1.7 Systematic encoder circuit for a direct parallel concatenation

Example 1.18 The two component codes considered in Example 1.17 are characterized by the same tern [5,2,3], and are equivalent, but they are associated to a pair of reciprocal generator polynomials. Clearly adopting the same code, which means two equal component codes, the effect of increasing the minimum distance is impossible. A benefit can be obtained, also with two equal codes, by introducing a proper interleaver.

A systematic encoder able to produce a direct parallel concatenation of two codes is that schematically depicted in Fig. 1.7. From the upper shift register the information sequence $i(x)$ and the first control symbol sequence $c_a(x)$ are sent to the channel. From the lower shift register the second control symbol sequence $c_b(x)$, after a proper delay, is sent to the channel too. In front of the channel, there is a switch which is maintained, during the first $(k + r_a)$ symbol times, connected to the upper link, and, during the remaining r_b symbol times, connected to the lower link.

Very simple variations in the scheme of Fig. 1.7 can be introduced, in order to take into account the presence of an interleaver before the lower shift register. In this case, instead of a direct parallel concatenation, we will obtain a parallel concatenation with an interleaver. An interleaver is a device which exchanges the order in the sequence of binary symbols at its input. Its mathematical description can be provided by a permutation matrix (see Definition 1.2). Often, the use of an interleaver allows to employ efficiently also two equal component codes, as already observed.

Example 1.19 Assuming an interleaver which inverts the order of the information symbols, instead of the delay device in the scheme of Fig. 1.7, it is possible to use two equal component codes, both characterized by the polynomial $g_a(x)$, given in Example 1.16. So we obtain an overall code with the following four code words

$$00.000.000, \quad 01.011.110, \quad 10.110.011, \quad 11.101.101.$$

It is an equivalent of the overall code obtained in Example 1.16, since it has the same word weight distribution $A(0) = 1$, $A(5) = 2$, $A(6) = 1$. Not always the introduction of an interleaver of this type gives the same result as the assumption of two reciprocal polynomials without interleaver.

Definition 1.13 A *direct serial concatenation of single code words*, taken from two s.s. time-invariant codes, described respectively by the generator polynomials $g_a(x)$ and $g_b(x)$, leads to a resulting code word of the type

$$t(x) = x^{r_a+r_b}i(x) + x^{r_b}c_a(x) + c_b(x),$$

$$c_a(x) = R\left[\frac{x^{r_a}i(x)}{g_a(x)}\right], \quad c_b(x) = R\left[\frac{x^{r_a+r_b}i(x) + x^{r_b}c_a(x)}{g_b(x)}\right], \tag{1.10}$$

where $r_a = \deg[g_a(x)]$ and $r_b = \deg[g_b(x)]$.

The difference with respect to Definition 1.12 is in the fact that here the whole code word of the first code is taken as information sequence by the second code. Clearly the order in which a serial concatenation is made influences the result which is obtained. Remark the importance of adopting a second code in systematic form. In fact, with a non-systematic code, we would have an overall code associated to the generator polynomial $g_a(x)g_b(x)$, which is a completely different situation.

Example 1.20 Still taking the two generator polynomials $g_a(x)$ and $g_b(x)$ introduced in Example 1.16, their direct serial concatenation yields a code with the following four code words

00.000.000, 01.011.110, 10.110.001, 11.101.111.

Here the overall minimum distance is $d = 4$, and the word weight distribution is $A(0) = 1$, $A(4) = 1$, $A(5) = 1$, $A(7) = 1$. Exchanging the order of the two component codes, one obtains a different result, that is the four code words

00.000.000, 01.101.110, 10.111.001, 11.010.111,

with overall minimum distance $d = 5$, and word weight distribution $A(0) = 1$, $A(5) = 2$, $A(6) = 1$.

Since there is a difference in the result obtained, for a serial concatenation of two codes exchanging their order (differently from what occurs in a parallel concatenation), the following specification appears more appropriate, inducing us to abandon subscripts a and b.

Definition 1.14 In a serial concatenation (direct or with an interleaver), we define *outer code* the one acting firstly and *inner code* the one acting then. The outer code will be associated to the generator polynomial $g_o(x)$, with $\deg[g_o(x)] = r_o$, and the inner code will be associated to the generator polynomial $g_i(x)$, with $\deg[g_i(x)] = r_i$.

A general scheme for a systematic encoder in a serial concatenation of two s.s. time-invariant codes, with interleaver, is shown in Fig. 1.8, where I represents the

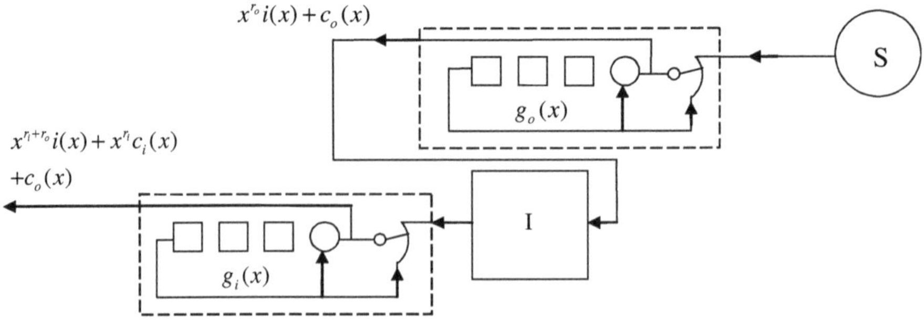

Fig. 1.8 Systematic encoder circuit for a serial concatenation with an interleaver

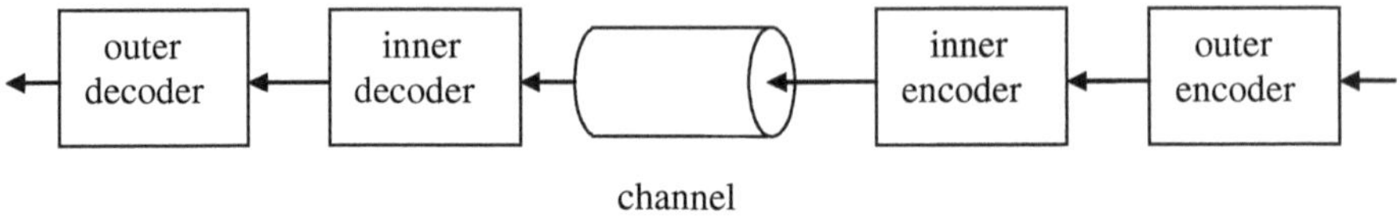

Fig. 1.9 Conceptual scheme of a serially concatenated coding

interleaver device. Remark that in a parallel concatenation (of single code words) the possible interleaver must exhibit a memory whose length is k. Here the possible interleaver must exhibit a memory whose length is $(k + r_o)$. The scheme for the case of a direct serial concatenation can be obtained from that reported in Fig. 1.8, by simply eliminating device I.

We can note that the meaning of inner and outer is referred to the position of the channel, as shown in Fig. 1.9, and not to the order in which the encoding procedure constructs the overall code word. The outer encoder operates firstly, the inner encoder then. As will be better explained in the following, this observation has a consequence in the construction of the resulting $\underline{G}$ matrix. In opposite sequence, at the output of the channel, the inner decoder operates firstly, the outer decoder then.

Since the inner decoder receives the signals directly from the channel, it can exploit a soft-decision decoding. The outer decoder often contributes with a simple hard-decision decoding. In certain situations, it is possible to arrange a unique soft-decision decoding algorithm for both (Hagenauer et al. 1996), typically adopting an iterative procedure which operates alternatively on the two component codes in search for a convergence. The best decoding algorithm must process the words of the overall code as they belong to a unique code, although with alternating efforts devoted to the two component codes. If an interleaver is adopted between the outer and the inner encoder, a proper de-interleaver will be set also between the inner and the outer decoder.

In order to stress the difference between parallel and serial concatenation, we can give the following typical comparative behaviour: parallel concatenation gives a modest increase in the minimum distance d, paid by a limited reduction in the code rate, whereas serial concatenation allows a stronger increase in d, accompanied by a remarkable lowering in the code rate. Clearly, in order to achieve these optimized results, different pair of codes, properly chosen depending also on the interleaver adopted, are to be considered. In fact, the first of (1.9) yields the new code rate in both schemes, when the same pair of component codes are employed.

Definition 1.15 Taking a linear block code, characterized by 2^k code words, whose length is n, it is possible to define a *subcode* of such code as the one collecting a subset of its code words. The new code has still code length n, and, due to linearity, it will exhibit 2^{k-b} code words, with $b < k$. The selection for obtaining 2^{k-b} code words from the total amount of 2^k code words previously available may be performed in different ways.

Lemma 1.16 *The overall code obtained from a code serial concatenation (direct or with interleaver) is a subcode of the inner code.*

Proof The selection mentioned in Definition 1.15 is represented here by the fact that any overall code word is also a word of the inner code, but associated to an information sequence which is a code word of the outer code (or a permuted version of it). The overall code length is equal to that of the inner code, but the number of code words is only 2^k, and not 2^{k+r_o}. Therefore, adopting the symbolism used in Definition 1.15, the reduction in the binary degrees of freedom in the subcode so obtained is expressed by $b = r_o$. □

Example 1.21 Considering again the direct serial concatenation introduced in Example 1.20, that is with $g_o(x) = x^3 + x + 1$ and $g_i(x) = x^3 + x^2 + 1$, we can observe that both these polynomials are factors of $(x^7 + 1)$. In particular, considering $g_i(x)$ and assuming $n = 8$, according to the situation here treated, among the 2^5 code words of such code, with n one unit greater than the length of the cyclic code, there will be a code word with weight $w = 2$ (see Sect. 1.3). This fact does not occur after concatenation, owing to the selection operated by the outer code.

In Appendix D many interesting even-distance binary block codes are listed, as derived from correspondent odd-distance block codes (see Definition D.21) after an increase by one unit of the minimum distance, at the expense of a single further control binary symbol. Here this operation will be better explained in the framework of a code concatenation process.

Definition 1.16 In a serial concatenation of two binary block codes, using a parity check code as the inner code, and an outer code characterized by odd minimum distance d, it is possible to obtain a new code, with minimum distance $(d + 1)$. This operation is called *binary code extension* of the outer code considered.

Example 1.22 Given a code characterized by the a generator polynomial $g_o(x) = x^3 + x + 1$, over the length $n = N = 7$ of its cyclic version, its extension leads to a generator matrix $\underline{G}$ which can be written as

$$\underline{G} = \begin{bmatrix} 1 & 0 & 1 & 1 & & & & \vline & 1 \\ & 1 & 0 & 1 & 1 & & & \vline & 1 \\ & & 1 & 0 & 1 & 1 & & \vline & 1 \\ & & & 1 & 0 & 1 & 1 & \vline & 1 \end{bmatrix},$$

in fact the remainder of the division of whichever polynomial with odd weight by $g_i(x) = x + 1$ is 1. The new code exhibits $d = 4$, as expected. It is a subcode of the parity check code, taken as the inner code. It is interesting to observe that, inverting the order of the two codes, we obtain a code whose minimum distance is only $d = 3$. In fact, for instance, the sequence 0010.1.100 (where dots separate the control symbols due to the two component codes), is a code word of the inner code.

We can conclude that binary extensions of odd-distance codes are subcodes of proper parity check codes. The following proposition summarizes the description of a procedure for the calculation of the $\underline{G}$ matrix in both parallel and serial concatenation without interleaver, using single code words.

Lemma 1.17 *In code concatenation (parallel or serial), without interleaver, the generator matrix of the overall code can be obtained in the following way. The submatrix on the left is that of the first component code or of the outer code. Any row is completed by attaching, on the right, the remainder of the division by the generator polynomial of the second code or of the inner code. For the parallel concatenation, this remainder is calculated just on the information sequence, for the serial concatenation on the entire outer code word.*

Proof These results follow directly from (1.8) and (1.10). We can see also Examples 1.17 and 1.22. □

Although very interesting composite codes can be obtained by means of a direct parallel or serial concatenation, using single code words, the best results are reached when an interleaver is adopted, and especially when more than one code word (proper ensemble of code words) are contemporary involved in this process. On such topic we will return, more in detail, in subsequent sections.

1.7 Code Puncturation: Effects on $\underline{G}$ Matrix

Code concatenation is a procedure whose goal is to increase the minimum distance of a certain code, at the expense of a reduction in its code rate. An opposite goal, that is to increase the code rate, at the expense of a reduction in the minimum distance, is what we will present here.

Definition 1.17 Given a linear block code, characterized by a generator matrix $\underline{G}$, it is possible to eliminate some control symbols, by suppression of the correspondent columns in $\underline{G}$. This procedure is called *code puncturation*. The word length is reduced from n to $(n - y)$, if y is the number of single puncturations performed (columns suppressed). The number of rows in $\underline{G}$ remains constant.

This description can be implicitly referred to a systematic version (at least in wide sense) of our code, where the positions of the control symbols are well identified. Nevertheless it can be generalized to consider also non-systematic versions, simply by speaking of a reduction in the quantity $(n - k)$ by y units, due to the suppression of as many columns in $\underline{G}$. After y single puncturation, the code rate from k/n grows to $k/(n - y)$. Clearly, although we can start from an s.s. time-invariant block code in $\underline{G}$, the punctured code so obtained will become time-varying in $\underline{G}$. Typically, in an s.s. time-invariant code described by (1.3), puncturation starts from the last position in the code word, and then proceeds from right to left.

Lemma 1.18 *The maximum number y of single code puncturations which can be performed is $(n - k - 1)$. Progressive code puncturation produces a reduction in the minimum distance. After $(n - k)$ single code puncturations the punctured code certainly becomes degenerate.*

Proof As long as $(n - k)$ is reduced by subsequent single code puncturations, the weight of the code words decreases. The code remains linear, therefore Theorem D.1 (invariance in the point of view) is still valid, and also the all 0-symbol code word is still present, therefore the minimum distance d of the punctured code decreases too. If one eliminates all the control symbols, that is operate $(n - k)$ subsequent single puncturations, matrix $\underline{G}$ is reduced to a $k \times k$ square matrix, so describing an additive linear transformation, that is a degenerate code. □

Equation (1.3) is no longer valid, even after the first single puncturation, and consequently the lower bound in (1.5) does not hold. So the condition characterized by $d = 1$ may be reached after less than $(n - k)$ single puncturations, that is before that the punctured code becomes a degenerate code. The process for reaching $d = 1$ is more or less rapid, depending on the type of polynomial $g(x)$ and on the precise positions where single puncturations are performed.

Example 1.23 Assuming $g(x) = x^3 + x + 1$, and an original code characterized by $k = 2$, the initial word weight distribution is $A(0) = 1$, $A(3) = 2$, $A(4) = 1$. After a number y of subsequent single puncturations (from the rightmost position and proceeding towards the left), we obtain the following modified word weight distributions

$$\begin{aligned} y = 0: &\quad A(0) = 1, \quad A(3) = 2, \quad A(4) = 1, \\ y = 1: &\quad A(0) = 1, \quad A(2) = 1, \quad A(3) = 2, \\ y = 2: &\quad A(0) = 1, \quad A(1) = 1, \quad A(2) = 1, \quad A(3) = 1. \end{aligned}$$

After substitution of the above $g(x)$ with its reciprocal, that is $g(x) = x^3 + x^2 + 1$, and proceeding exactly in the same way, we obtain instead

$$\begin{aligned} y = 0: \quad & A(0) = 1, \quad A(3) = 2, \quad A(4) = 1, \\ y = 1: \quad & A(0) = 1, \quad A(2) = 1, \quad A(3) = 2, \\ y = 2: \quad & A(0) = 1, \quad A(2) = 2, \quad A(4) = 1. \end{aligned}$$

The last code in this new sequence coincides with a binary parity check code, with $n = 3$, and is different from the last code of the previous group.

Definition 1.18 A very general punctured code word $\underline{t}'$ can be calculated by the use of a *puncturation matrix* $\underline{P}$ (for block codes), starting from the original code word $\underline{t}$. For a single puncturation in position j, $\underline{P}$ is obtained by the $n \times n$ identity matrix, in which the jth column is suppressed and the jth row is formed by all 0-symbols. In a multiple puncturation, this construction can be repeated for any position to be punctured. In any case, we can write

$$\underline{t}' = \underline{t} \cdot \underline{P}, \tag{1.11}$$

where $\underline{P}$ is an $n \times (n - y)$ matrix with y null rows.

Lemma 1.19 *The generator matrix of a punctured code can be obtained from that of the original code as follows*

$$\underline{G}' = \underline{G} \cdot \underline{P}. \tag{1.12}$$

Proof Linearity assures the validity of this equality, after (1.11). □

In conclusion, code puncturation maintains the property of linearity, but breaks the uniform behaviour of the various rows in the $\underline{G}$ matrix, when it was initially constructed by means of a unique $g(x)$ shifting in its rows. So, like in the case of a code concatenation, we obtain a time-varying code. Clearly the new code can be constructed employing an encoder circuit of the type shown in Figs. 1.3 or 1.5, and then properly eliminating the symbols placed in the positions to be punctured. Sometimes a single puncturation does not produce any reduction in the minimum distance, but, in this case, the original code necessarily was not well constructed.

Example 1.24 Assuming $g(x) = x^4 + x^3 + 1$, and an original code with $k = 2$, we start from this word weight distribution $A(0) = 1$, $A(3) = 2$, $A(4) = 1$. Applying a single puncturation to it, characterized by the puncturation matrix

$$\underline{P} = \begin{bmatrix} 1 & & & & \\ & 1 & & & \\ & & 1 & & \\ & & & & \\ & & & 1 & \\ & & & & 1 \end{bmatrix},$$

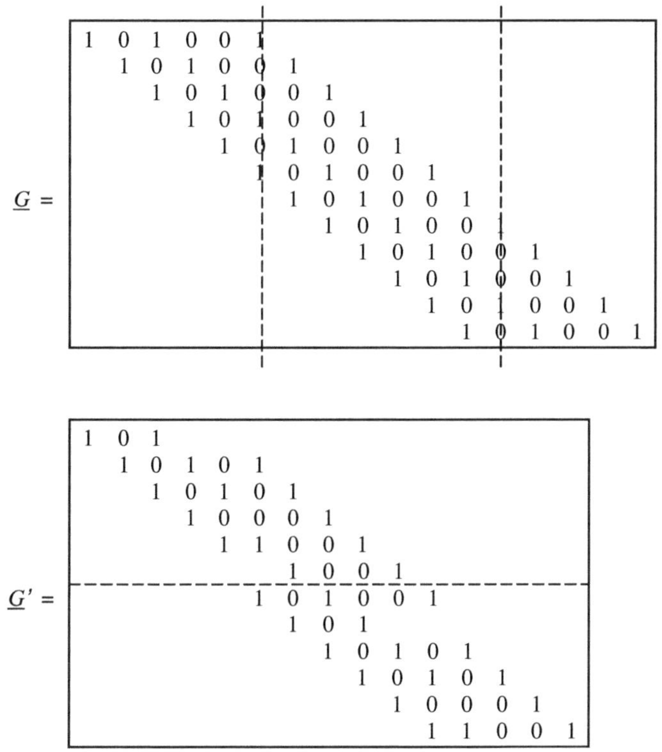

Fig. 1.10 Puncturation of a generator matrix as described in Example 1.25

the word weight distribution of the punctured code remains unchanged. The reason for this surprising behaviour is in the fact that the four code words of the original code, 000000, 011001, 101011, 110010, show the fourth position always null, that means however a bad code design. In this case, generally the position always null is suppressed.

After code puncturation, if made in some central positions of the code word, it is possible to rearrange the new generator matrix $\underline{G}'$, by proper ex-OR sums of its rows, in order to have the minimum number of different rows, except for a shift. A possibility is sometimes to maintain information symbols entering one at a time, that is an upper triangular form for the generator matrix. Otherwise, it is possible to set up a periodic structure with information symbols and control symbols properly alternated. A procedure of this type is particularly powerful in code concatenation and in convolutional code design.

Example 1.25 Assuming the generator polynomial $g(x) = x^5 + x^3 + 1$, and the original code characterized by $k = 12$, which means $n = 17$, after application of a double puncturation acting on the 6th and the 13th positions, we are in the situation shown in Fig. 1.10. In it the old $\underline{G}$ matrix is reported up, with the pair of eliminated

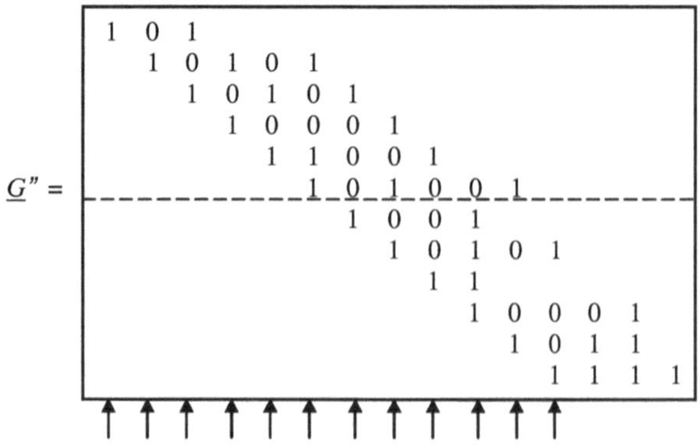

Fig. 1.11 Rearranged version of the punctured $\underline{G}$ matrix described in Example 1.25

columns represented by dashed vertical lines. The new generator matrix $\underline{G}'$ is then depicted down, putting in evidence that it is formed by two parts, characterized by 6 rows each (separated by a dashed horizontal line). The new code is time-varying and exhibits minimum distance $d' = 2$. If puncturation is made in the 6th and in the 12th positions, it becomes periodically time-varying, and its new generator matrix shows the repetition of the same ordered ensemble of five rows three times (but this choice would entail a minimum distance $d' = 1$). As already noted, in general, the new generator matrix is no longer upper triangular. In fact, as shown by Fig. 1.10 down, the 6th row exhibits its leftmost 1-symbol not aligned with the diagonal formed by all the other rows. In particular, two information symbols would appear both in the same position (the seventh one).

In order to avoid this fact, by rearranging the $\underline{G}'$ matrix (after ex-OR sums of rows and substitutions), it is possible to obtain a new version for such matrix, denoted $\underline{G}''$, which appears as shown in Fig. 1.11. It still exhibits two parts, with six rows each, but is not periodic, and the separation is suggested just for making interpretation easier. This rearrangement has given an upper triangular matrix, allowing to assign the positions of the $k = 12$ information symbols (denoted by small arrows) to as many first subsequent symbol times. This type of result is not always obtainable.

A completely different code structure is that in which the positions of control symbols are intentionally distributed, like occurs in convolutional codes, where information positions and control positions are properly alternated in a period. In a situation of this type also the puncturation matrix can be set in a periodic form. It is rather easy to predict the behaviour of a code so constructed, extrapolating what exhibited over the first few periods. Nevertheless, in such prediction, a fundamental role is played by how control symbols are distributed. This topic will be better investigated in Sect. 1.13 and in Chap. 3. A most exhaustive analysis of the positions occupied by control symbols requires the introduction of the parity check matrix, which gives a complementary description of linear codes, with respect to what discussed so far by means of the generator matrix. This alternative approach will be developed starting from Chap. 5.

1.8 Cyclic Block Codes

Theorem 1.2 *Given an s.s. time-invariant code, characterized by a generator polynomial g(x), in its cyclic form (that is with code word length n = N), all its code words exhibit the following property*

$$t'(x) = x^j t(x)\big|_{\mathrm{mod}(x^N+1)}, \quad j = 1, 2, \ldots N, \tag{1.13}$$

which means that, considered a certain code word t(x), any cyclic permutation obtained from it is still a code word t'(x).

Proof Since by definition the highest-order coefficient in $g(x)$ is non-null, it is possible to write $g(x) = x^r + \Delta g(x)$. Also $\Delta g(x)$ has a non-null highest-order coefficient, of degree smaller than r. As long as j increases, the remainder of the division of $x^j g(x)$ by $g(x)$ is null. Starting from $j = N - r$, we separate remainder calculation into two parts as follows

$$R\left[\frac{x^N + x^{N-r}\Delta g(x)}{g(x)}\right] = R\left[\frac{x^N}{g(x)}\right] + R\left[\frac{x^{N-r}\Delta g(x)}{g(x)}\right] = 0.$$

Polynomial $g(x)$ is factor of $(x^N + 1)$, then it is possible to write

$$R\left[\frac{x^N}{g(x)}\right] = R\left[\frac{1}{g(x)}\right].$$

Comparison with the above equality gives

$$R\left[\frac{x^{N-r}\Delta g(x)}{g(x)}\right] + R\left[\frac{1}{g(x)}\right] = R\left[\frac{x^{N-r}\Delta g(x) + 1}{g(x)}\right] = 0,$$

therefore the symbol sequence $x^{N-r}\Delta g(x) + 1$ is still a code word.

Extracting other non-null coefficients from $g(x)$, it is possible to complete a cyclic permutation. Linearity finally assures that this property holds for whichever code word of the considered cyclic code. □

After this Theorem, the denomination of cyclic code is better clarified. Not all the code words of a cyclic code exhibits $(N - 1)$ distinct cyclically shifted replicas. For instance, when the all-1-symbol code word belongs to our code, it has no replicas at all, that is it turns out to be the unique version of any possible cyclic shift. The same behaviour characterizes the all-0-symbol code word, which is always a code word, as known.

Definition 1.19 A sequence of 1-symbols, separated by 0-symbols, over a length b, within an N-length code word of a cyclic code, is called *pattern* of length b, with

$b \leq N$. It can be described by a polynomial, whose degree is $(b - 1)$ and is subjected to a cyclic shift invariance property, according to (1.13).

Definition 1.20 We will distinguish *regular patterns* as those having all the possible different cyclically permuted versions, and *degenerate patterns* as those having a smaller number of different cyclically permuted versions.

Clearly degenerate patterns are originated by particular symmetry properties.

Example 1.26 The cyclic code characterized by $g(x) = x^3 + x + 1$, has $N = 7$, that is $k = 4$. Its only degenerate pattern is 1111111, besides the all 0-symbol code word. All the other ones are regular. Therefore, being all the code words in number of 16, subtracting to this amount the code words 0000000 and 1111111, we are left with 14 code words, which means just 2 regular patterns (1011 and 11101).

Example 1.27 The cyclic code characterized by $g(x) = x^4 + x^3 + x + 1$, exhibits $N = 6$, that is $k = 2$. There are no regular patterns in this case. We have a unique degenerate pattern, besides the all 0-symbol code word. The three code words associated to this degenerate pattern are 011011, 110110, 101101 (two replicas besides the generator polynomial itself). As already noted in Example 1.15, in this cyclic code the all-1-symbol code word is not present.

Lemma 1.20 *In a cyclic code, if some degenerate patterns exist, besides the ones associated to the all-0-symbol and to the all-1-symbol code word, N must be not prime.*

Proof In order to complete the cyclic permutation, periodicity must be subdivided into the factors present in N. If integer N is the product of, say, two prime numbers, denoted n_1 and n_2, a pattern of the type $(x^{(n_1-1)n_2} + \ldots + x^{(n_1-1)} + 1)p(x)$ exhibits only n_2 cyclically permuted different versions. □

Example 1.28 The cyclic code described by $g(x) = x^4 + x^3 + 1$, is characterized by $N = 15$, that is $k = 11$. The all 1-symbols code word $\underline{1}$ is present in it. The degenerate patterns, listed placing $\underline{t}$ and $(\underline{1} + \underline{t})$ on the same row, are

000000000000000,	111111111111111,	(2 code words),
100100100100100,	110110110110110,	(6 code words),
100001000010000,	111101111011110,	(10 code words),
110011100111001,	110001100011000,	(10 code words),
101001010010100,	110101101011010,	(10 code words).

The remaining code words are $2^{11} - 38 = 2010$. Dividing by 15, we are left with 134 regular patterns. This is a [15,11,3] Hamming code, characterized by $A(3) = \frac{1}{3}\binom{N}{2} = 35$ (see Definition D.18 and subsequent considerations in Appendix D for details). There is just one degenerate pattern with $w = 3$ among those listed above, and it originates 5 code words. The remaining thirty 3-weight code words are to be subdivided into just two regular patterns. They are associated to 11001 and 101000001, corresponding respectively to $g(x)$ and to its square $g(x^2)$.

Definition 1.21 The co-factor of polynomial $g(x)$ in the factorization of $(x^N + 1)$ is called the *parity check polynomial*, and traditionally labelled $h(x)$. Its degree is $\deg[h(x)] = N - r$.

In Appendix B some examples of co-factors $\bar{g}(x)$ of as many polynomials $g(x)$ have been discussed. In the theory of channel coding, the parity check polynomial is useful in relation with a parity check matrix, traditionally labelled $\underline{H}$, which will be introduced in Chap. 5. Since now, however, we can expect that between polynomial $h(x)$ and matrix $\underline{H}$ holds a relation similar to that between polynomial $g(x)$ and matrix $\underline{G}$.

Definition 1.22 Given a cyclic code characterized by generator polynomial $g(x)$, factor of $(x^N + 1)$, the new cyclic code, having the same length N, but with polynomial $h(x)$ as its generator polynomial, is called the *dual code* of the original one.

Example 1.29 The binary parity check code is the dual of the binary repetition code on a same code word length $n = N$.

It is necessary to restrict the above definition of dual code pair to the only case of N odd. Only with this restriction, in fact, the dual of the dual turns out to be the original code. See for example the code characterized by $g(x) = x^4 + x^3 + x + 1$, factor of $(x^6 + 1)$. The co-factor of $g(x)$ and parity check polynomial of this code is $h(x) = x^2 + x + 1$, which is factor of $(x^3 + 1)$, hence the dual of the cyclic code characterized by $g'(x) = x^2 + x + 1$ is the parity check code, whose generator polynomial is $g''(x) = x + 1$, and not the original code. Under the above restriction, the following important statement holds.

Lemma 1.21 *For N odd, if (x + 1) is present in g(x), it will be absent in h(x), and vice versa. The dual of a code for which w[g(x)] is even is characterized by an odd-weight generator polynomial and vice versa.*

Proof If N is odd, there is just one binomial $(x + 1)$ in its factorization (see Lemma B.16 in Appendix B). On the other hand, from Lemma B.15 and Lemma B.20, necessary and sufficient condition for a polynomial to exhibit an even weight is that, among its factors, $(x + 1)$ is present. □

In Appendix D, to the purpose of easy identifying a linear code, as commonly adopted in the literature, the use of the tern $[n,k,d]$ is suggested. A more precise identification should require the knowledge of matrix $\underline{G}$, or, if the code is based on (1.3), of its generator polynomial $g(x)$. Nevertheless, for cyclic codes, the above tern can represent an acceptable identification, also because often many polynomials in the factorization of a certain binomial $(x^N + 1)$ give rise to equivalent cyclic codes.

Table 1.5 Examples of binary cyclic codes

Cyclic code	Tern	$g(x)$	$h(x)$	Possible equiv. versions
Parity check	$[n,n-1,2]$	$x+1$	$u_n(x)$	1
Repetition	$[n,1,n]$	$u_n(x)$	$x+1$	1
Hamming	[7,4,3]	1011	10111	2
	[15,11,3]	11001	111101011001	2
	[31,26,3]	100101	100101100111110001101110101	6
	.....			
BCH	[15,7,5]	111010001	11010001	2
	[15,5,7]	10100110111	101011	2
	[31,21,5]	11000001011	1111111011010110010111	6
	[31,16,7]	1001101110010101	10011001011000101	6
	.....			
Golay	[23,12,7]	10101110001	1010010011111	2
Simplex	[7,3,4]	10111	1011	2
	[15,4,8]	111101011001	11001	2
	[31,5,16]	100101100111110001101110101	100101	6
	.....			

Let us remember that in Definition 1.10 the minimum distance d_c is introduced, characterizing the cyclic parent of various codes derived from it, with $n < N$. We will consider here terns of the type $[N, N - r, d_c]$ for identifying just cyclic codes. Table 1.5 collects many interesting binary cyclic codes (see Appendix D for further details on them), and examples for their $g(x)$ or $h(x)$ polynomials. The parity check codes and the repetition codes, by themselves, express two families of codes, each characterized by any possible integer value for parameter $n = N$. Except for these codes, polynomials $g(x)$ and $h(x)$ have been written directly by means of the correspondent binary sequences, in order to have more compact expressions. The number of equivalent versions represents that of different generator polynomials able to originate the same tern and word weight distribution. Such number rapidly grows as N increases. Definition B.13 introduces the unitary polynomial $u_n(x)$.

We can observe that a dual pair is formed by simplex codes and Hamming codes, both briefly described in Appendix D (see also Examples 1.12 and 1.14 for more details on the [7,3,4] simplex code, several Examples on the [7,4,3] Hamming code, finally Example 1.28 on the [15,11,3] Hamming code). In simplex codes there is a unique regular pattern, besides the all-0-symbol code word.

Other interesting even-distance block codes, like the extended Hamming, BCH, and Golay, or the pth order Reed-Muller codes (see Appendix D for their description) cannot be obtained in cyclic form, but require other types of constructions, typically a code concatenation. The next section is devoted to relate the procedure for determining irreducible factors of $(x^N + 1)$ with N odd, to all the possible generator polynomials (and hence cyclic codes) having length N.

1.9 Enumeration of all the Possible Cyclic Codes of Length N

The decomposition of $(x^N + 1)$, with N odd, into irreducible factors, based on the theory of the search for the Nth root of unity is outlined in Appendix B. It leads to the results reported in Table B.1, for N odd, up to 63. From the irreducible polynomials written in the same row of such Table (associated to a precise N value), considering also possible products, we can take proper combinations to construct the generator polynomial of a cyclic code, whose length is N. The analysis of these codes can be made according to the following two propositions.

Lemma 1.22 *Taken the irreducible polynomials in the factorization of* $(x^N + 1)$, *with N prime and odd, the product of any combination of these polynomials, excluded that characterized by $(x + 1)$ alone, can be assumed as the generator polynomial of a cyclic block code having length N.*

Proof If N is a prime odd number, in the factorization of $(x^N + 1)$ there are only irreducible factors not present in the factorization of any $(x^{N'} + 1)$, $N' < N$, besides $(x + 1)$, which is always present. □

Lemma 1.23 *Taken the irreducible polynomials in the factorization of* $(x^N + 1)$, *with* $N = ab$, *a and b odd prime numbers, the product of any combination of these polynomials, excluded* $(x + 1)$ *and provided that they are chosen not only from the factors of* $u_a(x)$ *or from the factors of* $u_b(x)$, *can be assumed as the generator polynomial of a cyclic block code having length N.*

Proof If $N = ab$, in the factorization of $(x^N + 1)$, besides $(x + 1)$, also other polynomials, which belong to the factorization of $(x^a + 1)$ or of $(x^b + 1)$, are present. Furthermore a polynomial $Q_{a,b}(x)$ (see Lemma B.21 for more details) comes out. Polynomial $Q_{a,b}(x)$, even when not irreducible, contains factors not present in any binomial $(x^{N'} + 1)$, $N' < N$. In order to obtain the generator polynomial of a cyclic block code having length N, it is necessary to assume at least a factor in $Q_{a,b}(x)$, or one factor from $u_a(x)$ and one factor from $u_b(x)$. On the contrary, taking factors only from $u_a(x)$ or only from $u_b(x)$, we obtain a cyclic block code with length a or b respectively. □

Example 1.30 For $N = 15$, there are two [15,11,3] Hamming codes (whose generator polynomials are 11001 or 10011), two [15,7,5] BCH codes (whose generator polynomials are the product of one of the two preceding polynomials by 11111), two [15,5,7] BCH codes (obtained multiplying one time more by polynomial 111). Multiplying all the previous combinations one time more by polynomial 11, the following pairs of codes [15,10,4], [15,6,6], [15,4,8], are obtained. They are called even-distance subcodes of the previous ones. The last pair is also formed by simplex codes. Besides these well known codes, there are several other possibilities, for instance: a pair of [15,9,3] codes obtained multiplying 11001 or 10011 by 111, a single symmetric [15,9,4] code obtained multiplying 11111 by 111 (already introduced in Example 1.13), a single symmetric [15,7,3] code obtained as the product of the reciprocal pair 11001, 10011, coincident with $Q_{3,5}(x)$. To all the last groups of codes we can add also their subcodes, obtained multiplying one time more by polynomial 11. Finally there is the repetition code, whose generator polynomial is $u_{15}(x)$. On the contrary, 111 or 11111, taken alone, are not possible generator polynomials for a cyclic code having $N = 15$, but for a cyclic code having only $N = 3$ or $N = 5$ respectively.

It is possible to argue, from the very simple cases treated in Example 1.30, which refers to a value as small as $N = 15$, how incredibly high is the number of possibilities for obtaining cyclic codes with increasing N. A widening of Lemma 1.23, able to analyze the case in which N is the product of three or more prime odd numbers, is straightforward. A generalization of this treatment, in order to consider also the case in which N is even, say $N = 2b$, with b odd, can be made considering that here all the factors are double, and in a possible $g(x)$ polynomial we have to choose one factor from the non-trivial factors of the one ensemble and the other factor from the non-trivial factors of the other ensemble. All the above conditions are to be satisfied in order to guarantee that N is really the smallest integer expressing the length of the desired cyclic code. To this purpose, when N is the power of a prime number, say $N = a^2$, we have to choose at least one factor from the factorization of $u_a(x^a)$ (see Lemma B.24). Extension of this condition to higher powers is obvious.

Historically, among the very large number of possibilities to design and use cyclic codes having length N, a particular family of codes was preferred, owing to simplicity in their design and hard-decision decoding. Their origin is strictly related to the decomposition of $(x^N + 1)$, with N odd, into irreducible factors, based on the theory of the search for the Nth root of unity (see Appendix B). Subsequent roots, the associated cyclotomic cosets, and the correspondent irreducible polynomials are considered for selecting the factors to be progressively introduced in the generator polynomial $g(x)$, leading to an increasing maximum number t of correctable single errors. Such codes are called binary BCH codes from the first letters in the names of their inventors (Bose and Ray-Chaudhuri 1960; Hocquenghem 1959). Parameters and main performance of primitive binary BCH codes (those obtained with $N = 2^m - 1$) are briefly described in Appendix D. Non-binary BCH codes have been presented for the first time in Gorenstein and Zierler (1961). They will be treated in Sect. 1.21.

Definition 1.23 For all the values of N, such that, in the decomposition of $(x^N + 1)$, a primitive polynomial exists (see Definition B.17), it is possible to introduce a family of binary BCH codes, as cyclic block codes having length N. They are called *primitive binary BCH* when $N = 2^m - 1$, *non-primitive binary BCH* in all the other cases (provided that N is odd).

Theorem B.6 assures that, when $N = 2^m - 1$, a primitive polynomial, whose degree is m, certainly exists. In many other cases a primitive polynomial exists too, but its degree can vary widely.

The following statement, whose demonstration can be found in MacWilliams and Sloane (1977), is of basic importance for understanding the structure of the generator polynomial $g(x)$, for either primitive or non-primitive binary cyclic BCH codes, and their design, starting from the maximum number t of single correctable errors in a hard-decision decoding procedure. It is strictly referred to binary codes and odd values of the minimum distance $d = 2t + 1$ (see Lemma D.2). Its purpose is to guarantee a minimum value for t. A generalization of this proposition, able to consider also a non-binary alphabet and even or odd values for d, will be discussed in Sect. 1.21.

Given a binary BCH cyclic code, designed in such a way as to correct up to τ single errors, the product forming its generator polynomial $g(x)$ contains 2τ subsequent groups of Nth roots of unity, collected in proper irreducible polynomials, obtained starting from a certain primitive polynomial. The maximum number of correctable single errors becomes $t \geq \tau$. Equivalently the minimum distance so obtained is so lower bounded $d_c \geq 2t + 1$. The above statement is known as the BCH minimum distance bound for binary polynomial codes.

It is useful to remember that any irreducible polynomial collects a group of conjugate roots (see Definitions B.15 and B.16). The number of conjugated roots equals the degree of the associated polynomial, which, in turn, is related to a proper cyclotomic coset C_j, with j odd (remember that j = 2, 4, 6, … are not to be taken into account, since such roots are already present in C_1, C_3, … for a binary alphabet). Finally the root associated to (x + 1), conventionally denoted with j = 0, always present, is to be considered separately, and is useless in what follows, in fact we are looking for codes characterized by odd value for d.

The primitive polynomial will be denoted $M_1(x)$. It corresponds to the cyclotomic coset $C_1 = \{1, 2, 4, 8, \ldots\}$. The subsequent irreducible polynomials will be denoted $M_j(x)$, and will correspond to the cyclotomic cosets $C_j = \{j, 2j, 4j, 8j, \ldots\}$, j being subsequent odd numbers not already included in previous ensembles. In such expressions, the numbers collected by any C_j are to be taken mod(N). As anticipated, the number of elements in C_j (conjugate roots) equals the degree of the correspondent $M_j(x)$. A couple of identical polynomials are, say, $M_j(x)$ and $M_{2j}(x)$. Between them, just one has to be taken, conventionally $M_j(x)$. In order to avoid a double presence of the same polynomial in the generator polynomial to be synthesized (due to the above pairwise coincidence but also to possible further coincidences) the following expedient is adopted. Denoting with LCM[…] the least common multiple of what contained between square brackets, these expressions for the generator polynomials are to be taken, for subsequent integer values of the design parameter τ,

$$
\begin{aligned}
\tau = 1 : g(x) &= \ \mathrm{LCM}[M_1(x), M_2(x)] = M_1(x), \quad t \geq 1, \\
\tau = 2 : g(x) &= \ \mathrm{LCM}[M_1(x), M_2(x), M_3(x), M_4(x)] \\
&= M_1(x)M_3(x), \quad t \geq 2, \\
\tau = 3 : g(x) &= \mathrm{LCM}[M_1(x), M_2(x), M_3(x), M_4(x), M_5(x), M_6(x)] \\
&= M_1(x)M_3(x)M_5(x), \quad t \geq 3, \\
& \quad \ldots\ldots,
\end{aligned}
$$

where the second equality in each row is formally correct, but often the number of different polynomials is lower.

The degree of any $M_j(x)$, that is the number of conjugate roots collected in it, depends on the length of the period for reaching the condition $(s2^{n_s}) \bmod N = s$ (see Appendix B). The number of elements in LCM[ ], before the pairwise simplification, is equal to twice the design parameter τ. Therefore, after such simplification, they become τ, except when a particular further simplification occurs, for instance $M_1(x) = M_3(x)$. Situations of these types may allow that the true t is greater than τ, in agreement with the inequality $t \geq \tau$, stated by the binary BCH bound. The following examples will show how this fact can occur.

Example 1.31 Taking N = 15, from Example B.34, we have these cyclotomic cosets and correspondent irreducible polynomials

$$\begin{aligned}
C_0 &= \{0\}, & M_0(x) &= x+1,\\
C_1 &= \{1,2,4,8\}, & M_1(x) &= x^4+x^3+1,\\
C_3 &= \{3,6,9,12\}, & M_3(x) &= x^4+x^3+x^2+x+1,\\
C_5 &= \{5,10\}, & M_5(x) &= x^2+x+1,\\
C_7 &= \{7,11,13,14\}, & M_7(x) &= x^4+x+1.
\end{aligned}$$

The primitive BCH codes for N = 15 then show the following terns, true maximum numbers of correctable single errors and generator polynomials,

$$[15,11,3],\quad t=1,\quad g(x) = M_1(x) = x^4+x^3+1,$$

$$[15,7,5],\quad t=2,\quad g(x) = M_1(x)M_3(x) = (x^4+x^3+1)(x^4+x^3+x^2+x+1),$$

$$[15,5,7],\quad t=3,\quad g(x) = M_1(x)M_3(x)M_5(x) = (x^4+x^3+1)(x^4+x^3+x^2+x+1)(x^2+x+1),$$

$$[15,1,15],\quad t=7,\quad g(x) = M_1(x)M_3(x)M_5(x)M_7(x) = u_{15}(x).$$

The first is also a Hamming code, the last a repetition code. Remark that, although the designed value τ for the last one was 4, we obtain $t = 7$, by far greater. This is due to $M_7(x) = M_{11}(x) = M_{13}(x)$ and $M_3(x) = M_9(x)$.

Example 1.32 In the case of $N = 31$, as a consequence of Theorem B.7, which gives 30/5 = 6, we have six cyclotomic cosets all characterized by five elements (see Example B.35). Assuming $M_1(x) = x^5+x^2+1$ as the primitive polynomial, we obtain the following primitive BCH codes

$$[31,26,3],\quad t=1,\quad g(x) = M_1(x) = x^5+x^2+1,$$

$$[31,21,5],\quad t=2,\quad g(x) = M_1(x)M_3(x) = (x^5+x^2+1)(x^5+x^4+x^3+x^2+1),$$

$$[31,16,7],\quad t=3,\quad g(x) = M_1(x)M_3(x)M_5(x) = (x^5+x^2+1)(x^5+x^4+x^3+x^2+1)(x^5+x^4+x^2+x+1),$$

$$[31, 11, 11], \quad t = 5, \quad g(x) = M_1(x)M_3(x)M_5(x)M_7(x)$$
$$= (x^5 + x^2 + 1)(x^5 + x^4 + x^3 + x^2 + 1)$$
$$(x^5 + x^4 + x^2 + x + 1) \cdot (x^5 + x^3 + x^2 + x + 1),$$

$$[31, 6, 15], \quad t = 7, \quad g(x) = M_1(x)M_3(x)M_5(x)M_7(x)M_{11}(x)$$
$$= (x^5 + x^2 + 1)(x^5 + x^4 + x^3 + x^2 + 1)(x^5 + x^4 + x^2 + x + 1)$$
$$\cdot (x^5 + x^3 + x^2 + x + 1)(x^5 + x^4 + x^3 + x + 1),$$

$$[31, 1, 31], \quad t = 15, \quad g(x) = M_1(x)M_3(x)M_5(x)M_7(x)$$
$$\cdot M_{11}(x)M_{15}(x) = u_{31}(x).$$

In the [31,11,11] code there is an increase in t by 2 units, instead of the usual 1 unit. This is due to the fact that $M_5(x) = M_9(x)$. In the [31,6,15] code there is the same increase. Finally the last polynomial introduces an increase in t by 8 units.

Lemma 1.24 *In a primitive BCH code, that is characterized by $N = 2^m - 1$, the following inequality holds between the degree r of the generator polynomial g(x) and the maximum number t of correctable single errors*

$$r \leq mt. \tag{1.14}$$

Proof For $N = 2^m - 1$, a primitive polynomial having degree m is always present. The subsequent irreducible polynomials, originated from it, exhibit a degree not higher than its degree. Since we have to take at most t of them, the overall $\deg[g(x)] = r$ will satisfy (1.14). □

For Hamming codes, (1.14) is always satisfied with equality sign. For subsequent primitive BCH codes, we can have the true t greater than the designed one, or (when N is composite) there are some irreducible polynomials whose degree is smaller than that of the primitive one. The last fact is justified by an anticipated occurrence of the condition $(s2^{n_s}) \bmod N = s$. Remark that inequality (1.14) is taken in Definition D.19, as (D.22), in order to introduce primitive BCH codes, in the general survey on the most known block codes, mainly from a performance point of view, given in Appendix D. Nevertheless an extrapolation of the increase in $\deg[g(x)]$ with increasing t, based on (1.14) with equality sign, is typically an over-estimation of parameter r, more and more as long as r approaches N.

Let us consider now some examples of non-primitive BCH codes, firstly with N prime, and at last with N not prime. It is important to stress that here the first step is not a Hamming code, which does not exist in this case.

Example 1.33 A very particular case is that characterized by $N = 23$. Here the cyclotomic cosets are

$$C_0 = \{0\}, \quad C_1 = \{1, 2, 3, 4, 6, 8, 9, 12, 13, 16, 18\},$$
$$C_5 = \{5, 7, 10, 11, 14, 15, 17, 19, 20, 21, 22\}.$$

So we can expect two only polynomials, whose degree is 11. From Table B.1, we can assume $M_1(x) = x^{11} + x^9 + x^7 + x^6 + x^5 + x + 1$. In order to obtain a code with at least minimum number of correctable single errors $\tau = 2$, we will write

$$\begin{aligned} g(x) &= \text{LCM}[M_1(x)M_2(x)M_3(x)M_4(x)] \\ &= M_1(x) = x^{11} + x^9 + x^7 + x^6 + x^5 + x + 1. \end{aligned}$$

Remark that here we have also the particular further simplification $M_1(x) = M_3(x)$. In spite of the fact that τ was 2, and hence the designed distance was 5, the true minimum distance turns out to be $d_c = 7$. This code is known as binary Golay code [23,12,7], and is described also in Definition D.20, where its word weight distribution is reported (Golay 1949). There are two equivalent binary Golay codes, different only in their generator polynomials, the one is $g(x) = M_1(x)$, and the other is $g(x) = M_5(x)$, reciprocal each other. In Example D.9 its residual binary symbol error probability, with hard- and soft-decision decoding, is calculated. In Example D.10 the asymptotic coding gain is finally estimated in both situations. The binary Golay code, which is then also a non-primitive BCH code, satisfies Hamming bound (D.13), for an odd minimum distance, with equality sign.

Example 1.34 A situation rather similar, although much less significant, is that characterized by $N = 17$. Also in this case, we have only three cyclotomic cosets

$$C_0 = \{0\}, \quad C_1 = \{1, 2, 4, 8, 9, 13, 15, 16\},$$
$$C_5 = \{3, 5, 6, 7, 10, 11, 12, 14\}.$$

So we can expect two polynomials, whose degree is 8. From Table B.1, we observe that the factorization of $(x^{17} + 1)$ shows only two irreducible symmetric polynomials. We can assume $M_1(x) = x^8 + x^5 + x^4 + x^3 + 1$. In order to obtain a code with at least minimum number of correctable single errors $\tau = 1$, we will write

$$g(x) = \text{LCM}[M_1(x)M_2(x)] = M_1(x) = x^8 + x^5 + x^4 + x^3 + 1$$

In spite of the fact that the designed distance was 3, the true minimum distance turns out to be $d_c = 5$. This is a [17,9,5] code. An equivalent could be obtained assuming the other irreducible polynomial $M_1(x) = x^8 + x^7 + x^6 + x^4 + x^2 + x + 1$. The word weight distribution of both such codes is

$$\begin{array}{lllll} A(0) = 1, & A(5) = 34, & A(6) = 68, & A(7) = 68, & A(8) = 85, \\ A(9) = 85, & A(10) = 68, & A(11) = 68, & A(12) = 34, & A(17) = 1. \end{array}$$

This non-primitive BCH code has performance slightly better than that of the [15,7,5] primitive BCH code, owing to a code rate a little higher.

Example 1.35 The case of $N = 21$ is similar to that of $N = 15$, but with even more possible options. It is characterized by these cyclotomic cosets and correspondent irreducible polynomials

$$\begin{array}{ll} C_0 = \{0\}, & M_0(x) = x + 1, \\ C_1 = \{1, 2, 4, 8, 11, 16\}, & M_1(x) = x^6 + x^4 + x^2 + x + 1, \\ C_3 = \{3, 6, 12\}, & M_3(x) = x^3 + x^2 + 1, \\ C_5 = \{5, 10, 13, 17, 19, 20\}, & M_5(x) = x^6 + x^5 + x^4 + x^2 + 1, \\ C_7 = \{7, 14\}, & M_7(x) = x^2 + x + 1, \\ C_9 = \{9, 15, 18\}, & M_9(x) = x^3 + x + 1. \end{array}$$

In order to have a code with designed number of correctable single errors $\tau = 1$, following the traditional BCH approach, we have to take $g(x) = M_1(x) = x^6 + x^4 + x^2 + x + 1$, and so a [21,15,3] non-primitive BCH code. Nevertheless it is possible also to assume $g(x) = M_3(x)M_7(x) = x^5 + x + 1$, whose degree is smaller, obtaining a code not BCH, but still with $d_c = 3$. In order to reach $d_c = 5$, with the traditional approach, we obtain the non-primitive BCH code whose generator polynomial is

$$g(x) = M_1(x)M_2(x) = (x^6 + x^4 + x^2 + x + 1)(x^3 + x^2 + 1),$$

that is the sequence 1110110011. This is a rather good [21,12,5] code, whose word weight distribution results in

$$\begin{array}{llll} A(0) = 1, & A(5) = 21, & A(6) = 168, & A(7) = 360, \\ A(8) = 210, & A(9) = 280, & A(10) = 1008, & \\ A(11) = 1008, & A(12) = 280, & A(13) = 210, & \\ A(14) = 360, & A(15) = 168, & A(16) = 21, & A(21) = 1. \end{array}$$

1.10 Shortened Cyclic (SC) Codes

Coming back to the Tables 1.1, 1.2, 1.3 and 1.4, we can observe that, for a same $g(x)$, the word length and consequently the code rate can be finely adjusted properly varying k, with r maintained constant. It is convenient to distinguish two situations, separated by the condition of cyclic code $k = N - r$. In this section, we will discuss the one characterized by $k < N - r$.

Definition 1.24 Given a cyclic code, whose words are expressed by means of (1.3) or (1.6), with $\deg[g(x)] = r$, factor of $(x^N + 1)$, we define the code obtained from it, with $k < N - r$, a *shortened cyclic code* (or briefly *SC code*). More precisely, assuming $k = N - r - s$, that is after s single shortening operations, we will identify the resulting code as *s-time shortened* starting from its cyclic version.

It is useful to remember that in (1.3) polynomial $u(x)$ has, in general, degree at most equal to $(k - 1)$. When this degree is at most $(N - r - 1)$, that is $k = N - r$, we have the cyclic code condition. Now, a code s-time shortened, starting from the cyclic one, will show deg[$u(x)$] at most equal to $(N - r - s - 1)$. If our codes (the cyclic and the shortened one) are in systematic form, let us consider expression (1.6) for their code words. In this expression, the above constraint on the degree of polynomial $u(x)$ is transferred, identical, to the degree of polynomial $i(x)$.

Example 1.36 Starting from the [7,4,3] Hamming cyclic code, with generator polynomial $g(x) = x^3 + x + 1$, a single shortening gives a [6,3,3] SC code with the following 8 code words

$$000000, \quad 001011, \quad 010110, \quad 011101,$$
$$100111, \quad 101100, \quad 110001, \quad 111010.$$

It is possible to observe that property (1.13) is no longer true. A second single shortening gives a [5,2,3] SC code, already discussed in several Examples, whose four code words are

$$00000, \quad 01011, \quad 10110, \quad 11101.$$

A third shortening would yield a [4,1,3] code, with the following two only code words

$$0000, \quad 1011,$$

where a 0-symbol is always present in the second position, which is then useless. After its elimination we obtain a [3,1,3] repetition code.

Making an inspection of the 8 code words of the [6,3,3] shortened Hamming code, listed in Example 1.36, we can observe that the two only regular patterns of the cyclic parent code (1011 and 11101, as shown in Example 1.26) are still present. Also their cyclic permutation can be recognized, by assuming a 0-symbol in an imaginary position before the leftmost one in the code words. This observation allows to give an alternative definition for a single shortening, valid also for non-polynomial codes (MacWilliams and Sloane 1977), here included in the following proposition.

Lemma 1.25 *The [$N - 1$, $N - r - 1$, d] one-time shortened code, obtained from an $[N, N - r, d_c]$ cyclic parent code, can be constructed as follows. Taken one half of the total set of code words characterizing the cyclic parent code, in particular those having a 0-symbol in their first position, suppress this first symbol.*

Proof It is straightforward to demonstrate that, in a cyclic code based on (1.3) or (1.6), in any of the N positions of the code word, 2^{k-1} code words show a 0-symbol and the remaining 2^{k-1} show a 1-symbol. With reference to the leftmost positions, it will be occupied by a 1-symbol when the highest-order power coefficient of $u(x)$ or $i(x)$ is non-null. Excluded this possibility, we are left with a polynomial $u(x)$ or $i(x)$ having at most degree $(N - r - 2)$. □

The choice of the leftmost position is a pure convention, in order to allow subsequent shortening operations proceeding progressively towards the right. On the other hand, this choice is able to maintain the separation between the positions occupied by information symbols and those occupied by control symbols (whose number remains unchanged and equal to r).

In Example 1.36, the third single shortening of the [7,4,3] Hamming code produces two only code words both with a 0-symbol in their second position. A similar situation was that described in Example 1.24, where a puncturation, starting from a [6,2,3] code, gave an unchanged word weight distribution. The original [6,2,3] code was obtained, also in that case, after too many single shortening operations with respect to its cyclic parent [15,11,3] Hamming code, having length $N = 15$. In conclusion, the number s of subsequent single shortening operations which can be performed, in general, is so upper bounded

$$s \leq N - r - 1, \tag{1.15}$$

in fact, when equality is satisfied, k remains as small as 1. Nevertheless when s is high, the shortened code may exhibit a too low code rate, only in part balanced by the elimination of possible useless positions in its code words.

As regards the minimum distance d of the shortened code, the only property we can suggest is

$$d \geq d_c, \tag{1.16}$$

where equality holds when $\mathrm{w}[g(x)] = d_c$. If the last condition does not hold, that is $\mathrm{w}[g(x)] > d_c$, keeping in mind the results of Examples 1.11 and 1.13 (Tables 1.2 and 1.4), it is possible to predict some increase in d, with respect to d_c, but only accompanied by a strong code shortening. However, even when d is maintained equal to d_c, the number of code words at minimum distance $A(d_c)$ progressively decreases as s increases, often leading to a more rapid approach to the asymptotic coding gain. In particular, with increasing s, the long low-weight patterns progressively disappear (see for instance the case of the only two 3-weight patterns in the [15,11,3] Hamming code, analyzed in Example 1.28).

1.11 Lengthened Cyclic (LC) Codes

Let us consider now a situation opposite to that characterizing shortened cyclic codes. It is associated to the condition $k > N - r$.

Definition 1.25 Given a cyclic code, described by the generator polynomial $g(x)$, with $\deg[g(x)] = r$, factor of $(x^N + 1)$, we define the code obtained from it, still with expression (1.3) or (1.6) for its code words, but with $k > N - r$, a *lengthened cyclic code* (or briefly *LC code*). More precisely, assuming $k = N - r + y$, that is after y single lengthening operations, we will identify the resulting code as *y-time lengthened* starting from its cyclic version.

A code y-time lengthened, starting from the cyclic one, will show $\deg[u(x)]$ in expression (1.3) or $\deg[i(x)]$ in expression (1.6) at most equal to $(N - r + y - 1)$.

Example 1.37 Starting from the [7,4,3] Hamming cyclic code, with generator polynomial $g(x) = x^3 + x + 1$, a single lengthening gives an [8,5,2] LC code, characterized by the following 32 code words

00000000,	00001011,	00010110,	00011101,
00100111,	00101100,	00110101,	00111010,
01000101,	01001110,	01010011,	01011000,
01100010,	01101001,	01110100,	01111111,
10000001,	10001010,	10010111,	10011100,
10100110,	10101101,	10110000,	10111011,
11000100,	11001111,	11010010,	11011001,
11100011,	11101000,	11110011,	11111110.

They are written ordering the first five symbols (information symbols) as progressive integer numbers expressed in a binary alphabet. The word weight distribution is $A(0) = 1$, $A(2) = 1$, $A(3) = 10$, $A(4) = 11$, $A(5) = 4$, $A(6) = 3$, $A(7) = 2$. It is possible to observe the presence of just one 2-weight code word, associated to the pattern 10000001, whose polynomial expression is $(x^7 + 1)$. This fact was expected (see Sect. 1.6). It is evident that the two outermost positions (the 1st and the 8th) will be less protected than all the other ones. Another important observation regards the fact that, like in a shortened cyclic code, property (1.13) is no longer true.

Clearly, this is a situation in which the positions less protected are known, and then they could be protected, for instance, by adding an outer code in a serial concatenation scheme. Remark that, in cyclic or in SC codes, generally, all the positions are equally protected. In some particular applications, unequal protection against errors can be efficiently exploited, for instance when a different reliability is intentionally requested to proper ensembles of decoded symbols (Kuriata 2008).

Subsequent single code lengthening operations initially show an increase in the number of these 2-weight code words, by one unit any single lengthening. In other terms, a $[N + y, N - r + y, 2]$ code, with $y < N$, exhibits $A(2) = y$. At the end of a complete lengthening period, that is for $n = 2N$, or $y = N$, there will be $A(2) = N$, and all the positions will turn out to be equally protected again. This situation repeats for any further lengthening period characterized by an additional amount of N information symbols. In such situations, periodically repeated, the geometry come back to be uniform, as in the original cyclic code.

Lemma 1.26 *For any $(q-1)N$-time lengthened code $[qN, N - r + (q-1)N, 2]$, q integer, property* (1.13) *is true for $N' = qN$.*

Proof If the generator polynomial $g(x)$ of our LC code is factor of $(x^N + 1)$, it is certainly factor also of $(x^{2N} + 1), (x^{3N} + 1), \ldots$. □

Although property (1.13) is that characterizing cyclic codes, we cannot define the lengthened codes described in Lemma 1.26 properly as cyclic codes, because we have reserved this definition only to those codes, described by a certain generator polynomial $g(x)$, factor of the binomial $(x^N + 1)$, N being the minimum integer for which this property holds.

Lemma 1.27 *If $g(x)$ contains a factor $(x + 1)$, the LC code, for $n = qN$, has a word weight distribution $A(w)$, which is symmetric with q even, and not symmetric with q odd. Instead, if $g(x)$ does not contain any factor $(x + 1)$, the cyclic code $(n = N)$ has a symmetric $A(w)$ distribution, and the same is for all its completely lengthened versions ($n = qN$, for q either even or odd).*

Proof Lemma 1.11 states that the condition for reaching a symmetric word weight distribution is to have the all-1-symbol $u_n(x)$ as a code word. So, if $(x + 1)$ is factor of $g(x), u_n(x)$ can be a code word for $n = qN$, only with q even. Instead, if $(x + 1)$ is not factor of $g(x), u_n(x)$ can be a code word for any $n = qN$, with q either even or odd. □

Example 1.38 Let us consider a [3,1,3] repetition code, whose generator polynomial is the irreducible polynomial $g(x) = x^2 + x + 1$. We have the word weight distributions shown in Table 1.6, as n increases. It is possible to observe how, in any situation characterized by $n = qN$, the word weight distribution is symmetric.

Example 1.39 From the simplex cyclic code [7,3,4], whose generator polynomial $g(x)$ includes a factor $(x + 1)$, after a complete lengthening period, that is $y = N$, or $n = 2N$, we obtain a [14,10,2] lengthened code, whose word weight distribution is symmetric (see Table 1.3).

Example 1.40 Taken the cyclic code [6,2,4], whose generator polynomial is $g(x) = x^4 + x^3 + x + 1 = (x + 1)(x^3 + 1)$, factor of $(x^6 + 1)$, a sequence of its lengthened versions shows the word weight distributions reported in Table 1.7. Only the version characterized by $n = 2N = 12$ is symmetric (and is evidenced in the Table). That of the cyclic code is not symmetric, as well as that characterized by $n = 3N = 18$. The one characterized by $n = 4N = 24$ will be symmetric again, and so on.

Table 1.6 Word weight distribution of some lengthened versions of the [3,1,3] repetition code

n	*qN*	*A*(0)		*A*(2)	*A*(3)	*A*(4)	*A*(5)	*A*(6)	*A*(7)	*A*(8)	*A*(9)	*A*(10)	*A*(11)	*A*(12)
3	1*N*	*1*			*1*									
4		1		1	2									
5		1		2	4	1								
6	2*N*	*1*		*3*	*8*	*3*		*1*						
7		1		5	12	7	4	3						
8		1		7	18	15	12	9	2					
9	3*N*	*1*		*9*	*27*	*27*	*27*	*27*	*9*		*1*			
10		1		12	36	46	60	60	28	9	4			
11		1		15	48	74	112	126	80	37	16	3		
12	4*N*	*1*		*18*	*64*	*111*	*192*	*252*	*192*	*111*	*64*	*18*		*1*

Table 1.7 Word weight distribution of some lengthened versions of a [6,2,4] cyclic code

n	*qN*	*A*(0)	*A*(2)	*A*(4)	*A*(6)	*A*(8)	*A*(10)	*A*(12)	*A*(14)	*A*(16)	*A*(18)
6	1*N*	1		3							
7		1	1	5	1						
8		1	2	9	4						
9		1	3	15	13						
10		1	4	26	28	5					
11		1	5	42	58	21					
12	2*N*	*1*	*6*	*63*	*116*	*63*	*6*	*1*			
13		1	8	89	216	163	32	3			
14		1	10	123	380	375	122	13			
15		1	12	165	640	795	372	63			
16		1	14	223	1,001	1,615	982	241			
17		1	16	294	1,548	3,060	2,392	802	76	3	
18	3*N*	1	18	278	2,322	5,508	5,382	2,406	342	27	1

From the above Examples, an observation about the increase in coefficient $A(2)$ can also be derived. Such increase turns out to be made by one unit at a time as long as the y single lengthening operations is so bounded $0 < y \leq N$, by two units at a time for $N < y \leq 2N$, and so on. This observation can be generalized through the following statement, considering a certain qth lengthening period.

Lemma 1.28 *In a lengthened code, characterized by a length n so constrained* $qN < n \leq (q + 1)N$, *the total number of 2-weight code word is*

$$A(2,n) = \begin{cases} (n-N), & q = 1 \\ \binom{q}{2} N + (n - qN)q, & q \geq 2 \end{cases} \quad (1.17)$$

Proof When $q = 1$, the upper line in equality (1.17) has been already justified observing that $n - N = y$. When $q \geq 2$, the following considerations hold. In the condition $n = qN$, the number is exactly $A(2, qN) = \binom{q}{2} N$, because of all the possible couples of periods, before the last one. The last lengthening period contributes with the additional number $\Delta = (n - qN)q$. In fact the differential increment is q. Addition of the two terms yields the lower line in (1.17). □

Lemma 1.29 *Given a y-time lengthened code* $[N + y, N - r + y, 2]$, *obtained from a cyclic code whose length is N, for whichever number y of single lengthening operations, its code words t(x), expressed by* (1.3) *or* (1.6), *can be so decomposed*

$$t(x) = a(x)(x^N + 1) + t_c(x), \tag{1.18}$$

where $a(x)$ is a polynomial having degree at most $(y - 1)$ and $t_c(x)$ is a code word of the original cyclic code.

Proof Assuming expression (1.3), that is $t(x) = u(x)g(x)$, it is possible to separate, in polynomial $u(x)$, the following two terms

$$u(x) = h(x)u_1(x) + u_o(x),$$

where $deg[u_o(x)]$ is at most $(N - r - 1)$. These two contributions are linearly independent, in fact the parity check polynomial $h(x)$ (see Definition 1.21), has degree equal to $(N - r)$, so the first term is completely determined by a part not superimposed to the second term. Since $h(x)$ and $g(x)$ form a pair of co-factors, their product is $(x^N + 1)$, hence we remain with $u_1(x) = a(x)$ and $t_c(x) = u_o(x)g(x)$. □

The above Lemma has a fundamental importance for the evaluation of the weight distribution which characterizes the code words of our lengthened code, for weights smaller than the minimum distance d_c of the parent cyclic code.

Lemma 1.30 *In the weight distribution of an $[N + y, N - r + y, 2]$ LC code, for $1 \le y \le N$, or $N < n \le 2N$, that is during the first lengthening period, all the code words whose weight is smaller than the minimum distance d_c of the original cyclic code show a weight distribution which obeys to the following binomial law, with just even weights,*

$$A(2) = y, \quad A(4) = \binom{y}{2}, \ldots \quad A(2b) = \binom{y}{b}, \quad 2b = \begin{matrix} d_c - 1, & (d_c\,odd) \\ d_c - 2, & (d_c\,even) \end{matrix}. \tag{1.19}$$

Proof Such expressions enumerate how many possible polynomials $a(x)$ give rise to the same weight $2w[a(x)]$ in (1.18). The statement regarding $A(2)$ for $q = 1$ is included in the upper line of (1.17), considering that $n - N = y$. The statement regarding $A(4)$ takes into account the fact that, if $y = 1$, $A(4) = 0$, whereas, if $y = 2$, $A(4) = 1$, because there is just one possible combination. For higher values of y, combinatorial algebra assures the result, with $w = 4$, but also with higher values of w, if any. Nevertheless, if d_c is equal to 3 or 4, the above expression for $A(4)$ is no longer valid, because there are other code words $t(x) = t_c(x)$ in (1.18), due to the original cyclic code, having such weight. In general, the list of valid expressions is upper bounded by the $2b$-weight coefficient $A(2b)$, where $2b$ equals $(d_c - 1)$ when d_c is odd, $(d_c - 2)$ when d_c is even. □

Example 1.41 Let us consider the BCH code [15,5,7], whose generator polynomial is $g(x) = M_1(x)M_3(x)M_5(x) = x^{10} + x^9 + x^8 + x^6 + x^5 + x^2 + 1$. Table 1.8 shows the first half of the coefficients in the word weight distributions of the $[N + y, N - r + y, 2] = [15 + y, 5 + y, 2]$ lengthened codes obtained from it, in the interval

Table 1.8 Word weight distributions of the lengthened versions of the [15,5,7] BCH cyclic code, over the whole first lengthening period

y	n	$A(0)$	$A(2)$	$A(4)$	$A(6)$	$A(7)$	$A(8)$	$A(9)$	$A(10)$	$A(11)$	$A(12)$	$A(13)$	$A(14)$	$A(15)$	…
0	15	*1*				*15*	*15*							*1*	
1	16	1	1			22	23	8	7					2	
2	17	1	2	1		32	35	24	22	4	3			4	
3	18	1	3	3	1	46	53	54	51	18	15	2	1	8	
4	19	1	4	6	4	65	81	108	104	54	48	12	8	17	
5	20	1	5	10	10	94	121	192	209	140	120	48	32	38	
6	21	1	6	15	20	134	185	330	384	316	393	148	100	94	
7	22	1	7	21	35	184	283	560	685	656	647	384	289	248	
8	23	1	8	28	56	248	438	912	1,176	1,304	1,340	928	776	616	
9	24	1	9	36	84	340	662	1,436	1,982	2,436	2,668	2,124	1,892	1,532	
10	25	1	10	45	120	464	962	2,224	3,292	4,368	5,154	4,528	4,280	3,696	
11	26	1	11	55	165	640	1,370	3,392	5,310	7,552	9,630	9,152	9,338	8,576	
12	27	1	12	66	220	848	1,967	5,956	8,408	12,800	17,248	17,856	19,416	18,976	
13	28	1	13	78	286	1,120	2,763	7,424	13,063	21,120	30,132	33,536	38,836	40,512	
14	29	1	14	91	364	1,472	3,817	10,752	19,922	34,048	51,387	60,928	75,112	83,584	
15	30	*1*	*15*	*105*	*455*	*1,920*	*5,205*	*15,360*	*29,883*	*53,760*	*85,645*	*107,520*	*140,835*	*167,168*	

$1 \le y \le 15$ (whole first lengthening period), in comparison with that of the original cyclic code. This [15,5,7] code and the completely lengthened [30,20,2] code have been evidenced. It is possible to observe the growth in the number of 2-weight, 4-weight and 6-weight code words, according to Lemma 1.30.

A binary parity check code exhibits a word weight distribution (see Lemma D.6) which resembles that expressed by combinatorial coefficients, only for even weights, similar to those collected in (1.19). Keeping in mind this analogy, it is possible to interpret any $[n,n-1,2]$ binary parity check code as a lengthened version of a cyclic parent code [2,1,2], whose generator polynomial is $g(x) = x + 1$. We can say that, if one accepts $N = 1$ as a permitted value for applying the theory of cyclic codes, the whole family of binary parity check codes contains lengthened versions of the same parent cyclic code, but each one of them is obtained after a complete code lengthening process. This means that parameter q will coincide with $(n-1)$, and is confirmed by the fact that any $[n,n-1,2]$ parity check code, with $n > 2$, is cyclic, exactly as the parent code [2,1,2].

In this sense, the strongest analogy appears with the situation described in Lemma 1.28, where the lower line of (1.17), for $N = 1$, degenerates into the well known binomial expression for coefficient $A(2)$ in a binary parity check code. Similar combinatorial coefficients could be obtained for $A(4), A(6), \ldots$. At this point, an interesting question can be opened: in which condition a family of codes, existing only in lengthened versions, but whose number of control symbols r is greater than 1, could be defined? A first answer to such a question is given by what follows.

Definition 1.26 An s.s. time-invariant block code, whose generator polynomial is $g(x) = x^r + 1$, is defined *rth-order generalized parity check code* (or briefly *GPC code*). It is described by a tern of the type $[r + k,k,2]$.

Lemma 1.31 *Expression* (1.18), *for a code word of an rth-order GPC code, reduces to*

$$t(x) = a(x)(x^r + 1). \tag{1.20}$$

Proof In this case the k degrees of freedom in (1.18) are all collected in polynomial $a(x)$, whose degree is at most $(k-1)$. Therefore $u_o(x) = t_c(x) = 0$. □

Lemma 1.32 *For an rth-order GPC code, the following identities hold*

$$\begin{aligned} u(x) &= i(x) = a(x), && k \le r, \\ u(x) &= a(x), && k > r, \end{aligned} \tag{1.21}$$

that is only for $k \le r$ expressions (1.3) *and* (1.6) *coincide.*

Proof Comparison of (1.3) with (1.20) yields $u(x) = a(x)$ for any value of k. On the other hand, if in (1.6) we let

$$i(x) = x^r i_1(x) + i_o(x),$$

for simplicity with $\deg[i_1(x)]$ and $\deg[i_o(x)]$ both smaller than r, it is possible to obtain the following expression for $u(x)$

$$u(x) = Q\left[\frac{x^{2r} i_1(x) + x^r i_o(x)}{x^r + 1}\right] = x^r i_1(x) + [i_1(x) + i_o(x)],$$

which coincides with $i(x)$ only when $i_1(x) = 0$. In the above passages we have considered the trivial property $Q\left[\frac{x^r}{x^r+1}\right] = 1$. The case of $\deg[i_1(x)]$ and $\deg[i_o(x)]$ higher is characterized by an even more extended superimposition, contributing to confirm the difference between $i(x)$ and $u(x)$. □

Lemma 1.33 *An rth-order GPC code, described by the tern $[r + k,k,2]$, if $k \leq r$, exhibits the following word weight distribution*

$$A(w) = \binom{k}{w/2}, \quad w = 2, 4, \ldots 2k. \tag{1.22}$$

Proof Over k positions, associated to polynomial $a(x)$, we can have $\binom{k}{j}$ combinations of j 1-symbols (as occurs in a degenerate code). They originate as many code words having weigh $w = 2j$. The maximum weight is clearly $2k$. □

Either lengthened cyclic codes (LC codes) or their special case, consisting of generalized parity check codes (GPC codes), are not subjected to a precise requirement about the overall length n of their code words (minimum distance remains $d = 2$, although the number of 2-weight code words can increase). Nevertheless the structural characteristics of such codes exhibit a periodic behaviour with increasing k.

Definition 1.27 After any period (whose length is N for LC codes, r for rth-order GPC codes), we have a cyclic replica of our code. This period is called *lengthening period.* The first lengthening period ranges between N and $2N$ for LC codes, between r and $2r$ for GPC codes.

Lemma 1.34 *Along the first lengthening period, a GPC code shows some central positions in its code words forced to exhibit 0-symbols.*

Proof In Fig. 1.12 the generator matrix $\underline{G}$ of a GPC code is schematically shown, in the two cases $k < r$ or $k > r$. It is possible to observe that, when $k < r$, there are precisely $(r - k)$ central positions, where 1-symbols cannot be present, whichever superimposition of rows one chooses. □

As already suggested for strongly shortened codes, where a similar bad feature might be present, it is always possible to perform a proper elimination of these positions, reducing the new code to an r'th-order GPC code, with $r' = k$, and with all

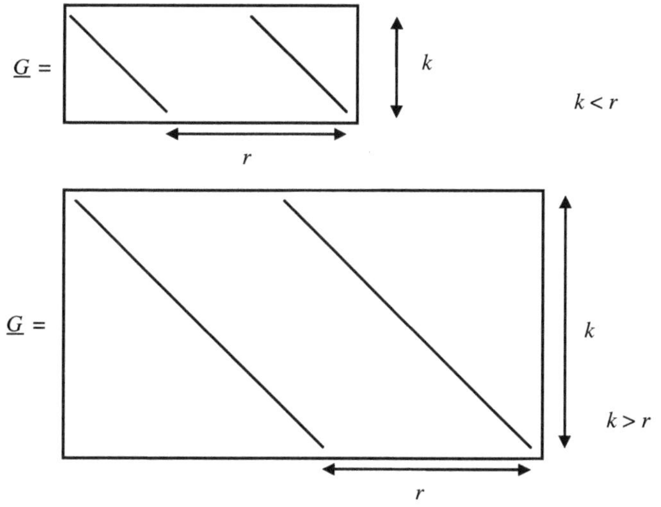

Fig. 1.12 Two examples of $\underline{G}$ matrix for GPC codes

the positions equally able to exhibit either 0- or 1-symbols. Nevertheless, in many situations, this is not advisable, since GPC codes represent only an efficient tool for interpreting the behaviour of more complicated code schemes, based on them.

In spite of their similitude, LC codes and GPC codes show also a remarkable difference, especially in their first lengthening period. A more interesting comparison can be made over a length of several lengthening periods, and also taking $r = N$. An LC code appears better, because, although characterized by the same word weight distribution in coefficient $A(2)$, it exhibits a higher code rate. Nevertheless GPC codes are interesting if they are employed as component codes in particular concatenation schemes. Their regular structure allows one to predict rather accurately the performance of the composite code so obtained.

Let us consider the following pair of comparable situations: $qN < n < (q + 1)N$ for an LC code and $qr < n < (q + 1)r$ for a GPC code. As already stressed, the codes characterized by $n = (q + 1)N$ and $n = (q + 1)r$ could be considered cyclic codes. Therefore our LC and GPC code can be interpreted as shortened versions of such cyclic codes respectively. The parity check polynomial $h(x)$, which can be identified also here as the co-factor of the generator polynomial, in the two cases, then becomes

$$h(x) = \frac{x^{(q+1)N} + 1}{g(x)}, \quad \text{(LC codes)}, \tag{1.23}$$

$$h(x) = \frac{x^{(q+1)r} + 1}{x^r + 1}, \quad \text{(GPC codes)}. \tag{1.24}$$

Lemma 1.35 *The parity check polynomial of an LC code or of a GPC code, characterized by lengths n within the qth lengthening period, is periodic, and can be written as follows*

$$h(x) = h_c(x)u_{q+1}(x^N), \quad \text{(LC codes)}, \tag{1.25}$$

where $h_c(x)$ *is the parity check polynomial of the cyclic parent code, or*

$$h(x) = u_{q+1}(x^r), \quad \text{(GPC codes)}. \tag{1.26}$$

Proof First we will face (1.26), which can be easily demonstrated by observing that in (1.24) it is possible to write $x^{(q+1)r} + 1 = (x^r + 1)u_{q+1}(x^r)$. In analogous way, assuming $g(x)\, h_c(x) = x^N + 1$, (1.25) follows from (1.23), taking into account that $x^{(q+1)N} + 1 = (x^N + 1)u_{q+1}(x^N)$. □

Example 1.42 Given the [7,4,3] Hamming code, with $g(x) = x^3 + x + 1$, and $h_c(x) = x^4 + x^2 + x + 1$, its lengthened version, up to the third lengthening period (q = 3), is characterized by parity check polynomial

$$h(x) = (x^4 + x^2 + x + 1)(x^{21} + x^{14} + x^7 + 1).$$

Definition 1.28 In the following, in order to identify cyclic codes, together with their shortened or lengthened versions, including them in a unique group, we will classify all such codes as *polynomial codes*. Since the generator polynomial is unique (in more general types of codes we can have different generator polynomials), we will precise this fact with the specification *s.s. time-invariant polynomial codes* in their $\underline{G}$ matrix (see also Definition 1.8).

1.12 Subcode of an s.s. Time-Invariant Polynomial Code

In Sect. 1.6, speaking about a direct serial concatenation of two block codes, we have introduced the definition of subcode of a linear block code (Definition 1.15). Here this definition will be particularized for s.s. time-invariant polynomial codes in $\underline{G}$.

Definition 1.29 Starting from an $[n,n - r,d]$ block code, whose code words are described by polynomial expression (1.3), we can obtain a subcode from it, by imposing some constraints to polynomial $u(x)$, maintaining the overall code length n unchanged. This *polynomial subcode* is still a linear code, with a reduced number of binary degrees of freedom, controlled by information symbols in $u(x)$.

Example 1.43 A primitive binary BCH code is a subcode of the correspondent Hamming code. In fact its generator polynomial is subjected to the constraint of having further irreducible factors in it. This feature can be interpreted also maintaining the same $g(x)$ and transferring the constraint to $u(x)$. For instance, the [15,7,5] BCH code characterized by $g(x) = g_H(x)(x^4 + x^3 + x^2 + x + 1)$, where $g_H(x)$ represents the generator polynomial of the Hamming code, shows code words which can be also written as

$$t(x) = u(x)g(x) = u'(x)g_H(x), \quad u'(x) = (x^4 + x^3 + x^2 + x + 1)u(x).$$

Definition 1.30 Given an odd-distance polynomial code, described by the tern $[n,k,d]$, multiplying its generator polynomial by $(x + 1)$, we obtain a so called *even-distance subcode* of the original code. It is described by the tern $[n,k - 1,d + 1]$.

Since the original code exhibits an odd minimum distance d, in its generator polynomial $g(x)$, the factor $(x + 1)$ is certainly absent. Multiplication of $g(x)$ by $(x + 1)$ will produce an increase in d to the even number immediately greater, $(d + 1)$, at the expense of a reduction, by just one unit, in the number k of the binary degrees of freedom present in $u(x)$. Hamming codes and BCH codes are characterized by interesting forms of even-distance subcodes. This possibility was already presented in Example 1.30, considering the case of $N = 15$.

Example 1.44 The [7,3,4] simplex code can be also interpreted as an even-distance subcode of the [7,4,3] Hamming code. This property holds just for $N = 7$.

Example 1.45 The [15,4,8] simplex code turns out to be also an even-distance subcode of the [15,5,7] BCH code. This property holds just for $N = 15$.

It is interesting to observe how, in a direct serial code concatenation, it is preferable to adopt expression (1.6) for describing a code word, and interpret the resulting overall code as a subcode of the inner code. On the contrary, here, considering a further factor added to the original $g(x)$, it is preferable to employ expression (1.3). In the latter situation, once modified $g(x)$, it is possible to still represent the code word $t(x)$ by means of the systematic expression (1.6). Instead, in the former case, to adopt a non-systematic expression like (1.3) is no longer possible, because a concatenated code generally is not an s.s. time-invariant polynomial code.

Another important way for obtaining a subcode of a certain s.s. time-invariant polynomial code is by selecting the times in which the source emits information symbols. This entails that polynomials $g(x)$ in the $\underline{G}$ matrix are still all equal, but they are not always separated by just one symbol time. As a consequence, control symbols are alternated with information symbols (if the code is systematic) or are alternated with the times in which the information source emits symbols (if it is non-systematic). This concept will be faced with more details in the next section.

Clearly a subcode is constructed from a certain original code, in order to increase its minimum distance, or, at least, to reduce the number of minimum-weight code words. Remark that a shortened code, although characterized by code words which, in their 1-symbol patterns, are the same as those of the original code, cannot be considered a subcode of it, because their lengths are different.

1.13 Modified Lengthened Cyclic (MLC) Codes

In order to insert distributed control symbols in a code word, it is necessary to force polynomial $u(x)$ to have 0-symbols in some prefixed positions. The resulting generator matrix $\underline{G}$ will show rows in which the same polynomial $g(x)$ shifts, but this is made occupying not all the possible positions, that is with a $\underline{G}$ matrix no longer upper triangular. In Fig. 1.13 two examples of such solution are reported: the first with one control symbol any two, and the second with two control symbols any three. Dashed small arrows denote the positions where control symbols will be inserted, and solid small arrows denote the positions devoted to information symbols (directly present if the encoder circuit is systematic, properly transformed if it is not). It is useful to remind the property $g_r = g_0 = 1$, which the generator polynomial coefficients must satisfy in order to assure a precise degree for $g(x)$ itself. Possible widening of this concept to more general situations, even non-periodic, can be easily imagined. Clearly the number of consecutive control symbols cannot be greater than r.

If the encoder circuit is non-systematic, the scheme of Fig. 1.3 can be still employed, with the only care of regulating the switch in such a way as to disconnect the information source in correspondence with the symbol times where a control symbol is to be inserted. The length of the code word may vary as one wishes. In order to increase the overall code rate, it is preferable to have a block length n larger than N, value which characterizes the cyclic version of our code. In this sense, such type of codes can be considered as derived from lengthened cyclic codes.

Definition 1.31 A block code, characterized by the generator polynomial $g(x)$, with block length $n > N$, for which expression (1.3) holds, but polynomial $u(x)$ is

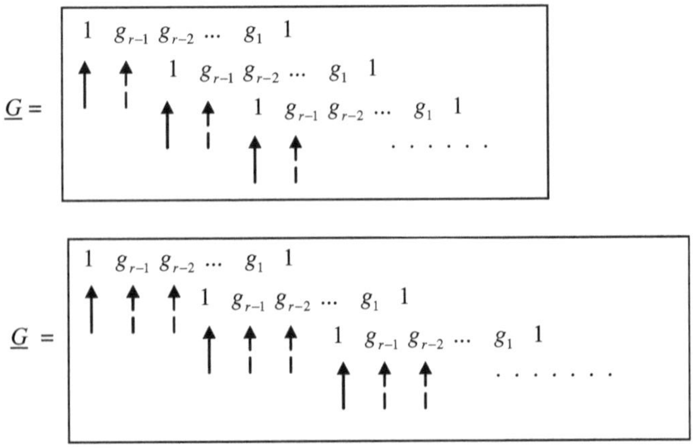

Fig. 1.13 Two examples of generator matrix in which periodically distributed control symbols have been inserted

constrained to show 0-symbols in proper positions, is called *modified lengthened cyclic code* (or briefly *MLC code*). It is a subcode of the correspondent LC code.

In order to have a well predictable behaviour, it is convenient to adopt a period, whose length is n_o, in which the first k_o positions are devoted to information symbols, and the remaining $r_o = n_o - k_o$ to control symbols. The choice $r_o \geq r$ is clearly useless, because it would mean to have separate code words.

Lemma 1.36 *The total length n of a code word in a periodic MLC code characterized by parameters r,* n_o*,* r_o*, is*

$$n = Zn_o + (r - r_o), \tag{1.27}$$

where Z is the number of periods where information symbols are inserted.

Proof The Zth period exhibits firstly k_o information symbols and then r_o control symbols, so the remaining control symbols taken from the last polynomial $g(x)$ are $(r - r_o)$. □

The entire last polynomial $g(x)$ is to be completely reproduced in order to correctly express (1.3). Only in this way, it is possible to have a uniform behaviour, in any part of the code word. Nevertheless sometimes few positions can remain with a 0-symbol forced to be present in them. In this case they can be eliminated, and (1.27) becomes an over-estimation of n by few units.

Lemma 1.37 *The code rate of a periodic MLC code, in the conditions of Lemma 1.36, results in*

$$R = \frac{Zk_o}{Zn_o + (r - r_o)}. \tag{1.28}$$

Proof The total number of information symbols is $k = Zk_o$. Then we calculate the ratio $R = k/n$, taking n from (1.27). □

A non-periodic MLC code can be classified as a time-invariant polynomial code in its $g(x)$, but not as an s.s. time-invariant code in the whole $\underline{G}$ matrix, because of a non-uniform behaviour along its length. On the contrary, a periodic MLC code is characterized by both the above properties. In the following, except when the contrary is expressly said, we will consider only periodic MLC codes.

Example 1.46 The [31,26,3] Hamming code, whose generator polynomial is $g(x) = x^5 + x^4 + x^3 + x + 1$, originates an MLC code, with $n_o = 2$, $k_o = 1$. This code exhibits the evolution of word weight distribution (with w up to 9), listed in Table 1.9, with increasing the number of periods Z in the code word (whose overall length n is also reported in the second column). Clearly, the minimum distance is $d = 5$. So the occurrence of 2-weight code words, and even of 4-weight code words, characteristic of the lengthened code, has been avoided. The increase in the

Table 1.9 Evolution of the word weight distribution for an MLC code derived from the [31,26,3] Hamming code

Z	n	$A(0)$		$A(5)$	$A(6)$	$A(7)$	$A(8)$	$A(9)$	...
1	6	1		1					
2	8	1		2	1				
3	10	1		3	3	1			
4	12	1		4	5	4	1		
5	14	1		5	7	8	5	1	
6	16	1		6	9	12	12	4	
7	18	1		7	11	16	20	17	
8	20	1		8	13	20	28	32	
9	22	1		9	15	24	36	48	
10	24	1		10	17	28	44	64	
11	26	1		11	19	32	52	80	
12	28	1		12	21	36	60	96	

coefficients $A(w)$ with small w values appears to be a linear function of the number of periods Z. The proportionality factors grow exponentially with the difference $(w - d)$.

Example 1.47 Let us consider now the repetition code [4,1,4] whose generator polynomial is $g(x) = x^3 + x^2 + x + 1$. Assuming here $n_o = 3$, $k_o = 2$, we obtain an MLC code, which shows the evolution of word weight distribution (with w up to 12) listed in Table 1.10, with increasing the number of periods Z in the code word. Also the overall code word length n and the number of information symbols k are reported in this Table. We have only even weights, as expected. It is possible to observe that here 2-weight code words remain, although one for each period, and without possibility of a binomial growth of $A(2)$ with increasing Z. This binomial growth is instead present in $A(4)$, and even a stronger growth law characterizes all the other coefficients. For $n = 8$ there is a symmetric word weight distribution.

There is also the possibility of constructing a systematic version for MLC codes. To this purpose, instead of the encoder circuit shown in Fig. 1.3, we can employ the encoder circuit depicted in Fig. 1.5. Also in this case, we have to modify the sequence of operations performed by the switch: there are Z periods in which, during the first k_o symbol times, it is connected to the information source, and during the remaining r_o symbol times to the feed-back, finally it maintains connection with the feed-back for the last $(r - r_o)$ symbol times. Any symbol time in which the switch is connected to the feed-back, a 0-symbol is inserted in the sequence representing $u(x)$.

Polynomial $u(x)$, in expression (1.3), taking into account also the constraint imposed to it by the MLC code word selection, can be interpreted as the evolution

Table 1.10 Evolution of the word weight distribution for an MLC code derived from the [4,1,4] repetition code

Z	n	k	A(0)	A(2)	A(4)	A(6)	A(8)	A(10)	A(12)	...
1	5	2	1	1	2					
2	8	4	1	2	10	2	1			
3	11	6	1	3	23	23	12	2		
4	14	8	1	4	38	84	89	32	8	
5	17	10	1	5	54	187	370	267	118	
6	20	12	1	6	71	324	999	1,322	937	
7	23	14	1	7	89	491	2,045	4,325	4,921	

of a state co-ordinate, which passes without changes through all the memory elements of the shift register, because it is organized in controller arrangement (see Appendix C). This property holds for either the non-systematic or for the systematic encoding procedure, exactly as occurs for a cyclic, shortened or lengthened code.

As for expression (1.6), it is not easy to find a condition on $i(x)$ valid for any combination of parameters k_o and r_o. In the particular case characterized by $n_o = 2$ and $k_o = r_o = 1$, we can observe that in polynomial $t(x)$ it is possible to distinguish odd and even positions, according to

$$t(x) = xi_o(x^2) + c_i(x^2), \tag{1.29}$$

where $i_o(x^2)$ contains the true information symbols, whereas $c_i(x^2)$ collects the distributed control symbols. The last polynomial can be constructed, taking the remainder of the polynomial division, one symbol any period. Finally (1.29) must be properly terminated by the addition of a certain number of final control symbols.

1.14 State Diagrams

Definition 1.32 In an encoder circuit like those depicted in Figs. 1.3 or 1.5, the contents of the r memory elements of the shift register form an r-symbol sequence, called its *state*. There are 2^r states on the whole. Each one of the r positions of the state is called *state co-ordinate.*

The evolution of the state, in a cyclic, shortened, lengthened code, and even in an MLC code, can be interpreted by means of a proper diagram, called the state diagram. It is a binary De Bruijn graph (1946), with 2^r nodes and 2^{r+1} branches, showing all the possible transitions from one state to another. The code words are represented by closed routes starting from and coming back to state $\underline{0}$. Remark that one state co-ordinate is enough for describing the state evolution, because all the state co-ordinates are characterized by the same state sequence, although time

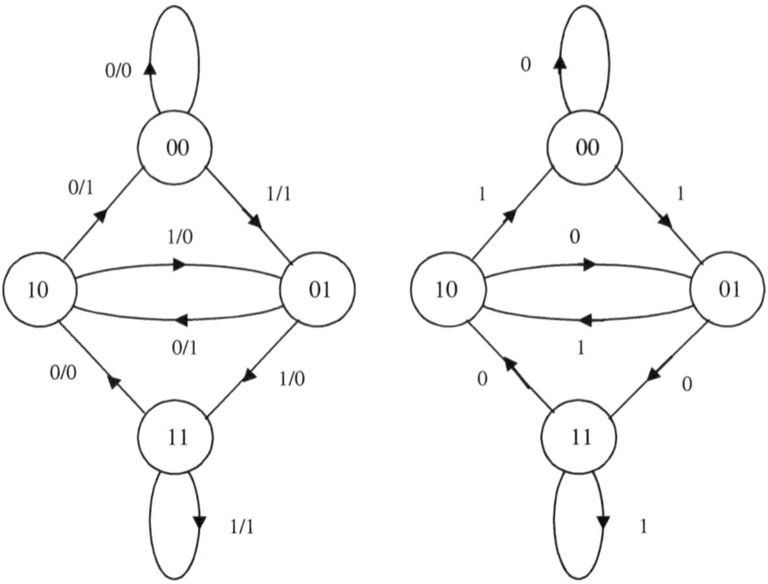

Fig. 1.14 State diagrams (non-systematic and systematic encoder), with $g(x) = x^2 + x + 1$

shifted. This unique state evolution is represented by polynomial $u(x)$, owing to the encoder structure, based on a shift register in controller arrangement. It is useful to remember that this property is no longer true for the encoder circuit shown in Fig. 1.6, where a polynomial remainder, instead of a polynomial quotient, is calculated.

Branch labelling is made in different ways, depending on the encoder circuit considered. If it is non-systematic, conventionally the input symbol and the correspondent output symbol are both represented, separated by a slash. If the encoder circuit is systematic, input and output symbol coincide, therefore it is enough to represent such unique symbol just one time.

Example 1.48 Taking the generator polynomial $g(x) = x^2 + x + 1$, which originates the [3,1,3] repetition cyclic code, as well as some lengthened or MLC versions of such code (code shortening is impossible), its state diagram is shown in Fig. 1.14. On the left we have the non-systematic form, on the right the systematic form. In the first, the convention of the type (input symbol)/(output symbol) has been adopted in labelling the branches. Here, the input symbol coincides with the entering state co-ordinate. Therefore the last two dummy input symbols, necessary for coming back to state $\underline{0}$, after the true information symbols, will be 0 and 0, since they directly fill the shift register. In general, r subsequent 0 input symbols assure satisfaction of this requirement. In the scheme on the right, such two last symbols are calculated by the circuit scheme of Fig. 1.5, after the final change in the position of the switch (for r = 2 last symbol times connected to the feed-back).

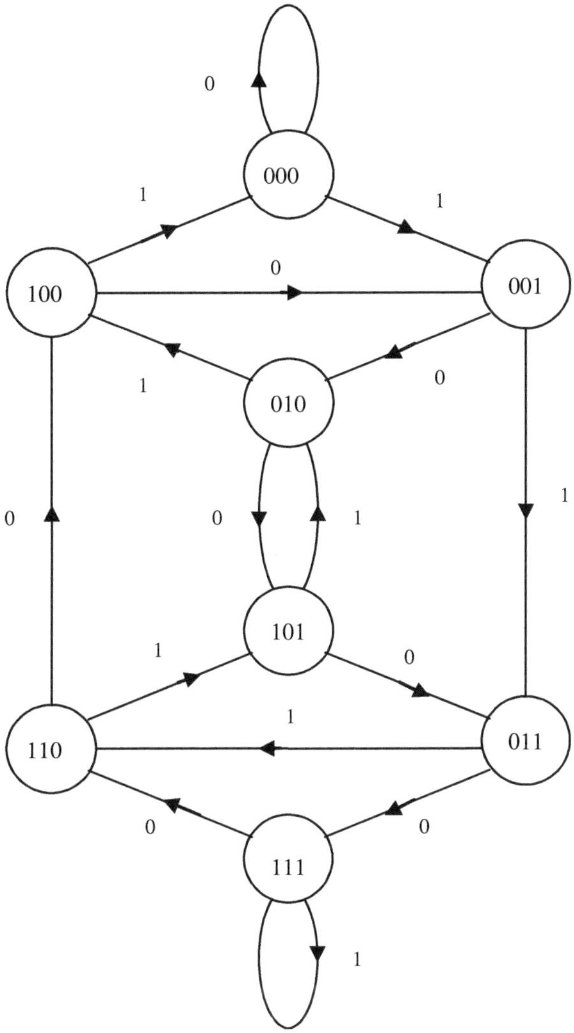

Fig. 1.15 Systematic state diagram for the polynomial codes associated to $g(x) = x^3 + x + 1$

All the structural properties of the code associated to a certain state diagram typically regard its output symbols, therefore in the following, when the contrary is not explicitly said, we will refer to the systematic version of state diagrams only. Output symbols denoting branches incoming in or outgoing from the same node are different.

Example 1.49 In Fig. 1.15 the systematic state diagram of the family of codes characterized by $g(x) = x^3 + x + 1$ is represented. It is associated to the codes deriving from the [7,4,3] Hamming cyclic code. It is possible to observe that, also here, two one-step cycles (closed-loop branches) are present, the first around state $\underline{0}$, the other around state $\underline{1}$. This property has general validity, although labelling of these closed-loop branches not always is like in the two cases treated so far. Precisely, state $\underline{0}$ always exhibits a 0-labelled closed-loop branch, whereas state $\underline{1}$ is characterized by a 0-labelled closed-loop branch too if $g(x)$ contains factor $(x + 1)$. This condition has effects also on the symmetry property of the state diagram with respect to a horizontal axis. On the contrary, the symmetry property with respect to a vertical axis is imposed by the symmetry characteristic of $g(x)$: if we replace $g(x)$ with its reciprocal, the state diagram will be subjected to a left-right mirror transformation in branch labelling.

As general comments, after the above two Examples, we can stress what follows. Any code word is represented by a closed route starting from and arriving at state $\underline{0}$, but a sequence of 0-symbols before or after this route is admitted, leading to as many shifted replicas of the same 1-symbol output pattern. In a cyclic, shortened or lengthened code, we have k steps imposed by information symbols, and then r steps for coming back to state $\underline{0}$. In MLC codes, the first part of the route shows an alternation between k_o steps imposed by information symbols and r_o steps in which coming back to state $\underline{0}$ just begins. At the end, state $\underline{0}$ is finally reached with r subsequent steps.

The shortest route around state $\underline{0}$, excluded the closed-loop branch, exhibits an $(r + 1)$-length symbol sequence representing $g(x)$. All the other routes are associated to as many linear combinations of it, according to (1.3), where $u(x)$ represents the state evolution before coming back to state $\underline{0}$. This property suggests to perform the enumeration of the output patterns characterized by lower weights directly in the state diagram. Sometimes this procedure is rather simple, as will be shown later.

There is a closed route, not comprehending state $\underline{0}$, whose length is N, characterized by N subsequent 0-symbols. After addition of two external 1-symbols, able to allow route beginning from and ending at state $\underline{0}$, we obtain the 2-weight code words of the lengthened code whose expression is $(x^N + 1)$.

It is convenient to adopt, as state identification, the sequence of r state co-ordinates employed for denoting the correspondent node in the state diagram. The state identifications of subsequent nodes along a route vary as follows: the leftmost symbol is eliminated, a new rightmost symbol is introduced, the remaining central symbols shift from right to left.

Lemma 1.38 *Not all the nodes of the state diagram are always touched by the routes representing code words. In a shortened, cyclic of even lengthened code, if we have $k < r$, the number of untouched nodes will be*

$$N_{ut} = \sum_{j=0}^{r-k-1} 2^{k-1+j}(r - k - j). \qquad (1.30)$$

They correspond to the states whose identification is characterized by a sequence of binary symbols where two external 1-symbols are separated by at least $(k - 1)$ symbols (either 0 or 1).

O	X	X	O

X	X	O	X

X	O	X	X

Fig. 1.16 Masks for the permitted states in the case of $r = 4$, with $k_o = 2$ and $r_o = 1$

Proof In the condition here treated, the state identification has a length r longer than the non-null sequences that $u(x)$ can represent. So the untouched states exhibit state identifications characterized by 1-symbol patterns, whose length is at least $(k + 1)$, and all 0-symbols in external positions. The shortest among these 1-symbol patterns has length exactly $(k + 1)$ and hence exhibits 2^{k-1} multiplicity, as the number of all possible different binary sequences of length $(k - 1)$. Remark that the two external 1-symbols are fixed. Furthermore it can shift in $(r - k)$ subsequent positions. Next we have a 1-symbol pattern whose length is $(k + 2)$, exhibits $2^{k-1+1} = 2^k$ multiplicity, and can shift in $(r - k - 1)$ subsequent positions. The longest 1-symbol pattern has length r, exhibits 2^{r-2} multiplicity, and can exist in just 1 position. The sum of all these contributions is collected in (1.30). There are no common terms, because the various groups of sequences are all distinct, since they are distinguished by a pair of external 1-symbols, whose separation progressively increases. □

Example 1.50 Assuming $r = 4$, $k = 2$, and the [6,2,4] cyclic code characterized by the generator polynomial $g(x) = x^4 + x^3 + x + 1$, the following states are untouched: 0101, 1010, 0111, 1110, 1001, 1011, 1101, 1111. Expression (1.30) gives two contributions: the first associated to a pair of 1-symbol patterns having length 3 (101, 111), in two possible shifted positions, and the second associated to four sequences having length 4 (1001,1011,1101, 1001), in a unique possible position. The total number is $N_{ut} = 4 + 4 = 8$, according to Lemma 1.38.

In an MLC code, untouched states are more difficult to be determined. Nevertheless periodicity can help. It is possible to imagine a mask shifting along the r memory elements of the shift register, or along the r co-ordinates of the state identification. For instance, assuming $r = 4$, with $k_o = 2$ and $r_o = 1$, this mask allows the only three types of permitted (touched) states shown in Fig. 1.16, where character X may represent either a 0-symbol or a 1-symbol, whereas character O always represents a 0-symbol.

An MLC code produces a strong selection in the permitted states, and hence in the 1-symbol output patterns forming low-weight code words. The enumeration of such patterns can be performed in an exact way, after direct inspection of the state diagram, to the point where the weight distribution coefficients $A(w)$, for w one or two units greater than the minimum distance d, can be accurately predicted. See, for instance, the coefficients $A(w)$ listed in Table 1.9 (Example 1.46). Coefficients $A(5)$ and $A(6)$ are obtained respectively from the unique output 1-symbol pattern 111011 of weight 5, and from the couple of output 1-symbol patterns 11010111, 1110001011 of weight 6. The first one is associated to $g(x)$ itself, the remaining two are associated to $(x^2 + 1)g(x)$ and to $(x^4 + 1)g(x)$ respectively. Their contributions to $A(5)$ and $A(6)$ produce an increase which is linear with the number of periods Z, by one unit per period in $A(5)$ and by two units per period in $A(6)$. Inspection of the state diagram of this code, taking into account the selection in the permitted states due to the choice $k_o = r_o = 1$, enables one to easily predict this behaviour, as well as those of all the other possible 1-symbol patterns characterized by at most $w = 9$. On the contrary, for $w \geq 10$, multiple passages through state $\underline{0}$ give rise to a binomial increase with Z and this analysis becomes less easy. Also the delay after which any linear growth starts, which depends on the length of the considered patterns, and hence on the degree of polynomial $u(x^2)$ originating them, can be exactly predicted.

All these properties are consequence of the adoption of an encoder circuit of the type shown in Fig. 1.5, in which the quotient of the division by $g(x)$ is calculated and expressed as the state evolution $u(x)$. The overall degree of $u(x)$ makes the code shortened, cyclic or lengthened, and its power coefficient selection describes the choice adopted in a possible MLC scheme. All these codes are s.s. time-invariant in their $\underline{G}$ matrix, since in any case they are constructed employing a unique generator polynomial $g(x)$, periodically shifting in its rows. In the next section this constraint will be removed.

Coming back to the general properties of a state diagram, either systematic or non-systematic, as already noted, when $(x + 1)$ is factor of $g(x)$, the closed-loop branch characterizing state $\underline{1}$ is 0-labelled (in output if the state diagram is non-systematic). This behaviour can be justified by the following considerations. Letting $g(x) = (x+1)g_o(x)$, any product with a unitary polynomial of weight z, $u_z(x)\,g(x) = (x^z+1)g_o(x)$, with $z \geq r$, is a word of our code (if N is not high enough, it is possible to take a lengthened version). This code word is formed as follows: state $\underline{1}$ is reached, after the state evolution consisting of $(r - 1)$ consecutive 1-symbols, corresponding to the output sequence $g_o(x)$, then in such state we remain for $(z - r)$ symbol times, and finally we will come back to state $\underline{0}$, after the same output sequence. The following noticeable result generalizes the above observation.

Lemma 1.39 *When, in the generator polynomial $g(x)$, a factor $p(x)$ is present, the state diagram shows a closed route, made by L consecutive branches, all 0-labelled in output, where L is factor of the exponent M in a proper binomial (x^M+1), in whose decomposition $p(x)$ is present.*

Proof It is possible to write $g(x) = p(x)g_o(x)$. Assuming $\bar{p}(x)$ as the co-factor of $p(x)$, with respect to a certain binomial, we can take this polynomial for representing the state evolution needed in order to reach our L-length 0-labelled closed route. In fact we can write $p(x)\,\bar{p}(x) = x^M + 1$. Then a product of the type

$$\begin{aligned} u_z(x^M)\bar{p}(x)\,g(x) &= [\bar{p}(x) + x^M\bar{p}(x) + \cdots + x^{(z-1)M}\bar{p}(x)]\,p(x)g_o(x) \\ &= (x^{zM}+1)g_o(x) \end{aligned}$$

is a word of our code, or of a lengthened version obtained from it. When N is prime and $g(x)$ is irreducible, we can have only closed routes of length N. Finally the state evolution for reaching our closed loop will be $u(x) = \bar{p}(x)$ or a possible delayed

Fig. 1.17 Systematic state diagram of the [4,1,4] repetition code

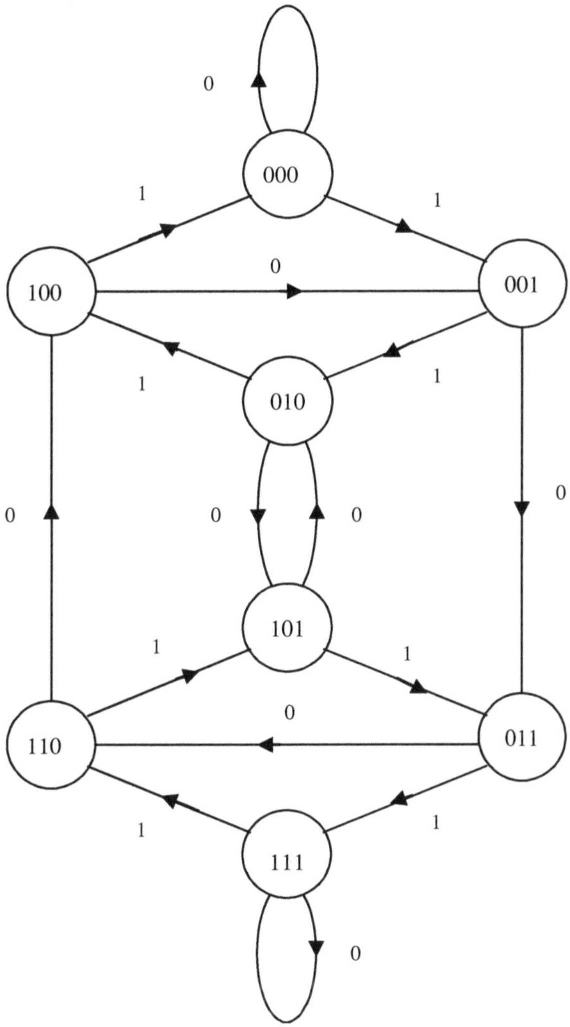

version $x^j\bar{p}(x)$. This confirms that, for $p(x) = x + 1$, the binomial is characterized by $M = r$, and the 0-labelled closed-loop branch is around state $\underline{1}$. □

Example 1.51 In Fig. 1.17 the systematic state diagram of the [4,1,4] repetition code is depicted. Its generator polynomial is $g(x) = x^3 + x^2 + x + 1 = (x + 1)(x^2 + 1)$. In such state diagram we can recognize the 0-labelled closed route of length 4 characterizing 2-weight code words of the type $(x^4 + 1)$ in the lengthened code. Then there is a 0-labelled closed route of length 2, associated to $g_o(x) = x + 1$, so $p(x) = x^2 + 1, \bar{p}(x) = 1$, and hence $L = M = 2$, with the code words $(x + 1)(x^4 + 1)$, $(x + 1)(x^6 + 1)$, … . The state evolution for reaching this closed route is $u(x) = x$. Finally there is the 0-labelled closed-loop branch around state $\underline{1}$, associated to factor $g_o(x) = x^2 + 1$, for which $p(x) = x + 1, \bar{p}(x) = x^2 + x + 1$, so $M = r = 3$, but $L = 1$, and the state evolution for reaching this 0-labelled closed-loop branch is $u(x) = x^2 + x + 1$. The last observation shows that condition $L = 1$ describes the already treated case of a $g(x)$ containing factor $(x + 1)$.

1.15 Direct Product Codes

In Sect. 1.6 we gave an anticipation about the possibility of performing code concatenation taking more than one code word from the first component code, before the action of the second component code. This scheme can be arranged in such a way as to maintain the two codes at the same level (neither inner nor outer), but with the overall code being a subcode of a proper ensemble of code words taken from the one code (or from the other). Such particular scheme of code concatenation exploits a sort of orthogonal relation between the two component codes.

Definition 1.33 Assuming a rectangular memory, with n_2 columns and n_1 rows, let us consider two linear block codes, respectively $[n_1,k_1,d_1]$ and $[n_2,k_2,d_2]$. The first will act on the rows, and the second on the columns, like shown in Fig. 1.18 (Elias 1954). The resulting $[n_1n_2, k_1k_2, d]$ code is defined *direct product code* between such two block codes.

In Fig. 1.18, the area occupied by control symbols is evidenced by means of a grey colour. The overall number of control symbols is $r = k_1r_2 + k_2r_1 + r_1r_2$, where we have set $r_1 = n_1 - k_1$ and $r_2 = n_2 - k_2$. The last term in the expression of r represents the so called controls on controls (MacWilliams and Sloane 1977). Remark that we have taken s.s. systematic forms for the two component codes, in order to simplify understanding, but the same result can be obtained also for w.w. systematic forms (see Definition 1.5), or even for non-systematic forms.

If we encode firstly along the upper k_1 rows, as many $[n_2,k_2,d_2]$ code words, and then along all the n_2 columns, as many $[n_1,k_1,d_1]$ code words, the remaining r_1 rows turn out to be correctly encoded as well. In fact linearity assures that an ex-OR sum of code words taken from a certain code is also a code word for the same code.

As a consequence of the above property, the roles of the two component codes are perfectly interchangeable. This fact is due to a sort of orthogonal property between the trajectories (horizontal or vertical) along which the two component codes operate. In principle, we could consider also a scheme for the encoder circuit of the type depicted in Fig. 1.8, where a pair of encoder circuits, in serial concatenation with interleaver, are employed. We would have to add only the specification that more than one code word are recorded in the rectangular interleaver. It has a size of $n_ok_i = n_1k_2$, is written by rows and encoded firstly by rows and then by columns. Even if the trajectories of the second component code are tilted by a certain angle with respect to the orthogonal direction, we obtain an equivalent of the direct product code described so far. Nevertheless it is preferable to stress the fact that the two component codes have an equivalent role. The conceptual scheme of

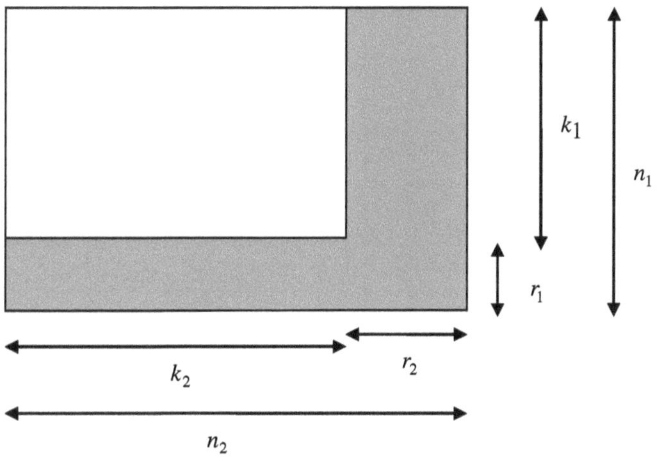

Fig. 1.18 Rectangular memory for encoding a direct product code

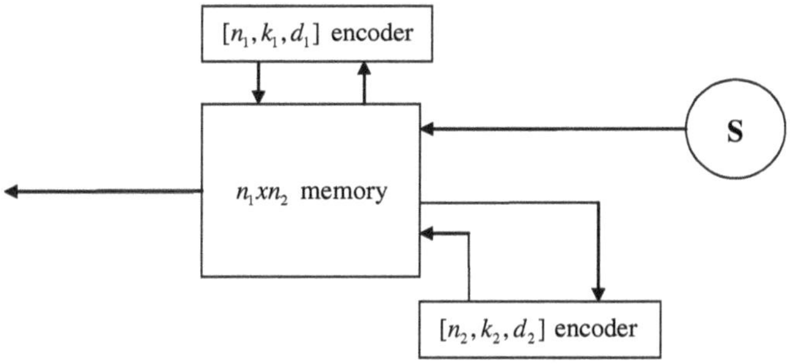

Fig. 1.19 Conceptual scheme of encoder circuit for a direct product code

encoder circuit most suited for this interpretation is then of the type depicted in Fig. 1.19. The definition direct product remarks that there is no interleaver, whose purpose would be to scramble the symbol sequence formed by the set of code words from one of the two codes, before being encoded by the other one, so leading to a different code scheme (product code with interleaver). In the following we will restrict our attention to a direct product code only.

Theorem 1.3 *Starting from the two word weight distributions* $A_1(w_1)$ *for the* $[n_1,k_1,d_1]$ *code and* $A_2(w_2)$ *for the* $[n_2,k_2,d_2]$ *code, the coefficients associated to weights* $w = w_1 w_2$ *in the direct product code between such two codes satisfy the condition*

$$A(w) \geq \sum_{S(w)} A_1(w_1) A_2(w_2), \quad (w = w_1 w_2), \tag{1.31}$$

where S(w) represents the ensemble of events leading to a code word having weight w.

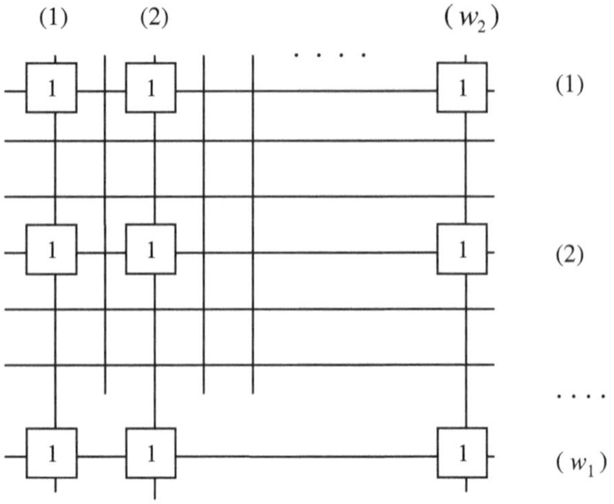

Fig. 1.20 Grid containing 1-symbols produced by two code words with weights w_1 and w_2

Proof In Fig. 1.20 the grid of 1-symbols produced by two code words, having weights w_1 and w_2 respectively, is shown. Such situation leads to a code word of the resulting direct product code with overall weight $w_1 w_2$. This is not the only possible mechanism for generating code words in the direct product code, but certainly it exists and gives a contribution $A_1(w_1)A_2(w_2)$. We have to enumerate all the events of this type, for a given $w = w_1 w_2$, varying the two factors w_1 and w_2, which are symbolically collected in the ensemble $S(w)$. Nevertheless there are situations not included in the model depicted in Fig. 1.20, justifying the inequality in (1.31). □

Expression (1.31) can be employed with increasing complication, as long as w grows. Furthermore there are also words, in the resulting direct product code, whose weight w is not given by a product $w_1 w_2$. Nevertheless small values of w can be easily analyzed. To this purpose, we have to note the following very interesting property.

Corollary a *The minimum distance d of the direct product code between two codes, characterized by minimum distances d_1 and d_2, is $d = d_1 d_2$, and $A(d) = A_1(d_1)\ A_2(d_2)$.*

Proof The two minimum weights are $w_1 = d_1$ and $w_2 = d_2$. Situations different from the one described in Fig. 1.20 cannot be present for the case of $w = d_1 d_2$. Furthermore a possible lower weight cannot occur in the resulting direct product code. The minimum value for w is then $d = d_1 d_2$. Therefore (1.31) has just one term, and holds with equality sign. □

Example 1.52 Let us take two [3,2,2] parity check codes, as component codes in a [9,4,4] direct product code. The 16 code words in this direct product code show the following word weight distribution: $A(0) = 1$, $A(4) = 9$, $A(6) = 6$. There are nine situations of the type shown in Fig. 1.20. They are reported in the upper row of Fig. 1.21, so verifying the result $A(4) = 9$, predicted by Corollary a. The remaining

$$\begin{array}{|c|}\hline 110\\ 110\\ 000\\ \hline\end{array}\;\begin{array}{|c|}\hline 110\\ 000\\ 110\\ \hline\end{array}\;\begin{array}{|c|}\hline 101\\ 101\\ 000\\ \hline\end{array}\;\begin{array}{|c|}\hline 101\\ 000\\ 101\\ \hline\end{array}\;\begin{array}{|c|}\hline 011\\ 011\\ 000\\ \hline\end{array}\;\begin{array}{|c|}\hline 011\\ 000\\ 011\\ \hline\end{array}\;\begin{array}{|c|}\hline 000\\ 110\\ 110\\ \hline\end{array}\;\begin{array}{|c|}\hline 000\\ 101\\ 101\\ \hline\end{array}\;\begin{array}{|c|}\hline 000\\ 011\\ 011\\ \hline\end{array}$$

$$\begin{array}{|c|}\hline 110\\ 101\\ 011\\ \hline\end{array}\;\begin{array}{|c|}\hline 101\\ 110\\ 011\\ \hline\end{array}\;\begin{array}{|c|}\hline 011\\ 101\\ 110\\ \hline\end{array}\;\begin{array}{|c|}\hline 110\\ 011\\ 101\\ \hline\end{array}\;\begin{array}{|c|}\hline 101\\ 011\\ 110\\ \hline\end{array}\;\begin{array}{|c|}\hline 011\\ 110\\ 101\\ \hline\end{array}$$

Fig. 1.21 Fifteen situations in the [9,4,4] direct product code considered in Example 1.52

6 situations, leading to $w = 6$, are reported in the lower row. They cannot be interpreted according to the scheme shown in Fig. 1.20, because w is not product of two weights in the component codes.

Example 1.53 Taking two [7,4,3] Hamming codes, we obtain a [49,16,9] direct product code, whose first two coefficients in the resulting word weight distribution turn out to be $A(9) = 49$ and $A(12) = 98$. The former one is due to Corollary a. The latter one derives from a term of the type (1.31), with equality sign, due to an ensemble $S(12)$ comprehending two events: the one characterized by $A_1(3)A_2(4) = 49$, and the other characterized by $A_1(4)A_2(3) = 49$ too. Remember that, in the word weight distributions of the two component codes, we have $A_1(3) = A_2(3) = 7$ and $A_1(4) = A_2(4) = 7$.

Example 1.54 Let us consider, as the two component codes in a direct product code, a [6,2,4] code (called C1), with generator polynomial $g(x) = x^4 + x^3 + x + 1$, characterized by $A_1(0) = 1$, $A_1(4) = 3$, and a [4,3,2] parity check code (called C2), characterized by $A_2(0) = 1$, $A_2(2) = 6$, $A_2(4) = 1$. The 64 code words of the [24,6,8] direct product code obtained from them exhibit the following word weight distribution

$$A(0) = 1,\ A(8) = 18,\ A(12) = 24,\ A(16) = 3 + 18 = 21.$$

Coefficient $A(8)$ follows from Corollary a, as $A(8) = A_1(4)A_2(2)$. The first term in the sum expressing coefficient $A(16)$ is the part due to a situation of the type shown in Fig. 1.20 as $A(16) = A_1(4)A_2(4)$. The second term in $A(16)$ comes out from the $\binom{4}{2}3$ different situations shown in Fig. 1.22 up, where two equal non-null code words of C1 are taken each time. Finally, code words having weight 12 are not predicted at all by (1.31). Coefficient $A(12) = 24$ comes out as 4!, and corresponds to the number of dispositions without repetition of the four code words of C1 (included the null one). In Fig. 1.22 down, six situations of this type are considered. In each one of them the 000000 code word of C1 can be inserted in 4 different columns. Those depicted in the figure refer to 000000 in the last column. There are three further configurations like the one here described.

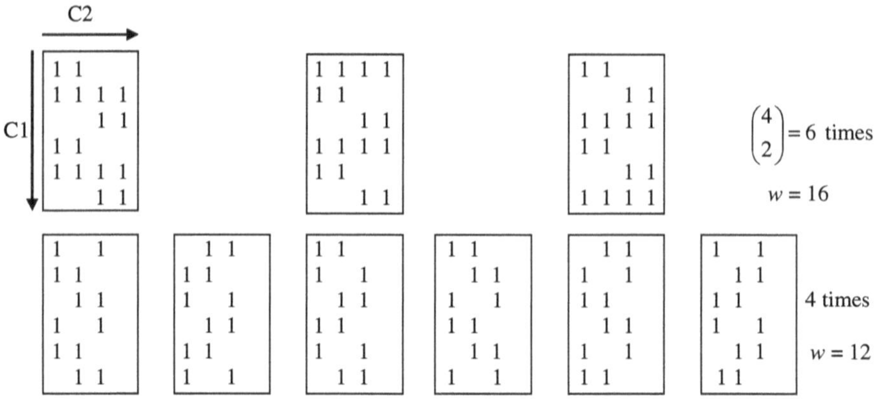

Fig. 1.22 Situations occurring in the evaluation of $A(16)$ and $A(12)$ for the direct product code [24,6,8] introduced in Example 1.54

Fig. 1.23 Ordered ensemble of n_1 code words in as many rows, for constructing a product code

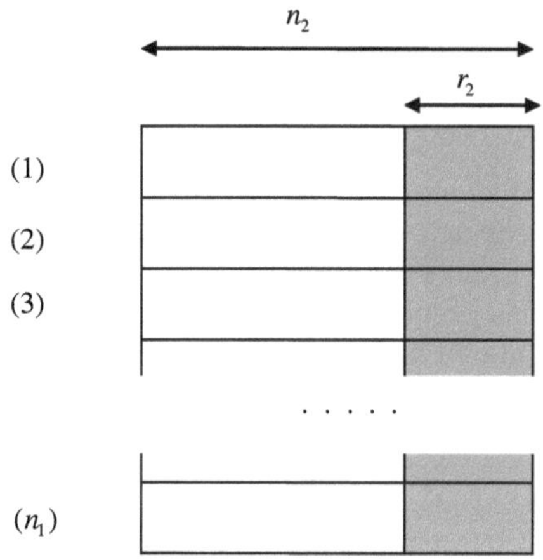

Definition 1.34 A set of code words, displayed as rows in a rectangular matrix, is called *ordered ensemble* of code words taken from the considered code.

In a direct product code, keeping in mind the organization shown in Fig. 1.18, before the application of the component code characterized by parameters n_1, r_1, we have an ordered ensemble of n_1 code words from the code characterized by parameters n_2, r_2. This ordered ensemble appears as shown in Fig. 1.23. The region occupied by control symbols is evidenced by means of a grey colour.

Let us consider the whole sequence of our code words in the ordered ensemble as a unique word of a new code, whose length is $n' = n_1 n_2$. Letting d_2 the minimum distance of the original code, it will be also the minimum distance of the new code, and the number of code words having this minimum distance is

$$A'(d_2) = n_1 A_2(d_2) \tag{1.32}$$

Higher-weight code words can be enumerated following a procedure similar to that employed in Theorem 1.3. For instance, we can write

$$A'(2d_2) \geq \sum_{S(2d_2)} \binom{n_1}{2} A_2(d_2), \tag{1.33}$$

where $S(2d_2)$ represents the ensemble of events leading to $w = 2d_2$ in the new code. The first factor in the terms of the summation represents all the possible couples of rows. The code words of the original code characterized by parameters n_2 and r_2, whose weight are directly $w = 2d_2$, are not included in (1.33). This fact justifies the inequality sign. We can interpret factor n_1 in (1.32) and factor $\binom{n_1}{2}$ in (1.33) as proper multiplicity factors, expressed by the combinatorial coefficients giving the number of 1-weight, 2-weight, … code words in a degenerate code $[n_1, n_1, 1]$. Such code can be thought as acting along the columns.

When another component code, characterized by parameters n_1 and r_1, substitutes the above degenerate code, still acting along the columns of the rectangular memory shown in Fig. 1.18, the direct product code is constructed. So a selection in the set of code words of the above ordered ensemble occurs. In this sense, the resulting direct product code can be considered as a subcode of the ordered ensemble of words from the first component code. In this situation, (1.32) and (1.33) are replaced by expressions of the type (1.31). Obviously the same conclusion can be drawn exchanging the roles of the two component codes.

1.16 Generator Matrix of a Direct Product Code

The generator matrix of an ordered ensemble of n_1 code words, taken from the component code characterized by parameters n_2, r_2, is depicted in Fig. 1.24. This matrix exhibits n_1 row layers of sub-matrices, each coincident with the generator matrix $\underline{G}_2$ of the considered code, and shifting by n_2 positions any row layer.

Theorem 1.4 *The generator matrix of a direct product code, obtained from two component codes, whose generator matrices are respectively $\underline{G}_1$ and $\underline{G}_2$, is the Kronecker product of them (see Definition A.21 in Appendix A)*

$$\underline{G} = \underline{G}_1 \otimes \underline{G}_2. \tag{1.34}$$

Proof Starting from the ordered ensemble of code words, whose generator matrix can be interpreted as a macroscopic $n_1 \times n_1$ identity matrix, made by sub-matrices $\underline{G}_2$, of the type shown in Fig. 1.24, we have to introduce the effect of the second

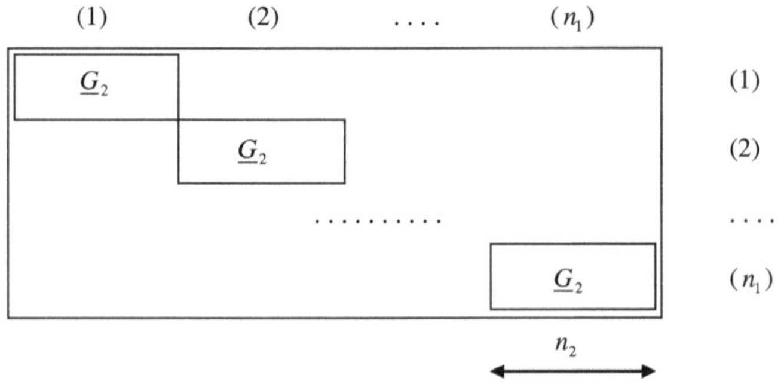

Fig. 1.24 Generator matrix of an ordered ensemble of code words

component code, acting along the columns, whose parameters are n_1, r_1. This effect can be considered by replacing the macroscopic $n_1 \times n_1$ identity matrix with a macroscopic $\underline{G}_1$ matrix. Such procedure is what expressed by the Kronecker product (1.34). □

Exchanging the order of the two component codes, we obtain an equivalent of the direct product code described by (1.34), as expected. In fact, assuming the component codes described by the terns [n_1,k_1,d_1] and [n_2,k_2,d_2], the generator matrix $\underline{G} = \underline{G}_1 \otimes \underline{G}_2$ or $\underline{G} = \underline{G}_2 \otimes \underline{G}_1$ anyway exhibits k_1k_2 rows, all linearly independent (see Lemma A.10), and n_1n_2 columns, leading to equivalent versions of the same code. We can imagine that the one generator matrix describes the microscopic structure of the overall $\underline{G}$, and the other one describes its macroscopic structure, but such two matrices superimpose their effects, leading to the same result also exchanging their roles.

Example 1.55 The direct product code between a [7,4,3] Hamming code, characterized by $g_1(x) = x^3 + x + 1$ and a [6,2,4] code, characterized by $g_2(x) = x^4 + x^3 + x + 1$, is a [42,8,12] code, whose generator matrix can be set in one of the two possible Kronecker forms shown in Fig. 1.25. In the first, four groups of two information symbols each are inserted in as many row layers. In the second, two groups of four information symbols each are inserted in as many row layers.

Theorem 1.4 has two important Corollaries. The first one regards the interpretation of any direct product code as an MLC code, and will be treated in Sect. 1.17. The second one deals with a GPC code, seen as a particular ordered ensemble of parity check code words, and the composite code arranged adding a second GPC code. The result is equivalent to the direct product of two parity check codes. This treatment will allow an easy enumeration of 4-weights code words in GPC codes, and will be presented in Sect. 1.18.

$$
\underline{G} = \left[\begin{array}{c|c|c|c|c|c|c}
11011 & & 11011 & 11011 & & & \\
11011 & & 11011 & 11011 & & & \\
\hline
 & 11011 & & 11011 & 11011 & & \\
 & 11011 & & 11011 & 11011 & & \\
\hline
 & & 11011 & & 11011 & 11011 & \\
 & & 11011 & & 11011 & 11011 & \\
\hline
 & & & 11011 & & 11011 & 11011 \\
 & & & 11011 & & 11011 & 11011
\end{array}\right]
$$

$$
\underline{G} = \left[\begin{array}{c|c|c|c|c|c}
1011 & 1011 & & 1011 & 1011 & \\
1011 & 1011 & & 1011 & 1011 & \\
1011 & 1011 & & 1011 & 1011 & \\
1011 & 1011 & & 1011 & 1011 & \\
\hline
 & 1011 & 1011 & & 1011 & 1011 \\
 & 1011 & 1011 & & 1011 & 1011 \\
 & 1011 & 1011 & & 1011 & 1011 \\
 & 1011 & 1011 & & 1011 & 1011
\end{array}\right]
$$

Fig. 1.25 Equivalent forms for the generator matrix of the [42,8,12] direct product code considered in Example 1.55

1.17 Direct Product Codes as MLC Codes

Corollary a *If the two component codes are characterized by the generator polynomials $g_1(x)$ and $g_2(x)$, their direct product code can be interpreted as an MLC code, whose generator polynomial is*

$$G_o(x) = g_1(x^{n_2})g_2(x), \tag{1.35}$$

with parameters $n_o = n_2$, $k_o = k_2$, and number of periods $Z = k_1$, or equivalently as an MLC code whose generator polynomial is

$$G_o(x) = g_2(x^{n_1})g_1(x), \tag{1.36}$$

with parameters $n_o = n_1$, $k_o = k_1$, and number of periods $Z = k_2$.

Proof In expression (1.35), x^{n_2} is adopted in polynomial $g_1()$ for taking into account shifts of as many positions. So it imposes the macroscopic structure of the overall $\underline{G}$ matrix. Polynomial $g_2(x)$ is replicated in coincidence with the above shifts. Finally the non-null coefficients of polynomial $g_1(x)$ establish where such replicas effectively must be set. The remaining equalities ($n_o = n_2$, $k_o = k_2$, $Z = k_1$) come out from a correspondence between the scheme introduced in Sect. 1.13 about MLC codes and the symbolism adopted here for direct product codes. Expression (1.36) follows from the possibility of exchanging the roles of the two component codes in the direct product. □

Example 1.56 The direct product code introduced in Example 1.55 can be interpreted according to Corollary a. The upper $\underline{G}$ matrix in Fig. 1.25 can be related to

(1.35), where the macroscopic structure of the matrix is assured by the Hamming generator polynomial $g_1(x) = x^3 + x + 1$. The lower $\underline{G}$ matrix, instead, can be related to (1.36), where the macroscopic structure of the matrix is imposed by the [6,2,4] code generator polynomial $g_2(x) = x^4 + x^3 + x + 1$.

In following chapters, we will come back to this interpretation of direct product codes as particular MLC codes, showing that they can be considered as not well designed in the framework of convolutional codes.

1.18 Interpretation of Particular Direct Product Codes by Means of GPC Codes

Let us consider an ordered ensemble of code words, of the type reported in Fig. 1.23, taking an $[n_2,n_2-1,2]$ parity check code as the code acting along the rows. This means $r_2 = 1$. The generator matrix of the overall code so obtained, according to Fig. 1.24, appears as shown in Fig. 1.26.

This structure is due to the fact that the information symbols are written by rows, from left to right, passing top–down from one row to the subsequent one. If one changes this assumption, and writes by columns, from top to down, passing left-right from one column to the subsequent one, a new $\underline{G}$ matrix is obtained, of the type shown in Fig. 1.27. Keeping in mind that, in the present situation, one of the two component codes is the $[n_1,n_1,1]$ degenerate code, the above exchange can be interpreted as the result of employing (1.36) instead of (1.35). Then the two codes so obtained form an equivalent pair.

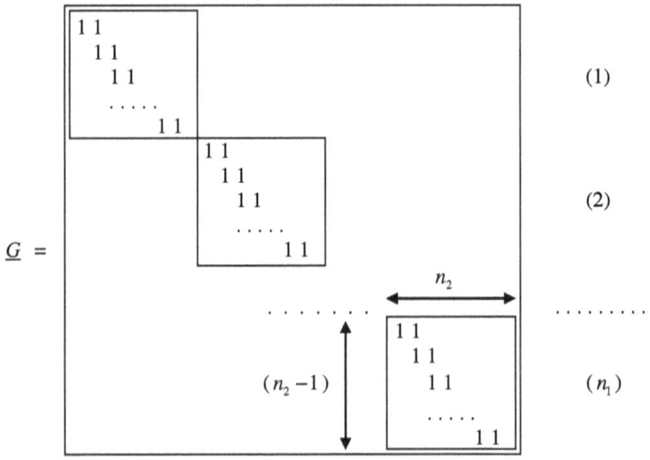

Fig. 1.26 Generator matrix of the ordered ensemble of n_1 words from the $[n_2,n_2-1, 2]$ code

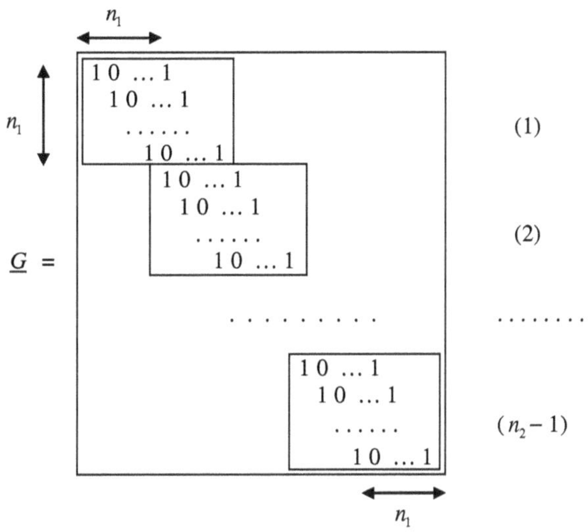

Fig. 1.27 $\underline{G}$ matrix of the same code as in Fig. 1.26, writing symbols by columns

Corollary b *The ordered ensemble of r code words, taken from a [q,q − 1,2] parity check code is equivalent to a unique rth-order GPG code, whose length n is exactly given by q lengthening periods.*

Proof Comparison between the generator matrix shown in Fig. 1.12 for an rth-order GPC code, in two situations ($k < r$ and $k > r$), and the generator matrix reported in Fig. 1.27, here for k_2 exactly equal to $(n_2 - 1)$, allows one to conclude that, after replacing n_1 with r and n_2 with q, they are identical. Therefore the two codes are equivalent. The unique specification to be added is that the GPC code has a length $n = qr$. □

Remark that in Fig. 1.27 it is impossible to have some central positions not protected by the code, as occurs in GPC codes having $k < r$. In fact the two 1-symbol diagonals are strongly overlapped. The number of code words having weight 2 in the considered ordered ensemble of code words, after (1.32), results in

$$A'(2) = rA(2) = r\binom{q}{2}. \tag{1.37}$$

In comparison with Lemma 1.28, we can observe that here (1.37) holds for precise values of k, that is $k = (q - 1)r$, or equivalently $n = qr$. Expression (1.37) can be interpreted as the multiplicity of patterns having just one 1-symbol in all the possible couples of periods.

In order to obtain the number of 4-weight code words, in our GPC code, it is possible to consider two ensembles of events: the first associated to the patterns having just one 1-symbol in all the possible sets of four periods, the other associated to a couple of 2-weight patterns in two different rows. These two terms allow a complete analysis and give us

$$A'(4) = r\binom{q}{4} + \binom{r}{2}\binom{q}{2}^2. \tag{1.38}$$

In the situation considered so far, regarding an rth-order GPC code, over q lengthening periods, taken alone, we can introduce now a second GPC code (or a second ordered ensemble of parity check codes), operating along the orthogonal direction in the encoding memory, so forming a direct product scheme. The previous code was made by an ordered ensemble of r code words taken from a $[q,q-1,2]$ parity check code. The new component code in the direct product scheme is an $[r,r-1,2]$ parity check code. So we obtain a subcode of the original GPC code, in which the selection of the code words is associated to a multiplication of the generator polynomial (x^r+1), shifting in its $\underline{G}$ matrix, as shown in Fig. 1.27, by the polynomial $(x+1)$. The resulting $G_o(x)$ polynomial, according to (1.36), is then

$$G_o(x) = (x+1)(x^r+1), \tag{1.39}$$

and the resulting generator matrix is of the type shown in Fig. 1.28 on the right. In conclusion, the direct product between two parity check codes, respectively described by the terns $[r,r-1,2]$ and $[q,q-1,2]$, whose encoding memory is reported in Fig. 1.28 on the left, can be interpreted as a proper subcode of a unique GPC code. Remark that (1.39) will shift in $(q-1)(r-1)$ rows of the overall generator matrix. One control symbol any r is introduced, as denoted by small dashed arrows in the figure.

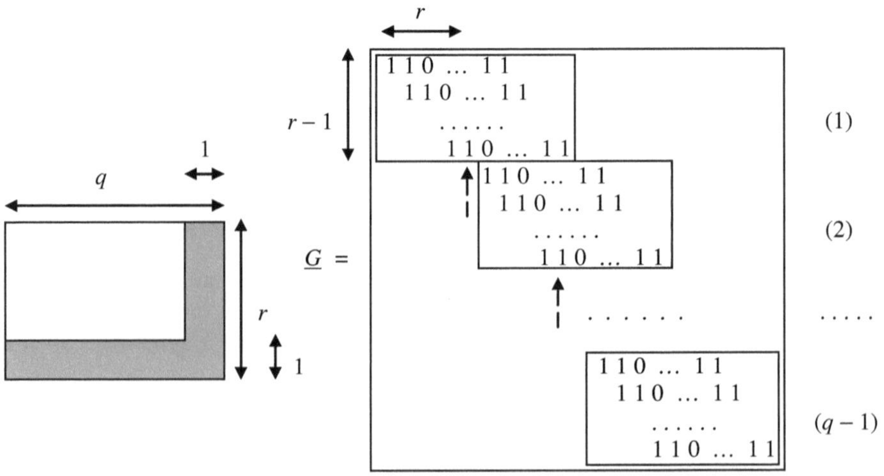

Fig. 1.28 Encoding memory and $\underline{G}$ matrix for the direct product code of two parity check codes

A perfectly equivalent description is that given by a generator polynomial

$$G_o(x) = (x+1)(x^q+1), \tag{1.40}$$

shifting in $(r-1)(q-1)$ rows of a new overall generator matrix. Here we have $(r-1)$ row layers, instead of $(q-1)$, and one control symbol any q is introduced.

Finally, the number of 4-weight code words in this direct product code can be calculated after Corollary a of Theorem 1.3. So we obtain

$$A''(4) = \binom{r}{2}\binom{q}{2}, \tag{1.41}$$

where the parameters of the two parity check codes are perfectly interchangeable, as expected. Comparison of (1.41) with (1.38) makes it clear how strong is the selection in the original amount of 4-weight code words. In fact the first term in (1.38) is now completely absent, and the second is reduced by a factor of $\binom{q}{2}$.

Example 1.57 Let us consider a pair of GPC codes having order respectively $r = 4$ and $q = 7$. The direct product code between the parity check codes [4,3,2] and [7,6,2] is equivalent to the serial concatenation, with a rectangular row-column interleaver, between such two GPC codes. The resulting composite code is characterized by the tern [28,18,4]. The number of its 4-weight code words turns out to be $A(4) = \binom{7}{2}\binom{4}{2} = 126$.

1.19 Cyclic and Pseudo-Cyclic Codes in a Non-binary Alphabet

In Appendix D, tables describing addition and multiplication in a finite field with $Q = 3$ or $Q = 4$ symbols are shown (see Fig. D.9). In a finite field, there is a non-null element, called primitive element of the field, which is able to generate, by subsequent integer powers, all the remaining non-null elements of the field (MacWilliams and Sloane 1977). This element is generally labelled α. From Fig. D.9, for $Q = 3$, taking $\alpha = 2$, we obtain $\alpha^0 = \alpha^2 = 1$, whereas for $Q = 4$, still taking $\alpha = 2$, we obtain $\alpha^0 = \alpha^3 = 1$, $\alpha^2 = 3$. This property will be useful in the following, but, we prefer to maintain distinguished the primitive element α, and its subsequent powers, from the Q-ary symbols s_j, $j = 1, 2, \ldots (Q-1)$, representing them, for simplifying understanding.

In general, we have $(Q-1)$ different binomials of the type the type $(x^N - s_j)$, where s_j represents one of the non-null symbols of the alphabet (in turn associated to a power of the primitive element α), therefore we will consider $j = 1, 2, \ldots (Q-1)$.

Example 1.58 Assuming $Q = 3$ and $N = 4$, with symbol 1 as the unitary element in multiplication, and $2 = -1$ as its opposite in addition, we can have two different binomials, whose factorization up to irreducible factors is

$$x^4 - 1 = (x+1)(x+2)(x^2+1), \quad x^4 - 2 = (x^2+2x+2)(x^2+x+2), \quad Q = 3.$$

Definition 1.35 The polynomial block codes, whose generator polynomial $g(x)$ is factor of $(x^N - 1)$, with degree r, are defined *cyclic codes in a non-binary alphabet*. Both expressions (1.3) and (1.6) are still valid for representing their code words, with the only care of considering, in all the polynomials, coefficients taken from our present Q-ary finite field.

Definition 1.36 In order to widen the concept of cyclic codes, we will introduce *pseudo-cyclic codes in a Q-ary alphabet*, as polynomial block codes whose generator polynomial $g(x)$ is factor of $\left(x^N - s_j\right)$, with degree r, for $j = 1, 2, \ldots (Q-1)$, being s_j one of the non-null symbols of the alphabet (Krishna and Sarwate 1990). Both expressions (1.3) and (1.6) are still valid for representing a code word, with the care already noted. Clearly cyclic codes turns out to be a particular case of pseudo-cyclic ones, characterized by $s_j = 1$. Finally in a binary field such two definitions coincide.

We could introduce other, more particular, definitions, like nega-cyclic codes, as those for which $s_j = -1$, or consta-cyclic codes, as those which are pseudo-cyclic but not cyclic (Bossert 1999). The only difference distinguishing such families of non-binary codes is in the transformation of the code words after a cyclic permutation.

Theorem 1.5 *Given a certain code word t(x) of a pseudo-cyclic code, whose generator polynomial g(x) is factor of* $\left(x^N - s_j\right)$, *its cyclic permutation multiplied by* s_j,

$$t'(x) = x^j t(x)\big|_{\mathrm{mod}(x^N - s_j)}, \quad j = 1, 2, \ldots N \tag{1.42}$$

is still a code word.

Proof Proceeding like for the demonstration of Theorem 1.2, we are left with a periodic polynomial division remainder, originating (1.42), but with a sort of super-period before coming back to the same code word. Along this super-period the non-null symbols of the field are progressively substituted. □

Example 1.59 Starting from the decomposition of the ternary binomial $(x^4 - 2)$ given in Example 1.58, we can take $g(x) = x^2 + 2x + 2$. The four code words forming a complete cyclic permutation (any time multiplying by 2) are 0122, 1220, 2202, 2021. These four code words belong to the same code, but they are not all its non-null code words. Just one further cyclic shift will give $2g(x) = 2x^2 + 1x + 1$, and hence the remaining 4 non-null code words 0211, 2110, 1101, 1012.

Conventionally, in order to distinguish true different polynomials from the simple multiplication of a given polynomial by s_j, all the polynomials are taken monic, that is with the highest-order coefficient equal to 1. It is possible to observe that, in Example 1.59, after one complete cyclic permutation, we reach $2g(x)$, which represents another code word of the same code, whereas in the decomposition of $(x^4 - 2)$ there is another factor, that is $(x^2 + x + 2)$, which cannot be obtained by simple cyclic shifts from $g(x) = x^2 + 2x + 2$.

Corollary a *Performing further cyclic permutations, of the type* (1.42), *still multiplying by the same s_j symbol, after a sort of super-cycle, whose length is* $(Q - 1)N$, *we reach again the original code word, provided that $s_j^2, s_j^3, \ldots, s_j^{Q-2}$ form all the remaining elements of our Q-ary finite field.*

Proof As already stressed, a primitive element α of the considered finite field is able to generate all the other symbols, assuming a proper integer power of it. More precisely, we can write $\alpha^0 = 1$, $\alpha^1 = \alpha \ldots$, up to the $(Q - 2)$th non-null symbol, being the next $\alpha^{Q-1} = \alpha^0 = 1$ again. So, taking $s_j = \alpha$, the above hypothesis is satisfied. At any N-length cycle, a new set of non-null code words comes out, but, at the end, owing to the closeness property of our finite field, related to subsequent powers of the primitive element α, it is possible to reach again the original code word. □

Example 1.60 For $Q = 3$ we can take 2 as the primitive element α, in fact $2^0 = 1$, $2^1 = 2$, $2^2 = 2^0 = 1$. For $Q = 4$, instead, we can assume either 2 or 3 as the primitive element α, in fact $2^0 = 1$, $2^1 = 2$, $2^2 = 3$, $2^3 = 2^0 = 1$, as well as $3^0 = 1$, $3^1 = 3$, $3^2 = 2$, $3^3 = 3^0 = 1$. See the tables in Fig. D.9 for explanation.

Example 1.61 The code introduced in Example 1.59 exhibits two sets of non-null code words, the one obtained from the other after a simple multiplication by 2. In general, the number of sets of non-null code words of this type is $(Q - 1)$ and they can be thought as obtained after what suggested in Corollary a. The considered code is a $[4,2,3]_3$ ternary Hamming code (see Appendix D). It is pseudo-cyclic. Remark that the same tern cannot be produced by a cyclic code, in fact $(x^4 - 1)$ for $Q = 3$ does not show any irreducible factor of degree 2 and weight 3 (see Example 1.58).

These observations are useful in order to extend to pseudo-cyclic codes the concept of cyclic code lengthening (see Sect. 1.11) . If one assumes a number of information symbols $k > N - r$, a lengthened version of our Q-ary code can be constructed. When it is cyclic, that is with $g(x)$ factor of $(x^N - 1)$, this binomial is a code word. When the code is pseudo-cyclic (but not cyclic), that is with $g(x)$ factor of $(x^N - s_j)$, for $s_j \neq 1$, the binomial $(x^N - s_j)$ is a code word. With increasing k, under the assumption of Corollary a of Theorem 1.5, all the possible binomials with degree integer multiple of N, become code words of this lengthened code, up to the binomial $(x^{(Q-1)N} - 1)$, which comes out at the end of the super-cycle.

In conclusion, we can state that any pseudo-cyclic code, comprehending in its definition also possible lengthened versions of it, becomes a cyclic code, over the length $(Q-1)N$ of its super-cycle.

Example 1.62 Q-ary repetition codes (see also Definition D.29 and Example D.28) are characterized by the tern $[N,1,N]_Q$. They are typically cyclic because factorization (B.33) is always valid for (x^N-1), but can be also pseudo-cyclic when a similar factorization holds for a different binomial. Taking $Q = 3$ and $N = 3$, and considering that $2 = -1$, the following binomial decompositions hold

$$x^3-1=(x^2+x+1)(x-1), \quad x^3-2=(x^2-x+1)(x+1), \quad Q=3.$$

From the former one, assuming $g(x)=x^2+x+1$, we obtain a $[3,1,3]_3$ cyclic repetition code, whose code words are 000, 111, 222. From the latter one, assuming $g(x)=x^2-x+1$, we obtain a $[3,1,3]_3$ pseudo-cyclic repetition code, whose code words are 000, 121, 212. In Example D.28, speaking about $[3,1,3]_3$ possible codes, we list the cyclic one, and then other two possible codes, respectively with the choices 000, 121, 212 and 000, 112, 221. The first one is pseudo-cyclic. The second one, instead, is not a polynomial code. Coming back to the two only permitted versions of the $[3,1,3]_3$ ternary repetition code, associated respectively to $g(x)=x^2+x+1$ (cyclic) and to $g(x)=x^2-x+1$ (pseudo-cyclic), let us perform one-time lengthening for both. We obtain the following sets of 9 code words: 0000, 0111, 0222, 1110, 2220, 1002, 2001, 1221, 2112 in the cyclic version, whereas 0000, 0121, 0212, 1210, 2120, 1001, 2002, 1122, 2211 in the pseudo-cyclic version. The minimum distance is $d = 2$, and we have $A(2) = 2$. In general $A(2) = Q - 1$ at the one-time lengthening. Remember that $-1 = 2$ and $-2 = 1$ in this ternary alphabet.

Example 1.63 Q-ary parity check codes (see also Definition D.30 and Example D.29) are characterized by the tern $[N,N-1,2]_Q$. For $Q = 4$ and $N = 3$, keeping in mind that negative sign is here useless, the following decomposition holds

$$x^3+1=(x+1)(x+2)(x+3), \quad Q=4.$$

Each one of these three factors can be taken as the generator polynomial $g(x)$ for a $[3,2,2]_4$ quaternary cyclic parity check code (which is also a Reed-Solomon code, as we will show in Sect. 1.22). Generator polynomials and code words are listed here

$$\begin{aligned} g(x)=x+1,\quad & 000,011,022,033,101,202,303,110,\\ & 220,330,123,132,213,231,312,321,\\ g(x)=x+2,\quad & 000,012,023,031,103,201,302,120,\\ & 230,310,111,132,213,222,321,333,\\ g(x)=x+3,\quad & 000,013,021,032,102,203,301,130,\\ & 210,320,111,123,222,231,312,333. \end{aligned}$$

The words associated to $g(x) = x + 1$ are reported also in Example D.29. For all, it is possible to obtain the same word weight distribution $A(0) = 1$, $A(2) = 9$, $A(3) = 6$, confirming that here not only even weights are present. Finally remark that, independently of the choice for $g(x)$, such three codes are all cyclic. This is consequence of the fact that all such $g(x)$ polynomials are factors of $x^3 - 1 = x^3 + 1$. It is important to stress that $(x^3 + 2)$ and $(x^3 + 3)$ are irreducible for $Q = 4$, therefore we cannot obtain parity check or repetition quaternary pseudo-cyclic codes of length 3 in this case. For $Q = 3$, instead, this is possible, say, assuming $N = 3$ and the factorization of $(x^3 - 2)$ presented in Example 1.62.

1.20 Q-ary State Diagrams

State diagrams have been introduced for describing state evolution and the relations between input and output symbols in encoder circuits. Nevertheless, independently of how encoding is performed, it could be interesting to represent the state diagram for a Q-ary code, in order to better understand its properties. We will consider systematic encoding, for simplicity, and the only case of parity check codes ($r = 1$), for $Q = 3$ and $Q = 4$. Particular attention will be devoted to the closed routes, not comprehending state $\underline{0}$, characterized by N consecutive 0-symbols, originating the code words of lengthened codes.

In Fig. 1.29 up, two state diagrams for $Q = 3$ are reported. They describe two possible families of ternary parity check codes. As a general comment, in comparison with the well known binary state diagrams (see Sect. 1.14), we can note here a further state, and three input or output branches for any state. On the left we have the choice $g(x) = x - 1 = x + 2$, which can be interpreted as a situation with $N = 1$, in the case of a cyclic choice. Therefore this state diagram describes also any LC code, for whichever $n > N$, like in a binary alphabet, being all these results as many complete lengthening operations. In fact 102, 1002, 10002 ... are all acceptable code words. On the right, instead, taking $g(x) = x - 2 = x + 1$, we consider a situation still with $N = 1$, but in the case of a pseudo-cyclic choice. Code lengthening is still possible for any $n > N$, but a complete lengthening can be performed only for $n = 3, 5, 7$ ..., in fact the only sequences 102, 10002, 1000002 ... are here acceptable code words. This difference is clearly explained considering the closed routes, not comprehending state $\underline{0}$, characterized by consecutive 0-symbols: they are two of length 1 in the cyclic state diagram and one of length 2 in the pseudo-cyclic state diagram. The property of having a longer 0-symbol closed route is a particular feature characterizing pseudo-cyclic codes.

In Fig. 1.29 down, two state diagrams for $Q = 4$ are shown. That on the left is constructed still for $N = 1$, assuming here $g(x) = x - 1 = x + 1$. It shows the same characteristics as the one considered above, included the 0-labelled closed-loop branch around any state (here four instead of three). This code exhibits complete lengthening for $n = 2, 3, 4$, ..., respectively leading to the terns $[2,1,2]_4$, $[3,2,2]_4$,

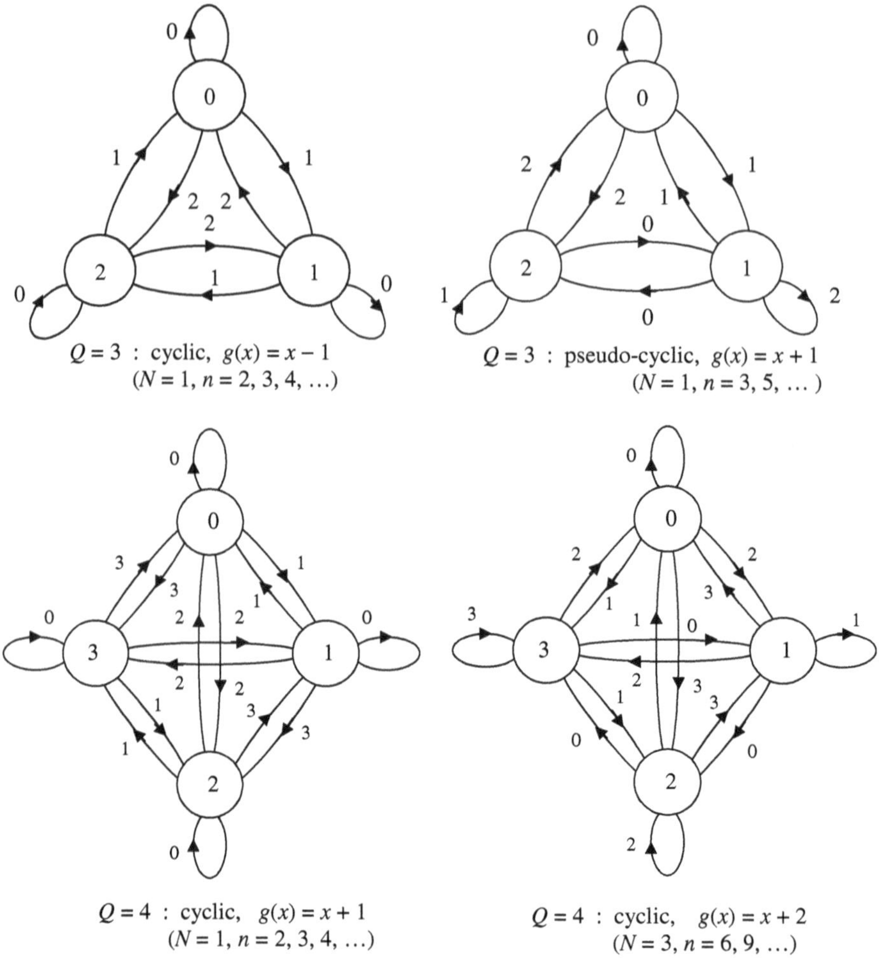

Fig. 1.29 State diagrams for parity check Q-ary codes, with $Q = 3$ and $Q = 4$

$[4,3,2]_4$, … . The state diagram down-right still holds for a cyclic code, but assuming $g(x) = x + 2$, with $N = 3$. In fact, the following equality holds in a quaternary alphabet $x^3 + 1 = (x+1)(x+2)(x+3)$ (see also Example 1.63), and any one of these three factors can be assumed as the generator polynomial in a parity check code. Nevertheless the natural length of such cyclic code is here $N = 3$. It is interesting to stress that, in general, N not always is 1 for a non-binary parity check code, and, when $N > 1$, complete lengthening is performed only over a length n integer multiple of N itself. The 0-symbol closed route in the state diagram reported down-right has length $N = 3$ for this reason.

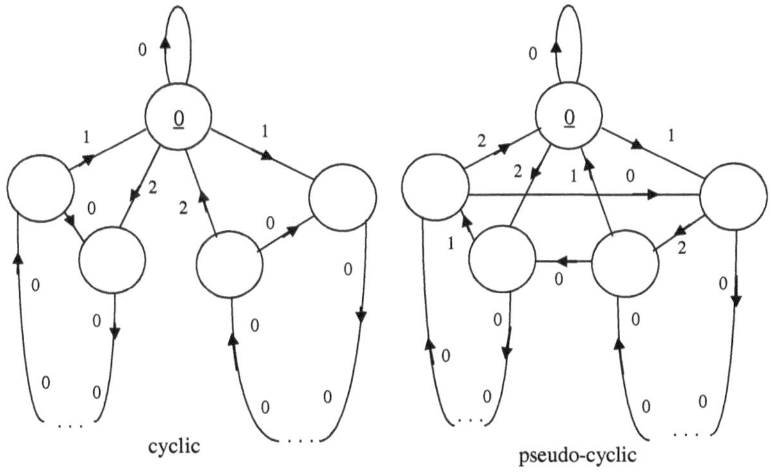

Fig. 1.30 Structures of the closed routes characterized by consecutive 0-symbols (2-weight words in lengthened versions) for $Q = 3$

Lemma 1.40 *The closed routes, not comprehending state $\underline{0}$, characterized by consecutive 0-symbols in Q-ary state diagrams have length N for cyclic codes, and up to $(Q - 1)N$ for pseudo-cyclic codes.*

Proof The first statement follows directly from Definition 1.35. For instance, taking $Q = 3$, we have two 0-labelled closed routes, separated by just one non-null symbol from state $\underline{0}$, like those qualitatively shown in Fig. 1.30 on the left. After Corollary a of Theorem 1.5 and the observation about lengthened versions of pseudo-cyclic codes, when s_j coincides with the primitive element of the field, also the second statement is demonstrated. This entails a unique 0-labelled closed route of the type qualitatively shown in Fig. 1.30 on the right. □

1.21 Main Families of Non-binary Block Codes

In the following, we will treat Hamming, simplex and BCH Q-ary codes, by means of as many Examples.

Example 1.64 Q-ary Hamming codes (see also Definition D.27) are characterized by the tern $\left[\frac{Q^r-1}{Q-1}, \frac{Q^r-1}{Q-1} - r, 3\right]_Q$. Let us consider the case of $Q = 3$ and $r = 2$. So we obtain $N = 4$. From the decompositions of $(x^4 - 1)$ and $(x^4 - 2)$ provided in Example 1.58, we can observe how just the second one admits two irreducible polynomials having weight $w = 3$. They both can be assumed as generator polynomials of as many equivalent $[4,2,3]_3$ ternary pseudo-cyclic codes. Remark that an equivalent cyclic version cannot exist. The code words in the two cases are

$$g(x) = x^2 + 2x + 2, \quad 0000, 0122, 0211, 1012, 1101, 1220, 2021, 2110, 2202,$$
$$g(x) = x^2 + x + 2, \quad 0000, 0112, 0221, 1021, 1120, 1201, 2022, 2120, 2210.$$

The code words of the first code are listed also in Example D.26. The word weight distribution, for both, turns out to be $A(0) = 1$, $A(3) = 8$. The first one of these two codes has been already treated also in Examples 1.59 and 1.61. Except the null code word, there are no code words in common.

Definition 1.37 Given a Q-ary cyclic code characterized by polynomial $g(x)$, factor of $(x^N - 1)$, the definition of its dual remains as that given for binary cyclic codes (Definition 1.22). If this code is pseudo-cyclic, characterized by polynomial $g(x)$, factor of $\left(x^N - s_j\right)$, its *dual code* can be defined as the code whose generator polynomial is the co-factor of $g(x)$ with respect to the binomial $\left(x^N - s_j\right)$.

The two $[4,2,3]_3$ ternary pseudo-cyclic Hamming codes, introduced in Example 1.64, are then a couple of dual codes.

Example 1.65 Q-ary simplex codes (see also Definition D.28) are characterized by the tern $\left[\frac{Q^{k-1}}{Q-1}, k, Q^{k-1}\right]_Q$. They are often pseudo-cyclic. A Q-ary simplex code is the dual of a Q-ary Hamming code, having the same length (Peterson and Weldon 1972). Decomposition of $\left(x^5 + 2\right)$ and of $\left(x^5 + 3\right)$ in a quaternary alphabet result in

$$x^5 + 2 = (x + 3)(x^2 + 2x + 2)(x^2 + x + 2), \quad Q = 4,$$
$$x^5 + 3 = (x + 2)(x^2 + 3x + 3)(x^2 + x + 3), \quad Q = 4.$$

In both, assuming $g(x)$ as one of the four possible irreducible polynomials with degree 2, we obtain a $[5,3,3]_4$ pseudo-cyclic quaternary Hamming code. Assuming $g(x)$ as the co-factor, we obtain its dual, that is a $[5,2,4]_4$ pseudo-cyclic quaternary simplex code. The one associated to the polynomial $g(x) = (x + 3)(x^2 + 2x + 2) = x^3 + x^2 + 3x + 1$ is characterized by the following 16 code words

00000, 01131, 02212, 03323, 10221, 11310, 12033, 13102, 20332, 21203, 22120, 23011, 30113, 31022, 32301, 33230, $Q = 4$.

All the 15 non-null code words have $w = 4$ confirming that, also in a Q-ary alphabet, the characteristic of a simplex code is to exhibit all non-null code words at the same distance from the null code word. The number of such non-null code words is $A\left(Q^{k-1}\right) = Q^k - 1$.

Remark that either the $[5,3,3]_4$ code or the $[5,2,4]_4$ code are MDS codes (see Definition D.31 and Fig. D.10). As already noted, Q-ary simplex codes can have also odd minimum distance d, differently from what occurs in a binary alphabet. For instance, the $[4,2,3]_3$ ternary Hamming code, which is also a simplex code, exhibits $d = 3$.

Also the $[5,3,3]_4$ quaternary Hamming code, as well as its dual $[5,2,4]_4$ simplex code, cannot be obtained in cyclic form, in fact the decomposition of $x^5 - 1 = x^5 + 1$ results in

$$x^5 + 1 = (x+1)(x^4 + x^3 + x^2 + x + 1), \quad Q = 4,$$

where the second factor is irreducible. On the other hand, these two factors still originate a parity check and repetition code pair, dual each other.

As already noted, Q-ary parity check codes and repetition codes are always available in cyclic form, since decomposition $x^{N-1} = (x-1)(x^{N-1} + \cdots + x + 1)$ is true for whichever N (see Lemma B.32). On the contrary, pseudo-cyclic versions of Q-ary parity check or repetition codes only sometimes exist. More precisely, these pseudo-cyclic versions are possible when they are assured by a decomposition of the type

$$x^N - s_j = (x - s_i)(x^{N-1} + \ldots + s_k), \tag{1.43}$$

where $s_i s_k = s_j$, and $s_j \neq 1$, with a latter factor having weight N and characterized by central terms all dropping out in couples after the product.

About the procedure for code extension, intended as a particular form of direct serial code concatenation, we can consider the following proposition.

Definition 1.38 The procedure of *code extension in a Q-ary alphabet* is performed by adding a further parity check symbol, so we obtain a direct serial concatenation, where the inner code is characterized by generator polynomial $(x - s_j)$. This procedure can be repeated, varying s_j, more than one time. The minimum distance too can increase here more than one time.

In order to allow a widening of the concept of BCH codes in a Q-ary alphabet, it is necessary to modify the BCH minimum distance bound, introduced in Sect. 1.9, as follows (MacWilliams and Sloane 1977).

Theorem 1.6 (BCH minimum distance bound for non-binary polynomial codes) *Given a Q-ary BCH cyclic code designed in such a way as to have minimum distance at least δ_c, the product forming its generator polynomial g(x) contains ($\delta_c - 1$) subsequent groups of Q-ary Nth roots of unity, collected in proper irreducible polynomials, which in turn are obtained starting from a certain primitive polynomial, in the non-binary finite field considered.*

The main difference with respect to the binary case is in the presence directly of the designed minimum distance δ_c, instead of the designed maximum number τ of correctable single errors. So here the true minimum distance d_c, as well as its designed value δ_c, can be either odd or even. Finally we have the inequality $d_c \geq \delta_c$, instead of the inequality $t \geq \tau$. Keeping in mind the relationship $\delta_c = 2\tau + 1$ and the fact that, in a binary field, there are always couples of conjugate roots, clearly the above non-binary version of BCH minimum distance bound includes the binary case as a particular situation.

Example 1.66 Given a certain Q, for which a finite field exists, a primitive Q-ary BCH code (Gorenstein and Zierler 1961) is associated to the tern $[Q^m - 1, Q^m - r - 1, d_c]$. The generator polynomial $g(x)$ of such code is the product of irreducible polynomials corresponding to $(\delta_c - 1)$ subsequent Q-ary roots of unity, being $d_c \geq \delta_c$, after the choice of a certain primitive polynomial, which always exists in this case. Since we are treating the Q-ary roots of unity, $g(x)$ will be factor of $(x^N - 1)$, and non-binary polynomial BCH codes will be cyclic.

Non-primitive Q-ary BCH codes comprehend codes for which N is not equal to $(Q^m - 1)$, but still exhibit a generator polynomial made by the product of irreducible polynomials. A strong difference with respect to the binary case is in the fact that here the equality $M_j(x) = M_{2j}(x) = M_{4j} = \ldots, j$ odd, is no longer valid, since it was consequence of being in a binary field. Therefore, in the expression

$$g(x) = \mathrm{LCM}[M_1(x), M_2(x), \ldots], \tag{1.44}$$

we can have the product of all the terms within square brackets, except those corresponding to subsequent powers of Q. This property makes useless to restrict our attention to only designed distances δ_c having odd values, as already stressed.

Comparison between the tern representing Q-ary Hamming codes introduced in Example 1.63 and the tern representing primitive Q-ary BCH codes allows one to conclude that the latter ones are not subcodes of the former ones (situation which occurs in a binary alphabet, as noted in Example 1.43). Lemma 1.24, which states the inequality $r \leq mt$, where t is the true maximum number of single correctable errors, and relates the performance of a binary Hamming code with that of a more general binary BCH code, is valid only with a binary alphabet. Here it is simply included in the inequality $d_c \geq \delta_c$.

Example 1.67 The quaternary primitive BCH code with $m = 2$ and hence $N = 4^2 - 1 = 15$, with designed distance $\delta_c = 3$, can be obtained assuming the following generator polynomial, factor of $(x^{15} + 1)$,

$$\begin{aligned} g(x) &= \mathrm{LCM}[M_1(x),\ M_2(x)] = M_1(x)M_2(x) \\ &= (x^2 + x + 2)(x^2 + x + 3) = x^4 + x + 1, \quad Q = 4. \end{aligned}$$

In it, two subsequent groups of roots, collected in the above pair of irreducible polynomials, have been taken. The resulting code can then be described by the tern $[15,11,3]_4$. The Hamming code characterized by $Q = 4$ and $r = 2$ is, instead, a $[5,3,3]_4$ code, already mentioned in Example 1.65, whose generator polynomial is one of the two possible irreducible factors of $(x^5 + 2)$ having degree $r = 2$.

Example 1.68 Still in a quaternary field, the following cyclotomic cosets and corresponding irreducible polynomials are available, for $N = 21$ (Wicker 1995),

$$\begin{array}{ll} C_0 = \{0\}, & M_0(x) = x + 1, \\ C_1 = \{1, 4, 16\}, & M_1(x) = x^3 + \alpha^2 x + 1, \\ C_2 = \{2, 8, 11\}, & M_2(x) = x^3 + \alpha x + 1, \\ C_3 = \{3, 12, 6\}, & M_3(x) = x^3 + x^2 + 1, \\ C_5 = \{5, 20, 17\}, & M_5(x) = x^3 + \alpha^2 x^2 + 1, \\ C_7 = \{7\}, & M_7(x) = x + \alpha, \\ C_9 = \{9, 15, 18\}, & M_9(x) = x^3 + x + 1, \\ C_{10} = \{10, 19, 13\}, & M_{10}(x) = x^3 + \alpha x^2 + 1, \\ C_{14} = \{14\}, & M_{14}(x) = x + \alpha^2, \qquad Q = 4, \end{array}$$

where α is the primitive element of this field (if $\alpha = 2$, then $\alpha^2 = 3$, or vice versa). For understanding how any cyclotomic coset is obtained, see (B.31) in Appendix B. It is possible to note the unique element in C_7, due to the result $(28)mod(21) = 7$, as well as in C_{14}, here due to the result $(56)mod(21) = 14$. The correspondent polynomials can be adopted for constructing as many parity check codes, besides the one associated to $M_0(x)$, all cyclic. If one is interested in a BCH code, with designed distance $\delta_c = 5$, it is necessary to take the generator polynomial as

$$\begin{aligned} g(x) &= \mathrm{LCM}[M_1(x), M_2(x), M_3(x), M_4(x)] = M_1(x)M_2(x)M_3(x) \\ &= (x^3 + \alpha^2 x + 1)\,(x^3 + \alpha x + 1)(x^3 + x^2 + 1), \quad Q = 4. \end{aligned}$$

since we have $M_4(x) = M_1(x)$. This generator polynomial has degree 9, therefore the code obtained is a $[21,12,5]_4$ quaternary BCH code, which is non-primitive since 21 is not an integer power of 4 minus one. The code considered in Example 1.67, instead, is a primitive quaternary BCH code.

With respect to the best binary codes, the quaternary BCH codes presented in Example 1.67 and Example 1.68 appear not particularly good. As a measure of their performance, we can assume the separation between r and d_c. On the contrary, a very efficient use of the number of control symbols is made in an important family of non-binary BCH codes, which will be introduced in the next section.

1.22 Reed-Solomon Codes and Other MDS Non-binary Codes

A particular situation, in the framework of Q-ary primitive BCH codes, is that characterized by $m = 1$, that is

$$N = Q - 1. \tag{1.45}$$

This means that here N equals the number of non-null symbols of our alphabet.

Lemma 1.41 *In the situation described by* (1.45) *it is always possible to express* $(x^N - 1)$ *as the product of N factors, all having degree 1. Such decomposition is then of the type*

$$x^N - 1 = \prod_{j=1}^{Q-1} (x - s_j). \tag{1.46}$$

Proof All the cyclotomic cosets have here only one element. Hence all the irreducible polynomials $M_j(x)$ have degree 1. Taking the primitive element α of this finite field, our binomial can be symbolically written as

$$x^N - 1 = (x - \alpha^0)(x - \alpha^1)(x - \alpha^2)\ldots(x - \alpha^{Q-2}), \tag{1.47}$$

which coincides with (1.46), except for symbolism (see also Lemma B.33). □

Lemma 1.42 *If we choose a subset of the set of* $(Q - 1)$ *factors in the product* (1.47), *following subsequent powers of the primitive element, it is possible to obtain a polynomial g(x) whose degree and weight satisfy the relationship* $w[g(x)] = deg[g(x)] + 1$.

Proof Let us consider all the possible products of field elements (proper integer powers of α) in expression (1.47). Following such subsequent powers of α, the addition between two terms with equal degree yields a new non-null term with the same degree. So the resulting product $g(x)$ will have degree equal to the number of subsequent factors considered, and weight greater by one unit. □

Example 1.69 Taking $Q = 4$ and hence $N = 3$, (1.47) yields

$$x^3 - 1 = (x - \alpha^0)(x - \alpha^1)(x - \alpha^2) = (x + 1)(x + 2)(x + 3),$$

where, being Q an integer power of 2, negative signs have been already changed into positive signs. Then the three possible products between two of the above factors result in

$$\begin{aligned}(x + 1)(x + 2) &= x^2 + 3x + 2,\\ (x + 2)(x + 3) &= x^2 + x + 1,\\ (x + 3)(x + 1) &= x^2 + 2x + 3.\end{aligned}$$

All these products exhibit weight $w = 3$. Assuming one of them as the generator polynomial, we obtain a quaternary BCH code with $N = d_c = 3$, which is also a $[3,1,3]_4$ quaternary cyclic repetition code.

Theorem 1.7 *After the procedure described in Lemma 1.41, it is possible to construct a* $[Q-1, k, Q-k]_Q$ *code, which is MDS, that is satisfies Singleton bound (see Definition D.31 and Theorem D.3) with equality sign.*

Proof The BCH bound, for a certain designed distance δ_c, assuming the generator polynomial

$$g(x) = (x-\alpha)(x-\alpha^2)\ldots(x-\alpha^{\delta_c-1}), \tag{1.48}$$

is able to guarantee that the true minimum distance d_c will be not lower than δ_c. On the other hand, after Lemma 1.42, w[$g(x)$] = deg[$g(x)$] + 1 = $Q - k = \delta_c$. Nevertheless Singleton bound states that $d_c \leq r + 1 = Q - k$. In conclusion, the BCH bound is to be taken with equality sign, otherwise Singleton bound could not be satisfied, and we are left with $d_c = \delta_c = Q - k$. □

Definition 1.39 The codes constructed according to Theorem 1.7 form a particular family of Q-ary primitive BCH codes, known as *Reed-Solomon codes*, briefly *RS* (see also Definition D.33) (Reed and Solomon 1960). In their polynomial form, since these codes derive from factorization (1.47), they are always cyclic.

In other terms, RS codes have $g(x)$ made by a product of the $(d_c - 1)$ generator polynomials taken from as many different Q-ary cyclic parity check codes. Among the three $[3,1,3]_4$ quaternary cyclic repetition codes described in Example 1.69, just one, rigorously, should be defined as an RS code. Nevertheless the other ones, as already stressed, are equivalent to it. This is justified by the possibility of taking different primitive elements.

Corollary a *The dual (according to Definition 1.37) of an RS code is also an RS code.*

Proof The co-factor of the product of a certain set of binomials $(x - s_j)$, associated to as many subsequent roots of $(x^N - 1)$, is the product of a set of binomials associated to as many subsequent roots of $(x^N - 1)$ as well. □

Example 1.70 The case characterized by $Q = 4$ and hence $N = 3$ has been treated in Example 1.69. The decomposition of the binomial $(x^3 - 1)$ gives rise to three possible $[3,1,3]_4$ quaternary repetition codes and to as many $[3,2,2]_4$ quaternary parity check codes, forming a dual code pair. The parity check codes have been already treated in Example 1.63. For higher values of Q, repetition and parity check codes would be considered trivial RS codes and separately treated.

Example 1.71 Taking $Q = 8$ and hence $N = 7$, we can have several non-trivial RS codes, depending on the choice of parameter k. Besides the trivial choices $k = 1$ and $k = 6$, leading to a repetition and a parity check code respectively, it is possible to list the following terns of RS codes: $[7,5,3]_8$, $[7,4,4]_8$, $[7,3,5]_8$, $[7,2,6]_8$. In Fig. 1.31 a map of these codes and other MDS codes in a $k \times r$ plane is shown. Similar maps are reported in Fig. D.10 for $Q = 2, 3, 4$. It is possible to note how the number of MDS codes is remarkably increased here. Grey small circles represent the above

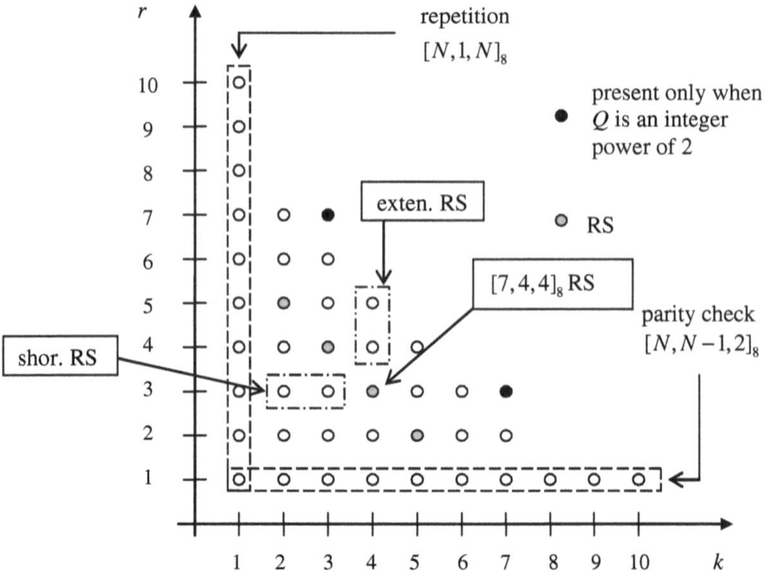

Fig. 1.31 Map of MDS codes in a $k \times r$ plane, for $Q = 8$

four RS codes with $N = 7$ and $Q = 8$. White small circles along the same columns towards up represent subsequent extensions of the above codes, able to increase the minimum distance. This is possible generally two times, so maintaining the property of being MDS codes. Only in the case of the $[7,3,5]_8$ code, this possibility is reached a third time, leading to the $[10,3,8]_8$ code (black small circle on the left).

A similar geometrical disposition of MDS codes can be drawn for any value of Q, able to guarantee the construction of a finite field (Q prime or power of a prime). Besides the row representing parity check codes and the column representing repetition codes, there is a triangular geometry, where pairs of dual codes occupy symmetric positions with respect to the bisector. The two codes out of this triangular geometry (for $Q = 8$, the $[10,7,4]_8$ and the $[10,3,8]_8$ code, denoted with black small circles) can be inserted in this ensemble of MDS codes just when Q is an integer power of 2 (MacWilliams and Sloane 1977).

Corollary b *A shortened version of an RS code is still an MDS code.*

Proof After s-time shortening, from a $[Q-1, k, Q-k]_Q$ RS code, we obtain a code described by the tern $[Q-1-s, k-s, Q-k]_Q$. The relationship between parameter r and the minimum distance d_c is unchanged, so the new code is MDS too. □

For instance, looking at Fig. 1.31, the shortened versions of the $[7,4,4]_8$ RS code are aligned along the same row (r remains constant) progressively on the left. The remaining part of this section is devoted to discuss binary codes equivalent to RS codes with Q equal to an integer power of 2.

Example 1.72 Let us describe how the $[7,5,3]_8$ RS code can be transformed into a binary code. We can follow Definition B.18 for obtaining all the elements of this finite field starting from a proper irreducible polynomial. Assuming $\alpha^0 = 1$ and $p(\alpha) = \alpha^3 + \alpha + 1$ as the irreducible polynomial, the following association comes out

$$\alpha^3 = \alpha + 1, \quad \alpha^4 = \alpha^2 + \alpha, \quad \alpha^5 = \alpha^2 + \alpha + 1, \quad \alpha^6 = \alpha^2 + 1.$$

The generator polynomial then is

$$g(x) = (x - \alpha)(x - \alpha^2) = x^2 + (\alpha^2 + \alpha)x + \alpha^3 = x^2 + \alpha^4 x + \alpha^3.$$

We can use the above association also for obtaining a binary equivalent of this code, through a suitable mapping of three binary symbols on any octal symbol employed so far. In order to do this, it is necessary to define a proper additive basis. For instance, taking $\alpha^0 = 001, \ \alpha^1 = 010, \ \alpha^2 = 100$ as a basis, we can write

$$g(x) = x^2 + (\alpha^2 + \alpha)x + (\alpha + 1) = (001)x^2 + (110)x + (011).$$

This generator polynomial exhibits binary weight $w = 5$. After multiplication of $g(x)$ by $\alpha^1 = 010, \ \alpha^2 = 100, \ldots$, the binary weight remains not lower than $w = 4$, as Table 1.11 clearly shows. Unfortunately further linear combinations of such polynomials lead to overall weights $w = 3$, up to the cyclic version of this code. In conclusion, the $[7,5,3]_8$ RS code here considered has been transformed into a [21,15,3] binary code.

Since the binary generator polynomial of the binary code obtained in Example 1.72 is $g(x) = x^6 + x^5 + x^4 + x + 1$, and the transformations after multiplication by all the other non-null symbols of the alphabet, are characterized by $w \geq 4$, at least a

Table 1.11 Binary expressions of the octal polynomials associated to $\alpha^j g(x)$ in Example 1.72

j	$\alpha^j g(x)$	Binary equivalent	Binary w$[\alpha^j g(x)]$
0	$g(x)$	$(001)x^2 + (110)x + (011)$	5
1	$\alpha g(x)$	$(010)x^2 + (011)x + (011)$	5
2	$\alpha^2 g(x)$	$(100)x^2 + (110)x + (111)$	6
3	$\alpha^3 g(x)$	$(011)x^2 + (111)x + (101)$	7
4	$\alpha^4 g(x)$	$(110)x^2 + (101)x + (001)$	5
5	$\alpha^5 g(x)$	$(111)x^2 + (001)x + (010)$	5
6	$\alpha^6 g(x)$	$(101)x^2 + (010)x + (100)$	4

[9,3,4] binary code can be obtained as a binary equivalent of the $[3,1,3]_8$ 4-time shortened version of the above RS code, after the binary symbol mapping adopted.

Comparison with codes directly designed in a binary alphabet can give a reference for evaluating the performance of the above RS octal codes. It is possible to assume, for instance, a [21,16,3] binary cyclic code, whose generator polynomial is $g'(x) = x^5 + x + 1$, or a [21,15,4] binary code, obtained as even-distance subcode of the previous one, that is adopting the generator polynomial $g''(x) = (x^5 + x + 1)(x + 1)$. In comparison with such two binary codes directly designed, the above transformation of the $[7,5,3]_8$ RS code appears characterized by a poorer performance. On the other hand, a direct product code, constructed employing two binary parity check component codes, yields the binary tern [9,4,4], clearly better than the [9,3,4] code previously obtained after 4-time shortening. Obviously different results could be obtained with another choice for binary symbol mapping on octal symbols, but not better than those reported here with the present binary symbol mapping.

In conclusion, typically the transformation of Q-ary codes, for Q equal to an integer power of 2, into binary codes, taking a number of binary symbols equal to the above power for replacing one original non-binary symbol, gives rise to codes with worse performance. Q-ary codes, especially if BCH, are generally preferred owing to the possibility of setting up an efficient hard-decision decoding with traditional procedures. Sometimes, however, it is possible to have a complete equivalence between a Q-ary code in its binary version and a certain binary code directly designed.

Example 1.73 Let us consider the case of $Q = 4$ and $N = 3$, already treated in Example 1.69, where we have analyzed three different $[3,1,3]_4$ cyclic quaternary repetition codes, associated to as many generator polynomials. Assuming the basis 1 = 01, 2 = 10, and hence 3 = 11, it is possible to rewrite these generator polynomials in a binary form respectively as

$$g(x) = (01)x^2 + (11)x + (10),$$
$$g(x) = (01)x^2 + (01)x + (01),$$
$$g(x) = (01)x^2 + (10)x + (11).$$

Two of them have binary weight 4. Proceeding as in Example 1.72, it is easy to verify that, also after multiplication of such generator polynomials by the two non-null remaining elements of this quaternary field, the binary weight remains $w = 4$. So there is a pair of [6,2,4] binary codes which can be obtained from such $[3,1,3]_4$ cyclic quaternary repetition code. It is possible to show that they are equivalent to the [6,2,4] binary cyclic code, constructed directly in a binary field, assuming the generator polynomial $g(x) = (x + 1)(x^3 + 1) = x^4 + x^3 + x + 1$, whose weight distribution is $A(0) = 1$, $A(4) = 3$ (see Example 1.15).

1.23 Trellis for an s.s. Time-Invariant Block Code Obtained from Its Generator Matrix

Let us come back to binary codes, in order to introduce a topic which is of fundamental importance in evaluating the computational complexity of possible soft-decision decoding applied to block codes. First of all, it is useful to remember that hard-decision decoding, performed by means of algorithms with acceptable computational complexity, is possible for BCH codes, up to their designed correction capability τ (Wicker 1995). Other algorithms, for instance majority-logic, error trapping or Meggitt decoding (Clark and Cain 1983), can be employed for particular codes, but give modest results being confined to small values of n and t. Therefore the possibility of soft-decision decoding is very attractive, but it is limited by computational complexity. In the following, a brief introduction to the most frequently proposed method for doing this, before the advent of modern coding (mainly related to turbo codes and LDPC codes), will be presented.

Given an s.s. time-invariant polynomial code (cyclic, shortened, lengthened, or modified lengthened), the temporal evolution of its state diagram, obtained with the encoder circuit reported in Fig. 1.3 (non-systematic version) or in Fig. 1.5 (systematic version), gives rise to a geometrical structure where a soft-decision decoding algorithm can be applied. It is based on the comparison between proper metrics in the states where two possible paths merge. Such decoding algorithm was initially proposed only for convolutional codes (Viterbi 1967), but it has been easily adapted also to the case of a block code, which will be treated here.

Definition 1.40 The temporal evolution of a state diagram obtained from polynomial $g(x)$, either in a block or in a convolutional code, is called *generator trellis*. The states where two paths merge are called *merge states*.

The specification that we are treating a generator trellis is due to the fact that also another type of trellis can be considered, the one associated to the parity check polynomial, which will be discussed in Chap. 5. The case of a convolutional code, still derived from a generator matrix approach, will be analyzed in Chap. 3.

Let k be the number of information symbols. For the moment, we suppose that they directly occupy (or at least influence) as many first positions in the code word, that is the polynomial block code considered has code words which can be expressed by (1.6) or at least by (1.3). The trellis shows firstly an expanding part, followed sometimes by a stationary part, and finally always by a contracting part. In fact a stage of states where output bifurcations occur is associated to a symbol time where one information symbol is inserted. Instead, a stage of states where pairs of input branches join can be associated to a symbol time where one control symbol is inserted, but also to a symbol time where the constraint due to parameter r forces two different paths to merge. This description refers to a binary state diagram, and then to a binary trellis. A widening of this concept, in order to comprehend also non-binary codes, is easy, adopting a state diagram of the type treated in Sect. 1.20, and will be presented in Chap. 2.

The trellis of a block code whose word length is n shows $(n + 1)$ subsequent stages, starting from the stage characterized by the unique state $\underline{0}$ and ending at the stage characterized by the unique state $\underline{0}$. The trellis collects all the possible $\underline{0}$-to-$\underline{0}$ paths having length of n branches, originated by as many closed paths around state $\underline{0}$ in the state diagram.

Lemma 1.43 *Assuming a block polynomial code (shortened, cyclic or lengthened, but not MLC), characterized by generator polynomial $g(x)$, whose degree is r, with k information symbols in the first part of the code word, the stage with the maximum number of states contains 2^k states if $k < r$, contains 2^r states if $k > r$. The two situations coincide when $k = r$.*

Proof If $k < r$, there are states which cannot be touched (see Lemma 1.38 about touched states). The contracting part of the trellis begins immediately after its expanding part. Although the states can be identified in different ways, the maximum number of states per stage is reached at the $(k + 1)$th stage. This number is 2^k, because such stage follows k subsequent binary bifurcations. If $k > r$, the contracting part of the trellis does not begin immediately after its expanding part. Precisely there are $(k - r + 1)$ stages containing 2^r states each, forming a central stationary part. □

MLC codes cannot be treated in a way so simple, because they introduce distributed control symbols. The generator trellis of MLC codes, where a soft-decision decoding is possible up to a certain computational complexity, will be analyzed in Chap. 3, owing to their resemblance with convolutional codes. The following propositions better describes the structure of the generator trellis.

Lemma 1.44 *Under the same assumption as in Lemma 1.43, if $k < r$, the trellis exhibits $(r - k + 1)$ stages containing transit states (without branch joints or bifurcations) and then $(k - 1)$ final stages formed by merge states, whose number decreases as the powers of 2. If $k > r$, the trellis exhibits a central stationary part characterized by 2^r merge states in the stages $(r + 1)$th, $(r + 2)$th, ... kth. Also here the contracting part of the trellis shows stages whose number decreases as the power of 2.*

Proof If $k < r$, not all the states of the state diagram are reached, so originating some transit states after the expanding part of the trellis. They are associated to the insertion of the first control symbols. New states are touched but their total number per stage does not change. The number $(r - k + 1)$ of such stages comes out from the part of r exceeding k. On the contrary, if $k > r$, at the $(r + 1)$th stage, pairs of paths merge, due to the absence of possible further states, so leading to the stationary part of the trellis. For the whole length of this stationary part, any stage contains 2^r merge states. In both cases, in the contracting part of the trellis, the number of states per stage (all merge states) decreases after one halving at any stage. □

Example 1.74 In Fig. 1.32 the case of $k = 2$ and $r = 3$ is reported. It refers to the [5,2,3] SC code derived from the [7,4,3] Hamming cyclic code, whose $g(x)$ is

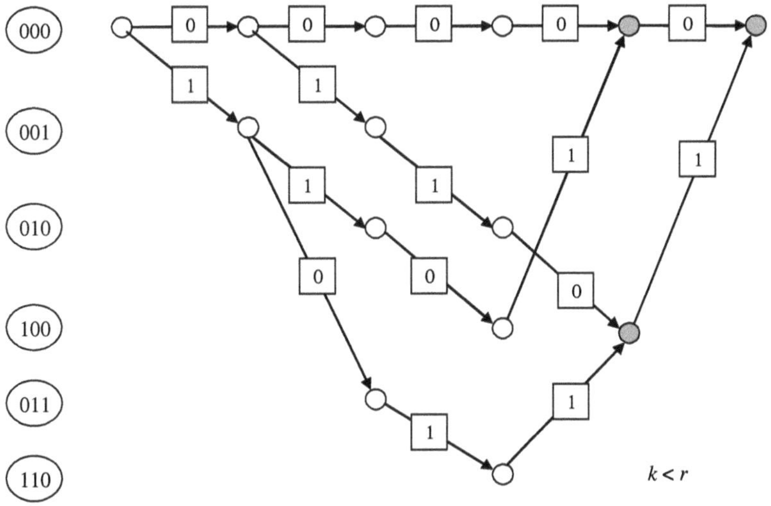

Fig. 1.32 Trellis obtained from the $\underline{G}$ matrix for the [5,2,3] SC code described in Example 1.74

associated to the sequence 1101. Merge states are denoted with solid small circles. It is possible to note the presence of $r - k + 1 = 2$ central stages containing transit states, and then a stage containing $2^{k-1} = 2$ merge states.

Example 1.75 In Fig. 1.33 a case with $k = 3$ and $r = 2$ is analyzed. It refers to the [5,3,2] two-time lengthened code derived from the [3,1,3] repetition cyclic code, whose $g(x)$ is associated to the sequence 111. Also here merge states are denoted with solid small circles. The stationary part of the trellis comprehends $k - r + 1 = 2$ stages. Except for the first one of them, any stage in general contains 2^r merge states. Here we have only one stage containing $2^r = 4$ merge states. It is easy to imagine what occurs for the [4,2,2] one-time lengthened code derived from the same [3,1,3] repetition code. In this case, characterized by $k = r$, we have only one stage with $2^r = 4$ states, and they are not merge states. Therefore the maximum number of merge states occurs in the subsequent stage, and this number is $2^{k-1} = 2^{r-1} = 2$, confirming what predicted in Lemma 1.43.

Remark that a precise identification of the states with the evolution of polynomial $u(x)$, as made in the trellises reported in Figs. 1.32 and 1.33, is possible only when we refer to an encoder circuit arranged as shown in Figs. 1.3 or in 1.5. It is not possible, although the structure of the trellises remains unchanged, if we refer to the systematic encoder circuit shown in Fig. 1.6, because its state evolution reports the current remainder and not the current quotient.

In order to evaluate the decoding computational complexity it is necessary to take into account the number of merge states and the number of stages where such merge states are placed. After Lemma 1.43 and Lemma 1.44, it is possible to summarize what follows. If $k \le r$, there are 2^{k-1} states in the first stage of the

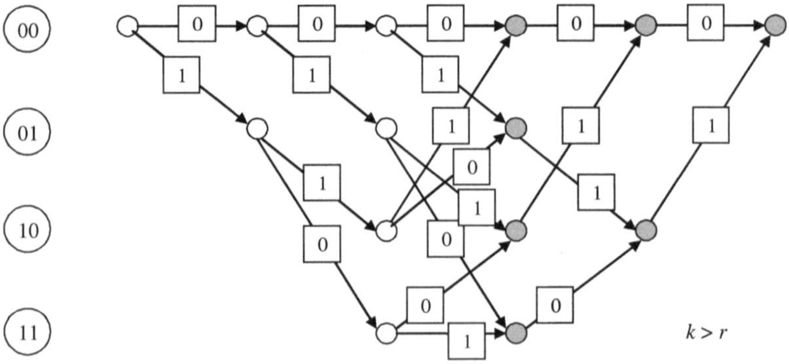

Fig. 1.33 Trellis obtained from the $\underline{G}$ matrix for the [5,3,2] LC code described in Example 1.75

contracting part of the trellis, and this stage is the one characterized by the maximum number of states. If $k > r$, there is at least one stage, along the stationary part of the trellis, where 2^r merge states are present.

In conclusion, the state computational complexity can be estimated as proportional to the quantity $\min\{2^{k-1}, 2^r\}$. When 2^r is the minimum number, we have to multiply also by the number of stages where this multiplicity occurs. The description given so far refers to polynomial block codes, but the same codes can be constructed also not in polynomial form, sometimes advantaging the decoding procedure. The problem of finding a minimal trellis (trellis with minimum number of merge states), in order to perform a soft-decision decoding of a certain block code, has been faced starting from late 1980. In Muder (1988) the above bound has been determined. In Forney (1988), Vardy and Be'ery (1994) it has been shown that, taking some equivalent of known polynomial s.s. time-invariant block codes, it was possible to reduce by far the maximum number of merge states with respect to this prediction. Such codes cannot be expressed in s.s. systematic form, and some control symbols are to be distributed along the code word. The resulting trellis turns out to be partitioned, leading to some independent sub-trellises, only converging in the final part of the structure. For instance, in Calderbank et al. (1998) a w.s. time-invariant version of the [24,12,8] extended Golay code has been proposed, with only 16 states per stage. These codes, and the soft-decision decoding algorithms performed in their simplified trellises, will be treated more in detail in Chap. 2.

References

Bose RC, Ray-Chaudhuri DK (1960) On a class of error correcting binary group codes. Inf Contr 3:68–79

Bossert M (1999) Channel coding for telecommunications. Wiley, Weinheim

Calderbank AR, Forney GD, Vardy A (1998) Minimal tail-biting trellises: the Golay code and more. In: Proceedings of ISIT 1998, Cambridge, USA, p 255

Cancellieri G, Marchesani R (2000) Transmission method using variable length code. Eur Patent EP 0 998070 A2
Clark GC, Cain JB (1983) Error correcting coding for digital communications. Plenum Publishers, New York
De Bruijn NG (1946) A combinatorial problem. J London Math Soc 21:167–169
Elias P (1954) Error-free coding. IRE Trans Inf Theor 4:29–37
Forney GD (1988) Coset codes. Part II: binary lattices and related codes. IEEE Trans Inf Theor 34:1152–1187
Golay MJE (1949) Notes on digital coding. Proc IRE 37:657
Gorenstein DC, Zierler N (1961) A class of error correcting codes in p^m symbols. J SIAM 9:207–214
Hagenauer J, Offer E, Papke L (1996) Iterative decoding of binary block and convolutional codes. IEEE Trans Inf Theor 42:429–445
Hocquenghem A (1959) Codes correcteurs d'erreurs (Error correcting codes). Chiffres 2:147–156
Krishna A, Sarwate DW (1990) Pseudocyclic maximum-distance separable codes. IEEE Trans Inf Theor 36:880–884
Kuriata E (2008) Creation of unequal error protection codes for two groups of symbols. Int J Appl Mat Comp Sci 18:251–257
MacWilliams FJ, Sloane NJA (1977) The theory of error correcting codes. North-Holland Publishing Company, New York
Muder DJ (1988) Minimal trellises for block codes. IEEE Trans Inf Theor 34:1049–1053
Peterson WW, Weldon EJ (1972) Error-correcting codes. MIT Press, Boston
Reed IS, Solomon G (1960) Polynomial codes over certain finite fields. J SIAM 8:300–304
Vardy A, Be'ery Y (1994) Maximum likelihood soft decision decoding of BCH codes. IEEE Trans Inf Theor 40:546–554
Viterbi AJ (1967) Error bounds for convolutional codes and an asymptotically optimum decoding algorithm. IEEE Trans Inf Theor 13:260–269
Wicker SB (1995) Error control systems for digital communications and storage. Prentice-Hall, Englewood Cliffs

Chapter 2
Wide-Sense Time-Invariant Block Codes in Their Generator Matrix

2.1 Periodically Time-Varying Generator Matrix

In Chap. 1 we have considered s.s. time-invariant block codes in their generator matrix $\underline{G}$, where a unique generator polynomial $g(x)$ shifts along the rows of $\underline{G}$, which is upper triangular. Here we will study a wider class of block codes, whose $\underline{G}$ matrix is still upper triangular.

Definition 2.1 A generator matrix, where some generator polynomials $g_j(x)$, $j = 1, 2, \ldots q_o$, among which at least two are different, occupy as many consecutive rows, forming periods of length q_o and shifting by one position at a time, is called a *periodically time-varying generator matrix*.

Examples of time-varying generator matrices, employing some different generator polynomials, can be found in Sect. 1.6 for concatenated block codes and in Sect. 1.7 for punctured block codes. It is important to stress the periodic nature of the time-varying generator matrix introduced here. The treatment of the linear block codes, characterized by a generator matrix of the type described in Definition 2.1, is greatly simplified if none of the polynomials $g_j(x)$ has its highest-order coefficient in correspondence with the highest-order coefficient of another one. This statement follows from the conditions for having a univocal polynomial interleaved division, as discussed in Appendix B. From another point of view, this choice corresponds to a generator matrix which is upper triangular (see Definition 1.7), and implies that all the rows of $\underline{G}$ are linearly independent. In the following, when the contrary is not explicitly said, we will consider only such type of $\underline{G}$ matrix.

Lemma 2.1 *The linear block code associated to a generator matrix constructed according to Definition* 2.1, *without overlapping of highest-order coefficients in its generator polynomials* $g_j(x), j = 1, 2, \ldots q_o$, *with* $\mathrm{w}[g_j(x)] \geq 2$, *may exhibit minimum distance* $d = 1$.

G. Cancellieri, *Polynomial Theory of Error Correcting Codes*,
Signals and Communication Technology, DOI 10.1007/978-3-319-01727-3_2

Proof In a cyclic block code, where a unique generator polynomial is present, owing to Theorem B.4 in Appendix B, the minimum distance, also after any possible code lengthening, is d = 2. Here we may have even lower values of d. Nevertheless at least d = 1 will remain, because of the choice of having highest-order coefficients in the generator polynomials without overlapping. □

Definition 2.2 A *wide-sense* (or briefly *w.s.*) *time-invariant linear block code, in its generator matrix*, with period q_o, is a code whose words are obtained as a polynomial interleaved product of the type

$$t(x) = x^{q_o-1}u_1(x^{q_o})g_1(x) + x^{q_o-2}u_2(x^{q_o})g_2(x) + \cdots u_{q_o}(x^{q_o})g_{q_o}(x), \qquad (2.1)$$

where $u_j(x), j = 1.2, \ldots .q_o$, are binary sequences related to information symbols, and polynomials $g_j(x)$ are chosen according to Definition 2.1.

For more details on interleaved product, see Appendix B. In (2.1) all the highest-order coefficients in any polynomial $g_j(x)$ must occupy different positions, furthermore each one of the q_o positions in a period contains an information symbol, so there are no positions available for control symbols. Only the last periods in the code word contain control symbols. In order to assure the above property, it is necessary to select the polynomial characterized by the maximum degree, and apply a proper shift factor x^{Δ_i} to any other polynomial. In this way, all the polynomials are forced to have the same degree (see Lemma B.28), condition required for making interleaved division univocal, and able to be interpreted as the inverse operation with respect to an interleaved product. In other terms, the generator polynomials are aligned, occupying subsequent positions in any period, on their left hand side (highest-order coefficients). This preliminary operation is assumed to be already performed, before writing (2.1).

Finally we can observe that expression (2.1) is proper for a non-systematic version of our code, that is with the information symbols not directly reproduced in the first positions of the code word. It can be considered as a generalization of expression (1.3), which holds for s.s. time-invariant codes. The possibility of obtaining a systematic counterpart of (2.1) regards the encoder circuit, and will be considered later.

Example 2.1 Taking the pair of generator polynomials $g_1(x) = (x^2 + x + 1)x$ and $g_2(x) = x^3 + x + 1$, matrix $\underline{G}$ appears as shown in Fig. 2.1 up. In it, 5 periods where information symbols are introduced have been considered, leading to an overall word length n = 13. In this computation the length of the last two generator polynomials is included. It is possible to conclude that there will be three control symbols placed in the final part of the code word. This code produces the evolution of word weight distribution up to weight w = 4, with increasing the period number z, reported down in the same Figure. We can observe that the first 1-weight code word appears in the second period. Since $q_o = 2$, any increase by one unit in the period number z implies a multiplication by four of the code word number.

$$
\underline{G} = \begin{array}{ccccccccccccc}
1 & 1 & 1 & & & & & & & & & & \\
 & 1 & 0 & 1 & 1 & & & & & & & & \\
 & & 1 & 1 & 1 & & & & & & & & \\
 & & & 1 & 0 & 1 & 1 & & & & & & \\
 & & & & 1 & 1 & 1 & & & & & & \\
 & & & & & 1 & 0 & 1 & 1 & & & & \\
 & & & & & & 1 & 1 & 1 & & & & \\
 & & & & & & & 1 & 0 & 1 & 1 & & \\
 & & & & & & & & 1 & 1 & 1 & & \\
 & & & & & & & & & 1 & 0 & 1 & 1
\end{array}
$$

n	z	$A(0)$	$A(1)$	$A(2)$	$A(3)$	$A(4)$
5	1	1			2	1
7	2	1	1	1	4	5
9	3	1	3	4	7	14

Fig. 2.1 Generator matrix $\underline{G}$ and word weight distribution for the code described in Example 2.1

Example 2.2 Assuming the following pair of generator polynomials $g_1(x) = (x^3 + x^2 + x + 1)x$ and $g_2(x) = x^4 + x^3 + 1$, we have the $\underline{G}$ matrix shown in Fig. 2.2 up. In it, still 5 periods where information symbols are introduced have been considered, leading to an overall word length $n = 14$, although the 13th position is useless. The evolution of the word weight distribution up to weight $w = 6$ results as reported down in the same Figure. Here $d = 2$ is the minimum distance. The first 2-weight code word appears in the second period.

$$
\underline{G} = \begin{array}{cccccccccccccc}
1 & 1 & 1 & 1 & & & & & & & & & & \\
 & 1 & 1 & 0 & 0 & 1 & & & & & & & & \\
 & & 1 & 1 & 1 & 1 & & & & & & & & \\
 & & & 1 & 1 & 0 & 0 & 1 & & & & & & \\
 & & & & 1 & 1 & 1 & 1 & & & & & & \\
 & & & & & 1 & 1 & 0 & 0 & 1 & & & & \\
 & & & & & & 1 & 1 & 1 & 1 & & & & \\
 & & & & & & & 1 & 1 & 0 & 0 & 1 & & \\
 & & & & & & & & 1 & 1 & 1 & 1 & & \\
 & & & & & & & & & 1 & 1 & 0 & 0 & 1
\end{array}
$$

n	z	$A(0)$	$A(1)$	$A(2)$	$A(3)$	$A(4)$	$A(5)$	$A(6)$
6	1	1			2	1		
8	2	1		1	6	5	2	1
10	3	1		3	12	17	16	9
12	4	1		5	24	42	56	58
14	5	1		8	40	89	160	256
16	6	1		12	60	171	380	618

Fig. 2.2 Generator matrix $\underline{G}$ and word weight distribution for the code described in Example 2.2

Table 2.1 Evolution of the word weight distribution for the code presented in Example 2.3

n	z	$A(0)$	$A(1)$	$A(2)$	$A(3)$	$A(4)$	$A(5)$	$A(6)$
6	1	1			1	1	1	
8	2	1			5	5	2	2
10	3	1			10	16	12	12
12	4	1			17	38	44	52
14	5	1			28	77	112	168
16	6	1		2	41	127	258	484
18	7	1		4	58	204	518	1148

Lemma 2.2 *Under the assumptions of Lemma* 2.1, *if all the generator polynomials have the same degree without any shift factor* x^{Δ_i}, *the minimum distance of the code cannot be smaller than* $d = 2$.

Proof In this case, besides the initial 1-symbol on the left, after the interleaved multiplication, we have always also, at least, a final 1-symbol on the right. □

Example 2.3 Assuming $g_1(x) = x^4 + x^3 + x^2 + x + 1$ and $g_2(x) = x^4 + x^3 + 1$, characterized by the same degree, without the need of any shift factor x^{Δ_i}, we obtain the evolution of the word weight distribution shown in Table 2.1.

In Fig. 2.3 the ex-OR sums of generator polynomials leading to the first two 2-weight code words, which appear together for $z = 6$, are shown. This result has been obtained by means of interleaved division (see Appendix B). The former sum exhibits a code word $(x^{14} + 1)$ with 1-symbols both in odd positions, whereas the latter sum exhibits the same code word with 1-symbols both in even positions. This fact justifies the appearance of a pair of 2-weight code words in the same period. As already noted, z expresses the number of periods where couples of information symbols are introduced. The overall number of periods is greater, because of the degrees of $g_1(x)$ and $g_2(x)$. In other terms, 8 overall periods are needed, since there are 4 control symbols in the last two periods. Finally we can remark that Lemma 2.2 represents only a sufficient (not necessary) condition for having $d \geq 2$, as Example 2.2 clearly shows.

```
1 1 1 1 1                          1 1 0 0 1
  1 1 0 0 1                          1 1 1 1 1
      1 1 0 0 1                        1 1 0 0 1
          1 1 0 0 1                          1 1 1 1 1
            1 1 1 1 1                            1 1 1 1 1
                1 1 1 1 1                              1 1 0 0 1
                  1 1 0 0 1
                    1 1 1 1 1
```

Fig. 2.3 Ex-OR sums of generator polynomials to give 2-weight code words, odd-to-odd position and even-to-even position, both on 8 overall periods

Definition 2.3 We define *distance-1 w.s. time-invariant block code in its $\underline{G}$ matrix* a code of the type described so far, which, after a certain length n, exhibits a minimum distance $d = 1$.

Clearly this characteristic makes such type of codes very bad, so they must be absolutely avoided. Fortunately the first 1-weight code word appears typically after few periods, allowing easy recognition of good codes and rejection of possible distance-1 codes. Remark that this behaviour resembles that of a punctured cyclic code, when code puncturation is too strong.

2.2 Quasi-Cyclic Codes (QC) as a Widening in the Concept of Cyclic Codes

Let us consider a w.s. time-invariant block code, not distance-1, whose code word length is of zq_o positions plus a certain number of positions able to collect the tails of the generator polynomials, which take place after the zth period. The number of periods, as well as the number of information symbols, can be as high as one wishes. After a certain length n, as in a cyclic code, the minimum distance (from a higher value if present) falls to $d = 2$. Nevertheless, many non-distance-1 codes reach minimum distance $d = 2$ originating a code word coincident with a binomial $(x^C + 1)$, with C not integer multiple of the period length q_o. Then the above binomial is characterized by 1-symbols in different positions of the period.

Example 2.4 Assuming the pair of polynomials $g_1(x) = (x^2 + x + 1)x^2$ and $g_2(x) = x^4 + x + 1$, the code obtained with them exhibits the evolution of the word weight distribution shown in Table 2.2.

In the third period, just one 2-weight code word appears. This means that such code word begins in a position of the period and ends in a different position of the period. Precisely, we have one 2-weight code word, from even position to odd position, of the type $(x^5 + 1)$, and another one 2-weight code word, from odd

Table 2.2 Evolution of the word weight distribution for the code presented in Example 2.4

n	z	$A(0)$	$A(1)$	$A(2)$	$A(3)$	$A(4)$	$A(5)$	$A(6)$
6	1	1			2	1		
8	2	1			5	5	2	2
10	3	1		1	11	15	11	11
12	4	1		3	20	36	42	48
14	5	1		5	36	75	112	175
16	6	1		7	58	145	268	495
18	7	1		11	84	249	586	1,227

```
1 1 1                                   1 0 0 1 1
  1 0 0 1 1                                   1 1 1
    1 1 1
      1 0 0 1 1
          1 0 0 1 1
            1 1 1
```

Fig. 2.4 Ex-OR sums of generator polynomials to give 2-weight code words, odd-to-even position and even-to-odd position, on different overall numbers of periods

position to even position, of the type $(x^9 + 1)$, as shown in Fig. 2.4, where the sums of generator polynomials leading to such code words are reported. It is interesting to note that the 1-symbol on the right in $(x^9 + 1)$ is not due to the last generator polynomial, but to the one preceding it. This fact occurs sometimes when the two generator polynomials have different degrees, and we are forced to apply a proper shift factor x^{Δ_i} in order to equate them.

In any case, Theorem B.8 assures that, starting from a dividend x^L, the interleaved division, if performed with an ordered set of equal-degree divisors, will lead to a remainder 1 in the same position of the period q_o, so inducing a large-scale periodicity. This behaviour resembles that occurring in a cyclic code. In the situation analyzed in Example 2.4, the code reaches minimum distance $d = 2$ before beginning a new cycle. In order to assure a large-scale periodicity (a cycle), on the contrary, the two 1-symbols in the 2-weight code word must belong to the same position in the period.

Definition 2.4 A *quasi-cyclic code* (or briefly *QC*) is a w.s. time-invariant block code in its $\underline{G}$ matrix, excluding distance-1 codes, which, on a certain length $n = N$, still exhibits $d > 2$, and, beyond this length, it shows $d = 2$, owing to the occurrence of 2-weight code words, coincident with the binomial $(x^N + 1)$, with N integer multiple of the period length q_o.

The code introduced in Example 2.4 is not a QC code, because its minimum distance falls to $d = 2$ forming a 2-weight code word between a pair of 1-symbols, in different positions of the period (n odd). On the other hand, the code introduced in Example 2.3 is a QC code when $z = 5$ is assumed, which means a word length $n = N = 14$. This QC code is then described by the tern [14,10,3]. In particular, the definition of QC code is reserved to the only code with word length exactly $n = N$. For smaller or higher values of n, it is possible to imagine that there are some shortened or lengthened versions of it, like occurs for cyclic codes. Such codes will be further investigated in following Sections. About the possibility of including in the class of QC codes also those of the type introduced in Example 2.4, it would be necessary to distinguish firstly a passage from odd positions to even positions and then a passage from even positions to odd positions, but we prefer to avoid this inclusion, in order to maintain the length N as that after which the first 2-weight code word appears.

It is important to note that Definition 2.4 is not the most widely adopted definition of quasi-cyclic codes (see for example Lin and Costello 2004). Here the choice of giving an immediate understanding of the similitude with cyclic codes has been preferred. Nevertheless, in the following, after a proper transformation of the $\underline{G}$ matrix, we will show the equivalence between such two interpretations.

Definition 2.5 In a QC code, the interleaved division of binomial $(x^N + 1)$ by the ordered set of generator polynomials $g_j(x), j = 1, 2, \ldots q_o$, gives rise to an exact interleaved quotient (see Definition B.21 and subsequent considerations in Appendix B). So we obtain an *ordered set of parity check polynomials* $h_j^{(i)}(x^{q_o}), j = 1, 2, \ldots q_o$, as the ordered quotients (see Lemma B.29) forming the above overall interleaved quotient, one for each position i of the period $(i = 1, 2, \ldots q_o)$.

Example 2.5 Assuming the code introduced in Example 2.3, that is characterized by $g_1(x) = x^4 + x^3 + x^2 + x + 1$ and $g_2(x) = x^4 + x^3 + 1$, from Fig. 2.3 it is easy to obtain the following pair of parity check polynomials. In position $i = 1$: $h_1^{(1)}(x^2) = x^{10} + x^4 + x^2 + 1$, $h_2^{(1)}(x^2) = x^{10} + x^8 + x^6 + x^2$. In position $i = 2$: $h_2^{(2)}(x^2) = x^{10} + x^8 + 1$, $h_1^{(2)}(x^2) = x^{10} + x^6 + x^4$. This result is better explained in Fig. 2.5. Remark the inversion of subscripts for $i = 2$. Superscripts refer to the position in the period where 1-symbols are placed in the code word $(x^N + 1)$, whereas subscripts are related to the generator polynomial taken in the sum. Remark also the importance of 0-symbols for a correct relative positioning of such polynomials.

Definition 2.6 The interleaved ex-OR sum of the ordered parity check polynomials corresponding to a same position in the period (same value for superscript i)

$$h_{\text{int}}^{(i)}(x) = \sum_{j=1}^{q_o} x^{q_o - j} h_j^{(i)}(x^{q_o}), \quad i = 1, 2, \ldots q_o, \tag{2.2}$$

is called the ith *interleaved parity check polynomial*. It is the interleaved quotient in the interleaved division of $(x^N + 1)$ by the ordered set of generator polynomials, each associated to the ith position in the period.

```
1 1 1 1 1                          1 1 0 0 1
  1 1 0 0 1                          1 1 1 1 1
      1 1 0 0 1                        1 1 0 0 1
          1 1 0 0 1                          1 1 1 1 1
            1 1 1 1 1                            1 1 1 1 1
                1 1 1 1 1                              1 1 0 0 1
                  1 1 0 0 1
                    1 1 1 1 1
```

$h_1^{(1)}(x^2) =$ 0 1 0 0 1 1 1 $\quad h_2^{(2)}(x^2) =$ 0 1 1 0 0 0 1

$h_2^{(1)}(x^2) =$ 0 1 1 1 0 1 0 $\quad h_1^{(2)}(x^2) =$ 0 1 0 1 1 0 0

Fig. 2.5 Determination of the parity check polynomials, over a QC code word length $N = 14$ (Examples 2.3 and 2.5)

Example 2.6 The QC code introduced in Example 2.3 is characterized by the following two interleaved parity check polynomials (see also Example 2.5 and Fig. 2.5)

$$h_{\text{int}}^{(1)}(x) = x^{11} + x^{10} + x^8 + x^6 + x^5 + x^3 + x^2 + x,$$
$$h_{\text{int}}^{(2)}(x) = x^{11} + x^{10} + x^9 + x^6 + x^4 + x.$$

Since the parity check polynomials of the same ordered sequence act on different positions, the weight of the correspondent interleaved parity check polynomial will be the sum of their weights. It is useful to remark that the following set of equations separately hold

$$\sum_{j=1}^{q_o} x^{q_o-j} h_j^{(i)}(x^{q_o}) g_j(x) = x^N + 1, \quad i = 1, 2, \ldots q_o. \tag{2.3}$$

Lemma 2.3 *A QC code, whose ordered generator polynomials $g_j(x)$, employed in an interleaved division, exactly divide the binomial $(x^N + 1)$, shows this property in each one of the q_o positions of the period.*

Proof As already stressed, N is multiple of q_o. Uniqueness of N is due to Theorem B.8, which assures the same large-scale periodicity in all the positions of the period. □

As a consequence of this property, the first period beyond the quasi-cyclic length N of our code will show a contemporary appearance of a number q_o of 2-weight code words, all of the type $(x^N + 1)$, one for each position of the period. For instance, the QC code introduced in Example 2.3, where $q_o = 2$, is characterized by the contemporary appearance of two 2-weight code words at the first period beyond the quasi-cyclic length $N = 14$. After the [14,10,3] code, addition of a further period gives a [16,12,2] code with $A(2) = 2$. This can be considered as a lengthened quasi-cyclic code.

Definition 2.7 Considering the q_o ordered generator polynomials in a QC code, let $(r_p + 1)$ be the number of consecutive periods they occupy. Parameter r_p is called *generator constraint length* of the considered QC code.

In order to simplify the treatment, we will suppose, for the moment, that all the q_o generator polynomials have the same degree and their ensemble occupy completely all the above $(r_p + 1)$ consecutive periods, say without positions to be eliminated because always null in the last period. This assumption is not strictly necessary in order to assure that the code is QC, and will be removed when a more general situation will be under consideration.

Lemma 2.4 *In a QC code, if $r_p q_o$ expresses the (equal) degree of all the ordered generator polynomials $g_j(x)$, the code word overall length n can be so expressed*

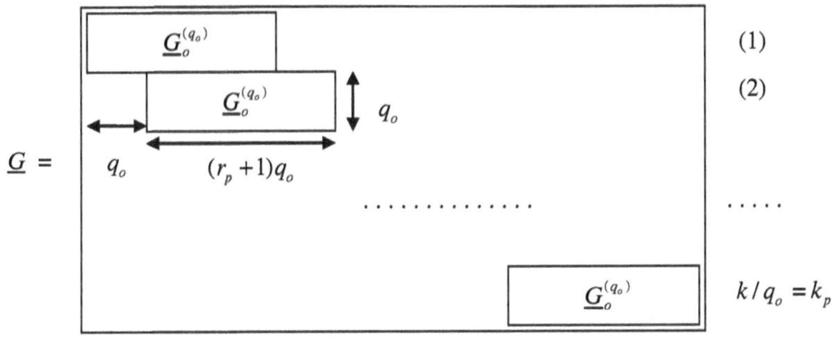

Fig. 2.6 Generator matrix of a QC code characterized by a generator sub-matrix $\underline{G}_o^{(q_o)}$

$$n = k + r_p q_o, \tag{2.4}$$

where k represents the total number of information symbols. In the QC condition, we have n = N.

Proof The number of columns in the $\underline{G}$ matrix, which equals the block length n, is formed by a first part due to the k information symbols (the number of rows of $\underline{G}$), plus a second part due to $r_p q_o$ positions necessary for terminating the last generator polynomials, with as many control symbols. In fact, being $\deg[g_j(x)] = r_p q_o$, any generator polynomial $g_j(x)$ in the last row layer of the $\underline{G}$ matrix exhibits its highest-order coefficient within the zth period, and its 0th-order coefficient within the $(z + r_p)$th period. A particular situation is that characterized by $n = N$, which is the length of the QC code. For $k < N - r_p q_o$ we have a shortened quasi-cyclic code. For $k > N - r_p q_o$, as already noted we have a lengthened quasi-cyclic code. □

Definition 2.8 In the $\underline{G}$ matrix of a QC code (and also shortened or lengthened), it is possible to identify a *generator sub-matrix* $\underline{G}_o^{(q_o)}$ with q_o rows and $(r_p + 1)q_o$ columns, containing its ordered set of generator polynomials, which shift by one period at a time along a sort a enlarged diagonal, as shown in Fig. 2.6.

Since both N and $(N - r_p q_o)$ are multiple of q_o, it is convenient to set

$$N - r_p q_o = k_p q_o, \quad N = p q_o, \tag{2.5}$$

so the code rate, in the QC condition, turns out to be

$$R = \frac{k_p}{p}, \quad \text{(QC code)}. \tag{2.6}$$

Both (2.5) mean that the number z of periods reaches the value $z = k_p$ in correspondence with the QC code length N. If the generator polynomials are characterized by different degrees, the crucial role in these considerations is played by the one showing the greatest degree. The meaning of parameter k and parameter k_p do not change. Sometimes, the true length of the code word could be a little shorter, and consequently the code rate a little higher, owing to elimination of some final useless positions.

Example 2.7 The code introduced in Example 2.3, characterized by $g_1(x) = x^4 + x^3 + x^2 + x + 1$ and $g_2(x) = x^4 + x^3 + 1$, exhibits the generator matrix $\underline{G}$ depicted in Fig. 2.7, where the separation into five sub-matrices $\underline{G}_o^{(q_o)}$ has been evidenced. Each sub-matrix spans $r_p + 1 = 3$ periods (separated by dashed vertical segments). This QC code is characterized by $N = 14$, $k = 10$, $q_o = 2$, and hence the number of generator sub-matrices $k_p = k/q_o$ is 5. There are no final useless positions.

Lemma 2.5 *Under the assumptions of Lemma* 2.4, *considering also* (2.5), *the following relationship holds among the parameters* p, k_p *and* r_p

$$p = k_p + r_p. \tag{2.7}$$

Proof Substitution of (2.5) into (2.4), for $n = N$, enables us to eliminate q_o and hence obtain (2.7). □

It is possible to consider cyclic codes as a particular sub-class of QC codes, characterized by $q_o = 1$. For them, equality $n = k + r$ replaces (2.4), and the condition $n = N$ is related to a particular value for k, exactly as here this particular value gives rise to (2.7).

Definition 2.9 A *cyclic permutation with steps of s over a length* $m = ps$, applied to a symbol sequence whose polynomial expression is $q(x)$, is represented by the following symbolic expression

$$q'(x) = x^{js} q(x)\big|^{(s)}_{\mathrm{mod}(x^m+1)}, \quad j = 1, 2, \ldots p. \tag{2.8}$$

Its meaning is to be intended as that of the traditional cyclic permutation in cyclic codes (see Sect. 1.8), with the only restriction on the shift to be integer multiple of parameter s.

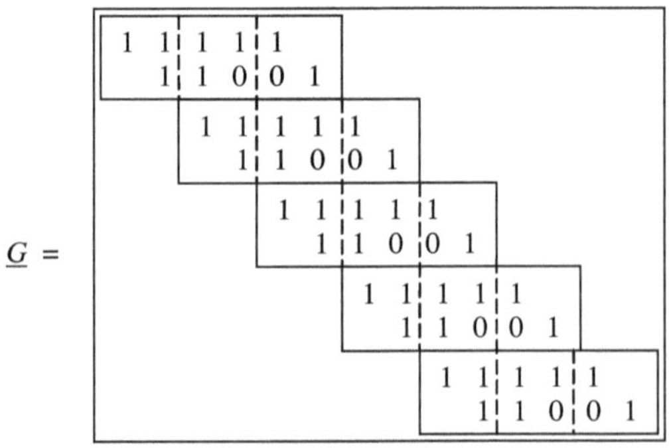

Fig. 2.7 Generator matrix of the QC code introduced in Example 2.3

Theorem 2.1 *In a QC code, characterized by period q_o, all the code words exhibit the following property*

$$t'(x) = x^{jq_o} t(x)|_{\mathrm{mod}(x^N+1)}^{(q_o)}, \quad j = 1, 2, \ldots p, \tag{2.9}$$

that is, given a certain code word t(x), any cyclic permutation with steps of q_o over a length $N = q_o$, obtained from it, is still a code word t'(x).

Proof This proof can be organized as that of Theorem 1.2 in Sect. 1.8, with the only difference of considering a polynomial interleaved division instead of a traditional polynomial division, and keeping in mind that, owing to (2.3), the remainder of the interleaved division of x^N by the ordered set of the q_o generator polynomials $g_j(x)$ is equal to 1. □

Corollary a *The parity check polynomials, and also the interleaved parity check polynomial, like in a cyclic code, are periodic with large-scale period N, but here with steps of q_o.*

Proof Any factor of $(x^N + 1)$ is also factor of $(x^{bN} + 1)$, with b representing whichever integer. Therefore the bth replica of this process, which interests a lengthened quasi-cyclic code, requires that the polynomials $h_j^{(i)}(x)$, as well as $h_{\mathrm{int}}^{(i)}(x)$, are prolonged according to a periodic behaviour, characterized by large-scale period N, with steps of q_o, up to the considered length (see also Lemma 1.35, which holds for cyclic codes). □

Example 2.8 Let us take the following pair a generator polynomials $g_1(x) = x^4 + x^3 + x^2 + 1$ and $g_2(x) = x^4 + x^2 + x + 1$. They give rise to a quasi-cyclic code, with length $N = 8$, whose generator matrix is

$$\underline{G} = \begin{bmatrix} 1 & 1 & 1 & 0 & 1 & & & \\ & 1 & 0 & 1 & 1 & 1 & & \\ & & 1 & 1 & 1 & 0 & 1 & \\ & & & 1 & 0 & 1 & 1 & 1 \end{bmatrix}.$$

It is easy to verify that, for instance, the ex-OR sum of the first three rows yields 10001110, which is a proper cyclic permutation with steps of 2 of the first row. The parity check polynomials associated to the original $\underline{G}$ matrix are obtained as shown in Fig. 2.8. Adding further rows to the above $\underline{G}$ matrix, one can verify the large-scale period, with length $N = 8$, exhibited by all these parity check polynomials.

It is possible to rearrange the rows of the above $\underline{G}$ matrix, by proper ex-OR sums, so obtaining a partitioned generator matrix $\underline{G}'$, equivalent to the previous one, which can be written as

```
11101              10111
 10111              10111
  11101              11101
    11101              10111
```

$h_1^{(1)}(x^2) = \; 0 \;\; 1 \;\; 1 \;\; 1$ $\qquad h_2^{(2)}(x^2) = \; 0 \;\; 1 \;\; 1 \;\; 1$

$h_2^{(1)}(x^2) = \;\; 0 \;\; 1 \;\; 0 \;\; 0$ $\qquad h_1^{(2)}(x^2) = \;\; 0 \;\; 0 \;\; 1 \;\; 0$

Fig. 2.8 Obtainment of the parity check polynomials for the code introduced in Example 2.8

$$\underline{G}' = \left[\begin{array}{ccccccc|c} 1 & 1 & 1 & 1 & 1 & 1 & 1 & 1 \\ & 1 & 1 & 1 & 0 & 0 & 0 & 1 \\ & & 1 & 0 & 1 & 1 & 0 & 1 \\ & & & 1 & 0 & 1 & 1 & 1 \end{array}\right].$$

Considering a [7,4,3] Hamming code, with generator polynomial $g(x) = x^3 + x + 1$, it is possible to recognize in $\underline{G}'$ the generator matrix of an extended version of this code (see Example 1.22). A vertical line separates the added parity check symbol, inserted in the final position. In the preceding seven positions there are four linearly independent code words of the [7,4,3] Hamming code, after an exchange between the first two columns on the left. Therefore the code introduced in Example 2.8 can be considered a QC equivalent of the [8,4,4] extended Hamming code.

2.3 Quasi-Cyclic Codes with Distributed Control Symbols Described with Their $\underline{G}$ Matrix

An alternative description of QC codes exists and is characterized by a word structure where the control symbols are distributed. Taking parameter q_o for the period length, $(r_p + 1)$ for the number of periods characterizing the generator sub-matrix $\underline{G}_o^{(q_o)}$, $N = pq_o$ for the overall length and $k = k_p q_o$ for the total number of information symbols, it is possible to construct a complete generator matrix $\underline{G}_{comp}$ as shown in Fig. 2.9. In it, the $q_o \times q_o$ square sub-matrices $\underline{g}_{(0)}, \underline{g}_{(1)}, \ldots \underline{g}_{(r_p)}$ are so related to the generator sub-matrix $\underline{G}_o^{(q_o)}$, introduced in Definition 2.8,

$$\underline{G}_o^{(q_o)} = \left[\underline{g}_{(r_p)} \;\middle|\; \underline{g}_{(r_p-1)} \;\middle|\; \cdots\cdots \;\middle|\; \underline{g}_{(1)} \;\middle|\; \underline{g}_{(0)} \right].$$

The last $r_p q_o$ rows (r_p row layers of sub-matrices) are obtained from the first $k_p q_o$, by means of proper cyclic shifts, with step of q_o, according to (2.9). They are linearly dependent on them. This redundant organization of the generator matrix of a QC code is needed in order to better understand the obtainment of a new version

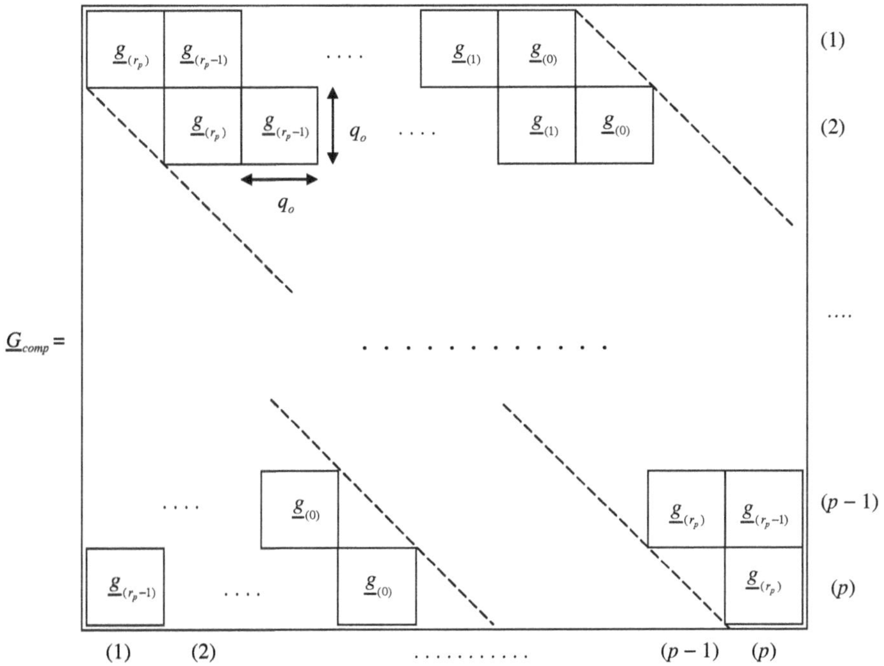

Fig. 2.9 Complete generator matrix as circulant of square blocks

of the generator matrix, in which control symbols are distributed along the code word. In order to develop this procedure some further definitions are required.

Definition 2.10 A *permutation circulant* is a square matrix which performs a permutation (of the type introduced in Definition 1.2), consisting of a cyclic shift applied to the input binary sequence.

Definition 2.11 A *circulant* is a square matrix formed by the ex-OR sum of a certain number of permutation circulants. Since the permutation circulants cannot have 1-symbol in the same position, the ex-OR sum of permutation circulants in the resulting circulant is the superimposition of them. The number of 1-symbols characterizing each one of its rows (or columns) is called *weight of the circulant.*

Example 2.9 In Fig. 2.10 up all the possible 5 × 5 permutation circulants are shown. They are characterized respectively by the shifts: 0 (identity matrix), 1, 2, 3, 4. The 5 × 5 circulant, of weight 3, obtained superimposing the permutation circulants having shifts 0, 1, 3, is reported in Fig. 2.10 down.

Clearly a permutation circulant is a circulant having weight 1. Definitions 2.10 and 2.11 can be widened also to comprehend square sub-matrices (called blocks), instead of 1-symbols. In this sense, the complete generator matrix $\underline{G}_{comp}$ in Fig. 2.9 is $N \times N$ square, but can be considered as a $p \times p$ circulant in the blocks $\underline{g}_{(i)}, i = 0, 1, \ldots r_p$. Each one of these blocks is $q_o \times q_o$ square.

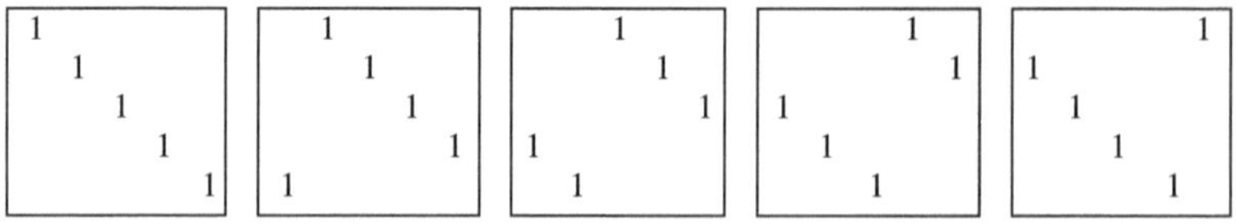

1	1	0	1	0
0	1	1	0	1
1	0	1	1	0
0	1	0	1	1
1	0	1	0	1

Fig. 2.10 The five possible 5 × 5 permutation circulants (shifts 0, 1, 2, 3, 4) and the 3-weight circulant constructed taking 0, 1, 3

Lemma 2.6 *It is possible to transform a $pq_o \times pq_o$ square matrix, which is a circulant of square $q_o \times q_o$ blocks, into a new $pq_o \times pq_o$ square matrix, which is a square block of $p \times p$* circulants, obtaining an equivalent of the original QC code.

Proof Let us take in sequence the columns occupying the 1st, the $(q_o + 1)$th, the $(2q_o + 1)$th, … positions, then the 2nd, the $(q_o + 2)$th, the $(2q_o + 2)$th, … positions, and so on. After this, let us take in sequence the rows occupying the 1st, the $(q_o + 1)$th, the $(2q_o + 1)$th, … positions, then the 2nd, the $(q_o + 2)$th, the $(2q_o + 2)$th, … positions, and so on. The resulting square matrix is subdivided into $p \times p$ circulants, in general all different. Since the operations performed consist of only changes in the order of columns or rows, after Lemma 1.6, the resulting code is equivalent to the original QC code. □

Theorem 2.2 *The generator matrix of a QC code with parameters $N = pq_o$, $k = k_pq_o$ and generator constraint length r_p, can be set in a form of the type shown in* Fig. 2.11, *that is a block of $k_p \times p$ truncated circulants (circulants in which the last r_p rows have been eliminated).*

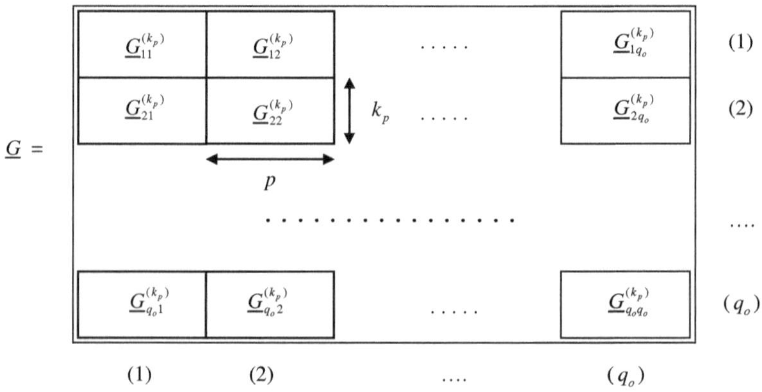

Fig. 2.11 Alternative structure for the generator matrix of a QC code

Proof Applying the transformation described in Lemma 2.6 to a QC code, from the complete matrix $\underline{G}_{comp}$ it is possible to obtain a new matrix, redundant as well, in which there are row layers of $p \times p$ circulants. Then we can eliminate r_p rows in any circulant, say the last ones, so remaining with k_p rows in any block $\underline{G}_{ij}^{(k_p)}$. In fact, from (2.7), we obtain $p - r_p = k_p$. The final result is a $k_p q_o \times p q_o$ generator matrix of the type shown in Fig. 2.11, with all linearly independent rows. □

The particular choice of eliminating the last r_p rows in any circulant enables one to set up a structural correspondence with the original $\underline{G}$ matrix, which was upper triangular. The new $\underline{G}$ matrix is no longer upper triangular, but the truncated circulants $\underline{G}_{ij}^{(k_p)}$ so obtained exhibit this property, taken one at a time. On the other hand, this choice is not the only possible one for obtaining the maximum number of linearly independent rows (that is a new $\underline{G}$ matrix having full rank).

In matrix algebra, the transformation adopted in Lemma 2.6 and Theorem 2.2 can be written symbolically as $\underline{M}_N \cdot \underline{G} \cdot \underline{M}_N^T$, where $\underline{M}_N$ is an $N \times N$ permutation matrix, in whose rows a solitary 1-symbol shifts with steps of q_o, and initial positions $1, 2, \ldots (q_o - 1)$. An example of this matrix is depicted in Fig. 2.12, for $q_o = 2$, $p = 3$ and hence $N = 6$. In particular, $\underline{G} \cdot \underline{M}_N^T$ represents column permutations.

It is important to stress that, in the new $\underline{G}$ matrix depicted in Fig. 2.11, the truncated circulants $\underline{G}_{ij}^{(k_p)}$ form layers, in general not cyclically invariant over the length of q_o block positions. Nevertheless the most important characteristic of this type of generator matrix is in the fact that control symbols are distributed along the code word. Precisely their number is r_p in each period.

Example 2.10 The QC code characterized by $g_1(x) = x^4 + x^3 + 1$ and $g_2(x) = x^4 + x^2 + x + 1$ exhibits $q_o = 2$, $N = 10$, and hence $p = 5$. Furthermore we have $r_p = 2$, so that $k_p = 3$. Therefore we are with a [10,6,3] QC code. Its generator matrix, containing three sub-matrices $\underline{G}_o^{(2)}$, its complete generator matrix $\underline{G}_{comp}$, the matrix $\underline{G}'_{comp}$ after column reordering, the matrix $\underline{G}''_{comp}$ after row reordering (both according to Lemma 2.6), and finally its new generator matrix, as block of truncated 3×5 circulants, are shown in Fig. 2.13.

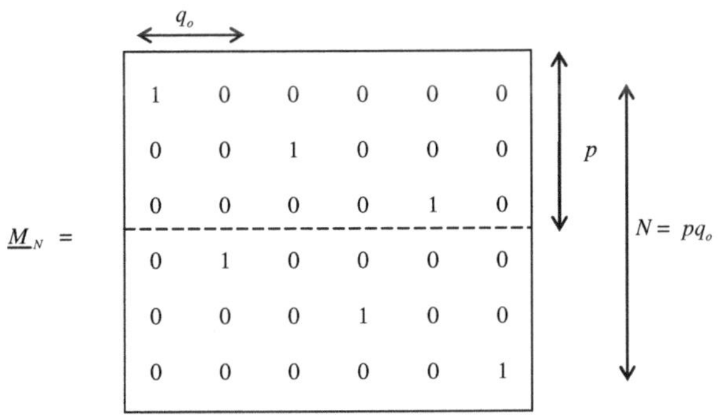

Fig. 2.12 Example of permutation matrix $\underline{M}_N$ for $q_o = 2$, $p = 3$ and hence $N = 6$

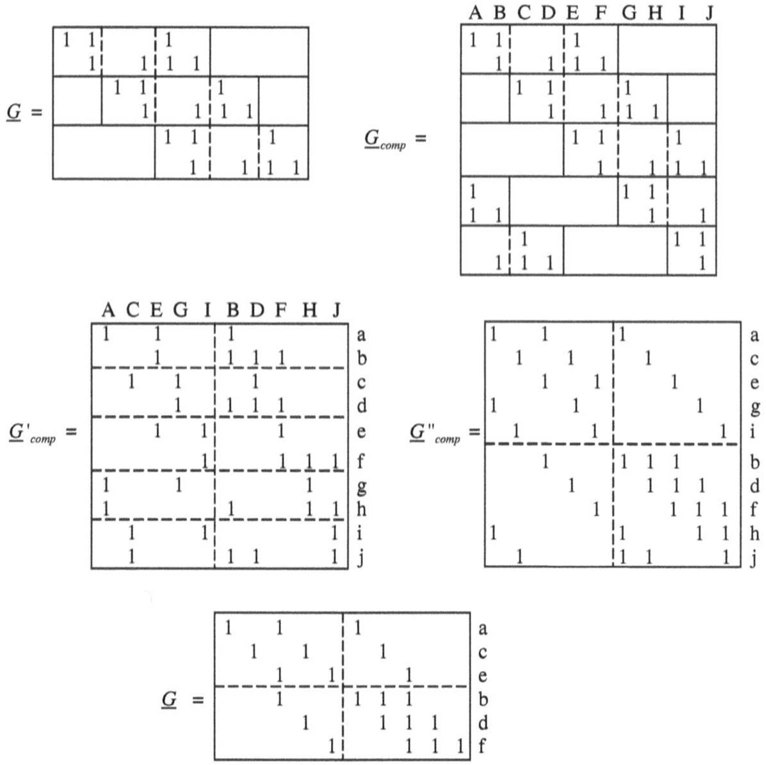

Fig. 2.13 Series of transformations on the generator matrix characterizing a [10,6,3] QC code

The $\underline{G}$ matrix so obtained shows periods whose length is p and generator sub-matrices having k_p rows. Keeping in mind how information and control symbols are distributed in the original code, it is reasonable to assume the first k_p positions in any period as devoted to information symbols, and the remaining $p - k_p = r_p$ positions to control symbols. A possible encoding procedure can be organized as follows: firstly encode (by means of an encoder circuit able to implement the $\underline{G}$ matrix shown in Fig. 2.6, which will be described in Sect. 2.6) and then reorder the positions in the code words so obtained. We can denote, as usual, the first k_p positions of any p-length period with solid small arrows to represent as many information symbols, and the remaining ones with dashed small arrows to represent as many control symbols. In Fig. 2.14, this association has been reported for a code word vector $\underline{t}$ characterized by $k_p = 3$, $r_p = 2$, $q_o = 2$ (the same parameters of the QC code introduced in Example 2.10). An organization of this type can be easily modified, increasing or decreasing parameter k_p, in order to obtain a lengthened quasi-cyclic code or a shortened quasi-cyclic code.

The advantage of a $\underline{G}$ matrix structure made by circulants consists in its ability to originate an important family of LDPC codes (Fan 2000) on which we will return in Sect. 2.7, and with more details in Chaps. 8 and 11. Sometimes further

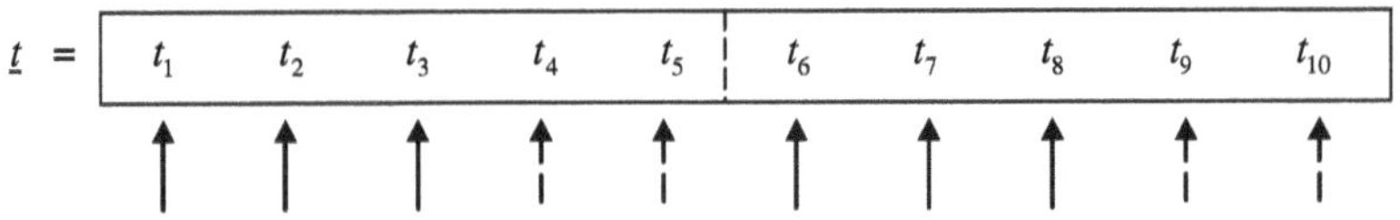

Fig. 2.14 Positions of information and control symbols after reordering for the QC code introduced in Example 2.10

manipulations can yield an upper triangular $\underline{G}$ matrix, although time-varying. In this case it is possible to have all the information symbols associated again to the first $k = k_p q_o$ positions.

Example 2.11 With reference to the $\underline{G}$ matrix reported down in Fig. 2.13, it is possible to sum its third, fourth and sixth row, obtaining the row 0000011111, substituted to the fourth, in such a way as to obtain the following equivalent form

$$\underline{G} = \left[\begin{array}{ccccc|ccccc} 1 & 0 & 1 & 0 & 0 & 1 & & & & \\ & 1 & 0 & 1 & 0 & 0 & 1 & & & \\ & & 1 & 0 & 1 & 0 & 0 & 1 & & \\ \hline & & & 1 & 0 & 0 & 1 & 1 & 1 & \\ & & & & 1 & 0 & 0 & 1 & 1 & 1 \\ & & & & & 1 & 1 & 1 & 1 & 1 \end{array}\right],$$

where the 4 control symbols are here located in as many positions at the end of the code word. In fact the new $\underline{G}$ matrix is upper triangular. Therefore an interleaved product could give a univocal result for constructing a code word in polynomial form, although time-varying (with three different generator polynomials).

So far we have studied just QC codes characterized by $q_o = 2$. It is however straightforward to widen this treatment in order to consider also greater values of q_o.

2.4 Representation of Known Block Codes as QC Codes with Distributed Control Symbols

Many good block codes, often available in s.s. time-invariant polynomial form, have an equivalent representation as QC codes, provided that their length $n = N$ is not a prime number. Typically they are directly expressed by a $\underline{G}$ matrix in a form of the type shown in Fig. 2.11 (block of truncated circulants), sometimes with layers characterized by different numbers of rows, especially when N and k have no common factors. In other situations, it is possible to have layers of complete square circulants (Chen and Peterson 1969).

Example 2.12 The $\underline{G}$ matrix obtained in Example 2.11 can be interpreted as subdivided into three layers of truncated circulants, the first with three rows, the second with two rows and the third with only one row.

Example 2.13 The cyclic [15,5,7] BCH code is characterized by a QC equivalent which can have either $q_o = 3$ (and $p = 5$) or $q_o = 5$ (and $p = 3$). Since the total number of information symbols is $k = 5$, the only possibility for having N and k with a common factor, is to choose $k_p = 1$, $q_o = 5$, and hence $r_p = 2$. In this situation, we obtain the following generator matrix

$$\underline{G} = \left[\begin{array}{ccccc|ccccc|ccccc}
1 & & & & & 1 & 1 & & 1 & & 1 & 1 & 1 & & \\
 & 1 & & & & & 1 & 1 & & 1 & & 1 & 1 & 1 & \\
 & & 1 & & & 1 & & 1 & 1 & & & & 1 & 1 & 1 \\
 & & & 1 & & & 1 & & 1 & 1 & 1 & & & 1 & 1 \\
 & & & & 1 & 1 & & 1 & & 1 & 1 & 1 & & & 1
\end{array}\right].$$

It is possible to observe that this matrix can be interpreted either in the form of Fig. 2.11, with just one sub-matrix $\underline{G}_o^{(5)}$, or in the form of a block of square circulants (having just one layer) with $p = 5$. Furthermore it is also in s.s. systematic form.

Example 2.14 Considering the QC equivalent of the [8,4,4] extended Hamming code introduced in Example 2.8, its complete generator matrix is as follows

$$\underline{G}''_{comp} = \left[\begin{array}{cccc|cccc}
1 & 1 & 1 & & 1 & & & \\
 & 1 & 1 & 1 & & 1 & & \\
1 & & 1 & 1 & & & 1 & \\
1 & 1 & & 1 & & & & 1 \\
\hline
 & & 1 & & 1 & 1 & 1 & \\
 & & & 1 & & 1 & 1 & 1 \\
1 & & & & 1 & & 1 & 1 \\
 & 1 & & & 1 & 1 & & 1
\end{array}\right].$$

Elimination of the last two rows in both layers yields

$$\underline{G} = \left[\begin{array}{cccc|cccc}
1 & 1 & 1 & & 1 & & & \\
 & 1 & 1 & 1 & & 1 & & \\
\hline
 & & 1 & & 1 & 1 & 1 & \\
 & & & 1 & & 1 & 1 & 1
\end{array}\right],$$

which is still upper triangular, but with two different polynomials in its first row layer and in its second row layer. Nevertheless it remains better interpreted as a 2×2 block of truncated circulants. In this sense, we have to pay attention to the following property: the same truncated circulant, having weight 3, passes from up-left to down-right, instead in going back from up-right to down-left, we have two different truncated circulants (although both with weight 1).

Example 2.15 There is also another possible organization for the complete generator matrix characterizing a QC equivalent of the [8,4,4] extended Hamming code. It exhibits a redundant form, before rearrangement, made by the following 8 × 8 matrix

$$\underline{G}_{comp} = \left[\begin{array}{cc|cc|cc|cc}
1 & & 1 & 1 & 1 & & & \\
 & 1 & 1 & 1 & & 1 & & \\
 & & 1 & & 1 & 1 & 1 & \\
 & & & 1 & 1 & 1 & & 1 \\
\hline
1 & & & & 1 & & 1 & 1 \\
 & 1 & & & & 1 & 1 & 1 \\
1 & 1 & 1 & & & & 1 & \\
1 & 1 & & 1 & & & & 1
\end{array}\right],$$

which is by itself in the form of a block of circulants. Elimination of the second row in each one of the four sub-matrices $\underline{G}_o^{(q_o)}$ here adopted gives us

$$\underline{G} = \left[\begin{array}{cc|cc|cc|cc}
1 & & 1 & 1 & 1 & & & \\
\hline
 & & 1 & & 1 & 1 & 1 & \\
\hline
1 & & & & 1 & & 1 & 1 \\
\hline
1 & 1 & 1 & & & & 1 &
\end{array}\right].$$

Now we have a unique polynomial shifting with step of 2 positions, and subjected to a cyclic permutation (going back from up-right to down-left without changes). The code can be then interpreted as s.s. time-invariant in its generator matrix, although the rows are cyclically shifted with steps of 2, so assuring four linearly independent rows.

2.5 Relation Between Some Binary QC-Codes and Cyclic or Pseudo-Cyclic Codes in a Q-Ary Alphabet

As already seen in Sects. 1.19, 1.20 and 1.21, in a non-binary alphabet, some polynomial codes exist, whose generator polynomial $g(x)$ is factor of $(x^N - s_j)$, s_j being a non-null symbol in the Q-ary considered alphabet, different from 1. Keeping in mind that, for cyclic codes, $g(x)$ is factor of $(x^N - 1)$, a so enlarged family of codes has been called pseudo-cyclic codes. Before introducing a correspondence between the behaviour of cyclic or pseudo-cyclic codes and that of certain QC codes, it is useful to report here a complete $\underline{G}$ matrix for a pseudo-cyclic code. Taking, for instance, the $[5,3,3]_4$ quaternary pseudo-cyclic Hamming code, whose code words have been listed in Example 1.65, we obtain

$$\underline{G}_{comp} = \begin{bmatrix} 1 & 1 & 3 & 1 & - \\ - & 1 & 1 & 3 & 1 \\ 3 & - & 1 & 1 & 3 \\ 2 & 3 & - & 1 & 1 \\ 3 & 2 & 3 & - & 1 \end{bmatrix},$$

where – denotes a null symbol. The particular characteristic of this behaviour is in the change which any symbol experiences in going back from up-right to down-left. We will call this behaviour pseudo-cyclic permutation. Clearly in a cyclic permutation, there is no change.

The total number of possible $p \times p$ circulants (of weight up to p) obtained in a binary alphabet is 2^p. They can be then formally associated to the symbols of a 2^p-ary alphabet (so assuming $Q = 2^p$). The same association can be constructed with reference to the truncated circulants obtained from them. The complete generator matrix $\underline{G}''_{comp}$, after reordering, sometimes shows a unique sub-matrix in any layer, except for its shift and cyclic or pseudo-cyclic permutation with steps of p. Such two possible types of permutation are made respectively changing or not changing the circulant which passes from the last column layer to the first column layer. In the former case we are treating a sort of 2^p-ary cyclic code, in the second case a sort of 2^p-ary pseudo-cyclic code (see Sect. 1.19 for better understanding this difference).

Remark that a correspondence of this type not always holds, since it requires that the same generator sub-matrix is present in any layer of $\underline{G}''_{comp}$, except for its cyclic or pseudo-cyclic shift and permutation. For instance, this is not for the QC code treated in Example 2.10 (Fig. 2.13). In other situations, such property may hold even directly for the complete generator matrix $\underline{G}_{comp}$, before reordering.

Example 2.16 The complete generator matrix $\underline{G}''_{comp}$, after reordering, obtained in Example 2.14 can be associated a 2^4-ary pseudo-cyclic code. It is possible to imagine an external super-cycle, whose length is greater than 1, before repeating the same 4-row sub-matrix.

Example 2.17 The complete generator matrix $\underline{G}_{comp}$, before reordering, treated in Example 2.15 can be associated to a 2^2-ary cyclic code, in fact no changes are needed in the cyclic permutations when a 2×2 circulant passes from the last column layer to the first column layer. The same is for the final $\underline{G}$ matrix, after reordering (with complete or truncated circulants).

The property of a given QC code in reordered form to be cyclic-like or pseudo-cyclic-like is important in order to fix the characteristics of its lengthening procedure (see Sect. 1.19 for Q-ary cyclic and pseudo-cyclic code lengthening).

Example 2.18 The [24,12,8] extended Golay code can be expressed in QC form employing the generator matrix shown in Fig. 2.15, where 4×8 blocks have been evidenced (Calderbank et al. 1998). We can observe one null block in any layer.

$$\underline{G} = \left[\begin{array}{cccccccc|cccccccc|cccccccc}
1&1&0&1&1&1&0&1&1&1&0&0&0&0&0&0&0&0&0&0&0&0&0&0\\
0&0&1&1&1&1&1&0&0&1&1&1&0&0&0&0&0&0&0&0&0&0&0&0\\
0&0&0&0&1&1&0&1&1&0&1&1&1&1&0&0&0&0&0&0&0&0&0&0\\
0&0&0&0&0&0&1&1&0&1&1&1&0&1&1&1&0&0&0&0&0&0&0&0\\
\hline
0&0&0&0&0&0&0&0&1&1&0&1&1&1&0&1&1&1&0&0&0&0&0&0\\
0&0&0&0&0&0&0&0&0&0&1&1&1&1&1&0&0&1&1&1&0&0&0&0\\
0&0&0&0&0&0&0&0&0&0&0&0&1&1&0&1&1&0&1&1&1&1&0&0\\
0&0&0&0&0&0&0&0&0&0&0&0&0&0&1&1&0&1&1&1&0&1&1&1\\
\hline
1&1&0&0&0&0&0&0&0&0&0&0&0&0&0&0&1&1&0&1&1&1&0&1\\
0&1&1&1&0&0&0&0&0&0&0&0&0&0&0&0&0&0&1&1&1&1&1&0\\
1&0&1&1&1&1&0&0&0&0&0&0&0&0&0&0&0&0&0&0&1&1&0&1\\
0&1&1&1&0&1&1&1&0&0&0&0&0&0&0&0&0&0&0&0&0&0&1&1
\end{array}\right]$$

Fig. 2.15 QC form of the generator matrix describing a [24,12,8] extended Golay code

This block and also the remaining two blocks shift along the three layers of such matrix. Inside each block, there are rows of different weights. So the blocks cannot be interpreted as truncated circulants. Nevertheless this structure can be imagined as obtained from a reordering procedure similar to that described in Fig. 2.11, but after a different selection of the rows to be eliminated as linearly dependent. In spite of this fact, we can consider such a code still a QC code, in an enlarged framework. The mechanism followed when a block passes from the last column to the first column resembles that of a cyclic code, in fact no changes occur. In this case, the generator polynomial is made by two only subsequent non-null 2^4-ary symbols (in which 1/2-rate redundancy is already present).

The advantage of organizing a generator matrix in such unusual way is in the possibility of arranging a soft-decision decoding algorithm, with reduced computational complexity (Forney 1988; Calderbank et al. 1998). Considering a matrix of the type reported in Fig. 2.15 for the [24,12,8] extended Golay code, it is possible to obtain four parallel independent trellises, having just $2^4 = 16$ states each, instead of a unique trellis with at least one stage with $2^{11} = 2048$ states, which would make any soft-decision decoding algorithm extremely heavy. This property holds also for the code treated in Example 2.15, although considering trellises by far simpler. The advantage of reducing computational complexity in the trellis of the QC version of known block codes will be investigated with more details in Sect. 2.11.

2.6 Encoder Circuits Based on the $\underline{G}$ Matrix for a QC Code

Figure 2.16 shows a non-systematic encoder circuit able to produce a w.s. time-invariant code in its $\underline{G}$ matrix of the type described by the generator matrix reported in Fig. 2.6, with $q_o = 2$. It is made by two shift registers in controller arrangement. The two generator polynomials $g_1(x)$ and $g_2(x)$ are characterized by coefficients g_{ji}, $j = 1, 2$, i variable from 0 to the degree of such polynomials. In this

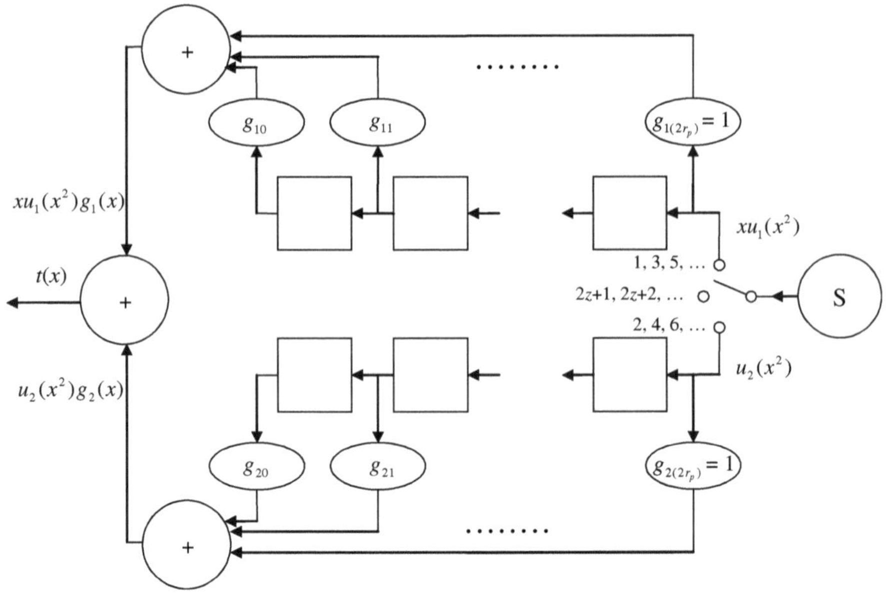

Fig. 2.16 Non-systematic encoder circuit for constructing a code word of the type (2.1) with $q_o = 2$

identification it is necessary to take into account also the convention introduced in Sect. 2.1 about the possible use of a shift factor, in order to make polynomial interleaved product and interleaved division a couple of reversible operations (see Appendix B and Sect. 2.1 for more details). Such convention gives rise to a unique degree r_p, so the highest-order coefficients $g_{1(2r_p)}$ and $g_{2(2r_p)}$ are both non-null, whereas, between g_{10} and g_{01}, it is enough that just one is non-null. We can justify the scheme reported in Fig. 2.16 by means of what discussed about Fig. C.10 in Appendix C.

The input switch is connected to the upper shift register during odd symbol times and to the lower shift register during even symbol times. The state evolutions $u_j(x^2), j = 1, 2$, directly feed the two shift registers, owing to the controller arrangement. Each one of these two alternated sequences acts like in a single MLC code, with $k_o = 1$ and $n_o = 2$. After the first $2z$ symbol times, the information source is disconnected, allowing that 0-symbols fill both shift registers. The encoding procedure regarding a code word ends with a series of 0-symbols in input to both shift registers, in order to reach state $\underline{0}$.

The state diagram, in general, is a 2^{q_o}-ary De Bruijn graph (De Bruijn 1946) The code words are represented in it by means of closed paths from state $\underline{0}$ to state $\underline{0}$, exactly as in the binary case. Nevertheless its structure resembles that of a Q-ary code. An example of the upper part of such diagram is shown in Fig. 2.17, for $q_o = 2$, $r_p = 3$, assuming $g_1(x) = x^6 + x^4 + x + 1$ and $g_2(x) = x^6 + x^4 + x^3 + x^2 + 1$. Branch and state labelling is as follows: the upper row describes input and

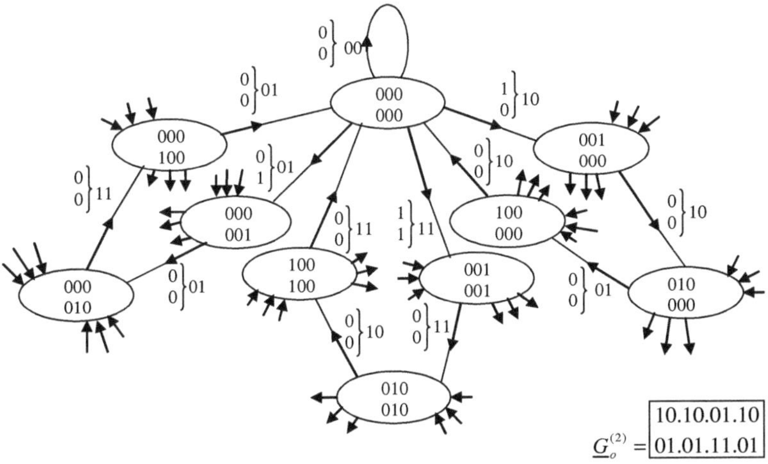

Fig. 2.17 *Upper* part of the state diagram for a non-systematic encoder circuit ($q_o = 2$, $r_p = 3$)

contents of the memory elements with reference to the upper shift register, the lower row the same with reference to the lower shift register, finally, after the graph bracket, a pair of output symbols is reported. Along the three closed paths around state $\underline{0}$, all four branches long, it is possible to recognize respectively $xg_1(x)$, $g_2(x)$ and $[xg_1(x) + g_2(x)]$. For any state, 4 branches go in and 4 branches go out. In general, this branch number is 2^{q_o}.

It is possible to arrange the encoder circuit described so far also in a systematic form. Following what discussed in Sect. 1.5 about the systematic encoder circuit in an s.s. time-invariant polynomial code, and in Appendix C for the polynomial interleaved division, we can organize a scheme of the type shown in Fig. 2.18. As long as information symbols are taken from the information source, they are subdivided into two fluxes: those placed in odd positions, described by $xi_1(x^2)$, and those placed in even positions, described by $i_2(x^2)$. This is assured by the input switch. The crossed feed-back wires are required because the current remainder is unique and is to be summed with the input sequences, but only when a proper quotient is to be calculated. To this purposes, the switch inserted on each one of such crossed feed-back wires separates even and odd symbols times. At the end, that is during the symbol times from $(2z + 1)$ to $2(z + r_p)$, the final remainder $c(x)$ is calculated, and attached to the information symbol sequence. Also owing to a possible shift factor, the degree of both generator polynomials is $\deg[g_1(x)] = \deg[g_2(x)] = 2r_p$, but the wire correspondent to the highest-order coefficient is not present, as usual in polynomial division.

After $2k_p$ symbol times (in general q_ok_p), the output switch attaches the control symbols collected in polynomial $c(x)$ in the final part of the code word. Such polynomial has degree at most $(2r_p - 1)$, being the final remainder of an interleaved division by a couple of polynomials whose degrees are $2r_p$.

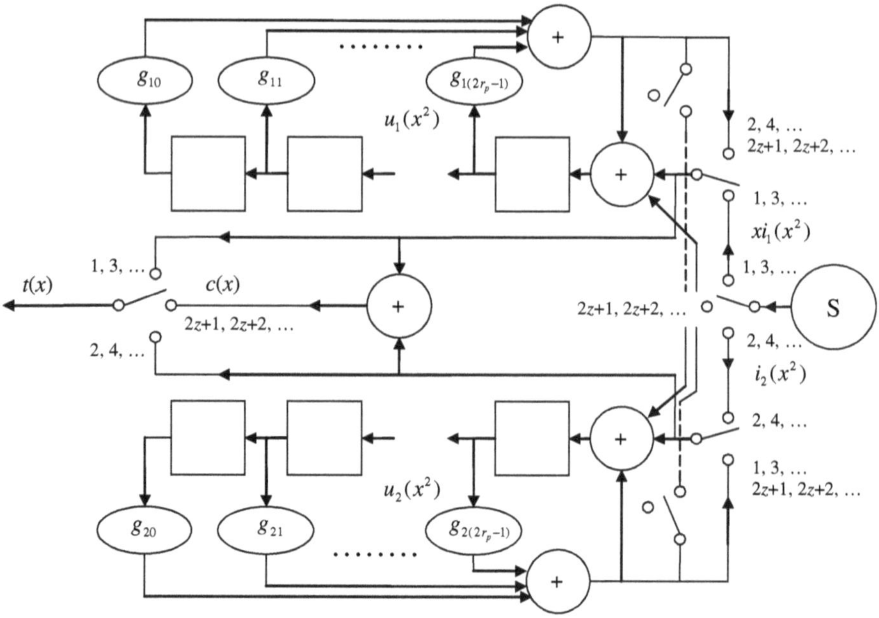

Fig. 2.18 Systematic encoder circuit for constructing a code word of the type (2.1) with $q_o = 2$

This encoder circuit is therefore able to transform a generator matrix of the type shown in Fig. 2.6 into its systematic equivalent. The quotients $u_1(x^2)$ and $u_2(x^2)$ are intermediate results of the calculation, and appear as state evolutions in the two shift registers. In this sense, they are characterized by one 0-symbols any two. This fact will be remarkably important in evaluating the computational complexity of a possible soft-decision decoding algorithm.

Example 2.19 Taken the [10,6,3] QC code introduced in Example 2.10, that is characterized by the two generator polynomials $g_1(x) = x^4 + x^3 + 1$ and $g_2(x) = x^4 + x^2 + x + 1$, a systematic encoder circuit of the type depicted in Fig. 2.18 is able to construct the following systematic generator matrix

$$\underline{G}_{sys} = \left[\begin{array}{cccccc|cccc} 1 & & & & & & 1 & 1 & 0 & 0 \\ & 1 & & & & & 1 & 0 & 0 & 1 \\ \hline & & 1 & & & & 0 & 0 & 1 & 1 \\ & & & 1 & & & 1 & 0 & 1 & 1 \\ \hline & & & & 1 & & 0 & 1 & 0 & 1 \\ & & & & & 1 & 0 & 1 & 1 & 1 \end{array}\right],$$

which can be interpreted as the systematic form of that reported in Fig. 2.13 up-left. The final remainder of each interleaved division is what reported in the last four columns of $\underline{G}_{sys}$. In Fig. 2.19 this particular encoder circuit is shown.

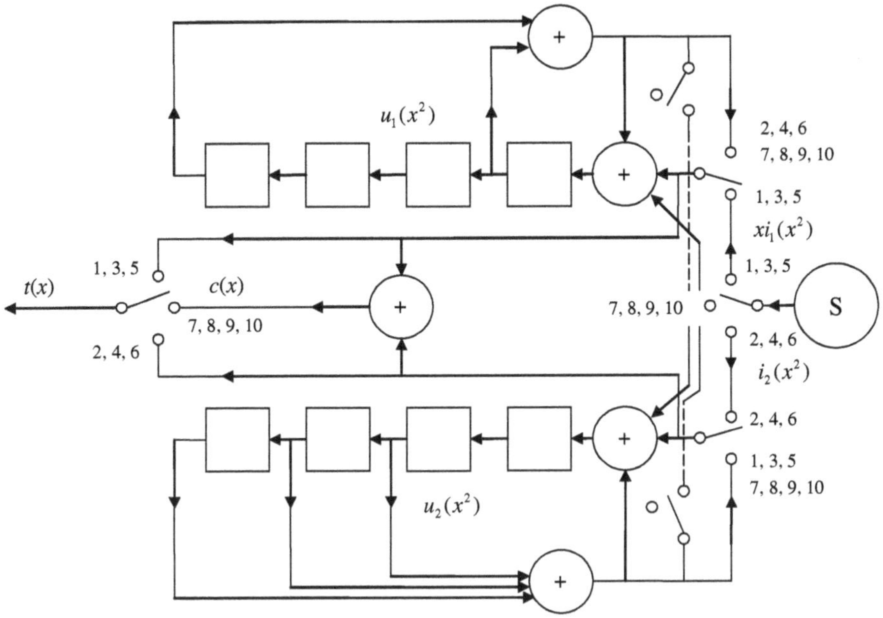

Fig. 2.19 Systematic encoder circuit for constructing the QC code described in Example 2.19

The generalization of the concept here exposed to the case of whichever $q_o > 2$ is obvious. It requires a circuit with q_o shift registers in controller arrangement and an input switch operating in such a way as to form q_o interleaved information symbol fluxes. In any case we have a unique final remainder in polynomial $c(x)$.

2.7 Shortened Quasi-Cyclic (SQC) Codes

Starting from a QC code, whose block length is $N = pq_o$, it is possible to assume a block length shorter than N (by a quantity multiple of q_o). This can be made in a way similar to what presented for cyclic codes. We will define the new code so obtained a shortened quasi-cyclic code. This form of code shortening can be arranged indifferently before or after the $\underline{G}$ matrix transformation described by Theorem 2.2. Furthermore, considering a wider class of QC codes, of the type characterized by a $\underline{G}$ matrix containing complete circulants, it is necessary to introduce a different form of code shortening.

Definition 2.12 Given a QC code, with parameters $p = k_p + r_p$, and q_o, it is possible to obtain an *s_o-time shortened version* of this code (briefly an *SQC code*), *with invariant generator sub-matrix* $\underline{G}_o^{(q_o)}$ (see Fig. 2.6), by operating a reduction in the word length and in the number of information symbols by the same quantity s_oq_o, with r_p and q_o constant.

Instead of varying k_p, which remains related to p and r_p through (2.7), it is preferable to vary the number k of information symbols, or equivalently the number z of periods where information symbols are inserted, taking $z < k_p$. It is useful to remember that, in (2.1), which holds for the QC code, the polynomials $u_j(x)$ have degree at most equal to $k_p - 1 = p - r_p - 1$. Now, an s_o-time shortened code, starting from its quasi-cyclic version, shows $\deg[u_j(x)]$ at most equal to $k_p - s_o - 1 = p - r_p - s_o - 1$. The word length, as a consequence, is reduced to $n = (p - s_o)q_o$.

If our considered codes (the quasi-cyclic and the shortened quasi-cyclic) are expressed in systematic form, we can imagine a polynomial expression for a code word, where the information symbols, represented by suitable polynomials $i_j(x)$, appear directly in its first part. The above constraint on $\deg[u_j(x)]$ is transferred, identical, to $\deg[i_j(x)]$. It is interesting to note what occurs in the reordered version of the generator matrix for an SQC code of this type.

Lemma 2.7 *An s_o-time shortened quasi-cyclic code with invariant generator sub-matrix $\underline{G}_o^{(q_o)}$ exhibits a $\underline{G}$ matrix in the reordered form represented by Fig.* 2.11, *in which any sub-matrix $\underline{G}_{ij}^{(k_p)}$ is transformed by eliminating its first s_o rows and s_o columns.*

Proof Any sub-matrix $\underline{G}_{ij}^{(k_p)}$ exhibits a number of columns equal to the number of column layers in the original $\underline{G}$ matrix (see Fig. 2.6). This number was p and is reduced here to $(p - s_o)$. Any sub-matrix $\underline{G}_{ij}^{(k_p)}$ exhibits a number of rows equal to the number of row layers in the original $\underline{G}$ matrix. This number was k_p and is reduced here to $(k_p - s_o)$. Such reductions are consequence of the elimination, in the original $\underline{G}$ matrix shown in Fig. 2.6, of s_o sub-matrices $\underline{G}_o^{(q_o)}$. The choice of eliminating the first rows and columns, instead of the last ones, has been made in analogy with what conventionally adopted for SC codes. □

The SQC code so obtained, for which property (2.9) does not hold, is characterized by the following expression for its code rate

$$R = \frac{k_p - s_o}{p - s_o}, \tag{2.10}$$

which is smaller and smaller with increasing s_o, as expected.

We can conclude that cyclic codes form a subset in the wider set comprehending QC codes. They are characterized by $\underline{G}_o^{(q_o)}$ consisting of just one row ($q_o = 1$), in which the unique generator polynomial $g(x)$ is present. Furthermore cyclic codes have $r_p = r$ and $p = N$. Finally, as regards code shortening, with the symbolism adopted in Chap. 1, SC codes are characterized by $s_o = s$.

Observing the generator matrix in its form of a block containing truncated circulants (Fig. 2.11), it is possible to say that this type of shortening does not modify the number of control symbols in any period, which remains r_p, but reduces the number of information symbols, which becomes $(k_p - s_o)$. The consequent elimination of rows and columns can be made from the original $\underline{G}$ matrix or from any truncated circulant in the reordered $\underline{G}$ matrix indifferently (the two codes are equivalent).

Example 2.20 Taking the [10,6,3] QC code introduced in Example 2.10, after elimination of $s_o = 1$ element in the ensemble containing the three $\underline{G}_o^{(2)}$ sub-matrices in its original generator matrix (up-left in Fig. 2.13), we obtain

$$\underline{G} = \left[\begin{array}{cc|cc|cc|cc} 1 & 1 & & & 1 & & & \\ & 1 & & 1 & 1 & 1 & & \\ \hline & & 1 & 1 & & & 1 & \\ & & & 1 & & 1 & 1 & 1 \end{array}\right].$$

In other terms, we have reduced z from 3 to 2, maintaining $\underline{G}_o^{(2)}$ unchanged. This is an [8,4,3] SQC code, whose generator matrix, in a form of the type shown in Fig. 2.11 (block of truncated circulants), turns out to be

$$\underline{G} = \left[\begin{array}{cccc|cccc} 1 & & 1 & & 1 & & & \\ & 1 & & 1 & & 1 & & \\ \hline & & 1 & & 1 & 1 & 1 & \\ & & & 1 & & 1 & 1 & 1 \end{array}\right].$$

It is interesting to note how the encoder circuits described in Sect. 2.6 can encode an SQC code as well, simply after a different sequence in the operations of the switches. Precisely the number of symbol times in which they assure connection with the information source is here reduced by $s_o q_o$ units. On the other hand, property (2.9) is no longer satisfied by SQC codes obtained after Definition 2.12, similarly to what occurs for cyclic and SC codes with respect to property (1.13).

A different class of QC codes is characterized by a generator matrix in the form of a block containing complete $p \times p$ square circulants, instead of truncated circulants, although maintaining the equality $N = pq_o$. Their redundancy is due to the number of row layers which is lower than that of column layers. Nevertheless some anomalous further rows can remain. Letting k the total number of information symbols, it is convenient to set

$$X = Q\left[\frac{k}{p}\right], \quad R_X = R\left[\frac{k}{p}\right]. \tag{2.11}$$

The quotient X represents the number of layers, where the same generator sub-matrix shifts, by one circulant size at a time, whereas the remainder R_X expresses the number of the above described anomalous rows, typically added in the lowermost part of the $\underline{G}$ matrix. When k is exactly divided by p, such set of separated rows is absent, and we can write $k = k_p p$. Otherwise a parameter like k_p cannot be defined here.

Example 2.21 The [15,5,7] BCH code in QC form introduced in Example 2.13, as already noted, can be interpreted according to the above description, that is directly with a $\underline{G}$ matrix in the form of a block of square 5×5 circulants, so that we have $k = p = 5$, $q_o = 3$, hence $X = 1$ and $R_X = 0$, and also $k_p = 1$. In the present case the generator sub-matrix coincides with the whole generator matrix.

Example 2.22 Let us consider the generator matrix shown in Fig. 2.20 up. It refers to a [16,9,4] QC code, characterized by parameters $n = 16$, $k = 9$, $q_o = 4$, $p = 4$, $X = 2$ and $R_X = 1$. We have the same 4×12 generator sub-matrix $\underline{G}_o^{(p)}$, made by 3 complete 4×4 circulants, shifting by a 4-length (in general p-length) period in the first two layers, plus one anomalous row. Its shortened version is reported down in the same Figure, and will be better explained in the following. The condition of maintaining the generator sub-matrix unchanged in this type of shortening will be adopted also for code lengthening (when the same unique generator sub-matrix will be repeated, properly shifted, a higher number of times). Finally we have a 9th row, consisting of a unitary polynomial having length $2p = 8$, which is not influenced by code shortening (and also by code lengthening).

The rows in the X layers, as well as the R_X final anomalous rows, exhibit the following property

$$t'(x) = x^j t(x)|_{\mathrm{mod}(x^p+1)}. \tag{2.12}$$

As usual, it means that, if $t(x)$ is a code word, also $t'(x)$ is a code word, but the operation mod(p) is to be performed separately in any p-length period. Remark that property (2.12) is equivalent to the description of the circulants, forming this type of $\underline{G}$ matrix, either in its X layers, or in its R_X anomalous rows (if present).

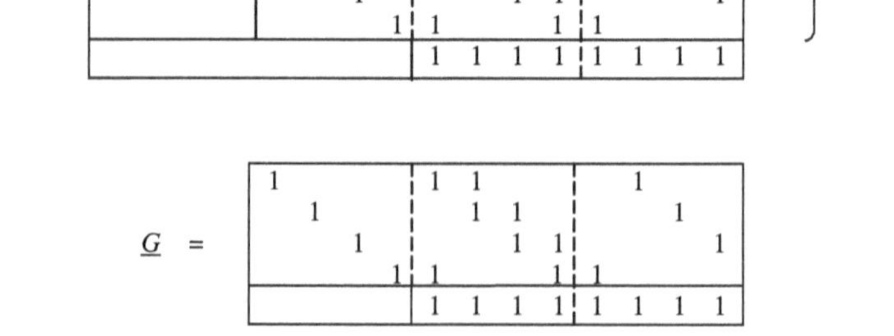

Fig. 2.20 Generator matrix of the [16,9,4] QC code and of its one-time shortened version, as described in Examples 2.22 and 2.24

In most literature, (2.12) is considered the basic property able to identify quasi-cyclic codes (Chen and Peterson 1969). Here we have preferred to introduce such codes starting from periodically time-varying polynomial codes in their generator matrix and then to apply the reordering procedure described in Lemma 2.6, so reaching the form of a block of circulants (complete or truncated), satisfying (2.12) as well. The forms of SQC codes constructed according to Definition 2.12, instead, do not satisfy property (2.12).

Lemma 2.8 *The code rate of a QC code constructed after* (2.11) *and* (2.12) *is*

$$R = \frac{Xp + R_X}{q_o p}. \tag{2.13}$$

Proof The number of information symbols (rows of the $\underline{G}$ matrix) is $k = Xp + R_X$. The number of columns of the $\underline{G}$ matrix is $n = N = q_o p$. □

A particular case is that characterized by $p = q_o$ (see Example 2.22). In Chaps. 8 and 11, codes so constructed (but better described by means of their parity check matrix) will be studied more in depth. They are known as array codes (Fan 2000) and, over lengths not longer than the one characterizing the QC condition, can be considered as very interesting LDPC codes. The number R_X of anomalous rows is related to the number of circulants spanned by the generator sub-matrix (Baldi et al. 2012).

For the class of QC codes whose $\underline{G}$ matrix is a block of complete $p \times p$ square circulants, forming X layers, plus possible R_X anomalous rows, it is useful to give the following alternative definition of code shortening, already anticipated in Example 2.22, for the particular case of a one-time shortening.

Definition 2.13 Given a QC code, with parameters p and q_o, whose $\underline{G}$ matrix, in the form of a block of complete circulants, contains X layers of $p \times p$ square circulants, plus possible R_X separated rows, we can obtain a *z_o-time shortened version* of this code (briefly an *SQC code*), *maintaining invariant parameter p*, after elimination of the first z_o row layers and z_o column layers of circulants from its $\underline{G}$ matrix.

It is possible to observe how parameter R_X is not involved in this type of code shortening. Furthermore such procedure can be adopted only when $X > 1$.

Lemma 2.9 *The code rate of a z_o-time shortened quasi-cyclic code with invariant parameter p, after the assumptions of Definition* 2.13, *turns out to be*

$$R = \frac{(X - z_o)p + R_X}{(q_o - z_o)p}. \tag{2.14}$$

Proof The number of rows of the $\underline{G}$ matrix before shortening was $k = Xp + R_X$. After shortening, it becomes $k' = (X - z_o)p + R_X$. The number of columns, from $n = q_o p$, becomes $n' = (q_o - z_o)p$. The ratio k'/n' gives the new code rate. □

Example 2.23 The QC code introduced in Example 2.21 cannot be shortened in the sense of Definition 2.13. In fact it is characterized by $X = 1$.

Example 2.24 Considering the QC code introduced in Example 2.22, elimination of the first row layer of square 4 × 4 circulants, together with the first column layer of 4 × 4 circulants enables us to produce a one-time shortening according to Definition 2.13. This is made maintaining invariant $p = 4$, and reducing X from 2 to 1, as well as q_o from 4 to 3. The resulting [12,5,4] SQC code is characterized by the generator matrix shown in Fig. 2.20 down. The code rate is reduced from 9/16 to 5/12.

In this particular case, elimination of one column layer and one row layer means the suppression of one generator sub-matrix, like in the shortening procedure performed according to Definition 2.12, except for the presence of the last row, which remains unchanged. About such sub-matrix labelling, it is advisable to substitute a symbol $\underline{G}_o^{(p)}$ to the symbol $\underline{G}_o^{(q_o)}$. Remark that both generator matrices in Fig. 2.20 are upper triangular. Finally, if one adopts Definition 2.13, property (2.12) is satisfied also in the SQC code.

SQC codes, either in the form introduced by Definition 2.12 or in the form introduced by Definition 2.13, exhibit a smaller number of low-weight code words and sometimes even a higher minimum distance with respect to their QC parents. This behaviour is similar to that obtained for SC codes with respect to their cyclic parents, and is obtained also here at the expense of a reduced code rate.

2.8 Lengthened Quasi-Cyclic (LQC) Codes

Starting from a QC code, in the form introduced by Definition 2.4, whose block length is $N = pq_o$, it is possible to assume a block length longer than N (by a quantity multiple of q_o). This can be made in a way analogous to what presented for cyclic codes, but, similarly to the case of QC code shortening, also for QC code lengthening, there are two possibilities: with invariant generator sub-matrix $\underline{G}_o^{(q_o)}$ or with invariant generator sub-matrix $\underline{G}_o^{(p)}$.

Definition 2.14 Given a QC code, with parameters $p = k_p + r_p$, and q_o, it is possible to obtain a *y_o-time lengthened version* of this code (briefly an *LQC code*), *with invariant generator sub-matrix* $\underline{G}_o^{(q_o)}$ (see Fig. 2.6), by operating an increase in the word length and in the number of information symbols by the same quantity $y_o q_o$, with r_p and q_o constant.

In comparison with what discussed about SQC codes, where polynomials $u_j(x)$ or $i_j(x)$ can have degree at most equal to $(p - r_p - s_o - 1)$, here, for LQC codes, such polynomials can have degree at most $(p - r_p + y_o - 1)$. In practice, this is a situation opposite to that of an SQC code, having as reference the QC condition, for which the above information polynomials have degree at most $(p - r_p - 1)$.

Lemma 2.10 *A y_o-time lengthened quasi-cyclic code with invariant generator sub-matrix* $\underline{G}_o^{(q_o)}$ *exhibits a* $\underline{G}$ *matrix in the reordered form represented by Fig.* 2.11, *in which any sub-matrix* $\underline{G}_{ij}^{(k_o)}$ *is transformed by adding y_o rows and y_o columns.*

Proof Any sub-matrix $\underline{G}_{ij}^{(k_p)}$ exhibits a number of columns equal to the number of column layers of the $\underline{G}$ matrix in Fig. 2.6. This number was p and is increased here up to $(p + y_o)$. Any sub-matrix $\underline{G}_{ij}^{(k_p)}$ exhibits a number of rows equal to the number of row layers of the $\underline{G}$ matrix in Fig. 2.6. This number was k_p and is increased here up to $(k_p + y_o)$. Such increases are consequence of the addition, in the original $\underline{G}$ matrix shown in Fig. 2.6, of y_o further sub-matrices $\underline{G}_o^{(q_o)}$. □

The code rate of the LQC code so obtained, turns out to be

$$R = \frac{k_p + y_o}{p + y_o}, \tag{2.15}$$

which is higher and higher with increasing y_o, as expected. It is important to observe that property (2.9) is no longer valid (except when n becomes equal to an integer multiple of N). In this fact we can found a strict resemblance with the behaviour of cyclic and LC codes with respect to property (1.13).

In lengthened quasi-cyclic codes, after Definition 2.14 (with invariant generator sub-matrix $\underline{G}_o^{(q_o)}$), there are 2-weight code words as already observed in Sect. 2.2. Along the code lengthening procedure, they appear in groups of q_o elements any single lengthening. A discussion on the origin of such 2-weight code words can be developed analogously to what made for cyclic codes in Sect. 1.11, with the only difference of taking here a q_o-length period. Cyclic codes represent, also in this respect, the particular situation characterized by $q_o = 1$.

Example 2.25 A QC equivalent of the [8,4,4] extended Hamming code has been introduced in Example 2.8, and a new version of it in Example 2.15. They both can be y_o-time lengthened, leading to the same word weight distribution, reported for y_o up to 4 in Table 2.3. Also its one-time shortened version has been reported for comparison.

Table 2.3 Word weight distribution of the family of codes described in Example 2.25

z	s_o	y_o	Code	$A(0)$	$A(2)$	$A(4)$	$A(6)$	$A(8)$	$A(10)$	$A(12)$	$A(14)$	$A(16)$
1	1		[6,2,4] SQC	1		3						
2	–	–	[8,4,4] QC	1		14		1				
3		1	[10,6,2] LQC	1	2	32	22	7				
4		2	[12,8,2] LQC	1	4	70	108	65	8			
5		3	[14,10,2] LQC	1	6	139	356	391	118	13		
6		4	[16,12,2] LQC	1	8	252	952	1,670	952	252	8	1

It is possible to observe the appearance of an increasing number of 2-weight code words. Remark that, for $z = 6$, that is after a complete lengthening period, the [16,12,2] code exhibits a symmetric word weight distribution, like the QC parent code [8,4,4].

We can argue, from similitude with cyclic and LC codes ($q_o = 1$), that in an y_o-time lengthened quasi-cyclic code, the number of 2-weight code words is $q_o y_o$, as long as $y_o \leq p$. This property is confirmed by the results collected in Table 2.3. Then a binomial increase in this number occurs. As in cyclic codes, it is important to avoid that the parity check polynomials are reproduced by the symbol sequences $u_j(x^{q_o})$, imposed by the information source. For w.s. time-invariant polynomial codes in $\underline{G}$, in order to assure this condition, it is enough that the degree of all the polynomials $u_j(x)$ is not greater than the bound $(p - r_p - 1)$. So the code must be QC or SQC. The addition of a suitable selection rule in the degrees of freedom expressed by polynomials $u_j(x^{q_o})$, so forming a subcode of our LQC code, can eliminate 2-weight code words, independently of the entity of y_o. This topic will be treated more in detail in following Sections, adopting the concept of modified lengthened codes as a reference.

Considering now the generator matrix in its form of a block containing truncated circulants (Fig. 2.11), it is possible to conclude that this type of code lengthening does not modify the number of control symbols in any p-length period (which remains r_p), but increases the number k of information symbols.

Example 2.26 Taking the [10,6,3] QC code introduced in Example 2.10, after addition of one $\underline{G}_o^{(2)}$ sub-matrix in its generator matrix, we obtain what shown in Fig. 2.21. The coefficients of two pairs of parity check polynomials are reported under the generator matrix. Their construction is made by polynomial interleaved division, as usual. It is easy to verify that they produce (if adopted as the sequence of information symbols) an even-to-even and an odd-to-odd code word of the type $(x^{10} + 1)$. Taking into account that parity check polynomials are periodic, with

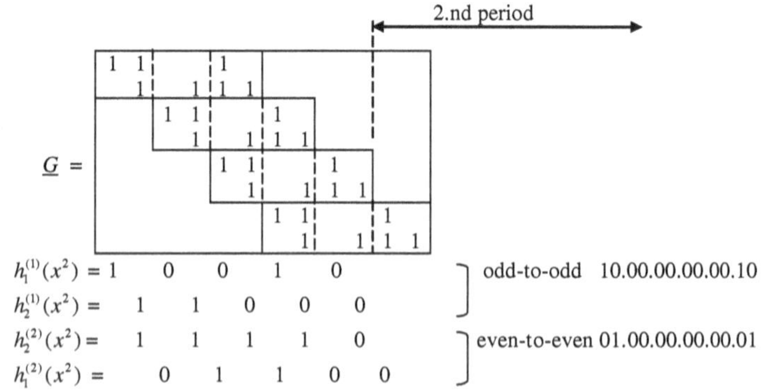

Fig. 2.21 Generator matrix and parity check polynomials for the [12,8,2] LQC code considered in Example 2.26

period N, like in LC codes (Corollary a of Theorem 2.1), in Fig. 2.21 the beginning of the second overall period has been evidenced. From these considerations, we can write le following periodic expressions

$$h_1^{(1)}(x^2) = (x^8 + x^2)(1 + x^{10} + x^{20} + \cdots),$$
$$h_2^{(1)}(x^2) = (x^8 + x^6)(1 + x^{10} + x^{20} + \cdots),$$
$$h_2^{(2)}(x^2) = (x^8 + x^6 + x^4 + x^2)(1 + x^{10} + x^{20} + \cdots),$$
$$h_1^{(2)}(x^2) = (x^6 + x^4)(1 + x^{10} + x^{20} + \cdots).$$

The resulting code is a [12,8,2] LQC code. Its generator matrix, in a form of the type shown in Fig. 2.11 (after reordering rows and columns according to Lemma 2.6) turns out to be as depicted in Fig. 2.22 up. The length of any truncated circulant is now 6, and the number of its rows is 4. Also in this case we can calculate two pairs of transformed parity check polynomials, able to produce here a transformed even-to-even and odd-to-odd pair of code words of the type $(x^5 + 1)$ and $(x^5 + 1)x^6$. Remark that even-to-even and odd-to-odd refer to the positions before the transformation.

Further lengthening operations on the same code will show truncated circulants properly scaled, but without changes in their structures. It is possible to interpret the new generator matrix as formed by two generator polynomials (in general q_o), the

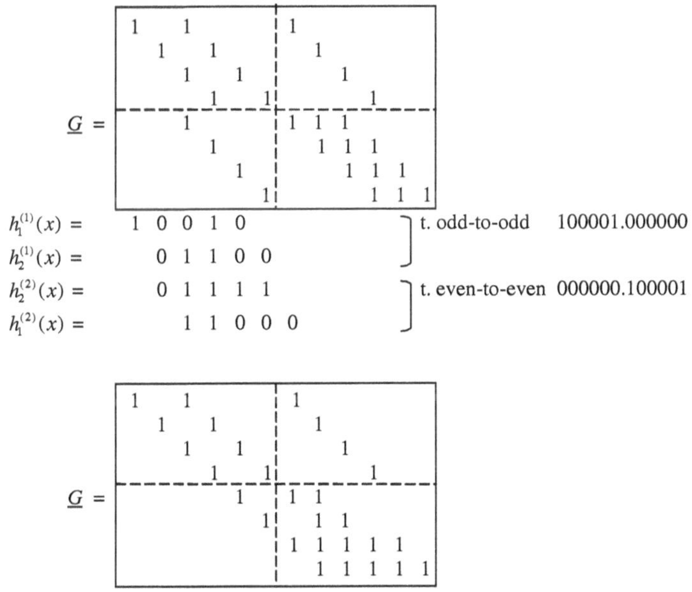

Fig. 2.22 Transformed generator matrices (the second upper triangular) and parity check polynomials for the [12,8,2] LQC code considered in Example 2.26

upper having weight 3 and the lower having weight 4. A proper ex-OR sum of rows yields a further equivalent form, shown in Fig. 2.22 down, which is upper triangular, but no longer formed by two only generator polynomials.

Definition 2.15 The q_o generator polynomials associated to as many layers in the generator matrix of a QC code in the form consisting of truncated circulants, will be called *truncated-circulant generator polynomials*, and denoted $G_{Tj}(x)$, $j = 1, 2, \ldots q_o$. Such polynomials, properly transformed, are present also in the SQC and the LQC versions of the original code.

Precisely an s_o-time shortening produces a reduction in both dimensions of the truncated circulants, say from p to $(p - s_o)$ in the horizontal dimension. Conversely, a y_o-time lengthening produces an increase, say from p to $(p + y_o)$ in the horizontal dimension. This transformation can be better interpreted looking at the truncated-circulant generator polynomials.

Example 2.27 The [10,6,3] code studied so far, in its original QC form, exhibits $G_{T1}(x) = x^5 + x^3 + 1$ and $G_{T2}(x) = x^5 + x^2 + x + 1$. The [8,4,3] one-time shortened quasi-cyclic code shows $G'_{T1}(x) = x^4 + x^2 + 1$ and $G'_{T2}(x) = x^4 + x^2 + x + 1$. In the [12,8,2] one-time lengthened quasi-cyclic code, we have $G''_{T1}(x) = x^6 + x^4 + 1$ and $G''_{T2}(x) = x^6 + x^2 + x + 1$. It is clear that the transformation consists of a progressive increment in the separation between the two parts of each polynomial, from the SQC up to the LQC version, passing through the QC parent code.

As already noted for SQC codes, the encoder circuits described in Sect. 2.4 are able to encode LQC codes as well, provided that a different sequence in the operations of the switches is adopted. Precisely the number of symbol times in which they assure connection with the information source are here increased by $y_o q_o$ units.

In the previous Section, we observed that a particular class of QC codes is characterized by a generator matrix in the form of a block containing complete $p \times p$ square circulants, in X layers, plus R_X separated rows. For them an alternative type of code shortening was introduced (see Definition 2.13). Now we will treat the correspondent type of code lengthening.

Code shortening is possible here by eliminating row layers and column layers, made by square circulants. Code lengthening, in opposite way, requires addition of row layers and columns layers. The last operation gives rise to code words of the type $(x^N + 1)$. As usual, we assume that a unique generator sub-matrix $\underline{G}_o^{(p)}$, made by square circulants, shifts along subsequent layers, except few possible R_X last rows. Code lengthening thus means addition of further $\underline{G}_o^{(p)}$ sub-matrices. In Chap. 6, treating the parity check matrix of such codes, in the framework of a similitude with some Q-ary cyclic codes, we will show that the condition for having a QC code of this type is $q_o = p$. This property will be verified here by means of an Example, adopting $q_o = p = 4$.

Definition 2.16 Given a QC code, with parameters p and q_o, whose $\underline{G}$ matrix contains X layers of complete $p \times p$ square circulants, plus possible R_X separated rows, we can obtain an *x_o-time lengthened version* of this code (briefly an *LQC code*), *maintaining invariant parameter p*, after addition of x_o row layers and x_o column layers of circulants in its $\underline{G}$ matrix.

Also here parameter R_X is not involved in this type of code lengthening. Clearly, the two operations of code shortening and code lengthening, according to Definitions 2.13 and 2.16 respectively, have sense only if a unique generator sub-matrix $\underline{G}_o^{(p)}$ is present in the X row layers of $\underline{G}$. On the other hand, if this assumption is not verified, the code would be no longer w.s. time-invariant, but completely time-varying. The above assumption will be always implicitly adopted in the following.

Lemma 2.11 *The code rate of an x_o-time LQC code with invariant parameter p, after Definition* 2.16, *turns out to be*

$$R = \frac{(X + x_o)p + R_X}{(q_o + x_o)p}. \tag{2.16}$$

Proof The number of row layers of the $\underline{G}$ matrix before lengthening was $k = Xp + R_X$. After lengthening, it becomes $k' = (X + x_o)p + R_X$. The number of columns, from $n = q_o p$, becomes $n' = (q_o + x_o)p$. Expression (2.16) has been written in a general form, but in it we can take $q_o = p$. □

Example 2.28 Let us consider the generator matrix shown in Fig. 2.20 (Example 2.22), where $X = 2$ layers are characterized by the same $\underline{G}_o^{(p)}$ sub-matrix, with $p = 4$. This [16,9,4] QC code can be one-time lengthened, maintaining $p = 4$ and increasing q_o from 4 to 5, as well as X from 2 to 3. The generator matrix of the one-time lengthened code so obtained is shown in Fig. 2.23. The resulting LQC code is a [20,13,2] code, with 4 code words of the type $(x^{16} + 1)$. The parity check polynomials $h_j^{(1)}(x^4), j = 1, 2, 3$ (the one for $j = 4$ is not needed in this interleaved division), able to produce the first of such 2-weight code words, are reported under the $\underline{G}$ matrix, together with a solitary parity check polynomial, associated to the anomalous generator polynomial $u_8(x)$ in the last row of $\underline{G}$, which has been denoted $h_o(x^4)$. In this construction, it is useful that polynomial $u_8(x)$ is placed over the rightmost two periods in the code word.

The word weight distributions, characterizing the shortened [12,5,4], the quasi-cyclic [16,9,4], the lengthened [20,13,2] and [24,17,2] codes, are shown in Table 2.4. We can observe the increase in the number of 2-weight code words, in the lengthened codes, with steps of $p = 4$. It is possible to note also a symmetry in the distribution for $n = N = 16$ (the QC code), and also for $n = 24$. The latter result is due to the fact that p is even. The next condition for having a symmetric distribution is $n = 2N = 32$, as usual.

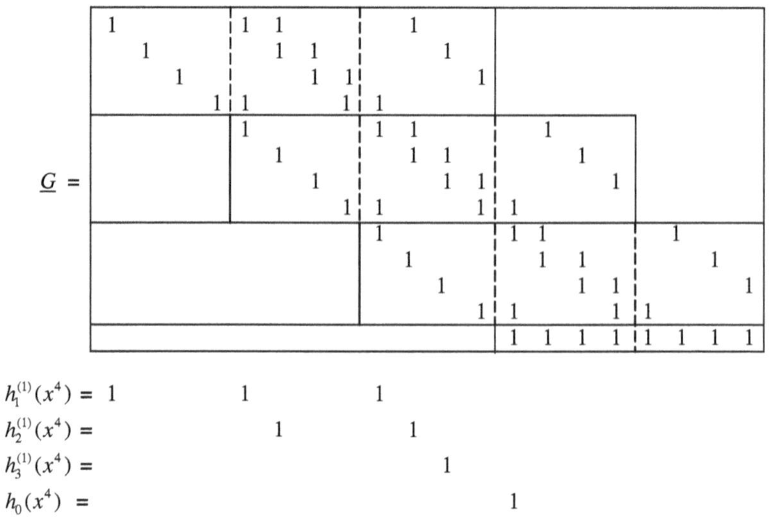

Fig. 2.23 Generator matrix of the [20,13,2] LQC code described in Example 2.28

Table 2.4 Word weight distributions of some codes introduced in Example 2.28

Code	A(0)	A(2)	A(4)	A(6)	A(8)	A(10)	A(12)	A(14)	A(16)	…
[12,5,4] SQC	1		6	16	9					
[16,9,4] QC	1		36	96	246	96	36		1	
[20,13,2] LQC ($x_o = 1$)	1	4	84	540	2,070	2,876	1,876	676	65	
[24,17,2] LQC ($x_o = 2$)	1	8	188	1,944	11,719	30,668	41,976	30,678	11,719	

2.9 Subcode of a w.s. Time-Invariant Polynomial Code

For w.s. time-invariant polynomial codes, in particular for QC, SQC and LQC codes, it is possible to extend the definition of polynomial subcode given in Sect. 1. 12 for s.s time-invariant polynomial codes (Definition 1.29).

Definition 2.17 Starting from an $[n, k, d]$ block code, whose generator matrix is w.s. time-invariant, as introduced in Sect. 2.1, with a code word polynomial expression given by (2.1), we can obtain a *polynomial subcode* from it, by imposing some constraints to the ordered polynomials $u_j(x^{q_o}), j = 1, 2, \ldots q_o$, maintaining the overall code length n unchanged. The subcode is still a linear code, with reduced number of degrees of freedom, controlled by information symbols in the polynomials $u_j(x^{q_o})$.

Lemma 2.12 *A code characterized by generator polynomials* $g'_j(x) = p_j(x^{q_o})\, g_j(x)$, $j = 1, 2, \ldots, q_o$, *where polynomial* $p_j(x^{q_o})$ *is present in the co-factor of* $g_j(x)$ *with respect to a binomial* $(x^N + 1)$, *being N the length of the original QC code, is a QC code itself, over the same length N, and turns out to be a subcode of the previous one.*

Proof Factor $p_j(x^{q_o})$ is transferred from the ordered parity check polynomials $h_j^{(i)}(x^{q_o})$ to the original generator polynomial. The result of their interleaved product remains $(x^N + 1)$. □

Example 2.29 In Fig. 2.13 up-left the generator matrix of the [10,6,3] QC code introduced in Example 2.10 is reported. Its four parity check polynomials are calculated in Fig. 2.21. They all are multiple of the common factor $(x^2 + 1)$ So multiplying both $g_1(x) = x^4 + x^3 + 1$ and $g_2(x) = x^4 + x^2 + x + 1$ by $(x^2 + 1)$, we respectively obtain $g'_1(x) = x^6 + x^5 + x^4 + x^3 + x^2 + 1$ and $g'_2(x) = x^6 + x^3 + x + 1$. This new QC code exhibits the generator matrix shown in Fig. 2.24. Under this $\underline{G}$ matrix, the new parity check polynomials are reported. This is a [10,4,4] QC code, which can be interpreted as a subcode of the previous one. On the right, the evolution of some low-weight $A(w)$ coefficients in the word weight distribution of the family of codes so obtained, is finally presented. We can find, for $z = 2$, the QC condition, and, for $z > 2$, the typical behaviour of a lengthened code over more than one lengthening period, with $A(2)$ increasing firstly by 2 units any single lengthening and then by 4 units, according to a binomial law. Down in the Figure the reordered $\underline{G}$ matrix, as a block of truncated circulants, is shown, for this [10,4,4] QC code.

In the above form of subcode construction, it is possible to recognize a resemblance with the obtainment of even-distance subcodes, starting from known cyclic codes (Definition 1.30), or with the design of binary BCH codes, starting

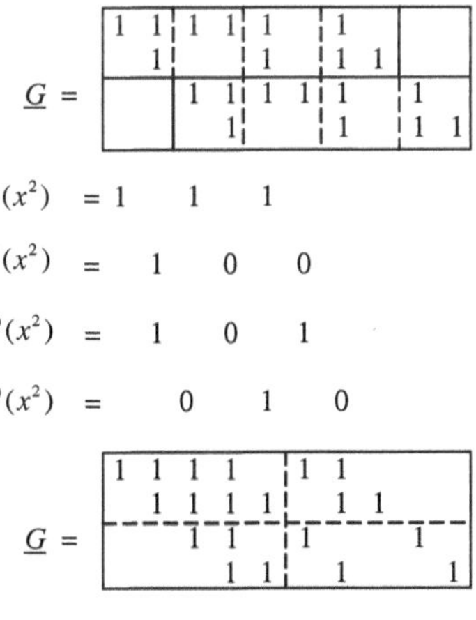

n	z	A(2)	A(4)	A(6)	
8	1		2	1	
10	2		5	10	(QC)
12	3	2	11	33	1.st length. period
14	4	4	24	100	
16	5	6	45	258	
18	6	8	80	592	
20	7	10	125	1240	(QC)
22	8	14	201	2328	2.nd length. period
24	10	18	303	4178	

Fig. 2.24 [10,4,4] QC code, subcode of the [10,6,3] QC code introduced in Example 2.10, and word weight distribution of the family of codes obtained from it (Example 2.29)

from the corresponding Hamming codes. Both these situations were assumed in Sect. 1.12 as examples of subcodes in the framework of s.s. time-invariant polynomial codes.

Another important way for obtaining a polynomial subcode of a given w.s. time-invariant polynomial code, following the analogy with what introduced for s.s. time-invariant polynomial codes, is by selecting the times in which the information source emits symbols, so distributing control symbols along the whole word length. Nevertheless, for QC codes (and SQC or LQC codes derived from them), in doing this we have a possible ambiguity, since there is an equivalent representation for such block codes already with distributed control symbols. It is made by a generator matrix constructed as a block of truncated circulants.

2.10 Modified Lengthened Quasi-Cyclic (MLQC) Codes

The definition of modified lengthened quasi-cyclic codes, starting from their parent quasi-cyclic code, can be given by proper widening of that presented in Sect. 1.13 for modified lengthened cyclic codes, starting from their parent cyclic code. We have only to substitute a $\underline{G}_o^{(q_o)}$ sub-matrix or a $\underline{G}_o^{(p)}$ sub-matrix to the unique generator polynomial $g(x)$ which characterizes s.s. time-invariant codes. Keeping in mind Definition 1.31, we can adopt here the following proposition in the case of a description based on $\underline{G}_o^{(q_o)}$.

Definition 2.18 Given an LQC block code described by a sub-matrix $\underline{G}_o^{(q_o)}$, with block length n greater than N by an integer multiple of q_o (the number of periods z in which information symbols are present is greater than k_p), its code words are still expressed by (2.1). If, in this representation, all the polynomials $u_j(x^{q_o})$ are constrained to show 0-symbols in proper positions, this code is called *modified lengthened quasi-cyclic code* (briefly *MLQC code*). It is a subcode of the original LQC code.

We can develop now considerations similar to those regarding an MLC code, but assuming here a super-period, whose length is N_o periods of the original QC code, that is N_oq_o positions on the whole. In it, the first K_oq_o positions are devoted to information symbols, and the remaining $R_oq_o = (N_o - K_o)q_o$ positions to control symbols. The choice of $R_o \geq r_p$, being $(r_p + 1)$ the number of periods spanned by $\underline{G}_o^{(q_o)}$, is useless, because it would entail separate code words.

Lemma 2.13 *The total length n of a code word in an* MLQC *code characterized by parameters* q_o, r_p, N_o, R_o, *is*

$$n = ZN_oq_o + (r_p - R_o)q_o, \tag{2.17}$$

where Z is the number of super-periods where information symbols are present.

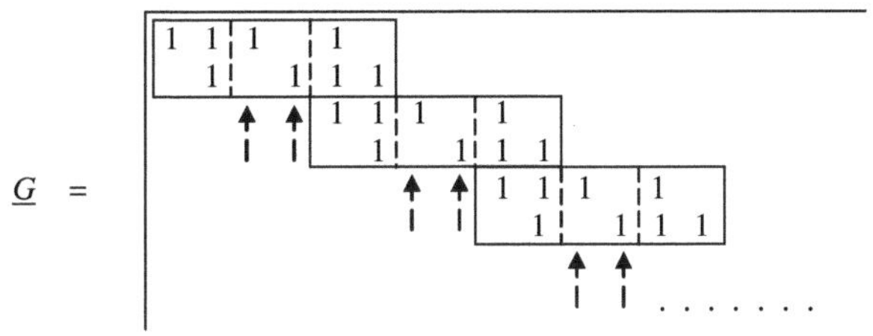

Fig. 2.25 *Top-left corner* of the generator matrix for the MLQC code described in Example 2.30

Proof Symbol Z has been adopted, in substitution of symbol z, for denoting that here we are treating proper super-periods, but also in analogy with what discussed in Sect. 1.13 about MLC codes. The Zth super-period exhibits firstly K_o periods entirely devoted to information symbols and then R_o periods entirely devoted to control symbols, so the number of the remaining control symbols taken from the last $\underline{G}_o^{(q_o)}$ sub-matrix is $(r_p - R_o)q_o$. □

Lemma 2.14 *The code rate of an MLQC code, in the conditions described by Lemma* 2.13, *results in*

$$R = \frac{ZK_o}{ZN_o + (r_p - R_o)}. \tag{2.18}$$

Proof The total number of information symbols is $k = ZK_oq_o$. Then we calculate the ratio $R = k/n$, in which n is taken from (2.17). Parameter q_o vanishes. □

Example 2.30 Assuming the QC code introduced in Example 2.8, which is an equivalent of the [8,4,4] extended Hamming code, we obtain the MLQC code characterized by $N_o = 2$ and $K_o = 1$, which shows the generator matrix depicted in Fig. 2.25, only in its top-left corner. Its number of columns n and number of rows k can be as large as one wishes (with steps of $q_o = 2$). Dashed small arrows denote the position where control symbols are placed.

The occurrence of 2-weight code words has been avoided, as Table 2.5, reporting the evolution of the word weight distribution, clearly shows. It is possible

Table 2.5 Word weight distribution of the MLQC code described by the $\underline{G}$ matrix in Fig. 2.19

Z	n	k	$A(0)$	$A(4)$	$A(6)$	$A(8)$	$A(10)$	...
1	6	2	1	3				
2	10	4	1	7	6	2		
3	14	6	1	11	16	27	8	
4	18	8	1	15	26	80	88	
5	22	10	1	19	36	150	278	
6	26	12	1	23	46	236	554	
7	30	14	1	27	56	338	910	

to observe a similarity of behaviour with respect to MLC codes. The linear increase with Z in the number of low-weight code words is a characteristic present both in MLC and in MLQC codes.

In the case of an MLQC code, like for a simple LQC code, there is still the possibility of reorganizing the $\underline{G}$ matrix in such a way as to assume the form of a block of truncated circulants. We can do this following two ways: still taking one column any q_o and one row any q_o, or taking one column any $N_o q_o$ and one row any $K_o q_o$. In order to put in evidence the relation with the QC or LQC original code, it is preferable to follow the first way.

Lemma 2.15 *An MLQC code obtained according to Definition* 2.18 *can be described also by a reordered generator matrix, constructed taking a row any* q_o *and a column any* q_o. *The new generator matrix exhibits rectangular* $a \times b$ sub-matrices, whose vertical size a and horizontal size b respectively are

$$a = ZK_o, \quad b = ZN_o + (r_p - R_o). \tag{2.19}$$

Proof The total number of rows is $ZK_o q_o$, whereas the total number of columns is given by (2.17). Division by q_o yields (2.19). □

Nevertheless sometimes there may be some null columns, either in the original matrix or in the reordered matrix, so they have to be eliminated. This can lead to a variation, even in the asymptotic code rate (with Z tending to infinity).

Example 2.31 Taking the [10,6,3] QC code introduced in Example 2.10, whose one-time lengthened version has been described in Example 2.26, we can obtain an MLQC code from it, assuming $N_o = 2$ and $K_o = 1$. With Z = 4, the original generator matrix is shown in Fig. 2.26 up. We have considered 10 periods on the whole, although the last is empty, for helping understanding. Furthermore it is possible to note some other null columns, denoted with dashed thick segments. The correspondent reordered generator matrix is shown in Fig. 2.26 down. After elimination of these useless columns, the asymptotic code rate turns out to be 2/3 instead of 2/4. In spite of this fact, the minimum distance remains d = 3, since no 2-weight code words appear.

Example 2.32 Let us consider the [10,4,4] QC code introduced in Example 2.29. It is characterized still by $q_o = 2$, but its generator sub-matrix exhibits $r_p = 3$. The MLQC version of such code, with $N_o = 2$ and $K_o = 1$ like in the previous Example, is described by the generator matrix, over Z = 4 periods, shown in Fig. 2.27 up. Here we have only one column to be eliminated, close to the end of the code word. The asymptotic code rate is 2/4 = 1/2, according to what has been fixed in the design. The correspondent reordered generator matrix is reported down in the same Figure.

A comparison between the reordered $\underline{G}$ matrices shown in Fig. 2.24 down and in Fig. 2.27 down makes it clear which type of modification occurs in the 4 sub-matrices after MLQC transformation. There is an increase, by a factor of 2, in the

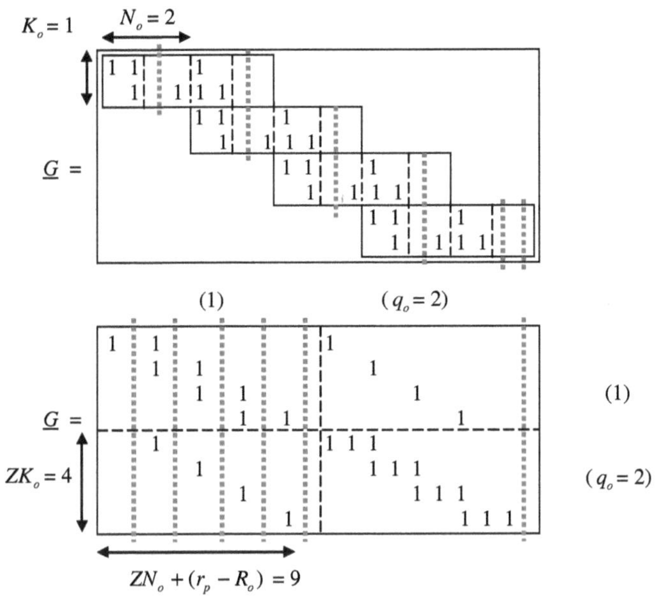

Fig. 2.26 Generator matrices of the MLQC code considered in Example 2.31, before and after reordering

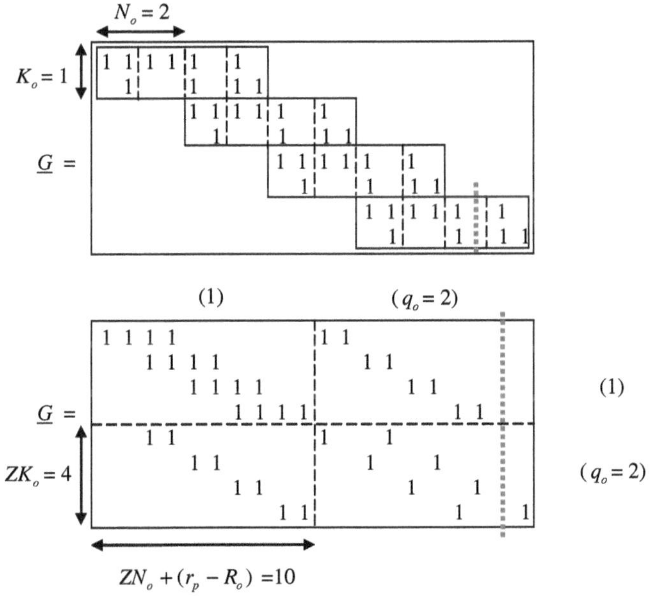

Fig. 2.27 Generator matrices of the MLQC code considered in Example 2.32, before and after reordering

horizontal shifts of their rows. In this situation such rectangular modules are no longer truncated circulants. A similitude can be found with the generator matrix of the [24,12,8] extended Golay code, reported in Fig. 2.15. We can imagine to have distributed control symbols also in these rectangular modules. Further details on the mechanisms responsible for the particular form assumed by the reordered generator matrix in MLQC codes will be given in Chaps. 3 and 4, taking advantage from their description as convolutional codes.

Starting from a QC code with a generator matrix formed by a block containing complete $p \times p$ square circulants, collected in X row layers, all with the same generator sub-matrix $\underline{G}_o^{(p)}$, plus R_X separated rows, we perform firstly a procedure of code lengthening as described in Definition 2.16. Then we are interested in constructing a subcode of this LQC code, operating a particular form of modified lengthening procedure, which acts only on the X row layers, maintaining unchanged the final R_X separated rows.

Definition 2.19 Given a QC code described by a generator matrix containing complete $p \times p$ square circulants, forming X layers, all with the same generator sub-matrix $\underline{G}_o^{(p)}$, plus possible R_X separated rows, the *modified lengthened version* of this code, *maintaining invariant parameter p*, can be constructed as follows. Assuming an MLQC super-period with length equal to N_p periods of the original QC code, that is an overall length $N_p p$, the first K_p periods are associated to $K_p p$ information symbols. Then we have $(N_p - K_p)p$ control symbols, to complete the super-period. Finally there are the R_X separated rows, which remain unchanged.

Example 2.33 Let us consider the generator matrix shown in Fig. 2.20 up (Example 2.22), where two layers are characterized by the same $\underline{G}_o^{(p)}$ sub-matrix, with $p = 4$, already treated also in Examples 2.24 and 2.28 (respectively for code shortening and code lengthening). This [16,9,4] QC code can produce an MLQC version, assuming $N_p = 2$ and $K_p = 1$, whose generator matrix is shown in Fig. 2.28, in the case of just two super-periods. This is a [20,9,4] MQLC code, where the last row $(R_X = 1)$ is unchanged. The positions devoted to control symbols have been denoted with dashed small arrows, as usual.

The evolution of the word weight distribution of the family of MLQC codes so obtained is reported in Table 2.6, for $Z = 2$ (the code described by the $\underline{G}$ matrix in Fig. 2.28), and $Z = 3, 4, 5$. Any 2-weight code word has been eliminated. The increase in the number of 4-weight (and even 6-weight) code words is proportional to the number of super-periods Z, as expected.

In conclusion, MLQC codes, as well as MLC codes, present distributed control symbols and also some additional final control symbols. A difference with respect to the case of cyclic codes is however in what follows. In an s.s. time-invariant context, distributed control symbols can be found just when an MLC arrangement is adopted. For a w.s. time-invariant $\underline{G}$ matrix, we can have distributed control symbols also in the original QC code, if such matrix is that made by truncated circulants (see Fig. 2.11). Nevertheless the element characterizing QC codes,

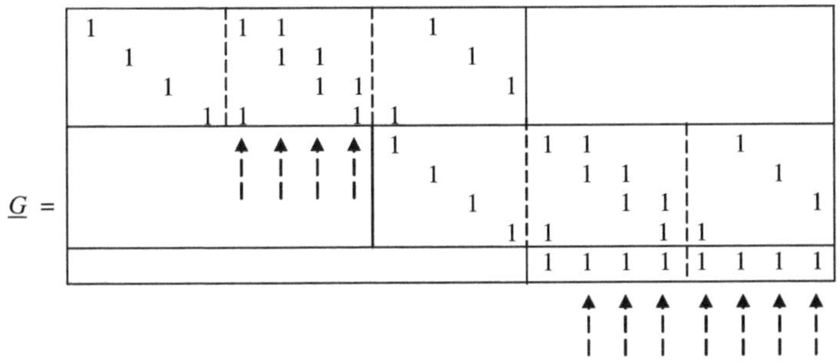

Fig. 2.28 Generator matrix of the [20,9,4] MLQC code described in Example 2.33

Table 2.6 Word weight distribution for the MLQC codes introduced in Example 2.33

Z	Code	A(0)	A(4)	A(6)	A(8)	A(10)	A(12)
2	[20,9,4]	1	10	28	114	192	138
3	[28,13,4]	1	14	36	194	736	1,800
4	[36,17,4]	1	18	44	287	1,240	4,666
5	[44,21,4]	1	22	52	397	1,828	8,090

although with distributed control symbols, remains the property of being constructed on a precise code word length N.

In Chap. 4, we will look for a correspondence between MLQC codes and a certain class of w.s. time-invariant convolutional codes. The limit in this approach is due to the condition that MLQC codes are subcodes of proper LQC codes (constructed at steps of q_o or p). In this sense, not all the possible w.s. time-invariant convolutional codes will be capable of having an MLQC code equivalent.

2.11 Trellis for a w.s. Time-Invariant Block Code Obtained from Its Generator Matrix

All the considerations developed in Sect. 1.23 about the generator trellis in an s.s. time-invariant block code can be repeated here as well, with the only difference of considering a higher number of states, owing to the presence of more than one shift register in the encoder circuit (see for instance Fig. 2.17, where a small part of the state diagram has been reported). In this sense, such state diagrams resemble those of Q-ary s.s. time-invariant block codes (Sect. 1.20), where, besides a state complexity, also a branch complexity is to be taken into account.

As anticipated at the end of Chap. 1 and also in Sect. 2.5, for QC codes with distributed control symbols, it is possible to reach a reduced number of merge states, and hence a reduced computational complexity in soft-decision decoding.

In such cases, the trellis is partitioned into some sub-trellises and more than one symbol is adopted for labelling branches, at least in some stages. It is particularly interesting to compare the traditional trellis for a certain polynomial code, or code obtained after a suitable concatenation of polynomial codes, and the partitioned trellis which can be arranged taking the QC equivalent of that code. As already noted, a QC equivalent exists when n in not prime, and its factors are used for proper partitioning the $\underline{G}$ matrix. This property is independent of the particular scheme of encoder circuit which could be adopted, but depends only on the structure of the new generator matrix. The QC equivalent of the [24,12,8] extended Golay code treated in Example 2.18 clearly represents a situation of this type.

A similar condition occurs for all the Reed-Muller codes (Forney 1988) and even for many BCH codes or extended BCH codes (Vardy and Be'ery 1994). As an introductory situation, we will present the case of the [8,4,4] 1st-order Reed-Muller code, or extended Hamming code. In Sect. 2.2 we presented some different QC equivalent versions of such code. In order to allow trellis partitioning a further equivalent version is needed. Precisely it is associated to the following sequence of information and control symbols *iiciccc*, where i denotes an information symbol position and c a control symbol position.

Example 2.34 A new equivalent version of the [8,4,4] extended Hamming code, useful for enabling reduced computational complexity in a soft-decision decoding procedure applied on a partitioned trellis, is described by this generator matrix

$$\underline{G} = \begin{bmatrix} 1 & 1 & 1 & 1 & 0 & 0 & 0 & 0 \\ 0 & 1 & 1 & 0 & 0 & 1 & 1 & 0 \\ 0 & 0 & 1 & 1 & 1 & 1 & 0 & 0 \\ 0 & 0 & 0 & 0 & 1 & 1 & 1 & 1 \end{bmatrix}.$$

It is possible to verify that all the 16 code words of this code are present in the partitioned trellis shown in Fig. 2.29. We have two sub-trellises, with two merge states per stage in each one of them. The branch complexity is here 2. Then we have a final merge state with branch complexity 4. Any branch collects two output symbols, so that two stages with 4 merge states each are present on the whole. We can remark the presence of a left-right symmetry in this structure, accompanied by a particular symmetry in the information and control symbol sequences *iiciccc*. The trellis in Fig. 2.29 is remarkably simpler than the traditional one, characterized by eight binary stages in which the largest one exhibits 8 merge states.

Now we will analyze how several BCH codes (primitive or non-primitive), or extended versions of BCH codes, whose n value is composite, can be arranged with an optimized partitioning of their co-ordinates (positions in the code word). First of all, we can observe that, if we exchange the second with the fourth row in the $\underline{G}$ matrix of the [8,4,4] code, discussed in Example 2.34, we obtain the following equivalent form

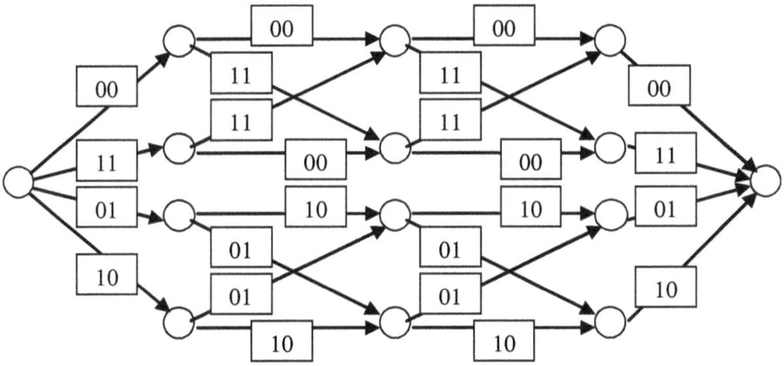

Fig. 2.29 Partitioned trellis for a proper QC version of the [8,4,4] extended Hamming code

$$\underline{G} = \begin{bmatrix} 1 & 1 & 1 & 1 & 0 & 0 & 0 & 0 \\ 0 & 0 & 0 & 0 & 1 & 1 & 1 & 1 \\ 0 & 1 & 1 & 0 & 0 & 1 & 1 & 0 \\ 0 & 0 & 1 & 1 & 1 & 1 & 0 & 0 \end{bmatrix},$$

which can be interpreted according to the partition scheme shown in Fig. 2.30 up-left. This scheme is a very simple case of a more general matrix partition, of the type described in Fig. 2.30 up-right (Kasami et al. 1990; Vardy and Be'ery 1994). Here we have a composite n, say $n = ab$, where a suitable co-ordinate reordering allows to obtain b equal codes, having code length a and k_a information symbols, each acting on a sequence of a symbols. The b sequences so obtained give rise to a partition in the ensemble of the co-ordinates of our original code. This operation has a sort of remainder $k_\Delta = k - bk_a$, where k represents the total number of information symbols in the original code. After the separation into b rectangles, as many as the component codes in the above partition, there is a large rectangle which collects the remaining rows of $\underline{G}$. Assuming, for instance, the [45,23,7] non-primitive BCH code, it is possible to take $b = 3$, $a = 15$, $k_a = 5$, in such a way as to employ three [15,5,7] primitive BCH codes as component codes. So we are left with $k_\Delta = 8$. The partition in the resulting $\underline{G}$ matrix which follows from this co-ordinate reordering is shown in Fig. 2.30 down. Certainly there is a further code, [45,8, d] with $d \geq 7$, but the condition for such decomposition to be valid is that any ex-OR sum of the rows in the generator matrix of this final component code with rows of the preceding component codes does not give weight lower than 7. This is effectively possible, so that the new code is equivalent to the original one.

A high number of codes admits this type of generator matrix decomposition. We can consider also even-distance subcodes of BCH codes, or extended codes obtained from BCH codes. In particular, the latter ones have overall length n which is an even number, and hence always composite. As a very short sample of possibilities, all the codes listed in Table 2.7 admit a decomposition whose parameters are collected on the right. In this list there are also the [8,4,4] extended Hamming

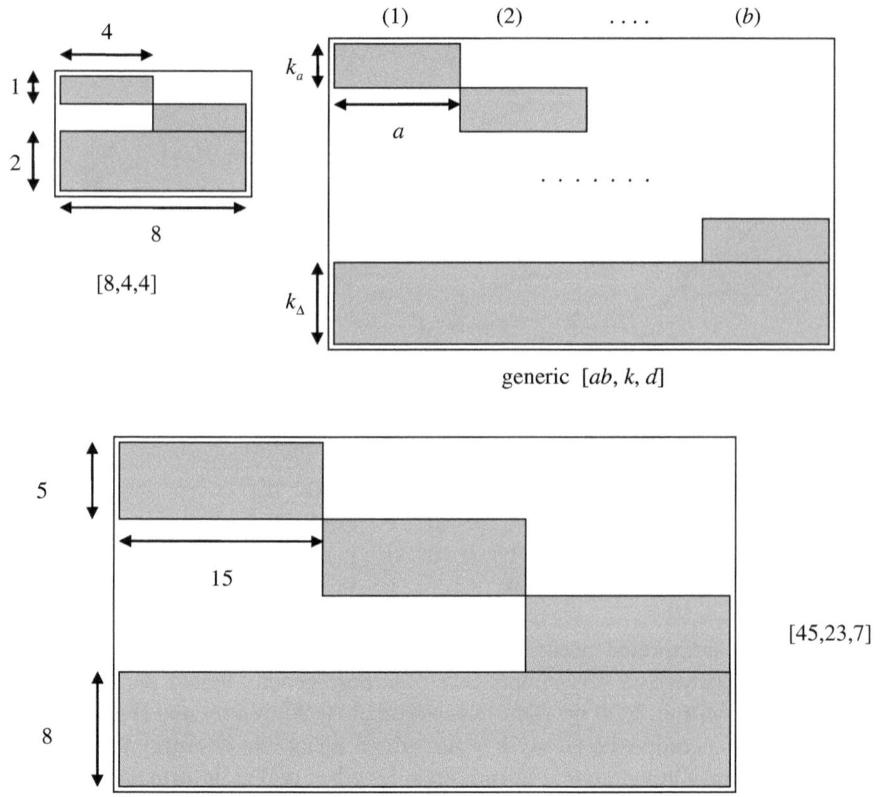

Fig. 2.30 Partition in the generator matrices, possible for certain codes derived from BCH codes with composite overall length n

Table 2.7 Some codes which admit proper decomposition of their $\underline{G}$ matrix

Code	Component codes	k_Δ	Final component code
[24,12,8]	Three [24,8,4]	–	
[8,4,4]	Two [8,2,4]	–	
[8,4,4]	Two [4,1,4]	2	[8,2,d], $d \geq 4$
[45,23,7]	Three [15,5,7]	8	[45,8,d], $d \geq 7$
[49,27,4]	Seven [7,3,4]	21	[49,21,d], $d \geq 4$
[75,31,7]	Five [15,5,7]	6	[75,6,d], $d \geq 7$
[93,53,7]	Three [31,11,7]	20	[93,20,7], $d \geq 7$
[32,16,8]	Two [8,1,8], two [16,4,8]	6	[32,6,d], $d \geq 8$
[16,7,6]	One [8,1,8], four [10,1,6]	2	[16,2,d], $d \geq 6$

code (in its first version, based on the trellis of Fig. 2.29) and the [24,12,8] extended Golay code, already mentioned, not needing any final component code.

The procedure described so far has been called direct sum of component codes. A variant of this procedure (Vardy and Be'ery 1994) has been also suggested, considering overlapped subsets of co-ordinates, instead of non-overlapped subsets, like in the partition requested for a direct sum. The result is a more involved decomposition of the $\underline{G}$ matrix, but substantially with the same properties. This further procedure has been called concurred sum, and many other known codes have been proposed for being so rearranged. Examples of cases in which a concurrent sum has been considered are the last two in Table 2.7. They are characterized by two types of component codes, whose co-ordinates are partially overlapped. All these structures are equivalent to suitable QC codes.

In all the decompositions that the proposed procedure makes possible, a partitioned trellis arises. The various sub-trellises so originated exhibit a number of states per stage and also a number of branches per stage which fix the overall decoding computational complexity. The advent of LDPC codes, for which decoding complexity is imposed by other considerations, has reduced the importance of such trellis complexity evaluations, from the practical point of view. Nevertheless these works remain of fundamental validity for understanding the structure of many textbook codes.

References

Baldi M et al (2012) On the generator matrix of array codes. In: Proceedings of Softcom 2012, Split, Croatia

Calderbank AR, Forney GD, Vardy A (1998) Minimal tail-biting trellises: the Golay code and more. In: Proceedings of ISIT 1998, Cambridge, USA, p 255

Chen CL, Peterson WW (1969) Some results on quasi-cyclic codes. Inf Contr 15:407–423

De Bruijn NG (1946) A combinatorial problem. J London Math Soc 21:167–169

Fan JL (2000) Array codes as low-density parity check codes. In: Proceedings of 2nd international symposium turbo codes, Brest, France, pp 543–546

Forney GD (1988) Coset codes—part II: binary lattices and related codes. IEEE Trans Inf Th 34:1152–1187

Kasami T, Takata T, Fujiwara T, Lin S (1990) Trellis diagram construction for some BCH codes. In: Proceedings of IEEE International Symposium Information Theory and Application, Honolulu

Lin S, Costello DJ (2004) Error control coding. Pearson, Prentice-Hall, Upper Saddle River

Vardy A, Be'ery Y (1994) Maximum likelihood soft decision decoding of BCH codes. IEEE Trans Inf Th 40:546–554

Chapter 3
Generator Matrix Approach to s.s. Time-Invariant Convolutional Codes

3.1 Traditional View of Non-systematic s.s. Time-Invariant Convolutional Codes

In the previous chapters, the following conclusion has been reached: starting from a code with minimum distance $d > 2$, in order to have a linear growth in the low-weight code word number with increasing the code word length, without appearance of 2-weight code words, it is necessary to have distributed control symbols. This behaviour occurs in some MLC or MLQC codes. On the other hand, this condition is not sufficient, as QC codes with reordered $\underline{G}$ matrix or direct product block codes have shown (they exhibit distributed control symbols, but not a linear growth in the low-weight code word number).

In order to assure performance independent of the code word length n, a linear growth in the low-weight code word number with n is needed, and is what linear convolutional codes are requested to guarantee. The traditional view of such linear codes (Elias 1955) starts from the model of a 1/2-rate non-systematic convolutional code. It is strictly related to the encoder circuit with feed-forward ex-OR sums, able to produce an encoded symbol sequence whose length is not prefixed. It is important to stress that the considered code is strict-sense time-invariant in $\underline{G}$, as will be better explained in the following.

Definition 3.1 A *binary linear non-systematic convolutional code having rate 1/2* is encoded by means of an encoder circuit, in which a shift register, in controller arrangement, with m memory elements, feeds two ex-OR sums. This circuit is depicted in Fig. 3.1. The wires connecting the shift register to the two ex-OR sums are imposed by the coefficients of two polynomials, $g^{(1)}(x)$ and $g^{(2)}(x)$, called *sub-generator polynomials*.

In Fig. C.2 of Appendix C a circuit, able to perform a polynomial product, is presented. It is based on a shift register, in controller arrangement, whose output is the polynomial product between an input binary sequence, represented by

G. Cancellieri, *Polynomial Theory of Error Correcting Codes*,
Signals and Communication Technology, DOI 10.1007/978-3-319-01727-3_3

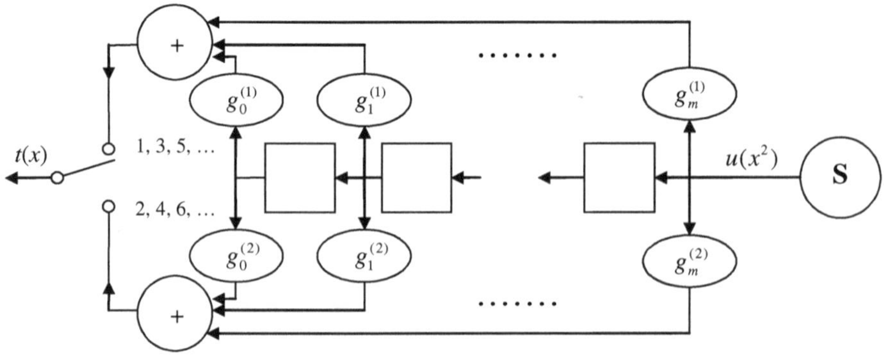

Fig. 3.1 Encoder circuit (in controller arrangement) for a 1/2-rate convolutional code

polynomial $u(x)$, and a wired polynomial $p(x)$. Here that scheme is doubled: a unique shift register, in which the input sequence is collected, feeds two sums. The switch at the encoder output selects, in the first symbol time, the upper sum output, and, in the second symbol time, the lower sum output. Then this cycle repeats.

We can observe a strict similitude with the circuit in Fig. 1.3, employed for a non-systematic s.s. time-invariant polynomial block code, except for the above mentioned doubling. The output of the information source is here taken one symbol time any two, as polynomial $u(x^2)$, representing here the input sequence, clearly expresses. Linearity is assured by the fact that all the circuit parts perform linear operations.

Lemma 3.1 *The output sequence* $t(x)$, *produced by an encoder circuit of the type depicted in* Fig. 3.1, *can be written as*

$$t(x) = u(x^2)\left[xg^{(1)}(x^2) + g^{(2)}(x^2)\right]. \tag{3.1}$$

Proof In odd symbol times the switch takes binary symbols from the upper sum, which collects the product $xu(x^2)g^{(1)}(x^2)$. In even symbol times it takes binary symbols from the lower sum, which collects the product $u(x^2)g^{(2)}(x^2)$. The degree of polynomial $u(x^2)$, as well as that of polynomial $t(x)$, is unconstrained, therefore the output sequence can be as long as one wishes. □

Definition 3.2 The output sequence, expressed by (3.1), is called *code frame*. This is a synonymous of code word, but stresses the fact that there is not a prefixed length for it. A correct frame termination is made in such a way as to complete expression (3.1), which means to fill all the m memory elements of the shift register with 0-symbols. Parameter m is called *generator constraint length* of the considered convolutional code.

Example 3.1 With $m = 1$, we may have four situations of interest, associated to as many schemes for the encoder circuit, depicted in Fig. 3.2, and correspondent

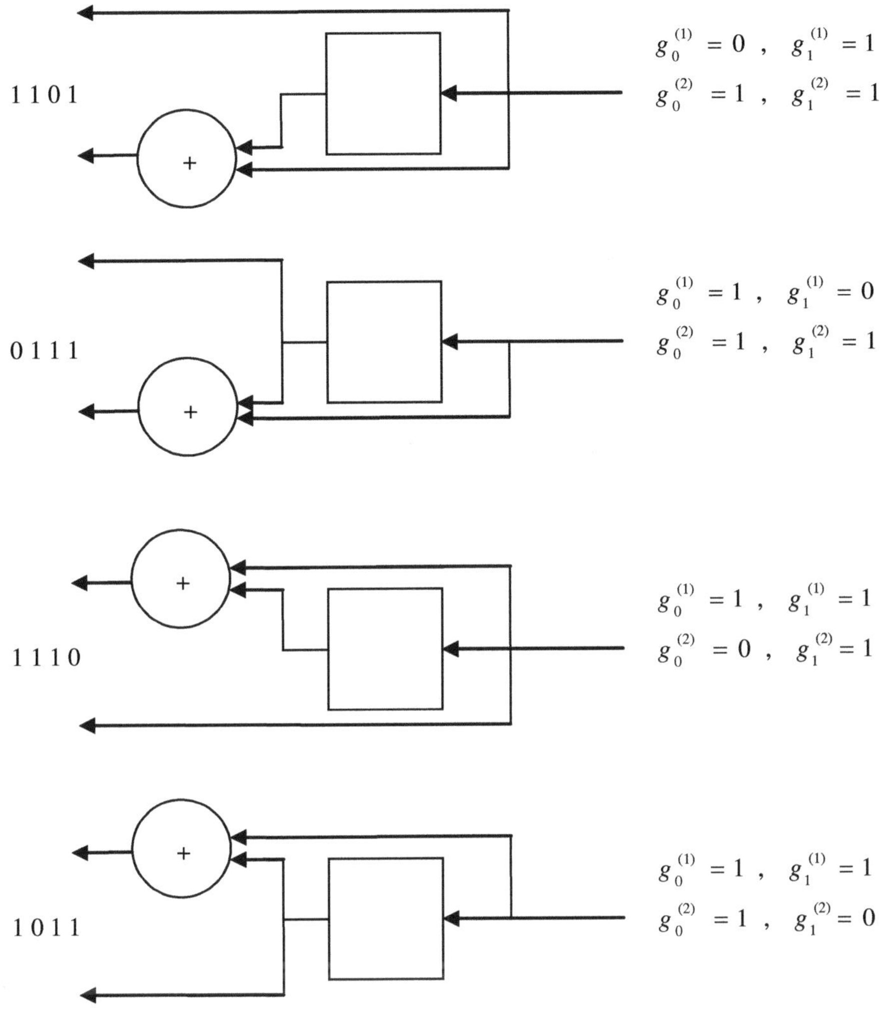

Fig. 3.2 Four possible situations characterized by $m = 1$

choices for parameters $g_j^{(i)}$, $i = 1, 2$, $j = 0, 1$. The output sequences represent the shortest possible frames, that is those obtained assuming just $u(x^2) = 1$.

In the following, we will show that a choice, characterized by sets of values for parameters $g_j^{(i)}$, $i = 1, 2$, $j = 1, 2$, out of the above four choices, does not represent a good design, although from different points of view. With less than three wires, there will be a minimum distance $d = 2$, considered too small. With four wires, a minimum distance as high as $d = 4$ can be achieved, but at the expense of a binomial increase in coefficient $A(4)$ with the frame length.

Since, for practical applications, the frame length is finite, the code rate 1/2 which identifies this class of convolutional codes is not rigorously exact. It could be more proper to define this parameter as an asymptotic code rate, so meaning the one achieved with an infinitely long code frame. Although the convolutional code described by (3.1) is non-systematic, we can interpret odd symbol times as those associated to information symbols and even symbol times as those associated to control symbols. Clearly, the case of an asymptotic code rate equal to 1/2 represents just a particular situation. It can be generalized as follows.

Definition 3.3 It is possible to introduce a *convolutional code period*, whose length is n_o (in the case treated so far $n_o = 2$) and a number of control symbols $(n_o - k_o)$ distributed in each period. They typically are placed in as many final positions of the period. Anyway the asymptotic code rate, in general, becomes k_o/n_o.

In order to encode a convolutional code with whichever asymptotic code rate k_o/n_o, sometimes it is necessary to employ more than one shift-register. In this case, we will still consider the unique parameter m as the length of the longest one of them. Coming back, for simplicity, to the case of a 1/2-rate convolutional code, we can observe that, in order to exploit the whole length m of the unique shift register, it is advisable to choose coefficients $g_0^{(1)}$ and $g_0^{(2)}$ in such a way as to have at least one of them non-null, and the same observation holds also about $g_m^{(1)}$ and $g_m^{(2)}$. This recommendation can be easily extended to a more general case, considering the longest shift register.

Lemma 3.2 *For a k_o/n_o-rate convolutional code, characterized by generator constraint length m, the overall frame length L (included the correct frame termination) of a finite-length frame, containing Z periods where information symbols are inserted, turns out to be at most*

$$L = Zn_o + mn_o. \tag{3.2}$$

Proof The last m periods are necessary for the correct frame termination, obtained by filling all the memory elements in each shift register (in particular in the longest one) with 0-symbols. For the encoder circuit shown in Fig. 3.1, expression (3.2) holds with $n_o = 2$. Its generalization to the case of whichever n_o and k_o should be made considering more complicated encoder circuits (and will be treated in subsequent sections), but L remains expressed by (3.2), independently of the choice for k_o, owing to the convention adopted in the meaning of parameter m. □

It is possible to observe that, sometimes, some final 0-symbols could be eliminated. For instance, in the situation described in Fig. 3.2, the frame termination involves just two symbols in the last period, but in the first and in the third case one of such two symbols is always 0, and can be eliminated (in the remaining two cases the same occurs in the first period). In this sense, (3.2) can be taken as an overestimation of the length L.

Lemma 3.3 *The true code rate of the convolutional code with parameters n_o, k_o, m, in a frame formed by Z periods where information symbols are inserted, is*

$$R = \frac{k_o Z}{n_o(Z+m)} = \frac{k_o}{n_o(1+\frac{m}{Z})}. \tag{3.3}$$

Proof The overall frame length L is given by (3.2), and the true code rate is $R = (k_o Z)/L$. Division yields (3.3). In (3.3) a possible overestimation of L, due to the possibility of eliminating few useless positions, can be present. It entails an underestimation of R. □

In Appendix D, Table D.3 collects the performance of the best 1/2-rate convolutional codes, taking m as an increasing parameter. For $m = 1$ the best minimum distance is $d = 3$. All the four codes in Fig. 3.2 exhibits $d = 3$, and also the same frame weight distribution. So they can be considered as equivalent versions of the same best convolutional code. With increasing parameter m, a proper choice of the two sub-generator polynomials $g^{(1)}(x)$ and $g^{(2)}(x)$ allows to obtain an increasing value for the best minimum distance. Nevertheless the computational complexity of the soft-decision decoding algorithm, which is based on a proper trellis (similar to that introduced in Sect. 1.23 for s.s. time invariant block codes), grows too with increasing parameter m.

Example 3.2 With $m = 2$, the best 1/2-rate convolutional code exhibits $g^{(1)}(x) = x^2 + x + 1$ and $g^{(2)}(x) = x^2 + 1$, or vice versa. The minimum distance so obtained is $d = 5$. Such two codes are clearly equivalent.

Lemma 3.4 *Given a 1/2-rate convolutional code, for any $m > 1$, the best performance is obtained when both $g^{(1)}(x)$ and $g^{(2)}(x)$ exhibit 0th-order and mth-order coefficients all not null.*

Proof From (3.1), these four coefficients are not subjected to possible cancellation in any ex-OR sum. So their non-null choice contributes to enlarge the minimum distance. The case characterized by $m = 1$ represents an exception, since, although $d = 4$ is reached, this good result is vanished, because of an improper increase in coefficient $A(4)$ with the frame length, as previously noted. □

Example 3.3 With $m = 3$, the best choice is associated to 1011 (or its reciprocal 1101) and 1111, taken indifferently as $g^{(1)}(x)$ or $g^{(2)}(x)$. The minimum distance of these four equivalent convolutional codes turns out to be $d = 6$. The choice of 1101 and 1011 (or their reciprocals) is rather good as well. The only difference is in coefficient $A(6)$, which in this case turns out to be double. On the contrary, the choice of 1101 and 1101 (or their reciprocals) is not good, because leads to $d = 4$, and furthermore coefficient $A(4)$ shows a binomial growth with the frame length.

3.2 State Diagram and Minimum Distance

In Sect. 1.14 we have introduced state diagrams for s.s. time-invariant block codes in the $\underline{G}$ matrix, encoded employing a shift register in controller arrangement. In that case the number of memory elements was r. Here we can proceed in similar way, with the only difference of assuming parameter m instead of r. Furthermore, since the encoder circuit is non-systematic, we will use branch labelling of the type (input symbol)/(output symbol), like the one depicted in Fig. 1.14 on the left, but here with two output symbols. A code frame is represented by a closed route in this state diagram, starting from and arriving at state $\underline{0}$. This last requirement is assured by the correct frame termination described in the previous section.

Example 3.4 The state diagrams associated to the four encoder circuits shown in Fig. 3.2 are depicted in Fig. 3.3. The four structures are identical. Just branch labelling changes. We can observe that the first state diagram shows a systematic behaviour. This property can be verified also from inspection of the correspondent encoder circuit in Fig. 3.2, where a wire connects the information source directly to the output. Nevertheless this is a particular situation, related to the fact that $m = 1$. For $m > 1$, the constraint of having a systematic encoder gives poor performance, as will be better explained in Sect. 3.3.

Example 3.5 In the case of $m = 2$, as already noted in Example 3.2, one of the two possible choices for obtaining the best 1/2-rate convolutional code is $g^{(1)}(x) = x^2 + x + 1$ and $g^{(2)}(x) = x^2 + 1$. The encoder circuit and the associated state diagram for such convolutional code are reported in Fig. 3.4. Its minimum distance is $d = 5$, as we can easily verify testing all the possible closed paths starting from and arriving at state $\underline{0}$.

Definition 3.4 In a 1/2-rate convolutional code, characterized by sub-generator polynomials $g^{(1)}(x)$ and $g^{(2)}(x)$, the polynomial

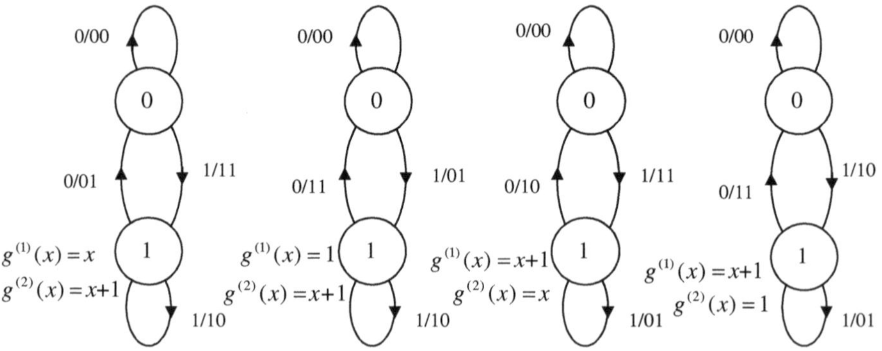

Fig. 3.3 State diagrams for $m = 1$ (in the four situations described in Fig. 3.2)

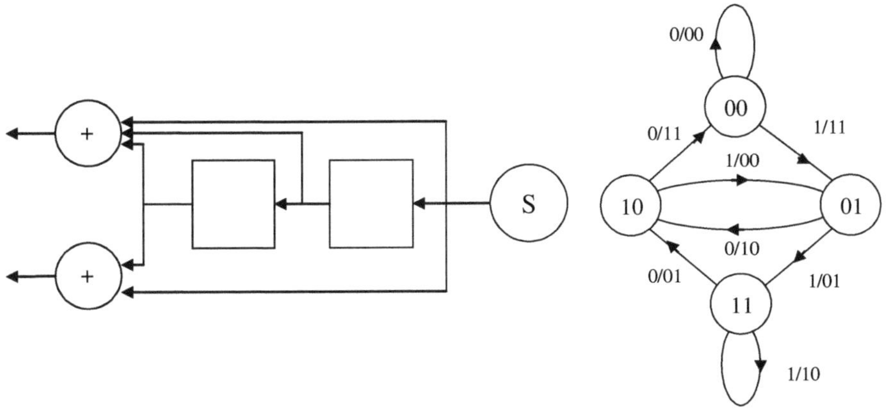

Fig. 3.4 Encoder circuit and state diagram for the convolutional code described in Example 3.5

$$g_{\text{int}}(x) = xg^{(1)}(x^2) + g^{(2)}(x^2) \tag{3.4}$$

is defined *interleaved generator polynomial for a 1/2-rate convolutional code*. Expression (3.4) is associated to a convolutional code, which, in turn, can be denoted as *strict-sense (s.s.) time-invariant in its* $\underline{G}$ *matrix*, since such matrix is characterized by a unique interleaved generator polynomial, shifting along its rows, by two positions any period.

Example 3.6 For the convolutional code described in Examples 3.2 and 3.5, we have $g_{\text{int}}(x) = x^5 + x^4 + x^3 + x + 1$, that is the sequence 111011. Its generator matrix, for $Z = 4$ periods in which information symbols are inserted, turns out to be

$$G = \begin{bmatrix} 1 & 1 & 1 & 0 & 1 & 1 & & & & & & \\ & & 1 & 1 & 1 & 0 & 1 & & & & & \\ & & & & 1 & 1 & 1 & 0 & 1 & 1 & & \\ & & & & & & 1 & 1 & 1 & 0 & 1 & 1 \end{bmatrix}.$$

A very important parameter of convolutional codes is their minimum distance d, sometimes called also minimum free distance (Proakis 1983), in order to stress that it is independent of the frame length. The number $A(w)$ of frames having low weights w, and in particular $A(d)$, are significant figures too. This fact follows from the possibility of adopting a soft-decision decoding procedure, whose performance are imposed by coefficients $A(w)$, as discussed in Appendix D. To this purpose, see, for instance, expression (D.7) on the binary symbol error probability after decoding. The number of different 1-symbol (output) patterns, forming low-weight frames, imposes the entity of the correspondent coefficients $A(w)$.

In Sect. 1.14 we noted that, in a state diagram, along the shortest closed route around state $\underline{0}$, it is possible to recognize, in output, the generator polynomial $g(x)$. Here, the same property holds, provided that we replace $g(x)$ with the interleaved polynomial $g_{int}(x)$. Furthermore, along a certain set of short closed routes around state $\underline{0}$, further low-weight 1-symbol patterns appear. In order to determine the frame weight distribution in a convolutional code, this heuristic approach can be very efficient, up to the weight $w = 2d - 1$, provided that the convolutional code considered is the result of a good design.

Definition 3.5 A convolutional code is *well designed* if the number of frames having a certain weight w increases linearly with the number Z of periods forming the frame, up to $w \leq 2d - 1$.

The above upper bound is due to the fact that, for $w \geq 2d$, a progressive appearance of new 1-symbol patterns is unavoidable, because of multiple passages through state $\underline{0}$. A similar phenomenon occurs in cyclic codes (or codes derived from them), with $k > r$.

Example 3.7 The 1/2-rate convolutional code, characterized by $m = 1$, and $g^{(1)}(x) = g^{(2)}(x) = x + 1$, is not well designed. In fact, although its minimum distance is $d = 4$, $A(4)$ exhibits a binomial growth with increasing the number Z of periods. From an easy inspection of its state diagram, which is shown in Fig. 3.5 on the left, it is possible to obtain the evolution of coefficient $A(4)$ reported in the same Figure on the right. This code exhibits non-null weight distribution coefficients just for w multiple of 4, and appears as a sort of widening in the concept of parity check codes.

Another example of not well designed 1/2-rate convolutional code is that characterized by $m = 3$ and two equal sub-generator polynomials 1101 (or their reciprocals), already mentioned at the end of Example 3.3. It is characterized by the interleaved generator polynomial 11.11.00.11 (or its reciprocal), which can be

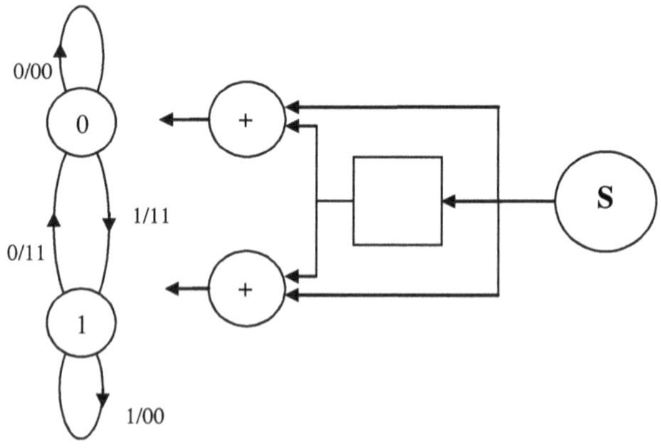

Z	A(4)
1	1
2	3
3	6
4	10
5	15
6	21
7	28
8	36
9	45

Fig. 3.5 Encoder circuit and state diagram of a not well designed 1/2-rate convolutional code

interpreted as a sort of twice repetition of a [7,4,3] Hamming code, in lengthened version. The co-factor 11101 (or its reciprocal), taken as an information sequence, clearly leads to a 4-weight frame, whose multiplicity increases binomially with the number Z of periods. These considerations suggest that it is not advisable to adopt a pair of equals sub-generator polynomials, and even polynomials having a common factor are to be treated with some cares, as will be better explained in the following.

Lemma 3.5 *The coefficients $A(w)$, for $w \leq 2d - 1$, of the weight distribution characterizing a well designed convolutional code, can be written as*

$$A(w) = B(w)Z - \delta(w), \tag{3.5}$$

where $B(w)$ is a proportionality constant, expressing a normalized weight distribution coefficient, whereas $\delta(w)$ takes into account the delay, after which a certain 1-symbol pattern, whose weight is w, appears.

Proof The first term in (3.5) expresses a proportionality with the number of periods Z in the frame, which accounts for the possibility of remaining in state $\underline{0}$ before and after the true closed route, owing to the 0-labelled closed-loop branch. All the distinct closed routes around state $\underline{0}$ then identify as many output patterns, which are ordered on the basis of their weights. So the distribution $A(w)$ can be calculated, up to $w \leq 2d - 1$, ignoring situations characterized by multiple passages through state $\underline{0}$. Nevertheless some patterns appear in the first periods, and other in some subsequent periods, owing to their longer lengths. Constant $\delta(w)$ is devoted to take into account such delays, properly cumulated. □

Example 3.8 The convolutional code, with $m = 2$, described in Example 3.5, exhibits the evolution of the frame weight distribution shown in Table 3.1. There is

Table 3.1 Frame weight distribution of the best 1/2-rate convolutional code with $m = 2$

Z	A(0)	A(5)	A(6)	A(7)	A(8)	A(9)
1	1	1				
2	1	2	1			
3	1	3	3	1		
4	1	4	5	4	1	
5	1	5	7	8	5	1
6	1	6	9	12	12	6
7	1	7	11	16	20	17
8	1	8	13	20	28	32
9	1	9	15	24	36	48
10	1	10	17	28	44	64
11	1	11	19	32	52	80
12	1	12	21	36	60	96

just one 5-weight pattern (11.10.11), the 6-weight patterns are two (11.01.01.11, 11.10.00.10.11), finally the 7-weight patterns are four (11.01.10.01.11, 11.10.00.01.01.11, 11.01.01.00.10.11, 11.10.00.10.00.10.11). Higher-weight patterns are less important, although their linear growth with Z is easily recognizable up to $w = 9$. Simple expressions, able to allow extrapolations over any number of periods Z, for the above three first non-null coefficients, according to (3.5), are

$$A(5) = Z, \quad A(6) = 2Z - 3, \quad A(7) = 4Z - 12.$$

We can observe that the two sub-generator polynomials act on different positions, so their weights are summed in $\mathrm{w}[g_{\mathrm{int}}(x)]$. The inequality $\mathrm{d} \leq \mathrm{w}[g_{\mathrm{int}}(x)]$ holds, like in polynomial block codes (see Lemma 1.8), after substitution of $g_{\mathrm{int}}(x)$ to $g(x)$, as usual. A difference between $\mathrm{w}[g_{\mathrm{int}}(x)]$ and d may arise when the sub-generator polynomials have high weights and the lowest-weight pattern appears after few periods. This is the case of the best code with $m = 3$, whose interleaved generator polynomial is 11.11.01.11 (or its reciprocal), already mentioned in Example 3.3. Its minimum distance is $d = 6$, in spite of the fact that $\mathrm{w}[g_{\mathrm{int}}(x)] = 7$. Here the lowest-weight code frames occur starting from $Z = 2$.

In most literature, a model, called convolutional code transfer function (Mason and Zimmerman 1960), is frequently adopted for determining the normalized weight distribution coefficients $B(w)$. Since in this model state $\underline{0}$ is halved, any possible extrapolation made beyond the bound $w \leq 2d - 1$, sometimes reported, must be taken with great cares.

An estimate of the average binary symbol error probability, after a soft-decision decoding, can be approximately calculated, by means of expression (D.7), from the knowledge of the only lowest-weight coefficient $A(d)$. This estimate is tighter and tighter with increasing the *SNR*, but it can be taken as a reference in order to compare the performance of different convolutional codes. A use of (D.7) with some further coefficients $A(w)$, after having determined their evolution according to (3.5), is possible, in order to obtain results over a wider range of possible values for the *SNR* (see Appendix D for details). Nevertheless a precise calculation of the break-even point, where the curves of the uncoded and of the coded error probability intersect, is not possible, because of the poor accuracy at low values of the *SNR* represented by (D.7). So the results of a computer simulation are preferable.

Since at the denominator of any term in the summation (D.7), there is the overall length n of the code word or code frame, being here this length proportional to the period number Z, from (3.5) we can obtain a performance which is practically independent of the frame length (or of the period number Z), confirming what expected for well designed convolutional codes.

The best convolutional codes, either at code rate 1/2 or more generally at code rates $1/n_o$, were discovered by attempts in late 1950 and during all the subsequent decade (Elias 1955; Forney 1970; Larsen 1973), with values of m up to some tenths, searching for the combination of sub-generator polynomials able to give the maximum d, having fixed m and the code rate. Finally, among different codes characterized by an equal value for d, the one with minimum $B(d)$ is preferred.

As anticipated, a soft-decision decoding procedure applied to a convolutional code, of the type here described, is made in a trellis, which represents the temporal evolution of the state diagram (see also Sect. 1.23). At any step of the decoding procedure, elimination of half the number of the possible paths in the state diagram is made according to Viterbi algorithm (Viterbi 1967), which was expressly proposed for being applied to such codes. From what discussed in Sect. 1.23, we can describe this type of trellis observing that it always shows three parts: two external transients, expanding and contracting respectively, and a very long stationary central part, where all the states in any stage are merge states. So 2^m can be considered a measure for the computational complexity of the above soft-decision decoding algorithm. The length of the framework is not considered, also because a sliding window decoding procedure is typically employed, as will be better explained in the following. Further details on the structure of the trellis will be given in Sect. 3.11.

Before the advent of Viterbi algorithm, some less efficient algorithms have been proposed (see also Appendix D). Sometimes they are able to face a number of states even higher, with acceptable computational complexity, but with poorer performance (in terms of reliability of the decoded binary symbol error probability obtained). In conclusion, values greater than $m = 7$ are rarely employed for practical applications, because of the decoding computational complexity or of the sub-optimal performance.

Table 3.2 shows the best convolutional codes characterized by code rate 1/2, that is with $k_o = 1$ and $n_o = 2$, up to $m = 7$. It collects the following parameters: generator constraint length m, minimum distance d, interleaved generator polynomial $g_{\text{int}}(x)$, normalized weight-distribution $B(w)$ up to $w = 2d - 1$ (for small values of d), or up to $w = d + 3$ (for high values of d). It is possible to note that, for $m = 3, 5, 7$, the minimum distance is $d = 6, 8, 10$, although $\text{w}[g_{\text{int}}(x)] = 7, 9, 11$ respectively. The interleaved generator polynomials suggested in the Table are not unique. In fact, if there is a convolutional code associated to an interleaved polynomial $g_{\text{int}}(x)$, certainly there is also an equivalent convolutional code associated to its reciprocal $g_{\text{int}}{}^*(x)$. The same property holds also exchanging the roles of the two sub-generator polynomials $g^{(1)}(x)$ and $g^{(2)}(x)$.

The first code is one of the four equivalent convolutional codes analyzed in Examples 3.1 and 3.4. The second code is that studied in Examples 3.2 and 3.5. The third code, as already noted in Example 3.3 and subsequent observations, is not the unique one with $m = 3$ and $d = 6$. It is characterized by $\text{w}[g_{\text{int}}(x)] = 7$ and $B(6) = 1$. Adopting 11.10.01.11 or 11.01.10.11 as $g_{\text{int}}(x)$, we obtain $\text{w}[g_{\text{int}}(x)] = 6$ and $B(6) = 2$.

We can observe that the codes associated to even-weight interleaved polynomials $g_{\text{int}}(x)$ exhibit just even-weight code frames (see for instance the case of $m = 6$), as expected. As long as m increases, there is a growing number of equivalent codes associated to different possible choices for $g_{\text{int}}(x)$. The number

Table 3.2 Best convolutional codes with $k_o = 1$, $n_o = 2$, and their $B(w)$ coefficients

m	d	$g_{\text{int}}(x)$	$B(3)$	$B(4)$	$B(5)$	$B(6)$	$B(7)$	$B(8)$	$B(9)$	$B(10)$	$B(11)$	$B(12)$	$B(13)$
1	3	11.01	1	1	1								
2	5	11.10.11			1	2	4	8	16				
3	6	11.11.01.11				1	3	5	11				
4	7	11.01.10.10.11					2	3	4	16			
5	8	11.01.10.11.10.11						1	8	7	11		
6	10	11.01.00.11.11.10.11								11	–	38	–
7	10	11.10.01.01.11.01.10.11								1	6	13	20

of codes having quite similar performance, for instance the same minimum distance but slightly different values for $B(d)$, grows even strongly. This topic will be analyzed more in depth in following sections, dealing with $1/n_o$-rate convolutional codes in general.

When d is rather high (of the order of 10 or more), its value could be somehow less important, whereas the multiplicity expressed by coefficient $B(d)$ could become more and more important. As known, such multiplicity influences how rapidly the asymptotic coding gain is reached (see Appendix D). For instance, with $m = 6$, the best code, reported in the Table, is characterized by $\mathrm{w}[g_{\mathrm{int}}(x)] = 10$, $B(10) = 11$ and $B(12) = 38$. There is another code, with $\mathrm{w}[g_{\mathrm{int}}(x)] = 9$, precisely $g_{\mathrm{int}}(x)$ associated to the sequence 11.01.10.01.01.10.11, which exhibits $d = 9$, but $B(9) = 3$, $B(10) = 3$, $B(11) = 9$, $B(12) = 19$, whose performance becomes better than that of the previous one for values of the binary symbol error probability after decoding greater than about 10^{-7}.

A consequence of the fact that control symbols are distributed along the frame is in the possibility of adopting a decoding procedure, able to give decoded information symbols before the complete processing of the whole frame. This feature is available only if the convolutional code is well designed.

Definition 3.6 In a decoding procedure performed determining the information symbols of the initial part of the frame before it is completely processed, there is a so called *sliding window*, shifting along the frame, in which the decoded information symbols are calculated step-by-step.

The length of the sliding window fixes a sort of step size in this step-by-step decoding procedure, which is an important parameter for evaluating the time latency of the error control system adopted. Considering Viterbi algorithm, it is possible to justify the finite length of such a sliding window, after the observation that all the surviving paths, at a certain stage, have a common initial part. We can expect that the length of the sliding window increases with m, but also with another quality parameter in the convolutional code design, that is the length of the longest pattern having a rather low weight ($w = d$ or a little more). The above considerations enable us to stress a further intuitive important property of a well designed convolutional code: the possibility of correcting more than one error pattern, provided that they are short and well separated (Wicker 1995). In this situation, the window relative to a first error pattern has to be already terminated at the moment in which a second error pattern occurs for being processed in another window.

On the contrary, not well designed convolutional codes cannot be decoded, in principle, employing a sliding window procedure, because two paths in the trellis, at small distance each other, may remain separated for the whole length of the frame, and merge only at its end. Most properties discussed so far, regarding 1/2-rate non-systematic convolutional codes, hold also for other types of convolutional codes, as will be better explained in following sections.

3.3 Systematic Convolutional Codes

The generator matrix $\underline{G}$ of a 1/2-rate convolutional code is characterized by the interleaved polynomial $g_{\text{int}}(x)$, given by (3.4), which shifts by two positions in each period, from one row to the subsequent one. This situation is shown in Fig. 3.6. Dashed small arrows denote the positions devoted to control symbols, whereas solid small arrows denote those where information symbols are emitted by the source forming polynomial $u(x^2)$. The binary symbols expressed by this polynomial directly enters the first memory element of the shift register, but generally do not appear in the output frame $t(x)$. A matrix $\underline{G}$ of the type shown in Fig. 3.6 is called semi-infinite, because the number of periods in the frame is not prefixed. A correct frame termination assures that the last polynomials $g_{\text{int}}(x)$ are completely inserted in its last rows, according to Lemma 3.2.

A matrix structure so organized cannot be systematic in strict sense, that is with all the information symbols in the first positions. It could be, instead, systematic just in wide sense, that is with information symbols appearing in some prefixed positions (see Definition 1.5). The most practical choice is to employ the first position of any period. In order to do this, keeping in mind the considerations developed after Lemma 1.5, describing the structure of a w.s. systematic $\underline{G}$ matrix, it is necessary that the columns where the interleaved polynomial $g_{\text{int}}(x)$ has its highest-order coefficient show the remaining elements all null.

Lemma 3.6 *In order to have a systematic 1/2-rate convolutional code, with generator constraint length m, the following condition on the first sub-generator polynomial is required*

$$g^{(1)}(x) = x^m \tag{3.6}$$

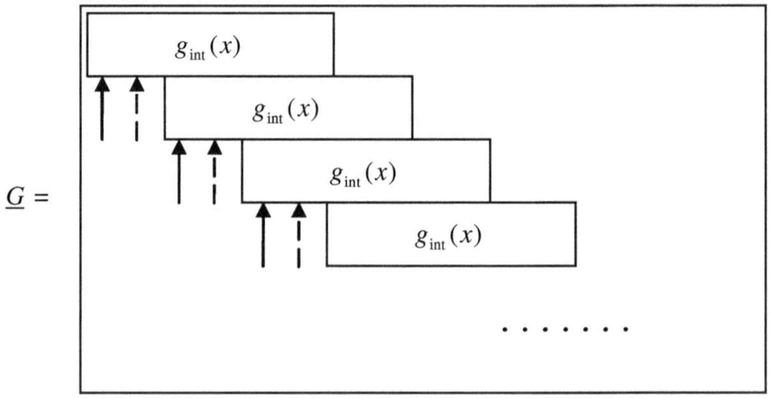

Fig. 3.6 Semi-infinite generator matrix of a 1/2-rate convolutional code

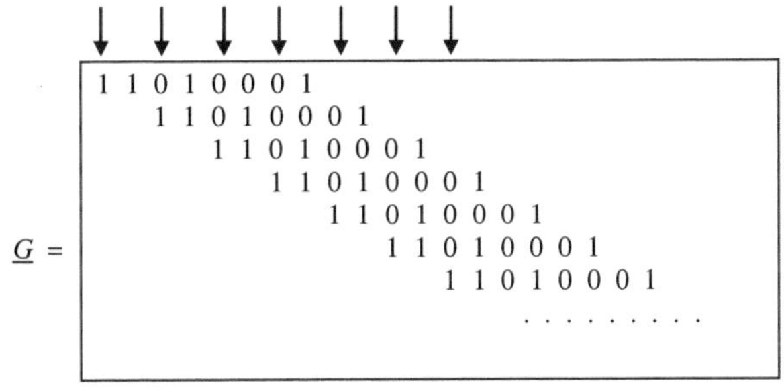

Fig. 3.7 Example of a systematic generator matrix for a 1/2-rate convolutional code

Proof If (3.6) holds, before the first memory element in the encoder circuit (of the type shown in Fig. 3.1), there is a wire connecting directly the information source to the encoder output. This wire takes the first symbol in any period. Since odd output symbol times can be occupied just by symbols coming from the upper ex-OR sum, the choice of $g^{(1)}(x)$ characterized by the only non-null $g_m^{(1)}$ coefficient assures that any information symbol in $u(x^2)$ reaches the encoder output without changing its identity. □

Example 3.9 In Fig. 3.7 a systematic semi-infinite $\underline{G}$ matrix is shown, for $m = 3$. The considered code exhibits $g^{(1)}(x) = x^3$, $g^{(2)}(x) = x^3 + x^2 + 1$, and a minimum distance $d = 4$. Solid small arrows denote the columns where information symbols are inserted: they correspond to coefficient $g_3^{(1)} = 1$, which expresses the highest-order coefficient of $g_{\text{int}}(x)$, and are characterized by all 0-symbols, except this 1-symbol.

Systematic convolutional codes show poorer performance with respect to non-systematic ones, for a same m, with $m > 1$. This fact is due to a not optimized exploitation of the weights characterizing the two sub-generator polynomials. As shown in Table 3.2, $\text{w}[g_{\text{int}}(x)]$, for the best 1/2-rate convolutional codes, is typically about 3/4 of the available overall weight, which in turn is equal to $(\deg[g_{\text{int}}(x)] + 1)$. Such a value cannot be obtained with one sub-generator polynomial having a weight as low as 1. A variant of Lemma 3.6 can be expressed as follows: in order to have a systematic convolutional code, it is enough to have $\text{w}[g^{(1)}(x)] = 1$ or $\text{w}[g^{(2)}(x)] = 1$. In this case a different position is to be considered for finding the information symbol, but the code performance is constrained exactly as in the case of (3.6) and cannot become better.

Fortunately it is always possible to construct a systematic encoder circuit, for whichever 1/2-rate convolutional code, for instance adopting a shift register in controller arrangement, with a suitable feed-back, as will be described in Sect. 3.6. Another possibility is based on a scheme with a shift register in observer arrangement, with ex-OR sums distributed along the shift register itself and a wire directly connecting the information source and the output. This alternative form is usually preferred when a different approach is followed, employing the parity check matrix instead of the generator matrix, and will be better analyzed in Chaps. 5 and 7.

Definition 3.7 A non-systematic convolutional code, in which an input sequence having an infinite weight produces an output sequence having finite weight, is defined *catastrophic convolutional code*.

Infinite, as usual, must be interpreted with a meaning of as high as one wishes. A code of this type produces the so called catastrophic propagation of single errors (Wicker 1995). In fact it can give a decoded sequence affected by much more errors than those present before decoding. In most literature catastrophic classification is reserved to the encoder circuit, and not to the code (Johannesson and Zigangirov 1999; Lin and Costello 2004). Nevertheless, adopting the concept that a code expresses the one-to-one correspondence between an information sequence and a code word, as made here so far, this characteristic appears to be correctly attributed also to the code itself. As an example of catastrophic convolutional code, we can take the one with $m = 1$, introduced in Example 3.7 (Fig. 3.5), and already described as not well designed. After an input consisting of all 1-symbols, it gives the output sequence 11.00.00.00 …, remaining in state $\underline{1}$ during all the length of the code frame. This fact is due to the closed loop characterizing state $\underline{1}$, whose output labelling is 00. There are other situations leading to catastrophic convolutional codes, associated to longer closed paths in the state diagram, whose branches are labelled by all 0-symbols in output. The following noticeable proposition holds.

Lemma 3.7 *A non-systematic 1/2-rate convolutional code with two sub-generator polynomials having a factor in common is catastrophic.*

Proof First of all, Lemma 1.39 states that, in a non-systematic state diagram describing a polynomial block code, a closed route, made by branches all labelled by 0-symbols in output, is present when $g(x)$ is not irreducible. This closed route is reached after a certain output sequence, and the same sequence is necessary for coming back to state $\underline{0}$. In the case of our convolutional code, we have two sub-generator polynomials which contribute independently to give 0-symbols or 1-symbols in output. If they have a common factor, it appears, squared, in the expression of $g_{\text{int}}(x)$. So a closed route made by branches labelled by pairs of 0-symbol in output takes place. It gives rise to code frames in which two equal non-null parts are separated by a sequence of 0-symbols long as one wishes. Precisely the length in output of this separation is multiple of twice the length of the closed route. □

Keeping in mind the structure of the code words discussed in the proof of Lemma 3.7, consequence of 0-labelled closed routes in the non-systematic state diagram, it is possible to conclude that the number of the correspondent code frames, all characterized by the same weight, grows binomially with the period number Z. This occurs for weights w lower than $2d$. In this sense, catastrophic convolutional codes are also not well designed convolutional codes. Nevertheless sometimes the binomial growth of a coefficient $A(w)$ is present only for w remarkably greater than d.

Example 3.10 The code characterized by $g^{(1)}(x) = x^3 + 1$ and $g^{(2)}(x) = x^3 + x^2 + x + 1$ is catastrophic. Its sub-generator polynomials exhibit factor $(x + 1)$ in common. Also in this case, we have state $\underline{1}$ with a closed-loop branch labelled 1/00. This situation is the most frequently encountered. Nevertheless the first weight distribution coefficient showing a binomial increase, due to the above closed-loop branch, is $A(10)$, whereas the minimum distance is $d = 6$.

Example 3.11 The 1/2-rate convolutional code characterized by sub-generator polynomials $g^{(1)}(x) = x^4 + x^3 + x + 1$ and $g^{(2)}(x) = x^4 + x^2 + 1$ is catastrophic. It exhibits $m = 4$, and an interleaved polynomial resulting in 11.10.01.10.11, leading to a minimum distance $d = 7$. Nevertheless the two polynomials have a factor $(x^2 + x + 1)$ in common. The closed route characterized by all 0-symbols in output has a length of two subsequent branches. The weight distribution coefficients are linear with Z up to $A(9)$ and become binomial only starting from coefficient $A(10)$. Precisely we have the following normalized coefficients: $B(7) = 2$, $B(8) = 2$, $B(9) = 2$. Comparison with the normalized coefficients characterizing the well designed code in the fourth row of the Table 3.2 shows substantially equivalent performance. In the following, still with $m = 4$, we will consider a catastrophic code even better, that is able to reach a minimum distance as high as $d = 8$.

A code, characterized by a certain minimum distance d, can be considered good or bad independently of the procedure by which information symbols are mapped on its code words. Unfortunately in a convolutional code, since the frames can be assumed to have whichever length, a bad mapping, as occurs for catastrophic convolutional codes, gives rise to a very high number of errors in the decoded information sequence, if we are operating just beyond its correction capability. In other terms, two code frames rather close each other (at the minimum distance or a little more) are associated to information symbols differing in a very great number of positions.

If the code is systematic, the information symbols, explicitly present in the frame, will be correct or not, after decoding, on average, in the same proportion as for all the symbols in the frame. The problem related to the possible presence of a catastrophic behaviour arises only in the typical operation condition of a non-systematic convolutional code, that is with very long frames, and in the presence of a strong lack of balance between the weight of information symbols and control symbols in a frame. It is absent in block codes, even if characterized by very long blocks, and also for convolutional codes employed with very short frames. On the other hand, the weight of the frame where catastrophic error propagation (or binomial number growth with Z) occurs is very important. If it is rather high (remarkably greater than d), the performance degradation is not so relevant.

Example 3.12 The convolutional code characterized by $m = 4$, and $g_{\mathrm{int}}(x)$ coincident with the sequence 11.01.11.10.11 (or equivalently 11.10.11.01.11), that is by the pair of sub-generator polynomials $g^{(1)}(x) = x^4 + x^2 + x + 1$ and $g^{(2)}(x) = x^4 + x^3 + x^2 + 1$ (or vice versa), is catastrophic if non-systematic, since the two sub-generator polynomials have $(x + 1)$ as a common factor, and remains not well designed even if it is transformed into an equivalent systematic version. Nevertheless it exhibits $d = 8$, all its low-weight code frames have weight multiple of 4, and the binomial growth with Z occurs only starting from coefficient $A(12)$. Precisely, after application of the input sequence $u(x^2) = x^6 + x^4 + x^2 + 1$, a frame characterized by the 1-symbol pattern 11.10.01.11.11.10.01.11 comes out. Its structure is made by two equal parts, which can be separated by pairs of 0-symbols, so originating a binomial growth in coefficient $A(12)$, like in a binary parity check code. On the other hand, the normalized coefficient $B(d)$ results $B(8) = 10$. In comparison with the fourth code reported in Table 3.2, which exhibits $d = 7$, for the same value of $m = 4$, this code seems to be very interesting.

Up to now we have discussed not well designed convolutional codes and catastrophic convolutional codes together, superimposing their limiting features. Nevertheless one of such features can be overcome. In fact the following proposition holds.

Lemma 3.8 *A systematic convolutional code cannot be catastrophic.*

Proof In the case of a systematic convolutional code, a fraction k_o/n_o of the whole amount of symbols in the frame are information (input) symbols. The above fraction is of the same order of magnitude of the total number of output symbols in the frame. So it is impossible to have a finite-weight output sequence in correspondence with an infinite-weight input (information) sequence. □

So, in principle, the catastrophic propagation of errors can be eliminated, by a proper systematic arrangement of the encoder circuit able to construct our code, but the binomial growth with Z in the number of some low-weight code frames and the consequent impossibility of applying a sliding window decoding procedure remain. In order to make in practice the above information mapping transformation, we have to consider that a de-mapping circuit with feed-back summations is needed (Massey and Sain 1968), so we can expect that the same structure will characterize also the desired systematic encoder circuit. In this sense, the solution based on a shift-register in observer arrangement is not available, but the one realized employing a shift register in controller arrangement with proper feed-back can be considered a valid solution.

3.4 Low-Rate Convolutional Codes

In Appendix D, where the main performance aspects regarding convolutional codes are summarized, low-rate convolutional codes are defined those characterized by code rate $1/n_o$. In Sect. 3.1, after the introduction of a non-systematic 1/2-rate convolutional code, some more general considerations have been carried out, about

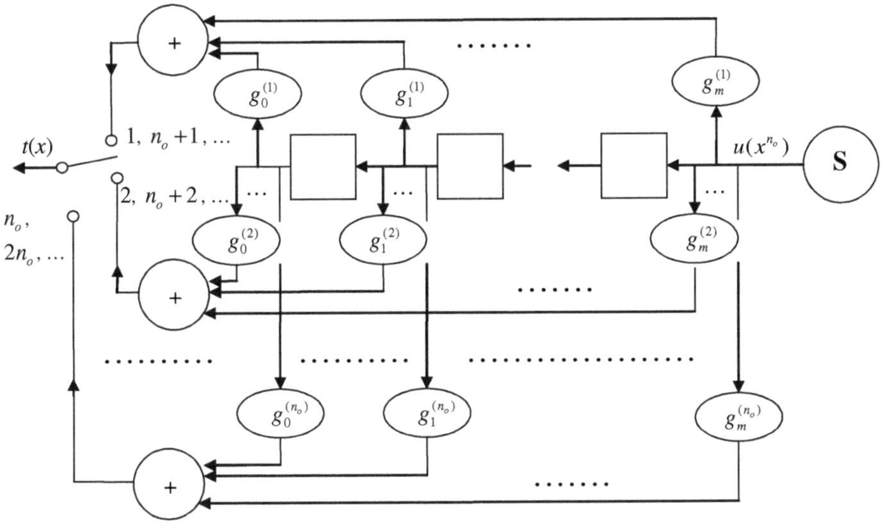

Fig. 3.8 Encoder circuit (in controller arrangement) for a $1/n_o$-rate convolutional code

the periods forming a frame. Any period is composed by n_o symbols, among which, the first k_o are related to as many information symbols. For low-rate convolutional codes, therefore, we can assume $k_o = 1$.

Definition 3.8 A *binary linear non-systematic convolutional code having rate $1/n_o$* is encoded by an encoder circuit, in which a shift register, in controller arrangement, with m memory elements, feeds n_o ex-OR sums. This circuit is depicted in Fig. 3.8. The wires connecting the shift register to these ex-OR sums are imposed by the coefficients of n_o sub-generator polynomials, $g^{(1)}(x)$, $g^{(2)}(x), \ldots, g^{(n_o)}(x)$.

Comparison with Definition 3.1 shows how this Definition includes the previous one as a particular case. We can observe that the scheme reported in Fig. C.2 of Appendix C, regarding a controller arrangement for operating a polynomial product, has been adopted here n_o times, exploiting a unique shift register, characterized by m memory elements. The output of the information source is taken one symbol time any n_o, as polynomial $u(x^{n_o})$ clearly expresses, representing the input sequence. An obvious difference with respect to the circuit shown in Fig. 3.1 regards the switch, which here exhibits n_o contacts, instead of only 2. The input symbol time is then subdivided into n_o output symbol times. The choice of maintaining the first sub-generator polynomial alone over the shift register will be useful for making a transformation of these circuits easier, in order to obtain systematic versions for them, as will be done in Sect. 3.6.

Lemma 3.9 *The output sequence t(x), that is the convolutional code frame, produced by an encoder circuit of the type depicted in* Fig. 3.8, *can be written as*

$$t(x) = u(x^{n_o})\left[x^{n_o-1}g^{(1)}(x^{n_o}) + x^{n_o-2}g^{(2)}(x^{n_o}) + \cdots + g^{(n_o)}(x^{n_o})\right]. \quad (3.7)$$

Proof We can follow a simple widening of the proof of Lemma 3.1. □

State diagrams can be introduced here as analogous widening of those presented in Sect. 3.2. They are still binary graphs (owing to the binary input symbols), but branch labelling has to be made with 1 input symbol and n_o output symbols.

Example 3.13 In the case of $m = 2$, there are various equivalent possibilities for obtaining the best 1/3-rate convolutional code. One of them consists of assuming $g^{(1)}(x) = x^2 + x + 1$, $g^{(2)}(x) = x^2 + x + 1$, $g^{(3)}(x) = x^2 + 1$. The encoder circuit and the correspondent state diagram for such code are reported in Fig. 3.9. The minimum distance turns out to be $d = 8$.

Definition 3.9 In a $1/n_o$-rate convolutional code, characterized by sub-generator polynomials $g^{(i)}(x)$, $i = 1, 2, \ldots, n_o$, the following polynomial

$$g_{\text{int}}(x) = x^{n_o-1}g^{(1)}(x^{n_o}) + x^{n_o-2}g^{(2)}(x^{n_o}) + \cdots + g^{(n_o)}(x^{n_o}) \quad (3.8)$$

is defined *interleaved generator polynomial for a* $1/n_o$*-rate convolutional code.*

This Definition is clearly a widening of Definition 3.4. We can recognize, in output, $g_{\text{int}}(x)$ along the shortest route around state $\underline{0}$, in the state diagram. This property is evident, for instance, in Fig. 3.9, where the interleaved polynomial turns out to be 111.110.111 (dots separates the three periods involved). Also for $1/n_o$-rate convolutional codes, provided that they are well designed, Lemma 3.5 holds. The mechanism after which a binomial increase in the number of code frames having weight greater than $(2d - 1)$ is still due to more than one passage through state $\underline{0}$.

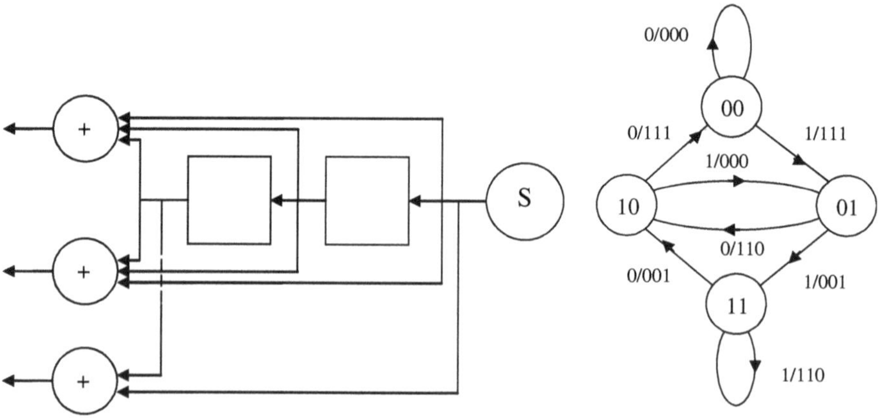

Fig. 3.9 Encoder circuit and state diagram for the 1/3-rate convolutional code described in Example 3.13

Table 3.3 Frame weight distribution of the best 1/3-rate convolutional code with $m = 2$

Z	$A(0)$	$A(8)$	$A(10)$	$A(12)$	$A(14)$
1	1	1			
2	1	3			
3	1	5	2		
4	1	7	6	1	
5	1	9	11	5	1
6	1	11	16	14	4
7	1	13	21	26	14
8	1	15	26	39	34
9	1	17	31	52	63
10	1	19	36	65	96
11	1	21	41	78	130
12	1	23	46	91	164

Example 3.14 The 1/3-rate convolutional code, with $m = 2$, described in Example 3.13, exhibits the evolution of the frame weight distribution shown in Table 3.3. Just even weights are present, since $w[g_{\text{int}}(x)]$ is even. There are two 8-weight patterns (111.110.111, 111.001.001.111), whereas the 10-weight patterns are five (111.110.000.110.111, 111.001.110.001.111, 111.110.000.001.001.111, 111 .001.001.000.110.111, 111.001.001.000.001.001.111). Higher-weight patterns are less important. The expressions, able to allow extrapolations over any number of periods Z, for the first three weight coefficients, according to (3.5), are

$$A(8) = 2Z - 1, \quad A(10) = 5Z - 14, \quad A(12) = 13Z - 65.$$

Table 3.4 shows the best convolutional codes characterized by code rate 1/3, that is with $k_o = 1$ and $n_o = 3$. It collects, as usual, the following parameters: total number m of memory elements (also here named generator constraint length), minimum distance d, interleaved generator polynomial $g_{\text{int}}(x)$, normalized weight-distribution $B(w)$ up to $w = 2d - 1$ (for $d = 5$), or up to a w value progressively reduced (for higher minimum distances d). In the second code, it is possible to recognize the code already described in Examples 3.13 and 3.14. The last three codes have just even-weight frames. We can observe a progressive reduction of the ratio d/m, already noted in Table 3.2, for 1/2-rate convolutional codes.

Table 3.5 shows the best convolutional codes with rate 1/4, that is characterized by $k_o = 1$ and $n_o = 4$. For all the codes, we have reported only the first five normalized coefficients $B(w)$. The last code is characterized by $B(15) = 0$ over any frame length. The progressive reduction of the ratio d/m is appreciable here only starting from higher values of m.

Table 3.4 Best convolutional codes with $k_o = 1$ and $n_o = 3$

m	d	$g_{int}(x)$	$B(5)$	$B(6)$	$B(7)$	$B(8)$	$B(9)$	$B(10)$	$B(11)$	$B(12)$	$B(13)$	$B(14)$	$B(15)$	$B(16)$
1	5	111.011	1	1	1	1	1							
2	8	111.110.111				2	–	5	–	13				
3	10	111.011.101.111						3	–	2	–	15		
4	12	111.011.101.011.111								5	–	3	–	13

Table 3.5 Best convolutional codes with $k_o = 1$ and $n_o = 4$

m	d	$g_{\text{int}}(x)$	$B(7)$	$B(8)$	$B(9)$	$B(10)$	$B(11)$	$B(12)$	$B(13)$	$B(14)$	$B(15)$	$B(16)$	$B(17)$
1	7	1111.0111	1	1	1	1	1						
2	10	1111.1110.1111				1	1	1	3	1			
3	13	1111.1110.0101.1111							2	1	–	3	1

The interleaved polynomials in Tables 3.4 and 3.5 can be considered as simple examples (Larsen 1973). In fact there are many possible equivalent codes, whose number rapidly grows with both m and n_o. Several properties among those listed in Sects. 3.2 and 3.3 for 1/2-rate convolutional codes still hold for $1/n_o$-rate convolutional codes. In particular, this is true for Lemma 3.6 about a condition on the first sub-generator polynomial, in order to have a systematic encoder circuit. Furthermore this is true for Lemmas 3.7 and 3.8 regarding catastrophic codes. The former one is to be extended here considering all the n_o sub-generator polynomials with a common factor.

About the convolutional codes called not well designed, for which Lemma 3.5 does not hold regarding some low-weight coefficients $A(w)$, we can observe that a trivial generalization for whichever n_o of the encoder circuit shown in Fig. 3.5 (Example 3.7) yields a code whose minimum distance is $d = 2n_o$. There are, also here, a binomial growth of $A(2n_o)$ and non-null weight distribution coefficient just for w multiple of $2n_o$. This family of codes appears as a further widening of the concept of parity check codes.

3.5 High-Rate Punctured Convolutional Codes

Many good high-rate convolutional codes are obtained by a proper code puncturation applied to the best 1/2-rate convolutional codes (Yasuda et al. 1984). In order to define univocally a punctured convolutional code, it is necessary to fix precisely which symbols of the frame are eliminated. In a systematic convolutional code, generally puncturation acts on control symbols only, in such a way as to maintain the character of being systematic. In non-systematic convolutional codes, such eliminated symbols can regard either information positions (typically the first ones in any period, and denoted i) or control positions (typically the last ones in any period, and denoted c). A special definition of puncturation matrix is to be given now, different from Definition 1.18, restricted to block codes only, because of the periodic behaviour here present.

Definition 3.10 In a 1/2-rate non-systematic convolutional code, a $2 \times Y$ matrix $\underline{P}$, called the *convolutional puncturation matrix*, has Y columns, as many as the periods on which it operates. In any column 1-symbols are placed in correspondence with the maintained symbols, 0-symbols are placed in correspondence with the eliminated symbols.

Example 3.15 Assumption of the convolutional puncturation matrix

$$\underline{P} = \begin{bmatrix} 1 & 0 & 0 & 0 & 1 \\ 1 & 1 & 1 & 1 & 0 \end{bmatrix},$$

means that, starting from a complete frame of the type *ic.ic.ic.ic.ic.* ...., we are left with a punctured frame of the type *ic.c.c.c.i.* ... . Nevertheless now symbols of the type i or c are no longer associated to information and control respectively, but only to the positions in the original complete frame. It is possible to calculate that this punctured convolutional code is characterized by asymptotic code rate 5/6.

Lemma 3.10 *The resulting asymptotic code rate of the punctured code, starting from a 1/2-rate convolutional code, after application of a puncturation matrix according to Definition 3.10, turns out to be*

$$R = \frac{Y}{w_p}, \tag{3.9}$$

where w_p is the total weight (overall number of 1-symbols) in the convolutional puncturation matrix.

Proof The resulting code rate will be the product of the initial rate, 1/2, by the ratio between the $2Y$ total positions in the puncturation matrix and its weight w_p. From this product, (3.9) comes out. □

As a check, if $w_p = Y$, all the control symbols are eliminated (or an equivalent situation occurs) and correctly the result is $R = 1$. At the opposite side, if $w_p = 2Y$, all the symbols are present and we are left with $R = 1/2$, as expected without any puncturation. It is important to stress that this puncturation matrix $\underline{P}$ has not to be multiplied by $\underline{G}$ in any way, but is useful only for fixing the choice of the surviving columns in the $\underline{G}$ matrix (with as many 1-symbols).

Example 3.16 To the 1/2-rate convolutional code characterized by $g^{(1)}(x) = x^2 + x + 1$ and $g^{(2)}(x) = x^2 + 1$, already discussed in Example 3.5, 3.6 and 3.8, we apply here the convolutional puncturation matrix

$$\underline{P} = \begin{bmatrix} 1 & 1 \\ 0 & 1 \end{bmatrix}.$$

The result is a code rate 2/3 and a reduction in the minimum distance d from 5 to 3. The new $\underline{G}$ matrix then becomes as shown in Fig. 3.10. An extended period comes out, whose length is 3 symbol times, the first and the second devoted to information symbols, and the third to a control symbol (dashed small arrows). An easy inspection of this generator matrix shows that it is periodically time-varying. There are two different generator polynomials, alternately present in its rows. We can introduce a generator sub-matrix $\underline{G}_o$ collecting them.

To summarize, in general, puncturation of an s.s. time-invariant convolutional code in the $\underline{G}$ matrix gives rise to a w.s. time-invariant convolutional code in the $\underline{G}$ matrix. An exhaustive analysis of this type of convolutional codes, also when not originated from puncturation, will be carried out in Chap. 4.

It is easy to argue that the decoding computational complexity in a punctured convolutional code remains the same as that characterizing the original 1/2-rate code (Cain et al. 1979). In fact, we can interpret symbol elimination after a puncturation like an erasure (see Definition D.35 in Appendix D for details). The

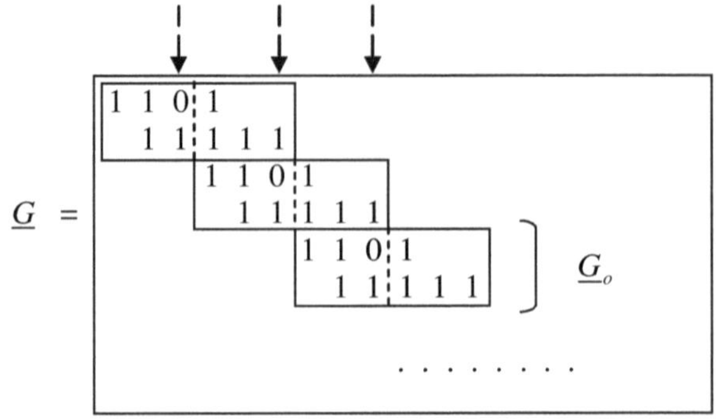

Fig. 3.10 Semi-infinite $\underline{G}$ matrix for the punctured code described in Example 3.16

decoding trellis is unchanged, and the metric adopted can be suitable for either a hard- or soft-decision decoding. Simply a value exactly on the threshold between the two signal domains (associated to the two symbols respectively) is taken to replace the punctured symbol in the decoding path comparison. In spite of this good feature, there are several risks that puncturation can introduce, such as: a too strongly reduced minimum distance, catastrophic error propagation, impossible information recover at the destination.

As discussed in Sect. 1.7, a strong puncturation may lead to not only 2-weight code frames, but even to 1-weight code frames. We can observe, in the generator matrix shown in Fig. 3.10, that 2-weight patterns do not occur, as well as 1-weight patterns. Fortunately these points of weakness have been prevented in the 2/3-rate punctured code treated in Example 3.16, owing to the code rate adopted (not too high) and also to the value of the generator constraint length m characterizing the original code (not too short).

Example 3.17 Application of whichever puncturation matrix of length $Y = 2$, with $w_p = 3$ (still leading to a code rate 2/3), to any of the four convolutional codes with $m = 1$ shown in Fig. 3.2 gives rise to a code characterized by minimum distance $d = 2$. If we assume a puncturation matrix of length $Y = 3$ and $w_p = 4$ (which means a code rate 3/4), even a code frame with weight 1 may appear. For instance, taking

$$\underline{P} = \begin{bmatrix} 1 & 1 & 0 \\ 0 & 1 & 1 \end{bmatrix},$$

where the eliminated symbols are one any three on average, applied to the code characterized by interleaved polynomial $g_{\text{int}}(x) = x^3 + x^2 + 1$, the resulting generator matrix is shown in Fig. 3.11. The extended period is here 4 positions long. In it, the first, second and fourth positions are devoted to as many information symbols, whereas the third is devoted to a control symbol (dashed small arrows). There are three different generator polynomials. One of them has weight 2, and one weight 1, therefore the minimum distance is $d = 1$. This code exhibits also another evident defect: there is no correlation between adjacent periods, making useless the concept of frame itself.

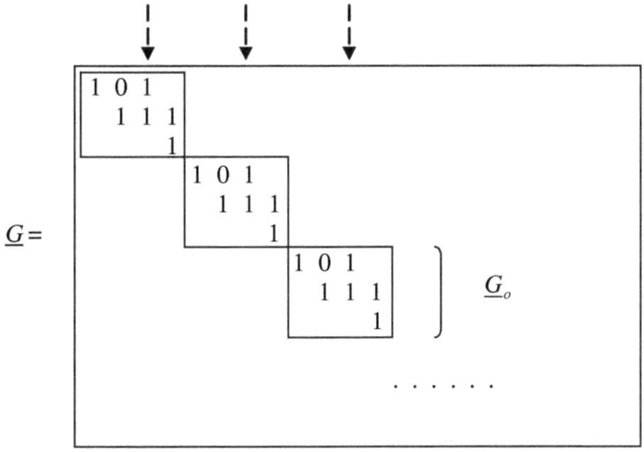

Fig. 3.11 Semi-infinite $\underline{G}$ matrix for the punctured code described in Example 3.17

Example 3.18 Assuming again the 1/2-rate convolutional code characterized by $g^{(1)}(x) = x^2 + x + 1$ and $g^{(2)}(x) = x^2 + 1$, with the following convolutional puncturation matrix

$$\underline{P} = \begin{bmatrix} 1 & 1 & 1 \\ 0 & 0 & 1 \end{bmatrix},$$

we obtain a 3/4-rate not well designed convolutional code (catastrophic if non-systematic). In fact the resulting generator matrix appears as shown in Fig. 3.12. In any generator sub-matrix, the first and the third row can form a chain of the type present in the 1/2-rate convolutional code introduced in Example 3.7, in which $A(4)$ grows binomially with the number Z of periods. The same result occurs here, although the minimum distance is now $d = 3$.

Let us denote with parameter $(m' + 1)$ the number of periods spanned by the $\underline{G}_o$ sub-matrix obtained after puncturation, taking, as a reference, the original generator constraint length m. In Examples 3.16 and in 3.18, from $m = 2$, after puncturation

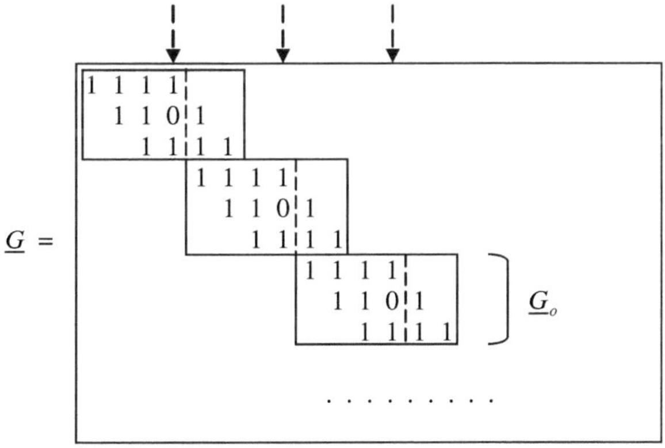

Fig. 3.12 Semi-infinite $\underline{G}$ matrix for the punctured code described in Example 3.18

we have obtained $m' = 1$. We can conclude that branch complexity have replaced state complexity (see Sect. 2.11). This phenomenon will be better investigated in Chap. 4, where w.s. time-invariant convolutional codes (like those obtained here after puncturation) will be analyzed. In Sect. 3.8 we will discuss, instead, the possibility of constructing high-rate s.s. time-invariant convolutional codes, that is with a unique interleaved generator polynomial, but with code rates higher than 1/2.

There is a constraint in the maximum possible number of single puncturations, in order to have a true code and not a degenerate code, that is a code without control symbols at all.

Lemma 3.11 *In order to prevent possible situations in which the punctured convolutional code turns out to be degenerate, for an original 1/2-rate convolutional code, it is necessary to satisfy the condition* $w_p \geq Y$ *in the puncturation matrix* $\underline{P}$ *described in Definition 3.10 and Lemma 3.10.*

Proof In the case of equality, we have an additive linear transformation (see Definition 1.1) expressed by a square $\underline{G}$ matrix. In a situation characterized by $w_p < Y$, information extraction is impossible, because we should have more constraints than variables (see Appendix A). □

Remark that the above obvious condition is only necessary in order to allow information recover. In fact, also when the punctured code is not degenerate, we have to add the well known requirement that the resulting $\underline{G}$ matrix admits a right-inverse. First of all, we can observe that the $\underline{G}$ matrix of the original (not punctured) convolutional code is always right-invertible. The possibility of not satisfying this condition may come out in punctured convolutional codes, when they are non-systematic, but also when they are systematic but puncturation is made on information symbols (Barbier et al. 2006). This risk typically occurs when there are too many subsequent 0-symbols in matrix $\underline{P}$. For instance, if an entire interleaved generator polynomial is suppressed, it is impossible to recover the information symbol destined to multiply it.

3.6 Recursive Systematic Convolutional (RSC) Codes

Starting from a non-systematic $1/n_o$-rate convolutional code, it is possible to obtain an equivalent systematic code. The objective of this transformation is also to maintain unchanged the number of memory elements m in the encoder circuit, which, as known, is related to the decoding computational complexity. In order to obtain a systematic equivalent of a certain $1/n_o$-rate convolutional code, it is necessary to introduce a feed-back in the encoder circuit, like occurs for block polynomial codes in transforming the encoder circuit of Fig. 1.3 into that of Fig. 1.5. The analysis of this problem is advantaged by the use of a symbolic polynomial expression for the generator matrix $\underline{G}$.

Definition 3.11 Maintaining separate the n_o sub-generator polynomials, expressed in their symbolic form (function of a formal variable x), it is possible to define a *symbolic polynomial generator sub-matrix* $\underline{G}_o(x)$, which, for a non-systematic convolutional code, results in

$$\underline{G}_o(x) = \left[g^{(1)}(x) \quad g^{(2)}(x) \quad \ldots \quad g^{(n_o)}(x) \right]. \tag{3.10}$$

Such matrix exhibits n_o elements, formed by one polynomial each. In a more general context, for a convolutional code with code rate k_o/n_o, we would have n_o columns, each containing k_o polynomials.

Example 3.19 Assuming the best non-systematic convolutional code at rate 1/3, already discussed in Example 3.13, that is with an interleaved generator polynomial associated to the sequence 111.110.111, its symbolic polynomial generator sub-matrix is

$$\underline{G}_o(x) = \left[(x^2 + x + 1) \quad (x^2 + x + 1) \quad (x^2 + 1) \right].$$

It is important to stress that $\underline{G}_o(x)$ does not shift along subsequent rows in the $\underline{G}$ matrix. It is simply an instrument for introducing systematic convolutional codes, and also for making easier the design of their encoder circuits, by proper modifications of the one associated to their non-systematic version, but does not replace, in any possible context, the interleaved polynomial $g_{\text{int}}(x)$.

Definition 3.12 A convolutional code is called *delay-free* if it exhibits a non-null m.th-order coefficient in its first sub-generator polynomial $g^{(1)}(x)$, that is it shows $g_m^{(1)} = 1$.

The condition for the considered convolutional code to be delay-free is due to the need of arranging a proper polynomial division, performed by a circuit having a general scheme. It is sufficient that one of the n_o sub-generator polynomials satisfies this property, in fact a proper permutation of the positions in the period can lead to an equivalent code with its first sub-generator polynomial so predisposed.

Definition 3.13 A symbolic polynomial generator sub-matrix is called *rational* if, in some of its symbolic columns, it shows polynomial divisions.

As detailed in Appendix C, it is possible to perform a division and a multiplication at the same time, and the linear shift register able to do this appears as depicted in Fig. C.8. In particular we are interested in the upper scheme, which is organized in controller arrangement.

Lemma 3.12 *In order to obtain a systematic equivalent of a given non-systematic $1/n_o$-rate convolutional code, provided that it is delay-free, we have to adopt a rational symbolic polynomial generator sub-matrix $\underline{G}_o(x)$, where the generator polynomials from $g^{(2)}(x)$ up to $g^{(n_o)}(x)$ are divided by $g^{(1)}(x)$.*

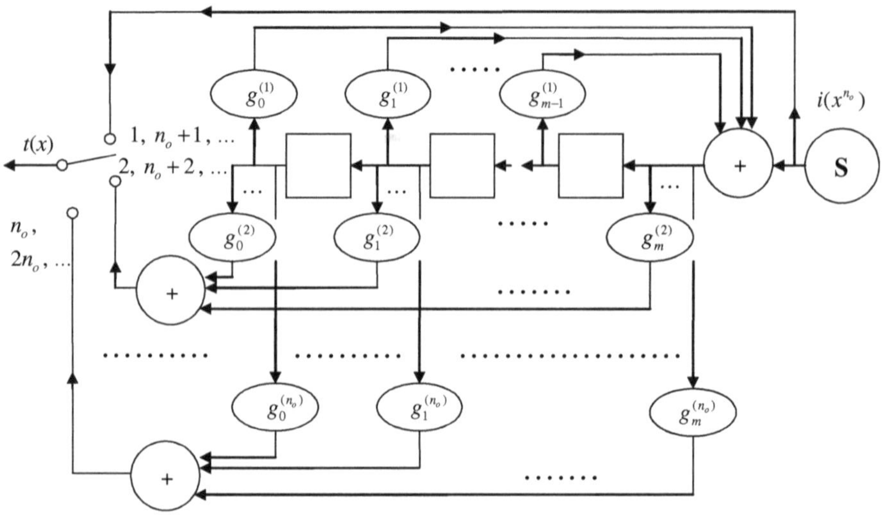

Fig. 3.13 Controller arrangement for an RSC code with rate $1/n_o$

Proof The symbolic polynomial generator sub-matrix of the desired code is

$$\underline{G}_o(x) = \begin{bmatrix} 1 & \frac{g^{(2)}(x)}{g^{(1)}(x)} & \cdots & \frac{g^{(n_o)}(x)}{g^{(1)}(x)} \end{bmatrix}. \tag{3.11}$$

It can be obtained from (3.10), performing a division by $g^{(1)}(x)$. The need of a solution of this type arises when $\mathrm{w}[g^{(1)}(x)] \geq 2$, otherwise the convolutional code is already systematic or easily suitable for being set up in systematic form. The validity of (3.11) is assured by the fact that we have a unique shift register. Since division, by means of a proper feed-back shift register, is possible only when the highest-order coefficient of the dividend is not null (see Appendix C), this entails the constraint of starting from a delay-free non-systematic convolutional code. □

Definition 3.14 A systematic $1/n_o$-rate convolutional code obtained with a rational symbolic polynomial generator sub-matrix of the type (3.11) is called *recursive systematic convolutional* (or briefly *RSC*), and its encoder circuit, in controller arrangement, is shown in Fig. 3.13.

A wire directly connects the information source to the first contact of the output switch, carrying the information symbol sequence, represented by polynomial $i(x^{n_o})$. All the wires imposed by $g^{(1)}(x)$ are connected to a feed-back ex-OR sum. It is useful to remember that the wire associated to $g_m^{(1)} = 1$ has to be ignored, as usual.

The weight distribution of a $1/n_o$-rate RSC code is coincident with that of the non-systematic convolutional code from which it has been obtained, so confirming that the two codes are equivalent. Nevertheless different mapping of the information symbols on the code frames are to be considered in the two cases. In particular, the

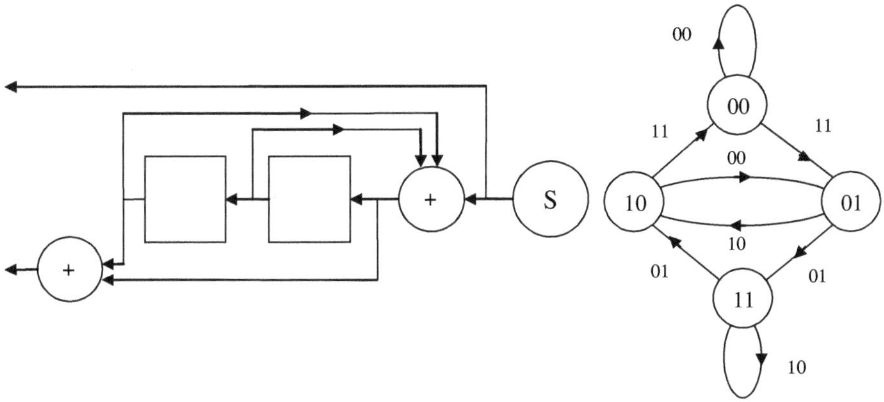

Fig. 3.14 RSC equivalent version of the code described in Fig. 3.4

following surprising property holds: short 1-symbol information patterns typically gives rise to long 1-symbol output patterns, and vice versa. The semi-infinite generator matrix of an RSC code is no longer characterized by an enlarged diagonal, but shows rows never-ending on their right hand side. These features will be better described in Chap. 9, where such codes will be considered as proper component codes in turbo concatenation schemes (Berrou et al. 1993).

Example 3.20 The RSC equivalent of the non-systematic 1/2-rate convolutional code characterized by sub-generator polynomials $g^{(1)}(x) = x^2 + x + 1$ and $g^{(2)}(x) = x^2 + 1$, described in Fig. 3.4, exhibits a systematic encoder circuit and a state diagram as shown in Fig. 3.14. In the state diagram it is no longer necessary to denote branches with input and output symbols.

It is important to note that RSC codes require a precise final sequence of m symbols emitted by the information source, in order to force the shift register to reach again state $\underline{0}$, so assuring a correct frame termination. Such symbol sequence depends on the state in which the shift register stops, after the action of the true information symbol sequence. Nevertheless this problem can be easily faced by means of a proper look-up table. The scheme of encoder circuit shown in Fig. 3.13 is valid also for a catastrophic convolutional code. In conclusion, catastrophic encoders turn out to be non-catastrophic in their RSC version, owing to the property which prevents a systematic code from being catastrophic (Lemma 3.8). Nevertheless the fact that the considered code is not well designed remains.

Example 3.21 The 1/2-rate convolutional code characterized by $g^{(1)}(x) = g^{(2)}(x) = x + 1$, presented in Example 3.7, in its RSC version can be encoded by the circuit shown in Fig. 3.15 on the left and exhibits the associated state diagram. It is described by the generator matrix reported on the right. So it turns out to be no longer catastrophic, in fact an infinite input sequence of 1-symbols is not required here for remaining in state 1, but it remains a not well designed convolutional code.

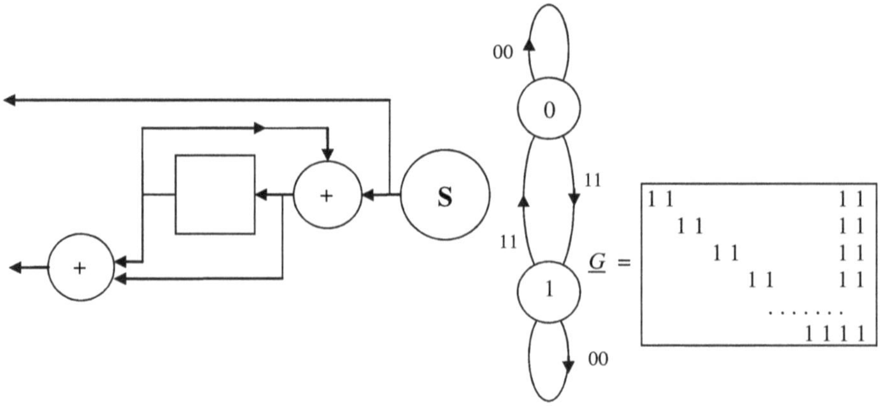

Fig. 3.15 RSC convolutional encoder for the code described in Example 3.7, state diagram and $\underline{G}$ matrix

Example 3.22 The 1/2-rate convolutional code characterized by $g^{(1)}(x) = x^4 + x^2 + x + 1$ and $g^{(2)}(x) = x^4 + x^3 + x^2 + 1$, presented in Example 3.12, in its RSC equivalent version is no longer catastrophic, although remains not well designed, but only in its 12-weight distribution coefficient. So it could be efficiently used, when a sliding window decoding procedure is not required.

3.7 Equivalence Between MLC Codes and s.s. Time-Invariant Convolutional Codes

Coming back to the description of MLC codes (Sect. 1.13), in comparison with the characteristics of s.s. time-invariant $1/n_o$-rate convolutional codes discussed so far, we can argue that there is an equivalence between them. In particular, the introduction of the interleaved polynomial $g_{\text{int}}(x)$ (Definition 3.9), for the convolutional code, allows an easy identification of such polynomial with the generator polynomial $g(x)$ of the cyclic code from which the correspondent MLC code derives.

Theorem 3.1 The equivalence between an MLC code, characterized by generator polynomial $g(x)$ whose degree in r, and an s.s. time-invariant convolutional code, described by an interleaved polynomial $g_{\text{int}}(x)$, both with code rate $1/n_o$, is supported by the following identities

$$r = (m + 1)n_o - 1, \quad g_{\text{int}}(x) = g(x). \tag{3.12}$$

Proof From (3.8) the degree of the interleaved polynomial describing a $1/n_o$-rate non-systematic convolutional code clearly is $\deg[g_{\text{int}}(x)] = (m + 1)n_o - 1$. More precisely, substitution of (3.8) into (3.7), for the same code, yields the output sequence

$$t(x) = u(x^{n_o})g_{\text{int}}(x). \tag{3.13}$$

This expression characterizes also the code word of an MLC code, in non-systematic version, for $k_o = 1$ and whichever n_o, provided that we replace $g_{\text{int}}(x)$ with the generator polynomial $g(x)$. So the structures describing the generator matrix of the two types of codes are coincident, as Figs. 1.13 and 3.6 clearly demonstrate. Also the block/frame termination is performed in the same way, that is assuring complete insertion of the last polynomials. □

The encoder circuits, either in non-systematic (Figs. 1.3 and 3.8) or in systematic form (Figs. 1.5 and 3.13), produce the same coded sequences, although with a different mechanism. This difference is evidenced by the state diagrams. In an MLC code there is a greater number of states. Nevertheless a fraction of the whole ensemble of such states consists of pure transit states, without branch joints in input or branch bifurcation in output. After suppression of such states and collection of the resulting subsequent branches into a single branch, labelled with the correspondent input/output symbol sequence, the two types of state diagrams turn out to be coincident. We have to note that not always the first of (3.12) is valid for a couple of parameters r and m both integer numbers. Given an integer m, the correspondent integer r can be calculated. So, in order to have a correspondence between integers, the convolutional code is to be constructed with all the sub-generator polynomials having degree m (condition for reaching the best performance). In the following, when the contrary is not explicitly said, we will consider this constraint always satisfied.

Example 3.23 About the coincidence in the state diagram, it is possible to compare the one characterized by $r = 3$, and $g(x) = x^3 + x + 1$, shown in Fig. 1.15 for an MLC Hamming code, with the one characterized by $m = 1$, $g^{(1)}(x) = x + 1$ and $g^{(2)}(x) = 1$, shown in Fig. 3.3 in the rightmost position (output symbols only) for the correspondent 1/2-rate convolutional code.

Example 3.24 The MLC code described in Example 1.46 and the 1/2-rate convolutional code described in Examples 3.5 and 3.8 exhibit the same word (frame) weight distribution, as Tables 1.9 and 3.1 clearly show, so confirming their equivalence.

At this point, one could argue that, being parameter r in the MLC code greater than parameter m in the equivalent $1/n_o$-rate convolutional code, for a soft-decision decoding applied on both, the computational complexity is lower in the latter one. This conclusion is not true.

Corollary a *For $1/n_o$-rate convolutional codes, the maximum number of merge states per stage in the trellis is equal to that of the correspondent MLC codes.*

Proof We remember that Definition 1.40 introduces merge states in a decoding trellis as the nodes where two branches join (and then where a choice between two decoding paths has to be made). A selection in the merge states of an MLC code occurs, owing to the insertion of distributed control symbols, leading to the transit

O	X	O	X	O

Fig. 3.16 Mask of the states where a choice between two decoding paths has to be made for $r = 5$ and code rate 1/2

states already mentioned. In fact they correspond to 0-symbols forced in the state evolution (see, for instance, the masks in Fig. 1.16). □

Let us take now a simple situation for which (3.12) gives a pair a equivalent codes (with r and m both integer numbers). For example, the mask corresponding to the merge states in the case of a 1/2-rate MLC code, with $r = 5$, is reported in Fig. 3.16, where, as usual, character X may represent either a 0-symbol or a 1-symbol, whereas character O expresses a position in which a 0-symbol is forced to be. The total number of states in this stage is $2^2 = 4$, exactly equal to the number 2^m of merge states in the correspondent stage of the trellis associated to the equivalent convolutional code. As a consequence, also the computational complexity of a soft-decision decoding is equivalent in the two families of codes. The above considerations hold for any code rate $1/n_o$, with $n_o \geq 2$.

Beyond the structural equivalence analyzed so far, we can observe that MLC codes appear to form a wider class of codes, in fact they can be arranged also in a high-rate form, assuming values greater than 1 for parameter k_o. They remain s.s. time-invariant also in this high-rate context, as will be better explained in the next section. On the contrary, starting from a 1/2-rate convolutional code, it is typically punctured for reaching higher code rates, so becoming a w.s. time-invariant convolutional code, as remarked in Sect. 3.5. Another practical difference regards the systematic versions of such two families of codes. RSC codes, as already noted, require a certain final sequence of m symbols emitted by the information source in order to force the shift register to reach again state $\underline{0}$. This procedure is not necessary for MLC codes, whose encoder circuit (see Fig. 1.5) requires simply that the switch remains connected to the feed-back for r consecutive symbol times.

Before concluding this section, let us come back to the interpretation of direct product codes as particular MLC codes discussed in Sect. 1.17. We can observe here that, when the rectangular memory for encoding a direct product code has one dimension much smaller that the other, encoding can be easily arranged by means of a proper unique shift register. Unfortunately, seen as a convolutional code (owing to the MLC code equivalence), such direct product code appears as a not well designed convolutional code. This leads to a binomial increase with Z (the longer dimension in the rectangular encoding memory) even for the lowest-weight coefficient $A(d)$, so making impossible any sliding window decoding procedure.

It is possible to verify the equivalence here summarized by proper interpretation of the code described in Examples 3.7 and 3.21 as the direct product between a [2,1,2] parity check code and a [Z + 1,Z,2] parity check code. In particular, the $\underline{G}$ matrix shown in Fig. 3.15 is clearly the Kronecker product of the generator matrices of such two codes, the latter one in systematic form.

3.8 Strict-Sense Time-Invariant High-Rate Convolutional (MLC) Codes

A wide-sense time-invariant high-rate convolutional code is characterized by different interleaved generator polynomials periodically inserted in the $\underline{G}$ matrix, as will be better explained in Chap. 4. As already noted in Sect. 3.5, a code of this type can be obtained, also after code puncturation, starting from a 1/2-rate s.s. time-invariant convolutional code. Alternatively, it can be designed and constructed with an encoder circuit employing more than one shift register, leading to better performance typically at the expense of a higher decoding computational complexity. Nevertheless, if we take a unique polynomial, repeated k_o times any period, this description degenerates into that of a strict-sense time-invariant high-rate convolutional code. The performance of such family of codes is generally worse than that of w.s. time-invariant convolutional codes, having the same parameters (total number of memory elements in the encoder circuit and code rate). This is due to the reduced number of degrees of freedom in code design.

On the other hand, s.s. time-invariant high-rate convolutional codes, owing to their equivalence with proper MLC codes, can be studied with the purpose of understanding how low-weight patterns can come out. We will consider only non-systematic codes and the interleaved polynomial will be denoted $g(x)$, for the sake of simplicity.

Lemma 3.13 *The output sequence of an s.s. time-invariant high-rate convolutional (MLC) code, characterized by parameters k_o, n_o, and generator polynomial $g(x) = g_{\text{int}}(x)$, can be written as*

$$t(x) = \left[x^{k_o-1}u_1(x^{n_o}) + x^{k_o-2}u_2(x^{n_o}) + \cdots + u_{k_o}(x^{n_o})\right]g(x). \tag{3.14}$$

Proof The input sequence can be thought as subdivided into k_o distinct interleaved fluxes, whose contribution can be considered taking one symbol from each of them in any position of the period. □

The structure of the generator matrix describing a 2/3-rate s.s. time-invariant convolutional code of this type is shown in Fig. 3.17. We have used a generator sub-matrix $\underline{G}_o$ like the one introduced in Sect. 3.5, in general with k_o rows, all containing here the same polynomial. This code exhibits a value for $r = \deg[g(x)]$ such as to occupy $(m + 1)$ subsequent periods. Dashed small arrows denote the positions where control symbols are inserted, as usual. Parameter m should represent the number of memory elements in each one of the k_o shift registers forming the convolutional encoder circuit, in controller arrangement, according to the traditional theory of convolutional codes. Nevertheless, since these shift registers should have all the same scheme, it is clearly preferable to employ an encoder circuit, with just one shift register, of the type shown in Fig. 1.3 (non-systematic) or Fig. 1.5 (systematic), so adopting the MLC code model. Only the switch operation has to be properly organized.

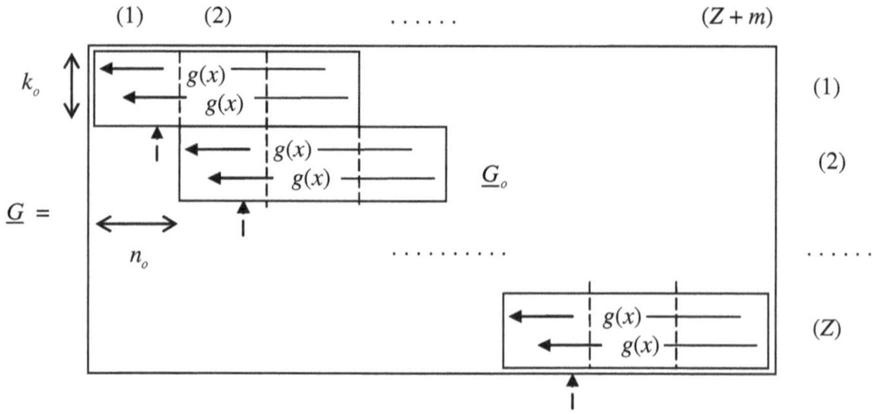

Fig. 3.17 Generator matrix of a 2/3-rate s.s. time-invariant convolutional code ($m = 2$)

Equality (3.12), which holds for $k_o = 1$, relating parameters m and r to the period length n_o, can be here so generalized

$$r = (m + 1)n_o - k_o, \tag{3.15}$$

where r represents the number of memory elements of the encoder circuit to be employed in the MLC code model. The difference between (3.12) and (3.15) regards only the number of rows in $\underline{G}_o$, with reference to the last period. The following proposition can be useful in order to understand how possible 2-weight code frames may be formed, and consequently how to exclude their presence.

Theorem 3.2 *Under the assumption of Lemma* 3.13, *if* $w[g(x)] > 2$, *the occurrence of 2-weight 1-symbol patterns is avoided, provided that the input sequence, within square brackets in expression* (3.14), *does not match the parity check polynomial* $h_c(x)$ *characterizing the cyclic code, parent of the considered MLC code.*

Proof For the parent cyclic code, the following equality holds $g(x)\, h_c(x) = x^N + 1$. If, exploiting the degrees of freedom present in the polynomial $\left[x^{k_o-1}u_1(x^{n_o}) + x^{k_o-2}u_2(x^{n_o}) + \cdots + u_{k_o}(x^{n_o})\right]$, imposed by the information symbol sequence, it is able to match polynomial $h_c(x)$, the above binomial will be formed as an output sequence. This is the only way for obtaining such a 2-weight code frame. Therefore, excluding it, the occurrence of this 1-symbol pattern will be prevented. □

Example 3.25 Considering a [7,4,3] Hamming code, characterized by the generator polynomial $g(x) = x^3 + x + 1$, it is possible to verify that, with $n_o - k_o = 1$ and $n_o \leq 3$, the minimum distance remains $d = 3$. Instead, with $n_o - k_o = 1$, but $n_o = 4$, that is with code rate 3/4, 2-weight code frames occur. More precisely, in Table 3.6 the evolution of the weight distribution for the 3/4-rate MLC code so obtained is reported.

Table 3.6 Frame weight distribution for the 3/4-rate code presented in Example 3.25

k	Z	$n = L$	$A(0)$	$A(2)$	$A(3)$	$A(4)$	$A(5)$	…
3	1	6	1		4	3		
6	2	10	1	1	10	13	12	
9	3	14	1	2	18	31	48	
12	4	18	1	3	26	55	118	

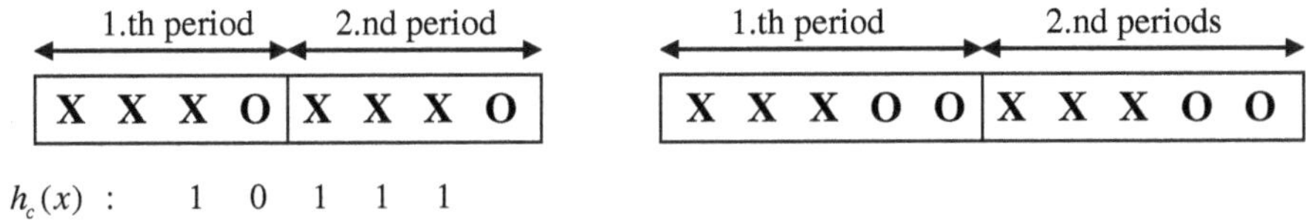

Fig. 3.18 State masks for two subsequent periods associated respectively to code rate 3/4 and 3/5 (Examples 3.25 and 3.26)

In this Table, k represents the total number of information symbols, $n = L$ the overall frame length (included termination), and Z, as usual, the number of periods where information symbols are inserted. Remark that we have an increase in the number of 2-weight code frames by one unit any period, starting from the second. It is due to the pattern $(x^7 + 1)$, shifting along the code frame, and, in turn, originated by the multiplication of $g(x)$ by $h_c(x) = x^4 + x^2 + x + 1$, which can be matched by suitable choices of polynomials $u_j(x^4)$ in expression (3.14). In fact the considered situation is described by a state mask, over two subsequent periods, of the type shown in Fig. 3.18 on the left.

It is interesting to observe that, starting from the 1/2-rate convolutional code with the same interleaved generator polynomial, and applying a puncturation matrix, as made in Example 3.17, the result turns out to be a 3/4-rate punctured code with minimum distance $d = 1$. Instead, the use of a high-rate s.s. time-invariant MLC code prevents this possibility, leading, in the worst case, to only 2-weight code frames. Unfortunately the decoding computational complexity is higher.

Example 3.26 Considering the same parent [7,4,3] Hamming cyclic code, as in Example 3.25, but here with $n_o - k_o = 2$, and $n_o = 5$, that is code rate 3/5, a different situation occurs. The evolution of the weight distribution for the 3/5-rate MLC code so obtained is reported in Table 3.7. Here 2-weight code frames are absent. This fact is due to a condition which prevents matching of $h_c(x) = x^4 + x^2 + x + 1$, by any choice of polynomials $u_j(x^5)$ in expression (3.14), as the state mask, depicted in Fig. 3.18 on the right, clearly shows. In this MLC code, interpreted as a convolutional code, the weight distribution exhibits the following normalized coefficients $B(3) = 4$, $B(4) = 7$, $B(5) = 12$.

Remark that the first code either in Table 3.6 or in Table 3.7 coincides with the [6,3,3] one-time shortened Hamming code. The [7,4,3] Hamming cyclic code could

Table 3.7 Frame weight distribution for the 3/5-rate code presented in Example 3.26

k	Z	$n = L$	$A(0)$	$A(3)$	$A(4)$	$A(5)$	…
3	1	6	1	4	3		
6	2	11	1	8	10	8	
9	3	16	1	12	17	20	
12	4	21	1	16	24	32	

be employed also for obtaining a 4/6-rate MLC code ($k_o = 4$, $n_o = 6$), without the occurrence of any 2-weight code frame, as it can be immediately argued from inspection of the correspondent state mask.

The risk that a high-rate s.s. time-invariant convolutional code becomes not well designed (see Definition 3.5) is stronger and stronger with increasing the code rate. In fact, although 2-weight code frames can be prevented, according to what suggested by Theorem 3.2, some even-weight 1-symbol patterns, like those characterizing an LC code when its cyclic parent code has a high minimum distance (see Sect. 1.11), may occur. Such code words typically show a more than linear increase with the number of periods Z, so leading here to a not well designed convolutional code. At least for the case of $w = 2$, a condition can be imposed, in order to prevent this bad behaviour.

Corollary a *Under the same assumptions as in Lemma 3.13 and Theorem 3.2, if coefficient $A(2)$ is non-null, in order to avoid that this coefficient grows binomially with the number Z of periods, it is sufficient that the convolutional period n_o is not factor of the parent cyclic code period N.*

Proof We remember that the periodic behaviour of the parity check polynomial in an LC code gives rise to different binomials, of the type $(x^{qN} + 1)$, being q an integer. The binomial growth of 2-weight code words comes out from the fact that code words of this type are originated by sums of properly shifted code words $(x^N + 1)$. Only if N is an exact multiple of n_o the above sums can occur. Remark that even when a common multiple between n_o and N exists, the whole periodic polynomial $h(x)$ given by (1.25) cannot be matched, and hence the binomial growth of $A(2)$ with Z is avoided. □

Example 3.27 Taking the cyclic code characterized by $g(x) = x^4 + x^3 + x + 1$, whose parity check polynomial is $h_c(x) = x^2 + x + 1$, their product gives $(x^6 + 1)$. With the choice $n_o = 4$ and $k_o = 3$, $h_c(x)$ can be matched. Nevertheless, since $n_o = 4$ is not factor of $N = 6$, $A(2)$ does not grow binomially with Z. The evolution of the frame weight distribution obtained with such code is reported in Table 3.8, up to $Z = 4$ periods, therefore well above the value $n = L = 12$, which expresses the least common multiple between n_o and N. It is possible to observe that $A(2)$ increases linearly with Z, with normalized coefficient $B(2) = 1$.

Table 3.8 Frame weight distribution for the 3/4-rate code presented in Example 3.23

k	Z	$n = L$	$A(0)$	$A(2)$	$A(4)$	$A(6)$	...
3	1	7	1	1	5	1	
6	2	11	1	2	20	30	
9	3	15	1	3	42	158	
12	4	19	1	4	69	434	

Table 3.9 Frame weight distribution for the 1/2-rate code presented in Example 3.28

k	Z	$n = L$	$A(0)$	$A(2)$	$A(3)$	$A(4)$	...
1	1	5	1		1		
2	2	7	1	1	2		
3	3	9	1	2	4	1	
4	4	11	1	3	8	3	
5	5	13	1	5	12	7	
6	6	15	1	7	18	14	
7	7	17	1	9	27	27	
8	8	19	1	12	36	36	

Example 3.28 Taking the cyclic code characterized by $g(x) = x^4 + x^2 + 1$, whose parity check polynomial is $h_c(x) = x^2 + 1$, their product still gives $(x^6 + 1)$. With the choice $n_o = 2$ and $k_o = 1$, the whole periodic polynomial $h(x)$ of the LC code can be matched here. This is due to the fact that $n_o = 2$ is factor of $N = 6$. Coefficient $A(2)$ exhibits a binomial increase with Z, as expected. In Table 3.9 the evolution of the frame weight distribution of this 1/2-rate code is reported. It is possible to observe a growing differential increase in the values of coefficient $A(2)$, according to a binomial law of the type $\binom{Int[Z/3]}{2}$. It is easy to verify that this code is equivalent to a lengthened repetition code. In fact its generator polynomial $g(x)$ can be interpreted simply as an enlarged 3-weight unitary polynomial. In Chap. 5, the family of codes so constructed will be named generalized repetition codes.

As already noted, expressions (3.2) and (3.3), respectively for the total length $n = L$ of the frame and for the code rate R, are approximated, since they consider the last period with its whole length. This is confirmed by direct inspection of the first three columns in the Tables from 3.6 to 3.9. In the case characterized by $k_o = 1$, it is convenient to take the degree $r = \deg[g(x)]$ and employ (3.12). In a more general case ($k_o \geq 1$), but still with a unique generator polynomial $g(x)$, it is possible to employ (3.15), and write

$r_o = 1$

$r =$	2	3	4	5	6	7	8	9	10	11	code rate
$k_o = 1$	2	2	2^2	2^2	2^3	2^3	2^4	2^4	2^5	2^5	1/2
$k_o = 2$	2	2^2	2^3	2^3	2^4	2^5	2^5	2^6			2/3
$k_o = 3$	2^2	2^2	2^3	2^4	2^5	2^5	2^6				3/4
$k_o = 4$	2^2	2^3	2^3	2^4	2^5	2^6					4/5

$r_o = 2$

$r =$	3	4	5	6	7	8	9	10	11	12	code rate
$k_o = 1$	2	2	2	2^2	2^2	2^2	2^3	2^3	2^3	2^4	1/3
$k_o = 2$	2	2^2	2^3	2^3	2^3	2^4	2^5	2^5			2/4
$k_o = 3$	2^2	2^2	2^3	2^4	2^5	2^5	2^5				3/5
$k_o = 4$	2^3	2^3	2^3	2^4	2^5	2^6					4/6

$r_o = 3$

$r =$	4	5	6	7	8	9	10	11	12	13	code rate
$k_o = 1$	2	2	2	2	2^2	2^2	2^2	2^2	2^3	2^3	1/4
$k_o = 2$	2	2^2	2^3	2^3	2^3	2^3	2^4	2^5			2/5
$k_o = 3$	2^2	2^2	2^3	2^4	2^5	2^5	2^5				3/6
$k_o = 4$	2^3	2^3	2^3	2^4	2^5	2^6					4/7

Fig. 3.19 Maximum number of merge states per stage in MLC codes for different values of r_o

$$n = L = n_o Z + (r - r_o), \quad R = \frac{k_o}{n_o\left(1 + \frac{r - r_o}{Z}\right)}. \tag{3.16}$$

where we have set $r_o = n_o - k_o$. In conclusion, matrix $\underline{G}$ in Fig. 3.17 is correct as regards the number of periods in the frame, but the last period is not always occupied by 1-symbols in its whole length. Expressions (3.16) fix more precisely frame length and code rate for the considered high-rate MLC codes. As already observed, this correction is almost negligible when Z is very high.

About the maximum number of merge states per stage in the trellis of a high-rate MLC code, which imposes the computational complexity of a soft-decision decoding algorithm, an exhaustive analysis leads to the main results summarized in Fig. 3.19. Such results are independent of the particular generator polynomial $g(x)$ adopted, in an encoder scheme of the type shown in Figs. 1.3 or 1.5, except for its degree r. Three values of r_o have been considered: $r_o = 1, 2, 3$. For each one of them, some increasing values for r and k_o are taken, obviously under the constraint $r > r_o$. The three tables, inside as many rectangles, report the maximum number of merge states per stage.

The first row of any table (characterized by $k_o = 1$) describes the case of a $1/n_o$-rate s.s. time-invariant code. From (3.12), the following relationships between m and r can be written, taking r_o as an increasing parameter,

$$m = \frac{1}{2}(r+1) - 1, \quad r_o = 1, \quad k_o = 1,$$
$$m = \frac{1}{3}(r+1) - 1, \quad r_o = 2, \quad k_o = 1,$$
$$m = \frac{1}{4}(r+1) - 1, \quad r_o = 3, \quad k_o = 1.$$

As already noted, only sometimes m turns out to be integer. For the values of r corresponding to integer values of m (less and less frequent with decreasing the code rate), the traditional estimate of the computational complexity for $1/n_o$-rate s.s. time-invariant convolutional codes, equal to 2^m, is confirmed. Otherwise we have new situations, but the resulting computational complexity, expressed as a function of parameter r, never is lower than the above mentioned cases, having fixed the minimum distance d. Inspection of the three tables reported in Fig. 3.19, even for higher values of k_o, allows to conclude that the computational complexity increases exponentially with parameter r (roughly at steps of r_o), more and more rapidly as parameter k_o grows. This behaviour could be expected, from an analysis of state masks like those reported in Figs. 3.16 and 3.18.

Following a traditional approach to convolutional codes, typically with different interleaved generator polynomials, it is possible to employ an encoder circuit with forward ex-OR sums, in controller or in observer arrangement, in order to encode a high-rate w.s. time-invariant non-systematic convolutional code of the type here described. This solution has the advantage of allowing the use of different interleaved generator polynomials, so reaching better performance. Externally, an encoder of this type will be characterized by k_o parallel input wires and n_o parallel output wires. The internal scheme can vary widely, depending on the proportion between k_o and n_o, consequently the most efficient choice can be adopted. For high-rate codes, typically the observer arrangement, based on a parity check matrix, yields a smaller overall number of memory elements (Johannesson and Zigangirov 1999), as will be better explained in Chap. 6.

A high-rate convolutional code, s.s. or even w.s. time-invariant in its $\underline{G}$ matrix, forced to be systematic (not in RSC code version), as described in Sect. 3.3 for 1/2-rate codes, typically shows poor performance, owing to the heavy constraint imposed on the coefficients of the interleaved polynomials. In Sect. 3.6, RSC codes have been introduced for $1/n_o$-rate s.s. time-invariant convolutional codes. The extension of such solution to high-rate convolutional codes is not easy, and will be faced in Chap. 4, considering also the w.s. time-invariant context. For s.s. time-invariant high-rate convolutional codes, a suitable version of RSC encoder arrangement is always possible, without increasing the total number of memory elements in the encoder circuit with respect to its non-systematic version. The w.s. time-invariant option, instead, may lead to rather different results, due to the right-inversion of the non-systematic generator matrix of the code (sometimes requiring a higher number of memory elements, sometimes impossible at all).

On the other hand, often it is preferable to take an 1/2-rate RSC code and puncture it, maintaining the information symbols. This allows to obtain a decoding computational complexity comparable with that of the original 1/2-rate code, and right-inversion of the generator matrix is always guaranteed. This choice is supported by the observation that the performance of properly punctured 1/2-rate convolutional codes is not far from that of the best high-rate w.s. time-invariant convolutional codes, obtained after encoder circuit optimized design, at least for moderate values of m. If one, in applying puncturation, eliminates also some information symbol, instead, the condition for allowing right-inversion of the resulting $\underline{G}$ matrix is not guaranteed, and must be carefully tested.

3.9 A First Bridge Between Cyclic Block Codes and s.s. Time-Invariant Convolutional Codes

Cyclic block codes and convolutional codes have been invented and developed separately. The attempt of constructing a connection between them has been proposed in the literature from different points of view. For instance, in Solomon and Van Tilborg (1979), it has been observed that, separating information and control symbols in an 1/2-rate convolutional code, it is possible to obtain a family of block codes, which can be interpreted as shortened versions of a cyclic code, whose generator polynomial has a degree as high as the number of periods Z in the frame. Furthermore quasi-cyclic codes can be encoded using typical convolutional encoding procedures. The tail-biting arrangement for a convolutional code, which will be introduced in the next section, gives the opportunity of closing a frame to form a wrapped block, enforcing the similitude between a QC code and a convolutional code (typically w.s. time-invariant). Often the goal of reducing the computational complexity in soft-decision decoding, owing to a partitioned trellis, has driven the search for the above similitude (Calderbank et al. 1998).

A further possibility of connection is provided by the equivalence, discussed in Sect. 3.7, between MLC codes and convolutional codes. It allows to recognize, in the interleaved generator polynomials $g_{\text{int}}(x)$ of the best $1/n_o$-rate s.s. time-invariant convolutional codes, the generator polynomials $g(x)$ of some well known cyclic block codes. The best time-invariant convolutional codes, which were obtained by attempts, are characterized by a very large number of possible choices for their sub-generator polynomials, leading to equivalent solutions. Tables 3.2, 3.4 and 3.5 (respectively for $n_o = 2$, $n_o = 3$, $n_o = 4$), collect only few examples of such choices, and the number of possible equivalent designs increases with both m and n_o. A large number of these possible choices can be associated to the tern of a well known parent cyclic code in an MLC code scheme. This correspondence, which appears

Table 3.10 Interpretation of the interleaved generator polynomials of some best $1/n_o$-rate s.s. time-invariant convolutional codes as generator polynomials of known cyclic block codes

$R = 1/n_o$	m	r	d	$g_{int}(x)$	Cyclic block code
1/2	1	3	3	11.01	Hamming [7,4,3]
	2	5	5	11.10.11	Hamming [31,26,5]
	3	7	6	11.11.01.11	Hamming [127,120,3]
	4	9	7	11.01.10.10.11	Hamming [511,502,3]
	5	11	8	11.01.10.11.10.11	Hamming [2047,2036,3]
	6	13	10	11.01.00.11.11.10.11	e.d. subcode H [4095,4083,3]
1/3	1	5	5	111.011	Hamming [31,26,3]
	2	8	8	111.110.111	[105,97,4]
	3	11	10	111.011.101.111	[868,857,4]
	4	14	12	111.011.101.011.111	–
1/4	1	7	7	1111.0111	Hamming [127,120,3]
	2	11	8	1111.1110.1111	[1905,1894,4]
	3	15	11	1111.1110.0101.1111	–

clear only looking at the interleaved generator polynomial $g_{int}(x)$, (and not maintaining the sub-generator polynomials separated) is summarized in Table 3.10.

At code rate 1/2, up to $m = 5$ ($r = 11$), we have just Hamming codes. They are associated to high-weight irreducible polynomials, with $w[g_{int}(x)]/\deg[g_{int}(x)]$ of the order of 3/4. With $m = 6$, the following factorization holds $g_{int}(x) = (x+1)(x^{12}+x^9+x^8+x^7+x^5+x^3+1)$, corresponding to the generator polynomial which characterizes the even-distance subcode of a [4095,4083,3] Hamming code. So we obtain a [4095,4082,4] overall code. In Sect. 3.2, we observed that, still with $m = 6$, a code substantially comparable in performance with the one above described is that characterized by $w[g_{int}(x)] = 9$ and precisely by 11.01.10.01.01.10.11, which is the generator polynomial of a [8191,8178,3] Hamming code.

At code rate 1/3, with $m = 1$, we still have the [31,26,3] Hamming code. With $m = 2$, the best convolutional code is associated to a cyclic code whose generator polynomial is $g(x) = (x+1)(x^4+x+1)(x^3+x+1)$, factor of $(x^{105}+1)$, and correspondent to a [105,97,4] cyclic code, being (x^4+x+1) and (x^3+x+1) the generator polynomials of a [15,11,3] and of a [7,4,3] Hamming code respectively. With $m = 3$, we can write $g_{int}(x) = (x+1)^3(x^5+x^4+x^3+x^2+1)(x^3+x^2+1)$. The last two polynomials are those of cyclic Hamming codes too. The resulting cyclic code has a length $N = 868$ and its minimum distance turns out to be $d = 4$. With $m \geq 4$, the equivalent best choices are too many, and their interpretation would be less significant.

At code rate 1/4, with $m = 1$, we still have the [127,120,3] Hamming code. With $m = 2$, we can write $g_{int}(x) = (x^4+x^3+1)(x^7+x^5+x^2+x+1)$, being such two factors the generator polynomials of a [15,11,3] and of a [127,120,3] Hamming code respectively. This cyclic code has a length $N = 1905$, and its minimum distance is $d = 4$. Also in this case, greater values of m are not so significant.

In order to give further insights into the structure of low-rate convolutional codes, it could be interesting to treat such codes all together. Instead of fixing the code rate $1/n_o$, and then varying parameter m, we will attempt to study, in a unitary way, the cases of some small values of m (say $m = 1$ and $m = 2$), and take n_o variable. This investigation is devoted to find possible regular behaviours, and show how large is the number of choices leading to an equivalent best performance. For m values of few units, the best performance is obtained with $g_{\text{int}}(x)$ characterized by both external periods having the maximum possible weight, that is $w = n_o$ (see also Lemma 3.4, valid for $n_o = 2$). In the particular case of $m = 1$, this choice has to be avoided, since, although able to give the maximum value for the minimum distance ($d = 2n_o$), it would lead to a family of not well designed convolutional codes. This conclusion was already remarked at the end of Sect. 3.4.

Lemma 3.14 *The best $1/n_o$-rate convolutional code with $m = 1$ shows $d = 2n_o - 1$. The number of possible equivalent choices is $2n_o$. The normalized weight distribution coefficients of such codes result in*

$$B(2n_o - 1) = 1, \quad B(2n_o) = 1, \quad \ldots \quad B(4n_o - 3) = 1.$$

Proof In Fig. 3.20, the state diagram characterizing this family of codes is depicted, with output weight labelling branches. An equivalent choice comes out after exchanging the roles between the two branches connecting state $\underline{0}$ and state $\underline{1}$. Polynomial $\alpha(x)$ represents a degree of freedom in this design. Clearly the best result, that is the highest possible value for d (excluded $2n_o$ after the need of obtaining a well designed convolutional code), is associated to the choice $\text{w}[\alpha(x)] = 1$. Equivalent choices can be obtained with all the possible expressions

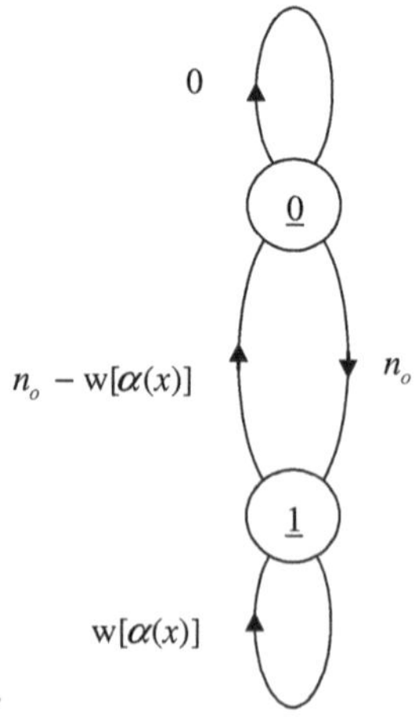

Fig. 3.20 State diagram for $m = 1$ with output weight labelling branches

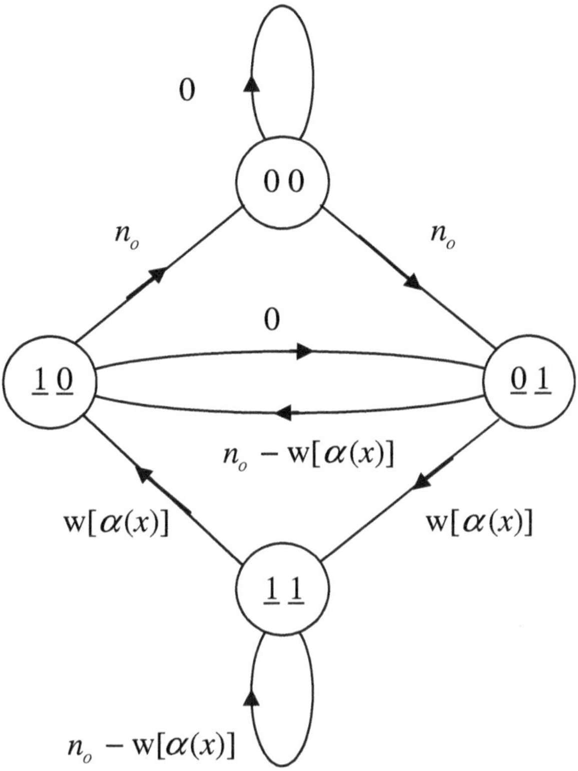

Fig. 3.21 State diagram for $m = 2$ with output weight labelling branches

$\alpha(x) = x^j$, $j = 0, 1, \ldots, (n_o - 1)$, multiplied by two, because of the possibility of exchanging the roles between the two branches connecting state $\underline{0}$ and state $\underline{1}$. The equivalence multiplicity is then $2n_o$. Finally the normalized weight distribution coefficients can be obtained enumerating all the possible routes from state $\underline{0}$ to state $\underline{0}$, without further passages through state $\underline{0}$. The normalized weight distribution coefficients reported above is obtained considering iterated passages through state $\underline{1}$. □

The case of $m = 2$ can be studied by inspection of the state diagram reported in Fig. 3.21, constructed in analogous way. Also here a unique degree of freedom is present, in the choice of w[α(x)], after the adoption of both external periods having the maximum possible weight, that is $w = n_o$.

We have to compare the two weights

$$\mathrm{w}[g(x)] = 3n_o - \mathrm{w}[\alpha(x)], \quad \mathrm{w}[(x^{n_o} + 1)g(x)] = 2n_o + 2\mathrm{w}[\alpha(x)], \tag{3.17}$$

respectively associated to $u(x^{n_o}) = 1$ and $u(x^{n_o}) = x^{n_o} + 1$. The best results obtained are summarized in Table 3.11, up to $n_o = 12$. It is possible to recognize a periodic behaviour with increasing parameter n_o, characterized by period 3. The equivalence multiplicity can be evaluated on the basis of combinatorial algebra,

Table 3.11 Results of the comparison between the two expressions (3.17) in order to obtain the best value for the minimum distance d, with $m = 2$, in low-rate convolutional codes

n_o	d	$w[\alpha(x)]$	$B(d)$	Equivalence multiplicity
2	$5 = 2n_o + \frac{n_o}{2}$	1	1	$\binom{2}{1} = 2$
3	$8 = 2n_o + \frac{n_o+1}{2}$	1	2	$\binom{3}{2} = 3$
4	$10 = 2n_o + \frac{n_o}{2}$	2	1	$\binom{4}{2}$
5	$13 = 2n_o + \frac{n_o+1}{2}$	2	1	$\binom{5}{3}$
6	$16 = 2n_o + \frac{n_o+2}{2}$	2	2	$\binom{6}{4}$
7	$18 = 2n_o + \frac{n_o+1}{2}$	3	1	$\binom{7}{4}$
8	$21 = 2n_o + \frac{n_o+2}{2}$	3	1	$\binom{8}{5}$
9	$24 = 2n_o + \frac{n_o+3}{2}$	3	2	$\binom{9}{6}$
10	$26 = 2n_o + \frac{n_o+2}{2}$	4	1	$\binom{10}{6}$
11	$29 = 2n_o + \frac{n_o+3}{2}$	4	1	$\binom{11}{7}$
12	$32 = 2n_o + \frac{n_o+4}{2}$	4	2	$\binom{12}{8}$

considering the number of different combinations of $(n_o - w[\alpha(x)])$ elements over n_o positions.

Similar considerations can be repeated also for $m = 3$ and even for $m = 4$. Higher values of m require a too long optimisation procedure. The state diagram exhibits 2^m states. So the number of possible paths, whose weights are to be compared, increases exponentially with m.

Before concluding the present section, it is interesting to remember that in Sect. 1.17 direct product codes have been interpreted as MLC codes. Now we can add the following observation about such codes (already outlined at the end of Sect. 3.7). Since they exhibit a binomial increase in the lowest-order weight distribution coefficient $A(d)$ with one of the two dimension of the rectangular encoding memory, they will be necessarily classified as not well designed, in the framework of s.s. time-invariant convolutional codes. Also this property contributes to enforce the conceptual bridge between block codes, in this case made by the concatenation of proper cyclic codes, and convolutional codes.

3.10 Tail-Biting s.s. Time-Invariant Convolutional Codes

Given the $\underline{G}$ matrix of an s.s. time-invariant convolutional code, whose parameters are k_o, n_o and m, after a certain number Z of periods, where information symbols are inserted, there are $k_o\,m$ interleaved generator polynomials in its last rows (last sub-matrices $\underline{G}_o$), to be completed in order to perform a correct frame termination. The final periods of such polynomials can be wrapped around, in such a way as to admit cyclic permutation with steps of n_o, over the length n_oZ.

Definition 3.15 An s.s. time-invariant convolutional code, whose last k_om interleaved generator polynomials $g(x)$ are wrapped around over the frame length n_oZ, in such a way as to produce a $k_oZ\mathrm{x}n_oZ$ overall $\underline{G}$ matrix, is called *tail-biting over the frame length* n_oZ.

We can adopt, for instance, the $\underline{G}$ matrix shown in Fig. 3.17, for $k_o = 2$, $n_o = 3$, $m = 2$ (that is $r = 7$ in the MLC equivalent form). Its tail-biting version is reported in Fig. 3.22. The above Definition holds indifferently for low-rate ($k_o = 1$, $n_o > 2$), 1/2-rate ($k_o = 1$, $n_o = 2$) or high-rate ($n_o > k_o > 1$) convolutional codes, provided that they are s.s. time-invariant. In the Figure, dashed small arrows denote the conventional positions for the insertion of control symbols, one any period (in general r_o any period). A suitable widening of Definition 3.15 will be given in Chap. 4 for w.s. time-invariant convolutional codes.

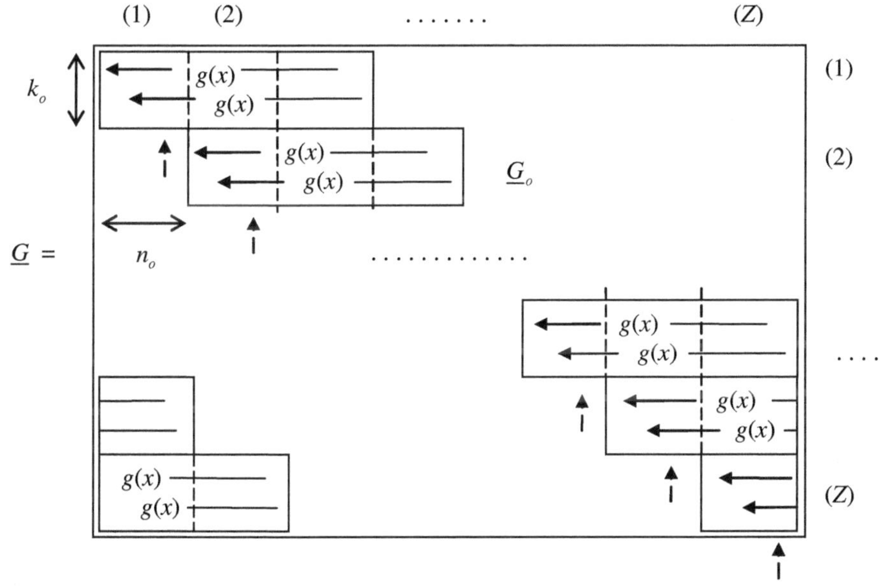

Fig. 3.22 Overall $\underline{G}$ matrix for a 2/3-rate tail-biting s.s. time-invariant convolutional code

The polynomial expression for a frame in a tail-biting s.s. time-invariant convolutional code, characterized by an interleaved generator polynomial $g(x)$, over a length of Z periods, is

$$t(x) = \left[x^{k_o-1}u_1(x^{n_o}) + x^{k_o-2}u_2(x^{n_o}) + \cdots + u_{k_o}(x^{n_o})\right]g(x)\Big|^{(n_o)}_{\mathrm{mod}(x^{n_o Z}+1)}. \qquad (3.18)$$

This expression can be taken as representative of any situation: with $k_o = 1$ for low-rate and 1/2-rate codes, with $n_o > k_o > 1$ for high-rate codes. Finally mod $(x^{n_o Z} + 1)$ operation assures wrapping around over a length $n_o Z$ with steps of n_o. We have adopted here a notation of the type employed in (2.9) for QC codes. The degree of any polynomial $u_j(x)$ is at most $(Z - 1)$. Nevertheless (3.18) can be considered only a formal expression, since the true positions of the information and control symbols, and even their numbers, depend on the structure of the convolutional code adopted.

We can observe a close resemblance with the $\underline{G}$ matrix of a QC code with distributed control symbols (see Sect. 2.3). The problem of attributing precise positions to the information and control symbols can be faced, taking advantage from this analogy. Nevertheless, first of all, it is necessary to understand if the operation of wrapping around might produce rows linearly dependent on the other ones, in the resulting $\underline{G}$ matrix.

Lemma 3.15 *The code rate of a tail-biting s.s. time-invariant convolutional code of the type assumed in Lemma 3.13, if it is well designed (see Definition 3.5), becomes*

$$R = \frac{k_o}{n_o}. \qquad (3.19)$$

Proof Expression (3.19) is valid only if all the rows of $\underline{G}$ are linearly independent. The binomial increase with Z of some coefficients $A(w)$, for $w < 2d$, typical in a not well designed convolutional code, implies the existence of code frames which, wrapped around over a length of any multiple of Z, are linearly dependent on the previous ones. If this fact does not occur, that is the considered convolutional code is well designed, a $\underline{G}$ matrix obtained after wrapping around has a full rank, and hence the true code rate is as expressed by (3.19). □

Remark that the above considerations do not regard coefficients $A(w)$, with $w \geq 2d$ (see Lemma 3.5 and consequent observations), because these coefficients are due to code frames that, by themselves, are obtained as a linear superimposition of shorter 1-symbol patterns (already present in $\underline{G}$). In conclusion, under the assumptions of Lemma 3.15, the frame length and code rate are to be modified with respect to what predicted by (3.16). In particular, the true code rate becomes equal to the asymptotic one. This benefit, which however is more and more negligible with increasing the number of periods Z, could be paid with some performance degradation and implementation problems.

$$\underline{G} = \begin{bmatrix} 1&1&1&0&1&1 \\ 1&1&1&1&1&0 \\ 1&0&1&1&1&1 \end{bmatrix} \quad Z = 3$$

A(0)	A(2)	A(3)	A(5)
1	3	1	3

$$\underline{G} = \begin{bmatrix} 1&1&1&0&1&1&& \\ &&1&1&1&0&1&1 \\ 1&1&&&1&1&1&0 \\ 1&0&1&1&&&1&1 \end{bmatrix} \quad Z = 4$$

A(0)	A(2)	A(3)	A(4)	A(5)	A(6)
1	2	4	1	4	4

$$\underline{G} = \begin{bmatrix} 1&1&1&0&1&1&&&& \\ &&1&1&1&0&1&1&& \\ &&&&1&1&1&0&1&1 \\ 1&1&&&&&1&1&1&0 \\ 1&0&1&1&&&&&1&1 \end{bmatrix} \quad Z = 5$$

A(0)	A(3)	A(4)	A(5)	A(6)	A(7)
1	5	5	6	10	5

$$\underline{G} = \begin{bmatrix} 1&1&1&0&1&1&&&&&& \\ &&1&1&1&0&1&1&&&& \\ &&&&1&1&1&0&1&1&& \\ &&&&&&1&1&1&0&1&1 \\ 1&1&&&&&&&1&1&1&0 \\ 1&0&1&1&&&&&&&1&1 \end{bmatrix} \quad Z = 6$$

A(0)	A(3)	A(4)	A(5)	A(6)	A(7)
1	2	9	12	13	18

Z	A(0)	A(3)	A(4)	A(5)	A(6)	A(7)
7	1		7	21	21	29
8	1		2	24	36	40
9	1			18	48	63
10	1			12	45	90
11	1			11	33	99
12	1			12	26	84

Fig. 3.23 Generator matrices and weight distributions of some tail-biting s.s. time-invariant convolutional codes with the same $g(x)$ with code rate 1/2

First of all, we have to note that the number of periods Z over which wrapping around takes place can influence the minimum distance d. Typically, if the original s.s. time-invariant convolutional code is well designed, there is a threshold value for Z above which d is maintained, but coefficient $A(d)$ can still be greater. For a value of Z remarkably higher, a normalized coefficient $B(d)$ can be introduced also for the tail-biting code, and its value matches the one characterizing the original code.

Example 3.29 The 1/2-rate convolutional code described by the interleaved generator polynomial $g(x) = x^5 + x^4 + x^3 + x + 1$ is arranged in tail-biting form, after Z = 3, 4, 5, 6 periods. The generator matrices and correspondent frame weight distributions are shown in the upper part of Fig. 3.23. It is possible to note that a minimum distance $d = 2$ characterizes the first two codes. Then we have $d = 3$ for $Z = 5, 6$. The table in the lower part of the Figure shows the evolution of the frame weight distribution for Z between 7 and 12. We have $d = 4$ for $Z = 7, 8$. Finally, for Z from 9 to whichever higher value, the minimum distance reaches the value $d = 5$, which is proper of such code without wrapping around (see Example 3.8). Also the normalized coefficient $B(d)$ is the same, resulting in $B(5) = 1$, and even with an equal delay constant $\delta(5) = 0$.

It is not possible to perform here a sliding window decoding procedure, because the first and the last part of the frame are not independent. In this sense, the tail-biting code is a block code under any point of view. The encoder circuit needs an additional memory, in which the whole code frame is stored, in order to allow that an ex-OR sum is made between its initial part and its final part. The decoding procedure can still be made adopting a soft-decision algorithm in a trellis, but this is circularly closed to form a ring. So it is necessary to perform, in general, 2^m separate decoding procedures, and then compare the results obtained, in order to choose the most likely one (Johannesson and Zigangirov 1999). Finally a precise association of the positions in the frame to information or control symbols is difficult. It is always possible to perform an encoding procedure able to give (3.18), and then wrap around, like in the reordered code words of a QC code, but information recovering can be not straightforward. It could be often preferable to locate information symbols before the control symbols acting on them, also in order to correctly construct a code concatenation scheme.

As regards the performance, we can observe that, if Z is high enough (several times m), and the original convolutional code is well designed, an expression of the type (3.5) for the weight distribution coefficients $A(w)$ is here still valid, with the same normalized coefficient $B(w)$, but with a delay constant $\delta(w)$ sometimes remarkably reduced, and in certain cases even with a negative sign.

Example 3.30 Assuming the same interleaved generator polynomial $g(x) = x^5 + x^4 + x^3 + x + 1$ as in Example 3.29, but with code rate 2/3 instead of 1/2, we obtain an MLC code (or high-rate s.s. time-invariant convolutional code), which can be arranged in tail-biting form as well. The two frame weight distributions (for the not tail-biting version on the left and for the tail-biting version on the right) are compared in Table 3.12. The tail-biting behaviour is characterized by coefficients showing unstable initial transients. 3-weight code frames disappear only after $Z = 10$ periods. The asymptotic expressions describing the first low-weight coefficients, for the not tail-biting code, are $A(4) = 3Z - 3$, $A(5) = 11Z - 22$. The tail-biting code approaches $A(4) = 3Z + 3$ only after a long oscillating transient. Oscillations in $A(5)$ go on, up to Z of the order of twenty. This characteristic is more evident for high code rates.

Example 3.31 In Example 3.12 a not well designed convolutional code has been introduced, whose interleaved polynomial is 11.01.11.10.11. As anticipated, such code exhibits $d = 8$, $B(8) = 10$, and only $A(12)$ growing binomially with Z. Table 3.13 shows a comparison between the not tail-biting and the tail-biting version of this 1/2-rate convolutional code. The number of information symbols k is smaller by one unit in the tail-biting version. This is due to the fact that one row is linearly dependent on the others. Furthermore, we can note in it some residual 4-weight code frames for $Z < 11$. Looking at coefficient $A(8)$ in the tail-biting version,

Table 3.12 Comparison between not tail-biting and tail-biting version of the same 2/3-rate s.s. time-invariant convolutional code introduced in Example 3.30

Z	k	Not tail-biting					Tail-biting						
		n	A(0)	A(4)	A(5)	A(6)	n	A(0)	A(2)	A(3)	A(4)	A(5)	A(6)
4	8	16	1	9	23	32	12	1	4	8	38	56	44
5	10	19	1	12	33	58	15	1	–	15	45	97	160
6	12	22	1	15	44	88	18	1	3	6	45	132	327
7	14	25	1	18	55	121	21	1		7	42	175	434
8	16	28	1	21	66	155	24	1		–	52	176	520
9	18	31	1	24	77	190	27	1		3	36	171	561
10	20	34	1	27	88	225	30	1			40	150	605
11	22	37	1	30	99	260	33	1			33	165	572
12	24	40	1	33	110	295	36	1			39	144	562

we can observe an oscillating behaviour, which tends, after about twenty periods, to the asymptotic expression $A(8) = 10Z + 10$, in comparison with the not tail-biting version, which exhibits $A(8) = 10Z - 30$. For $k = 7, 9, 11, 13, 15, \ldots$ symmetric weight distributions characterize the tail-biting arrangement, even when coefficient $A(4)$ is still present. This singular behaviour is consequence of the noticeable symmetry properties present inside this polynomial $g(x)$.

It is interesting to note that, for QC codes, the code word length N is a structural property, whereas in tail-biting convolutional codes the frame length (or the number of periods Z), over which wrapping around is performed, is a free choice. In other terms, the frame length of a tail-biting convolutional code is a scalable parameter. This is not for QC codes. The discussion about analogies and differences between such two classes of codes, both with distributed control symbols, will be continued in Chap. 4, after the introduction of w.s. time-invariant tail-biting convolutional codes.

3.11 Trellis of an s.s. Time-Invariant Convolutional Code Obtained from Its Generator Matrix

In several parts of the present chapter, we mentioned some properties about the generator trellis which characterizes an s.s. time-invariant convolutional code, resulting to be of fundamental importance for evaluating the computational complexity in its soft-decision decoding procedure. In this trellis a Viterbi algorithm is typically adopted, as already said. Also a modification of this algorithm, called SOVA (Soft-Output Viterbi Algorithm) can be employed, which is able to give an estimate of the reliability of the decoded information symbol sequence (Hagenauer and Hoeher 1989). This improved algorithm is efficiently used in concatenated code schemes.

Table 3.13 Comparison between not tail-biting and tail-biting version of the same 1/2-rate s.s. time-invariant convolutional code introduced in Example 3.31

Z	Not tail-biting							Tail-biting							
	k	n	$A(0)$	$A(8)$	$A(12)$	$A(16)$	$A(20)$	k	n	$A(0)$	$A(4)$	$A(8)$	$A(12)$	$A(16)$	$A(20)$
8	8	24	1	50	152	53		7	16	1	12	102	12	1	
9	9	26	1	60	266	179	6	8	18	1	9	171	75		
10	10	28	1	70	420	481	52	9	20	1	5	250	250	5	1
11	11	30	1	80	612	1,093	260	10	22	1		330	616	77	
12	12	32	1	90	836	2,185	948	11	24	1		375	1,296	375	
13	13	34	1	100	1,085	3,937	2,802	12	26	1		390	2,340	1,313	52
14	14	36	1	110	1,351	6,517	7,061	13	28	1		378	3,717	3,717	378
15	15	38	1	120	1,628	10,046	15,704	14	30	1		345	5,320	8,835	1,848
16	16	40	1	130	1,912	14,580	31,520	15	32	1		300	6,976	18,214	6,976
17	17	42	1	140	2,201	20,104	58,047	16	34	1		255	8,466	33,439	21,420

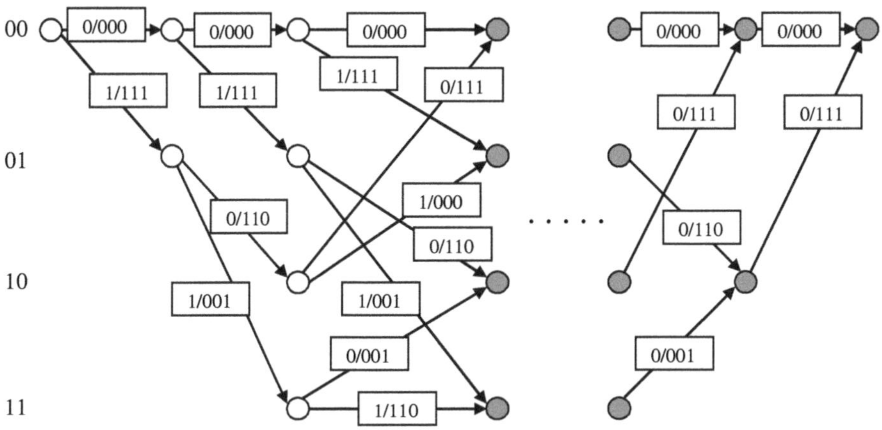

Fig. 3.24 Trellis for the 1/3-rate convolutional code characterized by interleaved generator polynomial 111.110.111

The traditional structure of the trellises for convolutional codes, having code rate $1/n_o$, resembles that of a block code (see Sect. 1.23), for k much greater than r, with the only difference of having n_o output symbols for labelling branches. It shows an initial expanding part, made by a series of subsequent bifurcations, and a final contracting part made by only merge states, in order to allow a correct frame termination. Such two external parts are separated here by a very long central stationary part, where the states per stage are 2^m and are all merge states.

Example 3.32 Let us consider the 1/3-rate code introduced in Example 3.13 (whose state diagram is shown in Fig. 3.9). It is characterized by the interleaved generator polynomial 111.110.111. The temporal evolution of this state diagram originates the trellis reported Fig. 3.24. As usual, merge states have been denoted with solid small circles. It is possible to recognize an expanding part containing the first two periods (in general m periods), a central stationary part, long as one wishes, and a contracting part, formed by the final two periods (in general m periods too).

If the code is well designed, a sliding window decoding procedure can be adopted without any performance penalization. The length of this sliding window depends on the particular structure of the considered code: if there are long low-weight patterns, such length increases. However rarely it exceeds four or five times mn_o (Bossert 1999).

In the case of code rates k_o/n_o, there are 2^{k_o} branches outgoing from any node. The same number is that of the branches entering any merge node in the stationary part of the trellis. A decoding computational complexity regarding the branches is to be added to the traditional one, which is referred to the number of states (or nodes). In fact the decoding algorithm, in a merge state, has to compare here 2^{k_o} different metrics (Johannesson and Zigangirov 1999), like in QC codes with distributed control symbols (see Sect. 2.11). In general, we can observe that the branches are labelled by k_o input symbols and by n_o output symbols (the former parameter fixing branch complexity). Since this problem regards also w.s. time-invariant convolutional codes, it will be analyzed more in depth in Chap. 4. It is however useful to remember that, when a high code rate is obtained after code puncturation, the decoding computational complexity remains that of the original convolutional code. Finally, for high-rate MLC codes, it is possible to note that they allow to adopt always binary trellises, as a consequence of the particular property shown by the encoder circuit of Figs. 1.3 or 1.5, in taking one binary symbol at a time from the source. Anyway one merge state of the traditional approach turns out to be equivalent to a series of subsequent binary merge states in this MLC code model, leading to a substantial equivalent overall computational complexity, as discussed at the end of Sect. 3.8.

Let us consider now a 1/2-rate s.s. time-invariant convolutional code, in tail-biting version. As anticipated, its trellis becomes a circularly closed structure forming a ring, where, in general, a sliding window decoding procedure cannot be applied. Here the decoding computational complexity increases, because of the choice among different possible solutions, associated to as many initial points. Although in principle their number is 2^m, many solutions can be immediately abandoned being clearly unlikely.

Example 3.33 Taking the 1/2-rate convolutional code whose generator polynomial is 10.11, its state diagram is the rightmost one in Fig. 3.3. The temporal evolution of this state diagram, closed in a ring, as needed for describing the tail-biting version of such code, with $Z = 10$ periods, is depicted in Fig. 3.25. The only output symbols have been reported (like in an RSC code arrangement).

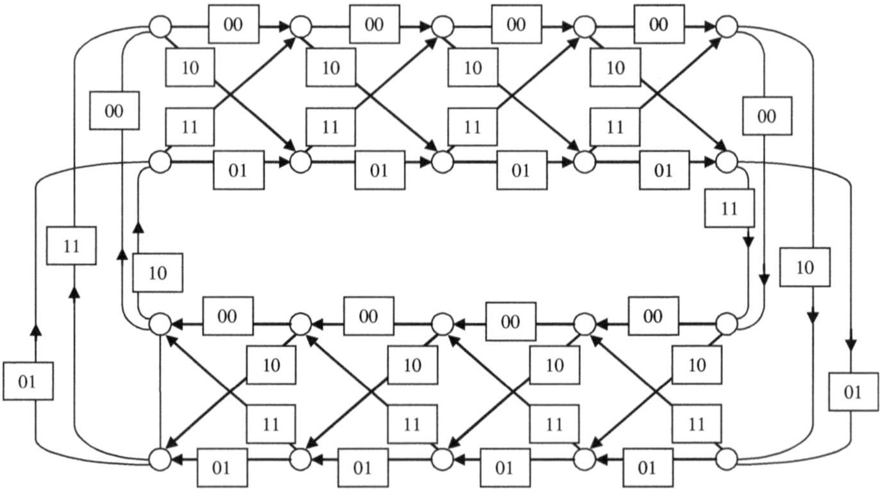

Fig. 3.25 Trellis for the 1/2-rate convolutional code characterized by interleaved generator polynomial 10.11 in tail-biting form

References

Barbier J, Sicot G, Houcke S (2006) Algebraic approach for the reconstruction of linear and convolutional error correcting codes. Int J Appl Math Comput Sci 2:113–118

Berrou C, Glavieux A, Thitimajshima P (1993) Near Shannon limit error correcting coding and decoding: turbo codes. In: Proceedings of IEEE international conference communications (ICC93), Geneva, pp 1064–1070

Bossert M (1999) Channel coding for telecommunications. Wiley, Weinheim

Cain JB, Clark GC, Geist JM (1979) Punctured convolutional codes at rate (n − 1)/n and simplified maximum likelihood decoding. IEEE Trans Inf Theor 25:97–100

Calderbank AR, Forney GD, Vardy A (1998) Minimal tail-biting trellises: the Golay code and more. In: Proceedings of ISIT 1998, Cambridge, USA, p 255

Elias P (1955) Coding for noisy channels. In: IRE Convention Rec Part 4, pp 37–46

Forney GD (1970) Convolutional codes I: algebraic structure. IEEE Trans Inf Theor 16:720–738

Hagenauer J, Hoeher P (1989) A Viterbi algorithm with soft-decision outputs and its applications. In: Proceedings of IEEE global telecom conference, Dallas, pp 1680–1686

Johannesson R, Zigangirov KS (1999) Fundamentals of convolutional coding. IEEE Press, New York

Larsen KJ (1973) Short convolutional codes with maximal free distance for rates 1/2, 1/3 and 1/4. IEEE Trans Inf Theor 19:371–372

Lin S, Costello DJ (2004) Error control coding. Pearson/Prentice-Hall, Upper Saddle River

Mason S, Zimmerman H (1960) Electronic circuits, signals and systems. Wiley, New York

Massey JL, Sain MK (1968) Inverses of linear sequential circuits. IEEE Trans Comput 17:330–337

Proakis JC (1983) Digital communications. McGraw Hill, Tokyo

Solomon G, Van Tilborg HCA (1979) A connection between block and convolutional codes. SIAM J Appl Math 37:358–369

Viterbi AJ (1967) Error bounds for convolutional codes and an asymptotically optimum decoding algorithm. IEEE Trans Inf Theor 13:260–269

Wicker SB (1995) Error control systems for digital communications and storage. Prentice-Hall, Englewood Cliffs

Yasuda Y, Kashiki K, Hirata Y (1984) High rate punctured convolutional codes for soft-decision Viterbi decoding. IEEE Trans Commun 32:315–319

Chapter 4
Wide-Sense Time-Invariant Convolutional Codes in Their Generator Matrix

4.1 Periodically Time-Varying Generator Matrix of a Convolutional Code

In Chap. 2 we have introduced and discussed periodically time-varying generator matrices in the framework of QC block codes, their shortened and lengthened versions (SQC and LQC) and even modified lengthened versions (MLQC). Distributed control symbols can be inserted also in the QC code structure (see, for instance, the case of a generator matrix whose form is a block of $k_p \times p$ truncated circulants, like that shown in Fig. 2.11). Nevertheless such forms of QC codes with distributed control symbols are characterized by a precise length N, therefore cannot be interpreted as convolutional codes in tail-biting arrangement. Instead MLQC codes, whose length can be as long as one wishes, forming the analogous of a frame, show the characteristic of convolutional codes, although they are only a particular class of the ensemble we are going to introduce now.

Definition 4.1 The generator matrix of a convolutional code, where some interleaved generator polynomials $g_{\text{int}}^{(i)}(x)$, $i = 1, 2, \ldots k_o$, among which at least two are different, form a period of length n_o in as many consecutive rows, can be considered that describing a *wide sense* (*w.s.*) *time-invariant convolutional code in its* $\underline{G}$ *matrix.*

The interleaved generator polynomials (here more than one) are originated by proper sub-generator polynomials in a way similar to what occurs for an s.s. time-invariant convolutional code. For QC codes and codes derived from them, in order to simplify the treatment (and make, for instance, interleaved multiplication and division unambiguous), it has been useful to add the following condition: the generator polynomials $g_j(x)$, $j = 1, 2, \ldots q_o$, must have their highest-order power coefficients in the jth position of the period (see Definition 2.2). This condition is no longer necessary for w.s. time-invariant convolutional codes in their $\underline{G}$ matrix.

G. Cancellieri, *Polynomial Theory of Error Correcting Codes*,
Signals and Communication Technology, DOI 10.1007/978-3-319-01727-3_4

A code frame, in the case of a non-systematic w.s. time-invariant convolutional code in its $\underline{G}$ matrix, whose parameters are described in Definition 4.1, can be expressed in polynomial form as follows

$$t(x) = \sum_{i=1}^{k_o} u_i(x^{n_o}) g_{\mathrm{int}}^{(i)}(x), \tag{4.1}$$

where polynomials $u_i(x^{n_o})$ represent the symbols outgoing from the information source, properly subdivided into k_o fluxes by means of a serial-to-parallel conversion. Expression (4.1) represents a generalization of (3.7), which holds for $k_o = 1$. A difference with respect to (2.1), which holds for quasi-cyclic codes and codes derived from them, is in the fact that (4.1) exhibits all the polynomials $u_i(x^{n_o})$ with their highest-order coefficients in the same position (not shifted). Polynomial product (4.1) is not a polynomial interleaved product, and hence, in general, cannot be inverted by means of a polynomial interleaved division. Sometimes it is possible to rearrange the rows of the $\underline{G}$ matrix so obtained, in order to derive an equivalent form with highest-order power coefficients in different positions. Often it is not possible. For this reason the availability of a right-inverse $\underline{G}_{right}^{-1}$ (see Definition A.23 in Appendix A) has to be verified.

A w.s. convolutional code requires a correct frame termination, exactly as an s.s. time-invariant convolutional code (Definition 3.2). The property of being well designed or not can be discussed, adopting also here Definition 3.5. In the following, in order to distinguish the generator matrix based on the interleaved generator polynomials from a symbolic polynomial generator matrix (Bossert 1999), introduced by the traditional approach to w.s. time-invariant convolutional codes (simple extension of what made for s.s. time-invariant convolutional codes in Definition 3.11), we will call the former one the true generator matrix.

4.2 Traditional Approach to w.s. Time-Invariant Convolutional Codes in Their $\underline{G}$ Matrix

The traditional approach to w.s. time-invariant convolutional codes in their $\underline{G}$ matrix is based on a symbolic polynomial generator sub-matrix. It expresses the input-output relations of the encoder circuit employing k_o shift registers in controller arrangement. In order to understand this approach, let us consider the scheme of encoder circuit depicted in Fig. 4.1, where a 2/3-rate code is assumed as an example. So we have $k_o = 2$ and $n_o = 3$. In this scheme, a serial-to-parallel converter, placed after the information source, feeding the two (in general k_o) shift registers, and an output parallel-to-serial converter, placed after the three (in general n_o) ex-OR sums, are not reported for the sake of simplicity. We have six sub-generator polynomials on the whole (in general k_on_o), whose superscripts denote input and output respectively. The maximum number of memory elements in each

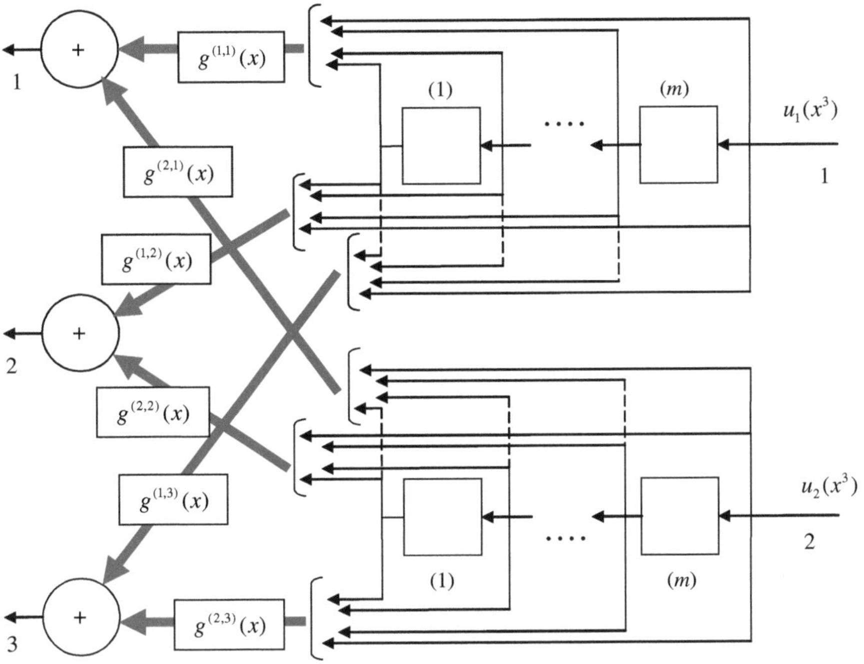

Fig. 4.1 Encoder circuit, in controller arrangement, for a 2/3-rate w.s. time-invariant convolutional code in non-systematic form

shift register is m. This is also the maximum degree of any input-output sub-generator polynomial $g^{(i,j)}(x)$, $i = 1, 2$ (in general up to k_o), $j = 1, 2, 3$ (in general up to n_o).

An important convention is that the input serial-to-parallel converter feeds, at the same symbol time, the two (in general k_o) shift registers. This choice has the consequence of producing expression (4.1), without any input shift. Nevertheless the serial disposition of the output symbols yields the following relationship between the ith interleaved generator polynomial and the input-output sub-generator polynomials, having the same input i,

$$g_{\text{int}}^{(i)}(x) = x^{n_o-1}g^{(i,1)}(x^{n_o}) + x^{n_o-2}g^{(i,2)}(x^{n_o}) + \cdots + g^{(i,n_o)}(x^{n_o}), \quad i = 1, 2, \ldots k_o. \tag{4.2}$$

The set of expressions (4.2) represents an extension of the single expression (3.8), which holds for s.s. time-invariant convolutional codes in their $\underline{G}$ matrix, that is characterized by a unique $g_{\text{int}}(x)$. A symbolic polynomial generator sub-matrix $\underline{G}_o(x)$ was introduced in Definition 3.11, with only one row ($k_o = 1$). Here this concept is extended to the case of $1 < k_o < n_o$.

Definition 4.2 For a non-systematic convolutional code characterized by input-output sub-generator polynomials $g^{(i,j)}(x)$, the *symbolic polynomial generator sub-matrix* $\underline{G}_o(x)$ can be written as

$$\underline{G}_o(x) = \begin{bmatrix} g^{(1,1)}(x) & g^{(1,2)}(x) & \dots & g^{(1,n_o)}(x) \\ g^{(2,1)}(x) & g^{(2,2)}(x) & \dots & g^{(2,n_o)}(x) \\ & \dots & & \\ g^{(k_o,1)}(x) & g^{(k_o,2)}(x) & \dots & g^{(k_o,n_o)}(x) \end{bmatrix}, \tag{4.3}$$

and exhibits n_o columns, formed by k_o elements each, in their turn consisting of proper polynomials, whose degree is at most m.

On the contrary, the true generator sub-matrix $\underline{G}_o$ is formed by k_o rows, where as many row vectors $\underline{g}_{\text{int}}^{(i)}$, correspondent to the interleaved polynomials $g_{\text{int}}^{(i)}(x)$, are placed. This is a $k_o \times (m+1)n_o$ matrix whose elements are binary symbols. The two sub-matrices $\underline{G}_o$ and $\underline{G}_o(x)$ are one-to-one related. In Fig. 4.2 the semi-infinite true generator matrix $\underline{G}$, where $\underline{G}_o$ horizontally shifts by n_o positions each row, and the correspondent symbolic polynomial generator sub-matrix $\underline{G}_o(x)$ are shown. The latter one is still formed by k_o rows, but has n_o columns and each position contains an input-output sub-generator polynomial. Furthermore it cannot be directly employed for shifting in the rows of $\underline{G}$.

As anticipated, only $\underline{G}$ is the true generator matrix. On the other hand, $\underline{G}_o(x)$ is useful because it simplifies right-inverse calculation and recursive systematic code arrangement. Finally it is straightforward to deduce it from the non-systematic encoder circuit scheme (in controller arrangement) adopted, and vice versa.

In Fig. 4.3 the generic structure of the true generator sub-matrix $\underline{G}_o$ is shown, assuming for $g^{(i,j)}(x)$ the following expressions

$$g^{(i,j)}(x) = g_m^{(i,j)}x^m + g_{m-1}^{(i,j)}x^{m-1} + \dots + g_1^{(i,j)}x + g_0^{(i,j)}, \\ i = 1, 2, \dots k_o, \quad j = 1, 2, \dots n_o, \tag{4.4}$$

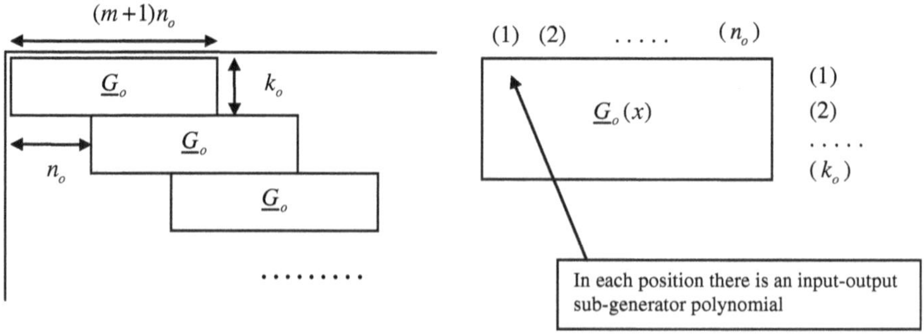

Fig. 4.2 Semi-infinite generator matrix and symbolic polynomial generator sub-matrix

$$\underline{G}_o = \begin{array}{|cccccccccc|} g_m^{(1,1)} & g_m^{(1,2)} & \ldots & g_m^{(1,n_o)} & g_{m-1}^{(1,1)} g_{m-1}^{(1,2)} & \ldots & g_{m-1}^{(1,n_o)} & \ldots\ldots & g_0^{(1,1)} \, g_0^{(1,2)} & \ldots \, g_0^{(1,n_o)} \\ g_m^{(2,1)} & g_m^{(2,2)} & \ldots & g_m^{(2,n_o)} & g_{m-1}^{(2,1)} g_{m-1}^{(2,2)} & \ldots & g_{m-1}^{(2,n_o)} & \ldots\ldots & g_0^{(2,1)} \, g_0^{(2,2)} & \ldots \, g_0^{(2,n_o)} \\ & & & & \ldots & \ldots & \ldots & \ldots & & \\ g_m^{(k_o,1)} & g_m^{(k_o,2)} & \ldots & g_m^{(k_o,n_o)} & g_{m-1}^{(k_o,1)} g_{m-1}^{(k_o,2)} & \ldots & g_{m-1}^{(k_o,n_o)} & \ldots\ldots & g_0^{(k_o,1)} \, g_0^{(k_o,2)} & \ldots \, g_0^{(k_o,n_o)} \end{array}$$

Fig. 4.3 Generic structure of the true generator sub-matrix

under the conditions $g_m^{(i,j)} = 1$, for at least one couple (*i*,*j*) and $g_0^{(i,j)} = 1$ for at least the same or another couple (*i*,*j*). The last specification confirms that parameter *m* maintains the meaning of generator constraint length (see Definition 3.2).

Example 4.1 The encoder circuit depicted in Fig. 4.4 gives rise to a 2/3-rate w.s. time-invariant convolutional code, whose symbolic polynomial generator sub-matrix is

$$\underline{G}_o(x) = \begin{bmatrix} (x^2 + x) & x & x^2 \\ 1 & x^2 & (x^2 + x + 1) \end{bmatrix}.$$

This code is characterized by $m = 2$. Nevertheless only the lower shift register exhibits two memory elements.

After (4.3) we obtain the two interleaved generator polynomials and hence the following true generator sub-matrix

$$\underline{G}_o = \left[\begin{array}{ccc|ccc|ccc} 1 & 0 & 1 & 1 & 1 & 0 & 0 & 0 & 0 \\ 0 & 1 & 1 & 0 & 0 & 1 & 1 & 0 & 1 \end{array}\right],$$

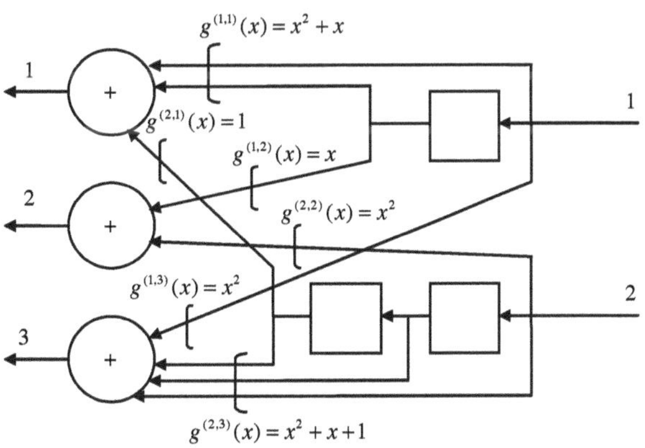

Fig. 4.4 Encoder circuit of the 2/3-rate w.s. time-invariant convolutional code described in Example 4.1

Table 4.1 Construction parameters of some good convolutional codes with $k_o = 2$ and $n_o = 3$

μ	m	d	$g^{(1,1)}(x)$ / $g^{(2,1)}(x)$	$g^{(1,2)}(x)$ / $g^{(2,2)}(x)$	$g^{(1,3)}(x)$ / $g^{(2,3)}(x)$	$g^{(1)}_{\text{int}}(x)$ / $g^{(2)}_{\text{int}}(x)$
2	1	3	10	01	11	101.011
			01	11	11	011.111
3	2	4	110	010	100	101.110.000
			001	100	111	011.001.101
4	2	5	011	100	101	011.100.101
			100	011	111	101.011.011
5	3	6	1100	1110	0110	110.111.011.000
			1101	0000	1111	101.101.001.101
6	3	7	0110	1101	1101	011.111.100.011
			1011	0010	1011	101.000.111.101

where the binary sequences corresponding to the two interleaved polynomials have been written with periods separated by vertical segments, in order to better distinguish them.

Definition 4.3 The *total number of memory elements* in all the k_o shift registers, in controller arrangement, forming the encoder circuit, is denoted μ. The following inequality holds $\mu \leq mk_o$.

In the case described in Example 4.1 (Fig. 4.4), we have $\mu = 3$. The state diagram associated to the encoder circuit characterizing this code, in controller arrangement, then will exhibit $2^3 = 8$ states. Furthermore any state has $2^2 = 4$ input branches and as many output branches.

The sub-generator polynomials and the correspondent interleaved polynomials of some good w.s. time-invariant convolutional codes, with code rate 2/3, are reported in Table 4.1 (Johannesson and Zigangirov 1999). They show the best minimum distance d, given a certain parameter μ, and have been obtained, as usual, by attempts. It is possible to note a rather slow increase in the minimum distance d, with increasing parameter μ. This parameter expresses only a partial measure of the decoding computational complexity. Another part is due to k_o, and regards branch complexity, as already observed.

The highest-order power coefficient of the two interleaved polynomials sometimes occupies different position in the period (like in QC codes or in codes derived from them), sometimes this is not. In the present case ($k_o = 2$), exchanging the order of the two interleaved polynomials we obtain an equivalent code. The same is after substitution of the two polynomials with their reciprocals. A further equivalent code can be constructed after substitution of one of the two interleaved polynomials in the $\underline{G}_o$ sub-matrix with their ex-OR sum. In this sense, there are many forms of $\underline{G}_o$, able to give possible equivalent codes. It is important to note that alignment is however made on the left hand side, implying that some 0-symbols can be placed on the right, in order to allow that Fig. 4.3 is consistent with expressions (4.1) and (4.2).

The second code in the Table is that introduced in Example 4.1. There is a family of codes even simpler than this: their encoder circuit exhibits just $\mu = 2$ memory elements (one for each shift register). Several members of this family of codes, like the one reported in the first row of the Table, are characterized by $d = 3$. They can be compared with the 2/3-rate punctured convolutional code, obtained starting from a 1/2-rate s.s. time-invariant convolutional code with $m = 2$. It is possible to verify that the one described in Example 3.16 is equivalent to that reported in the first row of Table 4.1 (the ex-OR sums of the two rows in $\underline{G}_o$ yields 110.100). Only in this case we obtain the same performance. For higher values of μ, code puncturation is not able to assure the same performance as high-rate w.s. convolutional codes designed on the basis of an encoder circuit like that shown in Fig. 4.1, owing to the larger number of degrees of freedom present in the last situation. Nevertheless the decoding computational complexity is lower in punctured convolutional codes, as already noted.

Example 4.2 Let us consider the 2/3-rate convolutional code, with $m = 1$ and $\mu = 2$, described by the following symbolic polynomial generator sub-matrix and true generator sub-matrix

$$\underline{G}_o(x) = \begin{bmatrix} (x+1) & 1 & (x+1) \\ 1 & x & x \end{bmatrix}, \quad \underline{G}_o = \left[\begin{array}{ccc|ccc} 1 & 0 & 1 & 1 & 1 & 1 \\ 0 & 1 & 1 & 1 & 0 & 0 \end{array}\right].$$

In Fig. 4.5 the state diagram of this code is depicted. It is characterized by $2^2 = 4$ (in general 2^μ) states, and by $2^2 = 4$ (in general 2^{k_o}) branches going into or coming from any state. In the correspondent trellis, it is possible to distinguish a state computational complexity and a branch computational complexity, in this case equal. As usual, a slash separates the input symbols and the output symbols. Closed paths around state $\underline{0}$, representing as many code frames, enable one to conclude that the minimum distance (calculated on the output sequences) is $d = 3$. It is possible to enumerate two 1-symbol output patterns having weight 3, therefore $B(3) = 2$, where $B(w)$, as usual, represents the normalized weight distribution coefficient associated to weight w, in an expression of the type (3.5). They are related respectively to the paths 00-01-00, 00-11-01-00, where pairs of symbols denote the touched states, and result in the code frames 011.100, 110.000.100.

All the codes reported in Table 4.1 are well designed. It is then possible to find, for each of them, a pair of parameters, the normalized coefficient $B(w)$ and the delay $\delta(w)$, in any weight distribution coefficient $A(w)$, provided that $w \leq 2d - 1$. A catastrophic propagation of single errors occurs if we adopt a w.s. time-invariant not well designed convolutional code in non-systematic form, exactly as in the s.s. time-invariant case.

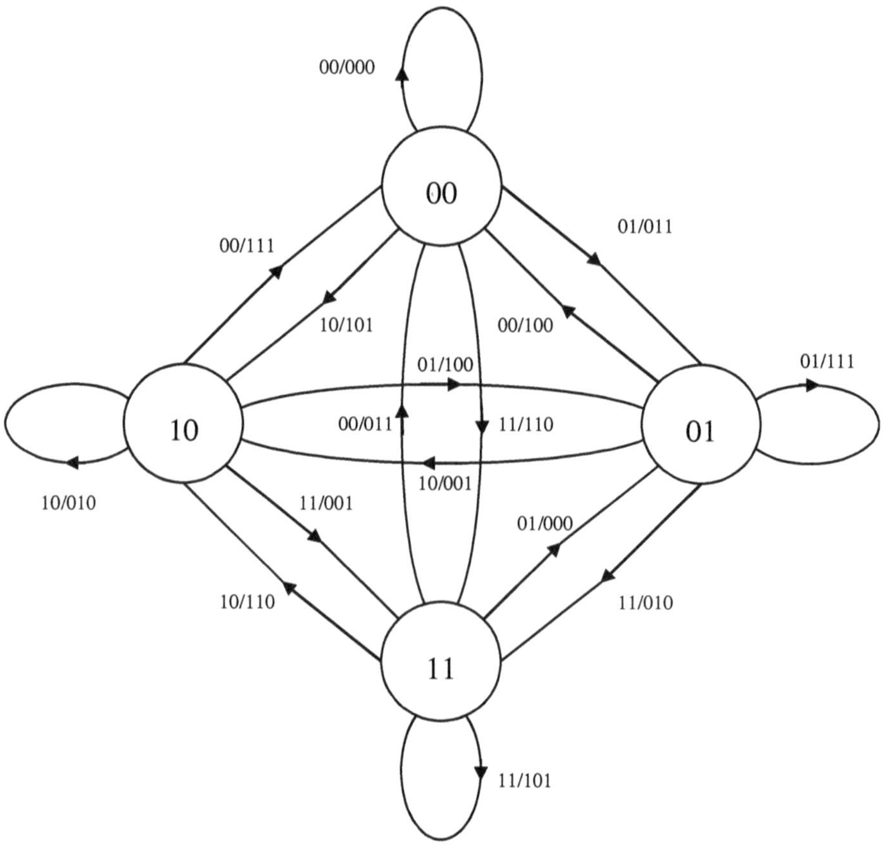

Fig. 4.5 State diagram of the 2/3-rate w.s. time-invariant convolutional code described in Example 4.2

Example 4.3 The evolution of the frame weight distribution of the 2/3-rate code introduced in Example 4.2 ($m = 1$, $\mu = 2$), whose state diagram is shown in Fig. 4.5, is reported in Table 4.2, with increasing the number of periods Z. It is possible to derive the following extrapolation laws

$$A(3) = 2Z - 1, \quad A(4) = 5Z - 8, \quad A(5) = 15Z - 39.$$

Therefore the normalized weight coefficients will result in $B(3) = 2$, $B(4) = 5$, $B(5) = 15$. Coefficient $A(6)$ will exhibit a binomial increase with Z, as expected.

Example 4.4 In Table 4.3 we report the interleaved generator polynomials describing two w.s. time-invariant convolutional codes, with $m = 1$ and $\mu = 2$, able to reach $d = 3$ at rate 2/3 as before, and the evolution of their frame weight distributions. The first one is a well designed convolutional code, and shows these

Table 4.2 Frame weight distribution of the code whose state diagram is shown in Fig. 4.5

Z	$A(0)$	$A(3)$	$A(4)$	$A(5)$
1	1	1	1	1
2	1	3	3	4
3	1	5	7	11
4	1	7	12	22
5	1	9	17	36
6	1	11	22	51
7	1	13	27	66

Table 4.3 Interleaved polynomials and frame weight distributions for two w.s. time-invariant 2/3-rate convolutional codes with $m = 1$ and $\mu = 2$

	Z	$A(0)$	$A(3)$	$A(4)$	$A(5)$
$g_{int}^{(1)}(x) = 101.011$	1	1	1	1	1
$g_{int}^{(2)}(x) = 011.111$	2	1	2	4	6
	3	1	3	8	15
	4	1	4	12	28
	5	1	5	16	42
	6	1	6	20	56
	7	1	7	24	70
$g_{int}^{(1)}(x) = 111.011$	1	1	1	1	1
$g_{int}^{(2)}(x) = 010.110$	2	1	2	5	4
(Not well designed)	3	1	3	10	13
	4	1	4	16	24
	5	1	5	23	37
	6	1	6	31	52
	7	1	7	40	69
	8	1	8	50	88
	9	1	9	61	109

weight distribution coefficients: $A(3) = Z$, $A(4) = 4Z - 4$, $A(5) = 14Z - 28$. So it appears better than that described in Examples 4.2 and 4.3, so becoming the best one with such set of parameters (it is reported in the first row of Table 4.1). The second code treated here is still characterized by $A(3) = Z$, but shows $A(4)$ which increases more than linearly with Z (the transient is certainly ended after $Z = 9$ periods). Therefore, on long frames, its performance will be poorer. It can be defined not well designed, although just for its $A(d + 1)$ weight distribution coefficient.

Example 4.5 The normalized weight distribution coefficients characterizing the w.s. time-invariant 2/3-rate convolutional code described in Example 4.1 ($m = 2$, $\mu = 3$), for $w = 4, 5, 6$, are as follows: $B(4) = 1$, $B(5) = 5$, $B(6) = 24$. As already noted, this is the best code with such values for parameters m and μ. There are no punctured convolutional codes able to reach a performance of this type, taking into account that the comparison is to be made with 1/2–rate original convolutional codes characterized by $m = 3$.

From the above Examples, in comparison with the best s.s. time-invariant $1/n_o$-rate convolutional codes, with a similar number of memory elements, described in Sects. 3.2 and 3.4, it is possible to note a stronger increase in the normalized weight coefficients $B(w)$ as the frame weight w grows.

We will give now few data regarding w.s. time-invariant convolutional codes in their $\underline{G}$ matrix characterized by $k_o = 3$ and $n_o = 4$, that is by a code rate 3/4. A list of precise polynomials for the best code design, obtained by attempts (Johannesson and Zigangirov 1999), is not significant, because of the very high number of possible equivalent results. Nevertheless it is useful to summarize the performance of the best ones. Assuming $m = 1$ and $\mu = 3$, a minimum distance as high as $d = 4$ can be obtained. With $m = 2$ and $\mu = 4, 5, 6$, it is possible to achieve the values $d = 4, 5, 6$ respectively. On the other hand, the number of next neighbour frames is so high to make such codes not particularly interesting. Furthermore the decoding computational complexity, especially due to branches, is remarkable, to the point where, in practice, punctured codes obtained from 1/2-rate s.s. time-invariant convolutional codes, having comparable performance, are typically preferred.

Also for w.s. time-invariant convolutional codes, provided that they are well designed, the decoding procedure is typically performed by means of a soft-decision decoding algorithm, with a sliding window, shifting along the frame, as described by Definition 3.6, for s.s. time-invariant convolutional codes.

It is interesting to note that w.s. time-invariant convolutional codes are not employed only for high code rates. It is possible to define codes having code rates in which the numerator and the denominator have a common factor (see, for instance, the MLQC code described by the $\underline{G}$ matrix in Fig. 2.25, whose code rate is 2/4). There are 4/8-rate w.s. time-invariant convolutional codes, well designed, with $m = 1$ and $\mu = 4$, which show $d = 8$. Therefore they are characterized by better performance with respect to the best 1/2-rate s.s. time-invariant convolutional codes, well designed too, with $m = 4$ (and hence with comparable state decoding complexity), whose best minimum distance is only $d = 7$. This better result is clearly paid by the presence of a branch decoding complexity to be added.

Example 4.6 A possible arrangement for a 4/8-rate w.s. time-invariant convolutional code able to reach $d = 8$, of the type mentioned above, is that characterized by the following $\underline{G}_o$ sub-matrix

$$\underline{G}_o = \left[\begin{array}{cccccccc|cccccccc} 1 & 1 & 0 & 1 & 1 & 1 & 0 & 1 & 1 & 1 & 0 & 0 & 0 & 0 & 0 & 0 \\ 0 & 0 & 1 & 1 & 1 & 1 & 1 & 0 & 0 & 1 & 1 & 1 & 0 & 0 & 0 & 0 \\ 0 & 0 & 0 & 0 & 1 & 1 & 0 & 1 & 1 & 0 & 1 & 1 & 1 & 1 & 0 & 0 \\ 0 & 0 & 0 & 0 & 0 & 0 & 1 & 1 & 0 & 1 & 1 & 1 & 0 & 1 & 1 & 1 \end{array}\right].$$

The sequence of information and control symbols in the period is of the type *icicicic* (this $\underline{G}_o$ sub-matrix exhibits rows which, although different, are characterized by polynomials having the same degree and shifting exactly by two positions any row). As anticipated, we have $m = 1$, $\mu = 4$, and $d = 8$. The normalized coefficient $B(d)$, with the above particular choice for $\underline{G}_o$, is as high as $B(8) = 49$. Such a code can be obtained, after a suitable procedure, from the QC version of the [24,12,8] extended Golay code, described in Example 2.18, as will be discussed with more detail in Sect. 4.8. An even better result can be achieved after a different design, to the point where $B(8) = 33$ can be reached (Johannesson and Zigangirov 1999).

4.3 Existence of the Inverse Linear Transformation

First of all, it is necessary to assure that, in the non-systematic w.s. time-invariant convolutional codes described so far, an inverse linear transformation exists, in such a way as to allow, from the encoded sequence, to recover the information sequence. This problem is related to the existence of the right-inverse of the generator matrix which describes our code. It is preferable to develop the discussion on the symbolic polynomial generator sub-matrix $\underline{G}_o(x)$, because possible realizations employing feed-back ex-OR sums will be straightforward by means of its rational polynomial version (see Definition 3.13).

Definition 4.4 A $k_o \times n_o$ symbolic polynomial generator sub-matrix $\underline{G}_o(x)$, with all rows linearly independent, represents a linear transformation, which is right-invertible if its *right-inverse matrix* $\underline{G}_o(x)^{-1}_{right}$ exists (see Definition A.23 in Appendix A). This is an $n_o \times k_o$ matrix. The relationship between $\underline{G}_o(x)$ and $\underline{G}_o(x)^{-1}_{right}$ is

$$\underline{G}_o(x) \cdot \underline{G}_o(x)^{-1}_{right} = \underline{I}_{k_o}, \tag{4.5}$$

where $\underline{I}_{k_o}$ represents a $k_o \times k_o$ identity matrix.

The following proposition can be constructed directly from Definition A.25 and expression (A.21) in Appendix A, about the Smith form of a matrix. The only difference is in the presence, here, of polynomial expressions.

Definition 4.5 It is always possible to decompose a $k_o \times n_o$ symbolic polynomial generator sub-matrix $\underline{G}_o(x)$ in the following product of three polynomial matrix factors

$$\underline{G}_o(x) = \underline{A}(x) \cdot \underline{\Gamma}(x) \cdot \underline{B}(x), \tag{4.6}$$

where $\underline{A}(x)$ is $k_o \times k_o$, $\underline{B}(x)$ is $n_o \times n_o$ and $\underline{\Gamma}(x)$ is $k_o \times n_o$, with non-null elements just in its leftmost $k_o \times k_o$ square sub-matrix. Expression (4.6) is called *Smith form of the polynomial matrix* $\underline{G}_o(x)$.

Lemma 4.1 *If the right-inverse* $\underline{G}_{o\,right}^{-1}(x)$ *of sub-matrix* $\underline{G}_o(x)$ *exists, using the Smith form of* $\underline{G}_o(x)$, *its expression will result in*

$$\underline{G}_{o\,right}^{-1}(x) = \underline{B}(x)^{-1} \cdot \underline{\Gamma}(x)_{right}^{-1} \cdot \underline{A}(x)^{-1}, \tag{4.7}$$

where $\underline{A}(x)^{-1}$ *and* $\underline{B}(x)^{-1}$ *are the inverse matrices respectively of* $\underline{A}(x)$ *and* $\underline{B}(x)$, *whereas* $\underline{\Gamma}(x)_{right}^{-1}$ *is an* $n_o \times k_o$ *matrix with non-null elements only in its uppermost* $k_o \times k_o$ *square sub-matrix, and is the right-inverse of* $\underline{\Gamma}(x)$.

Proof After substitution of (4.5) into (4.6), it is possible to obtain (4.7). The positions where $\underline{\Gamma}(x)_{right}^{-1}$ has its non-null elements are consequence of the positions where $\underline{\Gamma}(x)$ has its non-null elements. □

A first condition for the transformation described by our convolutional code to be invertible is that the non-null elements of $\underline{\Gamma}(x)_{right}^{-1}$ are just on the diagonal of its uppermost $k_o \times k_o$ square sub-matrix (see Appendix A). A further condition regards the possibility that such right-inverse is realizable by a circuit, in controller or observer arrangement. To this purpose, it is useful to remind that Definition 3.12 introduces s.s. time-invariant delay-free convolutional codes with the only trivial constraint to have $g_m^{(1)} = 1$ in the first sub-generator polynomial $g^{(1)}(x)$. Now this definition will be extended to w.s. time-invariant convolutional codes in their $\underline{G}$ matrix, looking at the structure of $\underline{\Gamma}(x)_{right}^{-1}$.

Definition 4.6 A $\underline{\Gamma}(x)$ matrix, characterized by non-null elements just on the diagonal of its uppermost $k_o \times k_o$ square sub-matrix, is called *delay-free* if, among all these non-null elements, at least one exhibits a non-null *m*th-order power coefficient. In this situation, the encoder circuit able to generate the considered code is defined delay-free too.

Equalities (4.6) and (4.7) are studied in Appendix A. Considering the symbolism introduced in Figs. A.5 and A.6, the elements on the diagonal of the leftmost $k_o \times k_o$ square sub-matrix in $\underline{\Gamma}(x)$ can be written $\gamma_i(x)$, $i = 1, 2, \ldots k_o$, and those on the diagonal of the uppermost $k_o \times k_o$ square sub-matrix in $\underline{\Gamma}(x)_{right}^{-1}$ can be written $\alpha_i(x)$, $i = 1, 2, \ldots k_o$.

Lemma 4.2 *The right-inverse* $\underline{G}_{o\,right}^{-1}(x)$ *exists and is realizable by a proper circuit if* $\underline{\Gamma}(x)$ *is delay-free.*

Proof In the inversion of the leftmost $k_o \times k_o$ square sub-matrix of $\underline{\Gamma}(x)$, to form the uppermost $k_o \times k_o$ square sub-matrix of $\underline{\Gamma}(x)^{-1}_{right}$, the elements $\gamma_i(x),\ i = 1, 2, \ldots k_o$, contribute at the denominator of the polynomial ratios $\alpha_i(x),\ i = 1,\ 2, \ldots k_o$. In order to allow polynomial division, at least one of them must be characterized by its mth-order coefficient non-null, as usual. Under this condition, a circuit of the type shown in Fig. C.8 of Appendix C is able to encode our code. □

Example 4.7 The w.s. time-invariant 2/3-rate convolutional code characterized by $m = 2$ and $\mu = 3$, introduced in Example 4.1, after the procedure described so far, shows a $\underline{\Gamma}(x)^{-1}_{right}$ matrix, whose uppermost square 2×2 sub-matrix is the identity matrix $\underline{I}_2$ multiplied by a polynomial. Therefore its $\underline{G}_o(x)$ sub-matrix is right-invertible, and, after long but straightforward calculations (Johannesson and Zigangirov 1999), its right-inverse turns out to be

$$\underline{G}_{o\,right}^{-1}(x) = \begin{bmatrix} x^4 + x^2 + 1 & x^4 + x^3 + x^2 \\ x^3 + 1 & x^3 + x^2 \\ x^3 + x + 1 & x^4 + x^2 \end{bmatrix}.$$

It is realizable by means of a circuit, in controller arrangement, characterized by three shift registers. The presence of the 4th-order power coefficient requires four memory elements in each of them. This inverse circuit is more complicated than the encoder circuit, where just $\mu = 3$ overall memory elements are needed. Anyway the decoding complexity remains that imposed by the encoder circuit, which is characterized by a state diagram with $2^3 = 8$ states.

The procedure needed for knowing if the considered w.s. time-invariant convolutional code admits an inverse transformation is required in order to avoid codes whose encoding operation is not reversible, but also in order to obtain an RSC version of the original code. If one chooses the interleaved generator polynomials in such a way as the highest-order power coefficient of each $g^{(i)}_{\text{int}}(x)$ occupies a different position in the period (like in QC codes or in codes derived from them), the above code selection is intrinsically performed. In fact, in encoding, we can calculate a polynomial interleaved product, and, in recovering the information symbols after decoding, we can proceed with the correspondent polynomial interleaved division. Nevertheless such choice, in principle, could limit the degrees of freedom in the code design, and prevent the obtainment of the best possible performance, like that permitted by an expression as general as (4.1).

Another important topic in the traditional approach to w.s. time-invariant convolutional codes in their $\underline{G}$ matrix regards the possibility of having different encoder circuits for encoding equivalent codes (Bossert 1999). Since the encoder state diagram influences the decoding computational complexity, there is a strong interest in finding the one characterized by the minimum number of memory elements, independently of the arrangement adopted, which can be neither completely in controller form nor completely in observer form (Ryan and Lin 2009).

Definition 4.7 Among the encoder circuits able to produce equivalent w.s. time-invariant convolutional codes, the one characterized by the minimum overall number μ of memory elements is called the *minimal basic encoder circuit.*

Given a very general $\underline{G}_o(x)$ sub-matrix, the procedure for searching the minimal basic encoder circuit for a convolutional code, equivalent to that described by this matrix, still takes advantage from the Smith form of $\underline{G}_o(x)$. Another important goal of the Smith form is in facing the problem of obtaining recursive systematic versions for the non-systematic w.s. time-invariant convolutional codes described so far. Finally, like for s.s. time-invariant convolutional codes, it is possible to use rational symbolic polynomial generator sub-matrices (see Definition 3.13) in order to construct the recursive systematic encoder circuit for whichever right-invertible $\underline{G}_o(x)$. Nevertheless the non-systematic minimal basic encoder circuit is often simpler than the encoder circuit of its RSC code version.

In the next section, employing the Smith form of $\underline{G}_o(x)$, we will treat how it is possible to construct, starting from (4.1), an RSC version for the most general type of w.s. time-invariant convolutional codes in their $\underline{G}$ matrix. The last specification regarding time-invariance in the $\underline{G}$ matrix, which was introduced in Definition 4.1, has not been repeated all the times for the sake of simplicity, since it could appear superfluous. The possible description of such codes with another type of matrix, the so called parity check matrix, will require to be more precise, in the following (starting from Chap. 5). Among the uses of the Smith form of $\underline{G}_o(x)$, there is finally the possibility of a straightforward obtainment of the parity check matrix describing our code. For this reason too, an exhaustive treatment will be carried out, so assuring a basis for further interpretations.

4.4 RSC Version of a w.s. Time-Invariant Convolutional Code

In order to assure that a k_o/n_o-rate convolutional code is systematic, it is necessary that, in its symbolic sub-matrix $\underline{G}_o(x)$, an identity matrix $\underline{I}_{k_o}$ is present as leftmost square sub-matrix. This feature must be set up in the symbolic sub-matrix $\underline{G}_o^{(RSC)}(x)$ describing the RSC version of our code.

Lemma 4.3 *Under the conditions for the existence of* $\underline{G}_{o\,right}^{-1}(x)$ *and for it to be realizable, it is possible to obtain a rational symbolic polynomial sub-matrix* $\underline{G}_o^{(RSC)}(x)$, *as the RSC version of the original* $\underline{G}_o(x)$.

Proof The Smith form of $\underline{G}_o(x)$ will be employed, like for the obtainment of $\underline{G}_{o\,right}^{-1}(x)$. Letting $\underline{T}(x)$ the leftmost $k_o \times k_o$ square sub-matrix present in the original $\underline{G}_o(x)$, if $\underline{\Gamma}(x)$ is delay-free, $\underline{T}(x)$ is non-singular, but its inverse $\underline{T}(x)^{-1}$ is typically a rational symbolic polynomial matrix. The RSC code version of the considered code is then described by the following rational symbolic polynomial generator sub-matrix

$$\underline{G}_o^{(RSC)}(x) = \underline{T}(x)^{-1} \cdot \underline{G}_o(x). \tag{4.8}$$

In fact product (4.8), considering the leftmost $k_o \times k_o$ square sub-matrix $\underline{T}(x)$ present in $\underline{G}_o(x)$, gives rise to the identity matrix $\underline{I}_{k_o}$ in the left hand part of $\underline{G}_o^{(RSC)}(x)$, as expected. □

In the situation characterized by $k_o = 2$, $n_o = 3$, the RSC symbolic polynomial generator sub-matrix appears so organized

$$\underline{G}_o^{(RSC)}(x) = \begin{bmatrix} 1 & 0 & \frac{g^{(1,3)'}(x)}{D(x)} \\ 0 & 1 & \frac{g^{(2,3)'}(x)}{D(x)} \end{bmatrix}, \tag{4.9}$$

where polynomials $g^{(1,3)'}(x)$, $i = 1, 2$, are suitable transformations of the previous correspondent sub-generator polynomials, whose degree is sometimes higher than the original ones, and $D(x) = \det[\underline{T}(x)]$. It is interesting to compare (4.9) with (3.11), which holds for s.s. time-invariant convolutional codes in the case $k_o = 1$. In this situation, characterized by just one row, at the denominator there is simply the first sub-generator polynomial $g^{(1)}(x)$.

Example 4.8 The w.s. time-invariant 2/3-rate convolutional code with $m = 1$ and $\mu = 2$, introduced in Example 4.2, is characterized by this RSC symbolic polynomial generator sub-matrix

$$\underline{G}_o^{(RSC)}(x) = \begin{bmatrix} 1 & 0 & \frac{x^2}{x^2+x+1} \\ 0 & 1 & \frac{x^2+1}{x^2+x+1} \end{bmatrix}.$$

Example 4.9 The w.s. time-invariant 2/3-rate convolutional code with $m = 2$ and $\mu = 3$, introduced in Example 4.1, is characterized by this RSC symbolic polynomial generator sub-matrix

$$\underline{G}_o^{(RSC)}(x) = \begin{bmatrix} 1 & 0 & \frac{x^3+x^2+x+1}{x^3+x^2+1} \\ 0 & 1 & \frac{x^3+1}{x^3+x^2+1} \end{bmatrix}.$$

In both the above Examples, the degrees of the polynomials in $\underline{G}_o^{(RSC)}(x)$ are increased with respect to those of the original non-systematic ones (which were referred to a minimal basic encoder form). This means more complicated encoder circuits, but just in a controller arrangement. In fact, adopting an observer

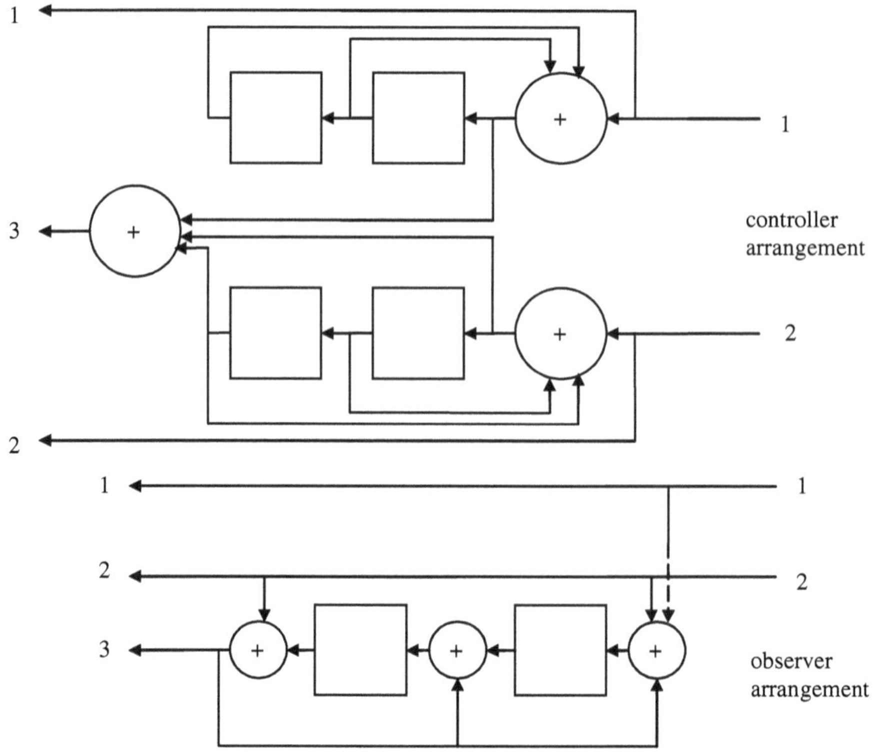

Fig. 4.6 Encoder circuits for the RSC version of the code described in Examples 4.2, 4.3 and 4.8 ($m = 1, \mu = 2$)

arrangement, we can exploit the advantage of having just one additional output line, carrying the unique control symbol, besides the two input lines directly connected to the output.

Example 4.10 Coming back to the w.s. time-invariant convolutional code in its $\underline{G}$ matrix, discussed in Examples 4.2, 4.3 and 4.8 ($m = 1, \mu = 2$), the non-systematic encoder circuit in controller arrangement able to generate this code has just one memory element in any of the two input shift registers. The RSC version of this circuit is shown up in Fig. 4.6. It is obtained from the description associated to the $\underline{G}_o^{(RSC)}(x)$ matrix, which has been reported in Example 4.8, and shows twice the number of memory elements of the non-systematic correspondent version. For this code, it is however also possible to set up an encoder circuit in observer arrangement, systematic as well. It has the structure depicted down in Fig. 4.6. This systematic observer arrangement is characterized by just two memory elements (as many as the non-systematic encoder in controller arrangement). The advantage in the observer arrangement is due to the unique shift register feeding the third output line. Remark that the input serial-to-parallel converter produces two information fluxes $i_1(x^3)$ and $i_2(x^3)$ in the two lines at the same symbol time, according to the convention adopted up to now, although in output a parallel-to-serial conversion is however due, in order to correctly reconstruct such two fluxes properly interleaved (with the addition of one control symbol any period).

The sub-generator polynomials at the numerator in the ratios occupying the third column of $\underline{G}_o^{(RSC)}(x)$ are responsible for the feed-forward wires, whereas the unique sub-generator polynomial at the denominators is responsible for the feed-back wires. We will come back to the encoder circuits in observer arrangement, here introduced, in Chaps. 7 and 8, where the parity check matrix of a convolutional code will be treated. Such matrix is better suited for this description.

Let us consider now an s.s. time-invariant high-rate convolutional code (high-rate MLC code), still characterized, say, by $k_o = 2$, $n_o = 3$. It shows the same interleaved polynomial $g_{\rm int}(x)$, simply shifted by one position in the two rows of $\underline{G}_o$. Employing the symbolic polynomial generator sub-matrix $\underline{G}_o(x)$, it always admits right inverse, as already noted.

Example 4.11 Assuming the s.s. time-invariant convolutional code in its $\underline{G}$ matrix (MLC code), characterized by $k_o = 2$, $n_o = 3$, and unique generator polynomial $g(x) = x^3 + x^2 + 1$, we have this true generator sub-matrix and correspondent symbolic polynomial generator sub-matrix

$$\underline{G}_o = \left[\begin{array}{ccc|ccc} 1 & 1 & 0 & 1 & 0 & 0 \\ 0 & 1 & 1 & 0 & 1 & 0 \end{array}\right], \quad \underline{G}_o(x) = \begin{bmatrix} (x+1) & x & 0 \\ 0 & (x+1) & x \end{bmatrix}.$$

Following the same procedure described so far, the RSC version of the latter sub-matrix is

$$\underline{G}_o^{(RSC)}(x) = \begin{bmatrix} 1 & 0 & \frac{x}{x^2+1} \\ 0 & 1 & \frac{x^2+x}{x^2+1} \end{bmatrix}.$$

Also in this very simple situation, we have an augment in the degree of the polynomials. It is then convenient to follow an approach based on the MLC code model, leading to a systematic encoder circuit of the type shown in Fig. 1.5. It is interesting to note that the overall length of the shift register in the traditional non-systematic encoder circuit, characterized by $m = 2$, is not completely exploited in this case (the last column of the true $\underline{G}_o$ sub-matrix is null). This is due to the fact that the interleaved polynomial degree $r = 4$ is not a good choice in this respect, with a 2/3 code rate.

On the other hand, an approach based on an upper triangular true generator sub-matrix $\underline{G}_o$ is able to guarantee, in any condition, the existence of an inverse transformation, as well as an easy design of a realizable encoder circuit. This consideration suggests to investigate more in detail the relation between MLQC codes and a particular class of w.s. time-invariant convolutional codes in their $\underline{G}$ matrix: those based on an upper triangular $\underline{G}_o$ sub-matrix.

4.5 Equivalence Between MLQC Codes and a Certain Class of w.s. Time-Invariant Convolutional Codes

Keeping in mind the equivalence between MLC codes and s.s. time-invariant convolutional codes, one could argue that a similar equivalence holds also between MLQC codes and w.s. time-invariant convolutional codes. This is not true, in fact MLQC codes have been introduced in Sect. 2.10, as subcodes of proper LQC codes. In particular, this implies the constraint of having a $\underline{G}_o^{(q_o)}$ sub-matrix shifting with steps of q_o, after Definition 2.18, or of having a $\underline{G}_o^{(p)}$ sub-matrix shifting with steps of p, after Definition 2.19. So there will be w.s. time-invariant convolutional codes which cannot be interpreted as MLQC codes. Furthermore expression (4.1) and the structure of $\underline{G}_o$ shown in Fig. 4.3 are more general and cannot be always matched by upper triangular generator sub-matrices.

Let us consider the class of MLQC codes introduced after Definition 2.18, where the most part of the control symbols are inserted along the code word (according to parameters K_o, N_o describing the super-period), and only a small number of control symbols remains at the end of the block. This behaviour resembles that of convolutional codes, with distributed control symbols and a certain number of final control symbols devoted to a correct frame termination. Nevertheless the only convolutional codes, w.s. time-invariant in their $\underline{G}$ matrix, for which an equivalence is possible are those characterized by asymptotic code rate (K_oq_o/N_oq_o) and by a mechanism for the insertion of distributed control symbols coincident with that adopted in the super-period. In this sense, let us remember the 2/4-rate MLQC code introduced in Example 2.30, whose semi-infinite generator matrix is shown in Fig. 2.25, and insertion of information and control symbols is made according to the scheme *iicc*. This MLQC code can be interpreted as a w.s. time-invariant convolutional code with $k_o = K_oq_o = 2$, $n_o = N_oq_o = 4$, $m = 1$, $\mu = 2$, and interleaved generator polynomials $g_{\text{int}}^{(1)}(x) = xg_1(x)$, $g_{\text{int}}^{(2)}(x) = g_2(x)$. Multiplication of the first by x is due to the different conventions about the meaning of such polynomials in the two models.

In spite of the above restriction, we can still make advantage from the theory of MLQC codes, in order to interpret the properties of this class of w.s. time-invariant convolutional codes. Sometimes, in performing MLQC transformation, some columns of the new $\underline{G}$ matrix can be eliminated, so obtaining an increase in the code rate. In other situations, it is possible to reveal the occurrence of 2-weight code words, or simplify the study of a not well designed convolutional code.

Example 4.12 The QC code characterized by $g_1(x) = x^4 + x^3 + 1$, $g_2(x) = x^4 + x^2 + x + 1$, with $q_o = 2$ and $N = 10$, introduced in Example 2.10, can be arranged in

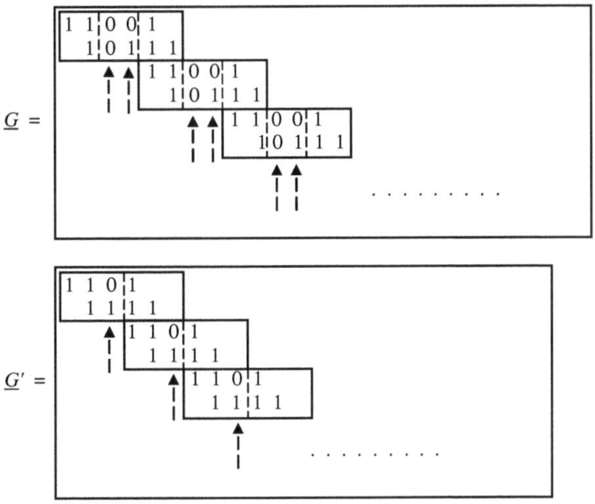

Fig. 4.7 Generator matrix of the MLQC or w.s. time-invariant convolutional code considered in Example 4.12 and its transformation after elimination of one useless control symbol

MLQC form with asymptotic code rate 2/4 (that is $N_o = 2$ and $K_o = 1$), adopting the semi-infinite generator matrix shown in Fig. 4.7 up, where control symbols are denoted with dashed small arrows as usual. The scheme adopted for information and control symbol insertion in any super-period is *iicc* also here. This situation was already treated in Example 2.31, in the case of Z = 4, with the main purpose of describing how the resulting generator matrix could be reordered (see Fig. 2.26). In doing this we discovered that the first control symbol in any period could be eliminated because always null.

After such elimination, instead of a code rate 2/4, we obtain a code rate 2/3, with interleaved generator polynomials $g^{(1)}_{\text{int}}(x) = x(x^3 + x^2 + 1)$, $g^{(2)}_{\text{int}}(x) = x^3 + x^2 + x + 1$. The new generator matrix $\underline{G}'$ is reported down in Fig. 4.7. The word (frame) weight distribution of this code, with increasing Z, is shown in Table 4.4 (where n is calculated after elimination of the useless control symbols). The following extrapolation expressions hold for such $A(w)$ coefficients

$$A(3) = 3Z{-}1, \quad A(4) = 7Z{-}9, \quad A(5) = 15Z{-}36,$$

demonstrating that this convolutional code is well designed. In this sense, we have obtained a good overall result by means of the unusual procedure here described.

Example 4.13 Let us consider the 2/4-rate MLQC code already treated in Example 2.30 (Fig. 2.25), starting from a QC equivalent of the [8,4,4] extended Hamming code. It can be interpreted as a 2/4-rate w.s. time-invariant convolutional code, characterized by $m = 1$ and $\mu = 2$, as already noted. Its frame weight distribution was shown in Table 2.5. From it, we can derive the following extrapolation expressions for the first two weight distribution coefficients: $A(4) = 4Z - 1$, $A(6) = 10Z - 14$. Here we have no useless control position to be eliminated, and the code rate remains 2/4. This code and that discussed in Example 4.12, both characterized

Table 4.4 Frame weight distribution for the 2/3-rate convolutional code in Example 4.12

k	Z	n	$A(1)$	$A(3)$	$A(4)$	$A(5)$
2	1	5	1	2	1	
4	2	8	1	5	5	2
6	3	11	1	8	12	10
8	4	14	1	11	19	24
10	5	17	1	14	26	39
12	6	20	1	17	33	54

by $m = 1$, do not completely exploit the second period of the $\underline{G}_o^{(2)}$ sub-matrix. A comparison between them shows substantially similar performance (the same asymptotic coding gain and a number of minimum-weight code frames not remarkably different).

In Example 2.32 the MLQC version of a subcode, taken from the previously considered [10,6,3] QC code, has been analyzed. This is a [10,4,4] QC code, where the minimum distance is increased from 3 to 4. The MLQC version obtained from it, adopting a code rate 2/4, surprisingly exhibits 2-weight code frames of the type $(x^{10} + 1)$, as it is possible to verify looking at the $\underline{G}$ matrix reported up in Fig. 2.27 (ex-OR sum of three consecutive rows gives this very bad result). Precisely, we obtain $A(2) = Z - 1$. So its performance is remarkably worse than that of the codes described in Examples 4.12 and 4.13, in spite of the fact that, interpreted as a w.s. time-invariant convolutional code, it is characterized still by $m = 1$ and $\mu = 2$, but exploits completely the second period of $\underline{G}_o^{(2)}$. In order to explain this poor behaviour, we can take advantage from the introduction of parity check polynomials, performed in Chap. 2 for QC codes, with the purpose of studying under which conditions 2-weight code frames may occur. In doing this, we can develop considerations similar to those presented in Theorem 3.2 regarding MLC codes and s.s. time-invariant convolutional codes, with the only difference that here we have more than one parity check polynomial.

Example 4.14 In Example 2.26 the parity check polynomials (either even-to-even or odd-to-odd) for the [10,6,3] QC code, already discussed here also in Example 4.12, have been calculated (see Fig. 2.21). Adopting an MLQC code with code rate 4/6 (that is $N_o = 3$ and $K_o = 2$), some 2-weight code frames associated to the binomial $(x^{10} + 1)$, due to the original LC code, are still present. The reason for this behaviour is in the sequence of information symbols, which, in this situation, is able to match $xh_1^{(1)}(x^2) + h_2^{(1)}(x^2)$. Here no useless control positions can be eliminated. In comparison with the 2/3-rate convolutional code described in Example 4.12, deriving from the same parent QC code, this 4/6-rate code is clearly worse.

In conclusion, the MLQC model according to Definition 2.18, although not able to comprehend all the possible w.s. time-invariant convolutional codes, allows easy prediction and interpretation of the properties of those convolutional codes, characterized by code rate of the type $(K_o q_o)/(N_o q_o)$ and by an upper triangular $\underline{G}_o$ sub-matrix. More in detail, especially when the code rate is rather high or the degrees of the generator polynomials not far from N, this model can help in preventing the occurrence of 2-weight code frames. As regards MLQC codes whose $\underline{G}$ matrix is characterized by complete square circulants, collected in a sub-generator matrix $\underline{G}_o^{(p)}$, according to Definition 2.19, they will be treated in following chapters, in the framework of LDPC convolutional codes.

4.6 Practical Importance of Punctured Convolutional Codes

The model discussed so far, regarding high-rate w.s. time-invariant convolutional codes, based on encoder circuits in controller arrangement, with more than one input wire, can be applied also to describe 1/2-rate time-invariant convolutional codes properly punctured, which is the solution most frequently adopted in practice. Often, they have rather good performance, which can be efficiently optimized, choosing the most suited puncturation matrix, without requiring an encoder circuit so complicated.

Example 4.15 The 1/2-rate convolutional code characterized by interleaved generator polynomial $g(x) = x^5 + x^4 + x^3 + x + 1$, punctured at code rate 2/3, has been treated in Example 3.16. It can be interpreted as a w.s. time-invariant 2/3-rate convolutional code with interleaved generator polynomials $g_{\text{int}}^{(1)}(x) = x(x^3 + x^2 + 1)$ and $g_{\text{int}}^{(2)}(x) = x^4 + x^3 + x^2 + x + 1$. Comparison with the best code with identical parameters (the first in Table 4.1, whose weight distribution is the former one in Table 4.3) shows that such two codes are equivalent, as already noted. Its frame weight distribution is then characterized by the following low-weight coefficients: $A(3) = Z$, $A(4) = 4Z - 4$, $A(5) = 14Z - 28$. Remark that, application of the same puncturation matrix to the 1/2-rate convolutional code characterized by $g(x) = x^5 + x^4 + x^3 + x + 1$ (the reciprocal of the previous one), gives rise to a not well designed code in coefficient $A(4)$, although d is still equal to 3, with $A(3) = Z$.

The above considerations show how critical can be the choice of the 1/2-rate convolutional code to be punctured, even within a set of equivalent codes, together with the proper puncturation matrix, and how the possibility of generating a not well designed convolutional code is always present. Stronger puncturation is possible, on the code described in Examples 3.16 and 4.15, maintaining the same principle based on elimination of control symbols only. We obtain $d = 3$ up to code rate 3/4, and, for higher code rates, a number of 2-weight code frames which becomes larger and larger. All these results can be predicted on the basis of a set of parity check polynomials, like those introduced for QC codes in Chap. 2, with the only difference of considering here distributed control symbols, whose fraction is lower and lower as the code rate increases.

From the practical point of view, it is possible to arrange an encoder circuit easily scalable in the code rate, simply adding a device able to eliminate an increasing number of control symbols. The decoding computational complexity, as already stressed, remains that of the 1/2-rate original code. After proper optimization of the puncturation matrices adopted for any selectable code rate, it is possible to obtain very good performance, but at the expense of more involved electronic circuits, in order to reach a true rate-adaptive behaviour (Cain et al. 1979).

In Sect. 3.5, a rough comparison between the generator constraint length m' of the punctured code and the generator constraint length m of the original 1/2-rate convolutional code has been reported. We observed that m' is generally smaller than m. This characteristic is more and more evident with increasing the final code rate, and, in a certain sense, contributes to balance the presence of a branch complexity which becomes heavier and heavier, if we adopt the general model of w.s. time-invariant convolutional code here introduced. As long as m and μ increase, as already noted, there is a growing number of high-rate convolutional codes, directly designed in the context of this model, whose behaviour is better than the correspondent punctured 1/2-rate convolutional codes. Nevertheless such optimized designs are typically of very difficult scalability.

4.7 Tail-Biting w.s. Time-Invariant Convolutional Codes

In Sect. 3.10 tail-biting s.s. time-invariant convolutional codes have been introduced. Also the case characterized by $k_o \geq 1$ has been treated, but only with a unique interleaved generator polynomial. Here we will extend this analysis to w.s. time-invariant convolutional codes. For the sake of simplicity, we will analyze the only class of codes having an upper triangular $\underline{G}_o$ sub-matrix.

Definition 4.8 A w.s. time-invariant convolutional code whose last $k_o m$ interleaved generator polynomials $g_{\text{int}}^{(i)}(x) = x^{k_o - i} g_i(x), i = 1, 2, \ldots k_o$, are wrapped around over the frame length $n_o Z$, in such a way as to produce a $k_o Z \times n_o Z$ overall $\underline{G}$ matrix of the type shown in Fig. 4.8, assuming $k_o = 2$, $n_o = 3$, $m = 2$, is called *tail-biting over the frame length* $n_o Z$.

The polynomials $g_i(x)$, $i = 1, 2, \ldots k_o$, (having equal degree after possible introduction of a shift factor x^{Δ_i}) show highest-order power coefficients in subsequent different positions, so satisfying the assumption that the $\underline{G}_o$ sub-matrix is upper triangular. This enables us to associate such positions to the information symbols, at least from a formal point of view (as already discussed).

Lemma 4.4 *The polynomial expression for a frame in a tail-biting w.s. time-invariant convolutional code, characterized by the interleaved generator polynomials* $g_{\text{int}}^{(i)}(x) = x^{k_o - i} g_i(x)$, $i = 1, 2, \ldots k_o$, *over a length of Z periods, is*

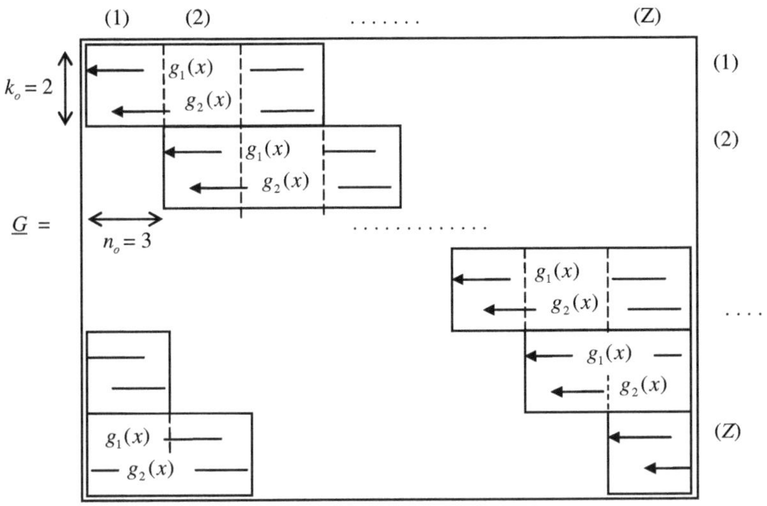

Fig. 4.8 Overall $\underline{G}$ matrix of a 2/3-rate tail-biting w.s. time invariant convolutional code

$$t(x) = \left[x^{k_o-1}u_1(x^{n_o})g_1(x) + x^{k_o-2}u_2(x^{n_o})g_2(x) + \cdots + u_{k_o}(x^{n_o})g_{k_o}(x)\right]\Big|^{(n_o)}_{\mathrm{mod}(x^{n_oZ}+1)}. \tag{4.10}$$

Proof Expression (4.1), under condition $g^{(i)}_{\mathrm{int}}(x) = x^{k_o-i}g_i(x)$, and after wrapping around, over the frame length n_oZ, becomes (4.10). The degree of any polynomial $u_j(x)$ is at most $(Z - 1)$, and mod $(x^{n_oZ} + 1)$ operation is exactly as in expression (3.18), holding for s.s. time-invariant codes in tail-biting form over a length of Z periods. In this sense, (4.10) represents an extension of (3.18) to comprehend also the case of w.s. time-invariant codes. □

Clearly, in a more general case, where (4.1) holds without the constraint introduced in Definition 4.8 on the interleaved generator polynomials, it is difficult to understand how information fluxes $u_j(x^{n_o})$ can act, and even a comparison with (3.18) risks to become impossible. Nevertheless, expression (4.10), as well as expression (3.18) for s.s. time-invariant tail-biting convolutional codes, is to be taken with some cares. In fact, as already noted, the true information symbols can be found rigorously only before the control symbols acting on them. This recommendation is particularly important in code concatenation schemes. In any case, (4.10) can be practically implemented, employing a traditional encoder circuit plus a memory able to store the whole code frame, so assuring suitable ex-OR sums of the its most external parts.

$$\underline{G} = \left[\begin{array}{ccc|ccc|ccc}
1&0&1&1&1&1&&& \\
0&1&1&1&0&0&&& \\
\hline
&&&1&0&1&1&1&1 \\
&&&0&1&1&1&0&0 \\
\hline
1&1&1&&&&1&0&1 \\
1&0&0&&&&0&1&1
\end{array}\right]$$

$Z = 3$

A(0)	A(2)	A(3)	A(4)	A(5)
1	3	13	15	15

$$\underline{G} = \left[\begin{array}{ccc|ccc|ccc|ccc}
1&0&1&1&1&1&&&&&& \\
0&1&1&1&0&0&&&&&& \\
\hline
&&&1&0&1&1&1&1&&& \\
&&&0&1&1&1&0&0&&& \\
\hline
&&&&&&1&0&1&1&1&1 \\
&&&&&&0&1&1&1&0&0 \\
\hline
1&1&1&&&&&&&1&0&1 \\
1&0&0&&&&&&&0&1&1
\end{array}\right]$$

$Z = 4$

A(0)	A(2)	A(3)	A(4)	A(5)
1	2	16	31	48

Z	A(0)	A(2)	A(3)	A(4)	A(5)
5	1		15	45	96
6	1		14	45	132
7	1		14	42	147
8	1		16	42	144
9	1		18	45	144

Fig. 4.9 Generator matrices with $Z = 3$, 4 and frame weight distributions up to $Z = 9$, for the 2/3-rate tail-biting code analyzed in Example 4.16

About the overall length of the frame in a w.s. time-invariant convolutional code, an expression of the type (3.2) is still valid. For fixing ideas, we can assume that the number of periods in the frame correctly terminated, and hence in the overall $\underline{G}$ matrix, is $(Z + m)$. So the whole frame length becomes $n_o(Z + m)$, except for the possible elimination of few useless columns in the final part of $\underline{G}$ made by all 0-symbols. On the contrary, for the tail-biting version of the above w.s. time-invariant convolutional code, the frame length results exactly in n_oZ, as Fig. 4.8 clearly shows, and consequently the code rate is k_o/n_o, as given by (3.19). The last property however holds only if the convolutional code is well designed. In fact, as in s.s. time-invariant convolutional codes, the tail-biting version of a not well designed convolutional code shows rows in its $\underline{G}$ matrix linearly dependent on the other ones, which are to be eliminated.

Example 4.16 Taking the 2/3-rate w.s. time-invariant convolutional code characterized by $m = 1$, $\mu = 2$, introduced in Example 4.2, it has been arranged in tail-biting form, over $Z = 3$ and $Z = 4$ periods. The generator matrices, all having full rank, are shown in Fig. 4.9 on the left. The correspondent frame weight distributions (for the above values of Z and then up to $Z = 9$) are reported on the right. It is possible to note that 2-weight code frames are no longer present when $Z > 4$. For $Z \geq 7$ the increase in the number of 3-weight code frames is proportional to Z, with a normalized coefficient $B(3) = 2$, like in the original not tail-biting code (see Table 4.2). The overall performance, after about twenty periods, is very close to that of the original code, as described in Example 4.3. Such results follow from the property of the present code to be well designed.

Example 4.17 Let us take the second 2/3-rate w.s. time-invariant convolutional code introduced in Example 4.4 (lower part of Table 4.3), characterized by interleaved generator polynomials $g_{\text{int}}^{(1)}(x) = 111.011$ and $g_{\text{int}}^{(2)}(x) = 010.110$, which is not well designed, but just in its $A(d + 1) = A(4)$ weight distribution coefficient. We can observe the following behaviour, as regards a tail-biting arrangement for this code. The last row in the wrapped around $\underline{G}$ matrix is to be eliminated, since it turns out to be a linear superimposition of the previous ones. The evolution of the weight distribution in comparison with that of the not tail-biting version of the same code is shown in Table 4.5. After a short transient, the linear increase of $A(3)$ with Z comes out also in the tail-biting arrangement, with the same normalized coefficient $B(3) = 1$. The $A(d + 1) = A(4)$ coefficient, like in the not tail-biting arrangement, does not increase linearly with Z, but this behaviour becomes evident only after a strongly oscillating transient. Any 2-weight code frame disappears for $Z > 4$. The only appreciable worse performance regards the fact that in the tail-biting arrangement k is smaller by one unit.

In Sect. 4.5 we have described the interpretation of some w.s. time-invariant convolutional codes as MLQC codes. Now we would like to focus our attention on another similitude between the behaviours of QC codes and w.s. time-invariant convolutional codes. The reordered $\underline{G}$ matrix of a QC code with distributed control symbols and the tail-biting arrangement of a convolutional code generator matrix, as already noted, appear very similar. In fact, in any period of length $n_o = p$, $k_o = k_p$ information symbols are followed by $r_o = r_p$ control symbols in both, having let $k_o + r_o = k_p + r_p = n_o = p$.

Nevertheless, there are also the following main differences. First of all, in the tail-biting convolutional code arrangement the same generator sub-matrix $\underline{G}_o$ shifts along row layers, and finally is wrapped around. On the contrary, for QC codes, in general, there are different row layers (see Fig. 2.11), made by all independent $\underline{G}_{ij}^{(k_p)}$ rectangular blocks (which in turn are truncated circulants). Therefore analogy is restricted only to those QC codes having equal row layers, except for the shifts. In

Table 4.5 Comparison in the frame weight distribution, between not tail-biting and tail-biting version of the same 2/3-rate not well designed convolutional code (Example 4.17)

Z	Not tail-biting						Tail-biting					
	k	n	$A(0)$	$A(3)$	$A(4)$	$A(5)$	k	n	$A(0)$	$A(3)$	$A(4)$	$A(5)$
3	6	12	1	3	10	13	5	9	1	14	18	12
4	8	15	1	4	16	24	7	12	1	8	44	56
5	10	18	1	5	23	37	9	15	1	5	20	56
6	12	21	1	6	31	52	11	18	1	6	27	54
7	14	24	1	7	40	69	13	21	1	7	35	70

this type of codes, after association of the $\underline{G}_{ij}^{(k_p)}$ blocks to the symbols of a 2^p-ary alphabet, we can recognize a cyclic behaviour (instead of, more generally, a pseudo-cyclic behaviour). The last property makes only such QC codes similar to tail-biting convolutional codes. Furthermore the code word length $N = q_o p$ of a QC code is consequence of the fact that the interleaved generator polynomials are factors, in a polynomial interleaved division, of a precise binomial $(x^N + 1)$. On the contrary, the frame length $n_o Z$, after which wrapping around of the $\underline{G}$ matrix takes place in a convolutional code, is a free parameter, adopted only for forcing the code to admit a cyclic permutation property with steps of n_o, like that expressed by (4.10). This observation was already anticipated at the end of Sect. 3.10.

Some variants of QC codes with distributed control symbols, especially the QC versions of known block codes, described in Sect. 2.4, appear particularly well suited for satisfying the conditions enabling the expected similitude. In the layers of their $\underline{G}$ matrix, different generator polynomials are periodically alternated, but a same structure (forming the equivalent of the convolutional code generator sub-matrix) shifts along the row layers, from left to right, with steps of p. For instance, we can consider the block code described in Example 2.18, regarding a QC version of the [24,12,8] extended Golay code, or the two QC versions of the [8,4,4] extended Hamming code (Examples 2.14 and 2.15). After all these premises, the following property holds.

Theorem 4.1 *A tail-biting w.s. time-invariant convolutional code, provided that it is well designed, is equivalent to a QC block code described by a $\underline{G}$ matrix arranged as a square block of rectangular sub-matrices, provided that it has full rank and exhibits row layers in which the same generator sub-matrix shifts.*

Proof In both, we have distributed control symbols (in number of $r_o = r_p$ any period, whose length is $n_o = p$), separated by information symbols (in number of $k_o = k_p$ any period). The condition on the QC block code to have a unique $\underline{G}_o$ sub-matrix makes its $\underline{G}$ matrix exactly coincident with the wrapped around generator matrix of a convolutional code. The condition, on the convolutional code, to be well designed, allows that all the rows of its tail-biting $\underline{G}$ matrix are linearly independent, as imposed to the generator matrix of the equivalent QC code. □

The only difference remains in the fact that $q_o p = N$ is a precise length in QC codes, whereas tail-biting convolutional codes has a length Zn_o, which can be varied starting from a sort of minimum threshold value for assuring a certain asymptotic behaviour. Roughly speaking, Z cannot be too small with respect to the generator constraint length m. Keeping well in mind the last remark, it is possible to construct interesting w.s. time-invariant convolutional codes from well known QC block codes, as will be described in the next section.

4.8 Unwrapping QC Block Codes and Reordered Versions of Convolutional Codes

We are going to describe two procedures, which allow to construct a bridge between QC block codes and w.s. convolutional codes, based on their generator matrices. Such procedures operate in opposite directions: the first from QC to convolutional codes, the other vice versa.

Definition 4.9 Given a QC code whose reordered $\underline{G}$ matrix satisfies the condition fixed in Theorem 4.1, the semi-infinite generator matrix obtained from it, after elimination of the property of being wrapped around in its last row layers, is called *unwrapped generator matrix of the original QC code.*

The w.s. time-invariant convolutional code described by the unwrapped generator matrix, obtained according to Definition 4.9, exhibits properties strongly influenced by the original QC code. Since in tail-biting convolutional codes good performance is reached when Z is remarkably greater than m, we can expect a good w.s. convolutional code after unwrapping the reordered generator matrix of a QC code characterized by some null $k_p \times p$ sub-matrices in its row layers.

Example 4.18 From the QC version of the [24,12,8] extended Golay code introduced in Example 2.18 (Fig. 2.15), it is possible to obtain, by unwrapping its $\underline{G}$ matrix, the 4/8-rate w.s. time-invariant convolutional code whose $\underline{G}_o$ sub-matrix has been introduced in Example 4.6 (Johannesson and Zigangirov 1999). Like in its extended Golay parent QC code, the only non-null weight coefficients are $A(8)$, $A(12)$, $A(16)$, the first two increasing linearly with Z (although with very high normalized coefficients). In the convolutional code version, obviously, there are also other non-null higher-weight coefficients, but just for weights integer multiple of 4.

Example 4.19 In Example 2.15, a second QC version of the [8,4,4] extended Hamming code has been obtained, whose $\underline{G}$ matrix, in reordered form, satisfies the condition fixed in Theorem 4.1. Here we have a unique generator polynomial, associated to the sequence 10111. Then the 1/2-rate convolutional code derived from it, after unwrapping, is s.s. time-invariant in its $\underline{G}$ matrix. It is characterized by $d = 4$, and normalized weight coefficient $B(4) = 2$. This code can be interpreted also as a 1/2-rate MLC code, having the [7,3,4] simplex code as its parent cyclic code. Another possibility, for the same QC code, is to start from the following generator matrix

$$\underline{G} = \left[\begin{array}{cc|cc|cc|cc} 1 & 1 & 1 & 0 & 1 & & & \\ & 1 & 0 & 1 & 1 & 1 & & \\ \hline & & 1 & 1 & 1 & 0 & 1 & \\ & & & 1 & 0 & 1 & 1 & 1 \end{array}\right],$$

directly introduced in Example 2.8. Owing to Theorem 2.1 about the cyclic permutation with step of $q_o = 2$, it can be rewritten as

$$\underline{G}' = \left[\begin{array}{cc|cc|cc|cc} 1 & 1 & 1 & 0 & 1 & & & \\ & 1 & 0 & 1 & 1 & 1 & & \\ \hline 1 & & & & 1 & 1 & 1 & 0 \\ 1 & 1 & & & 0 & 1 & 0 & 1 \end{array}\right] ,$$

which is clearly, by itself, in a form with distributed control symbols, with a unique row layer structure, shifting by 4 positions at a time. Unwrapping this matrix yields the 2/4-rate MLQC code introduced in Example 2.30 and treated also in Example 4.13. This rather surprising result is due to the small integer values of parameters q_o, p, k_o, n_o and to some equalities among them.

Then unwrapping represents a procedure for obtaining convolutional codes starting from block codes, generally quasi-cyclic, but sometimes also simply cyclic. Let us consider now an opposite procedure, able to give a block code equivalent to a convolutional code (Solomon and Van Tilborg 1979). Although we can consider either the not tail-biting or the tail-biting form of our code, in order to favour an easy first understanding, we will begin with the latter one.

Definition 4.10 Starting from the generator matrix of an s.s. or w.s. time-invariant convolutional code, well designed and in tail-biting form over a frame length of Z periods, we take in sequence all the first, the second, … the n_oth columns, and then all the first, the second, … the k_oth rows. So we obtain a block code, with Zk_o information symbols, whose positions, in general, are not singly identified, but are expected to be all before the sequence of control symbols, whose total number is $Z(n_o - k_o) = Zr_o$. This reordered form of our tail-biting convolutional code is called *complete circulant form* of the original tail-biting convolutional code.

The above denomination derives from the fact that, in this reordered matrix, complete square circulants will appear. So the block code which will be obtained turns out to be a QC code, characterized by X layers of complete circulants, without anomalous rows, of the type introduced in Sect. 2.7. In understanding why anomalous rows are absent, the following proposition can help.

Corollary a *The $\underline{G}$ matrix obtained from the transformation described in Definition 4.10 exhibits k_o row layers and n_o column layers, forming $k_o n_o$ square circulants on the whole, whose size is $Z \times Z$. If the assumptions of Theorem 4.1 are satisfied, all the rows of the old and of the new generator matrix are linearly independent.*

Proof The overall length of the code is Zn_o in both cases, and the total number of rows in both $\underline{G}$ matrices is Zk_o. Theorem 4.1 assures that all these rows are linearly independent. Hence any further (anomalous) row cannot be present. □

We are treating a code which can be considered either tail-biting convolutional and well designed, or quasi-cyclic with $X = k_o$ row layers, made by square circulants without anomalous rows. After the transformation described in Definition 4.10, we will obtain a rectangular block of $k_o n_o$ complete square circulants, whose size is $Z \times Z$, owing to the linear independence of all the rows in the original $\underline{G}$ matrix. We can consider this transformation as a sort of inverse with respect to that described by Lemma 2.6. Nevertheless such block code exists only over this word length Zn_o (which does not coincide, in general, with a precise natural length N). The size Z of the circulants then remains a degree of freedom in the construction.

Example 4.20 Let us consider the 2/3-rate tail-biting convolutional code introduced in Example 4.16 (see Fig. 4.9). This code is characterized by $m = 1$. After reordering the $\underline{G}$ matrix with $Z = 4$, as described in Definition 4.10, we obtain the rectangular block of complete square circulants shown in Fig. 4.10. It exhibits two rows and three columns of 4×4 complete circulants (in general $Z \times Z$). The first 8 positions are occupied by information symbols, and the last 4 by control symbols, although this $\underline{G}$ matrix is not in an upper triangular form. So we can conclude that, even when we start from a $\underline{G}_o$ sub-matrix which is upper triangular, the same property is typically not verified in the transformed matrix of the equivalent QC code. Attribution of positions to information and control symbols has been made on the basis of their positions before reordering.

We can observe how the input-output sub-generator polynomials directly appear in the structures of the six circulants involved in the new representation. This is a general property, and an increase in Z will entail only a growth in the size of the circulants.

Example 4.21 Let us consider the s.s. time-invariant 1/2-rate convolutional code, whose tail-biting form has been analyzed in Example 3.29. This code is characterized by $m = 2$. Reordering its $\underline{G}$ matrix with $Z = 5$ (Fig. 3.23), we obtain the rectangular block of complete square 5×5 circulants shown in Fig. 4.11. This result is similar to the previous one, but here we have just one row layer and two column layers, because the original code was characterized by $k_o = 1$ and $n_o = 2$.

The constraint length m of the original convolutional code fixes an upper bound to the phase shifts characterizing the circulants. In Fig. 4.12 this upper bound is

$$\underline{G} = \left[\begin{array}{cccc|cccc|cccc}
1 & 1 & & & & 1 & & & 1 & 1 & & \\
 & 1 & 1 & & & & 1 & & & 1 & 1 & \\
 & & 1 & 1 & & & & 1 & & & 1 & 1 \\
1 & & & 1 & 1 & & & & 1 & & & 1 \\
\hline
 & 1 & & & 1 & & & & 1 & & & \\
 & & 1 & & & 1 & & & & 1 & & \\
 & & & 1 & & & 1 & & & & 1 & \\
1 & & & & & & & 1 & & & & 1
\end{array}\right]$$

Fig. 4.10 $\underline{G}$ matrix describing the complete circulant form of the tail-biting convolutional code introduced in Example 4.16 for $Z = 4$

$$\underline{G} = \left[\begin{array}{ccccc:ccccc} 1 & 1 & 1 & & & 1 & & 1 & & \\ & 1 & 1 & 1 & & & 1 & & 1 & \\ & & 1 & 1 & 1 & & & 1 & & 1 \\ 1 & & & 1 & 1 & 1 & & & 1 & \\ 1 & 1 & & & 1 & & 1 & & & 1 \end{array}\right]$$

Fig. 4.11 $\underline{G}$ matrix describing the complete circulant form of the tail-biting convolutional code introduced in Example 3.29 (Fig. 3.23) for $Z = 5$

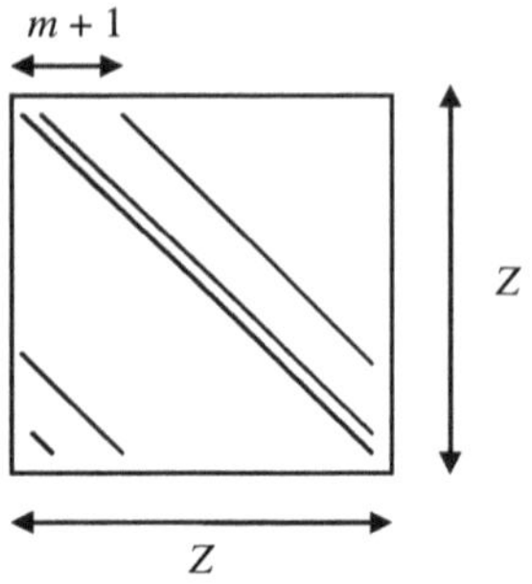

Fig. 4.12 Qualitative comparison between m and Z in a complete circulant

qualitatively depicted. The positions of 1-symbols have been represented by straight lines tilted at an angle of $-\pi/4$, and properly broken, in the right-down corner, by the circulant size. About the role of the comparison between m and the overall Z, given a certain value for m, as long as Z increases, better and better performance is expected, but, beyond a certain limit, such improvement reaches an asymptotic behaviour. This occurs when the original tail-biting code approaches the performance of the original not tail-biting code. Beyond this condition, any further increase in Z gives no benefit.

Example 4.22 The complete circulant form of the convolutional code described in Examples 4.16 and 4.20, with $Z = 4$ is characterized by $d = 2$. This is confirmed after inspection of the $\underline{G}$ matrix in Fig. 4.10, taking ex-OR sums of proper rows. If we assume $Z = 5$, any 2-weight code word is eliminated, as can be derived from the table in the lower part of Fig. 4.9. It is convenient to take Z of the order of, at least, some tenths, in order to reach an asymptotic behaviour.

In Sect. 2.4 QC codes having a structure of the type here under discussion, that is with layers of square circulants, sometimes with addition of R_X possible further anomalous rows, have been introduced (see Examples 2.13, 2.22, 2.24 and 2.28). For them, particular forms of code shortening, code lengthening and modified lengthening have been presented in subsequent sections of Chap. 2. Now we can observe that, when $R_X = 0$, such codes are equivalent to proper tail-biting convolutional codes. In particular, the QC equivalent of the [15,5,7] BCH cyclic code, described in Example 2.13, can be interpreted as an 1/3-rate convolutional code, in tail-biting version, with $Z = 5$, $m = 3$, and interleaved generator polynomial 111. 011.001.010. Proper scaling of parameter Z, provided that $Z > 5$, enables us to maintain $d = 7$, with $A(7) = Z$, $A(8) = Z - 1$.

It is possible to extend the procedure for reordering the $\underline{G}$ matrix described so far also directly to not tail-biting convolutional codes. Following this way, we will obtain codes rather different in their structure, although asymptotically they are expected to give the same performance.

Definition 4.11 Starting from the generator matrix of an s.s. or w.s. time-invariant convolutional code, not in tail-biting form, with Z row layers, we take in sequence all the first, the second, … the n_oth columns, and then all the first, the second, … the k_oth rows, in any period. So we obtain a block code, with Zk_o information symbols and by $(Zr_o + mn_o)$ control symbols, where m is the generator constraint length of the considered code. This reordered form of our convolutional code is called *truncated circulant form* of the original code *in its $\underline{G}$ matrix*.

The result will be a reordered $\underline{G}$ matrix where, instead of complete square circulants, we can find rectangular circulants. More precisely these sub-matrices can be interpreted as enlarged circulants, with size $(Z + m)$, in which the last m rows are eliminated, giving rise to proper $Z \times (Z + m)$ truncated circulants. Such codes are not requested to satisfy the condition of having a minimum value for Z, for avoiding the risk of poor performance. Nevertheless with an m value of the same order of magnitude as Z itself, they pay a strongly reduction in the code rate. Comparison between Definitions 4.10 and 4.11 can be made considering that we have adopted the convention after which in the tail-biting form the frame length is Zn_o, whereas in the not tail-biting form the frame length becomes $(Z + m)n_o$.

Example 4.23 The generator matrices discussed in Example 4.20 (Fig. 4.10) and Example 4.21 (Fig. 4.11), starting from the same codes, but not in tail-biting version (that is following Definition 4.11 instead of Definition 4.10), will appear as

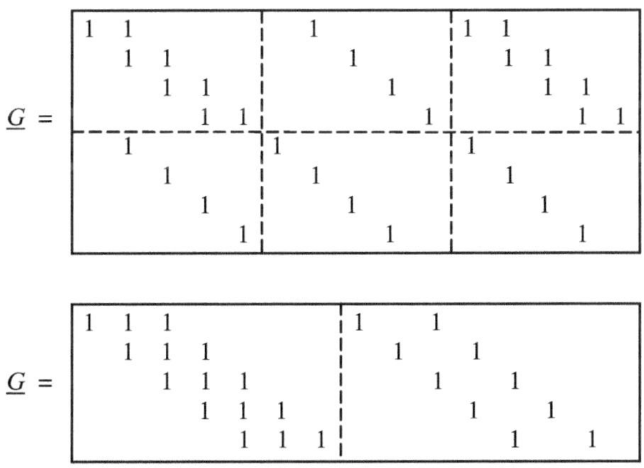

Fig. 4.13 $\underline{G}$ matrices describing the truncated circulant form of two not tail-biting original codes (Examples 4.20 and 4.21)

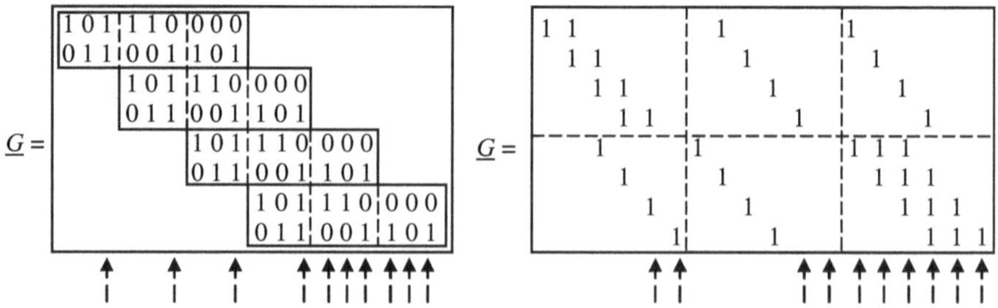

Fig. 4.14 Two equivalent forms of the $\underline{G}$ matrix for the same 2/3-rate convolutional code (Example 4.24)

shown in Fig. 4.13. We remember that the first code is characterized by code rate 2/3, $Z = 4$, $m = 1$, and the second code by code rate 1/2, $Z = 5$, $m = 2$.

Corollary b *In the truncated circulant form of a convolutional code, obtained after the transformation described in Definition 4.11, besides the final control symbols, there are also distributed control symbols. Precisely they occupy the last m positions in each column layer.*

Proof In the reordering procedure such positions derive from the positions of as many final control symbols in the original frame. Their total number is mn_o, as imposed by the requirement of a correct frame termination. So, since the number of column layers is n_o, they will be distributed in number of m any column layer, and they are placed in the last m positions of this column layer. □

In Fig. 4.13, the upper $\underline{G}$ matrix is characterized by $m = 1$ and $Z = 4$, so the fifth position, in the first and in the second column layers, will be occupied by a control symbol. The lower $\underline{G}$ matrix is characterized by $m = 2$ and $Z = 5$, so the sixth and seventh position, in the first column layer, will be occupied by a pair of control symbols. This position association is made, as usual, considering information and control symbols in the frame of the original convolutional code.

Example 4.24 The 2/3-rate convolutional code introduced in Example 4.1 is characterized by $m = 2$. A frame with $Z = 4$ periods where information symbols are introduced ($Z + m = 6$ overall periods) can be described either by the $\underline{G}$ matrix reported in Fig. 4.14 on the left, or by a reordered version (in truncated circulant form) of the same $\underline{G}$ matrix, which is reported on the right. Dashed small arrows denote control symbol positions, which are 10 on the whole. We have six 4×6 truncated circulants in two row layers. The input-output sub-generator polynomials are responsible for their structures. We can observe that in both matrices there is a null column which can be eliminated, so producing a contraction in the length of the correctly terminated code frame by one position.

Looking at the symbolic polynomial generator sub-matrix $\underline{G}_o(x)$ for a convolutional code, it is possible to interpret the (truncated) circulant form of the overall $\underline{G}$ matrix as a sort of its representation. The input-output sub-generator polynomials are mapped in the correspondent circulants. The code rate of a convolutional code in truncated circulant form (exactly as for the original convolutional code) reaches the value k_o/n_o only asymptotically (for Z much greater than m). We can conclude that convolutional codes, even in their truncated circulant form, are characterized by two types of control symbols: those contributing to impose the asymptotic code rate, and those related to the correct frame termination. The last ones typically have a much smaller number.

If the generator matrix reordering described so far is made on an MLQC code, instead of on a QC code, we obtain a reordered $\underline{G}$ matrix of the type introduced in Lemma 2.15 (see Fig. 2.27). Nevertheless, owing to adoption of a transformation based on the unique parameter q_o, instead of on the couple $K_o q_o$, $N_o q_o$, we cannot recognize truncated circulants in the reordered matrices so obtained. On the other hand, such transformation enables us to obtain a scheme for the sequence of information and control symbols of the type *icic* in any column layer.

Definition 4.12 A *symbolic representation of a permutation circulant* can be made considering the shift that it exhibits with respect to an identity matrix (symbolically represented by the permutation circulant $\underline{0}$). A *symbolic representation of a circulant* is made as the sum of their component permutation circulants.

Example 4.25 The generator matrices shown in Figs. 4.10 and 4.11 (and collected in Fig. 4.13 for their truncated circulant versions) can be symbolically written respectively as

$$\underline{G} = \begin{bmatrix} (\underline{0}+\underline{1}) & \underline{1} & (\underline{0}+\underline{1}) \\ \underline{1} & \underline{0} & \underline{0} \end{bmatrix}, \quad \underline{G} = [\,(\underline{0}+\underline{1}+\underline{2}) \quad (\underline{0}+\underline{2})\,],$$

where, according to Definitions 2.10 and 2.11, an ex-OR sum of permutation circulants gives a circulant with weight higher than 1. These representations are independent of the circulant size. Furthermore they can be referred either to $Z \times Z$ complete circulants or to $Z \times (Z + m)$ truncated circulants.

Lemma 4.5 *The integer number J in the* $\underline{\mathrm{J}}$ *symbols, adopted for the circulant symbolic representation of the reordered generator matrix describing a w.s. convolutional code, are so upper bounded*

$$\mathrm{J} \leq m, \tag{4.11}$$

where m represents the generator constraint length of the code.

Proof Inequality (4.11) follows from the above considerations regarding a comparison between m and Z, qualitatively summarized in Fig. 4.12. □

4.9 A First Bridge Between Quasi-cyclic Block Codes and w.s. Time-Invariant Convolutional Codes

In Sect. 3.9 a conceptual bridge between convolutional codes and cyclic block codes has been discussed. It regarded s.s. time-invariant codes in their $\underline{G}$ matrix. All the s.s. time-invariant convolutional codes in their $\underline{G}$ matrix have an equivalent MLC code. Unfortunately not all the w.s. time-invariant convolutional codes in their $\underline{G}$ matrix have an equivalent MLQC code. Nevertheless many other new analogies and correspondences can be carried out here, owing to the presence of parameter q_o, which for cyclic codes was equal to 1. We will start from the last observation for attempting to construct here a bridge between QC codes and w.s. time-invariant convolutional codes, still based on their $\underline{G}$ matrix. Complementary bridges will be presented in subsequent chapters, regarding the parity check matrix.

First of all, good well known QC codes, under the assumptions of Theorem 4.1, can be unwrapped, according to Definition 4.9, leading to w.s. (sometimes s.s.) time-invariant convolutional codes. These codes can be then adopted, over any frame length, with the same asymptotic code rate k_p/p characterizing the original QC code. They typically have information symbols interleaved with control symbols also within any period. See, for instance, the 4/8-rate convolutional code obtained unwrapping the [24,12,8] extended Golay code. The correspondent convolutional code is characterized by $k_o = 4$, $n_o = 8$, and the positions of information and control symbols obey to the scheme *icicicic*. Its structure, after unwrapping, can be prolonged over any frame length. Figure 4.15 up shows a picture where the ensembles of codes involved in this correspondence are related. The subset denoted Theorem 4.1 represents the QC codes whose reordered $\underline{G}$ matrix satisfies the assumption listed in Theorem 4.1. On the other side, the w.s. time-invariant convolutional codes so obtained form a subset of all the w.s. time-invariant convolutional codes.

Another correspondence can be summarized as follows. The tail-biting form of the generator matrix describing a w.s. time-invariant convolutional code (provided that it is well designed) can be reordered in such a way as to produce k_o row layers and n_o column layers, made by complete square circulants (Definition 4.10). The size of such complete circulants is Z (Corollary a of Theorem 4.1), and the overall structure of the generator matrix so obtained resembles that of a particular class of QC codes. Precisely those whose $\underline{G}$ matrix is a rectangular block containing complete $p \times p$ square circulants, plus some possible further anomalous rows. In the case of the correspondence here considered such anomalous rows are absent. This ensemble correspondence is depicted in the middle of Fig. 4.15. The input-output sub-generator polynomials of the original w.s. time-invariant convolutional codes appear, as proper phase shifts, in the circulants of the new representation.

The same reordering transformation of the $\underline{G}$ matrix proposed for tail-biting codes can be applied also to not tail-biting codes, leading to a block of truncated circulants (Definition 4.11). Precisely this block contains k_o row layers and n_o column layers, made by $Z \times (Z + m)$ truncated circulants (Corollary b of Theorem

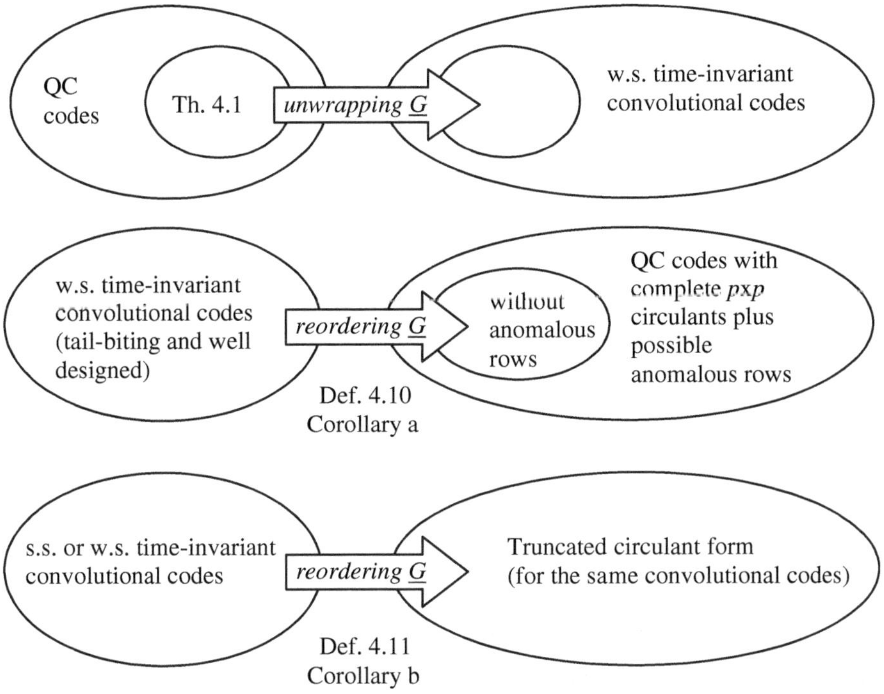

Fig. 4.15 Main ensemble correspondences in the bridge between QC codes and w.s. time-invariant convolutional codes, based on their $\underline{G}$ matrix

4.1). Such ensemble correspondence is reported in Fig. 4.15 down. In this situation the assumption that we start from a well designed convolutional code is no longer necessary, in fact it was requested in the framework of Corollary a, because the tail-biting form of a not well designed convolutional code exhibits a lower number of information symbols, owing to linear dependence of some rows in its $\underline{G}$ matrix.

Considering the role that the number of periods Z plays in a convolutional code, the following observation can be made, about the new form of the $\underline{G}$ matrix, obtained after reordering (with either complete or truncated circulants). Once having chosen the macroscopic structure of the layers (parameters k_o and n_o) and of the circulants (for instance in symbolic form according to Definition 4.12), we can vary Z. Beyond a certain minimum value, we will obtain performance independent of Z. This fact is consequence of the presence of distributed control symbols in the original convolutional code. The need of a minimum value for Z is due to the risk of low-weight code frames in the tail-biting arrangement, whereas it is due to a sacrificed code rate in the not tail-biting arrangement.

Coming back to MLQC codes (after Definition 2.18), it is useful to remember that they have a $\underline{G}$ matrix characterized by the same structure as that of a certain class of w.s. time-invariant convolutional codes (Sect. 4.5). Precisely this class comprehends convolutional codes whose period is integer multiple of q_o and whose

$\underline{G}_o$ sub-matrix is upper triangular. In such codes, the overall number of information symbols then turns out to be $k_oZ = q_oK_oZ$. Instead of applying a transformation of the type suggested for QC codes (with steps of q_o both horizontally and vertically, like made in Fig. 2.27), we can proceed exactly as for w.s. time-invariant convolutional codes, that is according to Definition 4.10 or 4.11. So we obtain a block characterized by q_oK_o row layers and q_oN_o column layers, made by circulants. In the complete circulant form, we have $Z \times Z$ square circulant, as usual. In the truncated circulant form, we have $Z \times (Z + m')$ rectangular sub-matrices, as truncated circulants, where the new generator constraint length m' can be calculated, as a function of parameter r_p, imposing the equality $(m' + 1)N_o = r_p + K_o$, derived in analogy with (3.15).

Example 4.26 In Example 2.31 the MLQC code obtained from the [10,6,3] QC code, assuming $N_o = 2$ and $K_o = 1$, has been analyzed (see Fig. 2.26). Reordering of the $\underline{G}$ matrix has been made according to Lemma 2.6 and Theorem 2.2, that is with steps of q_o for both columns and rows. We can repeat now the same operation, but according to Definition 4.11, that is with steps of 2 vertically and with steps of 4 horizontally. The new $\underline{G}$ matrix so obtained can be partitioned into two row layers, each made by four 4×5 truncated circulants, where the input-output sub-generator polynomials of such equivalent convolutional code can be immediately recognized. The third column layer is null and can be eliminated, so leading to a resulting code rate 2/3, instead of 2/4, as already noted. This is the final result in symbolic form

$$\underline{G} = \begin{bmatrix} (\underline{0}+\underline{1}) & \underline{0} & - \\ \underline{1} & (\underline{0}+\underline{1}) & \underline{0} \end{bmatrix}.$$

About the possibility of obtaining an MLQC code according to Definition 2.19, that is starting from a $\underline{G}$ matrix already organized with layers of circulants, we can observe what follows. The presence of the circulants in this rectangular block, by itself, means that the considered code is equivalent to a tail-biting convolutional code. Addition of an external MLQC structure, like in Example 2.33, introduces a further redundancy, leading to an overall code which can be interpreted as a doubly convolutional code. On this topic, we will come back in Chaps. 6 and 12.

4.10 Trellis of a w.s. Time-Invariant Convolutional Code Obtained from Its Generator Matrix

Assuming a code rate k_o/n_o, for a code not obtained after puncturation, the generator trellis is characterized by 2^{k_o} branches outgoing from any node, except in its final contracting part. In the stationary part of this trellis and in its contracting part, there will be also 2^{k_o} branches entering each node (always a merge node). The number of nodes per stage, instead, can be calculated as 2^{μ}, where μ represents the overall number of memory elements in the k_o shift registers (in controller arrangement) employed in the encoder circuit.

Example 4.27 Let us imagine the temporal evolution of the state diagram shown in Fig. 4.5 (Example 4.2), which describes a 2/3-rate convolutional code characterized by $m = 1$ and $\mu = 2$. We have a trellis with an expanding part, consisting of only one 4-fold bifurcation. Then there is the stationary part, with $2^2 = 4$ states (nodes) in any stage. Each state collects $2^2 = 4$ entering branches. Finally there is the contracting part, consisting of the collection in state $\underline{0}$ of the last four branches. Any branch is labelled with 3 output symbols. If the code is non-systematic also 2 input symbols are needed in labelling branches. The metrics are calculated as departures of the received signal from the output symbol permitted sequences. We can consider a branch decoding complexity and a state decoding complexity, in this case, coincident and equal to 4.

Typically a good trade-off is to maintain such two contributions to the overall computational complexity comparable. Nevertheless, as already said, the same code can be encoded also by different encoder circuits, for instance those based on shift registers in observer arrangement. When the code rate is high ($k_o > n_o$) this solution appears typically preferable, as will be better explained in Chaps. 6 and 8, where the parity check matrix of the considered convolutional codes will be taken into account. For values of the code rate near 1/2, it is of some interest an accurate search for the minimal basic encoder circuit (see Definition 4.7), in order to minimize μ, especially when the code rate is a ratio between two numbers having a common multiple (for instance 4/6, 6/9, 8/12, all equal to 2/3). In such cases, the branch complexity often turns out to be the strongest constraint. About the codes obtained after unwrapping starting from some known block QC codes, they typically derive from a co-ordinate reordering and $\underline{G}$ matrix partitioning, of the type described in Sect. 2.11. In this situation, the decoding computational complexity of the resulting convolutional code is the same as that of the original block code.

As a final remark, considering convolutional codes with very high rate, of the type $(n_o - 1)/n_o$, the encoder circuit based on a unique shift register in observer arrangement and the construction of a proper code puncturation system, starting from a certain 1/2-rate parent code, can show comparable behaviours, both as a trade-off between performance and decoding computational complexity. The advantage of using a punctured convolutional code can be in a more direct possibility of arranging a rate-adaptive encoder-decoder couple (Cain et al. 1979).

On the other hand, modern convolutional codes are often based on a different structure, which can be understood only considering an approach based on the parity check matrix, and consequently on a special decoding algorithm. They belong to the family of LDPC convolutional codes (Gallager 1962; Tanner 1981; Jimenez-Felstrom and Zigangirov 1999; Costello et al. 2007) and will be treated in Chaps. 10–12. Also for them, it is possible to set up rate-adaptive solutions.

References

Bossert M (1999) Channel coding for telecommunications. Wiley, Weinheim

Cain JB, Clark GC, Geist JM (1979) Punctured convolutional codes at rate (n − 1)/n and simplified maximum likelihood decoding. IEEE Trans Inf Th 25:97–100

Costello DJ, Pusane AE, Jones CR et al (2007) A comparison of ARA-and protograph-based LDPC block and convolutional codes. Proc Inf Th Appl Workshop, La Jolla (CA), 111–119

Gallager RG (1962) Low-density parity-check codes. IRE Trans Inf Th 8:21–28

Jimenez-Felstrom AJ, Zigangirov KS (1999) Time-varying periodic convolutional codes with low density parity-check matrix. IEEE Trans Inf Th 45:2181–2191

Johannesson R, Zigangirov KS (1999) Fundamentals of convolutional coding. IEEE Press, New York

Ryan WE, Lin S (2009) Channel codes: classical and modern. Cambridge University Press, Cambridge

Solomon G, Van Tilborg HCA (1979) A connection between block and convolutional codes. SIAM J Appl Math 37:358–369

Tanner RM (1981) A recursive approach to low complexity codes. IEEE Trans Inf Th 27:2966–2984

Part II
Parity Check Matrix

Chapter 5
Parity Check Matrix Approach to Linear Block Codes

5.1 Parity Check Matrix of a Linear Block Code

Given a linear block code $[n, k, d]$, its description can be made by means of a generator matrix $\underline{G}$, whose number of rows is k. Alternatively, its description can be provided also by an $(n - k)$-row matrix, expressing parity check equations among the symbols in its code words. In the following, except when the contrary is explicitly said, we will refer to binary codes only.

Definition 5.1 Letting $\underline{t}$ the row vector representing a code word, its n elements satisfy $(n - k)$ linear equations (parity check equations), which can be expressed as

$$\underline{t} \cdot \underline{H}^T = \underline{0}, \tag{5.1}$$

where the $(n - k) \times n$ matrix $\underline{H}$ is called *parity check matrix* of the considered code, and $\underline{0}$ is a row vector having $(n - k)$ null elements. Superscript T means matrix transposition (see Appendix A).

More precisely, (5.1) represents $(n - k)$ ex-OR sums equated to zero, all regarding the same vector $\underline{t}$. The jth one of such equations is a product between $\underline{t}$ and the jth column of $\underline{H}^T$. The above equations turn out to be as many linear constraints acting on the n degrees of freedom present in vector $\underline{t}$, so reducing them to those present in vector $\underline{u}$ introduced by (1.1). In this sense, (5.1) can be interpreted as a homogeneous linear system, with $(n - k)$ equations in the n unknowns consisting of all the symbols in the code word. Since we are in a binary finite field, the multiplicity of the possible solutions is 2^k, included the solution consisting of all 0-symbols (see Appendix A for details).

Lemma 5.1 *If a column of* $\underline{H}$ *is made by all 0-symbols, the corresponding position in the code word is not protected by the code.*

G. Cancellieri, *Polynomial Theory of Error Correcting Codes*,
Signals and Communication Technology, DOI 10.1007/978-3-319-01727-3_5

Proof This result follows directly from the definition of matrix product row by column (see Appendix A). □

Lemma 5.2 *All the rows of* $\underline{H}$ *must be linearly independent, that is this matrix must have full rank (see Appendix A for details).*

Proof If there is a row linearly dependent on the other ones, the number of constraints is reduced by one unit, and the number of code words is no longer 2^k, but becomes 2^{k+1}. □

Lemma 5.3 *After a change in the order of the rows in* $\underline{H}$*, we obtain again the original code.*

Proof The number and the structures of constraints do not change. □

Lemma 5.4 *After a change in the order of the columns in* $\underline{H}$*, we obtain an equivalent of the original code.*

Proof The new code exhibits code words equal to those of the original code, except for a change in their symbol positions. □

Lemma 5.5 *Replacing a row with an ex-OR sum of it and other rows, provided that* $\underline{H}$ *maintains rank* $(n - k)$*, yields again the original code.*

Proof The number of independent linear constraints expressed by the equations collected in $\underline{H}$ does not change. □

Example 5.1 Let us consider the $\underline{H}$ matrix shown in Fig. 5.1 up, where 0-symbols have not been represented. It has full rank equal to 3. Since the number n of column is 4, the code has just $k = 1$ degree of freedom, that is just one information symbol. Its transpose $\underline{H}^T$ is depicted in Fig. 5.1 down. It is possible to multiply a vector $\underline{t}$ by this transpose matrix, according to (5.1), and obtain the following result. Independently of the position where this information symbol is set, if there is one 1-symbol, all the remaining 3 symbols are 1, if there is one 0-symbol, all the remaining 3 symbols are 0. In other terms, the only two code words are 0000 and 1111.

In vector $\underline{t}$, sometimes, there are precise positions where information symbols can be found, according to Definition 1.5 about systematic codes, in strict sense or in wide sense. In particular, for a strict-sense systematic code, the information symbols occupy the first k positions. We remember also that the systematic form of the $\underline{G}$ matrix is characterized by a $k \times k$ identity matrix $\underline{I}_k$ in its left hand side part (Lemma 1.5). The definition of a systematic form for the correspondent parity check matrix $\underline{H}$ is somehow different, and regards the final $(n - k)$ positions, when such positions are devoted to control symbols.

Definition 5.2 A parity check matrix $\underline{H}$, showing an $(n - k) \times (n - k)$ identity submatrix $\underline{I}_{n-k}$ in its right hand side part, is called *systematic*. In this type of $\underline{H}$ matrix, each control symbol, when placed in one of the last $(n - k)$ positions of the code word, does not depend directly on other control symbols.

$$\underline{H} = \begin{array}{|cccc|} \hline 1 & 1 & & \\ & 1 & & 1 \\ & & 1 & 1 \\ \hline \end{array}$$

$$\underline{H}^T = \begin{array}{|ccc|} \hline 1 & & \\ 1 & 1 & \\ & & 1 \\ & 1 & 1 \\ \hline \end{array}$$

Fig. 5.1 Parity check matrix and its transpose for the code described in Example 5.1

Example 5.2 Let us still consider the code described in Example 5.1, whose parity check matrix is depicted in Fig. 5.1. Taking proper ex-OR sums of some rows in such matrix, but maintaining three rows all linearly independent, we obtain

$$\underline{H} = \begin{bmatrix} 1 & 1 & & \\ 1 & & 1 & \\ 1 & & & 1 \end{bmatrix},$$

which expresses the systematic form of the considered matrix. 0-symbols have not been represented. On the right, sub-matrix $\underline{I}_3$ is clearly evident. The new code is, in general, equivalent to the previous one. Nevertheless, in this particular situation, characterized by $k = 1$, equivalence implies coincidence.

Definition 5.3 A parity check matrix $\underline{H}$ of the type depicted in Fig. 5.2, that is with all 0-symbols over the diagonal of its $(n - k) \times (n - k)$ right sub-matrix (expressed by all 1-symbols), is called *lower triangular*. The example, where a, b, … can be either 0-symbols or 1-symbols, refers to the case of $n - k = 4$.

Remark that the systematic form of $\underline{H}$ is a particular case of a lower triangular $\underline{H}$. In a lower triangular form of a parity check matrix, the control symbols, if placed in the final $(n - k)$ positions of the code word, are calculated in sequence. This property contributes to make the encoding procedure easier. We can compare Definition 5.3 with Definition 1.7, which holds for an upper triangular $\underline{G}$ matrix. A situation more particular is that characterized by a unique pattern of 1-symbols which shifts along any row of $\underline{H}$.

Example 5.3 The code treated in Examples 5.1 and 5.2 can be described also by another form of $\underline{H}$, which exhibits a unique 1-symbol pattern shifting along its rows. This matrix turns out to be

$$\underline{H} = \begin{array}{|cccccc|} \hline & a & 1 & & & \\ & b & c & 1 & & \\ \ldots\ldots\ldots & d & e & f & 1 & \\ & g & h & i & j & 1 \\ \hline \end{array}$$

Fig. 5.2 Lower triangular form for an $\underline{H}$ matrix for $n - k = 4$

$$\underline{H} = \begin{bmatrix} 1 & 1 & & \\ & 1 & 1 & \\ & & 1 & 1 \end{bmatrix}.$$

It is possible to observe an analogy in the properties here introduced for the parity check matrix with respect to those discussed in Sect. 1.2 about the generator matrix $\underline{G}$. In particular, all Chap. 1 deals with s.s. time-invariant block codes in their generator matrix, Definition 2.2 introduces w.s. time-invariant block codes still in their generator matrix. For convolutional codes, respectively in Chaps. 3 and 4, the same definitions are adopted too. Here we will propose a similar classification, but referred to the parity check matrix $\underline{H}$.

Definition 5.4 A parity check matrix in which a unique 1-symbol pattern shifts along its rows, by one position each time, characterizes a block code which is defined *strict-sense* (or briefly *s.s.*) *time-invariant in its parity check matrix*.

Remark that a parity check matrix satisfying Definition 5.4 certainly has full rank, that is all its rows are linearly independent. Many codes are s.s. time-invariant both in their generator matrix and in their parity check matrix. Sometimes this is not. In particular, a different behaviour may characterize convolutional codes having code rate $1/n_o$, necessarily s.s. time-invariant in $\underline{G}$ and often not in $\underline{H}$, or having code rate $(n_o - 1)/n_o$, which can be expected to be necessarily s.s. time-invariant in $\underline{H}$, but generally not in $\underline{G}$. A different behaviour comes out also for shortened or punctured block codes.

In certain situations, it is convenient to eliminate columns on the left hand side in an s.s. time-invariant $\underline{H}$ matrix, so obtaining a time-varying parity check matrix, but still lower triangular. This occurs for a shortened cyclic code, as will be discussed in following Sections. A different case is that characterized by a periodically time-varying behaviour in the rows of $\underline{H}$. This feature originates wide-sense time-invariant codes in $\underline{H}$, which will be treated in Chap. 6.

Example 5.4 Taking the 1-symbol pattern 11001, it is possible to construct an s.s. time-invariant $\underline{H}$ matrix of the type shown in Fig. 5.3 up. This is a 5×9 parity check matrix, with full rank, therefore it refers to a linear block code, with $n = 9$ and $n - k = 5$, that is $k = 4$. It is possible to eliminate columns on its left hand side. A first elimination gives rise to a code characterized by $n = 8$ and $k = 3$ (matrix $\underline{H}'$). A second elimination makes the first row useless (matrix $\underline{H}''$). In fact a parity check equation with just one 1-symbol forces the position where this symbol is placed to be 0, independently of the information symbol sequence. So this row is eliminated, and the new code exhibits $n = 7$ but still $k = 3$.

Lemma 5.6 *Fixed a certain value for n, the maximum number of rows in $\underline{H}$, all linearly independent, is $(n - 1)$.*

$$\underline{H} = \begin{array}{|ccccccccc|} 1 & 1 & 0 & 0 & 1 & & & & \\ & 1 & 1 & 0 & 0 & 1 & & & \\ & & 1 & 1 & 0 & 0 & 1 & & \\ & & & 1 & 1 & 0 & 0 & 1 & \\ & & & & 1 & 1 & 0 & 0 & 1 \end{array}$$

$$\underline{H}' = \begin{array}{|cccccccc|} 1 & 0 & 0 & 1 & & & & \\ 1 & 1 & 0 & 0 & 1 & & & \\ & 1 & 1 & 0 & 0 & 1 & & \\ & & 1 & 1 & 0 & 0 & 1 & \\ & & & 1 & 1 & 0 & 0 & 1 \end{array}$$

$$\underline{H}'' = \begin{array}{|ccccccc|} 0 & 0 & 1 & & & & \\ 1 & 0 & 0 & 1 & & & \\ 1 & 1 & 0 & 0 & 1 & & \\ & 1 & 1 & 0 & 0 & 1 & \\ & & 1 & 1 & 0 & 0 & 1 \end{array}$$

Fig. 5.3 Parity check matrices of the codes described in Example 5.4

Proof If the rows of $\underline{H}$ are all linearly independent, when their number is $(n - 1)$, we are left with a unique degree of freedom, therefore the code is characterized by $k = 1$. Addition of a further row, linearly independent of the other ones, gives rise to a degenerate situation, with $k = 0$, with a unique code word $\underline{t} = \underline{0}$, which is clearly unacceptable. □

Example 5.5 The code introduced in Example 5.1 (treated also in Examples 5.2, 5.3) exhibits $k = 1$. It is possible to recognize in this code the [4,1,4] binary repetition code (see Definition D.16 and Sect. 1.4). As already noted, it is always at least w.s. systematic.

5.2 Parity Check Matrix and Hard-Decision Decoding

The parity check matrix is directly related to the procedure for hard-decision decoding. In order to introduce such procedure, it is useful to define error row vectors.

Definition 5.5 A row vector having n element, all null except one, in the jth position, $j = 1, 2, \ldots n$, is called *single-error row vector* and denoted $\underline{e}_j$.

A multiple-error row vector $\underline{e}$ turns out to be the ex-OR sum of a certain number of distinct single-error row vectors. Its weight (number of non-null symbols) will be equal to the number of distinct single-error row vectors considered.

Definition 5.6 The jth row vector of the transpose $\underline{H}^T$ of a parity check matrix $\underline{H}$ is defined *single-error* syndrome, and denoted $\underline{s}_j$.

Example 5.6 The $\underline{H}$ matrix in Example 5.3 exhibits these single-error syndromes: [100], [110], [011], [001]. They are all distinct. The same property holds also for other equivalent forms of this parity check matrix (Examples 5.1 and 5.2).

Lemma 5.7 *Letting* $(\underline{t} + \underline{e}_j)$ *a code word affected by one single error, in position j, its multiplication by* $\underline{H}^T$ *yields the jth single-error syndrome of* $\underline{H}$*. In other terms, we have* $\underline{s}_j = \underline{e}_j \cdot \underline{H}^T$.

Proof In the product $(\underline{t} + \underline{e}_j) \cdot \underline{H}^T$, the former term is null owing to (5.1). □

Lemma 5.8 *In order to correct any single error, it is necessary and sufficient that all the single-error syndromes in the parity check matrix of the considered code are non-null and distinct.*

Proof When the product $(\underline{t} + \underline{e}_j) \cdot \underline{H}^T$ is null the decoder decides, as the most likely choice, that no errors are present (only errors of weight not smaller than the minimum distance d can produce this situation). When the above product is different from $\underline{0}$, if the result coincides with one of the single-error syndromes, and it is unique, the correspondent position is where the decoder estimates, most likely, that the single error occurred. If all the single-error syndromes are distinct, any single error can be corrected. □

Example 5.7 Assuming the $\underline{H}$ matrix of Example 5.3, independently of the code word $\underline{t}$, if we have a single-error row vector [0010], the result of the product $(\underline{t} + \underline{e}_3) \cdot \underline{H}^T$ is $\underline{s}_3 = [011]$. Since the single-error syndromes of the considered code are all distinct, the decoder estimates that a single error occurred in the third position of the code word.

A hard-decision decoder, first of all, calculates the product $(\underline{t} + \underline{e}) \cdot \underline{H}^T$, where both $\underline{t}$ and $\underline{e}$ are separately unknowns (but not their ex-OR sum, which is the n-tuple to be decoded). Then it attempts to calculate $\underline{e}$. In general, the overall $\underline{e}$ is the superimposition of various single errors. The decoder will search single errors up to its error correction capability (see Appendix D about parameter t, expressing the maximum number of correctable single errors). Once determined the most likely vector $\underline{e}$, the correspondent vector $\underline{t}$ is obtained from a trivial ex-OR sum. The last procedure is possible just in a binary alphabet.

From the above discussion, we can conclude that a column of $\underline{H}$ with all 0-symbols is useless. Furthermore equal columns represent a point of weakness for the considered code.

Lemma 5.9 *In a parity check matrix with all non-null single-error syndromes, if there are two equal single-error syndromes, the code is characterized by minimum distance at most* $d = 2$*, and the two single errors in the corresponding positions are not distinguishable.*

Proof Letting $\underline{e}_j$ and $\underline{e}_i$ the two single errors, corresponding to two equals columns of $\underline{H}$. Both products $(\underline{t}+\underline{e}_j)\cdot\underline{H}^T$ and $(\underline{t}+\underline{e}_i)\cdot\underline{H}^T$ give the same result. The double error $\underline{e}=\underline{e}_j+\underline{e}_i$ exhibits, in turn, a null result after product by $\underline{H}^T$. Hence the code admits a code word $\underline{t}=\underline{e}_j+\underline{e}_i$, and its minimum distance then is, at most, $d = 2$. □

Definition 5.7 The procedure performed by the decoder, in order to not correct errors, but simply reveal their presence, is called *error detection*.

Error detection is briefly described also in Appendix D, for even minimum distance codes, and also for codes whose purpose is in part devoted to error detection, besides direct error correction. Error detection procedures can be interpreted also in the standard array table (see Definition D.12). When, in a same row of the standard array table, there are more than one n-tuple with equal minimum weight, the decoder can only reveal that an error having this weight occurred.

Lemma 5.10 *A code, whose minimum distance is d, when devoted to detect errors only, can detect all the error patterns having up to* $(d-1)$ *single errors.*

Proof When the whole minimum distance d is entirely devoted to detect errors, it is enough that the n-tuple $(\underline{t}+\underline{e})$ is different from whichever code word for allowing error detection. The weight of $\underline{e}$, in order to do this, can then be so upper bounded $w[\underline{e}] \le d-1$. □

It is possible to separate the two functions of error correction and error detection (Klove 2007). This is typical when d is even. After Lemma D.3, a code characterized by an even minimum distance d can correct up to $t = (d-2)/2$ single errors, and detect up to $d/2$ single errors. Just when all the minimum distance d is devoted to detect errors only, the bound expressed by Lemma 5.10 is reached. Many intermediate situations can be chosen, as long as d increases, in which the functions of error detection and error correction are properly balanced. On the other hand, for a code characterized by $d = 2$, it is possible only to detect single errors, although often some particular errors can be corrected too.

Example 5.8 A binary parity check code $[n, n-1, 2]$ can detect only single errors and cannot correct at all. Its parity check matrix is made by just one row, having all 1-symbols. Therefore all its syndromes are equal. Once an n-tuple with odd weight is to be decoded, the decoder can take n possible original code words, with equal confidence, obtained varying just one symbol in the n-tuple. To be more precise, a binary parity check code is able to detect any odd-weight error. Nevertheless the events of 3, 5, … contemporary single errors are much less probable than the events of one error (of the order of p^3 and p^5, with respect to the single-error probability p). Therefore they are ignored by the decoder.

On the other hand, the binary parity check code is not the unique code characterized by $d = 2$. For instance, a y-time LC code (see Definition 1.25), with

$N < y < 2N$, shows an increasing number of external positions in its code words where single errors can be only detected. In this case, there are y situations of two-fold ambiguity. In the remaining central positions the code can correct up to the correction capability of the original cyclic code. On the contrary, a binary parity check code is characterized by all the n positions in the same situation, and whichever ambiguity is n-fold. We can interpret the measure of this ambiguity as the number of code words at the same distance from the n-tuple to be decoded.

Example 5.9 A binary Hamming code $[2^r - 1, 2^r - 1 - r, 3]$ can correct only single errors. Its parity check matrix exhibits $(2^r - 1)$ different columns, as many as all the possible different sequences of r binary elements, except that made by all 0-symbols.

Definition 5.8 Given a multiple-error row vector $\underline{e}$, independently of the code word $\underline{t}$ affected by it, we can define a *multiple-error syndrome* as follows

$$\underline{s} = (\underline{t} + \underline{e}) \cdot \underline{H}^T = \underline{e} \cdot \underline{H}^T. \tag{5.2}$$

In obtaining the second equality of (5.2), Eq. (5.1) has been employed, like in the proof of Lemma 5.7. Remark that only single-error syndromes $\underline{s}_j$ are rows in $\underline{H}^T$. Multiple-error syndromes are proper ex-OR sums of them. The fundamental goal of a hard-decision decoding algorithm is to separate the components of a multiple-error syndrome, related to as many single errors. This goal is reached by decomposing $\underline{s}$ into the minimum number of single-error syndromes.

Keeping in mind the concepts of standard array table and decoding hyper-spheres (see Definition D.12, Figs. D.3 and D.4 in Appendix D), it is possible to say that there is a one-to-one correspondence between the lowest-weight n-tuples of any row of the standard array table and the multiple-error syndromes (MacWilliams and Sloane 1977). Just when this lowest-weight n-tuple is unique, error correction is possible. When this is not, error detection is sometimes possible, with a certain degree of ambiguity, related to the number of lowest-weight n-tuples in the considered row.

For a Hamming code, we have only single-error syndromes, and as many rows in their standard array tables. For a BCH code able to correct up to t single errors, there are 1-error, 2-error, ... t-error syndromes, all distinct. Typically many other distinct multiple-error syndromes, beyond the correction capability of the code, originate further rows in the standard array table. Just codes able to satisfy Hamming bound (D.13) with equality sign (see Appendix D) exhibit exactly up to t-error syndromes, all distinct, and as many rows of their standard array table, with unique lowest-weight n-tuple, without any further row. The exceeding rows are related to a certain number of possible error corrections beyond the limit t, and often to an even larger number of possible error detections. Nevertheless this form of exhaustive error correction-detection is possible in practice only when the standard array table is characterized by an acceptable number of elements, that is only for small block codes. In the following, we will restrict our attention to error correction up to the limit t, with particular attention to BCH codes, for which a powerful hard-decision decoding algorithm is available, starting from the multiple-error syndrome $\underline{s}$ given by (5.2).

The operation of decomposition of $\underline{s}$ can be summarized as follows. Letting

$$\underline{H}^T = \begin{bmatrix} \underline{s}_1 \\ \underline{s}_2 \\ \cdots \\ \underline{s}_n \end{bmatrix},$$

we obtain

$$\underline{s} = \sum_{j \in S(e)} \underline{s}_j, \tag{5.3}$$

where the ensemble $S(e)$ comprehends only the single-error syndromes in the positions where single errors occurred. The purpose of the decoding procedure is to find the minimum number of distinct $\underline{s}_j$'s, able to satisfy (5.3).

Clearly the maximum weight w[$\underline{e}$], for which the above decomposition is correct, coincides with the maximum number t of single errors which the code can correct, in turn related to the code minimum distance d through (D.3) for d odd or through (D.4) for d even. If $w[\underline{e}]$ exceeds t, (5.3) is still true, but generally gives rise to a wrong decoding, because the resulting ensemble $S(e)$ is no longer associated to the true error pattern, but to another error pattern $\underline{e}'$, with $w[\underline{e}'] \leq t$.

Example 5.10 In Fig. 5.4 the $\underline{H}$ matrix of a [15,7,5] BCH code is reported, for which we have $t = 2$. After reception of the 15.tuple $(\underline{t} + \underline{e}_3 + \underline{e}_5)$, we obtain the double-error syndrome $\underline{s} = [00111000]$. It is not a row of $\underline{H}^T$, but can be so decomposed into the superimposition of the minimum number of single-error syndromes (rows of $\underline{H}^T$)

$$\underline{s} = [01100000] + [01011000] = \underline{s}_3 + \underline{s}_5.$$

$$\begin{array}{c} \\ \\ \\ \underline{H} = \\ \\ \\ \\ \end{array}
\begin{array}{|ccccccccccccccc|}
\hline
 & & \downarrow & & \downarrow & & & & & & & & & & \\
1&1&0&1&0&0&0&1& & & & & & & \\
 &1&1&0&1&0&0&0&1& & & & & & \\
 & &1&1&0&1&0&0&0&1& & & & & \\
 & & &1&1&0&1&0&0&0&1& & & & \\
 & & & &1&1&0&1&0&0&0&1& & & \\
 & & & & &1&1&0&1&0&0&0&1& & \\
 & & & & & &1&1&0&1&0&0&0&1& \\
 & & & & & & &1&1&0&1&0&0&0&1 \\
\hline
\end{array}$$

Fig. 5.4 $\underline{H}$ matrix of the [15,7,5] BCH code and a pair of single-error syndromes

The columns of $\underline{H}$ corresponding to the transposes of such two single-error syndromes are denoted with small arrows. Their identification is univocal, because the situation considered is within the error correction capability of the code.

For binary BCH codes a suitable algorithm allows the obtainment of the solution to the problem of finding the positions of the single errors, but just up to the designed value τ, imposed during the code synthesis, as noted in Sect. 1.9. The design of such codes starts from the symbolic construction of the generator polynomial $g(x)$ as the product of irreducible factors, related to groups of roots α^j (see Sect. 1.9 and Appendix B). The same construction is followed in order to arrange, among all the possible equivalent versions of $\underline{H}$, the one best suited for decoding (Wicker 1995). Also when the BCH code is not in polynomial form, such an $\underline{H}$ matrix must be properly arranged. Such alternative version of $\underline{H}$ is different from the one characterized by the property of being s.s. time-invariant (and hence also lower and upper triangular). An example of this arrangement will be given in Sect. 5.11.

So we have to distinguish when the designed maximum number of correctable single errors τ equals the true value t obtained after BCH code construction, with respect to the cases for which $t > \tau$. The latter situation, for instance, is that of the [23,12,7] binary Golay code, which is also a BCH non-primitive code (see Example 1.33). It is characterized by $\tau = 2$, but several ad hoc algorithms, able to exploit its whole true minimum distance $d = 7$, have been proposed (Clark and Cain 1983).

For binary BCH codes, the firstly proposed decoding algorithm has been suggested by Peterson (1960). This algorithm exhibits a computational complexity growing with the square of t. A simpler algorithm was proposed few years later by Berlekamp (1965). Here the computational complexity grows only linearly with t. In both these algorithms, the computational complexity is related to the block length n as $n\lg(n)$ (MacWilliams and Sloane 1977).

For non-binary BCH codes, after having located the single errors, they must be properly processed and the right symbols have to be substituted to the wrong symbols. In Gorenstein and Zierler (1961) an extension of Peterson algorithm, suitable for non-binary BCH codes, was presented. In Massey (1969) an extension of Berlekamp algorithm was finally proposed, showing also a similar reduced computational complexity. About the growth in the computational complexity with the number Q of symbols in the considered alphabet, we can observe that it strongly depends on the structure of the code. For example, Reed-Solomon codes (Definition 1.39) admit more straightforward procedures, like the so-called Euclid's algorithm or the frequency-domain algorithm, which are remarkably simpler (Sweeney 2002). Typically their computational complexity grows only as $n\lg(n)$ (MacWilliams and Sloane 1977).

Finally, it is important to stress that all the algorithms for hard-decision decoding of BCH codes, although invented for codes whose length n is that of their cyclic code version, can be employed also for shorter lengths as well, that is for shortened BCH codes. Often this is a useful degree of freedom to be exploited for adapting channel coding systems to the particular situations where they are requested to operate, allowing fine scalability, either in the code rate or in the block length. From the practical point of view, when obtained the $\underline{H}$ matrix in the proper form for the parent cyclic code, it is enough to eliminate those of its columns where code shortening has been performed (typically the leftmost ones).

5.3 Relations Between the Parity Check Matrix and the Generator Matrix

Theorem 5.1 *Given a certain linear block code, its* $\underline{G}$ *matrix and its* $\underline{H}$ *matrix are so related*

$$\underline{G} \cdot \underline{H}^T = \underline{0}, \tag{5.4}$$

where $\underline{0}$ *is a* $k \times (n - k)$ *matrix having all null elements.*

Proof The generator matrix of a code collects k code words (linearly independent) as rows. Each of them can be assumed as represented by a different vector $\underline{t}$ in (5.1). Since the result of the product (5.1) is a row $(n - k)$ vector $\underline{0}$, the product on the left hand side in (5.4) will give a $k \times (n - k)$ null matrix. □

It is important to observe that (5.4), in general, is not a univocal relationship, in the sense that, given $\underline{G}$, there are various possible $\underline{H}$ matrices, able to satisfy (5.4), and vice versa. More precisely, given one of the two matrices, the code is unique, but it can be represented by a form of the other matrix, in which changes in the row order, or substitutions of rows with linear superimpositions of other rows are always possible.

We remember that equivalent codes (see Definition 1.6) have the same tern of parameters $[n, k, d]$ and word weight distribution. They can be originated by different matrices $\underline{G}$ (or $\underline{H}$). A typical situation is that regarding the same code, but characterized by different associations between information symbols and code words. Sometimes we can have even a different structure, although equivalent. This is the case, for instance, of codes having $g^*(x)$ instead of $g(x)$ or $h^*(x)$ instead of $h(x)$. Furthermore this occurs when we apply a reordering in the columns of the two matrices.

Corollary a *If* $\underline{G}$ *is given in systematic form, that is*

$$\underline{G} = [\underline{I}_k | \underline{E}], \tag{5.5}$$

where $\underline{I}_k$ *is a* $k \times k$ identity matrix and $\underline{E}$ is a $k \times (n - k)$ matrix, (5.4) is able to give univocally $\underline{H}$, in systematic form as well, and resulting in

$$\underline{H} = [\underline{E}^T | \underline{I}_{n-k}], \tag{5.6}$$

where $\underline{I}_{n-k}$ is an $(n-k) \times (n-k)$ identity matrix. This property is reversible.

Proof Let us write

$$\underline{t} = [i_1 \quad i_2 \quad \ldots \quad i_k \quad c_1 \quad c_2 \quad \ldots \quad c_{n-k}] = [\underline{i}|\underline{c}],$$

where $\underline{i}$ and $\underline{c}$ denote proper row sub-vectors. So, from (5.6), we have

$$\underline{t} \cdot \underline{H}^T = \underline{t} \cdot \left[\frac{\underline{E}}{\underline{I}_{n-k}}\right] = \underline{0},$$

which implies $\underline{c} = \underline{i} \cdot \underline{E}$, and hence

$$\underline{t} = \underline{i} \cdot [\underline{I}_k | \underline{E}]. \tag{5.7}$$

From (1.1), with $\underline{u} = \underline{i}$, because we are in the situation characterized by a systematic generator matrix, (5.7) demonstrates (5.5). Clearly also the inverse deduction procedure can be followed, starting from (5.5) and arriving at (5.6). □

Example 5.11 The [4,1,4] repetition code, introduced in Example 5.1, is characterized by a systematic parity check matrix, in which sub-matrix $\underline{E}^T$ is a column vector formed by three elements, all 1-symbols (see Example 5.2). From (5.5), its generator matrix, in systematic form as well, turns out to be

$$\underline{G} = [1 \quad | \quad 1 \quad 1 \quad 1],$$

where the leftmost 1-symbol can be interpreted as $\underline{I}_1$ and the remaining row vector [111] is the transpose of the above mentioned column vector. In this sense, we can conclude that the [4,1,4] code has a $\underline{G}$ matrix which is always in systematic form. This property holds for all the $[n,1,n]$ binary repetition codes.

Example 5.12 Assuming a [7,4,3] Hamming code characterized by the following systematic generator matrix

$$\underline{G} = \left[\begin{array}{cccc|ccc} 1 & & & & 1 & 0 & 1 \\ & 1 & & & 1 & 1 & 1 \\ & & 1 & & 1 & 1 & 0 \\ & & & 1 & 0 & 1 & 1 \end{array}\right],$$

from (5.6), its parity check matrix, in systematic form as well, becomes

$$\underline{H} = \left[\begin{array}{cccc|ccc} 1 & 1 & 1 & 0 & 1 & & \\ 0 & 1 & 1 & 1 & & 1 & \\ 1 & 1 & 0 & 1 & & & 1 \end{array}\right].$$

It is possible to recognize in the two parts of both matrices, separated by a vertical segment, the same sub-matrix $\underline{E}$, except for a transposition. Finally we can note how, as columns in $\underline{H}$, there are all the possible different non-null sequences of 3 binary symbols, confirming the property described in Example 5.9. This property, however, characterizes any possible equivalent form of the $\underline{H}$ matrix in a Hamming code.

Univocal obtainment of $\underline{H}$, when $\underline{G}$ is given in upper triangular form, is generally not possible, as well as univocal obtainment of $\underline{G}$, when $\underline{H}$ is given in lower triangular form. Just when the two matrices are s.s time-invariant (in a code s.s time-invariant both in $\underline{G}$ and in $\underline{H}$), univocal obtainment of the one, given the other, is possible, as will be better explained in the next Section. In order to obtain a $\underline{G}$ matrix for a certain code, whose $\underline{H}$ matrix is known, without the requirement that the former one is unique, it is possible to proceed by attempts, in searching for a structure made by k rows, all linearly independent, each arranged as described in the following proposition.

Lemma 5.11 *A row of the generator matrix $\underline{G}$ will exhibit 1-symbols in correspondence of a cluster of single-error syndromes having null ex-OR sum.*

Proof A row so constructed satisfies all the constraints in (5.1), imposed by $\underline{H}$. So it is a code word, and can be assumed as a row of $\underline{G}$. □

A problem, however, has to be faced, and regards the requirement of finding k linearly independent rows in $\underline{G}$. Among the $\underline{G}$ matrices which can be obtained following the procedure described in Lemma 5.11, one is certainly able to assure linear independence of the rows, and is that based on the choice of an upper triangular form for $\underline{G}$. In a construction of this type for the $\underline{G}$ matrix, it is not rare to discover this situation: although $\underline{H}$ is s.s. time-invariant, it is impossible to construct an s.s time-invariant $\underline{G}$. Even the opposite situation may occur: although $\underline{H}$ is time-varying (or at least w.s. time-invariant), it is possible to construct an s.s. time-invariant $\underline{G}$.

In Definition 1.22 the dual code of a linear block code has been introduced, restricting the attention to polynomial block codes only. Now this concept can be widened, referring it to the a pair of matrices $\underline{H}$, $\underline{G}$.

Definition 5.9 Given a linear block code characterized by an $\underline{H}$ matrix (or by a $\underline{G}$ matrix), the new code described by $\underline{G}' = \underline{H}$ (or by $\underline{H}' = \underline{G}$) is called the *dual code* of the original one.

Given a linear block code $[n, k, d]$, its dual will be an $[n, n - k, d']$ code, and vice versa. In fact, the code length is the same, because $\underline{H}$ and $\underline{G}$ have the same number of columns. The number of information symbols (rows of $\underline{G}$) becomes that of control symbols (rows of $\underline{H}$) and vice versa. In the next Section, we will discuss under which conditions Definitions 1.22 and 5.9 can be considered contemporary valid. With reference to the vector spaces associated to proper rectangular matrices (see Appendix A), we can observe how matrix $\underline{G}$ allows to say that a certain code represents the vector space defined by $\underline{G}$ itself. In this respect, the dual of a code represents the null space of the previous vector space (Definition A.30). In fact (5.4) coincides with (A.23), after the substitutions $\underline{G} = \underline{A}$ and $\underline{H}^T = \underline{A}_\perp^T$.

Before concluding the present Section, it is useful to remark how a row in $\underline{G}$ or in $\underline{H}$, made by just one 1-symbol, is not advisable, although from different points of view. When the row with just one 1-symbol belongs to $\underline{G}$, it means that the minimum distance of the code is $d = 1$, and, in particular, the correspondent $\underline{H}$ matrix will show a null column in the considered position, demonstrating how it is not protected at all (see Lemma 5.1). When the row with just one 1-symbol belongs to $\underline{H}$, it means that in the considered position all the code words exhibit a 0-symbol, and the correspondent $\underline{G}$ matrix will show a null column in that position (which, as already observed, can be eliminated). This behaviour confirms duality.

5.4 Polynomial Description of the Parity Check Matrix in a Linear Block Code

The parity check polynomial $h(x)$ of a cyclic code has been introduced in Definition 1.21 as the co-factor of the generator polynomial $g(x)$ in the factorization of $(x^N + 1)$. So we have

$$g(x)h(x) = x^N + 1. \tag{5.8}$$

Keeping in mind property (1.13), characterizing all the code words in a cyclic code, and considering that $g(x)$ is itself a code word, we can write

$$x^j g(x)h(x)|_{\text{mod}(x^N+1)} = (x^N + 1)|_{\text{mod}(x^N+1)} = 0. \tag{5.9}$$

Lemma 5.12 *For a cyclic code, it is possible to express the parity check matrix* $\underline{H}$ *in an* s.s. *time-invariant form, where the reciprocal of the parity check polynomial,* $h^*(x)$*, shifts by one position each row.*

Proof Property (5.9) assures that (5.1) is satisfied. Equating to zero all the terms in the above polynomial product (which expresses a temporal convolution, as shown by Theorem B.1) entails to have null result for a number of vector products equal $\deg[g(x)] = r$. These r equations are those needed for a sequential determination of all the control symbols c_i, $i = 1, 2, \ldots r$. Taking the reciprocal of $h(x)$ is necessary because in a temporal convolution one of the two sequences is reversed, whereas in a vector product the two sequences are both ordered from left to right. □

Remark that, if we accept Definition 5.9 for code duality, it would be necessary to substitute $h^*(x)$ to $h(x)$ in Definition 1.22, since in the rows of $\underline{H}$ matrix, as established by Lemma 5.12, polynomial $h^*(x)$ shifts. Nevertheless, since the codes obtained after replacing $g(x)$ with $g^*(x)$ or $h(x)$ with $h^*(x)$ are equivalent, this correction will be ignored.

$$\underline{G} = \left[\begin{array}{ccccccccccccccc}
1&0&1&0&0&1&1&0&1&1&1&&&& \\
&1&0&1&0&0&1&1&0&1&1&1&&& \\
&&1&0&1&0&0&1&1&0&1&1&1&& \\
&&&1&0&1&0&0&1&1&0&1&1&1& \\
&&&&1&0&1&0&0&1&1&0&1&1&1
\end{array}\right] \qquad g(x) = 10100110111$$

$$\underline{H} = \left[\begin{array}{ccccccccccccccc}
1&1&0&1&0&1&&&&&&&&& \\
&1&1&0&1&0&1&&&&&&&& \\
&&1&1&0&1&0&1&&&&&&& \\
&&&1&1&0&1&0&1&&&&&& \\
&&&&1&1&0&1&0&1&&&&& \\
&&&&&1&1&0&1&0&1&&&& \\
&&&&&&1&1&0&1&0&1&&& \\
&&&&&&&1&1&0&1&0&1&& \\
&&&&&&&&1&1&0&1&0&1& \\
&&&&&&&&&1&1&0&1&0&1
\end{array}\right] \qquad h(x) = 101011$$

Fig. 5.5 Generator matrix and parity check matrix of the [15,5,7] BCH code

Example 5.13 The [15,5,7] BCH code, whose generator and parity check polynomials respectively are

$$g(x) = x^{10} + x^8 + x^5 + x^4 + x^2 + x + 1, \; h(x) = x^5 + x^3 + x + 1,$$

is described by the generator matrix and by the parity check matrix shown in Fig. 5.5. We can observe that $g(x)$ shifts in the rows of $\underline{G}$, whereas $h^*(x)$ shifts in the rows of $\underline{H}$.

After property (1.13), in the $\underline{G}$ matrix of a cyclic code, any cyclic permutation of the generator polynomial $g(x)$ can be assumed as a row, provided that such a row is linearly independent of the other ones.

Lemma 5.13 *In the $\underline{H}$ matrix of a cyclic code, any cyclic permutation of $h^*(x)$ can be assumed as a row, provided that such row is linearly independent of the other ones.*

Proof In (5.9), it is possible to consider x^j either applied to $g(x)$, or applied to $h(x)$, or even in part to the former and in the remaining part to the latter. □

Lemma 5.14 *Also for shortened cyclic codes (introduced in Sect.* 1.10, *and briefly called SC codes), property* (5.9) *holds, although some positions in the code word are eliminated. Instead, for lengthened cyclic codes (introduced in Sect.* 1.11, *and briefly called LC codes), property* (5.9) *is no longer true.*

Proof An SC code is characterized by r control symbols, exactly as its parent cyclic code. Therefore the null temporal convolution, expressed by (5.9), remains valid for assuring satisfaction of (5.1). In particular, for a one-time shortening, while the

$\underline{G}$ matrix looses one row (and one column), in the $\underline{H}$ matrix all the rows are still present, and one column is to be eliminated (this behaviour will be better explained in Sect. 5.8). For an LC code, owing to the presence of 2-weight code words of the type $(x^N + 1)$, $h^*(x)$ in the $\underline{H}$ matrix will exhibit periodic replicas, and its product with $g(x)$ does not admit cyclic permutations (see Lemma 1.35). □

5.5 Encoder Circuit Based on the Parity Check Polynomial and Its State Diagram

Lemma 5.15 *In Fig.* 5.6 *an electronic circuit, able to perform systematic encoding of a cyclic code characterized by the parity check polynomial* $h(x)$ *is shown. The switch is maintained for* $k = N - r$ *symbol times connected to the information source, and then, for* r *symbol times, to the feed-back wire* (MacWilliams and Sloane 1977).

Proof This circuit makes in practice the r equations present in (5.1), under the assumption that the $\underline{H}$ matrix exhibits the s.s. time-invariant structure described in Lemma 5.12. More precisely, it allows calculation of the control symbols c_i, $i = 1, 2, \ldots r$, in sequence. □

As usual, polynomial $i(x)$ contains information symbols and polynomial $c(x)$ control symbols. The absence of the wire denoted $h_{N-r} = 1$ is due to the fact that this coefficient multiplies firstly the current information symbol outgoing from the source, and then the current ex-OR sum of the preceding control symbols.

Definition 5.10 In an encoder circuit of the type depicted in Fig. 5.6, the contents of the $(N - r)$ memory elements of the shift register form an $(N - r)$-symbol sequence, called its *state*. There are 2^{N-r} states on the whole. Each one of the $(N - r)$ positions of the state is called *state co-ordinate*.

Remark the analogy with Definition 1.32, given for an encoder circuit constructed exploiting the generator polynomial, and in particular able to calculate a

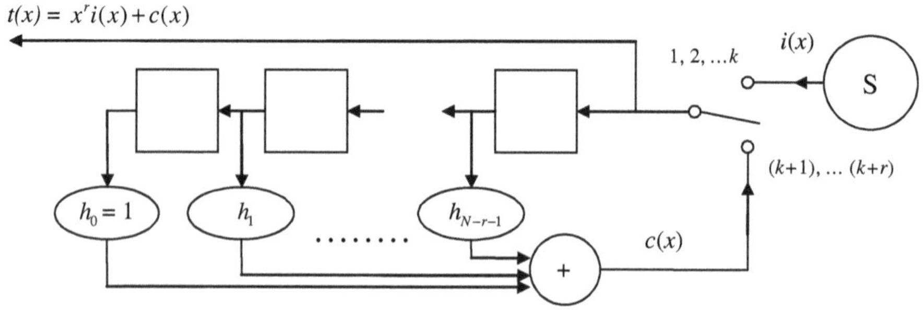

Fig. 5.6 Systematic encoder circuit based on a shift register constructed exploiting the parity check polynomial for a cyclic code (and also other codes derived from it)

polynomial division quotient. Also here the associated state diagram is a De Bruijn graph (De Bruijn 1946), showing all the possible transitions from one state to another. Here too, one state co-ordinate is enough for describing this evolution, because all the state co-ordinates show the same state sequence, although time shifted. In this respect, we could define this scheme as a controller arrangement. The apparent only difference is in the fact that here the numbers of nodes and branches are respectively 2^{N-r} nodes and 2^{N-r+1} branches. In all these considerations, since we are treating cyclic codes only, parameter k coincides with $(N - r)$. Nevertheless there is another important difference: here in a state co-ordinate we can found directly the whole code word, in fact branch labelling imposes, by itself, the last state co-ordinate in any state touched by the path in the state diagram.

Starting from state $\underline{0}$, after having reached a certain state, owing to polynomial $i(x)$, the remaining part of the code word follows a path internal to the state diagram without coming back to state $\underline{0}$. The structure of a certain state diagram (included branch labelling) can be employed for all the codes characterized by parity check polynomials having the same degree $(N - r)$, although different. The path in the state diagram, instead, depends on the particular parity check polynomial. Closed routes, internal to the state diagram, represent the code word patterns and then identify the considered code. Such routes are closed just because we are treating, for the moment, cyclic codes only. Finally it is possible to verify that it is not necessary to start from state $\underline{0}$. In fact reaching of any state after a proper $i(x)$ is possible not only from state $\underline{0}$, but also from any other state. So we prefer to imagine that it is reached starting from a state already in the closed route associated to the code word considered, k symbol times before, in such a way as to form the complete closed route. In other terms, the encoder circuit in Fig. 5.6 is able to begin any encoding procedure starting from whichever state.

Example 5.14 In Fig. 5.7 the state diagram related to a systematic encoder of the type described so far, associated to the whole class of cyclic codes with $\deg[h(x)] = 3$, is shown. We will describe the encoding procedure for the [7,3,4] simplex code (see Example 1.12), which can be performed in this state diagram. Such code is characterized by one between two possible parity check polynomials, reciprocal each other, $h(x) = x^3 + x + 1$ or $h(x) = x^3 + x^2 + 1$. The corresponding generator polynomials respectively result in $g(x) = x^4 + x^2 + x + 1$ or $g(x) = x^4 + x^3 + x^2 + 1$ (see Table 1.5). For this pair of codes, it is particularly simple to recognize the closed routes followed by the 7 non-null code words, since they all derive from a unique pattern, coincident with $g(x)$, after a complete cyclic permutation. The three information symbols present in $i(x)$, starting initially from state $\underline{0}$, enables one to reach whichever non-null state. Then the subsequent four branches are chosen following one of two possible closed routes (one the reciprocal of the other, and denoted by dashed or dotted lines in the Figure). These two patterns are both regular patterns of length 7 (see Definition 1.20). Remark that any state can be reached from any other, after 3 symbol times, so subsequent code words can be encoded without the need of passing through state $\underline{0}$ from one to another.

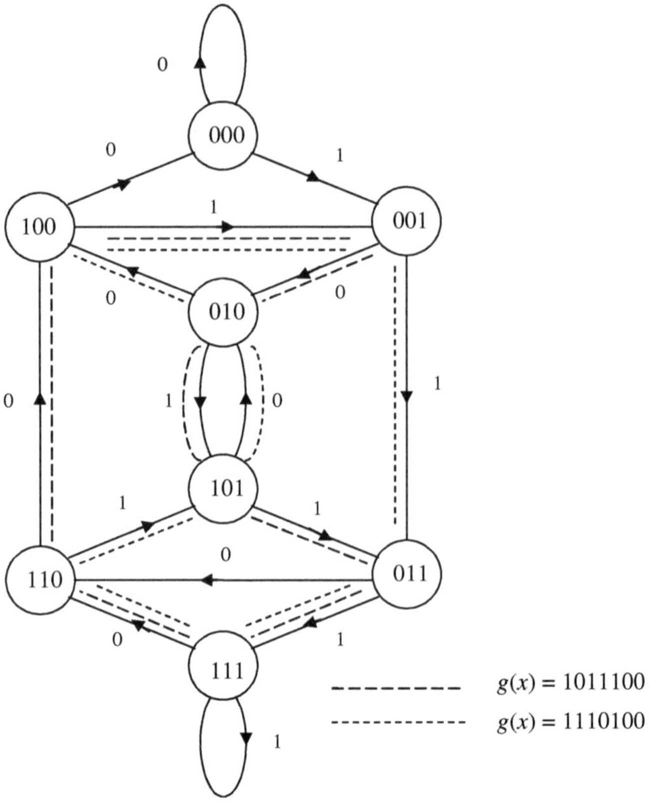

Fig. 5.7 State diagram for the whole class of cyclic codes with deg[$h(x)$] = 3, and closed routes giving the 7 non-null words of a [7,3,4] simplex code (two possible parity check polynomials)

Example 5.15 In Fig. 5.8 the same state diagram is employed now for describing the encoding procedure in the case of the [4,3,2] parity check code, whose generator polynomial is $g(x) = x + 1$. Here we have one regular pattern (1100), and two degenerate patterns (1010 and 1111). They correspond to closed routes of length respectively 4 (in dashed line), 2 (in dotted line), and 1 (in dashed-dotted line). All the 7 non-null code words are so obtained. Remark that, also here, we can begin the path either from state $\underline{0}$ or from whichever state along the closed route. This property is independent of the particular code considered.

If N is composite, it is possible to have degenerate closed routes (whose lengths are factors of N) corresponding to code words associated to degenerate patterns. This is the case, for instance, treated in Example 5.15.

In studying the state diagram associated to the systematic encoder circuit, based on the generator polynomial $g(x)$, in Sect. 1.14, we observed that closed routes exist, having length N, whose branches are denoted by as many consecutive 0-symbols. Typically there are several closed routes of this type. They originate

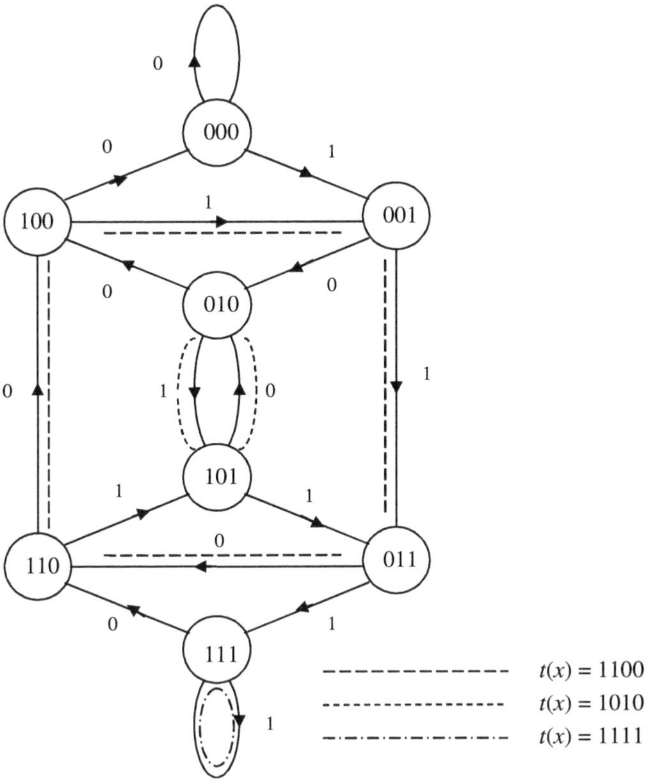

Fig. 5.8 State diagram for the whole class of cyclic codes with $\deg[h(x)] = 3$, and closed routes expressing the 7 non-null words of the [4,3,2] parity check code

code words whose polynomial form is $a(x)(x^N + 1)$, characterizing an LC code, and correspond to paths starting from and arriving at state $\underline{0}$ in the state diagram based on $g(x)$, as usual. About the state evolution on such 0-labelled closed routes now we can add the following property.

Lemma 5.16 *Closed routes of length N, characterized by N consecutive* 0*-symbols, in the state diagram of a systematic encoder circuit based on* $g(x)$, *exhibit a state evolution which expresses the code words of the dual of the considered cyclic code.*

Proof We remember that polynomial $u(x)$ in (1.3) coincides with the state evolution in a state diagram based on $g(x)$. So the state evolution on a closed route characterized by N consecutive 0-symbols, turns out to be the division of $a(x)(x^N + 1)$ by $g(x)$. Therefore it will be $a(x)h(x)$. This is an expression of the type (1.3) for a code word in the dual of the considered code, where $a(x)$ replaces $u(x)$ and $h(x)$ replaces $g(x)$. Any cyclic permutation of $a(x)h(x)$ is still a code word of this dual code, owing to the character of being a cyclic code. This property holds also for degenerate closed routes. □

Theorem 5.2 *In the state diagram of a systematic encoder circuit based on h(x), the closed paths where the code words are formed coincide with the closed paths made by N consecutive 0-symbols (or by a factor of N when degenerate) in the state diagram of an encoder circuit constructed with the generator polynomial g(x) of the dual of the considered code.*

Proof After Lemma 5.16, on such closed routes the code words of the dual of the considered code are formed. Then, if the state diagram refers to the dual code, on such closed routes $u(x)g(x)$ will appear. The dual of the dual of a code coincides with it. On the other hand, the state evolution along a path in a state diagram based on polynomial $h(x)$ is equal to the correspondent branch label sequence. Therefore such closed paths are those related to the N consecutive 0-symbols in the state diagram based on the generator polynomial of the dual of the considered code, taking into account also possible pattern degenerations in both. □

Example 5.16 For a [7,4,3] Hamming code, the state diagram of a systematic encoder constructed with $g(x)$ shows a unique 7-length closed route where the state evolution is $h(x)$ (see Fig. 1.15). Its dual, that is the [7,3,4] simplex code, is characterized by the same closed route, where its unique pattern (regular) forms all the non-null code words of such code, as discussed in Example 5.14 (Fig. 5.7), treating a state diagram based on $h(x)$.

Example 5.17 For a [4,1,4] repetition code, we have a state diagram based on polynomial $g(x)$, showing three closed routes with length respectively 4, 2, 1 (see Fig. 1.17). The first is regular and gives rise to $t(x) = 10001$, the remaining two are degenerate, and precisely correspond to $t(x) = 110011$ and $t(x) = 1010101$. Respectively we have $a(x) = 1$, $(x + 1)$, $(x^2 + 1)$. The correspondent result has been obtained for its dual, the [4,3,2] parity check code, in Example 5.15 (Fig. 5.8), exploiting its state diagram based on $h(x)$. Here the code words are 0011, 1011, 1100, 0110 on the first, 1010, 0101 on the second, 1111 on the third.

The code considered so far are in their cyclic form. For punctured cyclic codes (see Definition 1.17), the following statement holds.

Corollary a *Considering a single code puncturation, performed on the last control symbol of the code word characterizing the original cyclic code, the procedure for obtaining it, by means of the encoder circuit shown in Fig. 5.6, consists of stopping control symbol calculations after the $(r - 1)$th control symbol. In the correspondent state diagram, what described in Theorem 5.1 occurs as well, but the closed route selected by the information sequence is not completely covered, and the last branch is ignored.*

Proof A punctured code, characterized by $n = N - 1$ and still $k = N - r$, of the type here considered, exhibits all code words of the original code with the last symbol suppressed. The procedure for obtaining the control symbols, by means of the systematic encoder circuit shown in Fig. 5.6 is the same as that for the parent cyclic code, but the switch remains connected to the feed-back wire one symbol time less.

$$\underline{G} = \begin{array}{|ccccccc|} 1 & 1 & 1 & 0 & 1 & & \\ & 1 & 1 & 1 & 0 & 1 & \\ & & 1 & 1 & 1 & 0 & 1 \end{array} \qquad \underline{H} = \begin{array}{|ccccccc|} 1 & 0 & 1 & 1 & & & \\ & 1 & 0 & 1 & 1 & & \\ & & 1 & 0 & 1 & 1 & \\ & & & 1 & 0 & 1 & 1 \end{array}$$

$$\underline{G} = \begin{array}{|ccccc|} 1 & 1 & 1 & 0 & 1 \\ & 1 & 1 & 1 & 0 \\ & & 1 & 1 & 1 \end{array} \qquad \underline{H} = \begin{array}{|ccccc|} 1 & 0 & 1 & 1 & \\ & 1 & 0 & 1 & 1 \end{array}$$

Fig. 5.9 Generator matrix and parity check matrix of the [7,3,4] simplex code (*up*) and of the two-time punctured [5,3,2] code obtained from it (*down*)

In the state diagram, this means to perform an incomplete run of the closed route reached by $i(x)$. In this incomplete path the last branch is lost. □

Clearly the above procedure can be repeated for more than one subsequent single puncturation, provided that puncturation is performed and iterated starting from the end of the code word. The lost branches will be as many as the single puncturations performed.

Example 5.18 In Example 5.14 the [7,3,4] simplex code has been treated. We will consider here the generator polynomial $g(x) = x^4 + x^3 + x^2 + 1$ and the parity check polynomial $h(x) = x^3 + x^2 + 1$, that is the closed route denoted with a dotted line in the state diagram shown in Fig. 5.7. The generator matrix and the parity check matrix of such code are shown up in Fig. 5.9. A two-time punctured code, starting from it, is a [5,3,2] code, whose generator matrix and parity check matrix are reported down in the same Figure. It is possible to verify that the three independent code words in the new $\underline{G}$ matrix are respectively: the one starting from state 100, the one staring from state 010, the one starting from state 101, all of length 5, on same closed route. Comments on the transformation occurring in the parity check matrices, after puncturation, will be given in Sect. 5.7.

In conclusion, code puncturation can be directly performed during the encoding procedure, by means of the systematic encoder circuit shown in Fig. 5.6, since $k = N - r$ remains constant. In code shortening and code lengthening, k is to be changed, reducing or increasing it with respect to the cyclic reference value $k = N - r$. Code shortening can be also performed by means of the same encoder circuit. Code lengthening, instead, is impossible with the same encoder circuit. The impact on $\underline{H}$ matrix of code shortening and code lengthening will be described in Sects. 5.8 and 5.9, together with the possibility of employing the encoder circuit scheme shown in Fig. 5.6.

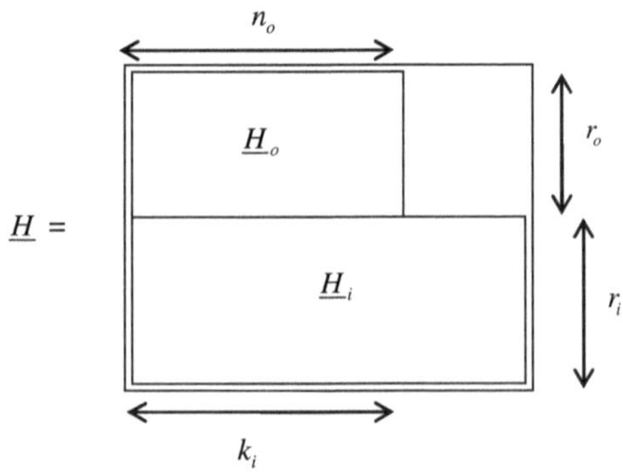

Fig. 5.10 Parity check matrix of the resulting code in a direct serial concatenation

5.6 Code Concatenation: Effects on $\underline{H}$ Matrix

In Sect. 1.6 we introduced different types of code concatenation (parallel, serial, direct, with interleaver, …). Our attention was devoted to evaluate the effects of these procedures on the generator matrix. Here we will consider the effects on the parity check matrix. Let us analyze firstly direct serial concatenation.

Lemma 5.17 *In a direct serial concatenation (see Definitions* 1.13 *and* 1.14) *of two block codes, characterized by the parity check matrices* $\underline{H}_i$ *(for the inner code) and* $\underline{H}_o$ *(for the outer code), the parity check matrix* $\underline{H}$ *of the concatenated code is made by adding the rows of* $\underline{H}_i$ *to the rows of* $\underline{H}_o$*, as depicted in Fig.* 5.10.

Proof We have $k_i = n_o$, so the inner code is requested to add its control symbols at the end of the code word. All the r_i rows of $\underline{H}_i$ are to be added to all the r_o rows of $\underline{H}_o$. Considering that the rows of $\underline{H}_i$ act on the whole length of the words of the resulting concatenated code, the matrix arrangement shown in the Figure appears the most intuitive possible. □

Example 5.19 The procedure of code extension, which is a direct serial concatenation with the inner code consisting of an $[n_i, n_i - 1, 2]$ parity check code (see Definition 1.16), can be obtained by means of the addition of just one row, containing a unitary polynomial $u_{n_i}(x) = u_{n_o+1}(x)$. For instance, the following parity check matrix describes an [8,4,4] extended Hamming code, having assumed $h_o(x) = x^4 + x^2 + x + 1$,

$$\underline{H} = \left[\begin{array}{cccccccc} 1 & 1 & 1 & 0 & 1 & & & \\ & 1 & 1 & 1 & 0 & 1 & & \\ & & 1 & 1 & 1 & 0 & 1 & \\ \hline 1 & 1 & 1 & 1 & 1 & 1 & 1 & 1 \\ & & & & & & & \end{array}\right].$$

$$\underline{H} = \left[\begin{array}{ccccccccccccccccccccc}
1&0&0&1&1&0&1&0&1&1&1&1&&&&&&&&& \\
&1&0&0&1&1&0&1&0&1&1&1&1&&&&&&&& \\
&&1&0&0&1&1&0&1&0&1&1&1&1&&&&&&& \\
&&&1&0&0&1&1&0&1&0&1&1&1&1&&&&&& \\
\hline
1&1&1&1&0&1&1&1&0&0&1&1&0&1&0&1&&&&& \\
&1&1&1&1&0&1&1&1&0&0&1&1&0&1&0&1&&&& \\
&&1&1&1&1&0&1&1&1&0&0&1&1&0&1&0&1&&& \\
&&&1&1&1&1&0&1&1&1&0&0&1&1&0&1&0&1&& \\
&&&&1&1&1&1&0&1&1&1&0&0&1&1&0&1&0&1& \\
&&&&&1&1&1&1&0&1&1&1&0&0&1&1&0&1&0&1
\end{array}\right]$$

$$\underline{G} = \left[\begin{array}{ccccccccccccccc|cccccc}
1&1&0&0&1&&&&&&&&&&&0&0&1&1&0&1 \\
&1&1&0&0&1&&&&&&&&&&1&0&1&1&1&1 \\
&&1&1&0&0&1&&&&&&&&&1&1&1&1&1&0 \\
&&&1&1&0&0&1&&&&&&&&0&1&1&1&1&1 \\
&&&&1&1&0&0&1&&&&&&&1&0&0&1&1&0 \\
&&&&&1&1&0&0&1&&&&&&0&1&0&0&1&1 \\
&&&&&&1&1&0&0&1&&&&&1&0&0&0&0&0 \\
&&&&&&&1&1&0&0&1&&&&0&1&0&0&0&0 \\
&&&&&&&&1&1&0&0&1&&&0&0&1&0&0&0 \\
&&&&&&&&&1&1&0&0&1&&0&0&0&1&0&0 \\
&&&&&&&&&&1&1&0&0&1&0&0&0&0&1&0
\end{array}\right]$$

Fig. 5.11 Parity check matrix and generator matrix for the direct serial concatenation described in Example 5.20

In Example 1.22 the generator matrix of the same code has been obtained. Remark that the [8,4,4] extended Hamming code cannot be organized in s.s. time-invariant polynomial form, neither in $\underline{G}$ nor in $\underline{H}$. It is possible to construct, instead, several equivalent quasi-cyclic forms for this code, as shown in Sects. 2.2 and 2.4.

Example 5.20 The direct serial concatenation of an outer Hamming code [15,11,3], characterized by generator polynomial $g_o(x) = x^4 + x^3 + 1$, that is 11001, with an inner [21,15,4] code characterized by generator polynomial $g_i(x) = (x^3 + x + 1)(x^3 + x + 1)(x + 1) = x^6 + x^4 + x + 1$, that is 1010011, yields a [21,11,4] code, whose parity check matrix turns out to be the one reported up in Fig. 5.11. It is possible to recognize, in this matrix, the reciprocal of the parity check polynomials associated to the two cyclic codes considered, the outer one in the first four rows, and the inner one in the last six rows. In particular, they are calculated as co-factors with respect to the two binomials $(x^{15} + 1)$ and $(x^{21} + 1)$ respectively. The generator matrix of such concatenated code, instead, obtained as suggested in Sect. 1.6, results in the matrix reported down in Fig. 5.11. It is useful to remember that, in the right hand side part of this matrix, the remainders of the division by $g_i(x)$, calculated for each one of the preceding polynomials, are collected.

After the above Examples of direct serial code concatenation, we can interpret the subsequent rows in a parity check matrix, even for a single code, as a direct serial concatenation of as many parity check codes, acting on a selection of the symbol positions in the overall code word. This interpretation explains the denomination of parity check matrix. The last 1-symbol of each row denotes the

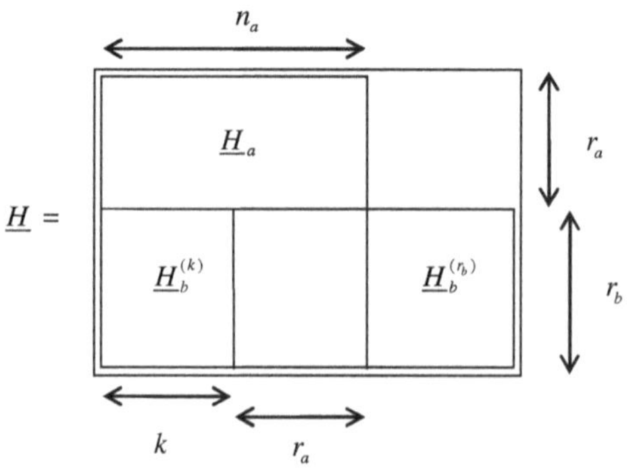

Fig. 5.12 Parity check matrix of the resulting code in a direct parallel concatenation

position where the correspondent control symbol is expected to be. For this reason it is preferable that the parity check matrices, when possible, are taken in a lower triangular form. We can observe that, if this condition is satisfied, for either the $\underline{H}_i$ or the $\underline{H}_o$ matrix, the resulting $\underline{H}$ matrix describing the direct serial concatenation is lower triangular as well.

Lemma 5.18 *Considering a direct parallel concatenation of two block codes (see Definition* 1.12), *respectively characterized by* $n_a = k + r_a$, *and by* $n_b = k + r_b$, *the resulting parity check matrix will appear as shown in Fig.* 5.12, *where the parity check matrix* $\underline{H}_b$ *of the second code is subdivided into two parts: that on the left,* $\underline{H}_b^{(k)}$, *which is an* $r_b \times k$ *matrix, and that on the right,* $\underline{H}_b^{(r_b)}$, *which is a square* $r_b \times r_b$ *matrix. These two parts are separated, in the lower layer of the overall* $\underline{H}$, *by an* $r_b \times r_a$ *null matrix.*

Proof The second code has to ignore the control symbols introduced by the first one. Its control symbols, contained in the $\underline{H}_b^{(r_b)}$ sub-matrix, then, are reported at the end of the code word, as in the case of the serial concatenation. □

Example 5.21 Let us consider the two equivalent cyclic versions of the same [7,4,3] Hamming code, respectively associated to $g_a(x) = x^3 + x + 1$ and to $g_b(x) = x^3 + x^2 + 1$. In Fig. 5.13 the $\underline{G}$ matrix and the $\underline{H}$ matrix for the composite code obtained from their parallel concatenation are reported. The resulting code is a [10,4,4] code. The 4 × 3 sub-matrix on the right hand side of the $\underline{G}$ matrix has been

$$\underline{G} = \left[\begin{array}{cccc|ccc|ccc} 1&0&1&1& & & &1&0&0\\ &1&0&1&1& & &1&1&0\\ & &1&0&1&1& &1&1&1\\ & & &1&0&1&1&1&0&1 \end{array}\right] \qquad \underline{H} = \left[\begin{array}{cccc|ccc|ccc} 1&1&1&0&1& & & & & \\ &1&1&1&0&1& & & & \\ & &1&1&1&0&1& & & \\ \hline 1&0&1&1& & & &1& & \\ &1&0&1& & & &1&1& \\ & &1&0& & & &1&1&1 \end{array}\right]$$

Fig. 5.13 Generator matrix and parity check matrix for the code in direct parallel concatenation, described in Example 5.21

obtained from the remainder of the division by $g_b(x)$ of the only information symbol sequences. Also in this case, the resulting $\underline{H}$ matrix, constructed as described in Fig. 5.12, turns out to be in lower-triangular form.

All the parity check matrices discussed in the previous Examples, although obtained from codes which are s.s. time-invariant in their $\underline{H}$ matrix, will be, in general, time-varying. The schemes reported in Figs. 1.7 and 1.8, respectively for a parallel concatenation and for a serial concatenation can be employed also here. The only difference is in replacing the encoder circuits of the type depicted in Fig. 1.5, with the ones here proposed starting from the parity check polynomial (Fig. 5.6). Both such encoder circuits are systematic.

So far we have considered only the concatenation of single code words, avoiding to treat the case of ordered ensembles of code words. Such treatment, with particular attention to product codes, direct or with an interleaver, will be faced in Sect. 5.16.

5.7 Code Puncturation: Effects on $\underline{H}$ Matrix

Let us consider a polynomial block code in its cyclic version. It is s.s. time-invariant in the $\underline{H}$ matrix, that is with a unique reciprocal parity check polynomial $h^*(x)$ shifting along the r rows of $\underline{H}$, and also s.s. time-invariant in the generator matrix. In fact $h(x)$ corresponds to a unique generator polynomial $g(x)$, representing its co-factor in the decomposition of the binomial $(x^N + 1)$. This generator polynomial shifts along the $(N - r)$ rows of $\underline{G}$. In this situation $\underline{H}$ is lower (and upper) triangular, whereas $\underline{G}$ is upper (and lower) triangular. Information and control symbols have positions in the code word which can be easily identified: respectively the first $(N - r)$, looking at the $\underline{G}$ matrix, and the last r, looking at the $\underline{H}$ matrix. Furthermore, the systematic encoder circuits shown in Fig. 1.5 or in Fig. 5.6 can be indifferently employed.

In order to maintain the property that $\underline{G}$ is upper triangular, in code puncturation, it is convenient to eliminate final rows of $\underline{H}$ (and the correspondent columns), like shown in Fig. 5.9 (Example 5.18), where a two-time punctured code has been described. Code puncturation has been introduced in Sect. 1.7 showing that the resulting $\underline{G}$ matrix necessarily becomes time-varying. In Sect. 5.5, Corollary a of Theorem 5.2 deals with punctured cyclic codes, and establishes that they can be encoded by means of a systematic encoder circuit of the type shown in Fig. 5.6, simply after interruption of the control symbol collection from the feed-back wire. This fact has two main important consequences: it is congruent with the choice of eliminating the last control symbols (last columns in $\underline{G}$), and it maintains $\underline{H}$ in an s. s. time-invariant form, lower (and upper) triangular.

Anyway elimination of an external symbol in a single puncturation, that is the first or the last row in $\underline{H}$, gives rise to the same $[N - 1, N - r, d]$ one-time punctured code, often with $d < d_c$, where d_c is the minimum distance value for the parent

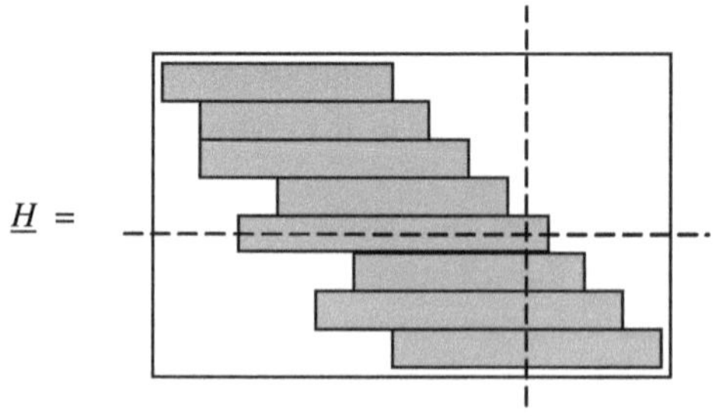

Fig. 5.14 Single code puncturation by elimination of one row and one central column in the $\underline{H}$ matrix

cyclic code. Nevertheless a quite different result comes out when, after application of a puncturation matrix $\underline{P}$ having a very general form (see Sect. 1.7 for further details), a central symbol is to be eliminated. In order to have, also in this case, the tern $[N - 1, N - r, d]$, a row and a column must be eliminated together in the original $\underline{H}$ matrix. It is advisable to eliminate the column where the control symbol of the considered row is located. This is possible when $\underline{H}$ is in lower triangular form (not necessarily time-invariant in $\underline{H}$). In fact, taking into account the observation, reported in the previous section, about the last 1-symbol of each row, where the correspondent control symbol can be assumed to be placed, we are induced to eliminate the column in that position.

Such choice is qualitatively described in Fig. 5.14 for an $\underline{H}$ matrix originally in lower triangular form, but not time-invariant. We have considered rows of different lengths (depicted as grey rectangles), one of which is suppressed, together with the column correspondent to the rightmost 1-symbol of the considered row. Row and column suppressions are represented by dashed lines. This type of code puncturation, when made on an s.s. time-invariant code in its $\underline{H}$ matrix, gives rise to a time-varying code in $\underline{H}$ (and also in $\underline{G}$).

The code puncturation made eliminating a central control symbol, like the one described above, although made on an s.s. time-invariant code in its $\underline{H}$, cannot be constructed by means of an encoder circuit of the type shown in Fig. 5.6. In fact elimination of a central column of $\underline{H}$ changes several rows, so that the invariance of $h^*(x)$, which is a fundamental assumption for allowing circuit validity, is no longer true. It is always possible, however, to encode the whole code word of the cyclic code and then eliminate the symbols in the punctured positions.

A completely different type of code puncturation is possible, still regarding a central position in the code word. It is performed imposing that the new code is characterized by the same value for the code word length (N in the cyclic case, n in more general situations), and by a reduced number of control symbols. It influences central rows of $\underline{H}$, but the number of information symbols is forced to increase, in order to compensate the reduction in the number of control symbols.

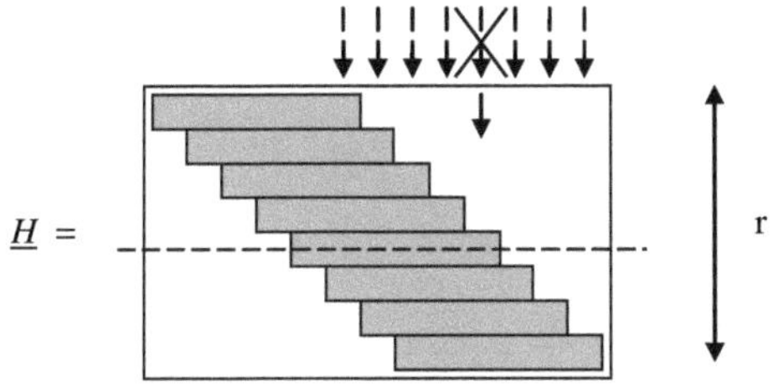

Fig. 5.15 Constant-length puncturation and change in symbol type according to Lemma 5.19

Definition 5.11 The elimination of one central row in an s.s. time-invariant $\underline{H}$ matrix, without eliminating any column, is defined *constant-length puncturation* for a code which is *s.s. time-invariant in its parity check matrix.*

The last specification is important, since, in Chap. 6, we will extend this Definition to the case of a code which is w.s. time-invariant in its parity check matrix. Nevertheless, in the following, we will use the procedure so described also for more general types of codes.

Lemma 5.19 *When a constant-length puncturation is applied to a polynomial code whose length is n, the new code is represented by the tern* $[n, n - r + 1, d]$. *If the original* $\underline{H}$ *matrix was lower triangular, the position occupied by the rightmost 1-symbol in the eliminated row, instead of being a control symbol position, becomes an information symbol position. If the original code is* s.s. *time-invariant in* $\underline{H}$, *the systematic encoder circuit shown in Fig.* 5.6 *is able to encode the new code, simply by turning the switch on the wire connected to the information source during the correspondent symbol time.*

Proof In Fig. 5.15 an s.s. time-invariant $\underline{H}$ matrix is depicted, on which the considered operation has to be performed. The eliminated row is denoted by a dashed line. The position where control symbols were located before the constant-length puncturation are denoted by dashed small arrows. The position in which one of them is changed into an information symbol is denoted by a cross and by the addition of a solid small arrow. Up to this point, the assumption that the original code was s.s. time-invariant in $\underline{H}$ has been not exploited, being sufficient that its $\underline{H}$ matrix was lower triangular. Assumption of s.s. time-invariance in $\underline{H}$ allows the possibility of encoding our code by means of the circuit shown in Fig. 5.6. Since the new parity check matrix remains s.s. time-invariant, this encoder circuit, based on a unique polynomial $h(x)$, will be able to encode also the new code. Simply the switch, in correspondence with the symbol time where the change in the type of symbol has occurred, is to be connected to the information source. Since any state of the state diagram associated to such encoder circuit is acceptable for beginning a new closed route, as discussed in Sect. 5.5, the encoding procedure can go on without any break. □

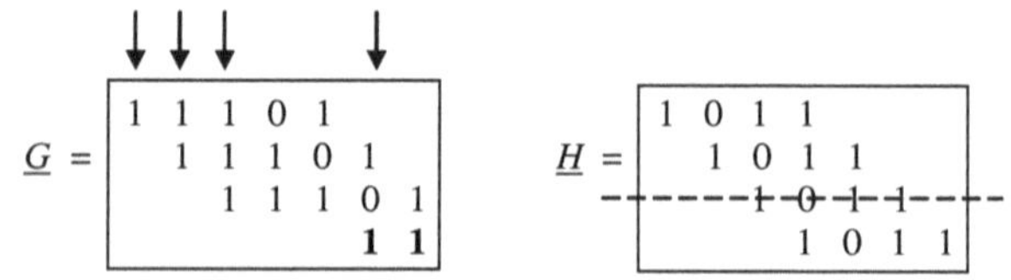

Fig. 5.16 Constant-length puncturation applied to a [7,3,4] simplex code

Nevertheless, if the switch does not remain connected to the feedback wire during all the r subsequent symbol times after a constant-length puncturation, the information symbol so introduced will be not protected as the preceding ones. This is the case qualitatively described in Fig. 5.15 as an example. So the minimum distance may decrease by far. In order to assure, also to this new information symbol, the same protection as that guaranteed to the other ones, it is necessary to increase the number of control symbols at the end of the code word with respect to the previous situation (and hence increase parameter n). This problem resembles that of a correct frame termination in a convolutional code (see Definition 3.2). In the following Sections we will show that such resemblance is truly justified.

Example 5.22 In Example 5.18 we treated a [7,3,4] simplex code, and a [5,3,2] two-time punctured code obtained from it, operating puncturation on the last two symbols. Now we apply a constant-length single puncturation to the same original simplex code, obtaining a [7,4,2] code, whose parity check and generator matrix become as shown in Fig. 5.16. The eliminated row in $\underline{H}$ is the third one, whereas the added row in $\underline{G}$ is the last one, whose 1-symbols have been written in bold characters. The positions of the four information symbols are denoted by small solid arrows. The 2-weight code word, reported as a fourth row in $\underline{G}$, comes out immediately from a procedure for searching clusters of single-error syndromes having null ex-OR sum (according to Lemma 5.11), applied to the last two columns of $\underline{H}$. This fact makes not well protected the symbols close to the end of the code words, as anticipated. Nevertheless also a code word 0100100 will appear, after ex-OR sum of the last three rows in $\underline{G}$, showing a further point of weakness for such code.

In spite of these considerations, constant-length puncturations, when performed periodically with frequency not too high, can give rise to a good trade-off between code rate and performance, as will be better explained in following Sections. It represents an alternative approach to convolutional codes, starting from cyclic block codes, besides that offered by MLC codes.

5.8 Shortening Cyclic Codes: Effects on $\underline{H}$ Matrix

In Sect. 1.10 code shortening has been introduced with reference to the $\underline{G}$ matrix. When $\underline{G}$ is s.s. time-invariant, such operation implies elimination of external rows, so maintaining the property of being s.s. time-invariant for such matrix (besides

$$\underline{G} = \begin{bmatrix} 1 & 0 & 1 & 1 & & \\ & 1 & 0 & 1 & 1 & \\ & & 1 & 0 & 1 & 1 \end{bmatrix} \qquad \underline{H} = \begin{bmatrix} 1 & 1 & 0 & 1 & & \\ 1 & 1 & 1 & 0 & 1 & \\ & 1 & 1 & 1 & 0 & 1 \end{bmatrix} \qquad [6,3,3]$$

$$\underline{G} = \begin{bmatrix} 1 & 0 & 1 & 1 & \\ & 1 & 0 & 1 & 1 \end{bmatrix} \qquad \underline{H} = \begin{bmatrix} 1 & 0 & 1 & & \\ 1 & 1 & 0 & 1 & \\ 1 & 1 & 1 & 0 & 1 \end{bmatrix} \qquad [5,2,3]$$

Fig. 5.17 Generator matrix and parity check matrix for the [6,3,3] and [5,2,3] SC Hamming codes

upper and lower triangular). An s-time shortened cyclic code or briefly an SC code (Definition 1.24) is an $[N - s, N - r - s, d]$ block code, whose $\underline{G}$ matrix exhibits s external rows (and columns) eliminated with respect to that of the $[N, N - r, d_c]$ parent cyclic code. It is preferable to eliminate the first rows and columns in order to maintain $\underline{H}$, which necessarily becomes time-varying, in a lower triangular form. This choice, for instance, has been made in Example 5.4 (Fig. 5.3), although not starting from a cyclic parent code. Precisely the original code was a [9,4,4] code, obtained from a [15,4,8] simplex code, after six single puncturations. The first single shortening, associated to the matrix $\underline{H}'$, gives an [8,3,4] code.

Example 5.23 In Example 1.36, starting from the [7,4,3] Hamming code characterized by the generator polynomial $g(x) = x^3 + x + 1$, we considered its shortened versions [6,3,3] and [5,2,3]. In Fig. 5.17, the generator matrix and the parity check matrix for such two codes are shown.

Lemma 5.20 *In cyclic code shortening, if the first information symbols are suppressed, it is possible to employ the systematic encoder circuit depicted in* Fig. 5.6, *provided that the switch is connected to the information source just for k symbols times, instead of* $(N - r)$ *symbol times, with* $k < N - r$.

Proof As already noted in the proof of Lemma 5.19, any state of the state diagram associated to such encoder is acceptable for beginning a new closed route. □

Lemma 5.21 *The dual of an SC code is a punctured cyclic code, and vice versa.*

Proof The modification in the $\underline{H}$ matrix induced after code puncturation (external row elimination), is the same as the modification in the $\underline{G}$ matrix induced after code shortening (external row elimination). As a consequence, in both, there is an external column elimination. □

The only difference is in the fact that, preferably, in the $\underline{G}$ matrix, column elimination owing to code puncturation is made on the right hand side (in order to maintain $\underline{G}$ upper triangular), whereas, in the $\underline{H}$ matrix, column elimination owing to code shortening is made on the left hand side (in order to maintain $\underline{H}$ lower triangular).

Example 5.24 In Fig. 5.18 up the $\underline{G}$ and $\underline{H}$ matrix for a [7,4,3] Hamming code and a [6,3,3] shortened Hamming code are reported. In Fig. 5.18 down the $\underline{H}$ and the $\underline{G}$ matrix for a [7,3,4] simplex code and a [6,3,3] punctured simplex code are

$$\underline{G} = \begin{array}{|ccccccc|} 1 & 0 & 1 & 1 & & & \\ & 1 & 0 & 1 & 1 & & \\ & & 1 & 0 & 1 & 1 & \\ & & & 1 & 0 & 1 & 1 \end{array} \qquad \underline{H} = \begin{array}{|ccccccc|} 1 & 1 & 1 & 0 & 1 & & \\ & 1 & 1 & 1 & 0 & 1 & \\ & & 1 & 1 & 1 & 0 & 1 \end{array} \qquad [7,4,3]\ \text{H.}$$

$$\underline{G} = \begin{array}{|cccccc|} 1 & 0 & 1 & 1 & & \\ & 1 & 0 & 1 & 1 & \\ & & 1 & 0 & 1 & 1 \end{array} \qquad \underline{H} = \begin{array}{|cccccc|} 1 & 1 & 0 & 1 & & \\ 1 & 1 & 1 & 0 & 1 & \\ & 1 & 1 & 1 & 0 & 1 \end{array} \qquad [6,3,3]\ \text{sh. H.}$$

$$\underline{H} = \begin{array}{|ccccccc|} 1 & 0 & 1 & 1 & & & \\ & 1 & 0 & 1 & 1 & & \\ & & 1 & 0 & 1 & 1 & \\ & & & 1 & 0 & 1 & 1 \end{array} \qquad \underline{G} = \begin{array}{|ccccccc|} 1 & 1 & 1 & 0 & 1 & & \\ & 1 & 1 & 1 & 0 & 1 & \\ & & 1 & 1 & 1 & 0 & 1 \end{array} \qquad [7,3,4]\ \text{simp.}$$

$$\underline{H} = \begin{array}{|cccccc|} 1 & 0 & 1 & 1 & & \\ & 1 & 0 & 1 & 1 & \\ & & 1 & 0 & 1 & 1 \end{array} \qquad \underline{G} = \begin{array}{|cccccc|} 1 & 1 & 1 & 0 & 1 & \\ & 1 & 1 & 1 & 0 & 1 \\ & & 1 & 1 & 1 & 0 \end{array} \qquad [6,3,3]\ \text{punc. simp.}$$

Fig. 5.18 Generator and parity check matrix before and after shortening/puncturing of a pair of dual codes (Example 5.24)

reported. It is possible to observe the difference between the $\underline{H}$ matrix of the shortened Hamming code (lower triangular) and the $\underline{G}$ matrix of the punctured simplex code (upper triangular). This difference is unessential, and is useful just for having a clear knowledge about the positions of information and control symbols. The two [6,3,3] codes are equivalent, and hence they can be interpreted as a unique self-dual code.

When, in code shortening, instead of an external row, a central row of $\underline{G}$ is eliminated, without suppressing any column, that is maintaining the same code word length, the result we can expect is the addition of a central control symbol, and is what occurs in MLC codes (see Sect. 1.13). This behaviour, however, is possible just after a proper code lengthening. In analogous way, in code puncturation, instead of an external row, a central row of $\underline{H}$ can be eliminated, without suppressing any column, giving rise to a constant-length puncturation (see Definition 5.11). This operation, if made after a procedure which can be considered as the dual of code lengthening, can originate very interesting codes, described in their parity check matrix. This concept will be better investigated in Sects. 5.13 and 5.15.

5.9 Lengthening Cyclic Codes: Effects on $\underline{H}$ Matrix

A lengthened cyclic code (briefly LC code) has been introduced in Sect. 1.11, with reference to the generator matrix $\underline{G}$ of the original cyclic code. It is characterized by some further rows in $\underline{G}$, where the same generator polynomial $g(x)$ still shifts. Lemma 1.26, treating a q-time completely lengthened code, whose tern is

$[qN, N - r + (q - 1)N, 2]$, states that it can be considered cyclic again (although with an enlarged meaning). Therefore an LC code with n so constrained

$$(q - 1)N < n < qN, \tag{5.10}$$

can be interpreted as a shortened version of the $[qN, N - r + (q - 1)N, 2]$ cyclic code. Expression (1.25), obtained in Lemma 1.35, gives the parity check polynomial for an LC code in the condition described by (5.10). It is characterize by q repetitions of the parity check polynomial $h_c(x)$ of its cyclic parent code. These observations enable us to construct the parity check matrix of an LC code.

Lemma 5.22 *The parity check matrix of an LC code whose block length n satisfies* (5.10) *can be expressed by r rows, where the polynomial* $h^*(x) = h_c^*(x)u_q(x^N)$ *shifts by one position at a time. Truncation of such polynomial on the left hand side allows to have the resulting* $\underline{H}$ *matrix in a lower triangular form.*

Proof Considering the $[qN, N - r + (q - 1)N, 2]$ cyclic code, its parity check matrix can be schematically shown as depicted up in Fig. 5.19, for $q = 3$. This conclusion follows from Lemma 1.35. In Sect. 5.8 the $\underline{H}$ matrix for a shortened version of a cyclic code has been obtained by eliminating some of its columns. So, in the present case, we are left with the parity check matrix $\underline{H}'$ reported down in Fig. 5.19. Elimination of the columns on the left hand side enables us to maintain for $\underline{H}'$ a lower triangular form, as usual. Such elimination entails polynomial truncation on the left hand side. □

Example 5.25 Let us consider a [7,4,3] cyclic Hamming code, whose parity check polynomial is $h_c(x) = x^4 + x^3 + x^2 + 1$. The [18,15,2] LC code obtained from such parent cyclic code exhibits a parity check matrix so arranged

$$\underline{H} = \left[\begin{array}{cccc|ccccccc|ccccccc} 1 & 1 & & & 1 & 0 & 1 & 1 & 1 & & & 1 & 0 & 1 & 1 & 1 & & \\ 1 & 1 & 1 & & & 1 & 0 & 1 & 1 & 1 & & & 1 & 0 & 1 & 1 & 1 & \\ 0 & 1 & 1 & 1 & & & 1 & 0 & 1 & 1 & 1 & & & 1 & 0 & 1 & 1 & 1 \end{array}\right].$$

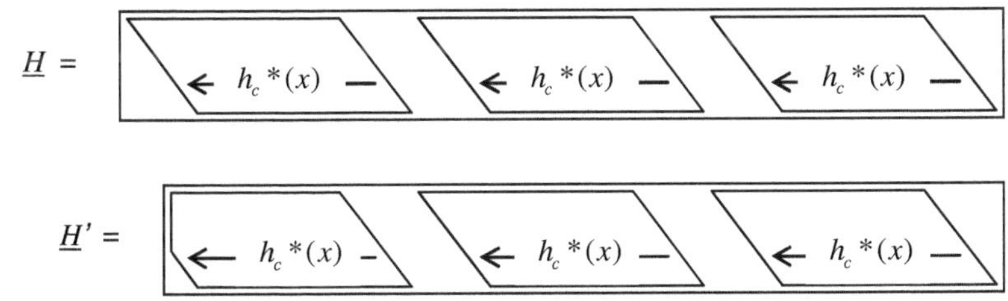

Fig. 5.19 Parity check matrix for a completely lengthened (cyclic) code, and for an LC code obtained by shortening it

On its left hand side three columns have been eliminated, starting from the parity check matrix of the completely lengthened (cyclic) [21,18,2] code. Vertical segments separate the three cyclic periods, whose length is $N = 7$, characterizing such LC code.

It is not convenient to employ the encoder circuit shown in Fig. 5.6 for an LC code. In fact the shift register would contain all the periodic parity check polynomial. In spite of this fact, this is what has been done in Example 5.15, treating the [4,3,2] parity check code. As already noted, it can be considered a lengthened code, starting from the [2,1,2] parent cyclic code characterized by $g(x) = x + 1$. Clearly it is much simpler to adopt an encoder circuit based on this generator polynomial, which is characterized by just one memory element.

5.10 MLC Codes (s.s. Time-Invariant Convolutional Codes in $\underline{G}$) and Their $\underline{H}$ Matrix

Modified lengthened cyclic codes (briefly MLC codes) have been introduced in Sect. 1.13, with reference to the generator matrix $\underline{G}$ of the original cyclic code. They are characterized by a length $n > N$ but exhibit proper central rows in $\underline{G}$ which have been periodically suppressed. In other terms, the same generator polynomial $g(x)$ shifts along the rows of $\underline{G}$, so maintaining the property to be s.s. time-invariant, but by more than one position at a time, according to a certain periodic behaviour. In Sect. 3.7 such codes have been demonstrated to be equivalent to s.s. time-invariant convolutional codes.

Fixing an MLC code period, whose length is n_o, in the last r_o positions of such period, polynomial $u(x)$ is forced to show as many 0-symbols (see Definition 1.31). In correspondence with such positions, distributed control symbols are inserted along the code word. In Sect. 1.14, interpreting $u(x)$ as the state evolution of the encoder circuit based on polynomial $g(x)$, the occurrence of 2-weight code words has been related to the possibility that $u(x)$, although constrained by the MLC code period, matches the parity check polynomial of the original cyclic code. Elimination of central rows in the $\underline{G}$ matrix without reducing code word length resembles analogous elimination of central rows in the $\underline{H}$ matrix without reducing code word length as well, operation introduced in Sect. 5.7 with the name of constant-length puncturation (see Definition 5.11). Then it could be useful to give a name also to the above operation on the $\underline{G}$ matrix, stressing duality.

Definition 5.12 The elimination of one central row in an s.s. time-invariant $\underline{G}$ matrix, without eliminating any column, is defined *constant-length shortening* for a code which is *s.s. time-invariant in its generator matrix.*

Constant-length shortening then gives rise to a subcode of the original code, whose tern exhibits a reduced overall parameter k, maintaining fixed n. Remark that traditional code shortening produces a reduction in k and in n of the same quantity.

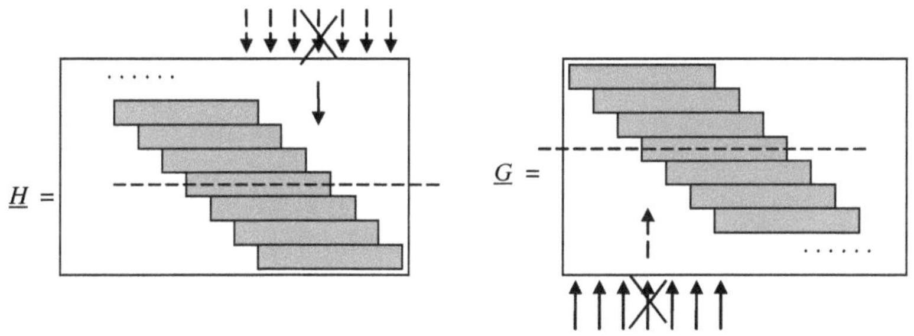

Fig. 5.20 Constant-length puncturation and constant-length shortening with transformation of a control symbol into an information symbol and vice versa

Finally we can observe that, when Definitions 5.11 and 5.12 are referred to an s.s. time invariant polynomial code (in $\underline{H}$ and in $\underline{G}$ respectively), it remains s.s. time-invariant in the matrix where this operation is made, and generally becomes time-varying in the other one. Furthermore the following statement holds.

Lemma 5.23 *The dual of an s.s. time-invariant code whose $\underline{H}$ matrix is subjected to a constant-length puncturation is an s.s. time-invariant code whose $\underline{G}$ matrix is subjected to a constant-length shortening, and vice versa.*

Proof The modification in the $\underline{H}$ matrix induced after code constant-length puncturation (central row elimination) is the same as the modification in the $\underline{G}$ matrix induced after constant-length shortening (central row elimination). No columns are eliminated in both, as a consequence of such operations. □

After elimination of a central row in $\underline{G}$, an information symbol is changed into a control symbol, in the position associated to the highest-order power coefficient in polynomial $g(x)$ (see Sect. 1.13 and Fig. 1.13). Instead, after elimination of a central row in $\underline{H}$, a control symbol is changed into an information symbol, in the position associated to the lowest-order power coefficient in polynomial $h^*(x)$ (see Lemma 5.19 and Fig. 5.15). These two behaviours are compared in Fig. 5.20 for matrices which are both s.s. time-invariant. The parts interested by the above operations are the right in the $\underline{H}$ matrix and the left in the $\underline{G}$ matrix. Solid small arrows denote information symbol positions and dashed small arrows control symbol positions, as usual. We can observe how the two new matrices are no longer upper or lower triangular, although they remain s.s. time-invariant. Encoding is possible by means of an encoder circuit based on polynomial $h(x)$ for $\underline{H}$ or on polynomial $g(x)$ for $\underline{G}$ respectively. During the symbol time in which the symbol nature is changed the input switch of the encoder circuit connects or disconnects the information source.

As already stressed, an MLC code is a subcode of the correspondent LC code. The cause of this fact is in constant-length code shortening performed on its $\underline{G}$ matrix. Hence the construction of a subcode requires to eliminate a certain number of code words from the original code. Each change of an information symbol into a control symbol contributes to reduce the number of code words by a factor of 2. This operation is sometimes called code word expurgation (MacWilliams and Sloane 1977).

Coming back to an MLC code (equivalent to an s.s. time-invariant convolutional code), we can state that its $\underline{G}$ matrix is characterized by an MLC code period. This, in turn, has length n_o, with the first k_o symbol times occupied by information symbols, and the remaining $r_o = n_o - k_o$ symbol times occupied by control symbols. How to obtain the correspondent $\underline{H}$ matrix is not easy to be investigated, at the moment. We can anticipate that, for $r_o = 1$, provided that the equivalent convolutional code is well designed, the $\underline{H}$ matrix is still s.s. time-invariant, but with a parity check polynomial different from $h_c(x)$. The $\underline{H}$ matrix of a convolutional code will be studied in Chap. 7, restricting our attention to the case of s.s. time-invariance in $\underline{H}$, that is with a unique parity check polynomial. Chapter 6 will treat block codes which are w.s. time-invariant in $\underline{H}$. Chapter 8 is finally devoted to discuss convolutional codes which are w.s. time-invariant in $\underline{H}$.

5.11 Some Further Considerations About $\underline{H}$ Matrix in Polynomial Codes

Let us remember that the $\underline{G}$ matrix of a certain polynomial code, s.s. time-invariant in its generator matrix, can be rewritten assuming linear combinations of a unique $g(x)$ polynomial (see property *iii* in Lemma 1.6). In this sense, the k rows of such new matrix are linearly independent and can have $g(x)$ as their greatest common divisor.

Example 5.26 Given the $\underline{G}$ matrix

$$\underline{G} = \begin{bmatrix} 1 & 1 & 1 & 0 & 1 & 0 & 0 \\ 0 & 1 & 1 & 1 & 0 & 1 & 0 \\ 0 & 0 & 1 & 1 & 1 & 0 & 1 \\ 1 & 1 & 1 & 1 & 1 & 1 & 1 \end{bmatrix},$$

it is possible to recognize in the code so described the [7,4,3] Hamming code, characterized by $g(x) = x^3 + x + 1$. See, for instance, the ex-OR sum of the first (or second or third) row and the last one. The rows of the above $\underline{G}$ matrix exhibit $g(x)$ as greatest common divisor, but none of them coincides with $g(x)$ itself. In spite of this fact, the code, except for a possible different association between information symbols and code words, is the same.

We can observe that the $\underline{G}$ matrix considered in Example 5.26 is not time-invariant (neither in strict sense, nor in wide sense), although it represents a code whose description can be also given by means of a generator matrix, in whose rows a unique $g(x)$ shifts.

Lemma 5.24 *Given an s.s. time-invariant polynomial code in its party check matrix, with parity check polynomial $h_c(x)$, an alternative representation for it can be provided by another $\underline{H}$ matrix, having full rank, with r rows, no longer time-invariant. The rows of such equivalent matrix, although different, must have a unique greatest common divisor, coincident with $h_c{}^*(x)$.*

Proof Lemma 5.5 assures the validity of such proposition. Alternatively, we can observe that the dual of the considered code exhibits a $\underline{G}$ matrix, whose rows are characterized by the same property, as discussed above. □

Definition 5.13 When a cyclic, punctured or SC code, s.s. time-invariant in $\underline{H}$, is described by a parity check matrix where a unique polynomial $h_c{}^*(x)$ (except for the possible truncations due to elimination of columns on its left hand side, because of code shortening) shifts along its rows by one position at a time, such an $\underline{H}$ matrix is called to be in *minimal form for a cyclic, punctured or SC code.*

This denomination follows from the property that $h_c{}^*(x)$ is the shortest 1-symbol pattern to be adopted. Remark that an analogous definition could be given also with reference to the $\underline{G}$ matrix regarding $g(x)$. The systematic forms of $\underline{G}$ and $\underline{H}$ are examples of matrices not in minimal form, because of 1-symbols which are more spread along each row. However there are few situations for which a $\underline{G}$ matrix not in minimal form is required. On the contrary, an $\underline{H}$ matrix not in minimal form is often employed with precise purposes, even abandoning the possibility of an easy identification of control symbol positions, like occurs when it exhibits different rows with rightmost 1-symbols in the same position.

Example 5.27 Among the possible alternative structures of $\underline{H}$ for a cyclic code, not in minimal form, there are those obtained after a cyclic permutation of polynomial $h_c{}^*(x)$. Let us consider the [7,4,3] cyclic Hamming code. In Fig. 5.21 two forms for its parity check matrix are shown. The former one is s.s. time-invariant, lower triangular and in minimal form. In its rows the same polynomial $h_c{}^*(x) = x^4 + x^3 + x^2 + 1$ shifts by one position at a time. The latter one, obtained after ex-OR sum of the last two rows, and substitution of the result to the second, is no longer lower triangular. In particular, it exhibits, in the second row, a cyclically permuted version of the same polynomial. Furthermore its last two rows have their rightmost 1-symbol in the same 7th position.

Example 5.28 The parity check matrix of a [15,5,7] cyclic BCH code can be written in minimal form, assuming the parity check polynomial $h_c(x) = (x+1)(x^4+x+1) = x^5+x^4+x^2+1$. This $\underline{H}$ matrix is shown up in Fig. 5.22.

$$\underline{H} = \begin{array}{|ccccccc|} 1 & 1 & 1 & 0 & 1 & & \\ & 1 & 1 & 1 & 0 & 1 & \\ & & 1 & 1 & 1 & 0 & 1 \end{array} \begin{array}{l} \\ \longleftarrow a \\ \longleftarrow b \end{array} \qquad \underline{H} = \begin{array}{|ccccccc|} 1 & 1 & 1 & 0 & 1 & 0 & 0 \\ 0 & 1 & 0 & 0 & 1 & 1 & 1 \\ 0 & 0 & 1 & 1 & 1 & 0 & 1 \end{array} \begin{array}{l} \\ \longleftarrow a+b \\ \\ \end{array}$$

Fig. 5.21 Two forms for the parity check matrix of the [7,4,3] cyclic Hamming code

$$\underline{H} = \left[\begin{array}{ccccccccccccccc}
1&0&1&0&1&1\\
&1&0&1&0&1&1\\
&&1&0&1&0&1&1\\
&&&1&0&1&0&1&1\\
&&&&1&0&1&0&1&1\\
&&&&&1&0&1&0&1&1\\
&&&&&&1&0&1&0&1&1\\
&&&&&&&1&0&1&0&1&1\\
&&&&&&&&1&0&1&0&1&1\\
&&&&&&&&&1&0&1&0&1&1
\end{array}\right] \qquad h_c{}^*(x) = x^5 + x^3 + x + 1$$

$$\underline{H} = \left[\begin{array}{ccccccccccccccc}
1&0&0&1&1&0&1&0&1&1&1&1\\
&1&0&0&1&1&0&1&0&1&1&1&1\\
&&1&0&0&1&1&0&1&0&1&1&1&1\\
&&&1&0&0&1&1&0&1&0&1&1&1&1\\
1&1&0&0&0&1&1&0&0&0&1&1\\
&1&1&0&0&0&1&1&0&0&0&1&1\\
&&1&1&0&0&0&1&1&0&0&0&1&1\\
&&&1&1&0&0&0&1&1&0&0&0&1&1\\
1&1&0&1&1&0&1&1&0&1&1&0&1&1\\
&1&1&0&1&1&0&1&1&0&1&1&0&1&1
\end{array}\right]
\begin{array}{l}
\left.\right\} h^*(x) = x^{11} + x^8 + x^7 + x^5 + x^3 + x^2 + x + 1\\
\left.\right\} h^*(x) = (x+1)(x^{10} + x^5 + 1)\\
\left.\right\} h^*(x) = (x+1)(x^{12} + x^9 + x^6 + x^3 + 1)
\end{array}$$

Fig. 5.22 Two possible forms for the parity check matrix of the [15,5,7] cyclic BCH code

Another form for $\underline{H}$, describing the same code, is that reported down, where the first 4 rows are characterized by polynomial

$$\frac{x^{15}+1}{x^4+x^3+1} = x^{11} + x^{10} + x^9 + x^8 + x^6 + x^4 + x^3 + 1,$$

then other 4 rows are characterized by polynomial

$$\frac{x^{15}+1}{u_5(x)} = (x+1)(x^{10} + x^5 + 1),$$

finally the last 2 rows are characterized by polynomial

$$\frac{x^{15}+1}{u_3(x)} = (x+1)(x^{12} + x^9 + x^6 + x^3 + 1).$$

It is easy to verify that the above three polynomials are respectively the parity check polynomials of the [15,11,3] cyclic Hamming code, of the [15,11,2] LC code whose cyclic parent is the [5,1,5] repetition code, and of the [15,13,2] LC code whose cyclic parent is the [3,1,3] repetition code. These three polynomials have $h_c(x)$ as greatest common divisor. The [15,5,7] BCH code is a subcode of all these three codes. In particular, in order to apply traditional hard-decision decoding procedures, based on syndrome evaluation, suitable for BCH codes, an $\underline{H}$ matrix not in minimal form is preferable, like that shown, for instance, in Fig. 5.22 down for the considered code.

A sort of crossed views of the single error positions can be so arranged. In fact, at the denominator in the division performed on the binomial $(x^{15}+1)$, there are the irreducible polynomials $M_1(x)$, $M_3(x)$, $M_5(x)$ respectively, which are employed for constructing the generator polynomial of the this code, according to BCH design rules (see Sect. 1.9). Considering a symbolic construction of $\underline{H}$, made taking subsequent powers of the primitive element α, representing as many roots collected in the co-factors of the irreducible polynomials forming $g(x)$, it is possible to express the parity check matrix for the [15,5,7] BCH code as

$$\underline{H} = \begin{bmatrix} 1 & \alpha & \alpha^2 & \alpha^3 & \alpha^4 & \alpha^5 & \alpha^6 & \alpha^7 & \alpha^8 & \alpha^9 & \alpha^{10} & \alpha^{11} & \alpha^{12} & \alpha^{13} & \alpha^{14} \\ 1 & \alpha^3 & \alpha^6 & \alpha^9 & \alpha^{12} & 1 & \alpha^3 & \alpha^6 & \alpha^9 & \alpha^{12} & 1 & \alpha^3 & \alpha^6 & \alpha^9 & \alpha^{12} \\ 1 & \alpha^5 & \alpha^{10} & 1 & \alpha^5 & \alpha^{10} & 1 & \alpha^5 & \alpha^{10} & 1 & \alpha^5 & \alpha^{10} & 1 & \alpha^5 & \alpha^{10} \end{bmatrix}$$

Each row of this symbolic matrix, in which any power of α represents a different non-null column vector, is associated to a different cyclotomic coset. The first row follows from C_1, in turn related to the irreducible polynomial $M_1(x) = x^4 + x^3 + 1$. All the possible sequences of 4 non-null binary elements are collected in this first row layer, as occurs in the Hamming code. The second row, still with column vectors having 4 elements, follows from C_3, related to $M_3(x) = x^4 + x^3 + x^2 + x + 1$. Finally the third row, with column vectors having only 2 elements, follows from C_5, related to $M_5(x) = x^2 + x + 1$. In this way, all the possible errors, of weight up to 3, give rise to a multiple-error syndrome (Definition 5.8) easily decomposable into the single-errors syndromes originating it (Sweeney 2002).

Another possible important application of $\underline{H}$ matrices not in minimal form is in the design of LDPC codes (Ryan and Lin 2009), as will be better explained in Chaps. 10–12.

Lemma 5.25 *Considering a cyclic code whose code word length is N, and whose generator polynomial has degree r, its parity check matrix in minimal form shows the maximum possible number of linearly independent rows under the above constraint, and then is unique.*

Proof After definition of $\underline{H}$ in minimal form, the rows of such matrix result in proper shifted replicas of $h_c{}^*(x)$, by one position at a time. Polynomial $h_c{}^*(x)$ has degree $(N-r)$, therefore, with exactly r rows, the entire word length N is covered. Any further possible row to be added will be a linear superimpositions of the rows already considered. In fact the full rank of $\underline{H}$ is equal to r. □

The procedure of code puncturation in an external positions of the code word, and even in its version regarding a central position in the code word without elimination of any column (constant-length puncturation according to Definition 5.11), is characterized by a unique $h^*(x)$, as in the cyclic version of our code. On the other hand, code puncturation performed in a central position, with elimination also of the column correspondent to its rightmost 1-symbol, produces a contraction in the whole matrix, which becomes time-varying, but, also in this case, the remaining rows exhibit all distinct positions for their rightmost 1-symbols. The procedure of code lengthening, as shown in Sect. 5.9, requires a modification of the parity check polynomial, introducing its periodic repetition. After Lemma 5.22, this case can be treated as that of a shortened code (with reference to the cyclic one with length qN). Also here we have different positions associated to the rightmost 1-symbols of all the rows.

Lemma 5.26 *Given a polynomial block code, whose description is made by means of a very general parity check matrix $\underline{H}$ (not in minimal form and even time-varying), if such matrix exhibits rows characterized by rightmost 1-symbols occupying different positions, each one of such positions represents that of the control symbol associated to the correspondent row.*

Proof In writing the $(n - k)$ parity check equations collected in $\underline{H}$, we adopt as many 1-symbol patterns, in general different. For all of them, the rightmost 1-symbol represents the non-null coefficient multiplying the control symbol so introduced. This allows a one-to-one correspondence between any row of $\underline{H}$ and the control symbol associated to it. □

Remark that the condition expressed by Lemma 5.26 is by far more general than that of a lower triangular $\underline{H}$ matrix. For instance, it includes the case of distributed control symbols, as occurs for convolutional codes, and also for QC codes in the reordered form obtained after Lemma 2.6.

Definition 5.14 An $\underline{H}$ matrix with rows whose rightmost 1-symbols occupy $(n - k)$ different positions is defined in *minimal form* (*for any type of polynomial codes*) when it is characterized by the shortest lengths of the 1-symbol patterns in all its rows.

This Definition is not restricted to cyclic, punctured or SC codes, like Definition 5.13 is (but includes that Definition as a particular case). It holds under the same assumptions of Lemma 5.26, that is, for instance, in the case of distributed control symbols and even for time-varying non-periodic parity check matrices.

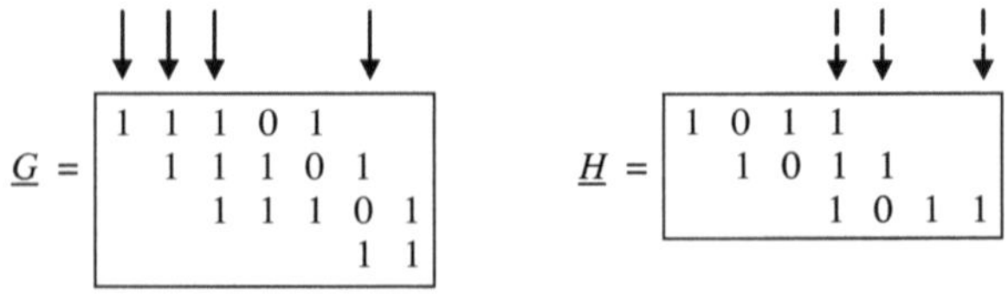

Fig. 5.23 $\underline{G}$ and $\underline{H}$ matrix of the [7,4,2] code obtained from constant-length puncturation of a [7,3,4] simplex code

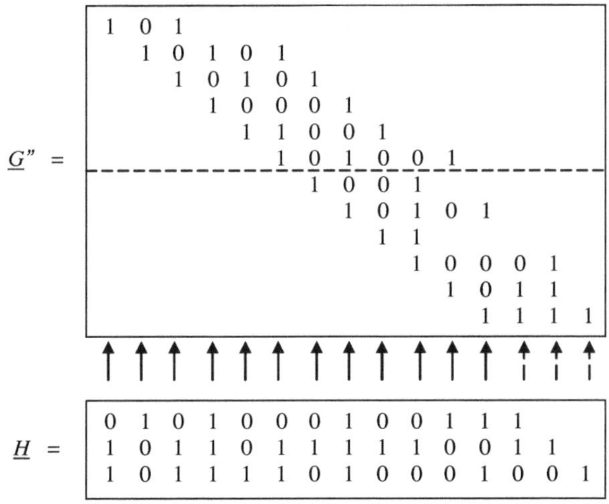

Fig. 5.24 Rearranged version of the punctured $\underline{G}$ matrix described in Example 1.25 and correspondent $\underline{H}$ matrix

Example 5.29 In Example 5.22, a constant-length puncturation has been applied to a [7,3,4] simplex code, obtaining a [7,4,2] code (see Fig. 5.16). In Fig. 5.23 the $\underline{G}$ matrix and the $\underline{H}$ matrix of the resulting code are compared. The latter one is in its minimal form (and is still s.s. time-invariant as expected). Solid small arrows denote information symbol positions, where the leftmost 1-symbols in the four rows of $\underline{G}$ are placed. Dashed small arrows denote control symbol positions. Remark that a lower triangular form is not possible for this parity check matrix, because of the 6th position, which is occupied by an information symbol.

A situation so simple like that presented in Fig. 5.23 is due to the fact that we are dealing with a constant-length puncturation. Some more difficulties can arise for traditional puncturation, especially if it regards information symbol positions.

Example 5.30 In Example 1.25 the puncturation of a [17,12,3] shortened Hamming code has been performed, eliminating, in its $\underline{G}$ matrix, the 6th column and the 13th column. After a proper rearrangement, an upper triangular generator matrix $\underline{G}''$ has been constructed and reported in Fig. 1.11. Such time-varying generator matrix is depicted here again in Fig. 5.24 up. One possible correspondent $\underline{H}$ matrix is shown in Fig. 5.24 down. Small arrows denote positions of information and control symbols, as usual. This parity check matrix is lower triangular and time-varying. The ex-OR sum of the last two rows and substitution to the third row enable us to obtain a third row with a 1-symbol pattern remarkably shorter. The new $\underline{H}$ matrix so obtained can be considered the one in minimal form for this code. Nevertheless, in comparison with that reported in the Figure, the improvement is rather poor.

In order to admit an $\underline{H}$ matrix in minimal form with 1-symbol patterns in the rows remarkably shorter than the overall code word length n, a code must be characterized by periodically distributed control symbols. We can imagine a sort of constraint length acting on the parity check matrix, similarly to what occurs for the generator matrix. This topic, which typically regards convolutional codes, will be faced in the next section.

5.12 Two Types of Rows in Polynomial Code $\underline{H}$ Matrix

Let us consider codes characterized by periodically distributed control symbols in their code words, with particular attention to the possibility of obtaining, also for them, a parity check matrix in minimal form. In order to fix ideas, it is possible to take an MLC code, for which there are positions devoted to control symbols, alternated with positions where information symbols are placed (explicitly or not, depending on the fact that the code is systematic or not). In the rows of the $\underline{G}$ matrix, which is s.s. time-invariant but no longer upper triangular, polynomial $g(x)$ shifts, occupying not all the possible positions.

Lemma 5.27 *The reciprocal of the parity check polynomial of the lengthened code, from which a certain MLC code has been derived, can be assumed in the rows of the* $\underline{H}$ *matrix of the MLC code considered.*

Proof The MLC code is a subcode of its parent LC code, characterized by the same polynomial $g(x)$. After Lemma 1.35, the parity check polynomial of such LC code will be of the type $h(x) = h_c(x)(1 + x^N + x^{2N} + \cdots)$. Its reciprocal $h^*(x)$ forms all the r rows in the LC code parity check matrix. Then polynomials $g(x)$ in the rows of the MLC generator matrix and the parity check matrix row $h^*(x)$ are a valid couple for satisfying constraint (5.4) in any condition. So $h^*(x)$ turns out to be a possible row for $\underline{H}$, independently of the position where it is placed. □

On the other hand the control symbols associated to the above periodic parity check polynomials are only the last ones, whose number is r. For distributed control symbols we have to add further rows in $\underline{H}$. A good choice can be that made in such a way as to assure that the rightmost 1-symbol of each row can be found in correspondence with the associated control symbol. Such rows can remain unbounded on their left hand side.

Example 5.31 The MLC code (or convolutional code), with $n_o = 2$ and $k_o = 1$, obtained from the cyclic code with $g(x) = x^3 + x^2 + 1$ and $h_c(x) = x^4 + x^3 + x^2 + 1$, exhibits a $\underline{G}$ matrix whose final part is reported up in Fig. 5.25. Taking parity check polynomials constructed by proper truncation, on the right hand side, of the periodic polynomials $h_c^*(x)(1 + x^7 + x^{14} + \cdots)$ can be a valid opportunity. Nevertheless, not all the possible rows so arranged (reported in the middle of the Figure) are compatible with all the rows of $\underline{G}$. On the other hand, all these parity check equations would be too many, considered the total amount of control symbol positions available. It is possible to verify that a proper matching with the rows of $\underline{G}$ is reached by means of the $\underline{H}$ matrix whose final part is reported down in the Figure, where the rightmost 1-symbol in any

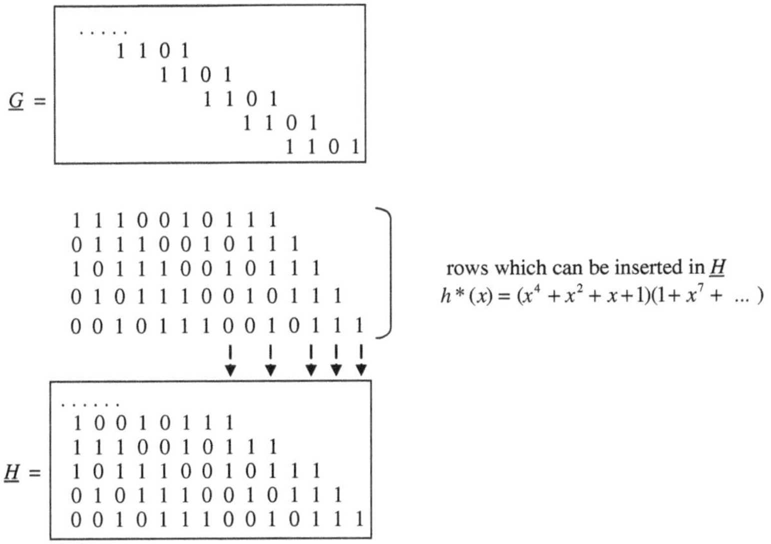

Fig. 5.25 Possible rows of $\underline{H}$, compatible with the rows of $\underline{G}$, and down-right corner of the $\underline{H}$ matrix for the convolutional code described in Example 5.31

row occupies a different position among those devoted to control symbols. Such rows are all linearly independent. Dashed small arrows are employed, as usual, for denoting control symbol positions. With the code rate adopted, they are distributed one in any period, plus a certain number at the end, for assuring the correct frame termination.

Clearly an $\underline{H}$ matrix so arranged exhibits rows which are unbounded on their left hand side, and hence not in minimal form. In order to construct an efficient $\underline{H}$ matrix of an MLC code (or of correspondent s.s. time-invariant convolutional code), it is convenient to search for its minimal form. This problem will be faced in Chap. 7. Now it is possible to anticipate that shorter 1-symbol patterns can be obtained by proper ex-OR sums of those obtained here. If the convolutional code is well designed, they can be eliminated at all, and the resulting rows of $\underline{H}$ will be all non-periodic and bounded either on the right or on the left hand side. It is important to remember that well designed convolutional codes are those whose low-weight code words, up to the weight $w = 2d - 1$, have coefficients $A(w)$ proportional to the period number Z (see Definition 3.5). Non-systematic convolutional codes, when not well designed, exhibit also a catastrophic behaviour (see Definition 3.7). Surviving of unavoidable periodic parity check polynomials, over the entire length of the code frame, is related to the characteristic of the considered convolutional code to be not well designed. This characteristic is not necessarily bad. For instance, a direct product code, which has been interpreted as a particular MLC code (and hence convolutional code), as discussed in Sect. 1.17, exhibits such property, and for this reason a sliding window decoding procedure cannot be applied on it. Nevertheless direct product codes remain a very interesting family of block codes.

In conclusion, we can summarize that, in the most general case, included that of time-varying codes, like the ones considered at the end of the previous Section (Examples 5.29 and 5.30), two types of rows can be present in the parity check matrix of a polynomial code: those bounded on both sides (certainly non-periodic) and those unbounded on their left hand side or unbounded at all (generally periodic).

5.13 H-Extended Cyclic (HEC) Codes

Let us remember how LC codes have been introduced (Definition 1.25). Such code lengthening operation is constructed on the $\underline{G}$ matrix. In this sense, they could be better identified as G-lengthened cyclic codes. However there is no ambiguity in the definition, also without such specification. It is important to stress, instead, that the $\underline{G}$ matrix from which we start is s.s. time-invariant. Now will introduce a similar operation (addition of rows continuing the uniform progressive shift of the unique polynomial in all the positions), but based on the $\underline{H}$ matrix, s.s. time-invariant as well. Such operation is similar to code extension, but with the difference that code extension is made employing an inner binary parity check code in direct serial concatenation. Here the rows to be added are those already present in the $\underline{H}$ matrix of our cyclic code, that is they are expressed by properly shifted reciprocal parity check polynomials $h_c{}^*(x)$. This operation can be repeated as many times as one wishes.

Definition 5.15 Given a cyclic code, characterized by the parity check polynomial $h_c(x)$, with $\deg[h_c(x)] = N - r$, factor of $(x^N + 1)$, with reference to its $\underline{H}$ matrix, s. s. time-invariant in minimal form, the code obtained from it, with a parity check matrix characterized by a number of rows $r' > r$, still with $h_{c^*}(x)$ shifting in these rows, is defined *H-extended cyclic code* (or briefly *HEC code*). More precisely, assuming $r' = r + j$, that is after j single H-extensions, we will call the resulting code *j-time H-extended* starting from its cyclic version.

A code which is j-time H-extended, starting from the parent cyclic code, will show a $\underline{G}$ matrix still with $(N - r)$ rows linearly independent. On the other hand, the code word length is $n = N + j$. The number of rows in the new $\underline{H}$ matrix becomes $(r + j)$. The operation of H-extension, we are considering now, then progressively reduces the code rate.

Example 5.32 The [7,4,3] Hamming cyclic code, whose parity check polynomial is $h_c(x) = x^4 + x^3 + x^2 + 1$, originates a three-time HEC code described by the following parity check matrix

$$\underline{H} = \begin{bmatrix} 1 & 0 & 1 & 1 & 1 & & & & & \\ & 1 & 0 & 1 & 1 & 1 & & & & \\ & & 1 & 0 & 1 & 1 & 1 & & & \\ & & & 1 & 0 & 1 & 1 & 1 & & \\ & & & & 1 & 0 & 1 & 1 & 1 & \\ & & & & & 1 & 0 & 1 & 1 & 1 \end{bmatrix}, \ h_c{}^*(x) = x^4 + x^2 + x + 1.$$

This code exhibits $n = 10$, and still a number of information symbols $k = 4$, as its parent cyclic code. The code rate, from 4/7, is then reduced to 4/10 = 2/5.

Lemma 5.28 *The dual of a y-time LC code is a y-time HEC code and vice versa.*

Proof Addition of y rows in the $\underline{G}$ matrix, where $g(x)$ still shows a uniform progressive shift, is the operation to be performed in a y-time lengthened code from a certain parent cyclic code. The same operation, made on the $\underline{H}$ matrix, using polynomial $h_c{}^*(x)$, is that after which a y-time H-extended code is obtained starting from the same parent cyclic code. □

Lemma 5.29 *The generator polynomial of an HEC code is periodic with period N.*

Proof This property follows from Lemma 5.28 and the part of Lemma 1.35 regarding LC codes. In other terms, once performed a number of single H-extensions, enough for reaching an overall word length so constrained $qN < n < (q + 1)N$, it is possible to write

$$g(x) = \frac{x^{(q+1)N} + 1}{h_c(x)}, \text{ (HEC codes)}, \tag{5.11}$$

where $h_c(x)$ is the parity check polynomial of the original cyclic code. Comparison with (1.23) clearly shows a duality property. □

In conclusion, the expression of the generator polynomial for an HEC code, after a number of single H-extensions able to reach a word length n between qN and $(q + 1)N$, is

$$g(x) = g_c(x)u_{q+1}(x^N), \text{ (HEC codes)}, \tag{5.12}$$

where $g_c(x)$ represents the generator polynomial of the original cyclic code. A possible truncation of the last replica of $g_c(x)$ is to be made on the right hand side of the $\underline{G}$ matrix, in order to maintain for such matrix an upper triangular form. Expression (5.12) can be compared with expression (1.25).

Example 5.33 The generator matrix of the HEC code introduced in Example 5.32, whose parent cyclic code is described by the generator polynomial $g_c(x) = x^3 + x^2 + 1$, is characterized by the following generator matrix

$$\underline{G} = \begin{bmatrix} 1 & 1 & 0 & 1 & & & & 1 & 1 & 0 \\ & 1 & 1 & 0 & 1 & & & & 1 & 1 \\ & & 1 & 1 & 0 & 1 & & & & 1 \\ & & & 1 & 1 & 0 & 1 & & & \end{bmatrix},$$

where some 0-symbols have been omitted, in order to favour immediate recognizing of the 1-symbol patterns. Remark that this $\underline{G}$ matrix, also after further possible H-extensions, will remain in an upper triangular form.

Considering that $\underline{G}$ is preferable in upper triangular form whereas $\underline{H}$ is preferable in lower triangular form, Lemma 5.28, about duality, could be a little modified, but this specification is not essential: the two codes remain a dual couple. In correspondence with word lengths $n = (q + 1)N$, where q is an integer, the HEC code satisfies property (1.13), hence it can be considered cyclic again, and is the dual of a cyclic LC code of equal length (see also Sect. 5.9). In the interval $qN < n < (q + 1)N$, the HEC code can be considered as a punctured version of the above cyclic HEC code (see also Lemma 5.21).

Lemma 5.30 *In a j-time H-extended code, the minimum distance d cannot be lower than that of the original cyclic code d_c. Furthermore we have the following two results*

$$d \geq d_c + 1,\ j \geq N - r, \tag{5.13}$$

$$d \geq 2d_c,\ j \geq N. \tag{5.14}$$

Proof After the first H-extension one parity check equations is added to the r already present in the $\underline{H}$ matrix of the considered cyclic code. Therefore the minimum distance of the new code cannot be lower. Expression (5.14) follows from the fact that, in correspondence with the $(N - r)$th single H-extension, the first 1-symbol of the replica of $g_c(x)$ will appear also in the last row of the $\underline{G}$ matrix, on its right hand side. So there are no rows in $\underline{G}$ whose weight is smaller than $(d_c + 1)$. Expression (5.1) is consequence of the doubling of all the 1-symbol patterns of the original cyclic code. □

Example 5.34 Let us consider the HEC code already discussed in Examples 5.32 and 5.33. It is originated from the [7,4,3] Hamming cyclic code. Table 5.1 shows the evolution in the word weight distribution characterizing some of its HEC versions, from $n = 10$ (the code already considered) up to $n = 14$ (first complete H-extension). We can observe that the number of information symbols is $k = 4$ for all these codes. The word weight distribution of the first completely H-extended code, which has been evidenced, is that of the original cyclic code only properly scaled. All these weight distributions are symmetric, because the all 1-symbol code word belongs always to such codes and after Lemma 1.11. For $j = 4$, expression (5.13) is confirmed, in the second row of the Table. For $j = 7$ (that is $n = 2N = 14$), expression (5.14) is confirmed, in the last row.

Table 5.1 Word weight distributions for some HEC codes obtained from the [7,4,3] Hamming code

k	j	n	$A(0)$	$A(3)$	$A(4)$	$A(5)$	$A(6)$	$A(7)$	$A(8)$	$A(9)$	$A(10)$	$A(11)$	$A(12)$	$A(13)$	$A(14)$
4	3	10	1	1	3	6	3	1			1				
4	4	11	1		3	4	4	3				1			
4	5	12	1		1	4	4	4	1				1		
4	6	13	1			3	4	4	3					1	
4	7	14	*1*				7		7						*1*

Lemma 5.31 *After $(q-1)$ complete H-extensions, starting from an $[N, N-r, d_c]$ cyclic code, we obtain a $[qN, qN-r, qd_c]$ code, whose word weight distribution can be calculated from the one of the original parent cyclic code, characterized by coefficients $A(w)$, as*

$$A'(qw) = A(w). \tag{5.15}$$

Proof At the end of the first complete H-extension the word weight distribution is $A'(2w) = A(w)$, since all the 1-symbol patterns are doubled. Then the process goes on, and, after the $(q-1)$th complete H-extensions, (5.15) is satisfied. □

Lemma 1.29 gives the code word polynomial expression in a y-time lengthened cyclic code. According to (1.18), it is made by two independent parts: the one, in which a polynomial, whose degree is at most $(N-r-1)$, multiplies polynomial $g_c(x)$, and the other one, in which a polynomial $a(x)$, whose degree is at most $(y-1)$, multiplies (x^N+1).

Lemma 5.32 *The generator matrix of a y-time lengthened cyclic code can be written with its first y rows where (x^N+1) shifts by one position at a time, and with the last $(N-r)$ rows where $g_c(x)$ shifts by one position at a time.*

Proof This property follows from (1.18). In the proof of Lemma 1.29, polynomial $u_c(x)$ multiplies $g_c(x)$, and polynomial $a(x)$ multiplies (x^N+1). These two parts are linearly independent if we construct a $\underline{G}$ matrix in upper triangular form, and subdivided into two layers. In the upper layer there are y rows of the second type, with $a(x) = 1, x, \ldots x^{y-1}$. In the lower layer, there are $(N-r)$ rows of the first type, with $u_c(x) = 1, x, \ldots, x^{N-r-1}$. □

Example 5.35 The [7,4,3] Hamming code, with generator polynomial $g_c(x) = x^3 + x^2 + 1$, gives rise to a [10,7,2] LC code, whose generator matrix can be written as follows

$$\underline{G} = \begin{bmatrix} 1 & 0 & 0 & 0 & 0 & 0 & 0 & 1 & & \\ & 1 & 0 & 0 & 0 & 0 & 0 & 0 & 1 & \\ & & 1 & 0 & 0 & 0 & 0 & 0 & 0 & 1 \\ & & & 1 & 1 & 0 & 1 & & & \\ & & & & 1 & 1 & 0 & 1 & & \\ & & & & & 1 & 1 & 0 & 1 & \\ & & & & & & 1 & 1 & 0 & 1 \end{bmatrix}.$$

Now we will apply a similar partitioning to the $\underline{H}$ matrix of an H-extended cyclic code. To this purpose, we can observe that a linear superimposition of rows in the parity check matrix of an HEC code, after j single H-extensions, can be written as

$$h^*(x) = x^p(x^N + 1),\ 0 \leq p \leq j-1. \tag{5.16}$$

On the other hand, a row of the type

$$h^*(x) = x^i h_c{}^*(x),\ j \leq i \leq j + N - r - 1, \tag{5.17}$$

is already present in the $\underline{H}$ matrix of its parent cyclic code, therefore it can be maintained in the new $\underline{H}$ matrix, besides the j rows of the type (5.16).

Lemma 5.33 *An HEC code, after j single H-extensions, can exhibit a parity check matrix, subdivided into two layers: the first layer, with* $(N - r)$ *rows of the type* (5.17) *and the second layer, with j rows of the type* (5.16).

Proof After duality, we can construct the $\underline{H}$ matrix of our H-extended code in the same way as discussed in Lemma 5.32 about the $\underline{G}$ matrix of a lengthened code. Also in this case, the $(N - r + j)$ overall resulting rows in the above construction are all linearly independent. In other terms, taken any value for integer i in (5.17) or any value for integer p in (5.16), we obtain linearly independent expressions. The only care to be taken into account is that, in order to obtain an $\underline{H}$ matrix having lower triangular form, the rows of the type (5.17) are to be placed before and the rows of the type (5.16) after. □

Example 5.36 Assuming a [7,4,3] Hamming cyclic code, whose parity check polynomial is $h_c(x) = x^4 + x^3 + x^2 + 1$, after j = 3 single H-extensions, it is possible to write the following parity check matrix

$$\underline{H} = \begin{bmatrix} 1 & 0 & 1 & 1 & 1 & & & & & \\ & 1 & 0 & 1 & 1 & 1 & & & & \\ & & 1 & 0 & 1 & 1 & 1 & & & \\ 1 & 0 & 0 & 0 & 0 & 0 & 0 & 1 & & \\ & 1 & 0 & 0 & 0 & 0 & 0 & 0 & 1 & \\ & & 1 & 0 & 0 & 0 & 0 & 0 & 0 & 1 \\ & & & & & & & & & \end{bmatrix}, \quad h_c{}^*(x) = x^4 + x^2 + x + 1.$$

This $\underline{H}$ matrix is not in minimal form. The [10,4,3] HEC code here considered coincides with the one introduced in Example 5.32, where the minimal form of $\underline{H}$ was employed. Finally we can observe how the $\underline{H}$ matrix obtained in Example 5.36 clearly comes out from the direct serial concatenation of two codes (see Sect. 5.6). In this sense, we can conclude that an H-extended code is a subcode of a generalized parity check code (see Definition 1.26). In this code concatenation scheme, the inner code is the GPC code, and the outer code is the original cyclic code considered. Let us take into account now the dual of an LC code.

Example 5.37 In order to obtain the dual of the LC code described by the $\underline{G}$ matrix in Example 5.35, it is enough to adopt such $\underline{G}$ matrix as the new $\underline{H}$ matrix (applying Definition 5.9). A different procedure is that performed starting from the dual of the original code and then apply three single H-extensions, at least in principle. The dual of the original cyclic code is the [7,3,4] simplex cyclic code, whose parity check polynomial is $h_c(x) = x^3 + x^2 + 1$. After $j = 3$ single H-extensions, the resulting $\underline{H}$ matrix, according to Lemma 5.33, turns out to be

$$\underline{H} = \begin{bmatrix} 1 & 0 & 1 & 1 & & & & & & & \\ & 1 & 0 & 1 & 1 & & & & & & \\ & & 1 & 0 & 1 & 1 & & & & & \\ & & & 1 & 0 & 1 & 1 & & & & \\ 1 & 0 & 0 & 0 & 0 & 0 & 0 & 1 & & & \\ & 1 & 0 & 0 & 0 & 0 & 0 & 0 & 1 & & \\ & & 1 & 0 & 0 & 0 & 0 & 0 & 0 & 1 & \end{bmatrix}, \quad h_c{}^*(x) = x^3 + x + 1.$$

Comparison with the $\underline{G}$ matrix described in Example 5.35 enables us to draw the following conclusions. We have here the reciprocal 3-weight polynomials, as expected, but, since 10000001 can be obtained from a linear superimposition of rows where 1101 shifts or where 1011 shifts as well, such two matrices are equivalent, so confirming Lemma 5.28.

Before concluding the present Section, it is important to remark that a systematic encoder circuit able to provide an HEC code, starting from a certain cyclic code, for whichever number j of single H-extensions, is still that reported in Fig. 5.6. In its state diagram, once reached a certain closed route, corresponding to a code word of the original cyclic code (see Fig. 5.7 or Fig. 5.8), the circuit remains in that closed route for j further steps. This behaviour is responsible for the periodic replicas of the code words characterizing the original cyclic code, due to a $\underline{G}$ matrix, in which a periodic generator polynomial is present, as given by (5.12).

5.14 Discussion About Dual Polynomial Codes

An s.s. time-invariant code (in its $\underline{G}$ or in its $\underline{H}$ matrix), if represented by matrices not in minimal form (like those introduced in the previous Section), can appear as time-varying. Nevertheless its structure remains the same. This occurs for the $\underline{G}$ matrix in an LC code (of the type described in Lemma 5.32), or for the $\underline{H}$ matrix in an HEC code (of the type described in Lemma 5.33). The properties of Lemmas 1.26 and 5.29 can be collected into a unique proposition, regarding complete code lengthening or complete H-extension.

Lemma 5.34 *An LC code, with $(q - 1)N < n \leq qN$, becomes cyclic again for $n = qN$. An HEC code, with $(q - 1)N < n \leq qN$, becomes cyclic again for $n = qN$. Such two codes, in any condition, form a dual pair.*

Proof For LC codes, if $g(x)$ is factor of $(x^N + 1)$, it is also factor of $(x^{qN} + 1)$. In this case the parity check polynomial becomes periodic. For HEC codes, if $h(x)$ is factor of $(x^N + 1)$, it is also factor of $(x^{qN} + 1)$. In this case the generator polynomial becomes periodic. The restriction to the interval $(q - 1)N < n \leq qN$ is due to the convention for which a cyclic code has generator and parity check polynomials, co-factors of the same binomial with minimum degree. □

Definition 5.16 A block polynomial code characterized by $h(x) = x^N + 1$ is called *generalized repetition code* (or briefly *GR code*). Clearly its length n must be equal to $2N$ or greater.

The last specification is due to the need of avoiding central positions in the code words not protected. Let us come back to Definition 1.26, in which GPC codes have been introduced, in comparison with LC codes. Here the meaning of GR codes can be interpreted in comparison with HEC codes. In GPC codes the central positions, in the case of $k < r$, were forced to be 0 (see Lemma 1.34).

Lemma 5.35 *The generator polynomial of a GR code, whose code words have length n so constrained* $qN < n < (q + 1)N$, *is*

$$g(x) = u_{q+1}(x^N), \text{ (GR codes)}. \tag{5.18}$$

Proof From division of $(x^{(q+1)N} + 1)$ by $h(x) = x^N + 1$, (5.18) can be easily obtained. Nevertheless, on the right hand side of the $\underline{G}$ matrix, proper truncations of polynomials (5.18) have to be made, like in HEC codes, for values of n not integer multiple of N. □

We can look at (5.11) and (5.18) in comparison with (1.24) and (1.26). After Lemma 5.35, the tern which identifies a GR code is $[n, N, q]$. In fact $\underline{G}$ has always N rows. The minimum distance d increases from q to $(q + 1)$ only when n, increasing from $(qN + 1)$, becomes equal to $(q + 1)N$. Such codes can be imagined to exist only in an H-extended form.

Lemma 5.36 *Under the assumptions of Lemma* 5.35, *the code rate of a GR code progressively decreases from* $1/q$ *up to* $1/(q + 1)$.

Proof The number of information symbols is N, the total number n of symbols is so bounded $qN < n < (q + 1)N$. From the ratio between these two quantities the two code rate values above mentioned come out. □

The code here considered can be interpreted as proper code puncturation of the cyclic code $[(q + 1)N, N, q]$. Remark that in Definition 1.26 we discussed about rth-order GPC codes, being r the degree of their generator polynomial $g(x)$, that is the separation between the two 1-symbols present in it. In order to make comparison with GR codes easier, it is convenient now to replace r with N.

Lemma 5.37 *A GR code is the dual code with respect to a GPC code, over the same length* $n \geq 2N$, *and vice versa.*

Proof In Fig. 5.26 up, the $\underline{G}$ matrix and the $\underline{H}$ matrix of a GR code are shown. In Fig. 5.26 down, the same is made for a GPC code, in order to allow an easy

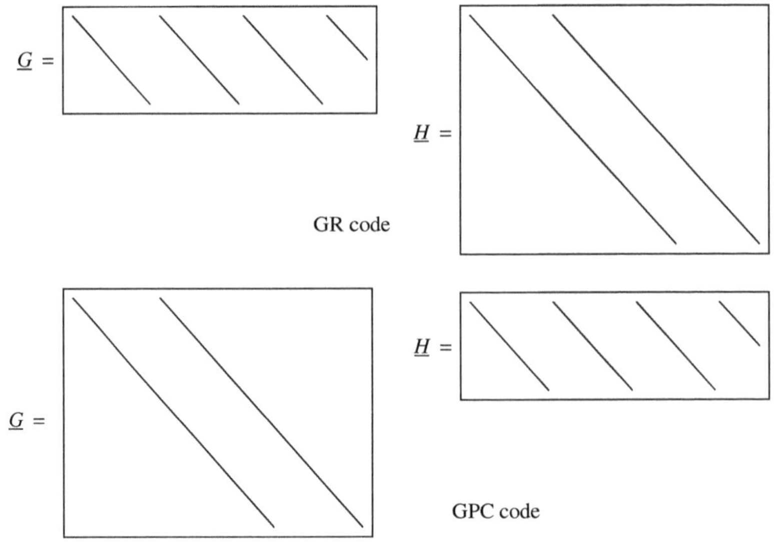

Fig. 5.26 Generator and parity check matrix of a GR code and of a GPC code

comparison. Clearly the $\underline{G}$ matrix of the one is the $\underline{H}$ matrix of the other one, and vice versa. Remark that the $\underline{H}$ matrix of a GPC code could be organized also in a lower triangular form, but we have avoided this option, in order to stress duality. □

For $n = 2N$, the families of GR codes and of GPC codes have an element in common. It is a self-dual code, and can be interpreted either as a GR or as a GPC code, at the same time. It can be imagined also by means of $2 \times N$ encoding memory, where an ordered ensemble of code words, taken from the [2,1,2] code, which is contemporary a parity check code and a repetition code, are listed. Then other codes of the GR family can be obtained substituting a $[q,1,q]$ repetition code to the [2,1,2] code, whereas other codes of the GPC family substituting a $[q,q-1,2]$ parity check code to the [2,1,2] code.

Let us consider a GR code with code word length n expressed by an integer multiple of N, say qN. Looking at its $\underline{G}$ matrix, it is easy to conclude that it can show only code words whose weight is integer multiple of q. Furthermore, the following proposition holds.

Lemma 5.38 *A GR code characterized by* $n = qN$ *is described by the tern* $[qN,N,q]$, *and exhibits this word weight distribution*

$$A(0) = 1, A(mq) = \binom{N}{m}, \quad m = 1,2,\ldots N. \tag{5.19}$$

Proof A code word of the considered code can be expressed by the polynomial

$$t(x) = i(x)u_q(x^N) \tag{5.20}$$

where $i(x)$ represents the information symbol sequence, whose degree is at most $(N - 1)$. So, over N positions, associated to polynomial $i(x)$, we can have $\binom{N}{m}$ combinations of m non-null symbols. Finally, from (5.20), the weight of this code word is $w = mq$. □

In conclusion, GR codes can exist only in an H-extended cyclic form, under condition always satisfied $k = N$, and their weight distribution is trivially given by (5.19). GPC codes, as already noted, can exist only in a lengthened cyclic form, under condition always satisfied $r = N$. Lemma 1.33 gives expression (1.22) for the weight distribution of a GPC code over the first lengthening period, that is with $k \le r$. Both such two families of codes have a common origin, in a code whose length is $n = 2N$. In this situation, (1.22) and (5.19) coincide, taking $k = N$, $q = 2$ and hence $w = 2m$.

Example 5.38 Assuming $k = N = 5$, the word weight distributions characterizing a family of GR codes, taken only with $n = qN$, are shown in Table 5.2, for $q = 2, 3, 4$. The code rate is $R = 1/q$, and the minimum distance is $d = q$. A correspondence with the family of binary repetition codes is clearly evident. Only the number of code words is greater than 2 (precisely it is 2^k). We can imagine a GR code as an ordered ensemble of k repetition codes. In a condition characterized by n between qN and $(q + 1)N$, we have a weight distribution characterized by a mixture of two adjacent non-null coefficients: those associated to mq and to $m(q + 1)$.

Another important pair of dual codes are Hamming codes and simplex codes on the same length N, equal to an integer power of 2 minus one. It is possible to construct the whole family of binary Hamming codes, forming a sort of chain with couples of such codes characterized by subsequent values of parameter r. The same is possible for the whole family of binary simplex codes taking subsequent values of parameter $(N - r)$.

Theorem 5.3 *Given a Hamming cyclic code, characterized by parameter r, an equivalent of the Hamming cyclic code, characterized by parameter $r' = r + 1$, can be obtained as follows: after one complete code lengthening procedure, one further control symbol is calculated as a parity check symbol acting on the last $N + 1 = 2^r$ symbols (additional control symbol included).*

Proof A complete lengthening procedure gives rise to a $[2N, 2N - r, 2]$ code, with a number N of 2-weight code words, of the type $(x^N + 1)$. Each of these code words exhibits one 1-symbol in the first set of N consecutive positions, and the other 1-symbol in the second set of N consecutive positions. Let us consider a parity check code acting on just one of these two sets of positions, say the second. So all the 2-weight code words, with the addition of the further control symbol, become 3-weight code words. The new code will be characterized by the tern

Table 5.2 Word weight distribution for the GR codes described in Example 5.38

q	n	k	$A(0)$	$A(2)$	$A(3)$	$A(4)$	$A(6)$	$A(8)$	$A(9)$	$A(10)$	$A(12)$	$A(15)$	$A(16)$	$A(20)$
2	10	5	1	5		10	10	5		1	(Coincident with the GPC code)			
3	15	5	1		5		10		10		5	1		
4	20	5	1			5		10			10		5	1

$$\underline{H} = \left[\begin{array}{ccccccc|cccccccc} 1&1&1&0&1&&&1&1&1&0&1&&& \\ &1&1&1&0&1&&&1&1&1&0&1&& \\ &&1&1&1&0&1&&&1&1&1&0&1& \\ \hline &&&&&&&1&1&1&1&1&1&1&1 \end{array}\right] \qquad \text{[15,11,3] Hamming code}$$

$$\underline{G} = \left[\begin{array}{ccccccc|cccccccc} 1&1&1&0&1&&&1&1&1&0&1&&& \\ &1&1&1&0&1&&&1&1&1&0&1&& \\ &&1&1&1&0&1&&&1&1&1&0&1& \\ \hline &&&&&&&1&1&1&1&1&1&1&1 \end{array}\right] \qquad \text{[15,4,8] simplex code}$$

Fig. 5.27 Parity check matrix for an equivalent of the [15,11,3] Hamming cyclic code and generator matrix for an equivalent of the [15,4,8] simplex cyclic code

$[2N + 1, 2N - r, 3] = [2^{r'} - 1, 2^{r'} - 1 - r', 3]$. So, owing also to Hamming bound with odd minimum distance d (Theorem D.4), such code must be necessarily an equivalent of the Hamming cyclic code with parameter r'. □

Example 5.39 Starting from a [7,4,3] Hamming cyclic code, with parity check polynomial $h(x) = x^4 + x^2 + x + 1$, one complete lengthening and addition of the further control symbol, according to Theorem 5.3, yield the $\underline{H}$ matrix depicted in Fig. 5.27 up. Dashed-line rectangles denote the two parts of the completely lengthened code. The new code is an equivalent of the [15, 11, 3] Hamming cyclic code.

Corollary a *Given a simplex cyclic code, characterized by parameter k, an equivalent of the simplex cyclic code, characterized by parameter* $k' = k + 1$, *can be obtained as follows: after one complete H-extension procedure, one further information symbol associated to a row in the* $\underline{G}$ *matrix, of the type* $u_{N+1}(x)$, *is added in correspondence with the* $(N + 1)$*th position of the new code word.*

Proof The same procedure described in Theorem 5.3 regarding the $\underline{H}$ matrix can be applied here on the $\underline{G}$ matrix. The new code will be represented by the tern $[2N + 1, k + 1, 2^k] = [2^{k'} - 1, k', 2^{k'-1}]$, so, owing also to duality, it must be necessarily an equivalent of the simplex cyclic code with parameter k'. □

Example 5.40 Starting from a [7,3,4] simplex cyclic code, with generator polynomial $g(x) = x^4 + x^3 + x^2 + 1$, one complete H-extension and addition of the further information symbol, according to Corollary a of Theorem 5.3, yield the $\underline{G}$ matrix reported in Fig. 5.27 down, with analogous convention in matrix partitioning as before. This code is an equivalent of the [15,4,8] simplex cyclic code.

5.15 Modified H-Extended Cyclic (MHEC) Codes

Starting from an HEC code, application of a periodic constant-length puncturation gives rise to a new family of codes, in which information symbols and control symbols are alternated, like in an MLC code.

Definition 5.17 Let us consider an s.s. time-invariant block code in its $\underline{H}$ matrix, characterized by the parity check polynomial $h(x)$, factor of $(x^N + 1)$, and having degree $(N - r)$. After several single H-extensions, leading to a block length n much greater than N, it is modified by periodic insertion of information symbols among the control symbols provided by the rows $h^*(x)$ in the original $\underline{H}$ matrix. This goal is reached by means of a series of single constant-length puncturations, and the code so obtained is called *modified H-extended cyclic code* (or briefly *MHEC code*).

According to Lemma 5.19, the position in the code word where a control symbol is transformed into an information symbol is that associated to the rightmost 1-symbol in the eliminated row, that is to the 0th-order coefficient in polynomial $h^*(x)$. It is important to stress that a periodic behaviour is to be adopted, in order to have efficient MHEC codes, so assuring a uniform performance in any part of the code word length. So, fixed a period, whose length is n_o, say the former k_o positions are associated to as many information symbols, and the latter $r_o = n_o - k_o$ positions to as many control symbols.

Example 5.41 Given the parity check polynomial $h(x) = x^4 + x^2 + x + 1$, and assuming $n_o = 3$, $k_o = 2$, $r_o = 1$, the MHEC code characterized by $n = 12$ is described by an $\underline{H}$ matrix of the type shown in Fig. 5.28 up. A 0-symbol has been maintained on the left of each polynomial $h^*(x)$, in order to have two complete periods, but its presence is clearly not influent. Such parity check matrix has been constructed after proper H-extension, constant-length puncturation (elimination of two central rows any three), and by a further traditional code shortening (external columns elimination), as depicted in Fig. 5.28 on the right. Eliminated rows are denoted by dashed horizontal lines, and eliminated columns by dashed vertical lines. Shortening has been performed on both sides of the parity check matrix in order to eliminate positions not protected as all the others, or useless. After this operation, we are left with three complete central rows and two external truncated

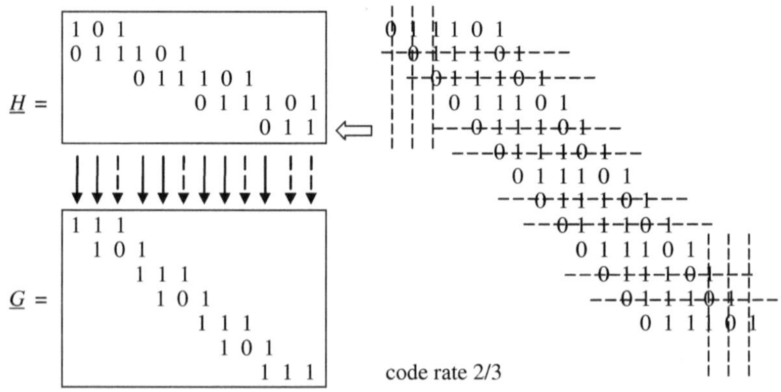

Fig. 5.28 Parity check matrix of the MHEC code described in Example 5.41, its construction and correspondent generator matrix

rows: five rows on the whole. They are all linearly independent. After imposing null ex-OR sum to proper clusters of syndromes (see Lemma 5.11), the correspondent $\underline{G}$ matrix is easily obtained, as shown in Fig. 5.28 down. This code is w.s. time-invariant in its $\underline{G}$ matrix. The minimum distance is $d = 2$. Solid and dashed small arrows denote information and control symbols respectively, as usual.

Looking at the generator matrix reported in Fig. 5.28, clearly we can note a close resemblance with that of a convolutional code, w.s. time-invariant in its $\underline{G}$ matrix, as described in Chap. 4. More precisely, we have here a true generator sub-matrix $\underline{G}_o$ (see Figs. 4.2 and 4.3), formed by two different interleaved generator polynomials: $g_{\text{int}}^{(1)}(x) = x(x^2 + x + 1)$ and $g_{\text{int}}^{(2)}(x) = x^2 + 1$. In the last row of matrix $\underline{G}$, there is $g_{\text{int}}^{(1)}(x)$ alone, since the overall code is a [12,7,2] code, leading to a not complete last period. Like in a convolutional code, we can assume whichever length n, provided that multiple of = 3, expecting a growth in the number of low-weight code words linear with the number Z of the periods where information symbols are inserted (if the code is well designed). The considered code has an asymptotic code rate of 2/3.

Example 5.42 Given the same parity check polynomial as before, that is $h(x) = x^4 + x^2 + x + 1$, but assuming here $n_o = 3$, $k_o = 1$, $r_o = 2$, the MHEC code so obtained can be better analyzed if we take n somewhat higher, say $n = 18$. The $\underline{H}$ matrix of this code is shown in Fig. 5.29 up, where one row any three has been already suppressed. This matrix has not full rank. Elimination of the 18th row (linearly dependent on the others) and of the 1st row (useless because characterized by only one 1-symbol) yields a code showing only 12 control symbols on the whole. Imposition to a proper cluster of syndromes to have a null ex-OR sum

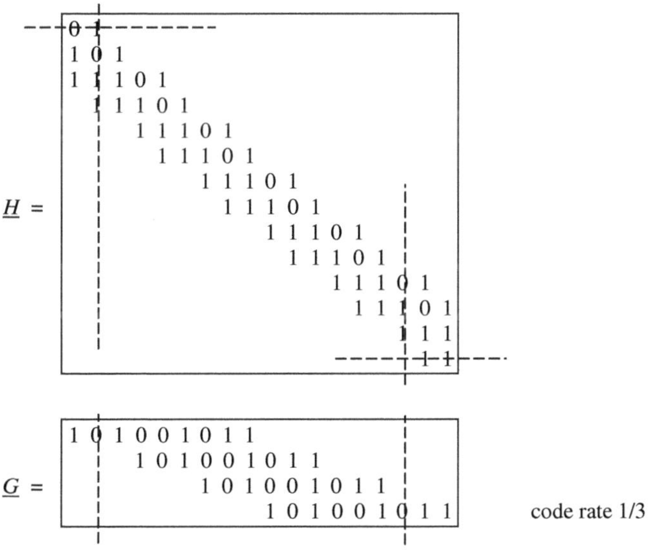

Fig. 5.29 Parity check matrix and generator matrix of the MHEC code described in Example 5.42

enables us to obtain the $\underline{G}$ matrix reported in Fig. 5.29 down. The 2nd and the 16th position in the code word are useless (always occupied by 0-symbols), and can be eliminated, allowing an interpretation for the eliminations of the 2nd and 16th column also in the $\underline{H}$ matrix. It is possible to consider the $\underline{G}$ matrix so obtained as s. s. time-invariant (except for the elimination of the above two positions in the overall code word, producing a sort of collapse in the outermost generator polynomials). The asymptotic code rate is 1/3, but, owing to a correct frame termination (and frame beginning), the tern which can be adopted for describing this code is [16,4,5]. The minimum distance is now $d = 5$. Remark that the present code, taken on the same number of periods, at least in the central part of the code words, appears as a subcode of that described in Example 5.41. All the code words having weight $w = 2, 3, 4$ have been eliminated, and just one with 1-symbol pattern having $w = 5$ remains. The unique interleaved polynomial is here $g_{\text{int}}(x) = x^8 + x^6 + x^3 + x + 1$. It forms a generator sub-matrix $\underline{G}_o$ characterized by just one row, horizontally shifting by three positions any period.

Example 5.43 As for the word weight distributions which characterize the MHEC codes introduced respectively in Examples 5.41 and 5.42, it is useful to adopt a normalized coefficient $B(w)$, of the type described by (3.5) for well designed convolutional codes. The code having asymptotic code rate 2/3 exhibits $B(2) = 1$ and $B(3) = 3$. The code having asymptotic code rate 1/3 is instead characterized by $B(5) = 1$ and $B(6) = 1$.

Let us remember that an MLC code is a proper subcode of the LC code from which it derives, but in that case the structure of the generator matrix does not change (except for the intentional periodic elimination of some central rows, owing to constant-length shortening). In an MHEC code, with respect to the H-extended code from which it derives, besides the elimination of some central rows in the parity check matrix, we have a transformation in the external parts of $\underline{H}$. Ignoring the last transformation, we could state that the HEC original code is a subcode of any MHEC codes obtained from it (remark the inversion in the roles of the two with respect to the previous situation). In analogous way, ignoring such external transformation, we could still conclude that an MHEC code is the dual of an MLC code.

Coming back to the subdivision of the possible types of rows in a parity check matrix (bounded on both sides or bounded only on their right hand side) described at the end of Sect. 5.12, we can observe that in the $\underline{H}$ matrices obtained for the MHEC codes in Examples 5.41 and 5.42, even the final rows, devoted to a correct frame termination, are bounded on both sides. This confirms that such MHEC codes, interpreted as convolutional codes, are well designed.

It is possible to choose the $\underline{G}$ matrix of an MLC code, and then obtain the correspondent $\underline{H}$ matrix. On the other hand, it is possible to choose the $\underline{H}$ matrix of an MHEC code, and then obtain the correspondent $\underline{G}$ matrix. Here, the last procedure has been developed by means of the null ex-OR sum of clusters of syndromes. The existence of a co-factor in the decomposition of a proper binomial $(x^N + 1)$ is of no help, in this case. Therefore it is necessary to set up a procedure by which, from the $\underline{G}$ matrix of an MLC code, we can derive the correspondent $\underline{H}$ matrix. It represents a sort of inverse procedure, with respect to that based on the null ex-OR sum of clusters of syndromes. This problem will be examined more in depth, speaking about the parity check matrix of s.s. time-invariant or w.s. time-invariant convolutional codes, in Chaps. 7 and 8 respectively.

Before concluding the present Section, it is important to stress that even MHEC codes can be encoded by means of the systematic encoder circuit shown in Fig. 5.6. There is, in fact, a unique parity check polynomial $h(x)$, which coincides with that of the cyclic code considered. The switch follows a periodic behaviour, remaining for k_o symbol times connected to the information source, and for r_o symbol times connected to the feed-back, any period. At last, for r consecutive symbol times, a proper symbol sequence outgoing from the information source assures a correct frame termination, like is due in convolutional codes. This procedure resembles that needed in an RSC encoder circuit. Elimination of few useless positions in the initial and in the final part of the code word (frame) can be performed after the true encoding procedure.

5.16 Direct Product Codes: Structure of Their $\underline{H}$ Matrix

Direct product codes have been introduced in Sect. 1.15 (Definition 1.33). Theorem 1.4 gives the $\underline{G}$ matrix of a direct product code as the Kronecker product of the two generator matrices describing the two component codes. As for the $\underline{H}$ matrix of this code, the following proposition is a straightforward consequence of its construction in the rectangular encoding memory which is adopted (Fig. 1.18).

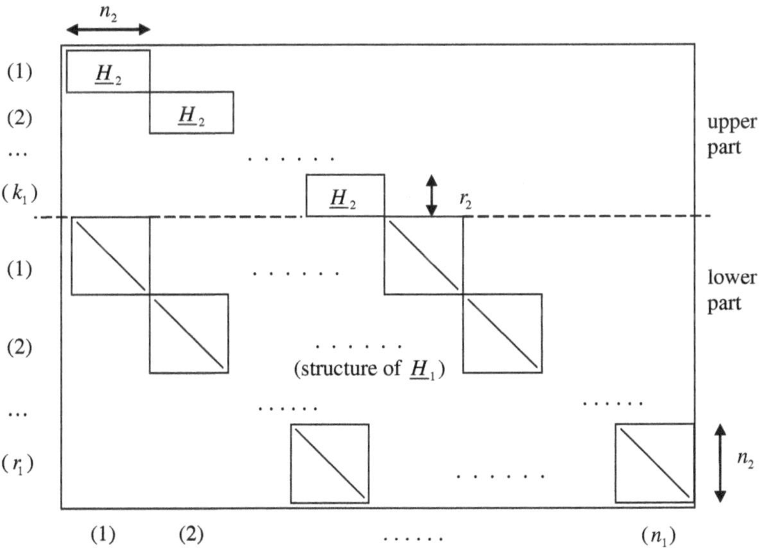

Fig. 5.30 Parity check matrix of a direct product code

Lemma 5.39 *A possible* $\underline{H}$ *matrix for a direct product code, whose parameters are those introduced in* Fig. 1.18, *exhibits two parts: an upper part, consisting of* k_1 *layers where as many* $\underline{H}_2$ *matrices shift forming an orthogonal set, and a lower part, consisting of the Kronecker product* $\underline{I}_{n_2} \otimes \underline{H}_1$, *where* $\underline{I}_{n_2}$ *represents the* $n_2 x n_2$ identity matrix. In Fig. 5.30 *a matrix of this type is schematically depicted.*

Proof The macroscopic structure of the lower part is imposed by the $\underline{H}_1$ matrix. It exhibits $n_2 x n_2$ identity matrices instead of 1-symbols. Therefore this lower part is made by r_1 row layers ($r_1 n_2$ rows on the whole), and represents the parity check equations acting on the columns in the encoding rectangular memory reported in Fig. 1.18. The upper part is the parity check matrix of the ordered ensemble of code words, assumed as rows in the same encoding rectangular memory, and is made by k_1 row layers ($k_1 r_2$ rows on the whole). In both parts it is possible to enumerate the columns: their number is $n_1 n_2$. □

There are several possible alternative versions for such parity check matrix. First of all, another version can be obtained exchanging the roles of the two component codes, so with an upper part, consisting of K_2 layers where as many $\underline{H}_1$ orthogonal matrices shift, and a lower part, consisting of the Kronecker product $\underline{I}_{n_1} \otimes \underline{H}_2$. Another pair of matrices can be obtained by prolongation of the first part up to n_1 (or n_2) layers. In this case, instead of $\underline{I}_{n_2}$ in Kronecker product, we have to consider $\underline{I}_{k_2}$ (instead of $\underline{I}_{n_1}$, we have to consider $\underline{I}_{k_1}$).

Example 5.44 Let us adopt the direct product code constructed with the [7,4,3] Hamming code, whose parity check polynomial is $h_1(x) = x^4 + x^2 + x + 1$, and the [6,2,4] code, characterized by $h_2(x) = x^2 + x + 1$. This code has been already treated in Example 1.55, analyzing its $\underline{G}$ matrix. Remark that the latter component code can be interpreted also as an HEC code, whose cyclic parent code is the [3,2,2] parity check code, after one complete H-extension process. The $\underline{H}$ matrix of this direct product code, according to Lemma 5.39, is shown in Fig. 5.31. We can observe $k_1 = 4$ layers in the upper part, with as many orthogonal matrices $\underline{H}_2$ (obtained exploiting the above interpretation as HEC code), and $r_1 = 3$ layers in the lower part, where the macroscopic structure of the $\underline{H}_1$ matrix is reproduced employing $\underline{I}_{n_2} = \underline{I}_6$ identity matrices instead of 1-symbols. Let us remember that, in $\underline{H}_1$, polynomial $h_1{}^*(x) = x^4 + x^3 + x^2 + 1$ shifts. The direct product code so obtained is described by the tern [42,8,12].

We can observe that, in the parity check matrix of a direct product code, there are rows bounded on both sides (those present in the upper part), producing distributed control symbols, and rows unbounded on their left hand side (those present in the lower part), producing final control symbols. This confirms the interpretation of direct product codes as particular MLC codes (Sect. 1.17). In the framework of convolutional codes, owing to the equivalence described in Sect. 3.7, because of the unbounded rows in the lower part of their $\underline{H}$ matrix, they have to be considered not well designed convolutional codes. On the other hand, the low-weight coefficients

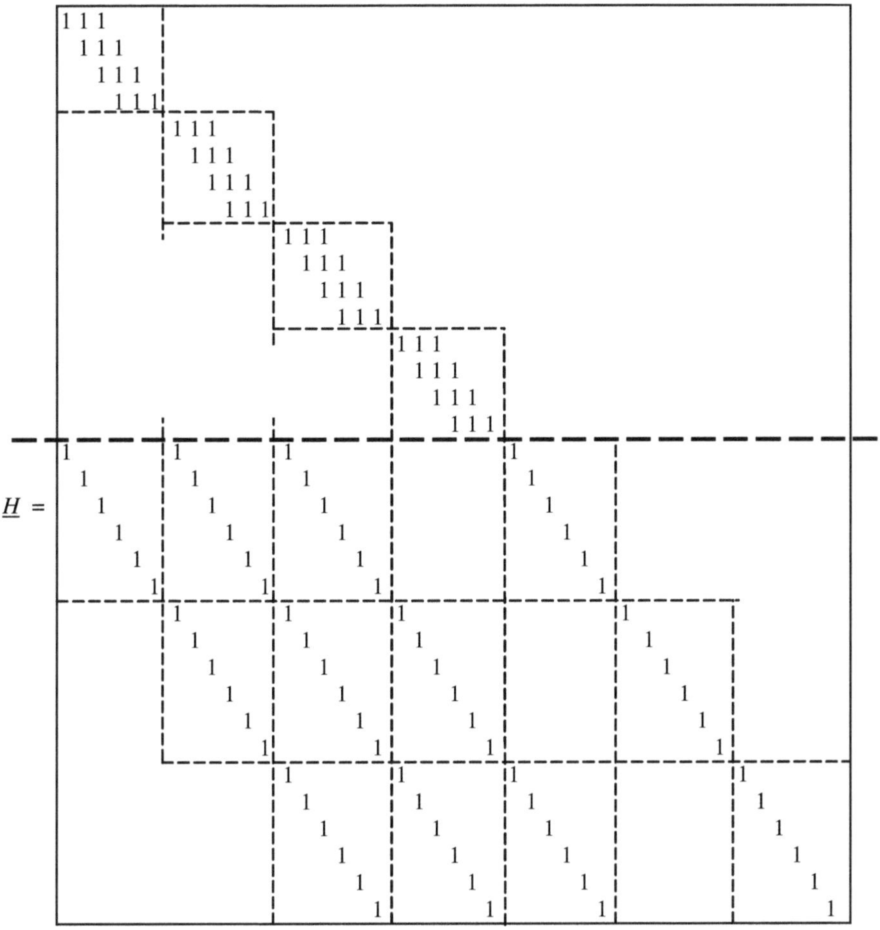

Fig. 5.31 Parity check matrix of the [42,8,12] direct product code described in Example 5.44

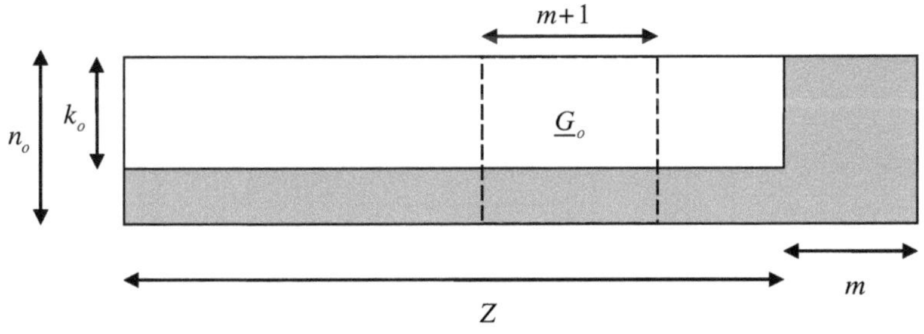

Fig. 5.32 Rectangular memory for interpreting the encoding process in a convolutional code

$A(w)$, in direct product codes, increase more than linearly with one of the two sizes of the rectangular memory employed for their encoding, so confirming this interpretation.

About the last analogy, it is possible to introduce a rectangular memory acting as an encoding memory also for convolutional codes. Since the number Z of periods in the frame is typically much greater than the period length n_o such rectangle will have one size much greater than the other. In Fig. 5.32 this rectangular memory is shown, adopting n_o as the vertical size and $(Z + m)$ as the horizontal size, m being the generator constraint length of our convolutional code (see Definition 3.2). The generator sub-matrix $\underline{G}_o$ covers $(m + 1)$ periods and is responsible for the final about mn_o control symbols, needed for a correct frame termination. It scans this rectangular memory, shifting by one column at any period, from left to right. Control symbols have been evidenced with a grey colour.

Comparison with Fig. 1.18 allows one to conclude that the particular case of a direct product code will be characterized by $n_o = n_1$, $k_o = k_1$, $Z = k_2$, $m = r_2$. In this rectangular memory we can consider perpendicular trajectories for the two component codes, in the direct product model. The generator sub-matrix of the equivalent not well designed convolutional code coincides with the Kronecker generator sub-matrix of our direct product code.

5.17 Composite Codes Based on GPC Codes

At the end of Sect. 1.18 the direct product code of two parity check codes was interpreted as a subcode of a GPC code. This view is confirmed by the $\underline{H}$ matrix of this

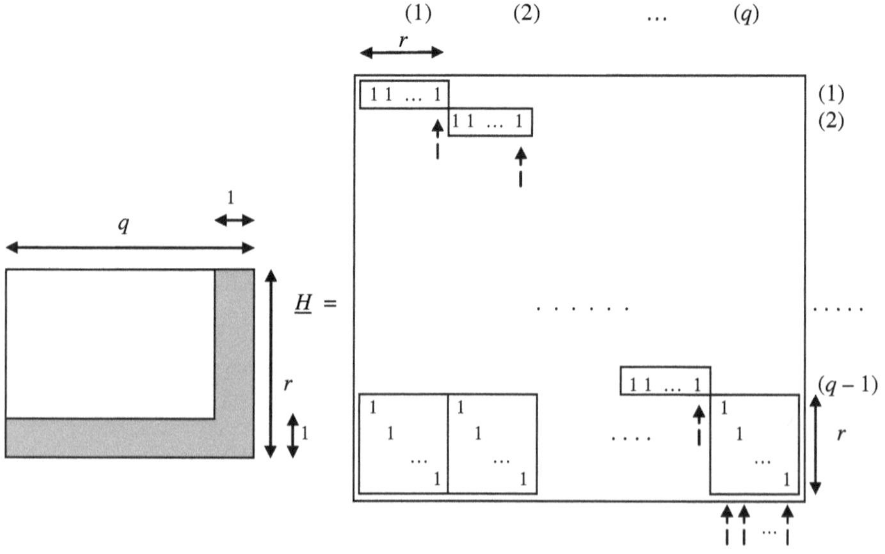

Fig. 5.33 Encoding memory and $\underline{H}$ matrix for the direct product code of two parity check codes

code, obtained according to the structure depicted in Fig. 5.30, which is shown in Fig. 5.33, together with the encoding memory to be employed. The two parity check codes here considered exhibit the terns $[q, q-1, 2]$, acting on the rows, and $[r, r-1, 2]$, acting on the columns. The tern $[qr, (q-1)(r-1), 4]$ describes this code. We can observe rows where unitary polynomials shift in the upper part of the $\underline{H}$ matrix, and a layer of $r \times r$ identity matrices I_r, having q elements, in its lower part. Clearly an equivalent parity check matrix can be obtained exchanging the roles between r and q. Dashed small arrows denote, as usual, the positions where control symbols are inserted. They are $(q + r - 1)$ on the whole, as many as the rows of the $\underline{H}$ matrix.

The lower part of the $\underline{H}$ matrix in Fig. 5.33 can be interpreted as due to a GPC code (see its parity check matrix in Fig. 5.26 down), which plays the role of an inner code in a direct serial concatenation, being the outer code represented by an ordered ensemble of code words taken from the same $[r, r-1, 2]$ parity check code (upper part in the overall $\underline{H}$ matrix). This $\underline{H}$ matrix, looking at the model shown in Fig. 5.32, can be interpreted also as an s.s. convolutional code characterized by $Z = q - 1$, $m = 1$, $n_o = r$, $k_o = r - 1$. It is a not well designed convolutional code, as a consequence of the layer of periodic unbounded rows in the lower part of $\underline{H}$. Such property holds, as already noted, for even more general direct product codes.

Considering now the $\underline{G}$ matrix describing the direct product code, constructed with a pair of parity check codes, following its interpretation as an MLC code, it is characterized by a unique generator polynomial, expressed by (1.39) or by (1.40), as obtained in Sect. 1.17. This generator polynomial is the product of two factors, the first coincident with the generator polynomial of a traditional parity check code, and the other one coincident with the generator polynomial of a GPC code. In other terms, we obtain $g(x) = (x+1)(x^q+1)$ in a generator sub-matrix with $(r - 1)$ rows, or $g(x) = (x+1)(x^r+1)$ in a generator sub-matrix having $(q - 1)$ rows.

Let us consider now the dual of this code, which is characterized by a $\underline{G}$ matrix coincident with the parity check matrix shown on the right in Fig. 5.33. It is a time-varying code in its generator matrix, in fact exhibits two different generator polynomials, both related to proper unitary polynomials: the first one is $u_r(x)$ and the second one is $u_q(x^r)$. The unitary polynomial $u_r(x)$ is the generator polynomial of a traditional binary repetition code. Polynomial $u_q(x^r)$ is the generator polynomial of a GR code, as given by (5.18). As usual, the roles of these two component codes can be exchanged.

Example 5.45 Let us consider the [9,4,4] direct product code obtained employing a pair of [3,2,2] parity check codes. Its dual is a [9,5,3] code, whose generator matrix results in

$$\underline{G} = \left[\begin{array}{ccccccccc} 1 & 1 & 1 & & & & & & \\ & & & 1 & 1 & 1 & & & \\ \hline 1 & & & 1 & & & 1 & & \\ & 1 & & & 1 & & & 1 & \\ & & 1 & & & 1 & & & 1 \end{array}\right],$$

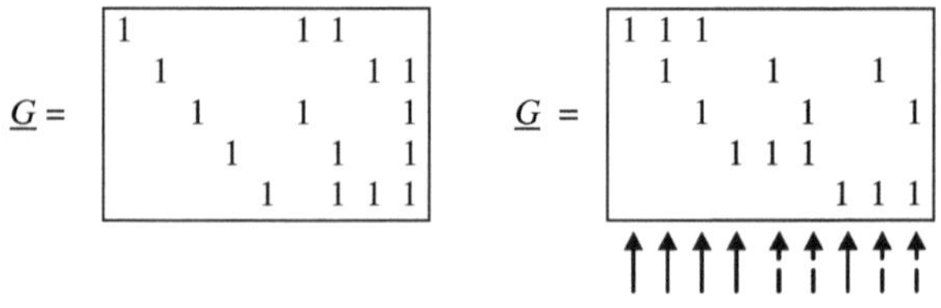

Fig. 5.34 Two alternative forms for the $\underline{G}$ matrix of the [9,5,3] code described in Example 5.45

This code is characterized by the following symmetric word weight distribution

$$A(0) = 1, A(3) = 6, A(4) = 9, \ \mathrm{A}(5) = 9, A(6) = 6, A(9) = 1$$

The above $\underline{G}$ matrix is not in upper triangular form. Its systematic form, which is shown in Fig. 5.34, on the left, appears strongly time-varying, and its obtainment has required exchange in column positions, besides linear superimpositions of rows. It is possible, instead, to arrange $\underline{G}$ in such a way as to allow recognition of the positions where information symbols are inserted, maintaining the only two types of rows used before. This further version for $\underline{G}$ is shown in Fig. 5.34 on the right. Solid small arrows denote information symbol positions, as leftmost 1-symbols in any row. By difference, dashed small arrows denote control symbols in the four remaining positions.

The concept of direct product code can be widened considering more than two orthogonal dimensions in the encoding memory, and as many component codes. The minimum distance turns out to be the product of all the single minimum distances. Unfortunately a drawback is in a progressively stronger reduction of the code rate. Use of high-rate parity check codes, as component codes, somehow limits this problem. Long block codes of this type have been proposed, together with an iterative soft-decision decoding procedure, up to five dimensions (Rankin and Gulliver 2001).

At this point it is useful to introduce a family of codes, obtained employing GPC codes with different orders r_j. Their correspondence with direct product codes is no longer valid when the number of component codes is greater than two. On the other hand, they do not exhibit a resulting minimum distance exponentially increasing with the number of component codes, as occurs for direct product codes.

Definition 5.18 The direct serial concatenation of M different GPC codes gives rise to a family of codes, which can be identified as the *direct concatenation of generalized parity check codes* (or briefly *DC-GPC code*).

Lemma 5.40 *Parity check matrix and generator matrix of a DC-GPC code are schematically shown in Fig.* 5.35, *assuming* r_j, $j = 1, 2, \ldots M$, *as the parameters of the component GPC codes.* $\underline{I}_{r_j}$ *represents an* $r_j \times r_j$ *identity matrix, as usual.*

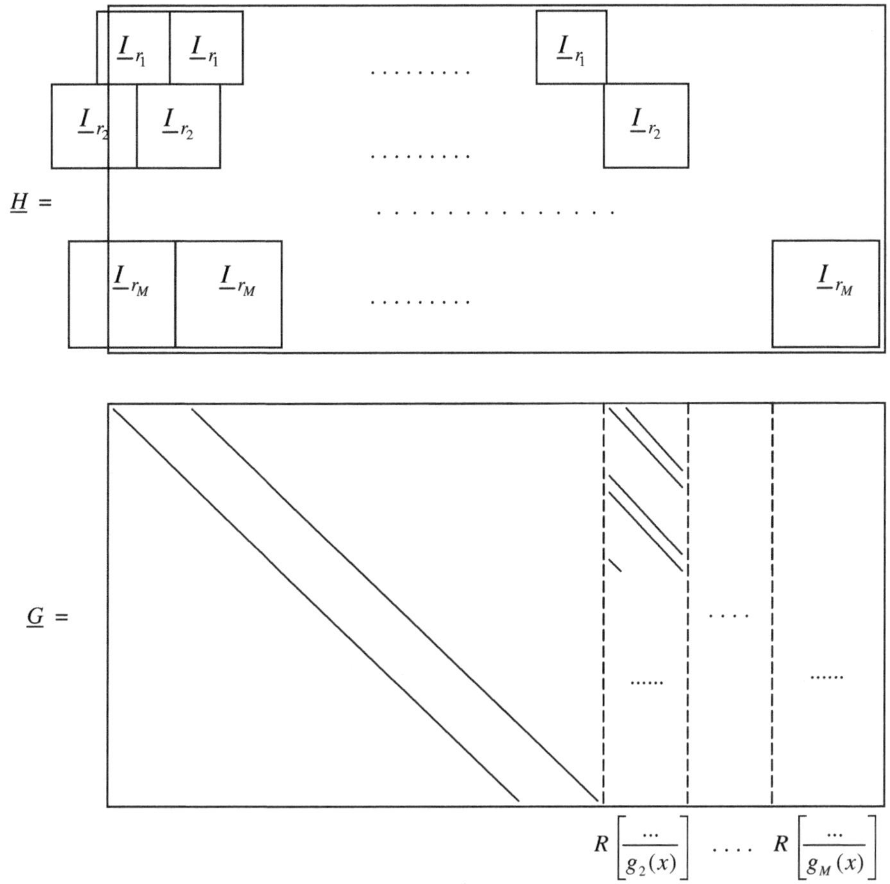

Fig. 5.35 Parity check matrix and generator matrix for a DC-GPC code with M component codes

Proof From Lemma 5.17 (Fig. 5.10), iteratively considered M subsequent times, it is possible to derive the structure of the parity check matrix reported in Fig. 5.35 up. In fact the parity check matrix of a GPC code, with generator polynomial $g_j(x) = (x^{r_j} + 1)$, consists of a layer where a unitary polynomial $u_w(x^{r_j})$ shifts (see Fig. 5.26 down). Any layer of the resulting parity check matrix has been organized in such a way as to form an overall lower triangular matrix. On the left hand side, identity matrices can be truncated as one wishes, in order to reach the desired total code length. Remark that the weight of such rows, is not constrained and can vary widely depending on the choice of the r_j parameters. The generator matrix, instead, is obtained following Lemma 1.17. In the leftmost part of this matrix, the generator matrix of the first component GPC code is reported. Then we have the remainders obtained from the division of what precedes by the polynomial generators (binomials) of the subsequent GPC codes. □

In the Figure, we have chosen a situation where the r_j parameters increase with increasing j, but this is not a forced choice, although it is preferable, in order to reach a higher code rate, as will be better explained in the following. It is important, however, that all the r_j are different, otherwise concatenation becomes useless, as already noted in Sect. 1.6, generally speaking about direct serial concatenation. From the properties of the polynomial division remainder, it is easy to conclude that the weights of all the rows of $\underline{G}$ are even and that they increase with increasing parameter M. The minimum distance d is then an even number as well.

Example 5.46 The parity check matrix of the direct concatenation of two GPC codes, characterized by $r_1 = 3$, $r_2 = 4$, and overall length $n = 16$, is

$$\underline{H} = \left[\begin{array}{cccccccccccccccc}
1 & & & 1 & & & 1 & & & 1 & & & & & & \\
& 1 & & & 1 & & & 1 & & & 1 & & & & & \\
& & 1 & & & 1 & & & 1 & & & 1 & & & & \\
\hline
1 & & & & 1 & & & & 1 & & & & 1 & & & \\
& 1 & & & & 1 & & & & 1 & & & & 1 & & \\
& & 1 & & & & 1 & & & & 1 & & & & 1 & \\
& & & 1 & & & & 1 & & & & 1 & & & & 1
\end{array}\right].$$

This code can be shortened by progressive elimination of columns on the left hand side of such $\underline{H}$ matrix, obtaining a structure where the leftmost square sub-matrices are not complete, so better resembling what depicted in Fig. 5.35 up. The minimum distance of the original code is $d = 4$. Such value remains unchanged after code shortening, but the number $A(4)$ of 4-weight code words progressively decreases, as expected.

Lemma 5.41 *In a DC-GPC code with $M = 2$, parameters r_1 and r_2, co-prime and characterized by $r_1 < r_2$, the overall length $n = r_1 r_2 + r_2$ is the maximum possible, before occurrence of 2-weight code words.*

Proof Let us take the case described in Example 5.46 for developing some general considerations. We can add only one further column on the left hand side of an $\underline{H}$ matrix of this type, and the condition $n = r_1 r_2 + r_2 + 1$ is verified. Such column is formed comprehending the last column of an $\underline{I}_{r_1}$ sub-matrix in its upper layer and the last column of an $\underline{I}_{r_2}$ sub-matrix in its lower layer. So we will obtain two equal columns, separated by $r_1 r_2$ positions, and hence the minimum distance will decrease from $d = 4$ to $d = 2$. This conclusion is independent of the particular values of r_1 and r_2 assumed, provided they are co-prime numbers. □

If r_1 and r_2 are not co-prime, minimum distance reduction occurs after a length n related to the common multiple between them. For this reason, since we are interested in reaching the maximum possible code word length, and hence code rate, when the contrary is not explicitly said, in the following all the r_j's will be considered co-prime. Coming back to Lemma 5.41, we can observe that, inverting the order of the two component codes, that is taking the one characterized by higher r_j

parameter as the first (outer) code, gives rise to a reduced maximum value for n. This conclusion holds because we are treating a direct serial concatenation. Hence the choice of an increasing set of co-prime values for the r_j parameters is the best for reaching the highest code rate possible with $d = 4$. The above property can be easily generalized to the case of whichever number M of component codes, as follows.

Lemma 5.42 *In a DC-GPC code, characterized by M component codes, with parameters r_j, $j = 1, 2, \ldots M$, increasing and all co-prime, the length $n = r_1r_2 + \sum_{j=2}^{M} r_j$ is the maximum possible, before occurrence of two equal and coincident vertical separations between a pair of 1-symbols in as many columns.*

Proof The four 1-symbols involved in the situation here considered form a rectangle whose vertical dimension equals the mentioned vertical separation. From proper iteration of the considerations developed in the proof of Lemma 5.41, the property about the maximum length of n is easily demonstrated. The most critical pair of layers (where the considered situation takes place earlier) remains that formed by the first two layers. □

About the reduction in the minimum distance d due to the above coincidence, it is possible to generalize what expressed by Lemma 5.41, saying that it produces an abrupt halving in the value of d, with respect to its maximum possible value. The situation described in Lemma 5.42 is very important, not only for this reduction in the minimum distance, but also in the perspective of a possible soft-decision decoding procedure, as LDPC code, to be applied to a code of this type (Ryan and Lin 2009). This concept will be better clarified in Chap. 10.

Definition 5.19 Considering $M = 2$ component codes, consisting of a pair of GPC codes with parameters r_1 and r_2 co-prime, and assuming an overall code word length $n = r_1r_2$, we can arrange an $\underline{H}$ matrix with two layers. The upper one will exhibit exactly r_2 square sub-matrices $\underline{I}_{r_1}$, and the lower one exactly r_1 square sub-matrices $\underline{I}_{r_2}$. This code is called *direct product concatenation of GPC codes* (or briefly *DP-GPC code*).

A DP-GPC code can be constructed by elimination of the last r_2 columns in the $\underline{H}$ matrix of a DC-GPC code, like that introduced in Definition 5.18 over its maximum possible length. Remark that, in the case of a DP-GPC code, the parity check matrix is no longer lower triangular. Furthermore one row is linearly dependent on the other ones, so the code turns out to be associated to the following tern $[r_1r_2, r_1r_2 - r_1 - r_2 + 1, d]$.

Example 5.47 In Fig. 5.36 on the right, a comparison between the $\underline{H}$ matrices in a DC-GPC code and in the correspondent DP-GPC code is shown, still for $r_1 = 3$ and $r_2 = 4$. On the left, we have the memories which can be adopted for encoding such codes. These memories are written by rows. Numbers in square borders denote the positions where the first component code acts (along as many columns), whereas numbers in round borders denote the positions where the second component code acts (along oblique trajectories). Any number expresses the correspondent row of $\underline{H}$. Dashed borders and a grey substrate denote control symbols, which, as usual, are taken as the rightmost 1-symbol in any row of $\underline{H}$. Up in the Figure, the direct concatenation solution is reported, which is associated to the $\underline{H}$ matrix already introduced in Example 5.46. There is an addition of $r_2 = 4$ positions with respect to the rectangle $r_1 \times r_2$, where the control symbols of the second component code are placed. This addition has been made prolonging the trajectories of the parity check equations due to the second component code. Down in the Figure, the direct product concatenation solution is described. Here the additional positions have been eliminated, and the control symbols of the second component code have been superimposed to those of the first component code. Two of such superimpositions can be resolved by transferring the control symbols denoted 5 and 6 in the rightmost position of the first and the second row. The third superimposition (denoted 7) is justified by the existence of one row in $\underline{H}$ linearly dependent on the other ones, as already noted.

In general, the trajectories of the parity check equations in the first component code are vertical lines. The trajectories of the parity check equations in the second component code are oblique and tilted by an angle which is smaller and smaller with increasing the difference $(r_2 - r_1)$. Except for the last property, the case treated in Example 5.47 can be considered completely general, maintaining $M = 2$. We can observe how the trajectories of two parity check equations, taken from different component codes, intersect just in one position. This property guarantees that the minimum distance remains $d = 4$. In fact, when two trajectories intersect in two positions, two equal columns appear in the $\underline{H}$ matrix, so reducing d from 4 to 2.

Lemma 5.43 *A DP-GPC code of the type introduced in Definition* 5.19, *with* $M = 2$, *is equivalent to the direct product code of two parity check codes.*

Proof Taken a rectangular encoding memory, whose sizes are r_1 and r_2, like that reported in Fig. 5.36 down, in the tilted trajectories (associated to the second component code), it is possible to exchange the positions of control symbols, in such a way as to align them all in the last column, as already noted. So we obtain a structure coincident with that characterizing a direct product code. The reordering procedure suggested above has the meaning of transforming a GPC code into its constituent parity check codes. □

Example 5.48 Let us consider the DP-GPC code whose encoding memory and parity check matrix are reported in Fig. 5.36 down, which has been constructed assuming $r_1 = 3$ and $r_2 = 4$. It is a [12, 6, 4] code, equivalent to a direct product code obtained with the [3,2,2] and [4,3,2] parity check code, taken as component codes (see also Sect. 1.18). The word weight distribution of such code is $A(0) = 1$, $A(4) = 18$, $A(6) = 24$, $A(8) = 21$. About $A(4)$, expression (1.41) is easily verified.

The equivalence stated by Lemma 5.43 holds for any difference $(r_2 - r_1)$, but is no longer true for $M \geq 3$. With increasing M, the two families of codes exhibit properties progressively diverging. The direct product code shows $d = 2^M$, at the expense of a strong reduction in the code rate. A DP-GPC code, with more than two

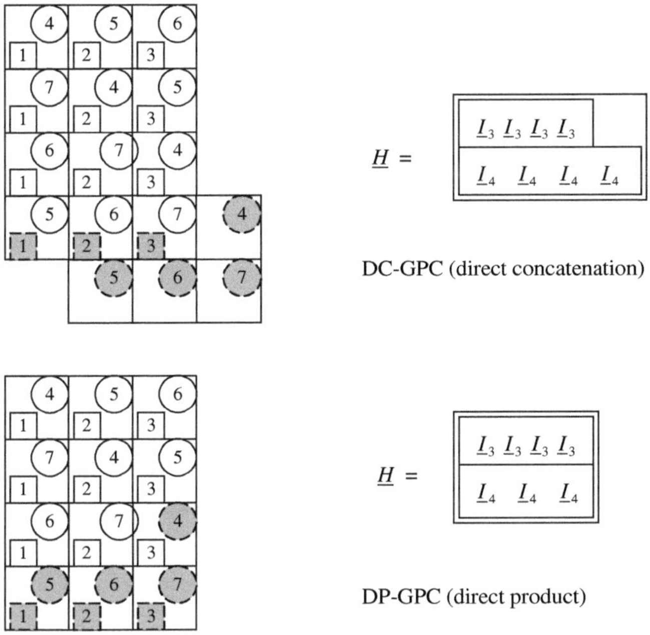

Fig. 5.36 Encoding memories for a DC-GPC code and for a DP-GPC code

component codes, is able to guarantee typically only smaller values for d, but its code rate can be maintained rather high, especially if one chooses r_j values not too small, and properly selected (Baldi et al. 2009, 1).

Lemma 5.44 *A DC-GPC code, of the type introduced in Definition* 5.18, *with* $M = 2$, *and parameters* r_1, r_2, *is equivalent to the direct product code between the pair of parity check codes characterized by parameters* $(r_1 + 1)$, r_2.

Proof Looking at the encoding memory reported in Fig. 5.36 up, it is possible to transfer the three control symbols of the second component code, denoted 5, 6, 7, in a fourth column, properly arranged, before the control symbols denoted 4, in exchanged order. The scheme so obtained coincides with that of a direct product code, like in the proof of Lemma 5.43, with the only difference of having parameter r_1 increased by one unit. □

Example 5.49 The DC-GPC code, whose encoding memory and parity check matrix are reported in Fig. 5.36 up, is a [16,9,4] code, equivalent to a direct product code obtained with two [4,3,2] parity check codes taken as component codes. It is possible to verify that, for both, the word weight distribution turns out to be $A(0) = 1$, $A(4) = 36$, $A(6) = 96$, $A(8) = 246$, $A(10) = 96$, $A(12) = 36$, $A(16) = 1$.

In Chap. 11 the family of DC-GPC codes and the family of PC-GPC codes will be better investigated, since they are very good LDPC codes. Also the minimum distance of such codes will be carefully analyzed, on the basis of the number of intersections exhibited by clusters of trajectories in their encoding memory, in turn produced by equalities relating the r_j's. Typical values for parameter M, in order to reach a good trade-off about code rate, word length n and minimum distance d, can vary between 4 and 6.

5.18 $\underline{H}$ Matrix for Non-binary Block Codes

The co-factor of polynomial $g(x)$ with respect to $(x^N - 1)$ can be assumed as the parity check polynomial $h(x)$ characterizing Q-ary cyclic codes. Rigorously, for Q-ary pseudo-cyclic codes, the longer binomial $(x^{(Q-1)N} - 1)$ should be taken into account. In the latter case, since we are interested anyway in the length of the first fraction of the whole super-cycle (except for LC codes), it is convenient to take the co-factor with respect to $(x^N - s_j)$ (see Sect. 1.19 for the symbolism). The reciprocal $h^*(x)$ of the parity check polynomial is finally adopted for the construction of the rows in the parity check matrix $\underline{H}$ of such non-binary block codes.

Definition 5.20 The parity check polynomial $h(x)$ of a Q-ary code turns out to be

$$\begin{aligned} h_c(x) &= \frac{x^N - 1}{g(x)}, \ (Q\text{–ary cyclic code}), \\ h_{pc}(x) &= \frac{x^N - s_j}{g(x)}, \ (Q\text{–ary pseudo-cyclic code}). \end{aligned} \tag{5.21}$$

The only difference is in the code lengthening procedure, which is to be performed, in the case of a pseudo-cyclic code with the above choice, after multiplication of any replica of $h_{pc}(x)$ by subsequent powers of the primitive element α, associated to the non-binary symbol s_j.

Example 5.50 Decomposition of binomial $(x^5 + 1)$, for $Q = 4$, gives

$$x^5 + 1 = (x + 1)(x^4 + x^3 + x^2 + x + 1), \ Q = 4,$$

in which also the second factor is irreducible. The $[5,4,2]_4$ quaternary parity check cyclic code characterized by $g(x) = x + 1$ exhibits the parity check polynomial $h_c(x) = x^4 + x^3 + x^2 + x + 1$.

Example 5.51 In Example 1.65 the decomposition of binomial $(x^5 + 2)$ for $Q = 4$ has been reported, giving

$$x^5 + 2 = (x + 3)(x^2 + 2x + 2)(x^2 + x + 2), \ Q = 4.$$

Assuming $g(x) = x^2 + 2x + 2$, which represents the generator polynomial of a $[5,3,3]_4$ quaternary Hamming pseudo-cyclic code, we are left with $h_{pc}(x) = (x+3)(x^2+x+2) = x^3 + 2x^2 + x + 1$.

In Sect. 5.11 we have introduced the minimal form for a parity check matrix (see Definition 5.13) in a cyclic, punctured or SC binary code. It is constructed by means of a unique reciprocal parity check polynomial, which, except for a possible truncation on its left hand side due to code shortening, shifts along the rows by one position at a time. This definition still holds here for both types of non-binary polynomial codes.

Example 5.52 Taking the $[5,3,3]_4$ quaternary Hamming pseudo-cyclic code introduced in Example 5.51, in Fig. 5.37 the generator matrices and the parity check matrices associated to its pseudo-cyclic, one-time shortened and two-time lengthened versions are compared. For the lengthened code, the periodic behaviour of the overall $h(x)$ is characterized by a multiplication by 2 in the first replica, furthermore the parity check matrix has been maintained in lower triangular form. All these parity check matrices can be considered in their minimal form, the last one also according to Definition 5.14, which is a widening of Definition 5.13.

The minimal form of the parity check matrix is useful for describing a convolutional code and sometimes an LDPC code. On the contrary, in order to perform hard-decision decoding of BCH codes, it is preferable to assume this matrix in a form based on subsequent powers of the primitive element α, of the type reported in Example 5.28 for the [15,5,7] binary BCH code. This is recommended for allowing to follow a procedure, based on the BCH minimum distance bound. A situation particularly interesting is that of Reed-Solomon codes.

Example 5.53 The symbolic construction of the $\underline{H}$ matrix, based on subsequent powers of the primitive element α, for a $[7,4,4]_8$ RS code (see Definition 1.39), gives

$$\underline{H} = \begin{bmatrix} 1 & \alpha^1 & \alpha^2 & \alpha^3 & \alpha^4 & \alpha^5 & \alpha^6 \\ 1 & \alpha^2 & \alpha^4 & \alpha^6 & \alpha^1 & \alpha^3 & \alpha^5 \\ 1 & \alpha^3 & \alpha^6 & \alpha^2 & \alpha^5 & \alpha^1 & \alpha^4 \end{bmatrix}.$$

This matrix is clearly not in minimal form, in fact all its rows have degree 6. Let us remember that the cycle of the subsequent powers of α has length $Q - 1 = 7$, as the number of non-null symbols of the considered alphabet, included $\alpha^0 = 1$. Comparison with what presented in Sect. 5.11 shows a strong analogy, but one difference: here also even values for d are admitted, as a consequence of the non-binary alphabet adopted. The decoding procedure is very straightforward, by means of a proper syndrome processing (Mac Williams and Sloane 1977). The equivalent minimal form of $\underline{H}$ should exhibit a unique $h_c{}^*(x)$ having degree 4, but it is useless for the above purpose.

	pseudo-cyclic	1-time shortened	2-time lengthened
$\underline{G}$	$\begin{bmatrix} 1 & 2 & 2 & & \\ & 1 & 2 & 2 & \\ & & 1 & 2 & 2 \end{bmatrix}$	$\begin{bmatrix} 1 & 2 & 2 & \\ & 1 & 2 & 2 \end{bmatrix}$	$\begin{bmatrix} 1 & 2 & 2 & & & & \\ & 1 & 2 & 2 & & & \\ & & 1 & 2 & 2 & & \\ & & & 1 & 2 & 2 & \\ & & & & 1 & 2 & 2 \end{bmatrix}$
$\underline{H}$	$\begin{bmatrix} 1 & 1 & 2 & 1 & \\ & 1 & 1 & 2 & 1 \end{bmatrix}$	$\begin{bmatrix} 1 & 2 & 1 & \\ 1 & 1 & 2 & 1 \end{bmatrix}$	$\begin{bmatrix} 2 & & 1 & 1 & 2 & 1 & \\ 3 & 2 & & 1 & 1 & 2 & 1 \end{bmatrix}$

Fig. 5.37 Generator and parity check matrix for codes related to the $[5,3,3]_4$ quaternary Hamming pseudo-cyclic code

Example 5.54 Considering the generator polynomial of the $[7,4,4]_8$ RS code already treated in Example 5.53, represented by its symbolic expression, after (1.48), we can write

$$g(x) = (x - a)(x - a^2)(x - a^3).$$

From (1.47), the correspondent parity check polynomial will be

$$h_c(x) = (x - a^0)(x - a^4)(x - a^5)(x - a^6).$$

The reciprocal of this polynomial can be employed for the construction of the parity check matrix of this code in its minimal and s.s. time-invariant form. It is possible to show that the three rows in the $\underline{H}$ matrix given in Example 5.53 are suitable linear superimpositions of the rows described here, after proper shift of $h_c{}^*(x)$. Precisely each of them is respectively the co-factor of one of the following symbolic binomials $(x - a)$, $(x - a^2)$, $(x - a^3)$, with respect to $(x^7 - 1)$. Remark the analogy with what described in Sect. 5.11, for the binary case.

About the possibility of repeated code extensions, in the perspective of gaining a progressive increase in the minimum distance d, we have to observe what follows. In Q-ary codes, whose code words have length N, the first code extension is performed by adding a final row in the $\underline{H}$ matrix where the unitary polynomial $u_{N+1}(x)$ takes place. This possibility always exists, since a factorization of the type $x^{N+1} - 1 = (x-1)u_{N+1}(x)$ is valid for any value of N. Repeating this operation with a second unitary polynomial $u_{N+2}(x)$ is useless, because the sum of such two rows gives rise to a row with just one 1-symbol in the last position. This conclusion, which is independent of Q, is due to the fact that multiplication of $u_{N+2}(x)$ by whichever non-null element of the field (in particular the addition opposite of 1) originates a row still valid. In conclusion, it is necessary to find another polynomial, not unitary, as co-factor of a binomial $(x - s_j)$, for $s_j \neq 1$, with respect to $(x^{N+2} - 1)$. Only at this point, it is possible to check if the minimum distance is increased.

Lemma 5.45 *In a Q-ary alphabet, the maximum number of single code extensions is* $(Q-1)$.

Proof The repeated operation described before can be performed with a different non-null element s_j any time. The same result can be reached considering the $\underline{G}$ matrix, and the remainders of the division by a proper binomial $(x - s_j)$. □

Example 5.55 Starting from the $[5,3,3]_4$ quaternary Hamming code, and using $u_6(x)$, we obtain a $[6,3,4]_4$ extended Hamming code. Nevertheless, although the addition of a further row in the $\underline{H}$ matrix, is possible, the minimum distance does not increase further.

5.19 Trellis of an s.s. Time-Invariant Block Code Obtained from Its Parity Check Matrix

In Sect. 1.23 the trellis obtained from polynomial $g(x)$ in a binary block code, s.s. time-invariant in its $\underline{G}$ matrix, has been introduced. It is based on the state diagram presented in Sect. 1.14, starting from the encoder circuit in Fig. 1.3 or Fig. 1.5. On the other hand, in Sect. 5.5 the state diagram based on the parity check polynomial, for a block code, s.s. time-invariant in its $\underline{H}$ matrix, has been discussed, starting from the encoder circuit in Fig. 5.6. Here we will consider the temporal evolution of this state diagram, that is an alternative form of trellis, for the same block polynomial code.

Definition 5.21 The temporal evolution of a state diagram obtained from polynomial $h(x)$, either in a block or in a convolutional code, is called *parity check trellis*. The states where two paths merge are called *merge states* (see also Definition 1.40).

Here the merge states, if present, should be characterized by the following property, directly deriving from the structure of the encoder circuit shown in Fig. 5.6. In any of such states, two paths merge only if they show the same $(N - r)$ final output symbol sequences, condition which imposes the identity of the state considered. On the other hand, the parity check trellis exhibits different behaviours for cyclic, shortened, punctured, H-extended and modified H-extended code versions. As noted in Sect. 5.9, it is not convenient to obtain an LC code from this encoder circuit. We will firstly analyze cyclic codes. Then it will be straightforward to extend this analysis also to other types of codes derived from them.

First of all, it is possible to note that a parity check trellis has a maximum number of states per stage given by 2^{N-r}, as many as the total number of states in the state diagram. Letting k the number of information symbols, since we are treating a systematic encoder, the first k stages are characterized by binary bifurcations. In the case of cyclic codes (for which we have exactly $k = N - r$), after the first k stages, we reach all the states. In the case of SC codes (for which we have $k < N - r$), we reach a subset of all the states. Next, the addition of the control symbols is made on closed routes inside the parity check state diagram (see Lemma 5.16 and Theorem 5.2). Such closed routes, being made by N consecutive 0-symbols (or by a factor of N consecutive 0-symbols) in the generator state diagram of the dual code, have no common nodes. In conclusion, the addition of the r final control symbols can contribute also to reach new states (as occurs for SC codes), but makes it impossible to have merge states.

The structure of a parity check trellis is schematically shown in Fig. 5.38. After an expanding part, made by the first k stages, there is a part containing the remaining r stages. Although, in this second part of the trellis, the overall number of touched states can be even higher than 2^k (if $k < N - r$), it is impossible to have more than 2^k states per stage, because there are no further bifurcations, being the information symbols all already used. On the other hand, since there are no merge states, as observed before, their number cannot become smaller than 2^k. In conclusion, the number of states per stage remains exactly equal to 2^k, and they are all transit states.

At this point, a unique comparison among the metrics of all the 2^k paths in the trellis should be made. This procedure resembles that occurring in non-binary trellises about branch complexity. Nevertheless, in order to better compare the present situation with that of a soft-decision decoding algorithm, performed in a binary generator trellis, we can add further 0-symbols, continuing the parity check trellis for k further stages, with the goal of reaching state $\underline{0}$. All these branches will be labelled with the same sequence of k subsequent 0-symbols, without modifying the metrics. Here a decreasing number of binary merge states per stage, according to the powers of 2, takes place. The first stage of this added part of the trellis (depicted with dashed line in Fig. 5.38) is formed by 2^{k-1} binary merge states.

Example 5.56 Let us consider the [7,3,4] cyclic simplex code, whose generator polynomial is $g(x) = 11101$. It is possible to see, from Fig. 5.39, that its parity check trellis exhibits no merge states in all its 7 subsequent stages. Only some paths have been labelled with branch symbols, since they can be recognized from the state evolutions. We can observe how the eight paths representing as many code words remain separate along the 4 stages associated to the $r = 4$ control symbols, confirming what anticipated for $k = N - r$. If such trellis is continued with 3 further dummy symbols (dashed branches), able to drive all the paths present in the state diagram to state 000, the first new stage exhibits $2^{N-r-1} = 4$ merge states, according to what anticipated in Fig. 5.38, in the case of a cyclic code.

Example 5.57 Let us consider the [5,2,3] two-time shortened Hamming code, whose generator polynomial is $g(x) = 1101$. It is possible to see, from Fig. 5.40, that its parity check trellis exhibits no merge states in all its 5 stages. We can observe how the four paths representing as many code words remain separate along the 3 stages associated to the $r = 3$ control symbols, confirming what anticipated with $k < N - r$. In this case the overall number of touched states is greater than $2^k = 4$, but there are no merge states, and hence they remain 4 states per stage. If such trellis is continued with 4 further dummy symbols (dashed branches), able to guide all the

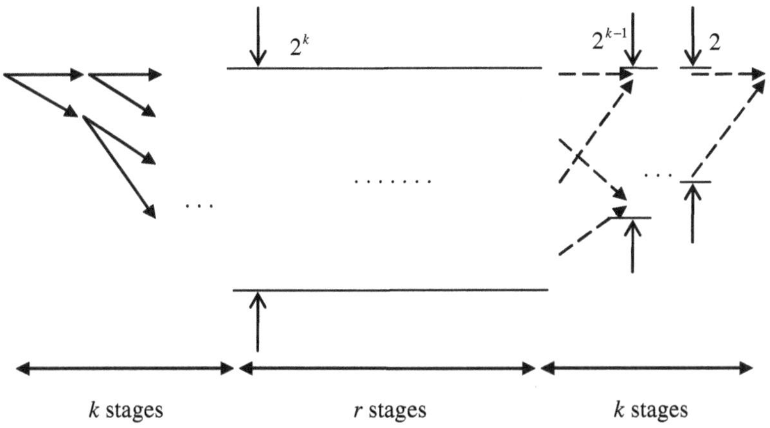

Fig. 5.38 Trellis obtained from a parity check polynomial $h(x)$

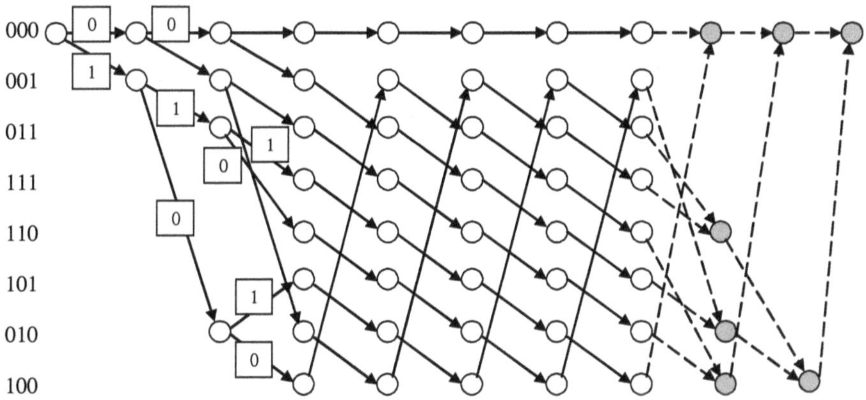

Fig. 5.39 Trellis associated to polynomial $h(x)$ for the [7,3,4] cyclic simplex code ($k = N - r = 3$)

paths present in the state diagram to state 0000, the first stage where branch merging occurs exhibits $2^{k-1} = 2$ merge states. Precisely, after the first 3 dummy 0-symbols, the paths in state 0111 and 1101 will merge in state 1000, whereas the paths in state 0000 and 1010 will merge in state 0000.

A punctured code can be interpreted as a cyclic code in which some final control symbols are suppressed. The structure of the trellis in Fig. 5.38 does not change if one replaces parameter r with a new parameter $r' < r$. In particular, the property of having 2^k separate paths along the second part of the trellis, made here by r' stages, is still true. An H-extended code, in turn, can be interpreted as a cyclic code with some further final control symbols, imposed by the same parity check polynomial $h(x)$. The structure of the trellis in Fig. 5.38 does not change if one replaces parameter r with a new parameter $r' > r$. In particular, the property of having 2^k

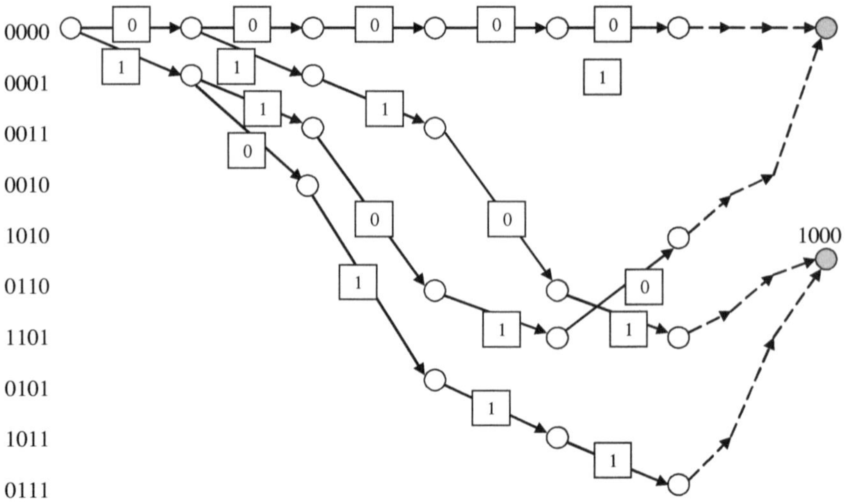

Fig. 5.40 Trellis associated to polynomial $h(x)$ for the [5,2,3] two-time shortened Hamming code ($k = 2$, $N - r = 4$)

separate paths along the central trellis segment, made by r' stages, is still true. A difference occurs only in MHEC codes, where, owing to distributed information symbols, some merge states will be present also along the code word (code frame) characterizing them. The same is for block codes whose reordered form leads to an alternation of information and control symbols.

In Sect. 1.23 we have discussed the computational complexity of a soft-decision decoding algorithm, applied to block codes, based on the generator trellis. This decoding procedure is possible for s.s. time-invariant polynomial codes in their $\underline{G}$ matrix. Here we can observe that, employing a parity check trellis, for an s.s. time-invariant polynomial block code in its $\underline{H}$ matrix, as made so far, it is possible to reach the value of 2^{k-1}, as the maximum number of binary merge states per stage. This limit coincides with that obtained in Sect. 1.23 for a generator trellis under the condition $k \leq r$. Otherwise it is convenient to employ the generator trellis, for which the number of merge states per stage is constrained to be 2^r.

Example 5.58 Let us consider the [3,2,2] cyclic parity check code, whose generator polynomial is $g(x) = x + 1$. in spite of the fact that the generator trellis would be more convenient, in Fig. 5.41 its parity check trellis has been reported. We can observe how the four paths representing as many code words remains separated in both the two stages associated to insertion of the unique control symbol. If such trellis is continued with 2 further dummy symbols (dashed branches), the first stage exhibits 2 merge states, exactly the same number of the solution obtained employing the generator trellis. Nevertheless we can imagine how this parity check trellis is to be transformed for a [4,3,2] cyclic parity check code. Now we will have a final stage with $2^3 = 8$ states, still without merge states. Also here we can add

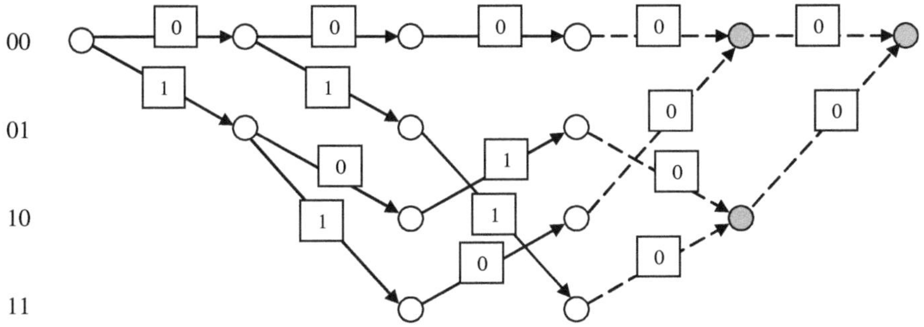

Fig. 5.41 Parity check trellis for the [3,2,2] cyclic parity check code ($k = N - r = 2$, $r = 1$)

some further stages, but in the first one of these additional stages, we will have 4 merge states, twice the number of merge states in the generator trellis, making this solution not convenient. Such conclusion is progressively enforced for all the [n, $n - 1$, 2] parity check codes, with $n \geq 4$.

In conclusion, in all the cases examined so far (characterized by all information symbols in the first part of the code word), there are no benefits in employing a parity check trellis. In the best condition, the computational decoding complexity equals that characterizing the generator trellis which can be used for the same code. A separate treatment is needed for an MHEC code, whose behaviour is that of a convolutional code described by means of its $\underline{H}$ matrix, which is still s.s. time invariant. Here merge states are produced by distributed information symbols, similarly to what described here after the addition of dummy symbols. In convolutional codes different possible encoding circuits are traditionally employed, like those already presented in Chaps. 3 and 4, exploiting shift registers in controller or in observer arrangement. The search for a minimal basic encoder circuit (Definition 4.7), as that which minimizes the decoding computational complexity in a proper trellis, can lead to a circuit which often is neither completely in controller arrangement nor completely in observer arrangement (Johannesson and Zigangirov 1999). Such considerations will be developed with more details in Chaps. 7 and 8.

References

Baldi M, Cancellieri G, Carassai A et al (2009) LDPC codes based on serially concatenated multiple parity—check codes. IEEE Comm Lett 13:142–144

Berlekamp ER (1965) On decoding Bose-Chauduri-Hocquenghem codes. IEEE Trans Inf Theory 11:577–580

Clark GC, Cain JB (1983) Error correcting coding for digital communications. Plenum, New York

De Bruijn NG (1946) A combinatorial problem. J London Math Soc 21:167–169

Gorenstein DC, Zierler N (1961) A class of error correcting codes in p^m symbols. J SIAM 9:207–214

Johannesson R, Zigangirov KS (1999) Fundamentals of convolutional coding. IEEE Press, New York

Klove T (2007) Codes for error detection. World Scientific, Singapore

MacWilliams FJ, Sloane NJA (1977) The theory of error correcting codes. North Holland, New York

Massey JL (1969) Shift-register synthesis and BCH decoding. IEEE Trans Inf Theory 15:122–127

Peterson WW (1960) Encoding and error-correction procedures. IRE Trans Inf Theory 6:459–470

Rankin DM, Gulliver TA (2001) Single parity check product codes. IEEE Trans Commun 49:1354–1362

Ryan WE, Lin S (2009) Channel codes: classical and modern. Cambridge University Press, Cambridge

Sweeney P (2002) Error control coding. Wiley, New York

Wicker SB (1995) Error control systems for digital communications and storage. Prentice-Hall, Englewood Cliffs

Chapter 6
Wide-Sense Time-Invariant Block Codes in Their Parity Check Matrix

6.1 Periodically Time-Varying Parity Check Matrix

Keeping in mind the description of w.s. time-invariant block codes in their generator matrix, developed in Chap. 2, it is possible to begin here the analysis of their counterpart, after the substitution of generator matrix with parity check matrix. It is convenient to start from the concept of time-varying $\underline{H}$ matrices. Examples of such parity check matrices, employing some different parity check polynomials, can be taken from Sects. 5.6 and 5.7, where concatenated codes and punctured codes have been described.

Definition 6.1 A parity check matrix, where the reciprocals of some parity check polynomials $h_1{}^*(x), j = 1.2, \ldots q_o$, among which at least two are different, occupy as many consecutive rows and shift along its length, forming a period of length q_o, is called a *periodically time-varying parity check matrix*.

Comparison with Definition 2.1 shows strong analogy. Nevertheless our attention is here to be focused on the rightmost one symbol (lowest-order coefficient) of such polynomials, where the associated control symbol is expected to be placed. The treatment of the linear block codes, with a parity check matrix of the type described in Definition 6.1, is greatly simplified if none of the polynomials $h_j{}^*(x)$ has its lowest-order coefficient in correspondence with the lowest-order coefficient of another one. A particular case of this situation is when the parity check matrix considered is lower triangular (see Definition 5.3), so assuring that all the control symbols are inserted in the final part of the code word. Finally it is useful to remember that only rows having at least weight 2 are valid as parity check rows.

Example 6.1 Taking $q_o = 2$ and $h_1{}^*(x) = x^2 + x + 1$, $h_2{}^*(x) = x^4 + x + 1$, in Fig. 6.1 a periodically time-varying code in $\underline{H}$, is described. On the left, its $\underline{H}$ matrix is reported, in which the polynomials are shifted in such a way as to form three periods in a lower triangular overall structure. On the right its $\underline{G}$ matrix is shown. It has been obtained applying the method based on the null ex-OR sum of clusters of

G. Cancellieri, *Polynomial Theory of Error Correcting Codes*,
Signals and Communication Technology, DOI 10.1007/978-3-319-01727-3_6

$$
\underline{H} = \begin{bmatrix}
 & 1 & 1 & 1 & & & & & \\
1 & 0 & 0 & 1 & 1 & & & & \\
 & & & 1 & 1 & 1 & & & \\
 & & 1 & 0 & 0 & 1 & 1 & & \\
 & & & & & 1 & 1 & 1 & \\
 & & & & 1 & 0 & 0 & 1 & 1
\end{bmatrix}
\qquad
\underline{G} = \begin{bmatrix}
1 & 0 & 0 & 0 & 1 & 1 & 1 & 0 & 1 \\
 & 1 & 1 & 0 & 0 & 0 & 1 & 1 & 1 \\
 & & 1 & 1 & 1 & 0 & 1 & 1 & 0
\end{bmatrix}
$$

Fig. 6.1 Parity check matrix and generator matrix for the code described in Example 6.1

syndromes (see Lemma 5.11) and imposing the constraint to have an upper triangular structure. This code is a [9,3,4] block code.

Lemma 6.1 *If an $\underline{H}$ matrix is constructed according to Definition* 6.1, *is lower triangular and has no null columns, the code that it describes cannot be a distance-*1 *code* (*Definition* 2.3).

Proof Since the procedure described in Lemma 5.11 for obtaining the correspondent $\underline{G}$ matrix requires at least a pair of columns, any row of $\underline{G}$ will have at least weight 2. Possible ex-OR sums of such rows cannot give a row having weight 1. Finally explicit exclusion of the possible presence of null columns, where, as already noted, 1-weight code words could be set, completes the treatment of all the possible origins for such code words. In conclusion, the minimum distance turns out to be $d \geq 2$.

Definition 6.2 A block code, whose $\underline{H}$ matrix has a form described by Definition 6.1, is lower triangular and has no null columns, will be called *wide-sense* (or briefly *w.s.*) *time-invariant in its parity check matrix.*

Such definition was anticipated in various circumstances, in order to make understanding of a possible code classification easier. After analogy with Definition 2.4, regarding a QC code introduced as w.s. time-invariant in its $\underline{G}$ matrix, we are now interested in searching for a code word length N, common to both descriptions (in the $\underline{G}$ matrix and in the $\underline{H}$ matrix). Nevertheless, after addition of further rows to our parity check matrix, according to the periodic behaviour expressed by Definition 6.1, we will maintain fixed the number of information symbols, and progressively increase that of control symbols, performing a sort of H-extension (see Definition 5.15).

The analogy in the behaviour with cyclic block codes allows one to expect that the generator polynomials are here periodic in their x-dependence. Coming back to Chap. 2, a certain form of parity check polynomials was introduced, in order to discover the existence of a natural length $N = pq_o$ for QC codes. This length is responsible for the cyclic permutation with steps of q_o, expressed by (2.9) for this class of codes. Such large-scale period of length N is evident in LQC codes, after Definition 2.14, where the parity check polynomials exhibit periodic expressions, like those described in Sect. 2.8. In this case the number of rows of $\underline{H}$ does not vary in code lengthening. After duality, this is expected to occur now for the rows of $\underline{G}$, whose number, as anticipated, remains unchanged, but whose form would be necessarily periodic with period N. On the other hand, this behaviour is not evident for the case treated in Example 6.1, because its length is too short (it can be considered strongly punctured with respect to the QC version).

Example 6.2 Given $q_o = 2$, $h_1{}^*(x) = x^2 + x + 1$, $h_2{}^*(x) = x^3 + x + 1$, and assuming a large number of periods, it is possible to obtain the pair of $\underline{H}$ and $\underline{G}$ matrices shown in Fig. 6.2. The latter one has been calculated from the former one by proper null ex-OR sum of clusters of syndromes. Clearly this is the H-extended form of a [4,2,2] quasi-cyclic code, which becomes [8,2,4], [12,2,6], after subsequent further complete H-extensions. So it is characterized by $N = 4$, $p = 2$. Remark that in the QC version of this code, the two generator polynomials show the following property $g_1(x) = h_2(x)$, $g_2(x) = h_1(x)$. This correspondence is evident only after having subdivided $\underline{G}$ into 4-length large-scale periods, and obviously represents a particular situation. In the H-extended version of this code, instead, we have

$$g_1(x) = (x^3 + x^2 + 1)(1 + x^4 + x^8 + \ldots),$$
$$g_2(x) = (x^2 + x + 1)(1 + x^4 + x^8 + \ldots).$$

which exhibit the periodic repetition of the two 1-symbol patterns, as anticipated.

Example 6.3 In Examples 2.8 and 2.15 two versions for a [8,4,4] QC code, equivalent to the extended Hamming code, have been discussed analyzing their generator matrices. For such codes, it is possible to adopt parity check matrices exactly coincident with the above generator matrices. So, we can speak of a self-dual code. A complete H-extension of the code introduced in Example 2.8 gives rise to a [16,4,8] code. In Fig. 6.3 up, the $\underline{H}$ matrix and the $\underline{G}$ matrix of such code are shown. It is possible to note two complete 8-length large-scale periods for the generator polynomial pair. Adding a proper unitary row $u_8(x)$ to this $\underline{G}$ matrix, we obtain a [16,5,8] code, equivalent to the 1st-order Reed-Muller code (see Definition D.26 and Example D.25 in Appendix D). This construction is shown in Fig. 6.3 down. Such non-conventional obtainment of 1st-order Reed-Muller codes resembles that proposed in Corollary a of Theorem 5.3, and Example 5.40, for simplex codes.

$$\underline{H} = \left[\begin{array}{cccccccc} 1 & 1 & 1 & & & & & \\ 1 & 0 & 1 & 1 & & & & \\ & & 1 & 1 & 1 & & & \\ & & 1 & 0 & 1 & 1 & & \\ & & & & 1 & 1 & 1 & \\ & & & & 1 & 0 & 1 & 1 \\ & & & & & & \ldots & \ldots \end{array}\right] \qquad \underline{G} = \left[\begin{array}{cccc|cccc|cccc|c} 1 & 1 & 0 & 1 & 1 & 1 & 0 & 1 & 1 & 1 & 0 & 1 & \ldots \\ & 1 & 1 & 1 & 0 & 1 & 1 & 1 & 0 & 1 & 1 & 1 & \ldots \end{array}\right]$$

Fig. 6.2 Parity check matrix and generator matrix for the code described in Example 6.2

$$\underline{H} = \left[\begin{array}{cccccccccccccccc}
1&1&1&0&1&&&&&&&&&&& \\
&1&0&1&1&1&&&&&&&&&& \\
&&1&1&1&0&1&&&&&&&&& \\
&&&1&0&1&1&1&&&&&&&& \\
&&&&1&1&1&0&1&&&&&&& \\
&&&&&1&0&1&1&1&&&&&& \\
&&&&&&1&1&1&0&1&&&&& \\
&&&&&&&1&0&1&1&1&&&& \\
&&&&&&&&1&1&1&0&1&&& \\
&&&&&&&&&1&0&1&1&1&& \\
&&&&&&&&&&1&1&1&0&1& \\
&&&&&&&&&&&1&0&1&1&1
\end{array}\right] \qquad \underline{G} = \left[\begin{array}{cccccccccccccccc}
1&1&1&0&1&0&0&0&1&1&1&0&1&&& \\
&1&0&1&1&1&0&0&0&1&0&1&1&1&& \\
&&1&1&1&0&1&0&0&0&1&1&1&0&1& \\
&&&1&0&1&1&1&0&0&0&1&0&1&1&1
\end{array}\right] \quad [16,4,8]$$

$$\underline{G} = \left[\begin{array}{cccccccccccccccc}
1&1&1&0&1&0&0&0&1&1&1&0&1&&& \\
&1&0&1&1&1&0&0&0&1&0&1&1&1&& \\
&&1&1&1&0&1&0&0&0&1&1&1&0&1& \\
&&&1&0&1&1&1&0&0&0&1&0&1&1&1 \\
&&&&&&&&1&1&1&1&1&1&1&1
\end{array}\right] \quad [16,5,8]$$

Fig. 6.3 Parity check and generator matrices for the codes described in Example 6.3

Up to now, we have considered just situations with $q_o = 2$ and characterized by particular symmetries. Generalization of these concepts to the case of higher values for q_o is easy. H-extension appears as a natural procedure in order to increase n, in search for the value $n = N = pq_o$, which characterizes the QC version of the considered family of codes, representing a sort of reference situation. To be more precise, this form of H-extension is a variant of that introduced by Definition 5.15, which takes into account only s.s. time-invariant codes in their $\underline{H}$ matrix. A proper widening in this definition will be given in Sect. 6.8. In the next Section we will discuss the cyclic permutation invariance with steps of q_o over the length N, which characterizes either the $\underline{G}$ matrix or the $\underline{H}$ matrix in the QC condition.

6.2 Parity Check Matrix of a Quasi-Cyclic Code

Quasi-cyclic codes have been introduced in Sect. 2.2 (Definition 2.4). Nevertheless Definition 2.4 is based on the existence of 2-weight code words of the type $(x^N + 1)$, with $N = pq_o$, in the lengthened version of a QC code. This property implies, for the QC code condition, that the generator polynomials satisfy an invariance law of the type (2.9), over the length N with step of q_o.

Lemma 6.2 *The parity check polynomials whose reciprocal appear as rows in the* $\underline{H}$ *matrix of a QC code, introduced after Definition* 2.4, *admit a cyclic permutation of the type described by Theorem* 2.1.

Proof Theorem 2.1, through expression (2.9), states that the code words of a QC code show the following property. Given a code word $t(x)$, any cyclic permutation with steps of q_o over the length $N = pq_o$, obtained from it, is still a code word $t'(x)$.

This implies that all the rows of $\underline{H}$ satisfies the same property, which is a structural characteristic of the code, and does not depend on the fact that it is described by means of $\underline{G}$ or by means of $\underline{H}$. The requirement of satisfying (5.1) is responsible for this characteristic. The same property, with step of $q_o = 1$, holds also for cyclic codes, where we have unique polynomials $g(x)$ an $h(x)$. □

In other terms, substitution of some rows of $\underline{H}$ with their cyclic permuted versions, with step of q_o, provided that the resulting parity check matrix is still of full rank, gives the same QC code. Starting from $\underline{H}$, it is possible to construct a correspondent $\underline{G}$ matrix, by the procedure which imposes the null ex-OR sum of clusters of syndromes (see Lemma 5.11). This way has been followed in the previous Section. We can perform also an inverse procedure: from $\underline{G}$ obtain $\underline{H}$. In any case, the rows of both matrices, if the code is taken over the length $n = N$ of its quasi-cyclic version, can be cyclically permuted, assuming a cyclic period N, with steps of q_o.

A same QC code (together with the lengthened or H-extended version derived from it) can be w.s. time-invariant in the $\underline{H}$ matrix and s.s. time-invariant in the $\underline{G}$ matrix, or vice versa. The more general case is that in which it is w.s. time-invariant both in $\underline{G}$ and in $\underline{H}$, but this behaviour is necessarily constrained to show the same value for the small-scale period q_o, as well as for the large-scale period $N = pq_o$. The parity check polynomials of a QC code are unbounded (in the sense discussed in Sect. 5.12). Nevertheless, like in cyclic codes, we can imagine to have 1-symbol patterns able to give a minimal form for it also in this case. Furthermore a lower triangular form is also advisable. Here such two forms of parity check matrix, in general, do not coincide.

The remaining part of this Section is devoted to justify the obtainment of a valid parity check matrix, starting from the $\underline{G}$ matrix of a certain QC code. As already noted, (5.4) cannot give a unique solution to this problem. Once noted that the ex-OR sum of cluster of syndromes is able to give a unique solution if $\underline{H}$ is lower triangular, imposing that $\underline{G}$ is upper triangular, we are searching for a sort of inverse procedure, able to give a unique lower triangular $\underline{H}$ starting from an upper triangular $\underline{G}$. A procedure of this type will be presented in Chaps. 7 and 8, treating convolutional codes, for which its introduction is simpler, because of the presence of non-periodic (and bounded) polynomials. It is valid also for QC codes (in turn including the particular case of cyclic codes, for $q_o = 1$), therefore it can be considered completely general. Nevertheless its presentation now would divert our attention to what we are mainly treating here.

In Chap. 2, an ordered set of parity check polynomials $h_j^{(i)}(x^{q_o}), j = 1.2, \ldots q_o$, has been introduced (see Definition 2.5). They are the ordered quotients, one for each position i of the period, of the binomial $(x^N + 1)$ divided by the generator polynomials $g_j(x)$'s. They show a periodic behaviour for $n > N$, after code lengthening with steps of q_o, but their reciprocal polynomials cannot be assumed directly as rows of the correspondent $\underline{H}$ matrix. The determination of such parity check polynomials has been anyway useful, in order to calculate N, always integer multiple of q_o, parameter $p = N/q_o$, and also its subdivision into k_p and r_p (see Sect. 2.2 for details).

Equation (2.3) are rewritten here considering also the positions in the period where any equality takes place

$$x^{q_o-i}\sum_{j=1}^{q_o} x^{q_o-j}h_j^{(i)}(x^{q_o})g_j(x) = x^{q_o-i}(x^N+1), \quad i=1,2,\ldots q_o. \tag{6.1}$$

It is possible to perform ex-OR sums, considering also the property of cyclic periodicity that any $h_j^{(i)}(x^{q_o})$ exhibits, in such a way as to separate the contributions due to the different $g_j(x)$'s. This solution is needed in order to satisfy a set of equalities directly deriving from (5.4). They can be expressed as

$$\underline{g}_j \cdot \underline{h}_b^T = 0, \quad j=1,2,\ldots q_o,\; b=1,2,\ldots r_p q_o, \tag{6.2}$$

where $\underline{h}_b^T$ is a column vector with N elements, representing a column of $\underline{H}^T$, but also any cyclic permutation of it, mod (x^N+1) with steps of q_o. The number of different $\underline{h}_i^T$ to be considered is $r_p q_o$. Often some of the p cyclic permutations of the same polynomial are linearly dependent. As long as the ex-OR sum leading to $\underline{h}_b^T$ and to all its cyclically permuted replicas is unique, the QC code, although w.s. time-invariant in $\underline{G}$, will be s.s. time-invariant in this particular form of its $\underline{H}$ matrix. In general more than one independent ex-OR sum are needed, and the QC code will be necessarily w.s. time-invariant also in its $\underline{H}$ matrix.

Example 6.4 In Examples 2.3 and 2.5, we treated a [14,10,3] QC code, with $q_o = 2$, $r_p = 2$, $g_1(x) = x^4+x^3+x^2+x+1$, $g_2(x) = x^4+x^3+1$, and the following polynomials

$$h_1^{(1)}(x^2) = x^{10}+x^4+x^2+1, \quad h_2^{(2)}(x^2) = x^{10}+x^8+1,$$
$$h_2^{(1)}(x^2) = x^{10}+x^8+x^6+x^2, \quad h_1^{(2)}(x^2) = x^{10}+x^6+x^4.$$

The pair of equalities (6.1) can be particularized here as follows

$$x[xh_1^{(1)}(x^2)g_1(x) + h_2^{(1)}(x^2)g_2(x)] = x(x^{14}+1),$$
$$xh_2^{(2)}(x^2)g_2(x) + h_1^{(2)}(x^2)g_1(x) = x^{14}+1.$$

A parity check matrix valid for this QC code can be arranged considering three possible rows: 11.11.00.01.10.00.10, 01.11.10.10.01.00.11, 01.01.01.01.01.01.01, with all their cyclically permuted versions with steps of 2. It is possible to verify that the ex-OR sum of the first with a cyclically permuted version of the second yields the third. Therefore, in arranging a valid $\underline{H}$ matrix, it is important to avoid rows which are linearly dependent, in such a way as to assure that $\underline{H}$ has full rank

$$
\underline{G} = \left[\begin{array}{cccccccccccccc}
1&1&1&1&1& & & & & & & & & \\
 &1&1&0&0&1& & & & & & & & \\
 & &1&1&1&1&1& & & & & & & \\
 & & &1&1&0&0&1& & & & & & \\
 & & & &1&1&1&1&1& & & & & \\
 & & & & &1&1&0&0&1& & & & \\
 & & & & & &1&1&1&1&1& & & \\
 & & & & & & &1&1&0&0&1& & \\
 & & & & & & & &1&1&1&1&1& \\
 & & & & & & & & &1&1&0&0&1
\end{array}\right]
$$

$$
\underline{H} = \left[\begin{array}{cccccccccccccc}
1&1&1&1&0&0&0&1&1&0&0&0&1&0\\
1&0&1&1&1&1&0&0&0&1&1&0&0&0\\
0&0&1&0&1&1&1&1&0&0&0&1&1&0\\
1&0&0&0&1&0&1&1&1&1&0&0&0&1
\end{array}\right]
$$

$$
\underline{H} = \left[\begin{array}{cccccccccccccc}
1&0&1&1&1&1&0&0&0&1&1&0&0&0\\
1&1&0&1&1&1&1&0&1&0&0&1&0&0\\
0&0&1&0&1&1&1&1&0&0&0&1&1&0\\
0&0&1&1&0&1&1&1&1&0&1&0&0&1
\end{array}\right]
$$

$$
\underline{H} = \left[\begin{array}{cccccccccccccc}
1&1&1&1&0&0&0&1&1&0&0&0&1&0\\
1&0&1&1&1&1&0&0&0&1&1&0&0&0\\
0&1&1&1&1&0&1&0&0&1&0&0&1&1\\
1&1&0&1&1&1&1&0&1&0&0&1&0&0
\end{array}\right]
$$

$$
\underline{H} = \left[\begin{array}{cccccccccccccc}
1&0&1&1&1&1&0&0&0&1&1&0&0&0\\
1&1&0&1&1&1&1&0&1&0&0&1&0&0\\
1&1&1&1&0&0&0&1&1&0&0&0&1&0\\
0&1&0&1&0&1&0&1&0&1&0&1&0&1
\end{array}\right]
$$

Fig. 6.4 Some different $\underline{H}$ matrices all valid for the [14,10,3] QC code, whose $\underline{G}$ matrix is reported on the *left* (Example 6.4)

$r_p q_o = 4$. Then, we can observe that the first two rows exhibit 7 permuted versions, and only four of them linearly independent. The last one is unique (any cyclically permuted version with steps of 2 gives unchanged result). It is possible to verify that all the four $\underline{H}$ matrices reported on the right in Fig. 6.4 are valid for this QC code, whose generator matrix is depicted on the left. The first corresponds to an s.s. time-invariant code in its $\underline{H}$ matrix. The second to an alternative version, which is w.s. time-invariant and lower triangular according to Definitions 6.1 and 6.2. The third is another version, which can be considered coincident with the previous one, except for a change in the order of the rows. Finally the fourth is completely time-varying.

After inspection of the $\underline{G}$ matrix, it is possible to conclude that, in any case, the last four positions in the code word are occupied by control symbols. In Fig. 6.5 the particular ex-OR sums leading to the two independent rows $\underline{h}_b$, previously described, are detailed. The upper one has weight 7 and employs polynomials $h_1^{(1)*}(x^2)$, $h_2^{(2)*}(x^2)$. The lower one has weight 8 and employs polynomials $h_1^{(1)*}(x^2)$, $h_2^{(1)*}(x^2)$. For both, the ex-OR sum of the two component rows gives the final $\underline{h}_b$ row reported above.

Example 6.5 The [8,4,4] QC code is self-dual, as already noted in Example 6.3. Taking the version of such code whose generator matrix is characterized by the pair of polynomials $g_1(x) = x^4 + x^3 + x^2 + 1$, $g_2(x) = x^4 + x^2 + x + 1$, the following parity check matrix can be adopted

$$\underline{h}_b \; : \; 11.11.00.01.10.00.10 \; = x^2 h_1^{(1)} * (x^2)\Big|^{(2)}_{\mathrm{mod}(x^{14}+1)} + x^8 h_2^{(2)} * (x^2)\Big|^{(2)}_{\mathrm{mod}(x^{14}+1)}$$

$$10.10.00.00.10.00.10 \; = x^2 h_1^{(1)} * (x^2)\Big|^{(2)}_{\mathrm{mod}(x^{14}+1)}$$

$$01.01.00.01.00.00.00 \; = x^8 h_2^{(2)} * (x^2)\Big|^{(2)}_{\mathrm{mod}(x^{14}+1)}$$

$$\underline{h}_b \; : \; 01.11.10.10.01.00.11 \; = x^{12} h_1^{(1)} * (x^2)\Big|^{(2)}_{\mathrm{mod}(x^{14}+1)} + x^8 h_2^{(1)} * (x^2)\Big|^{(2)}_{\mathrm{mod}(x^{14}+1)}$$

$$00.10.10.10.00.00.10 \; = x^{12} h_1^{(1)} * (x^2)\Big|^{(2)}_{\mathrm{mod}(x^{14}+1)}$$

$$01.01.00.00.01.00.01 \; = x^8 h_2^{(1)} * (x^2)\Big|^{(2)}_{\mathrm{mod}(x^{14}+1)}$$

$$h_1^{(1)} * (x^2) = 10.10.10.00.00.10.00$$

$$h_2^{(2)} * (x^2) = 01.00.00.00.01.01.00 \qquad h_2^{(1)} * (x^2) = 00.01.00.01.01.01.00$$

Fig. 6.5 Identification of some interleaved parity check polynomials in the two independent rows $\underline{h}_b$ having respectively weight $w = 7$ and $w = 8$ (Example 6.4)

$$\underline{H} = \begin{bmatrix} 1 & 1 & 1 & 0 & 1 & & & & \\ & 1 & 0 & 1 & 1 & 1 & & & \\ & & 1 & 1 & 1 & 0 & 1 & & \\ & & & 1 & 0 & 1 & 1 & 1 & \end{bmatrix},$$

which is exactly coincident with $\underline{G}$. In order to simplify identification some external 0-symbols have been ignored. Nevertheless also whichever parity check matrix obtained from the above one, after a cyclic permutation with step of $q_o = 2$, applied to any of its rows, provided that the resulting four rows are linearly independent, is a valid parity check matrix. For instance, the following one is acceptable as well

$$\underline{H} = \begin{bmatrix} 1 & 0 & 0 & 0 & 1 & 1 & 1 & 0 \\ 0 & 0 & 0 & 1 & 0 & 1 & 1 & 1 \\ 1 & 0 & 1 & 0 & 0 & 0 & 1 & 1 \\ 1 & 1 & 0 & 0 & 0 & 1 & 0 & 1 \end{bmatrix}.$$

Example 6.6 In Example 2.10 the [10,6,3] QC code characterized by $q_o = 2$, $r_p = 2$, $g_1(x) = x^4 + x^3 + 1$, $g_2(x) = x^4 + x^2 + x+1$, has been introduced. In this case we have two possible rows in the $\underline{H}$ matrix: 01.01.11.11.00 and 10.00.11.01.00. They both admit 5 cyclically permuted versions with steps of 2 all different. Only four of them, also here, are linearly independent. In Fig. 6.6, the generator matrix and four possible $\underline{H}$ matrices are reported for such QC code. The first one is s.s. time-invariant in $\underline{H}$. The second one satisfies Definition 6.1 but not Definition 6.2 (it is not lower triangular). The third one is time-varying, but is characterized by a lower triangular form. The fourth one satisfies both Definitions 6.1 and 6.2.

$$\underline{G} = \begin{array}{|cccccccccc|}\hline 1&1&0&0&1&&&&& \\ &1&0&1&1&1&&&& \\ &&1&1&0&0&1&&& \\ &&&1&0&1&1&1&& \\ &&&&1&1&0&0&1& \\ &&&&&1&0&1&1&1 \\ \hline\end{array}$$

$$\underline{H} = \begin{array}{|cccccccccc|}\hline 0&1&0&1&1&1&1&1&0&0 \\ 0&0&0&1&0&1&1&1&1&1 \\ 1&1&0&0&0&1&0&1&1&1 \\ 1&1&1&1&0&0&0&1&0&1 \\ \hline\end{array}$$

$$\underline{H} = \begin{array}{|cccccccccc|}\hline 1&0&0&0&1&1&0&1&0&0 \\ 0&1&0&1&1&1&1&1&0&0 \\ 0&0&1&0&0&0&1&1&0&1 \\ 0&0&0&1&0&1&1&1&1&1 \\ \hline\end{array}$$

$$\underline{H} = \begin{array}{|cccccccccc|}\hline 1&1&0&1&0&0&1&0&0&0 \\ 1&0&0&0&1&1&0&1&0&0 \\ 0&0&1&1&0&1&0&0&1&0 \\ 0&0&0&1&0&1&1&1&1&1 \\ \hline\end{array}$$

$$\underline{H} = \begin{array}{|cccccccccc|}\hline 1&1&0&1&0&0&1&0&0&0 \\ 0&1&0&1&1&1&1&1&0&0 \\ 0&0&1&1&0&1&0&0&1&0 \\ 0&0&0&1&0&1&1&1&1&1 \\ \hline\end{array}$$

Fig. 6.6 Generator matrix and some $\underline{H}$ matrices all valid for the [10,6,3] QC code, whose $\underline{G}$ matrix is reported on the *left* (Example 6.6)

6.3 Quasi-Cyclic Codes with Distributed Control Symbols Described with Their $\underline{H}$ Matrix

In Sect. 2.3 we have introduced a transformation of the $\underline{G}$ matrix, able to provide a new version of a quasi-cyclic code, characterized by distributed control symbols. The same procedure can be followed here for transforming the parity check matrix $\underline{H}$ discussed in the previous Sections in the same way. The only care to be considered is to start from an $\underline{H}$ matrix satisfying Definition 6.1, in which the period q_o is clearly identified, and having full rank. This matrix, which is referred to the QC condition, has $r_p q_o$ rows and $N = pq_o$ columns.

First of all, it is necessary to construct a complete matrix $\underline{H}_{comp}$, which is $N \times N$ square, and can be considered as a $p \times p$ circulant of $q_o \times q_o$ blocks. Then we apply Lemma 2.6, in order to obtain a new $pq_o \times pq_o$ square matrix, which is a square block of $p \times p$ circulants. Such Lemma can be referred clearly not only to the generator matrix (as made in Sect. 2.3), but also to the parity check matrix. The following proposition is a sort of dual version, with respect to Theorem 2.2.

Theorem 6.1 *The parity check matrix of a QC code with parameters* $N = pq_o$, $k = k_p q_o$ *and generator constraint length* r_p, *can be set in a form of the type shown in Fig. 6.7, that is a block of* $r_p \times p$ *truncated circulants (circulants in which the first* k_p *rows have been eliminated).*

Proof Applying the transformation described in Lemma 2.6 to a QC code with the above parameters, from the complete matrix $\underline{H}_{comp}$ it is possible to obtain a new matrix, redundant as well, in which there are row layers of $p \times p$ circulants. Starting from this, we can eliminate k_p rows in any circulant, say the first ones. This is a

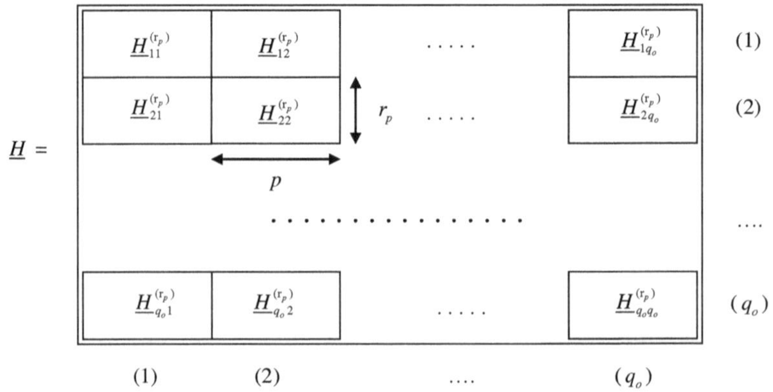

Fig. 6.7 Alternative structure for the parity check matrix of a QC code

choice complementary with respect to that adopted in Theorem 2.2, where we eliminated the last r_p rows. Remark that also a different choice could be considered, provided that the remaining rows in the resulting matrix are all linearly independent. Coming back to the above choice, we are left with r_p rows in any block $\underline{H}_{ij}^{(r_p)}$, which then becomes an $r_p \times p$ truncated circulant. The final result is an $r_p q_o \times p q_o$ parity check matrix of the type shown in Fig. 6.7, in which all the rows are linearly independent. □

Example 6.7 Taking the [10,6,3] QC code described in Example 2.10, and whose parity check matrix has been obtained in Example 6.6, we assume the two rows (10.00.11.01.00 and 01.01.11.11.00) alternated in such a way as to form a period with length $q_o = 2$. This parity check matrix is the second one listed in Fig. 6.6. Remark that it is not lower triangular, since this property is not necessary for application of Theorem 6.1. Thus we start with the parity check matrix $\underline{H}$ reported up-left in Fig. 6.8. Then the complete parity check matrix $\underline{H}_{comp}$, the matrix $\underline{H}_{comp}'$ obtained after column reordering, and the matrix $\underline{H}_{comp}''$ obtained after row reordering (both according to Lemma 2.6) are shown in proper sequence. Finally we have the new parity check matrix, whose form is a block of 2 × 5 truncated circulants. In this reordered parity check matrix, it is not clear, at a first view, where information and control symbols can be found.

The reordered $\underline{H}$ matrix obtained after Theorem 6.1 shows transformed periods whose length is p and parity check sub-matrices having r_p rows. Keeping in mind the positions of information and control symbols in the original code, also considering the structure of its $\underline{G}_o^{(q_o)}$ generator sub-matrix (see Definition 2.8), it is possible to assume the first k_p positions in any transformed period as devoted to information symbols, and the remaining $p - k_p = r_p$ positions as devoted to control symbols. The parity check sub-matrices, describing the various layers in this reordered parity check matrix, in general, are different.

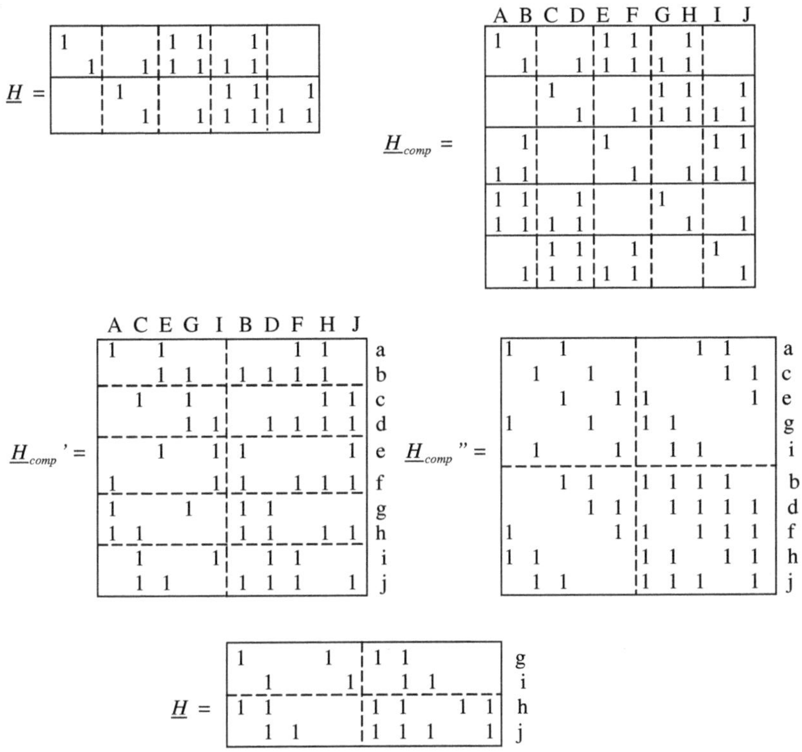

Fig. 6.8 Series of transformations on the parity check matrix characterizing a [10,6,3] QC code

Definition 6.3 When a QC code is characterized by a w.s. time-invariant parity check matrix $\underline{H}$, with all the control symbols at the end of the code word (that is before a possible reordering of the type described by Lemma 2.6), exploiting cyclic permutations with steps of q_o, we can obtain an equivalent form for its $\underline{H}$ matrix, in such a way as to span the smallest number of periods. It is called the *minimal form of* $\underline{H}$ *for a QC code.*

It is possible to compare this Definition with other ones previously given (Definition 5.13 for cyclic, punctured or SC codes only, and Definition 5.14 for more general codes, included the QC codes here treated in particular). The specification introduced now regards the possibility of cyclic permutations with step of q_o. It is devoted to evidence a series of subsequent 0-symbols, whose length should be as large as possible. This optimization, as usual, is made in order to find a reference representation for a certain matrix, also in the perspective of simplifying encoding procedures. Remark that a similar definition would be without sense for a reordered $\underline{H}$ matrix, in which the above series of subsequent 0-symbols have been dispersed.

Example 6.8 The [10,6,3] QC code treated in Example 6.6, for which several $\underline{H}$ matrices have been proposed (see Fig. 6.6), is characterized by more than one possible $\underline{H}$ matrix in minimal form (with the same number of spanned periods, equal to 4). In Example 6.7, before reordering, we took one of them (the second in Fig. 6.6). The lowermost one in Fig. 6.6, reported here

$$\underline{H} = \begin{bmatrix} 1 & 1 & 0 & 1 & 0 & 0 & 1 & & & \\ & 1 & 0 & 1 & 1 & 1 & 1 & 1 & & \\ & & 1 & 1 & 0 & 1 & 0 & 0 & 1 & \\ & & & 1 & 0 & 1 & 1 & 1 & 1 & 1 \end{bmatrix},$$

is in minimal form too, but also lower triangular (and even upper triangular, although this property is of no interest in the present treatment).

As already noted, if the QC code has an $\underline{H}$ matrix not in lower triangular form, it is however possible to attribute the last positions in its code word to control symbols, if the correspondent $\underline{G}$ matrix has an upper triangular form. This conclusion, drawn by difference, has been already stressed for the [10,6,3] QC code, whose reordering was presented in Example 6.7. Nevertheless a lower triangular form for the $\underline{H}$ matrix, besides of being also in minimal form, could represent an advantage in encoding and even in unwrapping processes. On the other hand, for hard-decision decoding (binary QC codes can be sometimes interpreted as proper non-binary BCH codes), or for assuming a structure suited for LDPC code behaviour (Ryan and Lin 2009), the desired parity check matrix of a QC code is often required to be neither in minimal form, nor in a lower triangular form.

6.4 Encoder Circuit Based on the $\underline{H}$ Matrix in a QC Code

A w.s time-invariant code in its $\underline{H}$ matrix, and in particular a QC code of the type described in Sect. 6.2, can be encoded by means of an encoder circuit which results a suitable widening of that presented in Sect. 5.5 (Fig. 5.6) for s.s. time-invariant codes. It is necessary that the considered parity matrix is lower triangular, so the control symbols will be added at the end of the code word, properly interleaved within the last r_p periods. When this requirement can be satisfied by an $\underline{H}$ matrix in minimal form, we will have also the minimum number of memory elements in the shift registers to be employed. In order to describe the structure of this encoder circuit, we will refer to the case characterized by $q_o = 2$, and by two parity check polynomials having the same degree $q_o k_p = q_o(p - r_p)$. Long but straightforward generalizations can be made in order to consider whichever q_o and a situation where the parity check polynomials have different degrees.

Under the above assumptions, a code word can be written in polynomial form as follows

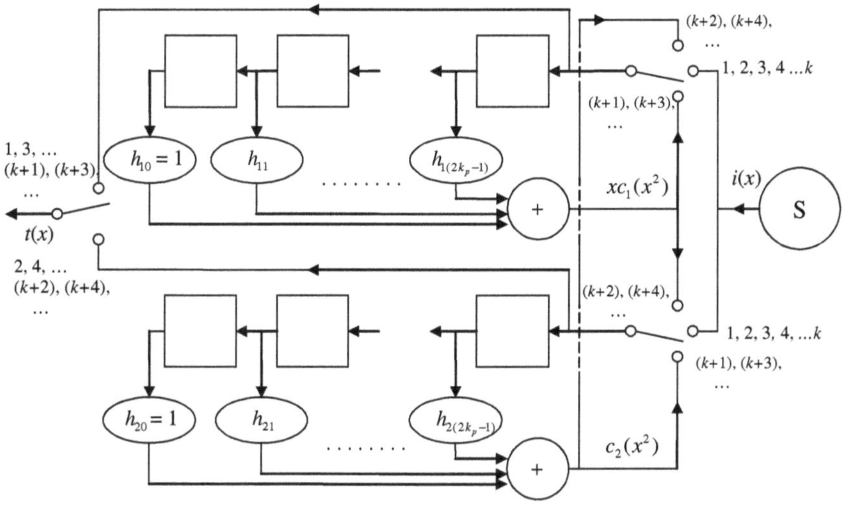

Fig. 6.9 Encoder circuit for a QC code with $q_o = 2$ and parity check polynomials having the same maximum degree $2k_p$

$$t(x) = x^{2r_p} i(x) + x c_1(x^2) + c_2(x^2), \tag{6.3}$$

where $i(x)$ represents the information symbol sequence, $c_1(x^2)$ the control symbols added by the rows of $\underline{H}$ in odd positions and $c_2(x^2)$ the control symbols added by the rows of $\underline{H}$ in even positions. Let such rows be described by polynomial $h_1{}^*(x)$, $h_2{}^*(x)$ respectively.

The encoder circuit to be considered has a scheme of the type reported in Fig. 6.9, where two shift registers, in controller arrangement, are employed, both with $2k_p$ memory elements (in general $q_o k_p$). The coefficients of polynomials $h_1(x)$ and $h_2(x)$ have been denoted respectively h_{1j} and h_{2j}, with $j = 0, 1, \ldots (2k_p\text{-}1)$, whereas by assumption we have $h_{1(2k_p)} = h_{2(2k_p)} = 1$, not present in the circuit, since these coefficients multiply the incoming symbols.

The analogy with the encoder circuit shown in Fig. 5.6 is evident. Also here, this solution can be adopted either for the QC code or for some shortened versions of it. So the number of information symbols has been assumed to be k, under the constraint that this parameter is even and satisfies the condition $k \le 2k_p$ (in general k is integer multiple of q_o and satisfies the condition $k \le q_o k_p$). Finally the degree of $i(x)$ turns out to be at most $(k - 1)$, as usual.

Remark that both shift registers are fed by all the information symbols collected in the sequence $i(x)$. This encoder circuit is not able to encode lengthened quasi-cyclic codes. Instead it is useful for possible H-extended and modified H-extended versions of a given QC code. We can note a strong resemblance with what discussed in Sect. 5.5, about the encoder circuit based on the unique parity check polynomial $h(x)$, and its applications.

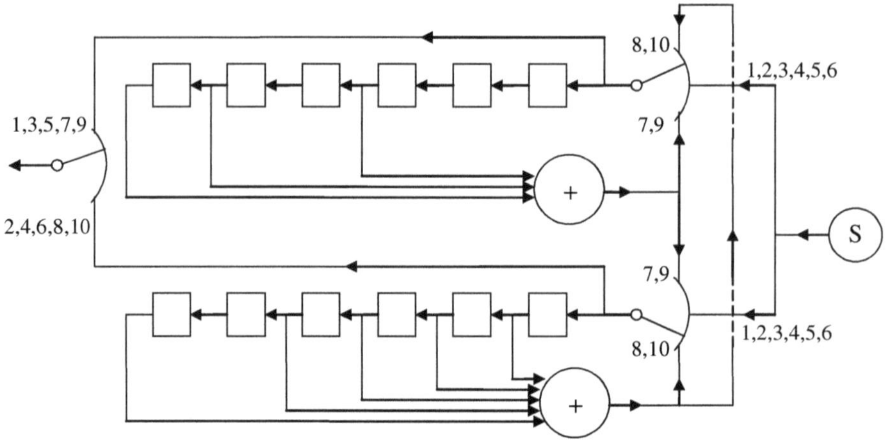

Fig. 6.10 Encoder circuit for the [10,6,3] QC code described in Example 6.9

Example 6.9 Assuming the parity check matrix in minimal form and lower triangular obtained in Example 6.8, with $h_1{}^*(x) = x^6 + x^5 + x^3 + 1$, $h_2{}^*(x) = x^6 + x^4 + x^3 + x^2 + x + 1$, which describes the well known [10,6,3] QC code introduced in Example 2.10, we obtain the encoder circuit shown in Fig. 6.10. In order to encode a shortened version of this QC code, we will maintain the two switches connected to the information source for a number of periods lower than $k_p = 3$. Remark that the $\underline{H}$ matrix adopted in Example 6.7, for starting the reordering process of the same QC code, although in minimal form as well, is not suited for being reproduced in an encoder circuit of the type here considered, because not in lower triangular form.

Lemma 6.3 *In the two shift registers of the encoder circuit depicted in Fig.* 6.9, *the same state evolution takes place.*

Proof Looking at the two switches it is possible to verify that they are connected to the same wires in the same symbol times. □

This proposition suggests the possibility of constructing a simplified scheme for this type of encoder circuit, which is based on a single shift register, still in controller arrangement, with feed-back wires dynamically controlled (periodically time-varying) (Jimenez-Felstrom and Zigangirov 1999). Anyway the conclusion expressed by Lemma 6.3 has great importance in the evaluation of the number of states, and hence of the decoding computational complexity characterizing a possible soft-decision decoding algorithm. This could be supported by a trellis, constructed on a state diagram, in turn associated to this encoder circuit, as will be better explained in Sect. 6.11. Such considerations, taking into account also what observed in Sect. 5.19, appear particularly important for H-extended (convolutional) schemes. On the other hand, the above construction of the encoder circuit is adopted also for LDPC codes, especially convolutional LDPC codes, but in this case the decoding algorithm is no longer based on a trellis, as will be discussed with more details in Chaps. 10 and 12.

6.5 Shortened Quasi-Cyclic (SQC) Codes and Their $\underline{H}$ Matrix

Shortened QC codes (briefly SQC) have been introduced in Sect. 2.5, looking at their generator matrix, in two different ways. Definition 2.12 refers to the version of QC codes characterized by a generator sub-matrix $\underline{G}_o^{(q_o)}$ (and also to the reordered form with distributed control symbols), whereas Definition 2.13 refers to the version of QC codes where the $\underline{G}$ matrix contains X layers of complete $p \times p$ square circulants $\underline{G}_{ij}^{(p)}$, plus possible R_X anomalous rows. We will treat here such two types of SQC codes, from the point of view of their parity check matrices.

It is possible to obtain the $\underline{G}$ matrix of an s_o-time shortened quasi-cyclic code, starting from the generator matrix of the original code, where the $\underline{G}_o^{(q_o)}$ sub-matrix shifts, by eliminating in it the first $s_o q_o$ rows (that is the s_o first $\underline{G}_o^{(q_o)}$ sub-matrices). So the number of periods in which information symbols are inserted turns out to be $(k_p{-}s_o)$, and the overall number of periods is $(p{-}s_o)$. Parameter r_p, instead, remains unchanged. Therefore the structure of the $\underline{G}$ matrix does not change: in it an invariant generator sub-matrix $\underline{G}_o^{(q_o)}$ still shifts by q_o positions at a time. On the contrary, in the $\underline{H}$ matrix, the first $s_o q_o$ columns (s_o column layers) are to be eliminated. A similar procedure holds also for cyclic codes in obtaining their shortened versions, simply with $q_o = 1$.

Example 6.10 Taking the [10,6,3] QC code described so far, its $\underline{G}$ matrix and its $\underline{H}$ matrix (the latter one in minimal form and lower triangular, as obtained in Example 6.8) are reported on the left in Fig. 6.11. A one-time shortened version of this code is an [8,4,3] SQC code, whose generator and parity check matrix,

$$\underline{G} = \left[\begin{array}{cc|cc|cc|cc|cc}
1 & 1 & 0 & 0 & 1 & & & & & \\
 & 1 & 0 & 1 & 1 & 1 & & & & \\
\hline
 & & 1 & 1 & 0 & 0 & 1 & & & \\
 & & & 1 & 0 & 1 & 1 & 1 & & \\
\hline
 & & & & 1 & 1 & 0 & 0 & 1 & \\
 & & & & & 1 & 0 & 1 & 1 & 1
\end{array}\right] \qquad
\underline{G}' = \left[\begin{array}{cc|cc|cc|cc}
1 & 1 & 0 & 0 & 1 & & & \\
 & 1 & 0 & 1 & 1 & 1 & & \\
\hline
 & & 1 & 1 & 0 & 0 & 1 & \\
 & & & 1 & 0 & 1 & 1 & 1
\end{array}\right]$$

$$\underline{H} = \left[\begin{array}{cc|cc|cc|cc|cc}
1 & 1 & 0 & 1 & 0 & 0 & 1 & & & \\
 & 1 & 0 & 1 & 1 & 1 & 1 & 1 & & \\
\hline
 & & 1 & 1 & 0 & 1 & 0 & 0 & 1 & \\
 & & & 1 & 0 & 1 & 1 & 1 & 1 & 1
\end{array}\right] \qquad
\underline{H}' = \left[\begin{array}{cc|cc|cc|cc}
0 & 1 & 0 & 0 & 1 & & & \\
0 & 1 & 1 & 1 & 1 & 1 & & \\
\hline
1 & 1 & 0 & 1 & 0 & 0 & 1 & \\
0 & 1 & 0 & 1 & 1 & 1 & 1 & 1
\end{array}\right]$$

Fig. 6.11 Generator and parity check matrix for the [10,6,3] QC code and for the [8,4,3] SQC code obtained from the previous one

$$\underline{G}' = \left[\begin{array}{cccc:cccc} 1 & & 1 & & 1 & & & \\ & 1 & & 1 & & 1 & & \\ \hdashline & & 1 & & 1 & 1 & 1 & \\ & & & 1 & & 1 & 1 & 1 \end{array}\right] \qquad \underline{H}' = \left[\begin{array}{cccc:cccc} & & & 1 & 1 & & & \\ 1 & & & 1 & 1 & 1 & & \\ \hdashline 1 & & & & 1 & & 1 & 1 \\ 1 & 1 & & & 1 & 1 & & 1 \end{array}\right]$$

Fig. 6.12 Reordered structures of the generator and parity check matrix for the [8,4,3] SQC code treated in Examples 6.10 and 6.11

$\underline{G}'$ and $\underline{H}'$, are shown on the right in Fig. 6.11. Matrix $\underline{H}'$ remains lower triangular, because we have eliminated the first column layer.

After reordering columns and rows, according to Lemma 2.6 and Theorem 6.1, for having distributed control symbols (see Fig. 6.7), the above procedure of code shortening gives rise to a new $\underline{H}$ matrix which is so transformed with respect to the one characterizing the original QC code: in any block of $r_p \times p$ truncated circulants, the first s_o columns are eliminated.

Example 6.11 The [8,4,3] SQC code, whose generator and parity check matrix are shown in Fig. 6.11 on the right, exhibits the new pair of reordered matrices $\underline{G}'$, $\underline{H}'$, reported in Fig. 6.12. We can observe the elimination of the first column in any transformed period. About the generator matrix, it was already obtained in Example 2.20. However, it is easy to verify that the null ex-OR sum of clusters of syndromes is satisfied between such two matrices.

Let us consider now a QC code constructed adopting, in its $\underline{G}$ matrix, X layers of complete $p \times p$ square circulants and R_X anomalous rows, as described in Sect. 2.7. For this type of $\underline{G}$ matrix, code shortening is made according to Definition 2.13 and Example 2.24. On the other hand, code lengthening is made according to Definition 2.16 and Example 2.28, so confirming that the QC condition is precisely the one characterized by code word length $n = N = p^2$. The codes considered in the above mentioned Examples have $\underline{G}$ matrices characterized by $X = 2$ layers of circulants and $R_X = 1$ anomalous row, the last one not being involved neither in shortening nor in lengthening procedures.

Typically the above class of QC codes are considered in the framework of LDPC code theory (Fan 2000). For them a soft-decision decoding algorithm is available, but it requires that 4-length short cycles of 1-symbols in $\underline{H}$ are not present (Tanner 1981). In Chap. 8, this particular type of LDPC codes will be better analyzed, and the general expressions for their $\underline{G}$ matrix and $\underline{H}$ matrix will be carried out. Now we can anticipate that the parity check matrix is characterized by layers of $p \times p$ permutation circulants. They are identified by proper phase shifts. When all the possible shifts have been used in any layer (over a length $N = p^2$), addition of one further column layer of permutation circulants produces a lengthened quasi-cyclic code and 2-weight code words as a consequence. Furthermore it introduces 4-length short cycles of 1-symbols in the new $\underline{H}$ matrix, which makes the decoding procedure, as LDPC code, useless. On the contrary, possible elimination of columns of permutation circulants, that is application of code shortening, can give rise to interesting families of rate-adaptive LDPC codes (Milenkovic et al. 2008).

Starting from the condition of QC code, that is with an $\underline{H}$ matrix characterized by p column layers of $p \times p$ permutation circulants, the parity check matrix of a z_o-time shortened code obtained from it (Definition 2.13) can be derived after elimination of z_o column layers in the original $\underline{H}$ matrix. Since the original parity check matrix is not in lower triangular form, it is equivalent to start elimination from the left hand side or from the right hand side. In order to better understand the phase evolutions along the row layers, a column layer made by all identity matrices is maintained on the left hand side, so starting elimination from the right hand side. Clearly the two procedures are equivalent and originate the same code. Finally it is important to stress that such parity check matrices are intentionally not in their minimal form.

Example 6.12 Let us consider the [16,9,4] QC code introduced in Example 2.22, whose one-time shortened version has been treated in Example 2.24, regarding its generator matrix. It is characterized by $p = 4$, and, in the QC version, has a length $N = p^2 = 16$. In Fig. 6.13 up the parity check matrix for this [16,9,4] QC code is reported. It is possible to verify that the $\underline{G}$ matrix shown in Fig. 2.20 up is a valid correspondent for this $\underline{H}$ matrix, by means of the procedure leading to a null ex-OR sum of clusters of syndromes (see Lemma 5.11). Performing, on this QC code, a one-time shortening procedure (starting from the right hand side as suggested above), we obtain a [12,5,4] SQC code, whose $\underline{H}$ matrix is reported down in Fig. 6.13. It is important to observe that one row in both these $\underline{H}$ matrices is linearly dependent on the others (their rank is 7 and not 8). This confirms that the rows of the $\underline{G}$ matrix are 9 and not 8, justifying the origin of the final anomalous row in it. Such row, consisting of a $u_8(x)$ unitary polynomial, in general a $u_{2p}(x)$ unitary polynomial, can be placed also in other positions, provided that it occupies exactly two column layers. In this sense, it can be also subdivided into two separate $u_4(x)$ unitary polynomials, as a rapid inspection of the $\underline{H}$ matrices in Fig. 6.13 can immediately confirm, owing to the null ex-OR sum of clusters of syndromes.

$$\underline{H} = \left[\begin{array}{cccc|cccc|cccc|cccc}
1 & & & & 1 & & & & 1 & & & & 1 & & & \\
 & 1 & & & & 1 & & & & 1 & & & & 1 & & \\
 & & 1 & & & & 1 & & & & 1 & & & & 1 & \\
 & & & 1 & & & & 1 & & & & 1 & & & & 1 \\
\hline
1 & & & & & 1 & & & & & 1 & & & & & 1 \\
 & 1 & & & & & 1 & & & & & 1 & 1 & & & \\
 & & 1 & & & & & 1 & 1 & & & & & 1 & & \\
 & & & 1 & 1 & & & & & 1 & & & & & 1 &
\end{array}\right] \quad [16,9,4]\ \text{QC}$$

$$\underline{H} = \left[\begin{array}{cccc|cccc|cccc}
1 & & & & 1 & & & & 1 & & & \\
 & 1 & & & & 1 & & & & 1 & & \\
 & & 1 & & & & 1 & & & & 1 & \\
 & & & 1 & & & & 1 & & & & 1 \\
\hline
1 & & & & & 1 & & & & & 1 & \\
 & 1 & & & & & 1 & & & & & 1 \\
 & & 1 & & & & & 1 & 1 & & & \\
 & & & 1 & 1 & & & & & 1 & &
\end{array}\right] \quad [12,5,4]\ \text{SQC}$$

Fig. 6.13 Parity check matrices of the [16,9,4] QC code and of its one-time shortened version

About the relationships between the rows in the parity check matrix reported up in Fig. 6.13 and the parity check polynomials $h_j^{(i)}(x^4)$, $i, j = 1, 2, 3, 4$, plus $h_o(x^4)$, needed (as quotient) in order to justify the presence of a 2-weight code word of the type $(x^{16} + 1)$ in Example 2.28, the following observations can be developed. In Fig. 2.23 down, only the first one of the four 2-weight code words due to one-time lengthening has been treated (that characterized by $i = 1$). It is convenient to maintain separate the $h_o(x^4)$ polynomial for the moment. Considering also the positions in the period, such polynomials can be written as

$$x^3 h_1^{(1)}(x^4) = x^3(x^8 + x^4 + 1), \quad x^2 h_2^{(1)}(x^4) = x^2(x^4 + 1), \quad x h_3^{(1)}(x^4) = x \cdot 1.$$

The remaining terns of polynomials (for $i = 2, 3, 4$) are simple shifts of the above ones, that is

$$x^2 h_2^{(2)}(x^4) = x^2(x^8 + x^4 + 1), \quad x h_3^{(2)}(x^4) = x \cdot (x^4 + 1), \quad h_4^{(2)}(x^4) = 1 \cdot 1,$$
$$x h_3^{(3)}(x^4) = x(x^8 + x^4 + 1), \quad h_4^{(3)}(x^4) = 1 \cdot (x^4 + 1), \quad x^3 h_1^{(3)}(x^4) = x^3 \cdot 1,$$
$$h_4^{(4)}(x^4) = 1 \cdot (x^8 + x^4 + 1), \quad x^3 h_1^{(4)}(x^4) = x^3 \cdot (x^4 + 1), \quad x^2 h_2^{(4)}(x^4) = x^2 \cdot 1.$$

To these constituent polynomials, it is necessary to add $x^{-1}h_o(x^4) = x^{-1} \cdot 1$, where also its position in the period has been taken into account. Since all these polynomials are symmetric, we can ignore reciprocal symbolism. For the rows in the first layer of the $\underline{H}$ matrix shown in Fig. 6.13 up, represented by enlarged unitary polynomials $u_4(x^4)$, it is easy to derive the following relationships

$$x^{4-j} u_4(x^4) = x^4 h_j^{(j)}(x^4) + h_j^{(j+2)\bmod 4}(x^4) \quad j = 1, 2, 3, 4.$$

We can note that all the constituent polynomials have been selected with the same subscript. Instead, for the rows in the second layer of $\underline{H}$, it is necessary to take constituent polynomials having different subscripts and employ $x^{-1}h_o(x^4)$ too. A lengthened version of this QC code is treated also at the end of the next Section, showing that four code words of the type $(x^{16} + 1)$ effectively occur after a one-time lengthening, as we can easily derive, now, from the only parity check matrix.

6.6 Lengthened Quasi-Cyclic (LQC) Codes and Their $\underline{H}$ Matrix

LQC codes have been introduced in Sect. 2.7 according to Definition 2.14, that is maintaining invariant the generator sub-matrix $\underline{G}_o^{(q_o)}$. The parity check matrix of a code so organized can be obtained simply by assuming the reciprocal of proper periodic parity check polynomials. The period has length $N = pq_o$.

$$\underline{G} = \left[\begin{array}{cccccccccccccc|cc}
1&1&1&1&1&&&&&&&&&&& \\
&1&1&0&0&1&&&&&&&&&& \\
&&1&1&1&1&1&&&&&&&&& \\
&&&1&1&0&0&1&&&&&&&& \\
&&&&1&1&1&1&1&&&&&&& \\
&&&&&1&1&0&0&1&&&&&& \\
&&&&&&1&1&1&1&1&&&&& \\
&&&&&&&1&1&0&0&1&&&& \\
&&&&&&&&1&1&1&1&1&&& \\
&&&&&&&&&1&1&0&0&1&& \\
\hline
&&&&&&&&&&1&1&1&1&1& \\
&&&&&&&&&&&1&1&0&0&1
\end{array}\right] \qquad
\underline{H} = \left[\begin{array}{cccccccccccccc|cc}
1&0&1&1&1&1&0&0&0&1&1&0&0&0&1&0 \\
1&1&0&1&1&1&1&0&1&0&0&1&0&0&1&1 \\
0&0&1&0&1&1&1&1&0&0&0&1&1&0&0&0 \\
0&0&1&1&0&1&1&1&1&0&1&0&0&1&0&0
\end{array}\right]$$

Fig. 6.14 Generator matrix and parity check matrix of the [16,12,2] one-time LQC code

Example 6.13 Taking the [14,10,3] QC code whose $\underline{G}$ matrix is shown on the left in Fig. 6.4, the generator matrix of its one-time lengthened version is reported here on the left in Fig. 6.14. A dashed rectangular border bounds the old $\underline{G}$ matrix. Adopting the second $\underline{H}$ matrix on the right in Fig. 6.4, the parity check matrix of the new code will appear as depicted in Fig. 6.14 on the right. It is possible to observe here the presence of two columns more, where a first periodic replica of the 4 rows present in the original $\underline{H}$ matrix begins. An analogous procedure can be applied also to all the other possible $\underline{H}$ matrices available for the original QC code. A pair of 2-weight code words comes out, as expected, having form $(x^{14} + 1)$ and $x(x^{14} + 1)$. Remark that lengthening has been performed here on the right hand side, in order to simplify a comparison with the behaviour exhibited by QC codes of the type treated in Example 6.12. On the other hand, a lower triangular form is suggested for simplifying the encoding procedure, by means of an encoder circuit of the type shown in Fig. 6.9, but this is anyway not advisable for lengthened codes.

When the $\underline{G}$ matrix is reordered in the version containing truncated circulants, where distributed control symbols are present, q_o truncated-circulant generator polynomials appear (see Definition 2.15). The LQC code will exhibit a greater number of rows in the $\underline{G}$ matrix, as described in Sect. 2.8, with progressively increasing separations between the two parts (in general q_o parts) of the truncated-circulant generator polynomials. The correspondent parity check polynomials turn out to be periodic in any rectangular sub-matrix of $\underline{H}$, showing a p-length period.

Example 6.14 In Example 2.27 the [10,6,3] QC code has been one-time lengthened. The [12,8,2] LQC code so obtained exhibits properly transformed truncated-circulant generator polynomials $G''_{T1}(x) = x^6 + x^3 + 1$, $G''_{T2}(x) = x^6 + x^2 + x + 1$, each one shifting in four subsequent rows of the reordered $\underline{G}$ matrix (see Fig. 2.22 down). Let us consider now this code from the point of view of its parity check matrix. In Example 6.8, we presented the parity check matrix of the original [10,6,3] QC code, in minimal form and lower triangular form. After one-time lengthening (on the right hand side like in example 6.13), we obtain the parity check matrix shown on the left in Fig. 6.15. Reordering yields the new $\underline{H}$ matrix reported on the right. Let us remember that the reordered version of this LQC code is characterized by 2-weight

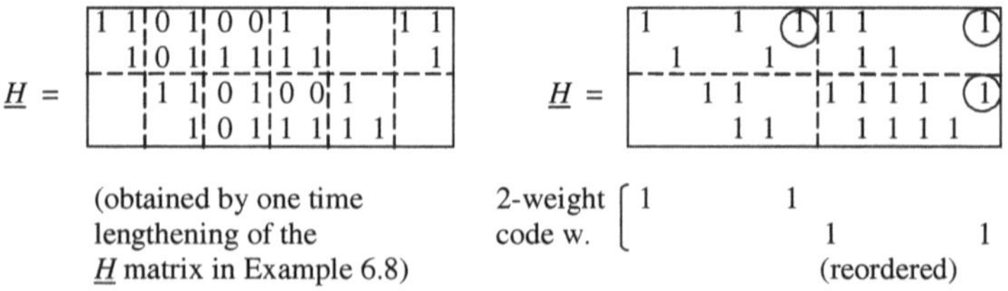

Fig. 6.15 Parity check matrices for the [12,8,2] LQC code treated in Example 6.14

code words of the type $(x^5 + 1)$, as already stressed in Example 2.26. Such code words are immediately recognizable in this $\underline{H}$ matrix, as proper null ex-OR sums of two columns (pairs of 1-symbols under the matrix). The three 1-symbols due to this code lengthening procedure have been denoted with small circles. Clearly they are responsible for the two 2-weight code words above discussed. Finally we can check that the last $\underline{H}$ matrix represents a valid correspondent for the generator matrix formed by polynomials $G''_{T1}(x)$, $G''_{T2}(x)$ in Fig. 2.22 up.

In Definition 2.16 the procedure for x_o-time code lengthening, maintaining invariant parameter p, has been introduced. It regards QC codes whose $\underline{G}$ matrix exhibits X layers of complete $p \times p$ square circulants and R_X possible anomalous rows. In this case the $\underline{G}$ matrix is modified adding x_o further layers of complete $p \times p$ square circulants and maintaining unchanged the R_X anomalous rows. The correspondent $\underline{H}$ matrix will show x_o further column layers, in which the permutation circulants to be added follow the shift sequence which characterizes the original QC code.

Example 6.15 Taking the [16,9,4] QC code already treated in Examples 2.22, 2.24 and 6.12, its one-time lengthened version is characterized by the $\underline{H}$ matrix shown in Fig. 6.16. The first layer is formed by the same unique permutation circulant, whereas the second layer shows two times the same permutation circulant (in particular a 4 × 4 identity matrix). This code is a [20,13,2] LQC code, with four 2-weight code words, confirming what expressed in Table 2.4 (Example 2.28).

A property of this type holds also when the layers of permutation circulants in the $\underline{H}$ matrix are more than two. In this case, as already noted, the generator polynomials forming the $\underline{G}$ matrix, by means of blocks of circulants, exhibit higher

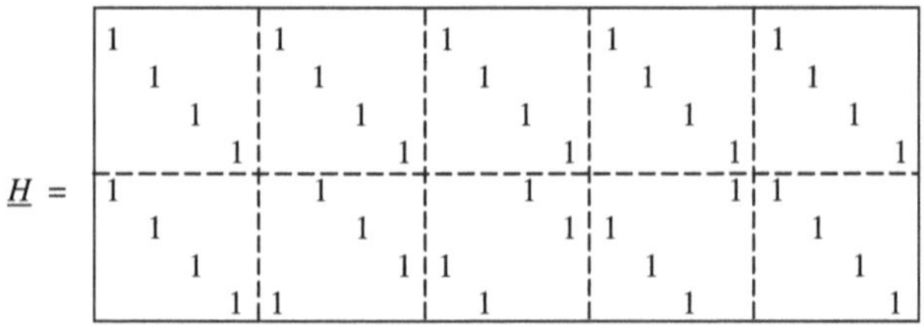

Fig. 6.16 Parity check matrix of the [20,13,2] LQC code described in Example 6.15

weight and degree. Nevertheless, besides the minimum distance, reduced to $d = 2$, another drawback occurs. Like in Fig. 6.16 (where two only row layers are considered), some 4-length short cycles of 1-symbols come out. This prevents the use of very efficient decoding algorithms for such codes, interpreted as LDPC codes (Fan 2000).

6.7 Punctured QC Codes

Punctured cyclic codes have been introduced in Sect. 1.7 regarding their $\underline{G}$ matrix (which becomes no longer time-invariant in general), and in Sect. 5.7 where, instead, we observed that the $\underline{H}$ matrix is characterized by progressive elimination of external rows (and columns), starting from the parity check matrix of the original code. Punctured QC codes will be introduced here, attempting to treat their $\underline{G}$ and $\underline{H}$ matrix at a same time (often exploiting duality). Although, in principle, it is possible to eliminate single rows (and columns) of $\underline{H}$, in order to maintain the property of w.s. time-invariance in $\underline{H}$, we will restrict ourselves to the elimination of complete periods.

Definition 6.4 Given a certain QC, SQC or LQC code, characterized by parameters q_o, p, r_p, whose generator matrix exhibits a $\underline{G}_o^{(q_o)}$ sub-matrix shifting in its rows, it is possible to perform subsequent puncturations in it by a progressive elimination of the last column layers in such $\underline{G}$ matrix. This corresponds to the progressive elimination of the last row layers (and column layers) in $\underline{H}$. The code so obtained is called *punctured QC, SQC or LQC code.*

As already noted for cyclic codes, code puncturation can lead to rows linearly dependent also when, before puncturation, they were all linearly independent. In this respect, starting from a $[pq_o, k_pq_o, d]$ QC code, after a single puncturation, we obtain a $[(p-1)q_o, k_pq_o, d']$ code, where typically the new minimum distance d' is smaller than d. Even a minimum distance $d' = 1$ may come out, if too many single puncturations are made. A situation similar to that here described occurs starting from an SQC or from an LQC code.

Example 6.16 Considering the [14,10,3] QC code already treated in Example 6.4, after one puncturation we obtain a [12,10,1] code, whose $\underline{G}$ and $\underline{H}$ matrix are shown in Fig. 6.17. The latter one has been obtained from the second $\underline{H}$ matrix shown in Fig. 6.4 on the right (which is the one in minimal form for this code), after elimination in it of the second pair of rows and the last pair of columns (dashed lines in the Figure). It is possible to verify that one 1-weight code word appears, represented by the position x^4, where the parity check matrix, after puncturation, shows a null column. The same conclusion can be reached also performing an ex-OR sum of the last three rows in the $\underline{G}$ matrix. Obviously cyclic permutations of the code words are here no longer possible. Nevertheless, if we perform a cyclic permutation before puncturation, in such a way as to set the upper row layer with its null column in the

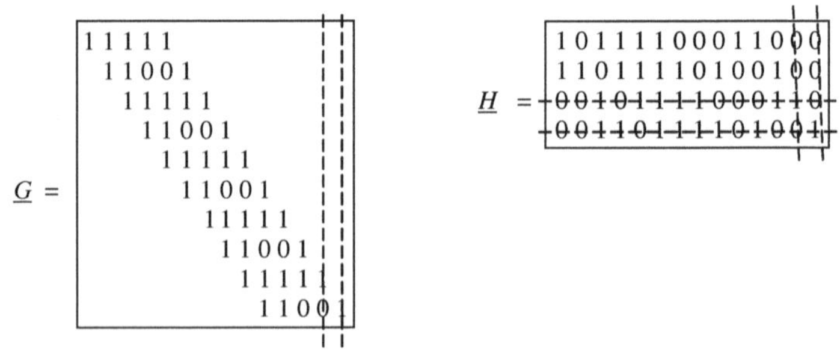

Fig. 6.17 Generator and parity check matrix of a [12,10,1] punctured QC code

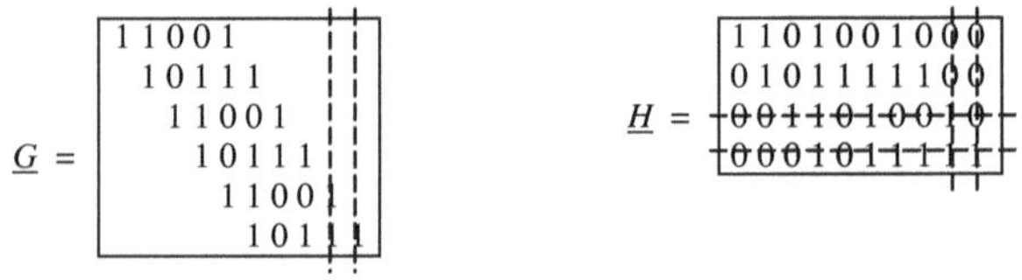

Fig. 6.18 Generator and parity check matrix of a [8,6,1] punctured QC code

first period, and then apply a one-time shortening, it is possible to obtain, after puncturation, a [10,8,2] code.

Example 6.17 Considering the [10,6,3] QC code already treated in Examples 2.10, 6.6, and also in some subsequent Examples, after one puncturation we obtain an [8,6,1] code whose $\underline{G}$ and $\underline{H}$ matrix are shown in Fig. 6.18. The latter one has been obtained from the parity check matrix in minimal form (see Example 6.8), after elimination in it of the second pair of rows and the last pair of columns (dashed lines in the Figure). Here the 1-weight code word occupies the position expressed by x^4. Clearly a so strong performance degradation is due to the fact that, like in Example 6.16, a puncturation produce control number halving. Another possibility for recovering an acceptable performance consists in the elimination of just the column where the above 1-weight code word is placed. So we obtain a [7,5,2] code. Clearly this new code looses the property of being derived from a QC code.

Lemma 6.4 *The dual of an SQC code is a punctured QC code, and* vice versa.

Proof We can proceed as in the proof of Lemma 5.21, with the only care of considering sets of q_o rows and integer periods at a time. □

For QC codes whose $\underline{G}$ and $\underline{H}$ matrix are organized in the form of block of truncated circulants, puncturation has the consequence of maintaining the structure of the latter one, except for a reduction in the row layers of its truncated circulants. On the contrary, the $\underline{G}$ matrix changes form, similarly to what described for the parity check matrix of SQC codes in Sect. 6.5.

At the end of the previous Section, codes characterized by layers of permutation circulants in their $\underline{H}$ matrix, with more than two layers, have been mentioned. Let Y be the number of layers. Varying parameter Y, it is possible to describe a family of codes called array codes (Fan 2000). A puncturation, in this context, is to be taken considering a whole layer at a time (constant-length puncturation of p rows). Nevertheless a matrix so organized cannot be expressed in its minimal form, without changing the LDPC nature of the code. Furthermore elimination of a column layer is equivalent to perform a code shortening. On these aspects we will return in Chaps. 8 and 11 with more details.

6.8 H-Extended Quasi-Cyclic (HEQC) Codes

Although, in Sects. 6.1 and 6.2, H-extension has been already introduced for QC codes, in order to classify possible $\underline{H}$ matrices with a number of rows greater than the number characterizing the QC condition, it is necessary to discuss here this topic more in detail. Duality properties will be widely exploited.

Definition 6.5 Given a QC code, whose length is $N = pq_o$ and whose generator constraint length is r_p, starting from its parity check matrix in minimal form (with $r_p q_o$ rows subdivided into r_p layers), the code obtained from it, increasing the number of row layers with the same set of q_o parity check polynomials, is defined *H-extended quasi-cyclic code* (or briefly *HEQC code*). More precisely, assuming a number of row layers $r_p' = r_p + j$, that is after j single H-extensions (regarding as many periods), we will define the resulting code as *j-time H-extended* starting from its quasi-cyclic version.

In the considered $\underline{H}$ matrix, q_o subsequent rows are collected in a proper sub-matrix, shifting along row layers with steps of q_o. This parity check sub-matrix is the basic element either for code puncturation or for code H-extension, in the first case reducing the number of row layers, in the second case increasing it, with respect to the QC condition. It is interesting to compare the above Definition with Definition 5.15, which regards cyclic codes. Also here, H-extension reduces the code rate, whereas code puncturation and code lengthening increase it. The following proposition expresses an important duality property.

Lemma 6.5 *The dual of a y-time LQC code is a y-time HEQC code and vice versa.*

Proof We can proceed as in the proof of Lemma 5.28, with the only care of considering sets of q_o subsequent rows and integer periods at a time. □

Lemma 6.6 *The generator polynomials of an HEQC code are periodic with period* $N = pq_o$.

Proof This property can be considered as a generalization of the one discussed in Lemma 5.29 about HEC codes. It follows directly from Lemma 6.5 and from the periodic behaviour of the parity check rows in an LQC code. □

The expressions of the generator polynomials for an HEQC code, after a number of single H-extensions able to reach a word length n, integer multiple of q_o, between qpq_o and $(q + 1)pq_o$, are

$$g_j(x) = g_j^{(QC)}(x)u_{q+1}(x^N), \quad j = 1, 2, \ldots q_o, \quad N = pq_o, \tag{6.4}$$

where $g_j^{(QC)}(x)$ represents the jth generator polynomial of the original quasi-cyclic code. A possible truncation of the last replica of the $g_j^{(QC)}(x)$'s is needed on the right hand side of the $\underline{G}$ matrix, in order to maintain it in upper triangular form. Remark the analogy with a correspondent property valid for HEC codes, as discussed in Sect. 5.13.

Example 6.18 In Example 6.3, the first complete $\underline{\text{H}}$-extension of the [8,4,4] QC code has been presented. In it no truncation of any generator polynomial is needed, because we have exactly $n = 2N = 16$. The $\underline{G}$ matrix so obtained (see Fig. 6.3 up-right) is upper triangular (and also lower triangular). Since the resulting [16,4,8] code is QC again, we can apply a cyclic permutation with step of $q_o = 2$ over its whole length $2N = 16$ to any rows of such generator matrix, provided that the four rows remain linearly independent. For instance, it is possible to adopt the following matrix

$$\underline{G} = \begin{bmatrix} 1&0&0&0&1&1&1&0&1&0&0&0&1&1&1&0 \\ 0&0&0&1&0&1&1&1&0&0&0&1&0&1&1&1 \\ 1&0&1&0&0&0&1&1&1&0&1&0&0&0&1&1 \\ 1&1&0&0&0&1&0&1&1&1&0&0&0&1&0&1 \end{bmatrix},$$

in which it is not easy to recognize where information symbols and to control symbols are placed.

Example 6.19 Taking the [14,10,3] QC code treated several times so far, it is possible to H-extend it, say three times, obtaining the pair of parity check and generator matrices shown in Fig. 6.19. The former one has been obtained starting from the second reported on the left in Fig. 6.4, which is the minimal form of $\underline{H}$ for

```
H =
1 0 1 1 1 1 0 0 0 1 1
1 1 0 1 1 1 1 1 0 1 0 0 1
    1 0 1 1 1 1 0 0 0 1 1
    1 1 0 1 1 1 1 1 0 1 0 0 1
        1 0 1 1 1 1 0 0 0 1 1
        1 1 0 1 1 1 1 1 0 1 0 0 1
            1 0 1 1 1 1 0 0 0 1 1
            1 1 0 1 1 1 1 1 0 1 0 0 1
                1 0 1 1 1 1 0 0 0 1 1
                1 1 0 1 1 1 1 1 0 1 0 0 1
```

```
G =
1 1 1 1 1                   1 1 1 1 1
  1 1 0 0 1                   1 1 0 0 1
    1 1 1 1 1                   1 1 1 1
      1 1 0 0 1                   1 1 0
        1 1 1 1 1                   1 1
          1 1 0 0 1                   1
            1 1 1 1 1
              1 1 0 0 1
                1 1 1 1 1
                  1 1 0 0 1
```

Fig. 6.19 Parity check and generator matrix of the [20,10,3] HEQC code treated in Example 6.19

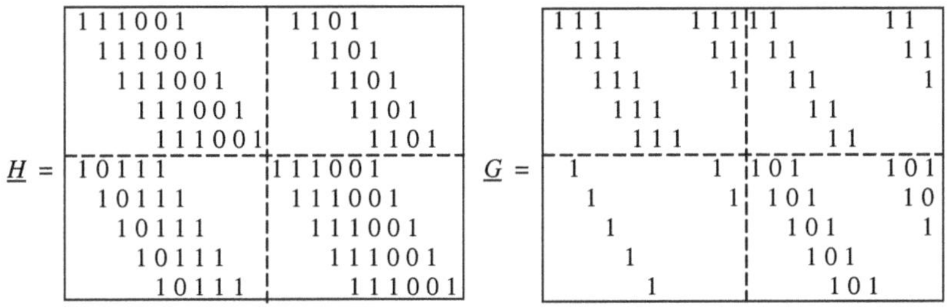

Fig. 6.20 Reordered $\underline{H}$ and $\underline{G}$ matrix of the [20,10,3] HEQC code described in Fig. 6.19

this code. The resulting parity check matrix is still lower triangular. The code so obtained is a [20,10,3] HEQC code. In order to have an increase in the minimum distance, several further single H-extensions have to be made, because some 3-weight truncated generator polynomials in the lower part of $\underline{G}$ remain.

Considering now the reordered QC codes, whose $\underline{G}$ and $\underline{H}$ matrix are in a form containing truncated circulants (see Figs. 2.11 and 6.7), H-extension has the following consequences. The parity check matrix exhibits still the form of a block of truncated circulants, but with increased size. Precisely they become $(r_p + j) \times (p + j)$ truncated circulants. In the $\underline{G}$ matrix, instead, any truncated circulant is transformed, by addition of further 1-symbols on the right hand side (a similar transformation is that shown by the $\underline{H}$ matrix in Fig. 6.15, where the case of a one-time lengthened QC code is treated).

Example 6.20 The [20,10,3] HEQC code analyzed in Example 6.19 can be reordered according to Lemma 2.6, so that its parity check and generator matrix assume respectively the forms shown in Fig. 6.20. Several 0-symbols have been eliminated in order to simplify identification of the 1-symbol patterns. Only the $\underline{H}$ matrix can be still interpreted as a true 2 × 2 block of 5 × 10 truncated circulants, in which the last five rows have been eliminated. The $\underline{G}$ matrix is characterized, instead, by the beginning of 1-symbol pattern replicas, in the up-right corner of any rectangular sub-matrix. The minimum distance is still $d = 3$, because of the last two rows of $\underline{G}$.

Finally about the QC codes constructed with complete square circulants in their generator and parity check matrix (array codes), we can observe that, since H-extension is to be made starting from a parity check matrix in minimal form, this procedure is not possible here. In fact such codes do not exhibit a parity check matrix with null circulants, property which is needed in order to identify the 1-symbol patterns shifting along its rows, when taken in minimal form. Nevertheless insertion of null circulants can be forced, leading to suitable variants in array codes, according to a procedure called masking (Lin and Costello 2004). Alternatively it is possible to introduce a procedure resembling a modified H-extension, although not rigorously coincident with it, as will be done in Sect. 6.10.

6.9 Modified Lengthened Quasi-Cyclic (MLQC) Codes and Their $\underline{H}$ Matrix

In Definition 2.18, MLQC codes have been described as proper subcodes of LQC codes. They can be interpreted as a particular class of w.s. time-invariant convolutional codes, according to what discussed in Chap. 4. Their $\underline{G}$ matrix is obtained by eliminating layers of consecutive q_o rows, in the generator matrix of the parent LQC code. The above layers of consecutive rows correspond to as many $\underline{G}_o^{(q_o)}$ sub-matrices. Nevertheless their $\underline{H}$ matrix is to be constructed independently of the original LQC parity check matrix, but simply as a valid counterpart of the $\underline{G}$ matrix considered, so satisfying constraint (5.4).

Keeping in mind the importance of the columns in $\underline{H}$ (syndromes), instead of grouping q_o rows of such matrix, and shifting this group from one row layer to the subsequent one (as for the $\underline{G}_o^{(q_o)}$ sub-matrices in the $\underline{G}$ matrix), it is preferable to introduce here a vertical shift of a proper sub-matrix, collecting layers of q_o subsequent columns of $\underline{H}$. First of all, it is however useful to widen the concept of constant-length shortening, introduced by Definition 5.12, in the framework of cyclic codes, in order to consider QC codes too.

Definition 6.6 The elimination of one central layer of q_o consecutive rows, which belong to the same $\underline{G}_o^{(q_o)}$ sub-matrix in a w.s. time-invariant $\underline{G}$ matrix, without eliminating any column, is defined *constant-length shortening* for a code which is *w.s. time-invariant in its generator matrix.*

Comparison with Definition 5.12 allows one to note that here we treat proper layers of q_o consecutive rows, instead of single rows. Both operations are made on the generator matrix. Clearly a periodic procedure of constant-length shortening, according to Definition 6.6, gives rise to an MLQC code, that is to a particular type of w.s. time-invariant convolutional code, described by means of its generator matrix.

Example 6.21 In Example 2.30 we started from the LQC version of the code characterized by $q_o = 2$, $r_p = 4$, $N = 8$, and polynomials $g_1(x) = x^4 + x^3 + x^2 + x + 1$, $g_2(x) = x^4 + x^2 + x + 1$. Adopting $N_o = 2$ and $K_o = 1$ as parameters for a periodic constant-length shortening, the semi-infinite generator matrix depicted in Fig. 2.25 comes out. Such $\underline{G}$ matrix and the correspondent $\underline{H}$ matrix, both semi-infinite, are shown in Fig. 6.21. There is a procedure for obtaining the latter one from the former one, but it will be described in following Chapters. For the moment, we can verify that this $\underline{H}$ matrix is valid for obtaining back the $\underline{G}$ matrix, by calculating proper null ex-OR sums of clusters of syndromes. It is possible to note the vertical shift of a proper sub-matrix, as anticipated, with steps of $q_o = 2$. The positions of information and control symbols have been denoted with small solid and dashed arrows respectively, as usual. They obey to the rule *iicc*. The original QC code was self-dual. This property has the consequence of producing the same polynomials both in $\underline{G}$ and in $\underline{H}$, also in this MLQC version of it.

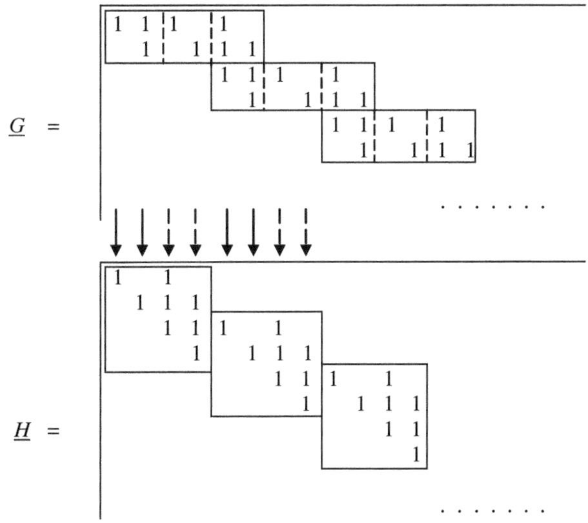

Fig. 6.21 *Top-left* corner of the generator and parity check matrix for the MLQC code described in Example 6.21

Nevertheless, a truncation of such polynomials on the left hand side originates two anomalous rows, whose weights are smaller than the weights of the remaining ones. The pair of matrices shown in Fig. 6.21, as already noted, can be considered semi-infinite, like in convolutional codes (see Sect. 3.3). A correct frame termination has to be performed, giving rise to a suitable down-right corner in the parity check matrix, where truncated parity check polynomials, similar to the ones occurred in the top-left corner, come out. This behaviour resembles that shown in Figs. 5.28 and 5.29 for MHEC codes, and has the purpose of assuring equal protection to all the positions in the frame.

Finally we can observe that the code treated in Example 6.21 is equivalent to a w.s. time-invariant convolutional code, whose asymptotic code rate is 2/4. It can be interpreted either as an MLQC code (following this way, it has been derived here), but also as a modified H-extended QC code, as will be better explained in the next Section. Nevertheless the case treated in Example 6.21 is very particular, since the original QC code was self-dual and also the periodic constant-length shortening adopted maintains this symmetry. In general, modified lengthening and modified H-extension, although both leading to convolutional codes, are rather different procedures.

6.10 Modified H-Extended Quasi-Cyclic (MHEQC) Codes and Their $\underline{H}$ Matrix

It is necessary, here, to widen the concept of constant-length puncturation, introduced by Definition 5.11 in the framework of cyclic codes, in order to consider QC codes too.

Definition 6.7 Starting from a w.s. time-invariant parity check matrix in its minimal form (whose rows are grouped in layers having q_o elements), the elimination of one central layer of q_o consecutive rows, without eliminating any column, is defined *constant-length puncturation* for a code which is *w.s. time-invariant in its parity check matrix.*

Comparison with Definition 5.11 allows one to note that proper layers of q_o rows substitute single rows. A periodic procedure of constant-length puncturation, according to Definition 6.7, is expected to produce a certain form of w.s. time-invariant convolutional code in the resulting $\underline{H}$ matrix. Remembering the derivation of MHEC codes from HEC codes in Sect. 5.15, the following steps can be listed.

Lemma 5.19 relates a periodic constant-length puncturation to the transformation of control symbols into information symbols for a cyclic code. A generalization of this Lemma, in order to take into account also QC codes, is obvious. The rightmost position of any eliminated row in the $\underline{H}$ matrix, also here, will be that where a control symbol is transformed into an information symbol. This is possible, however, only if the original $\underline{H}$ matrix is lower triangular (see also Lemma 5.26). Such transformation enables us to introduce a proper modification in the H-extension procedure applied to a QC code.

Definition 6.8 Let us consider a w.s. time-invariant block code in its $\underline{H}$ matrix. After several single H-extensions, starting from the minimal form of $\underline{H}$ in its QC version (whose code word length is *N*), we obtain a block length *n* remarkably greater than *N*. Such $\underline{H}$ matrix is modified by periodic insertion of information symbols among the control symbols associated to the rows of $\underline{H}$. This procedure is performed by means of proper constant-length puncturations (according to Definition 6.7), and the code so obtained is called *modified H-extended quasi-cyclic code* (or briefly *MHEQC code*).

Comparison with Definition 5.17, where MHEC codes have been introduced, shows the presence in both of a preliminary H-extension. In analogy with MLQC symbolism, we can take a super-period of length q_oN_o, whose first q_oK_o symbol times are the result of as many constant-length puncturations. This MHEQC code will exhibit an asymptotic code rate $(q_oK_o)/(q_oN_o)$.

Example 6.22 Let us consider the H-extended quasi-cyclic code treated in Example 6.19, starting from the parity check matrix of the [14,10,3] QC code, characterized by $q_o = 2$. In order to obtain the MHEQC code, equivalent to a w.s. time-invariant convolutional code with rate 2/4 (that is with $K_o = 1$, $N_o = 2$), it is possible to eliminate, from that parity check matrix, two rows any four (one row layer any two). The result is shown in Fig. 6.22. The new parity check matrix is depicted up, in semi-infinite form, already subdivided into proper sub-matrices vertically shifting. Six super-periods of length $q_oN_o = 4$ have been taken into account. The rectangular sub-matrices collect 4 columns, and vertically shift by two positions at a time. The correspondent $\underline{G}$ matrix is reported down, semi-infinite as well, where two generator polynomials form a $\underline{G}_o$ sub-matrix (the symbolism is that of convolutional codes w.s. time-invariant in $\underline{G}$), horizontally shifting by four positions at

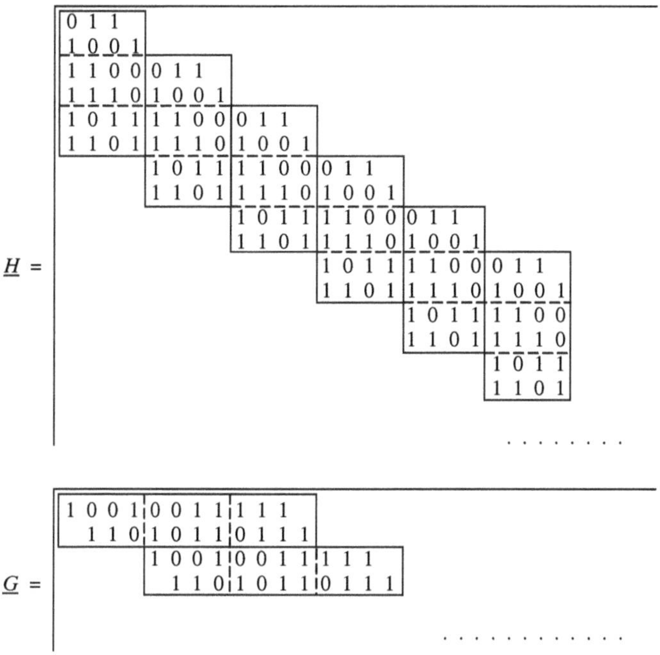

Fig. 6.22 Semi-infinite form of the parity check and generator matrix for the MHEQC code described in Example 6.22

a time. Such two generator polynomials have weights $w = 7$ and $w = 8$ respectively. The minimum distance of this code turns out to be $d = 6$, with low-weight coefficients $A(6) = Z-2$, $A(7) = 3Z-1$, where Z is the number of periods in which information symbols are inserted. After this number of periods, there are 2 periods more, with only control symbols, for assuring a correct frame termination. The above linear growth of coefficients $A(w)$ with Z confirms the convolutional nature of the code so obtained, which can be considered also well designed. Proper ex-OR sums of clusters of syndromes enable us to verify that the $\underline{G}$ matrix and the $\underline{H}$ matrix here discussed constitute a valid pair of matrices.

In spite of Lemma 6.5, where LQC codes and HEQC codes are recognized to form a dual pair of codes, we cannot state now, rigorously, that MLQC codes and MHEQC codes form a dual pair of codes, because of the different behaviour of some external periods. A similar conclusion has been derived in Sect. 5.15 (about MLC and MHEC codes). In this sense, we can consider the generator sub-matrix $\underline{G}_o$ of the convolutional code resulting after the above modified lengthening, in comparison with the generator sub-matrix $\underline{G}_o^{(2)}$ of the original QC code. It is possible to note how they are different (see Example 6.22), in polynomial degrees, but especially in polynomial weights. The minimum distance increases. On the other hand, the code rate has been reduced from 5/7 to an asymptotic value of 2/4.

A trivial transformation can be obtained organizing the $\underline{G}$ and $\underline{H}$ matrix of an MHEQC code in the form of blocks of truncated circulants. Here we will expect to have two sets of control symbols (corresponding to as many sets in the original code): a first set distributed along the length of any transformed period, and a second set at the end of any transformed period.

A quite different situation is that of QC codes showing complete square circulants (array codes). They have been introduced in Sect. 2.7 (see Example 2.22). In Sect. 6.5 their parity check matrix has been discussed, showing that the presence of the anomalous rows in $\underline{G}$ made by unitary polynomials is related to the existence of as many rows of $\underline{H}$ not linearly independent. In Sect. 6.9 we have ignored the problem of constructing the possible $\underline{H}$ matrix of MLQC versions for such codes. This choice was justified by the interest in the parity check matrix only, where it is possible to perform an efficient soft-decision decoding in an LDPC model. At the end of Sect. 6.8, we observed that traditional H-extension is impossible in such codes, since their $\underline{H}$ matrix is not in its minimal form. Therefore a particular type of modified H-extension is to be considered here, not made executing previously an H-extension and then a constant-length puncturation, but simply vertically shifting the entire block of the old parity check matrix as a proper sub-matrix.

Example 6.23 Taking the [16,9,4] QC code described by the parity check matrix shown in Fig. 6.13 up, and whose generator matrix was introduced in Fig. 2.20, it is possible to construct a particular MHEQC code by means of a modified semi-infinite $\underline{H}$ matrix of the type depicted up in Fig. 6.23. It is characterized by all linearly independent rows, and is obtained starting from a 2/3-rate shortened version of this code (see Fig. 6.13 down). The module so obtained vertically shifts by one row layer at a time. The correspondent $\underline{G}$ matrix, constructed by means of the procedure based on the null ex-OR sum of cluster of syndromes, is reported down in Fig. 6.23. It is possible to note that the anomalous row made by a unitary polynomial is no longer present, as anticipated. Furthermore the $\underline{G}_o$ sub-matrix, which horizontally shifts with step of 12 positions at a time, exhibits 8 rows, in two layers. The first layer is characterized by the same circulants which are present in the original QC code, whereas the second layer has a longer length, because it is placed across two super-periods. The overall asymptotic code rate is 8/12. Four dashed small arrows denote the positions occupied by as many control symbols in any super-period. The two matrices described so far can be written also in symbolic form, labelling the circulants according to Definition 4.12, that is employing shifts $\underline{J}$, J = 0, 1, … ($p-1$), with respect to an identity matrix (symbolically represented by the permutation circulant $\underline{0}$). This symbolic representation is depicted in Fig. 6.24.

The minimum distance of the w.s. convolutional code described in Example 6.23 is $d = 4$, provided that the frame is correctly terminated. This property holds as long as two equal columns in its $\underline{H}$ matrix are not present, like in the original QC block code. In order to guarantee that this condition is satisfied also for higher asymptotic code rates it is necessary to increase parameter p. The most interesting situation is that characterized by a higher number of row layers. A convolutional code of this type has been named doubly convolutional code in Sect. 4.9, owing to the

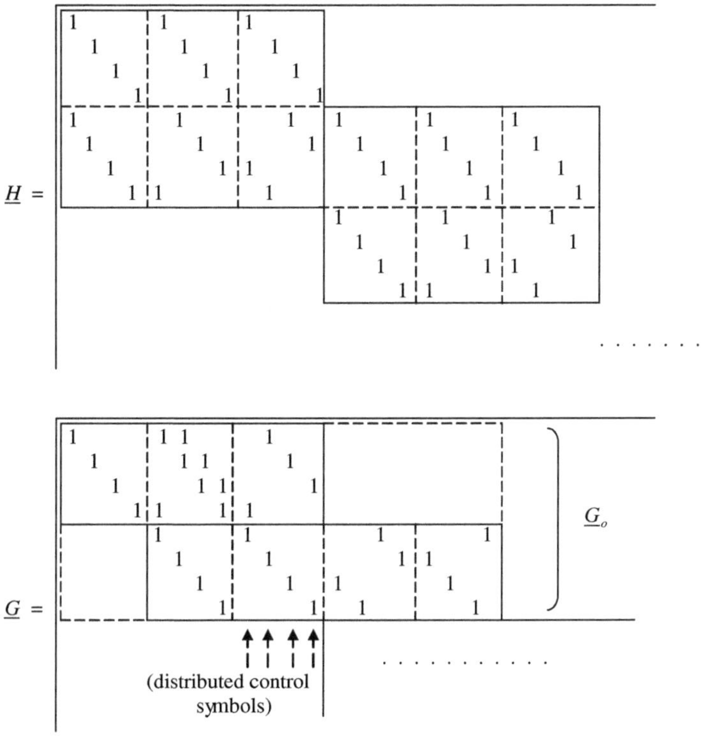

Fig. 6.23 A particular type of MHEQC code obtained starting from the [16,9,4] QC code treated in several previous Examples

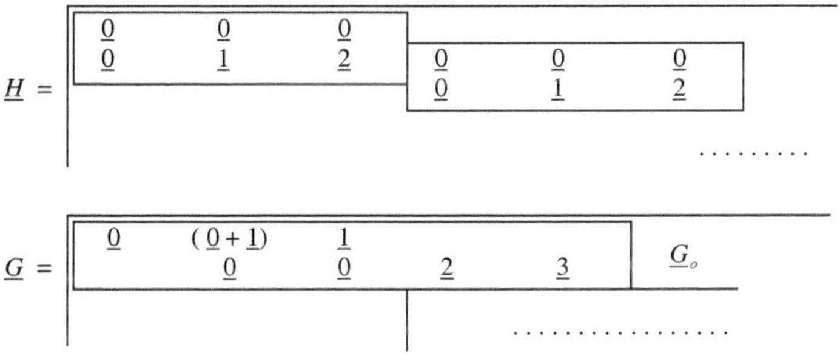

Fig. 6.24 Symbolic representation of the matrices reported in Fig. 6.23

tail-biting arrangement (represented by the circulants) and to the presence of a further convolutional code structure. These topics will be investigated more in detail in Chaps. 8 and 12.

6.11 Trellis of a w.s. Time-Invariant Block Code Obtained from Its Parity Check Matrix

It is useful to remember which types of w.s. time-invariant codes in their parity check matrix can be encoded adopting the scheme of encoder circuit in Fig. 6.9. They are the original QC code, any shortened version, H-extended version and even modified H-extended version. All these possibilities are assured simply by a different regulation of switch operation. Even punctured codes can be obtained, if one eliminates the control symbols subjected to code puncturation after a complete encoding procedure. As for modified H-extended codes, it is necessary to add a proper symbol sequence outgoing from the information source in order to assure a correct frame termination, as already noted for s.s. time-invariant codes at the end of Sect. 5.15.

In Sect. 5.19 the trellis associated to the encoder circuit depicted in Fig. 5.6 has been described, for an s.s. time-invariant block code in its $\underline{H}$ matrix. It is characterized by binary branch complexity. Here we have the encoder circuit, still based on the parity check matrix, reported in Fig. 6.9. As already observed, the same state evolution occurs in both the shift registers. The maximum number of states is then 2^{2k_p} (in general $2^{q_o k_p}$). We can expect a structure for this trellis, with first part characterized by subsequent bifurcations, followed by a second part made by only transit states. Not always the above overall number of states are contemporary present in the same stage, as already shown in other parity check trellises (see Figs. 5.39 and 5.40).

About the branch complexity, since we proceed at steps of q_o binary symbols at a time, it results in 2^{q_o}, like in the generator trellis. The trade-off between branch complexity and state complexity can suggest the best solution, taking into account also that equivalent versions of a same code can be found reordering the sequence of information and control symbols, as already noted in Chap. 2. Especially for MHEQC codes, where distributed control symbols introduce merge states along the frame, a particular analysis is to be made case by case, still in comparison with what could occur in the generator trellis for the same code, in order to choose the solution with minimum decoding computational complexity.

In spite of this discussion, the advent of soft-decision decoding algorithms in LDPC codes has completely changed the point of view. It is not so important to have a short parity check constraint length (or generator constraint length), because the decoding algorithm is no longer performed in a trellis. On the contrary, it is crucial that the whole $\underline{H}$ matrix of the considered code satisfies certain requirements, say a small overall 1-symbol number and absence of 4-length short cycles of 1-symbols. These considerations will be developed with greater attention in Chaps. 10–12.

References

Fan JL (2000) Array codes as low-density parity check codes. In: Proceedings of 2nd international symposium on turbo codes, Brest (France), pp 543–546

Jimenez-Felstrom AJ, Zigangirov KS (1999) Time-varying periodic convolutional codes with low density parity-check matrix. IEEE Trans Inf Th 45:2181–2191

Lin S, Costello DJ (2004) Error control coding. Pearson/Prentice-Hall, Upper saddle River

Milenkovic O, Kashyap N, Leyba D (2008) Shortened array codes of large girth. IEEE Trans Inf Th 52:3707–3722

Ryan WE, Lin S (2009) Channel codes: classical and modern. Cambridge University Press, Cambridge

Tanner RM (1981) A recursive approach to low complexity codes. IEEE Trans Inf Th 27:533–547

Chapter 7
Strict-Sense Time-Invariant Convolutional Codes in Their Parity Check Matrix

7.1 Syndrome Former Sub-matrix

In Sect. 5.2 we have introduced single-error syndromes (Definition 5.6), as the row vectors in $\underline{H}^T$. Now it is useful to employ the whole $\underline{H}^T$, known as syndrome former matrix, and, by partitioning it into proper layers, to put in evidence suitable sub-matrices. Such model is devoted to describe convolutional codes, either s.s. or w.s. time-invariant in their $\underline{H}$ matrix.

Definition 7.1 Letting n_o be the period of a convolutional code, described by its parity check matrix $\underline{H}$, a layer of n_o rows in $\underline{H}^T$ (columns in $\underline{H}$) is called *syndrome former sub-matrix*, and denoted $\underline{H}_o^T$. In order to have at least a w.s. time-invariant (periodically time-varying) parity check matrix, this sub-matrix has to be unique, except for a vertical shift along the columns of $\underline{H}$.

For the moment, we refer to a convolutional code, whose period n_o shows k_o information symbols in as many first positions of the period, and $r_o = n_o - k_o$ control symbols in as many subsequent positions. If this convolutional code is well designed, its rows are all bounded on both sides, and their rightmost 1-symbols identify the positions where the correspondent distributed control symbols are placed. In a polynomial description, following the classification given at the end of Sect. 5.12, such rows have been named non-periodic parity check polynomials, stressing the difference with respect to other types of rows, bounded only on their right hand side. The rightmost 1-symbol of the considered row will be then occupied by the 0th-order coefficient of a reciprocal parity check polynomial $h_j{}^*(x)$, $j = 1, 2, \ldots r_o$, having finite degree.

Definition 7.2 If the reciprocal parity check polynomials are all equal, and coincident with a certain $h^*(x)$, the considered code is called *strict-sense* (or briefly *s.s.*) *time-invariant convolutional code in its $\underline{H}$ matrix*.

G. Cancellieri, *Polynomial Theory of Error Correcting Codes*,
Signals and Communication Technology, DOI 10.1007/978-3-319-01727-3_7

This definition regards convolutional codes, whereas Definition 5.4 was referred to $\underline{H}$ matrices in general. Definition 6.2, introducing w.s. time invariance in $\underline{H}$, is to be considered restricted to block codes only, where the concept of syndrome former sub-matrix is not needed. In fact their parity check polynomials are bounded only on the right and periodic (over the natural length $N = pq_o$). Only at the end of Chap. 6, treating MHEQC codes, the existence of a rectangular sub-matrix vertically shifting in the columns of $\underline{H}$ has been outlined (see, for instance, Fig. 6.22). It is necessary to focus our attention firstly on the fact that this rectangular sub-matrix is able to completely describe the code, and then we will distinguish when it is constructed by means of a unique parity check polynomial (s.s. time-invariance in $\underline{H}$) or not. Anyway we will start from the case of a unique polynomial, for simplicity.

Lemma 7.1 *In a well designed convolutional code, s.s. time-invariant in its $\underline{H}$ matrix according to* Definition 7.2, *we can subdivide such parity check matrix into layers of n_o columns, where a same syndrome former sub-matrix $\underline{H}_o^T$ vertically shifts from top-left to down-right, with vertical steps of r_o.*

Proof The $\underline{H}_o^T$ sub-matrix exhibits a number of rows multiple of r_o. Subdividing it into layers, each formed by r_o consecutive rows, in such layers the same $h^*(x)$ polynomial (owing to s.s. time-invariance assumption) shifts, forming horizontal periods of length n_o. The assumption for the code to be well designed assures that there is no need of periodic polynomials in the $\underline{H}$ matrix. So $\underline{H}_o^T$ completely describes the whole code frame. □

Example 7.1 Let us consider the 1/3-rate convolutional code whose semi-infinite $\underline{H}$ matrix is shown in Fig. 7.1. Two equal polynomials $h^*(x) = x^4 + x^3 + x^2 + x + 1$ shift, occupying, with their 0th-order coefficients, the last two positions in any period, whose length is $n_o = 3$. The same unique $\underline{H}_o^T$ sub-matrix, characterized by 4 rows and 3 columns, is bounded by a rectangle, and vertically shifts with steps of $r_o = 2$. In the Figure, three periods have been considered, and as many $\underline{H}_o^T$ sub-matrices are represented. Solid small arrows denote the positions where information symbols are placed, dashed small arrows those devoted to control symbols. Proper termination of the frame is needed, where some final control symbols will be added. As a comparison, we can look at the $\underline{H}$ matrix in Fig. 6.22, where a w.s. time-invariant situation is considered, being characterized by two different parity check polynomials.

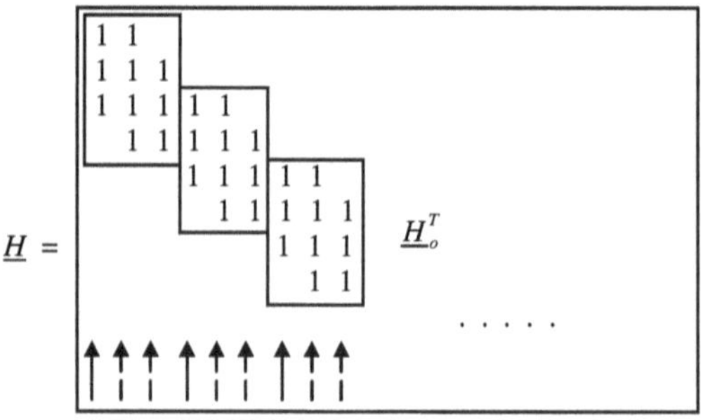

Fig. 7.1 Semi-infinite $\underline{H}$ matrix for the code described in Example 7.1

In Sect. 5.15 we have anticipated that MHEC codes, whose $\underline{H}$ matrix exhibits a structure of the type depicted in Fig. 7.1, can be interpreted as s.s. time-invariant convolutional codes in their $\underline{H}$ matrix. Nevertheless Lemma 7.1 can be extended to consider also w.s. time-invariant convolutional codes, for which a unique $\underline{H}_o^T$ is still present, although different rows are periodically repeated in its layers (this is the case of the MHEQC code whose parity check matrix is shown in Fig. 6.22).

Coming back to the case of s.s. time-invariance, it is useful to remember that MHEC codes are introduced starting from a precise parent cyclic code, described by the parity check polynomial $h_c(x)$. The generator polynomial $g_c(x)$ of this parent cyclic code is not present in the $\underline{G}$ matrix of the considered MHEC code. Obtainment of this generator matrix $\underline{G}$, compatible with the $\underline{H}$ matrix adopted, is possible only imposing the null ex-OR sum of clusters of syndromes (Lemma 5.11).

Example 7.2 After imposition of the null ex-OR sum for the shortest possible cluster of syndromes, a $\underline{G}$ matrix associated to the $\underline{H}$ matrix described in Example 7.1 is constructed. It is characterized by a $\underline{G}_o$ sub-matrix with just one row, where the interleaved generator polynomial $g_{\text{int}}(x) = x^8 + x^7 + x^5 + x^4 + x^3 + x + 1$ is present. As anticipated, this polynomial is not the co-factor of the parity check polynomial adopted. Although $\text{w}[g_{\text{int}}(x)] = 7$, this 1/3-rate convolutional code has minimum distance $d = 6$. Its behaviour is rather far from that of the best 1/3-rate s.s. time-invariant convolutional code with $m = 2$ (see Table 3.4). This poor result is consequence of the constraint of having two equal parity check polynomials.

As already stressed, it is possible to design a code, starting from its $\underline{G}$ matrix, and then obtaining a correspondent $\underline{H}$ matrix. Alternatively, it is possible to proceed in the reverse direction. Traditional convolutional codes are designed optimizing $\underline{G}$. On the contrary, LDPC convolutional codes (Jimenez-Felstrom and Zigangirov 1999; Tanner et al. 2004) are directly designed optimizing $\underline{H}$.

The concept of time-invariance for convolutional codes, when referred to the $\underline{H}$ matrix, can be somehow ambiguous. The convention adopted so far (Definition 7.2 and Lemma 7.1) is based on the presence of a unique polynomial $h^*(x)$, although with different positions in the period (when $r_o > 1$). A wider definition can be proposed as follows, still exploiting the concept of a unique syndrome former sub-matrix $\underline{H}_o^T$.

Definition 7.3 We have a *wide-sense (w.s.) time-invariant convolutional code in its* $\underline{H}$ *matrix*, when this matrix shows a unique syndrome former sub-matrix $\underline{H}_o^T$, but, at least, two different rows shifting in it.

Example 7.3 Let us consider the semi-infinite $\underline{H}$ matrix shown in Fig. 7.2 up. In it, two reciprocal parity check polynomials $h_1{}^*(x) = x^4 + x^3 + x^2 + x + 1$ and $h_2{}^*(x) = x^4 + x + 1$ horizontally shift by 3 positions at a time. Since this convolutional code has $n_o = 3$, $k_o = 1$, its $\underline{H}_o^T$ sub-matrix vertically shifts with steps of $r_o = 2$ each period. This code can be considered w.s. time-invariant, because the $\underline{H}_o^T$ sub-matrix is unique, but there are two different parity check polynomials alternately present in the rows of $\underline{H}$. On the contrary, the code treated in Examples

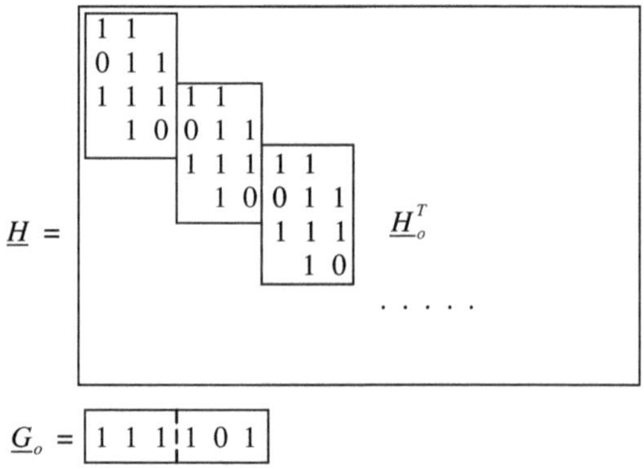

Fig. 7.2 Semi-infinite $\underline{H}$ matrix and $\underline{G}_o$ sub-matrix for the code described in Example 7.3

7.1 and 7.2, although the $\underline{H}_o^T$ sub-matrix vertically shifts in its $\underline{H}$ matrix with steps of $r_o = 2$ as well, turns out to be s.s. time-invariant (owing to the presence of a unique parity check polynomial). After imposition of a null ex-OR sum for the shortest possible cluster of syndromes, starting from the first position in any period, we obtain the generator sub-matrix $\underline{G}_o$ shown in Fig. 7.2 down, where the unique interleaved generator polynomial $g_{\text{int}}(x) = x^5 + x^4 + x^3 + x^2 + 1$ is present. Therefore this code is s.s. time-invariant in its $\underline{G}$ matrix. Finally we can observe that it represents an equivalent of the best 1/3-rate convolutional code, s.s. time invariant in $\underline{G}$, with $m = 1$ (see Table 3.4).

Lemma 7.2 *A convolutional code with asymptotic code rate* $(n_o - 1)/n_o$ is *forced to be s.s. time-invariant in* $\underline{H}$.

Proof In this case, the parity check matrix exhibits an $\underline{H}_o^T$ sub-matrix vertically shifting with steps of $r_o = 1$. So the reciprocal parity check polynomial $h^*(x)$ is the same in all the rows of $\underline{H}$, and then is unique. □

Example 7.4 In Fig. 7.3 up, the semi-infinite parity check matrix of a 2/3-rate convolutional code is reported. It is s.s. time-invariant, because the unique reciprocal parity check polynomial $h^*(x) = x^8 + x^7 + x^6 + x^4 + x + 1$ is present in all the rows of such $\underline{H}$ matrix. After imposition of a null ex-OR sum for two clusters of syndromes (starting from the first and the second position in any period), we obtain the generator sub-matrix $\underline{G}_o$ shown in Fig. 7.3 down.

The code treated in Example 7.4 is w.s. time-invariant in its $\underline{G}$ matrix, since its $\underline{G}_o$ sub-matrix collects two different rows, represented respectively by the interleaved generator polynomials $g_{\text{int}}^{(1)}(x) = x(x^4 + x^3 + x^2 + x + 1)$ and $g_{\text{int}}^{(2)}(x) = x^4 + x^3 + x^2 + 1$. It is characterized by minimum distance $d = 3$.

Convolutional codes, time-invariant in the $\underline{H}$ matrix only in wide sense, will be studied more in depth in Chap. 8. In Sect. 7.11, the equivalence between MHEC codes and s.s. time-invariant convolutional codes in their $\underline{H}$ matrix will be demonstrated, giving also some precise correspondences between the parameters of the two. MHEC codes are characterized by a unique parity check polynomial, although

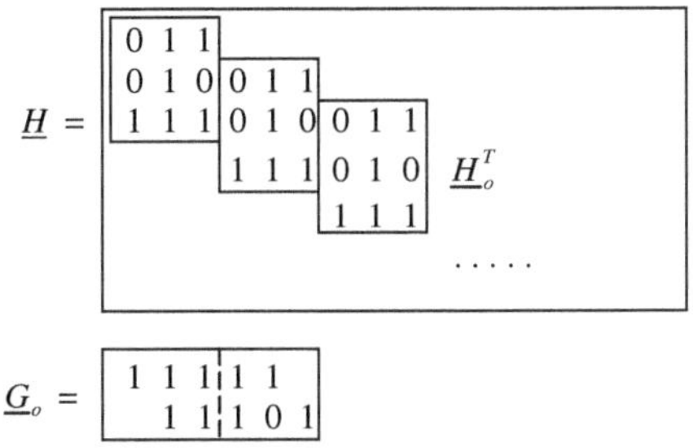

Fig. 7.3 Semi-infinite $\underline{H}$ matrix and $\underline{G}_o$ sub-matrix for the 2/3-rate convolutional code treated in Example 7.4

in the initial and in the final part of the code word (code frame), it is properly truncated. In this, we obtain a structure which is completely equivalent to that depicted in Fig. 7.1, here introduced for s.s. time-invariant convolutional codes in $\underline{H}$. Although, in such figure, the correct frame termination is not displayed, it is possible to argue, by symmetry, that it will result similar to frame beginning, that is with complete $\underline{H}_o^T$ sub-matrices.

Also for the $\underline{G}$ matrix we can widen the concept of time-invariance, with particular reference to convolutional codes. A concept of this type has been introduced in Definition 4.1, considering the rows of the whole $\underline{G}$ matrix. It is possible to make it more effective taking into account the generator sub-matrix $\underline{G}_o$.

Definition 7.4 We have a *wide-sense (w.s.) time-invariant convolutional code in its* $\underline{G}$ *matrix*, when this matrix shows a unique $\underline{G}_o$ sub-matrix, shifting along its rows with steps of n_o, in which, at least, two different rows are present.

Comparison with Definition 4.1 shows a substantial coincidence. Nevertheless here we have stressed the presence of a unique $\underline{G}_o$ sub-matrix, so showing a better correspondence with Definition 7.2, where we speak about a unique $\underline{H}_o^T$ sub-matrix in the parity check matrix. In Sects. 4.1 and 4.2, assuming expression (4.1) for the code frame $t(x)$, the most general situation has been considered, regarding the positions of the highest-order coefficients in the generator polynomials. Then great attention has been devoted to find the condition for admitting the inverse transformation, needed for obtaining an RSC version of our code. These preliminaries are necessary also for the construction of its $\underline{H}$ matrix. Nevertheless it is possible to develop a simplified theory, based on the assumption that $\underline{G}_o$ shows polynomials with highest-order coefficients all in different positions (and in particular with an upper triangular form). This assumption guaranties that the inverse transformation is always possible and that the $\underline{H}$ matrix can be easily calculated, especially if we adopt an $\underline{H}_o^T$ sub-matrix whose uppermost layer is lower triangular.

Lemma 7.3 *A convolutional code with asymptotic code rate* $1/n_o$ *is forced to be s.s. time-invariant in* $\underline{G}$.

Proof In this case, the generator matrix exhibits a $\underline{G}_o$ sub-matrix with just one row, where a unique interleaved generator polynomial $g_{\mathrm{int}}(x)$ is present. □

It is possible to consider several examples with $n_o = 3$ and $k_o = 1$ from Table 3.4, or with $n_o = 4$ and $k_o = 1$ from Table 3.5, both reported in Sect. 3.4. Finally, we can observe a clear duality property comparing Lemmas 7.2 and 7.3. Remark that in both, the condition considered here is not the only one for obtaining s.s. time-invariance (see, for instance, high-rate MLC codes or low-rate MHEC codes). Two complementary points of views are those based respectively on $\underline{G}_o$ and on $\underline{H}_o^T$. This consideration is particularly important when we would like to have a structure, which is scalable in the code rate. A separate case is that characterized by asymptotic code rate 1/2, that is by $n_o = 2$ and $k_o = r_o = 1$. The very singular properties of such a situation are consequence of the fact that a code of this type is s.s. time-invariant both in $\underline{G}$ and in $\underline{H}$. In other terms, this case is common to either Lemmas 7.2 or 7.3.

7.2 Construction of the Syndrome Former Sub-matrix for Low-Rate Convolutional Codes

It is convenient to introduce the procedure for calculating $\underline{H}_o^T$ from the knowledge of $\underline{G}_o$ gradually. First of all, we will consider $1/n_o$-rate codes, that is codes characterized by a $\underline{G}_o$ sub-matrix with just one row, where a unique interleaved generator polynomial is present. It shifts horizontally with steps of n_o. Keeping in mind the interpretation of such codes as low-rate MLC codes, their generator polynomial will be denoted $g_c(x)$. It coincides with the interleaved generator polynomial $g_{\mathrm{int}}(x)$ introduced in Chap. 3 according to a convolutional model. Now, we have to construct $(n_o - 1)$ interleaved reciprocal parity check polynomials $h_i^*(x)$, $i = 1, 2, \ldots (n_o - 1)$. In general they are different.

Let us consider a syndrome former sub-matrix $\underline{H}_o^T$ with a certain number of layers, each made by $(n_o - 1)$ rows, where the $h_i^*(x)$ polynomials horizontally shift with steps of n_o. Let the first layer be lower triangular, so that the last $(n_o - 1)$ positions of the period will be occupied by different 0th-order power coefficients $h_{i0}^* = 1$ of as many $h_i^*(x)$ polynomials. Precisely we adopt the following sequence: $i = 1$ in second position, $i = 2$ in third position, ... $i = n_o - 1$ in the last position. We are searching for non-periodic parity check polynomials, according to the classification given at the end of Sect. 5.12. The procedure we are going to describe is not suited for the periodic rows which have to be added in the $\underline{H}$ matrix of not well designed convolutional codes.

Under the sequence of binary symbols representing $g_c(x)$, we can construct the generic $h_i^*(x)$ polynomial, starting from $h_{i0}^* = 1$ in the $(i + 1)$th position of the leftmost period. In each step, the already constructed part of $h_i^*(x)$ shifts by one period to the right, and its present leftmost period is formed assuring that the number of coincidences of 1-symbols between $g_c(x)$ and $h_i^*(x)$ is even. This condition guarantees that multiplication of any row of $\underline{G}$ by the considered column of $\underline{H}^T$ gives a null result, according to (5.4). In Fig. 7.4 this procedure is schematically shown. Letting $(m + 1)$ the number of periods spanned by $g_c(x)$ (see Definitions 3.1 and 3.2), after construction of any polynomial $h_i{}^*(x)$, the ith row of any layer of the $\underline{H}_o^T$ will appear, properly shifted, under $g_c(x)$ itself.

Let $(m_h + 1)$ be the number of layers in $\underline{H}_o^T$, each made by $(n_o - 1)$ rows. A procedure of the type described so far gives rise to the ith row of any layer. Then the $(n_o - 1)$ polynomials so constructed are to be vertically interleaved, in order to obtain the overall $\underline{H}_o^T$. Let $\underline{H}_o^{T(i)}$ be the part of $\underline{H}_o^T$ comprehending the contribution of the ith row, that is the one associated to $h_i{}^*(x)$, and let $(m_h^{(i)} + 1)$ be the number of its rows, in such a vertically interleaved structure. The following constraint holds

$$m_h = \max\{m_h^{(i)}\}. \tag{7.1}$$

Often, some empty rows are present in the $\underline{H}_o^T$ sub-matrix, as a consequence of possible differences in the $m_h^{(i)}$ parameters. In Fig. 7.4, we have assumed $m_h > m$, but sometimes the opposite may occur, or such parameters may be equal.

Example 7.5 The best s.s. time-invariant convolutional code in $\underline{G}$, characterized by $n_o = 3$, $k_o = 1$, $m = 2$, is the one associated to the interleaved generator polynomial 111.110.111 (see Table 3.4). In Fig. 7.5 the constructions of $h_1{}^*(x)$, on the left, and of $h_2{}^*(x)$, on the right, are reported. The first is trivial, has weight $w = 2$, and occupies a single period. The second, instead, has weight $w = 5$, and requires three

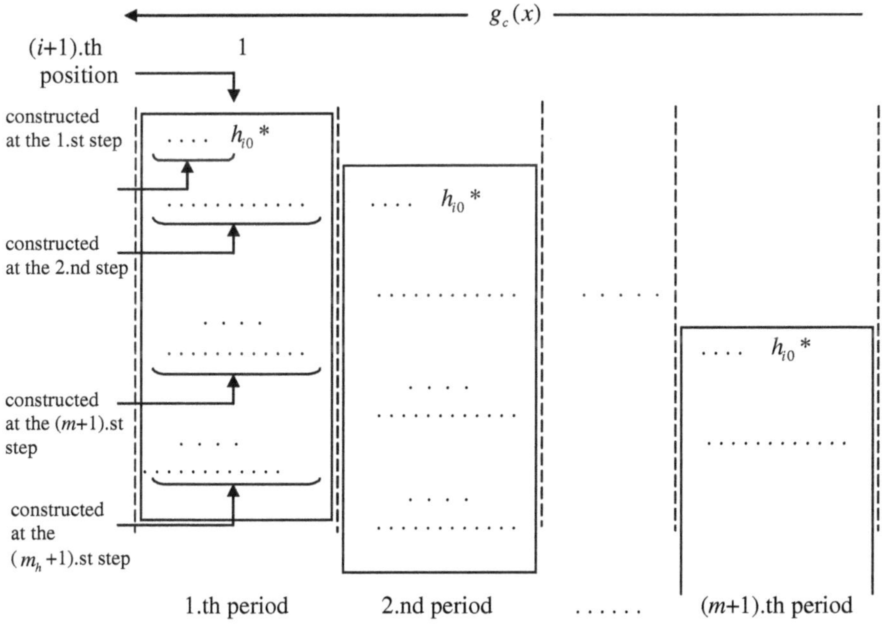

Fig. 7.4 Scheme of the procedure for the obtainment of $h_i^*(x)$ in $1/n_o$-rate convolutional codes

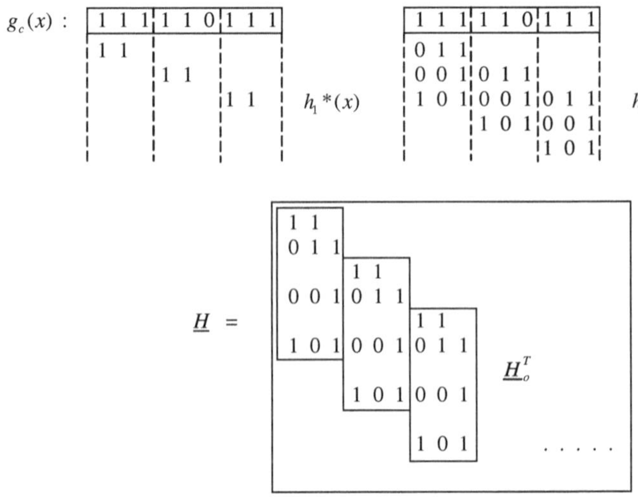

Fig. 7.5 Construction of the syndrome former sub-matrix for the 1/3-rate code described in Example 7.5

periods. So we can write $m_h^{(1)} = 0$, $m_h^{(2)} = 2$, and hence $m_h = 2$. Proper interleaving of these two polynomials gives rise to the overall semi-infinite parity check matrix depicted down in Fig. 7.5, where three $\underline{H}_o^T$ sub-matrices have been evidenced, vertically shifting with steps of $r_o = 2$.

Example 7.6 The best s.s. time-invariant convolutional code in $\underline{G}$, characterized by $n_o = 4$, $k_o = 1$, $m = 1$, is the one associated to the interleaved generator polynomial 1111.0111 (see Table 3.5). In Fig. 7.6 the constructions of $h_1{}^*(x)$, $h_2{}^*(x)$, $h_3{}^*(x)$ are shown. The resulting overall semi-infinite parity check matrix is depicted down, where two sub-matrices $\underline{H}_o^T$ have been considered, vertically shifting with steps of $r_o = 3$. In this case, we have $m_h = 1$, because of the presence of $h_1{}^*(x)$.

The construction of $\underline{H}_o^T$ presented so far, which proceeds by attempts, can give a not unique result. Different solutions for the same $h_i{}^*(x)$ are available, as a consequence of the multiplication by a polynomial function of x^{n_o}. In fact various solutions, all equal but superimposed and separated exactly by one period, are acceptable as well. In this sense, among the possible solutions $h_i{}^*(x)$, the one characterized by minimum overall parameter m_h is to be preferred.

Definition 7.5 When a convolutional code, s.s. or w.s. time-invariant in $\underline{H}$, is described by an $\underline{H}_o^T$ sub-matrix, formed by polynomials $h_i{}^*(x)$, parameter m_h, introduced according to (7.1) is defined *parity check constraint length*. The choice associated to the $\underline{H}_o^T$ sub-matrix characterized by minimum m_h is called *minimal form for the parity check matrix of this convolutional code.*

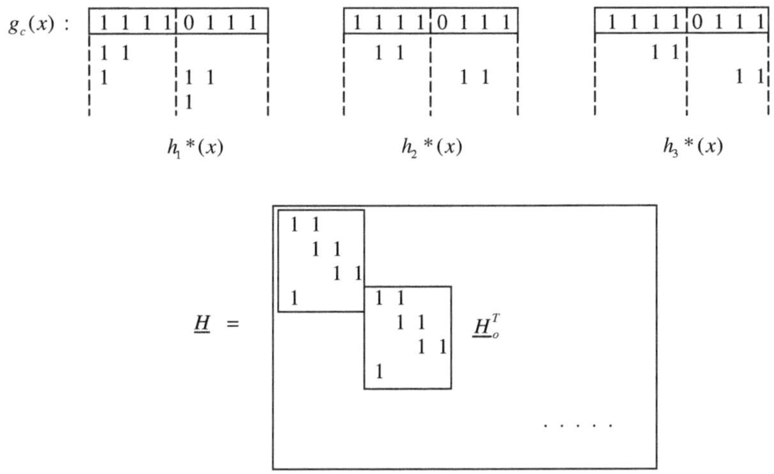

Fig. 7.6 Construction of the syndrome former sub-matrix for the 1/4-rate code described in Example 7.6

The concept of minimal form for $\underline{H}$ has been introduced in Definition 5.14 from a general point of view, in Definition 5.13 for cyclic or SC codes and revisited in Definition 6.3 for QC and SQC codes. Here such concept has been associated to the vertical size of the syndrome former sub-matrix $\underline{H}_o^T$ of a convolutional code, that is to its parity check constraint length.

It is important to note that the single $m_h^{(i)}$, involved in (7.1), can be related also to the degree deg $[h_i^*(x)]$ by means of the following double bound

$$m_h^{(i)} n_o < \deg[h_i^*(x)] < (m_h^{(i)} + 1) n_o. \tag{7.2}$$

Clearly parameters $m_h^{(i)}$, $i = 1, 2, \ldots r_o$, and the overall m_h as a consequence of (7.1), impose the number of memory elements in the shift registers employed for encoding the considered convolutional code, when encoding is based on the parity check matrix. In order to fix ideas, we can refer to the schemes described in Figs. 5.6 and 6.9, but there are also other schemes, properly organized for convolutional codes described through the parity check matrix. If the decoding algorithm is performed in the trellis deriving from this encoder circuit, its computational complexity is measured by the maximum number of states per stage and by the number of branches entering each merge state. To this purpose, depending on which parameter between m and m_h is smaller, it will be preferable to consider the generator matrix or the parity check matrix, taking into account also the number of shift registers needed. This topic will be investigated more in depth in following sections of the present chapter and in Chap. 8.

The possible results of the procedure presented so far, for obtaining $\underline{H}$ from the knowledge of $\underline{G}$, as already noted, can be distinguished because of a multiplication by different polynomials $p(x^{n_o})$. The solution associated to the minimal form of $\underline{H}$ is assumed to be that characterized by $p(x^{n_o}) = 1$. Nevertheless sometimes we can have also more than one minimal form of $\underline{H}$ for the same code, for instance after ex-OR sums of the various $h_i^*(x)$ polynomials, all in minimal form. Another important observation is that the different $h_i^*(x)$'s act separately, therefore one of them can be changed, without requiring any consequent change in the other ones, provided that they remain all linearly independent. These considerations are useful if one is looking for some particular form of the $\underline{H}$ matrix, say that free of 4-length short cycles of 1-symbols, as is needed in LDPC codes (Tanner 1981). Frequently this requirement is satisfied with an $\underline{H}$ matrix not in minimal form.

Lemma 7.4 *A convolutional code having code rate* 1/2, *provided that it is well designed, is s.s. time-invariant either in* $\underline{H}$ *or in* $\underline{G}$, *and, except for external period differences, it is also self-dual.*

Proof In this very particular situation, given a certain $g_c(x)$, assuming $h(x) = g_c(x)$, their product gives $h(x)\, g_c(x) = g_c(x^2)$, which turns out to be null in alternating positions, so satisfying (5.4). The two codes which can be so obtained are then one the dual code of the other, except for some external periods, where the $\underline{G}$ matrix exhibits complete $\underline{G}_o$ sub-matrices, whereas the $\underline{H}$ matrix is characterized by complete $\underline{H}_o^T$ sub-matrices. □

Example 7.7 The best 1/2-rate convolutional code characterized by $n_o = 2$, $k_o = 1$, $m = 3$, is the one associated to the interleaved generator polynomial 11.11.01.11 (see Table 3.2). In Fig. 7.7 the structure of an $\underline{H}$ matrix valid for this code is reported. It is possible to observe the property $h(x) = g_c(x)$, according to Lemma 7.4. It is reached with the choice of $h^*(x)$ characterized by minimum degree. The semi-infinite $\underline{H}$ matrix is directly constructed under $g_c(x)$. Four periods are shown, and as many syndrome former sub-matrices $\underline{H}_o^T$ have been reported. Finally it is possible to observe that, obviously, this is a case characterized by $m = m_h$.

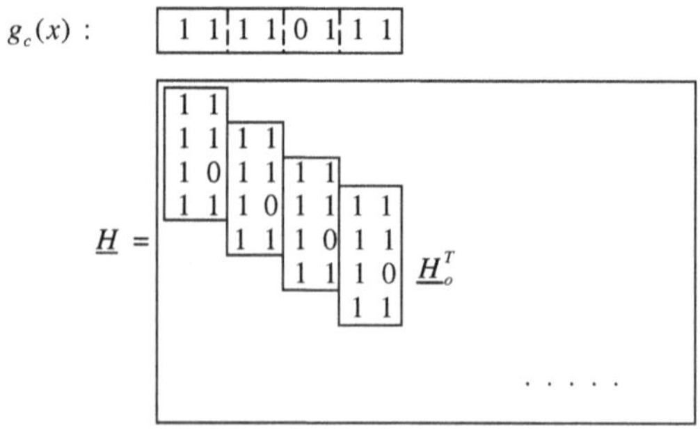

Fig. 7.7 Semi-infinite $\underline{H}$ matrix for the 1/2-rate convolutional code described in Example 7.7

In order to provide a correct frame termination, that is to assure that also the final symbols in the frame are protected as all the other ones, the last $\underline{G}_o$ sub-matrices and the last $\underline{H}_o^T$ sub-matrices should be entirely considered. Nevertheless there are situations in which some rows of the last syndrome former sub-matrices are to be eliminated (because null or made by just one 1-symbol). It is difficult to give a general rule, and a particular discussion is to be made depending on the considered case.

In Fig. 7.8, generator and parity check matrices for the codes described in Examples from 7.5 to 7.7 are compared, assuming $Z' = 5$ overall periods in all the cases. In the parity check matrix obtained for the 1/3-rate code, two rows have been eliminated (those characterized by all 0-symbols). In the parity check matrix of the 1/4-rate code, one row has been eliminated (that characterized by just one 1-symbol) and also one useless external positions in the frame (all 0-symbols in $\underline{G}$) has been suppressed. In the parity check matrix of the 1/2-rate code, no rows are eliminated and no positions in the frame are suppressed. As regards Z', this number of periods includes the correct frame termination, and is not to be confused with Z (number of periods where information symbols are inserted). Precisely we have $Z = 3$, $Z = 4$ and $Z = 2$ in the three cases examined. This difference comes out from the fact that we have different parity check constraint lengths.

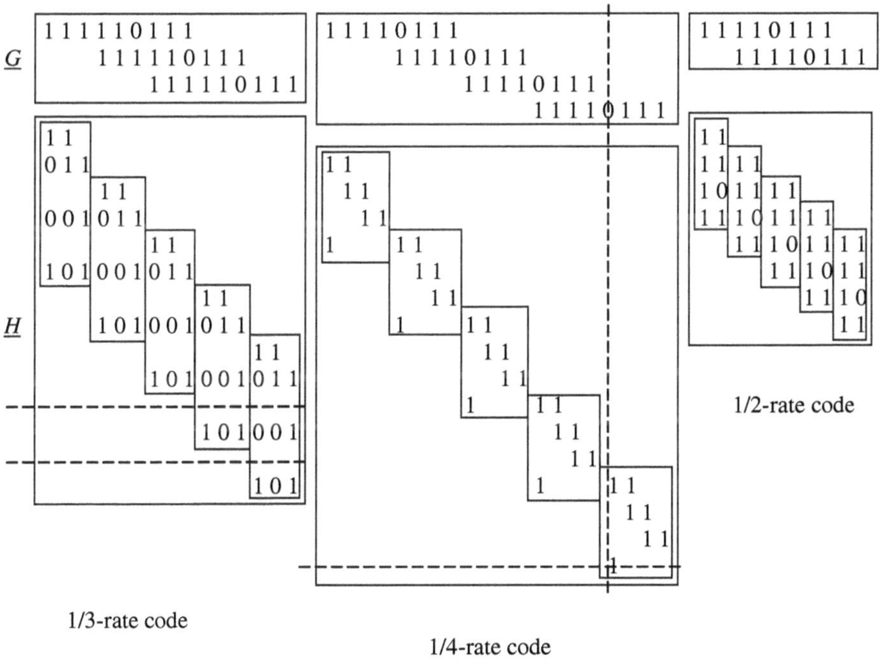

Fig. 7.8 Correct frame termination for the three codes described in Examples 7.5, 7.6 and 7.7

In all the above situations, it is possible to verify how the total frame length n results the sum of the row number in $\underline{G}$ and of the row number in $\underline{H}$ (as long as these matrices have full rank), that is the sum of the information symbol number and of the control symbol number. The number of periods is too small, so the true code rate turns out to be remarkably smaller than the asymptotic value $1/n_o$. More precisely the three codes so obtained can be represented respectively by the following terns: [15,3,8], [19,4,7], [10,2,6]. In the correct frame termination, important roles are played by the generator constraint length m and by the parity check constraint length m_h (in the above Examples we have respectively $m = m_h = 2$, $m = m_h = 1$, $m = m_h = 3$, but such result cannot be taken as a rule).

7.3 Extension of the Procedure for Obtaining the Syndrome Former Sub-matrix to High-Rate Convolutional Codes

In this section, the situation characterized by $k_o > 1$ will be investigated. We will analyze codes with $k_o = n_o - 1$, that is $r_o = 1$, which are forced to be s.s. time-invariant in the $\underline{H}$ matrix, but can be either s.s. time-invariant or w.s. time-invariant as regards the $\underline{G}$ matrix. The former choice can be related to MLC codes.

The procedure described in the previous section, in order to construct $\underline{H}_o^T$ from $\underline{G}$, in the case characterized by $k_o = n_o - 1$, even when different generator polynomials are present, is still valid. It has the goal of finding a unique reciprocal parity check polynomial $h^*(x)$ contemporary satisfying more than one constraint of the type imposed in the previous section. More precisely, the synthesis of this polynomial is performed with steps of n_o, entering on the left hand side and outgoing on the right hand side, in such a way as to contemporary assure that all the rows in the $\underline{G}_o$ sub-matrix show an even number of coincidences between the positions of their 1-symbols and those of the $h^*(x)$ polynomial under construction.

Example 7.8 Assuming $k_o = 2$ and $n_o = 3$, let us consider the same interleaved generator polynomial represented by the sequence 11001 on both rows of $\underline{G}_o$, except for a shift by one position. Interpreting this s.s. time-invariant convolutional code in $\underline{G}$ as an MLC code with asymptotic code rate 2/3, such a generator polynomial is factor of $(x^{15} + 1)$. On the other hand, it is an s.s. time-invariant convolutional code in $\underline{G}$ with generator constraint length $m = 1$, but also s.s. time invariant in its $\underline{H}$, owing to Lemma 7.2. In Fig. 7.9, on the left, the construction of the unique $h^*(x)$ is shown, under $\underline{G}_o$ sub-matrix, having here two rows. This reciprocal parity check polynomial is associated to the sequence 100.010.111. The $\underline{H}_o^T$ sub-matrix so obtained is characterized by $m_h = 2$, in its minimal form. In Fig. 7.9, on the right, both matrices $\underline{G}$ and $\underline{H}$ are reported, assuming $Z' = 5$ periods on the whole, and considering a correct frame termination. One row in $\underline{H}$ and one column in both matrices are eliminated, so the terminated code becomes described by the tern [14,8,3]. This is a case in which the generator constraint length and the parity check constraint length are different ($m = 1$ and $m_h = 2$).

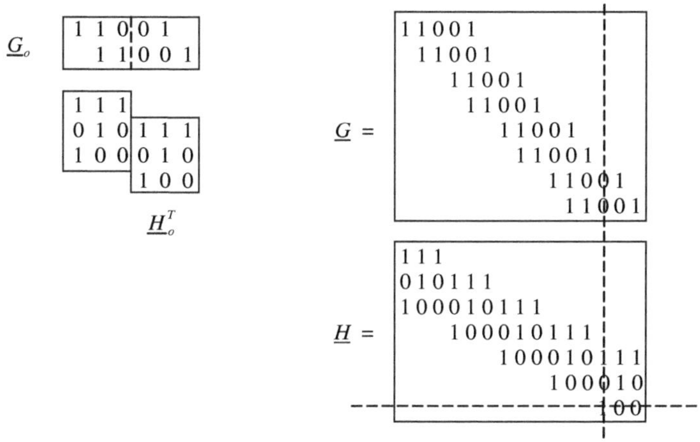

Fig. 7.9 $\underline{G}$ and $\underline{H}$ matrix for the s.s. time-invariant 2/3-rate convolutional code introduced in Example 7.8

It is possible to verify that a whole family of codes exists, for which, assuming $k_o = n_o - 1$, we can perform a scaling of the $\underline{H}_o^T$ sub-matrix obtained in Example 7.8, varying n_o. On this family of codes, all characterized by $d = 3$, we will return at the end of this section and in Chap. 12 (since they represent a simple case of rate-adaptive LDPC convolutional codes with fine scalability). For $n_o \geq 4$, such codes are no longer s.s. time-invariant in the $\underline{G}$ matrix.

Example 7.9 Still assuming $k_o = 2$ and $n_o = 3$, let us consider the couple of generator polynomials 11010 and 11111. A code of this type has been introduced in Example 3.16, and has been treated also in Example 4.15. It can be obtained by proper code puncturation of the best 1/2-rate time-invariant convolutional code with $m = 2$, and turns out to be the best w.s. time-invariant convolutional code with $m = 1$ and $\mu = 2$ at rate 2/3. In Fig. 7.10, on the left, following the usual procedure, a valid $\underline{H}_o^T$ sub-matrix is obtained. In it, the unique reciprocal parity check polynomial associated to the sequence 011.101.111 shifts. On the right, the resulting $\underline{G}$ and $\underline{H}$ matrix are shown, considering $Z' = 5$ overall periods and a correct frame termination. Here no rows and no columns are to be eliminated, and the overall code is represented by the tern [15,8,3].

We can observe that, in both the last two Examples, there is $Z' = Z + 1$, although in Example 7.8 the elimination of one position in the frame has been necessary. In general, it is possible to consider the following inequality, where the generator constraint length m is taken into account, for the overall frame length,

$$n \leq Z' n_o = (Z + m) n_o, \tag{7.3}$$

this upper bound being reached only when position eliminations are not needed and the whole length of the $\underline{G}_o$ sub-matrix is exploited.

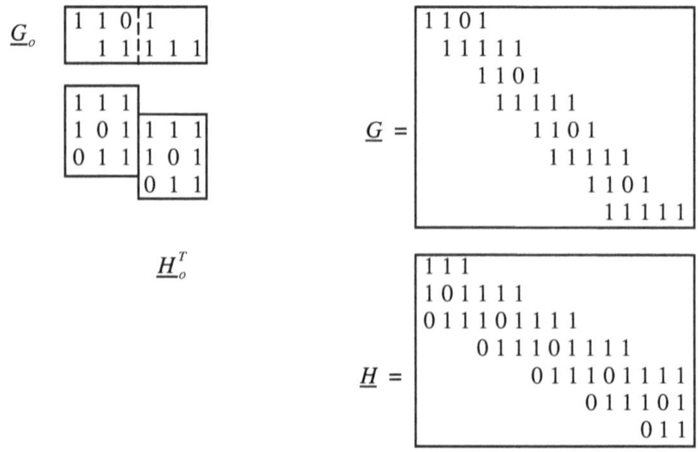

Fig. 7.10 Construction of the $\underline{H}$ matrix for a 2/3-rate w.s. time-invariant convolutional code in its $\underline{G}$ matrix (Example 7.9)

In the situation characterized by $k_o > 1$ the risk of having a not well designed convolutional code is greater, more and more with increasing the code rate. Not well designed convolutional codes exhibit some anomalous rows in $\underline{H}$, which are characterized by a periodic behaviour (see the classification at the end of Sect. 5.12). Their number is typically very small with respect to the non-periodic rows. In order to investigate how non-periodic parity check polynomials can be originated also in this context, taking into account that in an LC code we have only periodic parity check polynomials, we will consider the case of $k_o = n_o - 1$, in a high-rate MLC code.

Lemma 7.5 *A convolutional code, s.s. time-invariant in* $\underline{G}$, *characterized by* $k_o = n_o - 1$, *and by a unique generator polynomial* $g_c(x)$, *factor of* $(x^N + 1)$, *is described by an* $\underline{H}$ *matrix where a non-periodic polynomial* $h^*(x)$ *shifts. This polynomial is factor of the binomial* $(x^{N'} + 1)$, *where* $N' = \text{LCM}\{N, n_o\}$.

Proof Keeping in mind the procedure here introduced for obtaining $\underline{H}_o^T$, we can observe that a multiple of N is necessary for giving null result in the polynomial multiplication between $g_c(x)$ and $h_c(x)(1 + x^N + x^{2N} + \cdots)$, like in an LC code, of which our MLC code is a subcode. The non-periodic polynomial $h^*(x)$ derives from suitable ex-OR sums of periodic polynomials of the type $h_c{}^*(x)(1 + x^N + x^{2N} + \cdots)$. In conclusion, if n_o is factor of N, $h^*(x)$ must be itself factor of $(x^N + 1)$, if n_o and N are co-prime numbers, $h^*(x)$ must be factor of $(x^{Nn_o} + 1)$. Finally if they have some factors in common, these factors can be eliminated. To summarize, we can assert that $h^*(x)$ must be factor of the binomial $(x^{\text{LCM}\{N, n_o\}} + 1)$. □

Example 7.10 Assuming $g_c(x) = x^3 + x^2 + 1$, factor of $(x^7 + 1)$, we will analyze the situations regarding the code rates 2/4, 2/3, and 3/4. In the first situation, neither

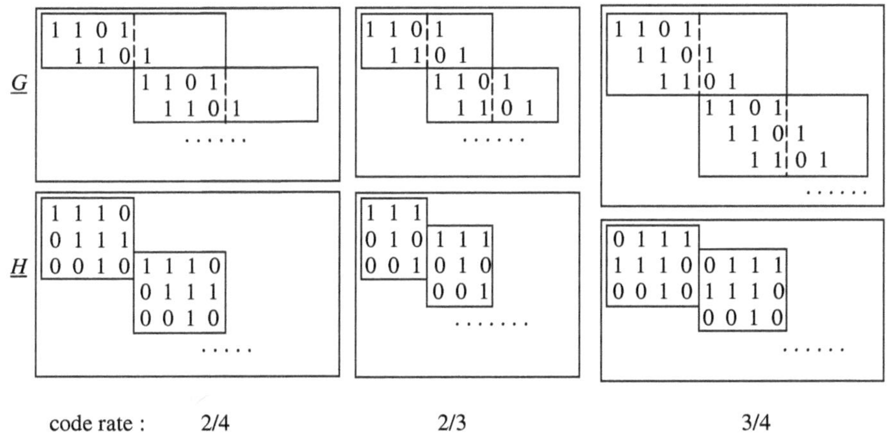

Fig. 7.11 $\underline{G}$ and $\underline{H}$ matrix for the code rates 2/4, 2/3, 3/4, all assuming $g_c(x) = x^3 + x^2 + 1$

Lemma 7.4, nor Lemma 7.5 can be applied, because the code rate is not of the type $(n_o - 1)/n_o$, like, in general, for any code rate of the type $P/(2P)$, being P an integer number. Nevertheless the procedure for obtaining the $\underline{H}_o^T$ sub-matrix from the $\underline{G}_o$ sub-matrix is valid also here. We obtain two different parity check polynomials, $h_1(x) = x^4 + x^3 + x^2 + 1$ and $h_2(x) = x^2 + x + 1$. The couple of semi-infinite $\underline{G}$ and $\underline{H}$ matrix are shown in Fig. 7.11, on the left. With $k_o = 2$, $n_o = 3$, instead, we obtain $h^*(x)$ associated to 001.010.111, which is factor of $(x^{21} + 1)$. With $k_o = 3$, $n_o = 4$, the result is $h^*(x)$ associated to 0010.1110.0111, which is factor of $(x^{28} + 1)$. Lemma 7.5 is verified in both the last cases, described on the right in Fig. 7.11, in fact LCM{7,3} = 21 and LCM{7,4} = 28.

Remark that also the last code treated in Example 7.10 (that with code rate 3/4) although with minimum distance $d = 2$, can be considered a well designed convolutional code, in fact it is characterized by a binomial growth of coefficient $A(4)$ with Z, but the minimum distance is $d = 2$, so $A(4) = A(2d)$. For this reason, it does not require any periodic additional row, and its description is complete employing the only syndrome former sub-matrix $\underline{H}_o^T$ obtained here.

Example 7.11 Starting from $g_c(x) = x^4 + x^3 + x + 1$, factor of $(x^6 + 1)$, and considering a code rate 3/4, the reciprocal non-periodic parity check polynomial $h^*(x)$, which can be obtained by the usual procedure, turns out to be that associated to the sequence 0001.1100.0111. It is factor of $(x^{12} + 1)$, because LCM{6,4} = 12, confirming Lemma 7.5. Nevertheless this is a not well designed convolutional code and requires the addition of a periodic parity check polynomial. It is easy to verify that $u_{z'}(x^2)$ can be adopted for this role.

Example 7.12 Assuming $g_c(x) = x^5 + x^3 + x^2 + 1$, factor of $(x^{12} + 1)$, we will analyze the situations regarding the code rates 2/3 and 3/4, both leading to not well

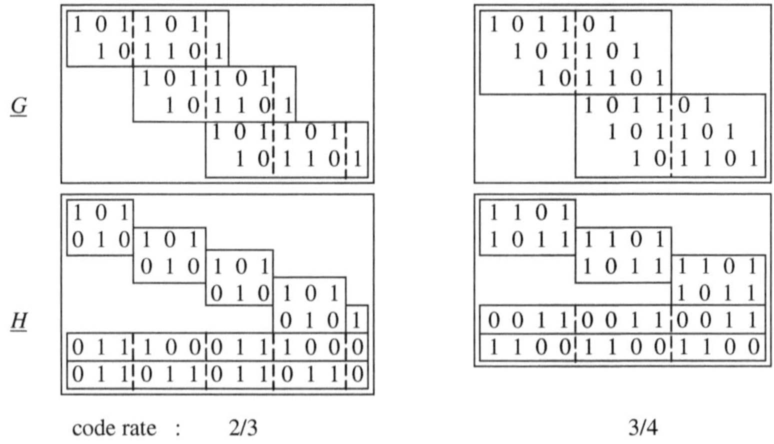

Fig. 7.12 $\underline{G}$ and $\underline{H}$ matrix for the code rates 2/3 and 3/4, assuming $g_c(x) = x^5 + x^3 + x^2 + 1$

designed convolutional codes. In Fig. 7.12 the $\underline{G}$ and $\underline{H}$ matrix for such codes are reported. Both parity check matrices exhibit two rows occupied by periodic polynomials, besides the usual rows characterized by the non-periodic polynomial $h^*(x)$, forming the $\underline{H}_o^T$ sub-matrices. The number of such sub-matrices equals Z', included those externally truncated. The periodic rows have been set in the lowermost positions for favouring their identification. We have considered $Z' = 4$ plus a fraction of one period, for the code rate 2/3, and exactly $Z' = 3$ for the code rate 3/4. The non-periodic rows contain a reciprocal parity check polynomial respectively associated to 010.101 (code rate 2/3) and to 1011.1101 (code rate 3/4), both factors of $(x^{12} + 1)$ confirming that LCM{12,3} = LCM{12,4} = 12, as predicted by Lemma 7.5.

Unfortunately Lemma 7.5 cannot be applied to convolutional codes which are only w.s. time-invariant in $\underline{G}$, although the procedure for obtaining $\underline{H}_o^T$ can still be employed (as made in Example 7.9 and in the first case treated in Example 7.10). In this situation, we cannot have an estimate of the binomial in whose factorization the non-periodic parity check polynomials can be found. A similar investigation could be developed in the framework of QC and MLQC codes, but is beyond the purpose of the present discussion.

Coming back to the obtainment of non-periodic parity check rows, we can observe that the condition on the $\underline{G}_o$ sub-matrix to be upper triangular (adopted so far) although not strictly necessary for employing the procedure here proposed, simplifies by far its application. At least when the code is well designed, it is convenient to adopt for $\underline{H}_o^T$ a form in which the first row layer is lower triangular. This choice has the advantage of providing clear positions to information and control symbols, so constructing an $\underline{H}$ matrix in minimal form without ambiguity. After this long series of considerations, it is possible to summarize the main characteristics of the procedure for the obtainment of $\underline{H}_o^T$ as follows.

Definition 7.6 Provided that the convolutional code considered is at least w.s. time-invariant in its $\underline{G}$ matrix, with upper triangular $\underline{G}_o$ sub-matrix, it is possible to obtain an $\underline{H}_o^T$ sub-matrix, whose first layer is lower triangular, by means of the procedure described by Fig. 7.4 and further generalizations. It will be called *column construction of the parity check matrix.*

The obtained sub-matrix $\underline{H}_o^T$ can collect a unique reciprocal parity check polynomial (when $r_o = 1$) or more than one reciprocal parity check polynomial (when $r_o > 1$). In the second case, each one of the $h_i^*(x)$'s acts separately from the other ones. In both cases, all the reciprocal parity check polynomials have to be properly matched with all the rows of $\underline{G}_o$. The inverse procedure, that is to obtain $\underline{G}_o$ from a certain number of adjacent $\underline{H}_o^T$ sub-matrices, is based on the imposition of having null ex-OR sum of clusters of syndromes. Both these procedures may have different results. Typically we are looking for the minimum possible values of the two constraint lengths, but sometimes it could be convenient to have a sparse $\underline{H}$ matrix (this occurs, for instance, with LDPC codes). In certain situations, there is also the possibility of having two equivalent codes, with different $\underline{G}$ and $\underline{H}$ matrix, although both obtained having the goal of minimizing the respective constraint length.

Example 7.13 In Fig. 7.9 (Example 7.8) the $\underline{H}$ matrix obtained after column construction is in its minimal form. It is possible to consider an equivalent of this 2/3-rate convolutional code, taking the reciprocal parity check polynomial $h^*(x)$ associated to the sequence 100.010.001.111. The semi-infinite parity check matrix shown in Fig. 7.13 up describes this code. The correspondent semi-infinite generator matrix, obtained after application of the null ex-OR sum of clusters of syndromes, is reported in Fig. 7.13 down. Here all the columns of $\underline{H}_o^T$ have weight $w = 2$. Furthermore parameter m_h is increased from 2 to 3. In spite of these differences, the two codes are equivalent (they show the same frame weight distribution), although a different mapping occurs between information symbols and code words. Finally here $\underline{G}_o$ is only w.s. time-invariant.

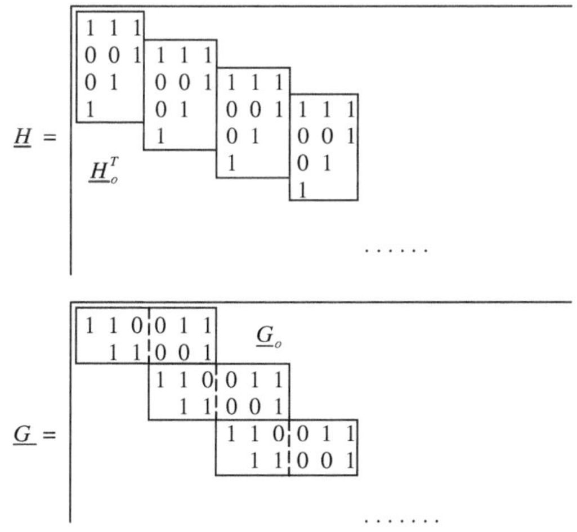

Fig. 7.13 Semi-infinite parity check and generator matrix for the code in Example 7.13

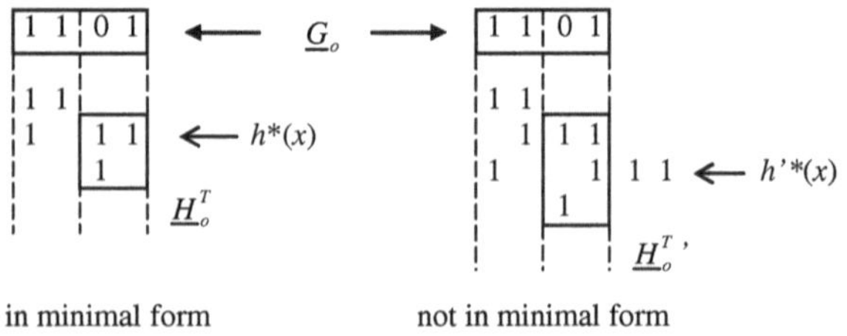

Fig. 7.14 Obtainment of two syndrome former sub-matrices for the same 1/2-rate s.s. time-invariant convolutional code in its $\underline{G}$ and $\underline{H}$ matrix

We can construct two families of codes which express a generalization of the ones having code rate 2/3 discussed in Example 7.8 and Example 7.13 respectively. They can have whichever asymptotic code rate $(n_o - 1)/n_o$. Such generalization is performed by prolonging the diagonal formed by the lower 1-symbols in $\underline{H}_o^T$. In the particular situation characterized by code rate 1/2, we have the two possible constructions described in Fig. 7.14, starting from the same generator polynomial $g_c(x) = x^3 + x^2 + 1$, obtained following Lemma 7.4. On the left we have an $\underline{H}_o^T$ sub-matrix in minimal form, associated to the polynomial $h^*(x) = x^3 + x + 1$. On the right, the new $\underline{H}_o^{T\prime}$ sub-matrix, not in minimal form, is associated to the polynomial $h'^*(x) = (x^2 + 1)(x^3 + x + 1) = x^5 + x^2 + x + 1$. It is possible to verify how, inversely, that is starting from the two parity check matrices, null ex-OR sum of clusters of syndromes gives rise anyway to the same $\underline{G}_o$ sub-matrix reported up in the Figure, so confirming coincidence.

The above two families of codes, at any code rate $(n_o - 1)/n_o$, give rise to equivalent codes, characterized by 3-weight normalized distribution coefficient

$$B(3) = \binom{n_o}{2}.$$

It is possible to verify that, with the family of codes characterized by all 2-weight columns, we obtain an $\underline{H}$ overall matrix having one row linearly dependent on the other ones (it is enough to perform the ex-OR sum of all the rows). Both such families of codes are suitable for being employed, as basic structures, for constructing LDPC convolutional codes.

7.4 Different Types of Not Well Designed Convolutional Codes

Let us remember that, according to Definition 3.5, a convolutional code is not well designed if, among the frame weight distribution coefficients $A(d)$, $A(d + 1)$, $A(d + 2)$, …, $A(2d - 1)$, at least one exhibits a binomial increase with the number of periods Z. The behaviour of a not well designed convolutional code is consequence of the presence, in its state diagram, of a closed route, not including state $\underline{0}$, characterized by s consecutive branches, labelled with sn_o 0-symbols in output.

Now we will follow the proof of Lemma 1.39, taking a $1/n_o$-rate convolutional code. One state of this closed route is reached, from state $\underline{0}$, by means of a state evolution, which was called $\bar{p}(x)$. Here polynomial $p(x)$ represents the common factor to all the sub-generator polynomials (see Lemma 3.7). In the expression of the interleaved polynomial we can then enucleate a factor $p(x^{n_o})$. Coming back to state $\underline{0}$, from the same or from another state of the closed path, is assured by means of an equal output symbol sequence $\bar{p}(x)$. This conclusion derives from the interpretation of our convolutional code as an MLC code. In fact the considered code frame (code word) belongs to the original LC code, being of the type (1.18) with $t_c(x) = 0$. It can be considered as a survivor, after the selection operated by the obtainment of a proper subcode, due to distributed control symbol insertion. The case of a w.s. time-invariant convolutional code in its $\underline{G}$ matrix is more complicated, but the mechanism is similar.

In Fig. 7.15 many possible situations have been reported, with reference to a state diagram based on the $\underline{G}$ matrix, having 2^m states, of the type associated to a convolutional code, encoded by means of a shift register in controller arrangement with m memory elements. The uppermost picture regards the case of $s = 1$, which interests the 0-labelled closed-loop branch around state $\underline{1}$. It can be present either with m even or with m odd. More precisely, it occurs when all the sub-generator polynomials exhibit $(x + 1)$ as a common factor. The situation described by the two central state diagrams refers to $s = 2$. Two different geometries are present for m even or m odd, regarding state labelling. Remark that such state labelling represents $\bar{p}(x)$. Finally the situation reported down in Fig. 7.15 refers to $s = 3$. Also in this case, two different geometries are to be considered, for m even or m odd. In this and in the previous case, common factors in all the sub-generator polynomials have degree greater than one (see Lemma 1.39).

Lemma 7.6 *In a not well designed convolutional code, the presence of a closed route in the state diagram whose length is s branches, all 0-labelled in output, is responsible for a part of the weight distribution coefficient related to the code frames passing through this closed route, which grows binomially with* Int[Z/s], *being Z the number of periods in the frame where information symbols are inserted.*

Proof In the proof o Lemma 1.39, we saw that the co-factor of the common factor represents the state evolution for reaching the closed path. Here we can assume $g_{\text{int}}(x) = p(x^{n_o})g_o(x)$. So the polynomial $g_o(x)$ will express, also in the present case, the output sequence needed for reaching our closed route. The longer the path for reaching this closed route, the higher the degree of the state evolution $\bar{p}(x)$ correspondent to the output sequence $g_o(x)$. For instance, taking the common factor $(x + 1)$, the co-factor is the unitary polynomial $u_m(x)$. For this reason the 0-labelled closed-loop branch, if present, is located around state $\underline{1}$. Anyway, the same symbol sequence $g_o(x)$ is needed for entering the closed route and for outgoing from it. In fact coming back to state $\underline{0}$ means that the shift register is progressively filled with

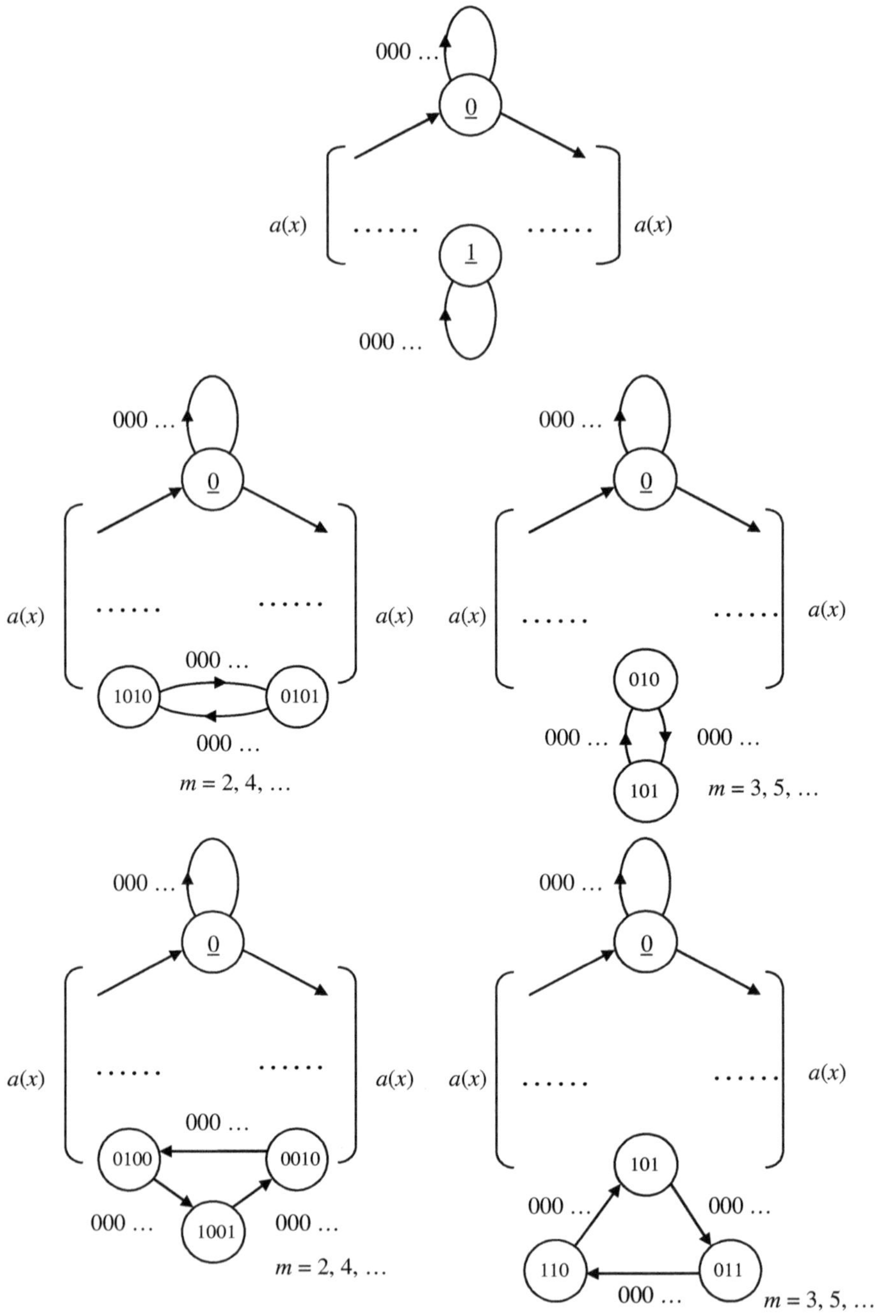

Fig. 7.15 Possible situations in the state diagrams of not well designed convolutional codes

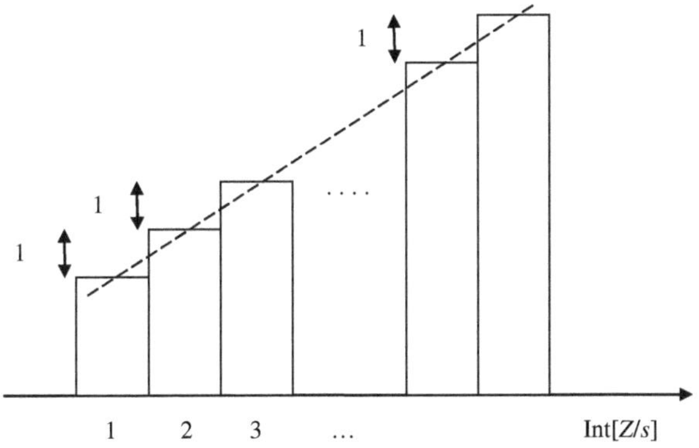

Fig. 7.16 Differential increments in the construction of the binomial part of coefficient $A(2w_a)$ with increasing Int[Z/s]

0-symbols. Letting $g_o(x) = a(x)$ for simplicity (and in analogy with the symbolism adopted in the LC code model), the code frames passing through the considered closed route can then be written as

$$t(x) = a(x)[x^{qsn_o} + 1], \qquad q = 1,\ 2,\ \ldots \tag{7.4}$$

In Fig. 7.16, it is possible to estimate the differential increments present in the construction of coefficient $A(2w_a)$, being $w_a = \mathrm{w}[a(x)]$: the increase occurs with steps of integer values for the ratio Z/s. Such triangular geometry is responsible for the binomial growth expected, according to (7.4), in the part of $A(2w_a)$ due to this anomalous behaviour. So we can adopt an expression of the type $A(2w_a) = \binom{\mathrm{Int}[Z/s] - \delta}{2}$, where δ represents a possible delay taking into account a transient behaviour. This term is always superimposed to the usual linear growth with Z of any $A(w)$. □

There are situations where the s-length closed route with all 0-symbols in output can be reached following a path, associated to polynomial $a(x)$ and left following a path associated to another polynomial $b(x)$. In this case, typically deg[$a(x)$] and deg[$b(x)$] are different, and their contribution to the binomial growth occurs in coefficient $A(w_a + w_b)$, being $w_b = \mathrm{w}[b(x)]$. Nevertheless the symmetric frame due to a same output path $a(x)$ is that always present, although it can be accompanied sometimes also by some asymmetric frames, with the same overall weight. This occurs especially when the coefficient $A(w)$, where binomial increase appears, is characterized by $w > d$. In this case, all the independent contributions must be added.

Example 7.14 The 1/2-rate convolutional code characterized by interleaved generator polynomial 11.11.11 exhibits a state diagram with four states, whose branches are labelled with pairs of output symbols as shown in Fig. 7.17, where we refer for simplicity to a systematic representation (branches labelled with only output symbols). It is clearly not well designed. The minimum distance is $d = 4$,

Fig. 7.17 State diagram of the 1/2-rate convolutional code treated in Example 7.14

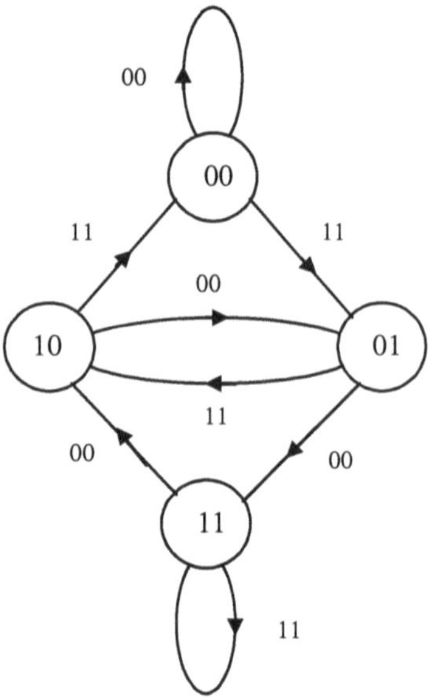

and the 4-weigth coefficient turns out to be approximately $A(4) \approx \binom{\mathrm{Int}[Z/3]}{2}$, which confirms what predicted by Lemma 7.6 with $s = 3$. The code frames having weight 4, of the type (7.4), can be written as 11.00.00.11, 11.00.00.00.00.00.11, and so on. We have here a polynomial $a(x)$ associated to the output sequence 11. It is possible to adopt a model, for describing the behaviour of this code, in which the LC code, constructed on the [3,1,3] repetition cyclic code, exhibits a word weight coefficient $A(2)$ showing the same dependence on the code length n.

Example 7.15 The 1/2-rate convolutional code characterized by interleaved generator polynomial 11.10.10.11 exhibits a state diagram with eight states, whose branches are labelled with pairs of output symbols as shown in Fig. 7.18. This code is characterized by $d = 6$. Here the binomial growth is present only in coefficient $A(10)$. This growth turns out to be approximately as $A(10) \approx \binom{Z}{2}$, justified by the fact that $s = 1$. The code frames having weight 10, of the type (7.4), can be written as 11.01.11.11.01.11, 11.01.11.00.11.01.11, 11.01.11.00.00.11.01.11, and so on. Clearly polynomial $a(x)$ is here associated to the output sequence 11.01.11.

Approximations in the above Examples are due to the initial transient and to the possible presence of further linear contributions in $A(w)$. As discussed in the proof of Lemma 7.6, the delay appears within the expression of the binomial, reducing by some units the contribution due to Z itself. In order to estimate when the binomial increase begins, it is convenient to analyze the evolution of $A(w)$ for various frame lengths (subsequent values of Z). For instance, with the convolutional code described in Example 7.15, the evolution of the frame weight distribution up to $w = 10$ is

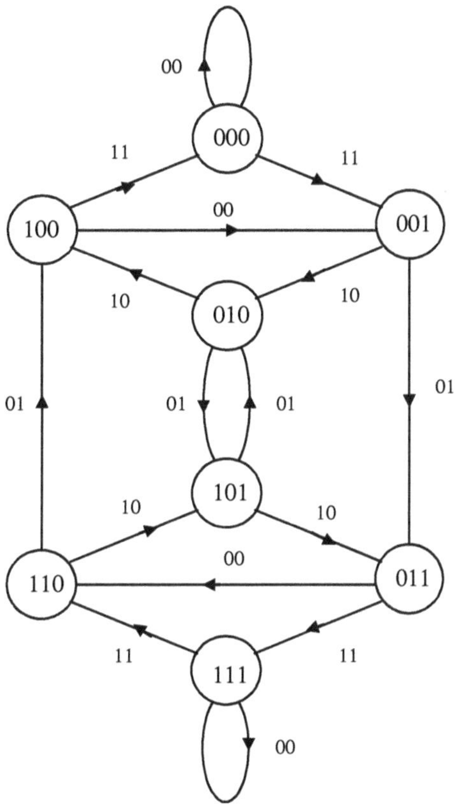

Fig. 7.18 State diagram of the 1/2-rate convolutional code treated in Example 7.15

reported in Fig. 7.19. Once reached the steady state, we can observe a differential increase which itself grows (by 1 unit at any unitary increment in Z). This binomial growth is regulated by parameters $s = 1$, $\delta = 3$. More precisely, we can adopt the extrapolation laws reported under the columns where the values of $A(w)$ are listed in Fig. 7.19. Remark that in $A(10)$ also a linear contribution is present, due to paths, from state $\underline{0}$ to state $\underline{0}$, not passing through the considered closed route.

As already noted, more than one path can be covered for reaching the closed route and for outgoing from it, say $a(x)$ and $b(x)$. When $w_a = w_b$ they both contribute to $A(2w_a)$, and also their combination, which leads to a pair of asymmetric code frames is to be taken into account. In this case the binomial law has to be multiplied by the number of all the possible path combinations.

Example 7.16 The convolutional code characterized by an interleaved generator polynomial associated to the sequence 11.01.11.10.11 has been demonstrated to be not well designed in the coefficient $A(12)$ (see Examples 3.12 and 3.22). It shows the evolution of the frame weigh distribution reported in Fig. 7.20. The state diagram of this code is characterized by $s = 1$ (00-labelled closed-loop branch around state 1111). The four code frames, responsible for the binomial increase with Z, are as follows

Z	$A(6)$	$A(8)$	$A(10)$	
1	1			
2	3			
3	5	1		
4	7	5	3	
5	9	12	7	
6	11	20	17	
7	13	28	38	
8	15	36	70	
9	17	44	108	
10	19	52	148	40
11	21	60	189	41
12	23	68	231	42
	$2Z-1$	$8Z-28$	$\binom{Z-3}{2}+34Z-213$ (with $Z>9$)	

Fig. 7.19 Evolution of the frame weight distribution for the convolutional code reported in Example 7.15 and extrapolation laws

11.10.01.11.11.10.01.11, 11.10.01.11.00.11.10.01.11, . . .

11.01.00.00.10.11.11.01.00.00.10.11, 11.01.00.00.10.11.00.11.01.00.00.10.11, . . .

11.10.01.11.11.01.00.00.10.11, 11.10.01.11.00.11.01.00.00.10.11, . . .

11.01.00.00.10.11.11.10.01.11, 11.01.11.00.10.11.00.11.10.01.11, . . .

The former two are symmetric and due to $a(x)$ associated respectively to the output patterns 11.10.01.11, 11.01.00.00.10.11. The latter two are asymmetric and due to $a(x)$ associated to 11.10.01.11, and $b(x)$ associated to 11.01.00.00.10.11. These output patterns justify the 4-fold multiplicity in the binomial part of the extrapolation law adopted under the column, where the values of $A(12)$ are listed in Fig. 7.20.

The binomial law discussed so far can be interpreted also in analogy with what expressed by (1.17) in Lemma 1.28, which deals with the binomial growth of 2-weight code words in LC codes. They give rise to even-weight code words, whose weight can be smaller than the minimum distance d_c of the parent cyclic code. The binomial growth of the weight distribution coefficient is typical in block codes (except for strongly shortened cyclic codes), therefore this property has not been stressed discussing them. In convolutional codes, this is instead the drawback which characterizes not well designed codes. Finally, the existence of 2-weight code frames, in the presence of generator polynomials having higher weights, is extremely negative, although only sometimes it entails a binomial growth of the weight distribution coefficient $A(2)$. The necessary condition for this occurrence is provided by Corollary a of Theorem 3.2 (N exact multiple of n_o). Typically this condition is

Z	$A(8)$	$A(12)$	
1	1		
2	3		
3	7		
4	13	2	
5	21	10	
6	30	32	
7	40	76	
8	50	152	
9	60	266	
10	70	420	
11	80	612	
12	90	836	
13	100	1085	
14	110	1351	
15	120	1628	
16	130	1912	] 289
17	140	2201	] 293
18	150	2494	] 297
19	160	2791	] 301
20	170	3092	
	$10Z-30$	$\binom{Z-4}{2}4+241Z-2208$ (with $Z>16$)	

Fig. 7.20 Evolution of the frame weight distribution for the convolutional code reported in Example 7.16 and extrapolation laws

much more restrictive than the characteristic of being simply a not well designed convolutional code, say in a coefficient $A(2w_a)$, with w_a rather high.

Example 7.17 Let us consider the interleaved generator polynomial 111.111 of the 2/3-rate convolutional code, whose semi-infinite $\underline{G}$ matrix is shown in Fig. 7.21 on the left. It can be interpreted as an MLC code with the same $g_c(x)$, which is factor of $(x^6 + 1)$. Its parity check polynomial is $h_c(x) = x + 1$, and can be matched by the information sequence (condition not satisfied in Example 7.14 with the same generator polynomial but with code rate 1/2). This convolutional code is not well designed directly in the coefficient $A(2)$, in fact we can write $6 = 2n_o$. The part of the state diagram of this MLC code, responsible for the 2-weight code frames, is reported in Fig. 7.21 on the right. This code is constructed starting from the [6,1,6] repetition cyclic code. The state diagram is that produced by an encoder circuit of the type shown in Fig. 1.5, that is with 2^r states, r being the degree of $g_c(x)$. In the present case $r = 5$. Clearly the 2-weight code frame originating the binomial growth of coefficient $A(2)$ is the sequence 1000001 itself (ex-OR sum of the two rows in $\underline{G}_o$). The binomial increase is related to Int[Z/2], in fact, in the model of 2/3-rate convolutional codes (with $m = 2$), we have here $s = 2$ and $n_o = 3$.

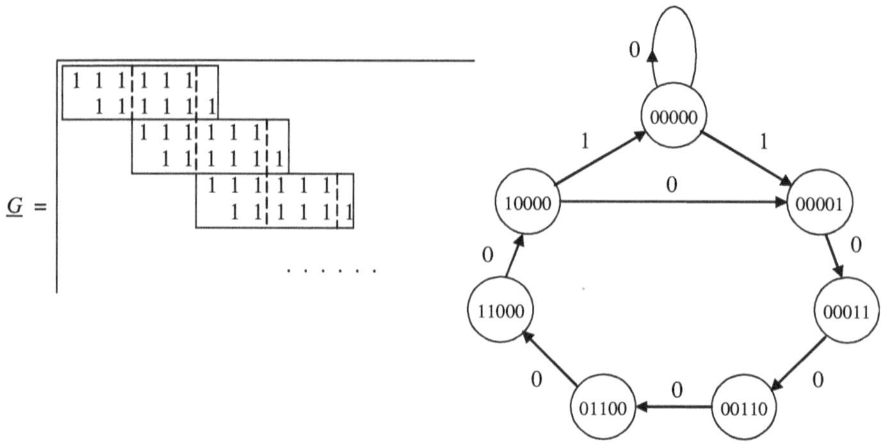

Fig. 7.21 Semi-infinite generator matrix and part of the state diagram of the not well designed convolutional code introduced in Example 7.17

Lemma 7.7 *In a convolutional code, which is s.s. time-invariant in both* $\underline{G}$ *and* $\underline{H}$*, if* $h_c(x)$ *can be formed after a proper information sequence, even when N is not an exact multiple of* n_o*, 4-weight code frames, whose minimum length is* $(N + n_o)$*, give a contribution to A(4) showing a binomial growth with Z.*

Proof A code frame of the type $(x^{n_o} + 1)(x^N + 1)$ occurs, owing to a pair of polynomials $h_c(x)$ formed by proper equal information sequences, separated by one period. This conclusion follows from a double application of Theorem 3.2. Therefore a family of code frames exists, whose expression is $t(x) = (x^N + 1)\ b(x^{n_o})$, where $b(x^{n_o})$ is a 2-weight polynomial, confirming the binomial growth of $A(4)$, provided that the frame length is greater than $(N + n_o)$. □

From another point of view, the rows of the $\underline{H}$ matrix describing an LC code are characterized by the periodic polynomial $h_c(x)(1 + x^N + x^{2N} + \cdots)$, responsible for the binomial increase in the number of 2-weight code words. For the MLC code, if $h_c(x)$ can be formed, but n_o is not factor of N, and hence the whole LC periodic parity check polynomial cannot be formed, 2-weight code words have a number which increases only linearly with Z. So the binomial increase regards just 4-weight code words.

In conclusion, there are two mechanisms enabling the occurrence of such anomalous low-weight code words (both with even weight w). The first one can be revealed in the state diagram of the convolutional code considered, and is associated to the presence of closed routes labelled with all 0-symbols in output, whose length is typically $s < N$ (see Lemma 7.6 for not well designed convolutional codes at any rate). The second one is due to proper linear combinations of $h_c(x)$, matched by the information symbol sequence (see Lemma 7.7, in which regular closed routes of length N are involved, like in LC codes) and typically comes out at high code rates.

A drawback characterizing not well designed convolutional codes, as already stressed, is the impossibility of performing a sliding window decoding procedure. Furthermore, in a tail-biting arrangement, some rows of $\underline{G}$ become linearly dependent on the other ones (see Example 4.17). Except for these limitations, the code remains acceptable for many practical applications, especially when its minimum distance d is high. On the contrary, the presence of 2-weight code frames represents, by itself, a severe limitation for the performance of a code. Therefore it has to be carefully avoided, also because it entails a binomial increase of coefficient $A(4)$ in any situation.

7.5 Interpretation of Direct Product Codes as Particular Not Well Designed Convolutional Codes

Direct product codes can be interpreted as not well designed s.s. time-invariant convolutional codes in their $\underline{G}$ matrix, as already discussed in Sect. 1.17 and at the end of Sect. 3.9. Nevertheless not all the s.s. time-invariant $\underline{G}$ matrices of not well designed convolutional codes are the Kronecker product of two matrices. In Sect. 5. 16, the parity check matrix of a direct product code has been presented. Besides an upper part with non-periodic rows, there is a lower part with rows that occupy the whole length of the code word, and can be considered as enlarged versions of the traditional reciprocal parity check polynomials characterizing one of the two component block codes. Looking at the parity check matrices shown in Fig. 7.12, for two not well designed convolutional codes, it is possible to observe a structure similar to that characterizing direct product codes.

Lemma 7.8 *Given an s.s. time-invariant not well designed convolutional code, with code rate* $1/n_o$*, only if all the sub-generator polynomials are equal (or null), the considered code is equivalent to a direct product code of two cyclic codes or codes derived from them.*

Proof Corollary a of Theorem 1.4 gives the condition for an MLC code for being equivalent to a direct product code. It must exhibit a generator polynomial having form $g_1(x^{n_o})g_2(x)$, where $g_1(x)$ and $g_2(x)$ are the generator polynomials of the two component codes. An s.s. time-invariant convolutional code, in order to be not well designed, must be characterized by an interleaved generator polynomial in which all the sub-generator polynomials show a common factor $p(x)$. In conclusion, not well designed s.s. time-invariant convolutional codes are characterized by an interleaved generator polynomial which can be written as

$$g_{\text{int}}(x) = p(x^{n_o})[x^{n_o-1}g^{(1)\prime}(x^{n_o}) + x^{n_o-2}g^{(2)\prime}(x^{n_o}) + \cdots + g^{(n_o)\prime}(x^{n_o})],$$

where we have set $g^{(j)}(x) = p(x)g^{(j)\prime}(x)$, $j = 1, 2, \ldots n_o$. Polynomial $g_{\text{int}}(x)$ is replicated k_o times, on as many rows of the generator sub-matrix $\underline{G}_o$. After comparison between $g_1(x^{n_o})g_2(x)$ and $g_{int}(x)$, we obtain $g_1(x) = p(x)$. Keeping in mind how the Kronecker matrix is constructed, the degree of the second factor in $g_{int}(x)$ will be equal to r_o. Then, after the assumption $k_o = 1$, it exhibits degree equal to $r_o = n_o - 1$, only if the polynomials $g^{(j)\prime}(x)$ are so constrained: $g^{(1)\prime}(x) = g^{(n_o)\prime}(x) = 1$, and the other ones equal to 0 or 1. If the considered code is not a direct product code, the degrees of polynomials $g^{(j)\prime}(x)$ can be also higher than 0, so leading to a more general situation. □

The case of $k_o > 1$ can be treated in a similar way, but requires to remember that the product code is s.s. time-invariant in its $\underline{G}$ matrix and then in the above expression of $g_{int}(x)$, maintaining the same value for the period length n_o, it is necessary to reduce the interval between the two external 1-symbols in the sequences of coefficients $g^{(j)\prime}(x)$. So, as long as $k_o = k_2$ increases, $g^{(1)\prime}(x)$, $g^{(2)\prime}(x)$, … will become null. In conclusion, Lemma 7.8, can be easily extended also to the case of $k_o > 1$. Besides the equality in the expressions of the unique generator polynomial, between a not well designed convolutional code and a direct product code, assured by Corollary a of Theorem 1.4 and consequent observations, also the constraint on the degree of factor $g_2(x)$ to be $r_2 = r_o$ has to be imposed. Among the codes in Examples 7.14, 7.15, 7.16, and 7.17, only the first one exhibits the behaviour described in Lemma 7.8, and hence can be considered a direct product code. Now we will study the parity check matrix of such code.

Example 7.18 The 1/2-rate not well designed convolutional code described in Example 7.14, which is characterized by an interleaved generator polynomial 11.11.11, that is $g_{int}(x) = (x^4 + x^2 + 1)(x + 1)$, shows the $\underline{G}$ matrix and $\underline{H}$ matrix reported in Fig. 7.22, on the left. $Z' = 5$ overall periods have been considered, This code can be interpreted as the direct product between the [2,1,2] parity check code and the [5,2,2] LC code obtained from the [3,1,3] repetition code, according to the rectangular encoding memory reported in Fig. 7.22 on the right, where control symbols have been denoted with a grey colour. In particular, the lower part of the $\underline{H}$ matrix is added taking a parity check polynomial of the type $(x + 1)(1 + x^3 + x^6 + \cdots)$, which is the one characterizing the [5,2,2] LC code considered. Two

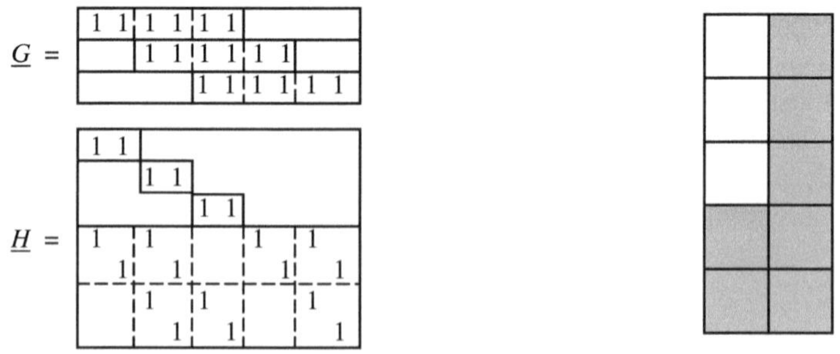

Fig. 7.22 Generator and parity check matrix for the code described in Examples 7.14 and 7.18, and its encoding memory, interpreted as a [10,3,4] direct product code

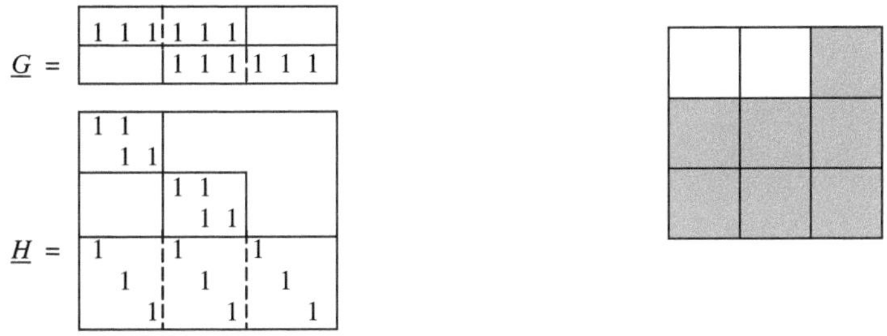

Fig. 7.23 Generator and parity check matrix for the code described in Example 7.19, and its encoding memory, interpreted as a [9,2,6] direct product code

further periods can be constructed prolonging the pseudo-diagonal structure characterizing the upper part of the matrix, by proper ex-OR sum of rows. In this case the number of rows of the lower part reduces. For this construction, it is possible to see Sect. 5.16, about the multiplicity of equivalent versions of the parity check matrix of a direct product code.

Example 7.19 The 1/3-rate not well designed convolutional code, characterized by an interleaved generator polynomial 111.111, that is $g_{\text{int}}(x) = (x^3 + 1)(x^2 + x + 1)$, over $Z' = 3$ overall periods, shows the $\underline{G}$ matrix and $\underline{H}$ matrix reported in Fig. 7.23, on the left. This code can be interpreted as the direct product code of the [3,1 3] repetition code and the [3,2,2] parity check code, whose square encoding memory is reported in Fig. 7.23 on the right. Here the parity check polynomial characterizing the macroscopic structure of the lower part of the overall $\underline{H}$ matrix is a unitary polynomial $u_3(x)$, as is due for the parity check code considered here.

The encoding memories to be adopted in order to construct longer frames, respectively for the two situations described in the above two Examples, are shown in Fig. 7.24. In the first case, the convolutional period length is $n_o = 2$, and the vertical size of the encoding memory has to be assumed long as one wishes, leading to an asymptotic code rate 1/2. In the second case, the convolutional period length is $n_o = 3$, and the horizontal size of the encoding memory has to be assumed long as one wishes, leading to an asymptotic code rate 1/3.

In the upper part of the $\underline{H}$ matrix of a direct product code, as already observed, there are orthogonal sub-matrices. This fact distinguishes them from not well designed convolutional codes of other type. The following Examples regard 1/2-rate convolutional codes, which, in spite of being not well designed, cannot be assumed as equivalent to proper direct product codes. For them, an expression of the type (1.35) is still valid, but with $g_2(x)$ having degree greater than r_o.

Example 7.20 The 1/2-rate convolutional code introduced in Example 7.15, whose interleaved generator polynomial is 11.10.10.11, over a frame length characterized by $Z' = 5$ overall periods, exhibits the generator and parity check matrix shown in Fig. 7.25. The parity check matrix shows just one periodic row, replacing the fourth row of the last $\underline{H}_o^T$ sub-matrix. In it, an enlarged unitary polynomial $u_5(x^2)$ occupies

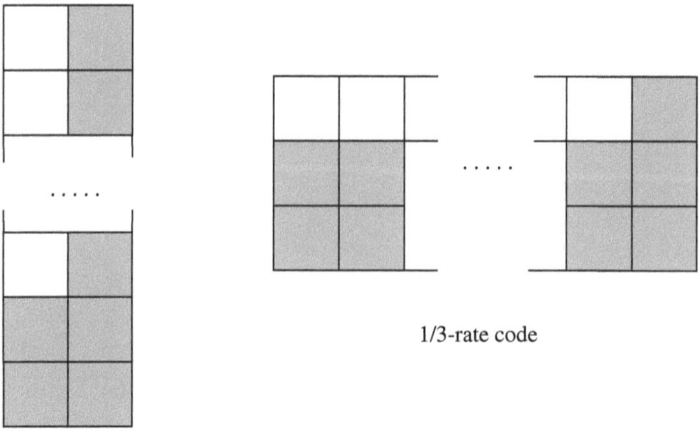

Fig. 7.24 Encoding memories for the not well designed convolutional codes (direct product codes) of Examples 7.17 and 7.18 with longer frames

with 1-symbols the second positions of all the periods. Alternatively it is possible to adopt the periodic polynomial $h^*(x) = xu_5(x^2)$, occupying with 1-symbols the first positions of all the periods. In the upper part of such matrix, we can recognize the property $h(x) = g(x)$, which holds after Lemma 7.4, but is not able here to allow the construction of the whole $\underline{H}$ matrix.

Example 7.21 The 1/2-rate convolutional code treated in Example 7.16, whose interleaved generator polynomial is 11.01.11.10.11, over a frame length characterized by $Z' = 5$ overall periods, exhibits the generator and parity check matrix shown in Fig. 7.26. Also in this case, we have only one periodic row, replacing the lowermost row of the last $\underline{H}_o^T$ sub-matrix. Remark that the periodic row is the same as before. This is due to the same common factor in the sub-generator polynomials, that is $p(x) = x + 1$.

The last two codes are characterized by binomial growths of coefficients $A(w)$ only for w remarkably greater than d. On the contrary, the condition expressed by Lemma 7.8 is responsible for the binomial growth with Z of coefficient $A(d)$, where

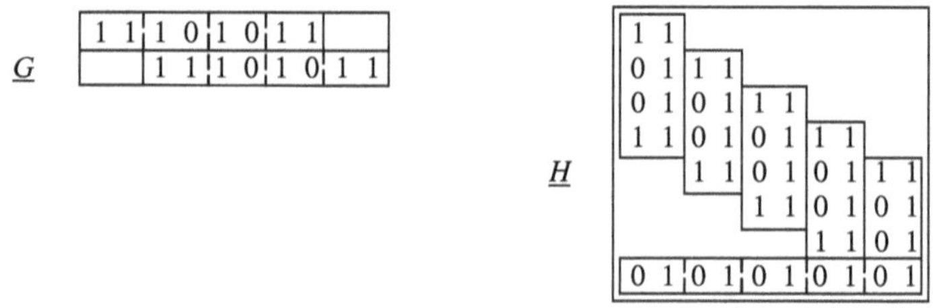

Fig. 7.25 Generator and parity check matrix for the code described in Examples 7.15 and 7.20

$$
\underline{G} = \begin{array}{cccccccccccc}
1 & 1 & 0 & 1 & 1 & 1 & 1 & 0 & 1 & 1 & & \\
 & & 1 & 1 & 0 & 1 & 1 & 1 & 1 & 0 & 1 & 1
\end{array}
$$

$$
\underline{H} = \begin{array}{cccccccccccc}
1 & 1 & & & & & & & & & & \\
1 & 0 & 1 & 1 & & & & & & & & \\
1 & 1 & 1 & 0 & 1 & 1 & & & & & & \\
0 & 1 & 1 & 1 & 1 & 0 & 1 & 1 & & & & \\
1 & 1 & 0 & 1 & 1 & 1 & 1 & 0 & 1 & 1 & & \\
 & & 1 & 1 & 0 & 1 & 1 & 1 & 1 & 0 & 1 & 1 \\
 & & & & 1 & 1 & 0 & 1 & 1 & 1 & 1 & 0 \\
 & & & & & & 1 & 1 & 0 & 1 & 1 & 1 \\
 & & & & & & & & 1 & 0 & 0 & 1 \\
0 & 1 & 0 & 1 & 0 & 1 & 0 & 1 & 0 & 1 & 0 & 1
\end{array}
$$

Fig. 7.26 Generator and parity check matrix for the code described in Example 7.21

d, in this situation, is the product of the minimum distances of the two component codes. A benefit of having a so large minimum distance is paid by this drawback.

To summarize, if one attempts to construct any type of $\underline{H}$ matrix by means of all non-periodic rows, in the case of not well designed convolutional codes, among such non-periodic rows, some ones will be not linearly independent or the total number of rows expected in $\underline{H}$ is not reached. In the first situation, an equal number of periodic rows must replace the linearly dependent rows (as occurs in Fig. 7.25 or 7.26). In the second situation, the periodic rows are to be added (as occurs in Fig. 7.12). The case of a direct product code is immediately recognizable, because the upper part of the matrix is characterized by orthogonal non-periodic sub-matrices ($m_h = 0$). A procedure able to find the additional periodic rows, in the most general situations, can be arranged as a proper variant of the column construction procedure introduced in Sects. 7.2 and 7.3. This will be done at the end of the next section, after a general discussion about the relationship between the generator constraint length m and the parity check constraint length m_h.

7.6 Various Situations for Well Designed and Not Well Designed Convolutional Codes Regarding Their Parity Check Matrix

Let us come back to well designed convolutional codes s.s time-invariant in $\underline{H}$, and interpret them as MHEC codes. They are characterized by just one parity check polynomial $h_c(x)$, whose reciprocal $h_c^*(x)$ shifts in the rows of $\underline{H}$. Starting from the parent HEC code, a period with length n_o is chosen, in which the first k_o rows are eliminated, and the last r_o rows show as many remaining $h_c^*(x)$ polynomials. Only close to the beginning and to the end of the frame there are truncated polynomials. As already observed, these codes can be encoded in systematic form by employing the encoder circuit shown in Fig. 5.6, after proper regulation of switch operation.

Remark that, from the above MHEC code construction of an s.s. time-invariant convolutional code, we obtain only non-periodic rows, all linearly independent.

Therefore not well designed convolutional codes, cannot be encoded by means of this encoder circuit. In fact their $\underline{H}$ matrix requires the addition of some final periodic rows, linearly independent of the previous non-periodic ones. In this sense, such codes should be rigorously defined as time-varying in $\underline{H}$ (see direct product codes as an example). On the other hand, encoding is possible here exploiting the $\underline{G}$ matrix, which remains at least w.s. time-invariant.

Equivalent versions of the same well designed convolutional code can be obtained assuming a certain generator sub-matrix $\underline{G}_o$, but different syndrome former sub-matrices $\underline{H}_o^T$, all obtained from the same polynomial $h_c{}^*(x)$ in minimal form, after multiplication by a certain polynomial $p(x^{n_o})$. They are no longer in minimal form and have not full rank. For instance, it is enough that few last reciprocal parity check polynomials (few last $\underline{H}_o^T$ sub-matrices) are set in minimal form for obtaining an equivalent behaviour, with the right number of $\underline{H}$ rows. This procedure cannot be considered rigorously as a simple correct frame termination, especially when polynomial $p(x)$ has a high degree. It leads to a time-varying code in $\underline{H}$, where the last periods exhibit a contracted structure, but does not change the convolutional nature of the code or the property of being well designed (see Example 7.13). The above two situations are similar, but the need of some periodic parity check polynomials is present only in the first one.

In general, better performance can be obtained with convolutional codes which are w.s. time-invariant in $\underline{G}$ (especially with high code rates), or which are w.s. time-invariant in $\underline{H}$ (especially with low code rates). For the moment, we can discuss the following important property, rather obvious, but not yet demonstrated in the present treatment.

Theorem 7.1 *A w.s. time-invariant convolutional code in its $\underline{G}$ matrix, if well designed, can be described by a parity check matrix which is s.s. or at least w.s. time-invariant. Vice versa, a w.s. time-invariant convolutional code in its $\underline{H}$ matrix, if in minimal form, can be described by a generator matrix which is s.s. or at least w.s. time-invariant.*

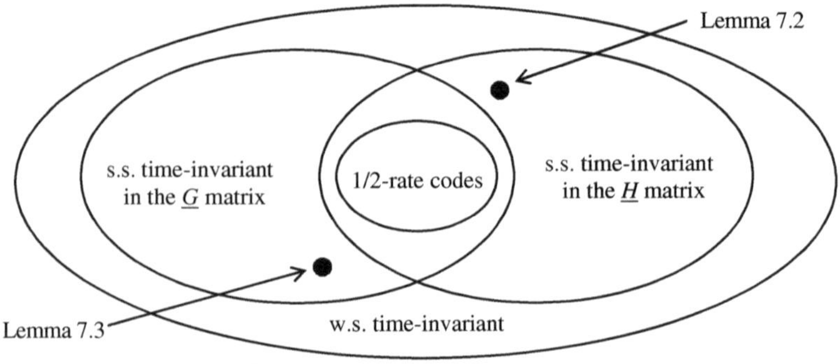

Fig. 7.27 Inclusion relationships among convolutional codes regarding their time-invariance properties (well designed convolutional codes only)

Proof Let us take a very general situation, characterized by $n_o = k_o + r_o$, with the last two parameters both greater than one (in fact Lemma 7.2 already treated the case of $r_o = 1$, and Lemma 7.3 that of $k_o = 1$). Sub-matrix $\underline{G}_o$ exhibits k_o rows, where, in general, different interleaved generator polynomials $g_{\text{int}}^{(j)}(x), j = 1, 2, \ldots k_o$, are present (see Sect. 4.1). If the considered code is w.s. time-invariant in $\underline{G}$, sub-matrix $\underline{G}_o$ is unique. It is not restrictive to assume that it is upper triangular. After the column construction of $\underline{H}_o^T$ (see Definition 7.6), such sub-matrix will be formed layer by layer, with r_o rows in each layer. The syndrome-former sub-matrix $\underline{H}_o^T$ in minimal form so obtained does not change passing from one period to the subsequent one, except for the usual correct frame termination. Furthermore there is no need of further periodic rows, owing to the assumption that the code is well designed. Hence the considered code is also at least w.s. time-invariant in $\underline{H}$. Conversely, let us start from an $\underline{H}_o^T$ sub-matrix, with layers having r_o rows each, unique as a consequence of the assumption that the code is at least w.s. time-invariant in $\underline{H}$ and in minimal form. Such parity check matrix, after Definition 7.3, will be characterized by different reciprocal parity check polynomials $h_i{}^*(x)$, $i = 1, 2, \ldots r_o$. It is not restrictive to assume that a lower triangular form characterizes the first layer of $\underline{H}_o^T$. After imposition of having null ex-OR sum of clusters of syndromes, we can obtain the correspondent $\underline{G}_o$ sub-matrix. It is necessary, also in this case, to choose the $\underline{G}_o$ sub-matrix with minimum value for the constraint length m. Otherwise we would have here a different code. The $\underline{G}_o$ sub-matrix so obtained, does not change passing from one period to the subsequent one, therefore the considered code is also at least w.s. time-invariant in $\underline{G}$. □

In conclusion, considering s.s. time-invariance as a property included in w.s. time-invariance, we can say that w.s. time-invariant convolutional codes both in $\underline{G}$ and in $\underline{H}$ constitute a closed ensemble. On the other hand, s.s. time-invariant convolutional codes in one of the two matrices can or cannot be s.s. time-invariant also in the other matrix. The case of 1/2–rate codes is only a special situation, where s.s. time-invariance regards both $\underline{G}$ and $\underline{H}$ (as stressed by Lemma 7.4). The inclusion relationships of the various ensembles of convolutional codes, with reference to their time-invariance properties, are shown in Fig. 7.27. Codes s.s. time-invariant only in one of the two matrices are those considered in Lemmas 7.2 and 7.3 (small black circles in the Figure). The diagram reported in Fig. 7.27 refers to the only codes for which Theorem 7.1 holds, therefore not well designed convolutional codes, say, are excluded. Some further comments are due about the impossibility of employing a longer generator constraint length, discussed at the end of the proof of Theorem 7.1. In principle, it is possible to proceed with a value of m greater than the minimum one, but, at the end of the frame, in this case, we cannot take the shortest non-periodic generator polynomials, because they would have no effect on the preceding periods. On the contrary, as a consequence of duality, some periodic polynomial whose length is that of the whole frame, are needed. Clearly this solution exceeds the concept of convolutional code. For this reason we have preferred to avoid the definition of minimal form for the $\underline{G}$ matrix.

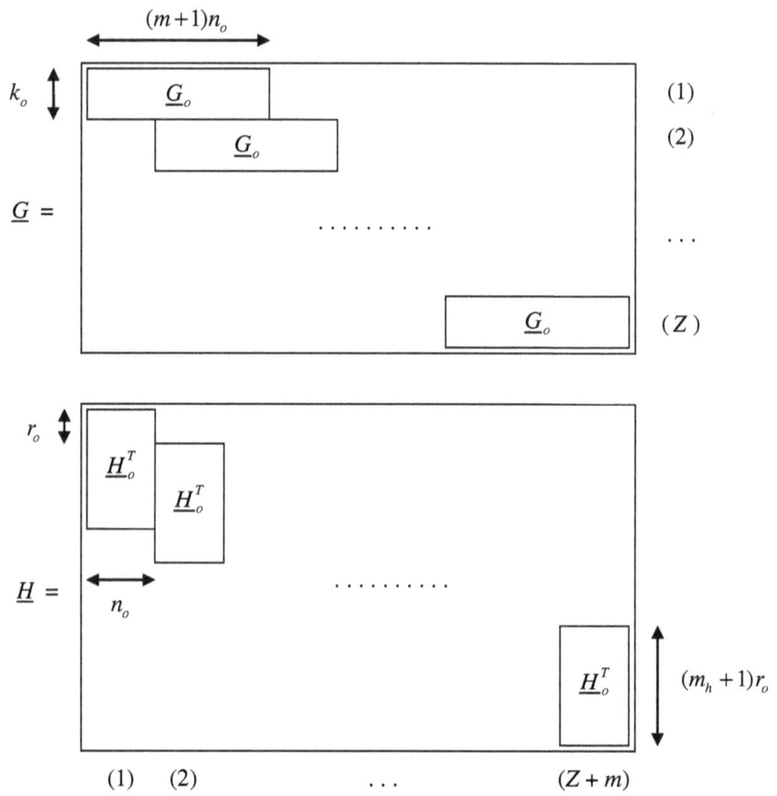

Fig. 7.28 Structures of the correctly terminated $\underline{G}$ and $\underline{H}$ matrices in a k_o/n_o-rate convolutional code

Let us consider the structures of the $\underline{G}$ and $\underline{H}$ matrix in a k_o/n_o-rate w.s. time-invariant convolutional code, schematically depicted in Fig. 7.28, where a correct frame termination has been taken into account in both. Furthermore, absence of periodic rows in $\underline{H}$ enables us to assert that we are treating the case of a well designed convolutional code. It is important to note a difference, regarding the external parts of such two matrices, already described and explained. After a proper comparison of the dimensions characterizing such two matrices, one could expect to find a relationship between the two constraint lengths m and m_h.

To this purpose we can start from the observation that the total frame length becomes $n = (Z + m)n_o$, whereas the total number of information symbols is $k = ZK_o$. On the other hand, the total number of control symbol (rows of $\underline{H}$) is $r = (Z + m)r_o + m_h r_o$. Assuming the relationship $n = k + r$, after brief calculations, we are left with

$$m_h r_o \approx m k_o, \tag{7.5}$$

where approximation is due to the possibility that not all the $(m + 1)n_o$ columns of the outermost $\underline{G}_o$ sub-matrices are occupied by 1-symbols and to the possibility that some initial or final rows in $\underline{H}$ are useless (made by just one 1-symbol or null at all). Although a first balance in these two processes is taken into account, often it is not sufficient for having an equality in (7.5), and, in certain cases it turns out to be rather far from validity.

It is exact for 1/2-rate codes (after Lemma 7.4). It is also exact for the particular codes having code rate 2/4 and 2/3 described in Fig. 7.11, with $m = 1$ and $m_h = 1, 2$, respectively. For the 2/3-rate code described in Example 7.8 (Fig. 7.9) equality (7.5) is still exact, with $m = 1$ and $m_h = 2$, although one row in $\underline{H}$ and one column in both matrices have been eliminated in order to perform a correct frame termination. The same is for the 2/3-rate code, only w.s. time-invariant in $\underline{G}$, analyzed in Example 7.9 (Fig. 7.10) without elimination of any column. For the 3/4-rate code described in the same Figure, with $m = 1$ and $m_h = 3$, instead, (7.5) is wrong by a factor of 2/3 in the estimation of parameter m_h, since a second column is to be eliminated in the last period. Finally the low-rate codes presented in Examples 7.5 and 7.6 (Fig. 7.8) are characterized by parameters for which (7.5) is wrong by factors greater than 1 in the estimation of parameter m. The reason for this strong departure from what expected is in the fact that the last row layers of $\underline{H}_o^T$ are occupied by non-null symbols only in a very small part. The above observations confirm that duality, from which expression (7.5) seems to derive, has to be taken often with some cares.

In order to calculate the periodic rows characterizing the $\underline{H}$ matrix of a not well designed convolutional code, it is possible to employ a variant of the column construction procedure. Such new procedure goes on by attempts as usual, but starts from the assumption of a certain pattern of 1-symbols, matching the 1-symbols of any row of the generator sub-matrix, in an even number of positions, interesting all the periods spanned by $\underline{G}_o$. It proceeds constructing the parity check row as usual, under the generator sub-matrix, and stops when the initial configuration is reached again, so assuring a large-scale periodicity (which typically is few times n_o). This procedure can be employed also for cyclic and quasi-cyclic codes giving N as the large-scale period. Some examples will clarify the concept (by treating previously these block codes), better than a long and articulate description.

Example 7.22 Assuming a [7,4,3] Hamming cyclic code, associated to the generator polynomial $g_c(x) = x^3 + x^2 + 1$, the column construction of its $\underline{H}$ matrix can be performed as shown in Fig. 7.29 on the left. Here we have a $\underline{G}_o$ sub-matrix consisting of just one row and the progressive shift of polynomial $h_c{}^*(x)$ under such row, from left to right, is made with steps of just 1 position. It is easy to verify that all the rows of this construction show an even number of coincidences with the generator polynomial reported above. Reading from down to top, we obtain the parity check row 0101110, which coincides with a cyclic permutation of polynomial $h_c{}^*(x) = x^4 + x^2 + x + 1$, characterizing this code. Here the syndrome former sub-matrix can be thought as to be degenerated into a periodic column vector.

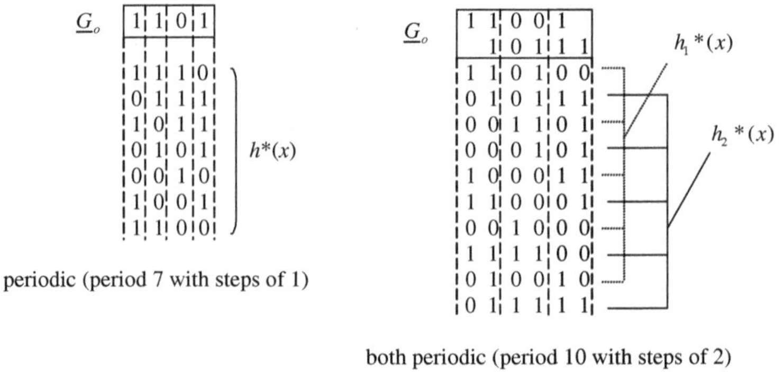

Fig. 7.29 Obtainment of the periodic parity check polynomials for a cyclic and a quasi-cyclic code

Example 7.23 Assuming the [10,6,3] QC code, introduced in Example 2.10 (and treated also in Examples 6.6, 6.7, and 6.8), which has $q_o = 2$ and is characterized by $g_1(x) = x^4 + x^3 + 1$, $g_2(x) = x^4 + x^2 + x + 1$, the column construction of its $\underline{H}$ matrix can be performed as shown in Fig. 7.29 on the right. Here we have a $\underline{G}_o$ sub-matrix consisting of two rows and the progressive shift of polynomials $h_i^*(x)$, $i = 1, 2$, under such rows, from left to right, is made with steps of $q_o = 2$ positions. The rows under $\underline{G}_o$ have been constructed already alternated, so obtaining the following pair of parity check sequences: 10.00.11.01.00 and 11.00.01.01.11. It is possible to recognize proper cyclic permutations of the periodic reciprocal parity check polynomials $h_1^*(x)$ and $h_2^*(x)$ already obtained (see, for instance, Example 6.8, where one possible $\underline{H}$ matrix in minimal form for this code has been discussed).

In general, the periodic parity check rows exhibit a large-scale period equal to $N = pq_o$, or to a factor of this quantity, maintaining q_o invariant. Some attempts are to be made before choosing the right initial distribution of 1-symbols in all the periods spanned by $\underline{G}_o$. Nevertheless, even when one starts from a wrong distribution, after a proper transient, the periodic behaviour can be reached as well. This can reduce by far the number of attempts. After this short description of a possible column construction for the parity check matrix in cyclic and QC codes, we will focus our attention on the application of such method for calculating the periodic rows in not well designed convolutional codes.

Example 7.24 In Example 7.12 we considered an MLC code, characterized by the cyclic generator polynomial $g_c(x) = x^5 + x^3 + x^2 + 1$, with asymptotic code rate 2/3 and 3/4. The overall parity check matrices have been reported in Fig. 7.12. In Fig. 7.30 on the left, the procedure applied to the 2/3-rate code is shown. Here we have two periodic polynomials, the one with period $2n_o = 6$, the other with period $n_o = 3$. In Fig. 7.30 on the right the case of the 3/4-rate code is examined. Here we have still two periodic polynomials, with period $n_o = 3$. They both are needed in order to complete the overall parity check matrix, after the non-periodic $\underline{H}$ rows, as

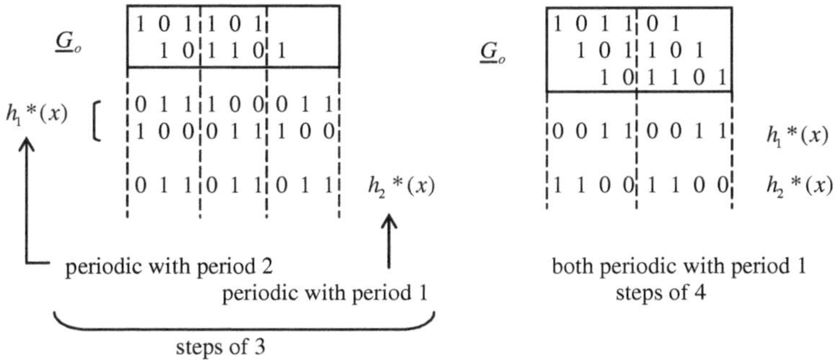

Fig. 7.30 Column construction of the periodic rows for the 2/3- and 3/4-rate MLC codes introduced in Example 7.12 (treated here in Example 7.24)

reported in Fig. 7.12 on the right. Alternatively their ex-OR sum (that is a unitary polynomial) could replace one of them. Both codes exhibit $d = 4$, and $A(4)$ itself growing binomially with Z. If we assume a code rate 4/5, the minimum distance is reduced to $d = 2$. In fact the parity check polynomial $h_c(\mathrm{x})$ of the cyclic original code is 10111101, sequence which, in this case, can be matched by a proper information sequence.

Example 7.25 Let us consider the code introduced in Example 7.15 and whose parity check matrix has been reported in Fig. 7.25, showing one periodic row, due to the common factor $(x + 1)$. It is a 1/2-rate convolutional code, whose interleaved generator polynomial is associated to the sequence 11.10.10.11. The column construction procedure for obtaining its periodic parity check row is depicted in Fig. 7.31 on the left.

Example 7.26 Let us consider now the code introduced in Example 7.16 and whose parity check matrix has been shown in Fig. 7.26. It is a 1/2-rate convolutional code, whose interleaved generator polynomial is associated to the sequence 11.01.11.10.11. Also here we have $(x + 1)$ as the common factor. The column construction procedure for obtaining its periodic parity check row is depicted in Fig. 7.31 on the right. The anomalous row is the same as before and is expressed by the enlarged unitary polynomial $u_{Z+m}(x^2)$. This is consequence of the common factor in the sub-generator polynomials, which is $(x + 1)$ in both. Furthermore it can be written with 1-symbols either in the first position of the period or in the second.

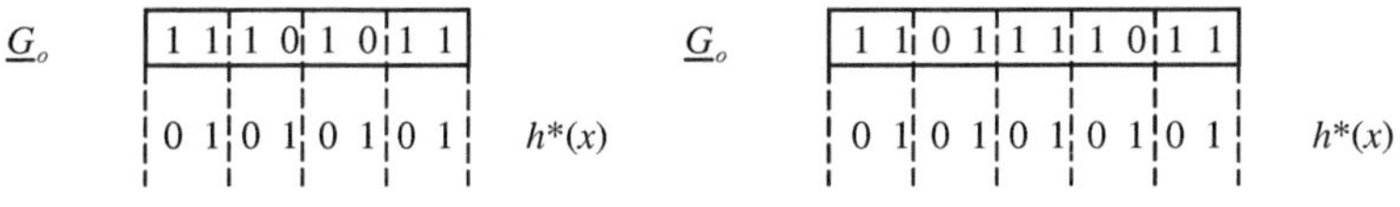

Fig. 7.31 Column construction of the periodic row for the 1/2-rate convolutional codes introduced in Examples 7.15 and 7.16 (treated here in Example 7.25)

Example 7.27 The 3/4-rate code whose semi-infinite $\underline{H}$ matrix is described in Fig. 7.11 on the right, exhibits $d = 2$, because the parity check polynomial of the parent cyclic code, which is associated to the sequence 10111, can be matched by the information sequence. The overall generator matrix and parity check matrix of this code, assuming a correct frame termination after 4 periods, are shown in Fig. 7.32 on the left. The last row in $\underline{H}$ is eliminated, because it exhibits just one 1-symbols. Also the correspondent column is eliminated. The whole code turns out to be described by the tern [15,10,2]. In fact the last period has length 3 instead of 4. The remaining rows of $\underline{H}$ are all linearly independent, confirming that this code, strictly speaking, cannot be classified as not well designed. In other terms, there is no need of periodic rows, besides the 5 parity check rows already obtained.

The origin of 2-weight code frames, of the type $(x^7 + 1)$ is due to the presence of two columns of $\underline{H}$ having just one 1-symbol in the same row. However coefficient $A(2)$ increases linearly with the period number Z, because $n_o = 4$ is not factor of $N = 7$ (see Corollary a of Theorem 3.2). Precisely we have $A(2) = Z - 1$. The 4-weight code frames originating a binomial increase in $A(4)$ have an expression of the type $(x^N + 1)b(x^{n_o}) = (x^7 + 1)b(x^4)$, where $b(x)$ is a binomial, due to proper information sequences (see the proof of Lemma 7.7). They are, say, the following two: 1000.1001.0001, 1000.0001.1000.001, respectively for $b(x^4) = x^4 + 1$ and $b(x^4) = x^8 + 1$. The latter one is ex-OR sum of two shifted versions of the former one, and exhibits a central passage through state $\underline{0}$. This shows that the origin of this

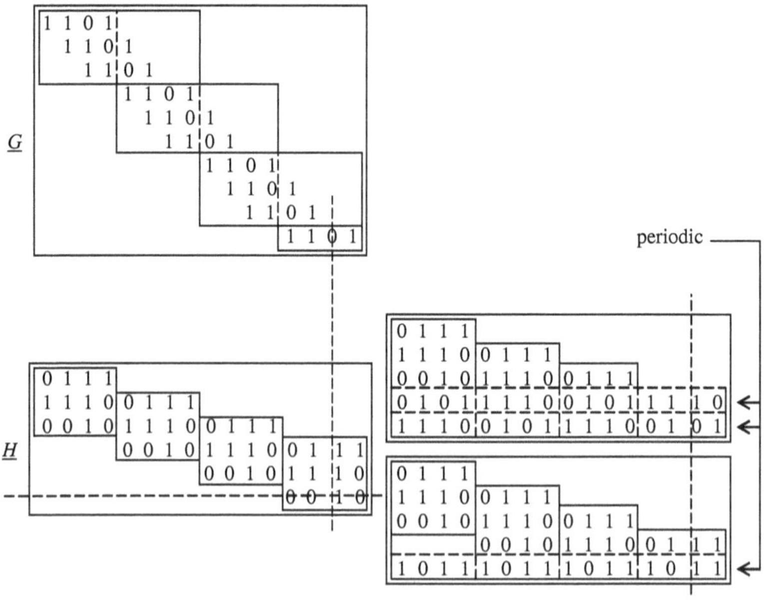

Fig. 7.32 $\underline{G}$ and $\underline{H}$ matrix for the 3/4-rate convolutional code with $d = 2$ described in Fig. 7.11 on the right, and alternative structures for $\underline{H}$

phenomenon is related to Lemma 7.7 only, and not to 0-labelled closed routes inside the state diagram.

Alternative structures for the parity check matrix of this code are reported on the right in Fig. 7.32. The former one exhibits two periodic rows, obtained after ex-OR sum respectively of the even rows and of the odd rows in the non-periodic part of $\underline{H}$. They both have period $2n_o = 8$. The latter one exhibits only one periodic row, obtained after ex-OR sum of all the rows in the non-periodic part of $\underline{H}$, and has period $n_o = 4$. The rows of the last $\underline{H}_o^T$ sub-matrices are progressively eliminated, in order to have five parity check independent equations on the whole. A similar construction of alternative forms for the parity check matrix (with few last periodic rows) can be adopted for any well designed convolutional code. Remaining in the framework of the model based on an MLC equivalent code, such periodic rows are those of the original LC code (see also Lemma 5.27), but their maximum number is upper bounded by the degree r of polynomial $g_c(x)$.

7.7 Systematic Encoder Circuit Based on a Unique Non-periodic Parity Check Polynomial

As already noted, in order to perform systematic encoding of a convolutional code, which is well designed and s.s. time-invariant in $\underline{H}$, that is with just one non-periodic parity check polynomial, it is possible to employ the scheme reported in Fig. 5.6. Here this possibility is better detailed. Figure 7.33 shows the encoder circuit which can be used with a k_o/n_o-rate convolutional code, in whose matrix the unique reciprocal parity check polynomial $h^*(x)$ shifts, being $D_h = \deg[h(x)] = \deg[h^*(x)]$. As usual, h_i represents the ith-order power coefficient of $h(x)$. The switch, during the first k_o symbol times, is connected to the information source, and, during the last r_o symbol times, to the feed-back. This cycle repeats at any period. Finally there are

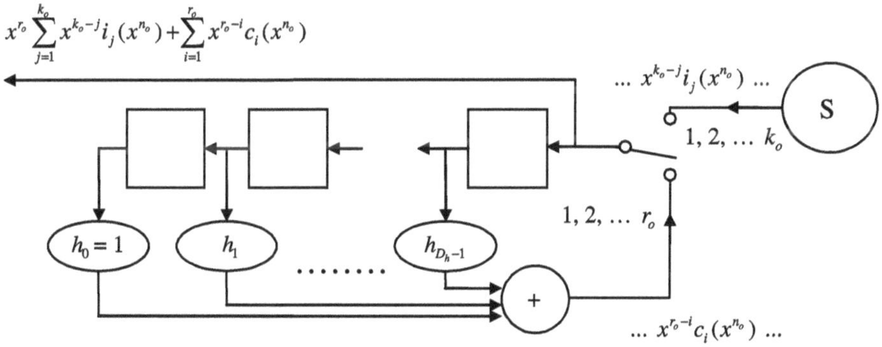

Fig. 7.33 Systematic encoder circuit based on polynomial $h(x)$, for a k_o/n_o-rate convolutional code, s.s. time-invariant in $\underline{H}$

some control symbols to be added, in order to perform a correct frame termination, depending on the particular $\underline{H}_o^T$ sub-matrix considered.

This result has been anticipated at the end of Sect. 5.15, speaking about MHEC codes, and can be better understood after the long description about frame termination developed in the last sections. From the circuital point of view, the scheme here considered is coincident with that reported in Fig. 5.6. Only the switch operates differently. We can observe that, when $r_o > 1$, the uppermost layer of the $\underline{H}_o{}^T$ sub-matrix is necessarily lower triangular. This can be a limitation in the degrees of freedom for the design of a convolutional code based on this model. In the next section, another scheme of encoder circuit will be presented, not subjected to such limitation. Furthermore there is always the alternative possibility of encoding our convolutional code by means of a circuital scheme based on its $\underline{G}$ matrix (typically this solution is more convenient for low-rate s.s. time-invariant convolutional codes).

Example 7.28 Assuming the reciprocal parity check polynomial $h_c{}^*(x)$ associated to the sequence 1101111, the semi-infinite parity check matrices shown in Fig. 7.34 can be obtained. They describe two MHEC codes with asymptotic code rate respectively 1/3 a 2/3. The correspondent semi-infinite generator sub-matrices are constructed down, after imposition of having null ex-OR sums of clusters of syndromes. In such generator matrices, two rather different sub-matrices $\underline{G}_o$ horizontally shift, by three position each time. They both are characterized by $n_o = 3$, but the first one exhibits a single row in $\underline{G}_o$, the second one exhibits two rows, and has

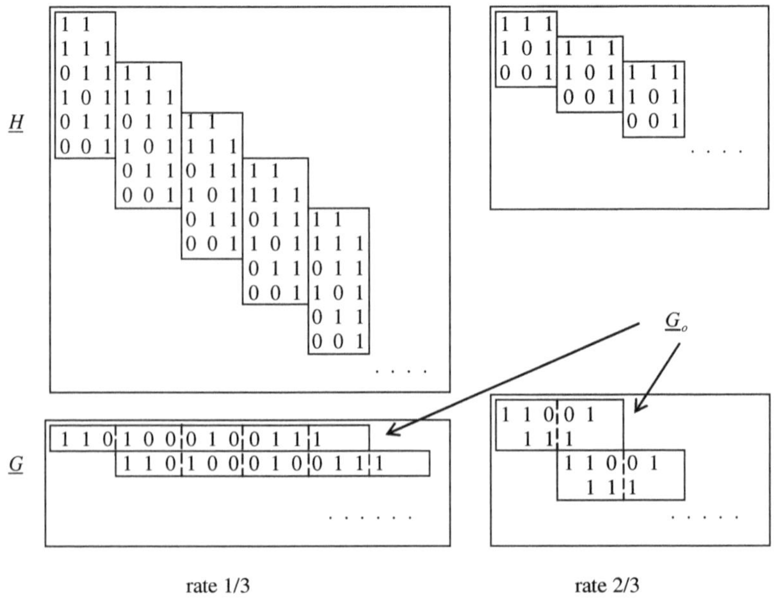

Fig. 7.34 Parity check and generator matrices for the MHEC codes described in Example 7.28

been constructed in upper triangular form. About the approximate relationship (7.5) between m and m_h, since we have $m_h = 2$ in both cases, it predicts $m = 1$ for the 2/3-rate code (exact) and $m = 4$ for the 1/3-rate code (exact too). We can finally observe that the overall generator constraint length m has not been fully exploited, since some 0-symbols are present at the end of the last period spanned by $\underline{G}_o$.

The systematic encoder circuit, based on $h_c(x)$, of the type schematically shown in Fig. 7.33, for the above pair of codes, is reported in Fig. 7.35 down. The table up in the same Figure contains low-weight coefficients in the frame weight distribution of such two codes. As for the correct frame termination, we have to observe that the 1/3-rate code requires 10 final control symbols, whereas the 2/3-rate code requires 2 only final control symbols. Such control symbols are generally produced forcing the information source to emit a sequence of symbols, taken from a proper look-up table (like in RSC encoders). The size of such table depends on the number of different states in which the encoder circuit can stop at the end of the periods, but, even in the case of the 1/3-rate code, this is not a problem in practice.

We can observe that the performance of the 1/3-rate code is rather poor, with respect to that of the best 1/3-rate convolutional codes, with the same value for parameter $m = 4$ (see Table 3.4). In particular, a minimum distance $d = 7$, although

Z	k	$A(0)$	$A(7)$	$A(10)$	$A(11)$	k	$A(0)$	$A(3)$	$A(4)$	$A(5)$
1	1	1	1			2	1	2	1	
2	2	1	2	1		4	1	5	5	2
3	3	1	3	2	1	6	1	8	12	10
4	4	1	4	4	3	8	1	11	19	24
5	5	1	5	6	6	10	1	14	26	39
6	6	1	6	8	9	12	1	17	33	54
			Z	$2Z-2$	$3Z-9$			$3Z-1$	$7Z-9$	$15Z-36$

rate 1/3 rate 2/3

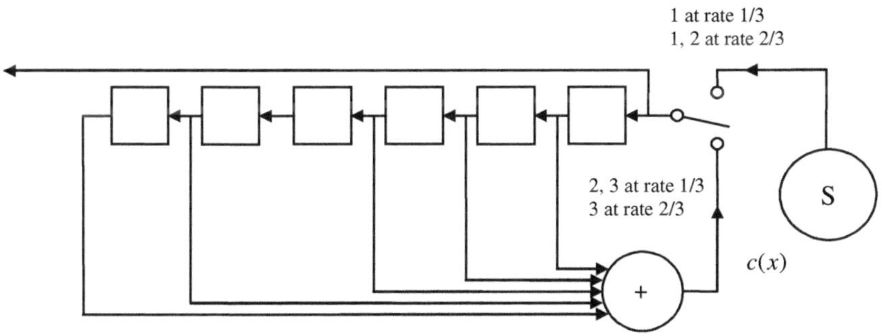

Fig. 7.35 Weight distributions for the two MHEC codes described in Fig. 7.34, and the systematic encoder circuit, based on the unique polynomial $h_c(x)$, which can be employed for both

higher weights are not smaller than 10, represents a severe performance limitation. This is due to the constraint of having a unique interleaved parity check polynomial $h_c(x)$, that is to be a low-rate MHEC code. Better performance can be achieved with two different parity check polynomials, that is employing a w.s. time-invariant convolutional code in its $\underline{H}$ matrix, as already noted. This topic will be faced in Chap. 8 with more details. Instead, the 2/3-rate code appears rather good. At this purpose, we can compare its extrapolation laws for coefficient $A(3)$ and coefficient $A(4)$ with those reported in Tables 4.2 and 4.3 for similar 2/3-rate codes ($m = 1$ and $\mu = 2$), and considering that here the generator constraint length is not fully exploited.

It could be of some interest to relate parameter m with the degrees of the generator polynomials and parameter m_h to the degrees of the parity check polynomials. It is useful to remember that, if the contrary is not explicitly said, we are treating interleaved polynomials, and not input-output polynomials typically employed in the traditional approach to convolutional codes. The following proposition holds for a w.s. time-invariant convolutional code, in the sense described by Theorem 7.1.

Lemma 7.9 *Letting* $D_g = \mathrm{Max}\{\deg[g_i(x)]\}$, *and* $D_h = \mathrm{Max}\{\deg[h_i^*(x)]\}$, *where* $g_j(x), j = 1, 2, \ldots k_o$, *are the generator polynomials, and* $h_i^*(x), i = 1, 2, \ldots r_o$, *are the reciprocal parity check polynomials, the following inequalities relate* D_g *to parameter* m, *and* D_h *to parameter* m_h,

$$(m-1)n_o + r_o + 1 \le D_g \le (m+1)n_o - 1, \tag{7.6}$$

$$(m_h-1)n_o + k_o + 1 \le D_h \le (m_h+1)n_o - 1. \tag{7.7}$$

Proof In Fig. 7.36 up, assuming that $\underline{G}_o$ is upper triangular, the conditions for having the shortest and the longest generator polynomial are reported. They correspond respectively to the lower and to the upper bound in (7.6). In Fig. 7.36 down, the uppermost layer of $\underline{H}_o^T$ being lower triangular, the conditions for having the shortest and the longest reciprocal parity check polynomial are described in analogous way. They correspond respectively to the lower and to the upper bound in (7.7). In all these considerations, we have adopted that $\underline{G}_o$ spans $(m + 1)$ consecutive periods, and $\underline{H}_o^T$ spans $(m_h + 1)$ consecutive row layers. □

Removing the assumptions that $\underline{G}_o$ is upper triangular and that the uppermost layer of $\underline{H}_o^T$ is lower triangular, (7.6) and (7.7) can be a little widened, but some drawbacks may arise. First of all, inversion is not guaranteed and must be checked (see Sect. 4.3). The minimal base encoder circuit, leading to the trellis with lowest computational complexity, is to be searched (see Definition 4.7). It can be constructed by shift registers not completely in controller or observer arrangement, often non-systematic. Since the benefit in removing the above assumptions are not

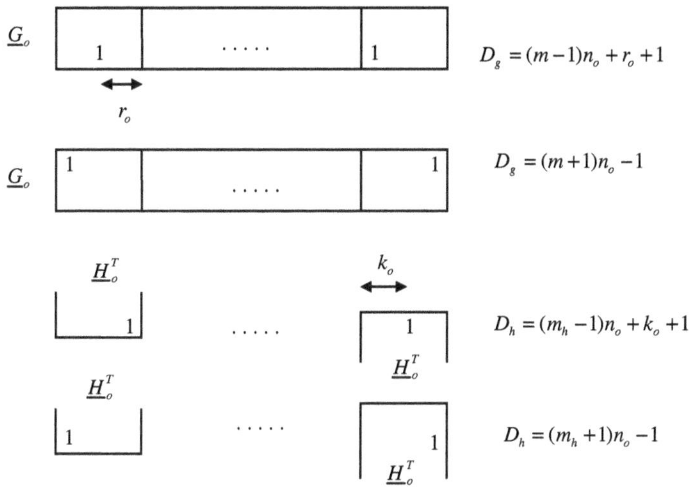

Fig. 7.36 Conditions for having the shortest and the longest polynomials

significant, we will continue the discussion of (7.6) and (7.7). Keeping in mind that m and m_h represent cost parameters, in terms of computational complexity of the soft-decision decoding procedure performed in the respective trellises, the best results are obtained when the upper bound of (7.6) and (7.7) are approached. The most efficient use of the shift register is, say, as follows: for an asymptotic code rate $1/n_o$ with an s.s. time-invariant $\underline{G}$ (according to Lemma 7.3), and for an asymptotic code rate $(n_o - 1)/n_o$ with an s.s. time-invariant $\underline{H}$ (according to Lemma 7.2). In these two cases it is convenient to adopt the choice $D_g = (m+1)n_o - 1$ and $D_h = (m_h + 1)n_o - 1$ respectively, so exploiting completely all the constraint lengths. Except in the last situation, when it becomes trivial, any possible transformation of (7.5) into a direct relationship between D_g and D_h becomes practically useless, because of the uncertainties in (7.5) and the variability in (7.6) or (7.7).

Example 7.29 Let us consider again the two MHEC codes, both described by the same reciprocal parity check polynomial $h_c{}^*(x)$, associated to the sequence 1101111, already treated in Example 7.28, to which (7.6) and (7.7) will be applied now. The one having code rate 1/3 is characterized by the following bounds: $12 \le D_g \le 14$, reaching only $D_g = 12$, and $5 \le D_h \le 8$, obtaining $D_h = 6$. The one having code rate 2/3 is characterized by the following bounds: $6 \le D_h \le 8$, reaching only $D_h = 6$, and $2 \le D_g \le 5$, obtaining $D_g = 4$. In spite of the fact that the last result is not optimum in shift register length exploitation, the trade-off between performance and costs is good, taking into account either decoding computational complexity or the number of final control symbols for the correct frame termination.

7.8 Another Type of Systematic Encoder Circuit for s.s. Time-Invariant Convolutional Codes in Their $\underline{H}$ Matrix

The RSC encoder in observer arrangement, shown in Fig. 4.6 down, can be interpreted as an encoder circuit based on the parity check matrix, although it was constructed starting from the generator matrix. This observation derives from uniqueness property for the sub-matrix $\underline{E}$ present both in the $\underline{G}$ and in the $\underline{H}$ matrix, except for a transposition, when they are expressed in systematic form (Corollary a of Theorem 5.1). The only care to be taken here is to consider symbolic rational polynomial matrices (see Definition 3.13). In a very general case (also for w.s. time-invariant convolutional codes in both $\underline{H}$ and $\underline{G}$), we can use the same polynomial sub-matrix $\underline{E}(x)$.

Lemma 7.10 *Given a w.s. time-invariant convolutional code, with parameters* $n_o = k_o + r_o$, *assuming its RSC polynomial generator sub-matrix in the form*

$$\underline{G}_o^{(RSC)}(x) = [\underline{I}_{k_o} | \underline{E}(x)], \tag{7.8}$$

where $\underline{E}(x)$ *is a* $k_o \times r_o$ *polynomial sub-matrix and* $\underline{I}_{k_o}$ *is a* $k_o \times k_o$ *identity matrix, it is possible to obtain the following symbolic expression of a systematic syndrome former sub-matrix*

$$\underline{H}_o^{T(RSC)}(x) = \begin{bmatrix} \underline{E}(x) \\ \underline{I}_{r_o} \end{bmatrix}, \tag{7.9}$$

where $\underline{I}_{r_o}$ *is an* $r_o \times r_o$ *identity matrix.*

Proof We can proceed as for the proof of Corollary a of Theorem 5.1, with the only difference of considering here proper symbolic polynomial matrices. □

Example 7.30 The 2/3-rate convolutional code characterized by $m = 1$ and $\mu = 2$, introduced in Example 4.2 and whose RSC symbolic polynomial generator sub-matrix has been obtained in Example 4.8, is w.s. time-invariant in $\underline{G}$ and is s.s. time invariant in $\underline{H}$. The last property follows from Lemma 7.2, in fact it is characterized by a code-rate of the type $(n_o - 1)/n_o$. The generator sub-matrix and the syndrome former sub-matrix describing this code, both in symbolic rational polynomial forms, are reported in Fig. 7.37.

The encoder circuit obtained in Fig. 4.6 down, for the code treated in Examples 4.2, 4.8, 4.10, and 7.30, can be better interpreted now as associated to a parity check matrix (and not to a generator matrix). It is straightforward to verify that the well known property (5.4) is satisfied here, giving

$$\underline{G}_o^{(RSC)}(x) = \begin{bmatrix} 1 & 0 & \frac{x^2}{x^2+x+1} \\ 0 & 1 & \frac{x^2+1}{x^2+x+1} \end{bmatrix} \qquad \underline{H}_o^{T(RSC)}(x) = \begin{bmatrix} \frac{x^2}{x^2+x+1} \\ \frac{x^2+1}{x^2+x+1} \\ 1 \end{bmatrix}$$

Fig. 7.37 Sub-matrix $\underline{G}_o^{(RSC)}(x)$ and sub-matrix $\underline{H}_o^{T(RSC)}(x)$ for the 2/3-rate convolutional code treated in Examples 4.2, 4.8, 4.10 and 7.30

$$\underline{G}_o^{(RSC)}(x) \cdot \underline{H}_o^{T(RSC)}(x) = \underline{0}. \tag{7.10}$$

where $\underline{0}$ is, in general, a $k_o \times r_o$ null matrix. Furthermore, transposition yields

$$\underline{H}_o^{(RSC)} = \left[\frac{h^{(1,1)}(x)}{h^{(3,1)}(x)} \quad \frac{h^{(2,1)}(x)}{h^{(3,1)}(x)} \quad 1 \right], \tag{7.11}$$

where $h^{(i,j)}(x)$, $i = 1, 2, 3$, $j = 1, 2$, are suitable input-output parity check polynomials.

In order to understand the structure of an encoder circuit in observer arrangement, and in particular the roles of the wires feeding the ex-OR sums, it can be useful to start from an $(n_o - 1)/n_o$-rate systematic convolutional code (in the sense of Definition 5.2, that is with control symbols not depending directly on other control symbols). In this case we have an $\underline{H}_o^T$ sub-matrix with columns associated to the control symbol positions (the n_oth positions in any period), with just one 1-symbol placed at the top. In Fig. 7.38 this situation is depicted for the case of $m_h = 2$ and $n_o = 3$. Clearly we have adopted the following notation, respectively for the symbolic polynomial and the true syndrome former sub-matrix,

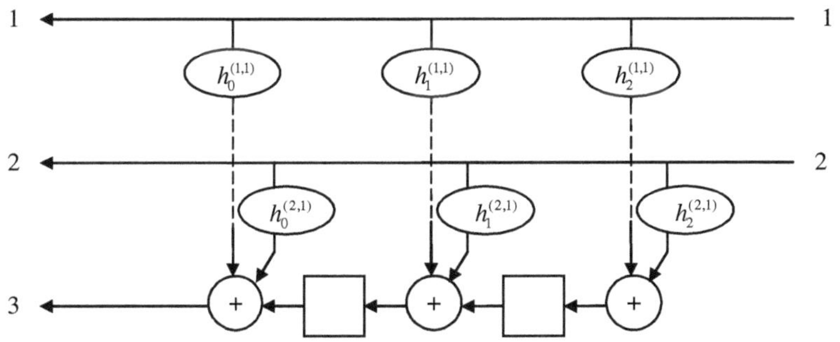

Fig. 7.38 General form for the encoder circuit in observer arrangement in the case of a systematic code with code rate 2/3

$$\underline{H}_o^T(x) = \begin{bmatrix} h^{(1,1)}(x) \\ h^{(2,1)}(x) \\ 1 \end{bmatrix}, \quad \underline{H}_o^T = \begin{bmatrix} h_0^{(1,1)} & h_0^{(2,1)} & 1 \\ h_1^{(1,1)} & h_1^{(2,1)} & 0 \\ h_2^{(1,1)} & h_2^{(2,1)} & 0 \end{bmatrix}.$$

A justification for coefficient disposition in the true $\underline{H}_o^T$ can be given observing that, adopting it, a row in the parity check matrix, associated to $h^*(x)$, turns out to be $[h_2^{(1,1)}\ h_2^{(2,1)}\ 0\ h_1^{(1,1)}\ h_1^{(2,1)}\ 0\ h_0^{(1,1)}\ h_0^{(2,1)}\ 1]$.

We have assumed the same convention as in Chap. 4 for labelling input and output wires. Subscripts regard subsequent periods, the first superscript refers to the output wire (or the position in the period), the second superscript is always 1, since we have only one control symbol in any period, placed in the last position.

Example 7.31 The code introduced in Example 7.8, whose parity check and generator matrix are shown in Fig. 7.9, is systematic in its representation by means of a parity check matrix. Its encoder circuit, according to the scheme shown in Fig. 7.38, is reported in Fig. 7.39. This encoder circuit is non-recursive. We can observe how sub-matrix $\underline{H}_o^T$, properly transposed, appears directly in the connection wires.

The following relationship holds between the symbolic polynomials $h^{(1,1)}(x)$, $h^{(2,1)}(x)$ and the unique parity check polynomial, whose reciprocal shifts in the rows of $\underline{H}$,

$$h(x) = x^2 1 + xh^{(2,1)}(x^3) + h^{(1,1)}(x^3), \quad \text{(systematic code).} \tag{7.12}$$

This situation resembles the one described by condition (3.6) for a $1/n_o$-rate systematic code in the representation based on its $\underline{G}$ matrix. Expression (7.12) allows to conclude that there is a third symbolic polynomial, which, in this description of a systematic code, degenerates into 1. A more general case, still for $n_o = 3$, is that characterized by the expression

$$h(x) = x^2 h^{(3,1)}(x^3) + xh^{(2,1)}(x^3) + h^{(1,1)}(x^3), \tag{7.13}$$

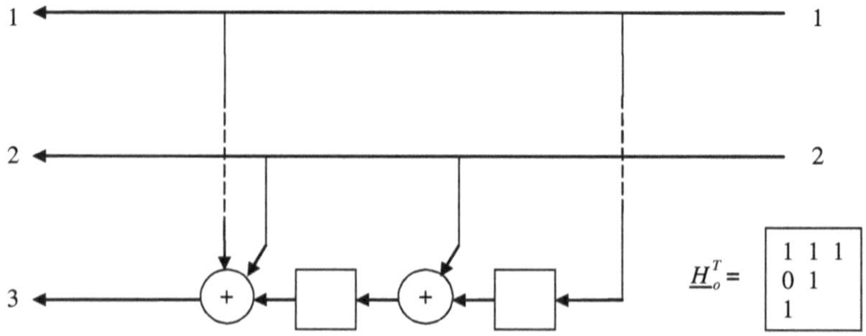

Fig. 7.39 Encoder circuit in observer arrangement for the systematic code treated in Examples 7.8 and 7.31

where we have introduced a third symbolic polynomial $h^{(3,1)}(x)$ for taking into account that the present control symbol can be function also of some preceding control symbols. Expression (7.13) is associated to the symbolic polynomial syndrome former sub-matrix

$$\underline{H}_o^T(x) = \begin{bmatrix} h^{(1,1)}(x) \\ h^{(2,1)}(x) \\ h^{(3,1)}(x) \end{bmatrix} = \begin{bmatrix} h^{(1)}(x) \\ h^{(2)}(x) \\ h^{(3)}(x) \end{bmatrix}. \tag{7.14}$$

where, in the latter expression, we have eliminated the second superscript, for simplifying notation.

Lemma 7.11 *When the original $(n_o - 1)/n_o$-rate convolutional code is non-systematic, in order to obtain a systematic equivalent of such code, it is necessary to adopt a recursive encoder circuit, in which the output of the wire associated to the control symbol (the n_oth symbol in the period) is properly taken for driving a feedback ex-OR sum.*

Proof Similarly to what occurs for the generator matrix of a 1/3-rate s.s. time-invariant convolutional code in its $\underline{G}$ matrix (see Lemma 3.12), where conventionally we divide by $g_1(x)$, here, conventionally as well, we will divide by $h^3(x)$, and obtain

$$\underline{H}_o^{T(RSC)}(x) = \begin{bmatrix} \frac{h^{(1)}(x)}{h^{(3)}(x)} \\ \frac{h^{(2)}(x)}{h^{(3)}(x)} \\ \\ 1 \end{bmatrix}, \tag{7.15}$$

or equivalently

$$\underline{H}_o^{(RSC)}(x) = \begin{bmatrix} \frac{h^{(1)}(x)}{h^{(3)}(x)} & \frac{h^{(2)}(x)}{h^{(3)}(x)} & 1 \end{bmatrix}. \tag{7.16}$$

So we can encode the RSC version of our code by means of a recursive encoder circuit, with just one shift register, in observer arrangement. See, for instance, the $\underline{H}_o^{T\,(RSC)}(x)$ sub-matrix shown on the right in Fig. 7.37, for which it is possible to employ the recursive encoder circuit in Fig. 4.6 down. Finally, in comparison with expression (4.9), we can find the following correspondences with the polynomials of the generator matrix description: $D(x) = h^{(3)}(x)$, $g^{(1,3)\prime}(x) = h^{(1)}(x)$, $g^{(2,3)\prime}(x) = h^{(2)}(x)$. The case characterized by whichever n_o can be easily obtained by proper generalization of the one treated so far, in particular, for $n_o = 3$. □

Definition 7.7 Taking a convolutional code characterized by code rate $(n_o - 1)/n_o$, a polynomial matrix, with n_o rows and just one column, describes the *circuit for*

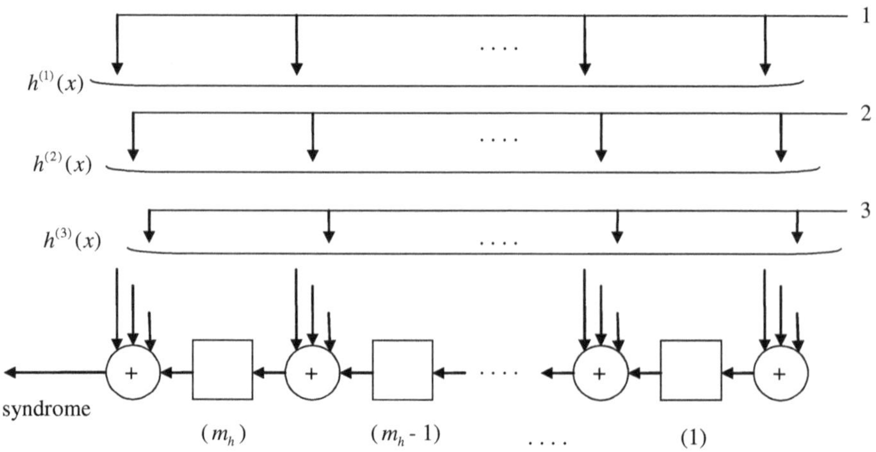

Fig. 7.40 Circuit for syndrome calculation in the case of a 2/3-rate convolutional code

syndrome calculation, shown in Fig. 7.40, for $n_o = 3$. In it m_h represents the maximum degree of polynomials $h^{(i)}(x)$, $i = 1, 2, 3$.

A circuit of this type could be employed in the framework of a hard-decision decoding algorithm (MacWilliams and Sloane 1977), at the destination of our coded symbol sequence. Sometimes it is used for immediately checking if a received frame is correct or not. Nevertheless its importance is especially in the possibility of generalizing what here discussed to the case characterized by whichever pair of parameters k_o and r_o (Johannesson and Zigangirov 1999; Ryan and Lin 2009). The meaning of input-output parity check polynomials for the $h^{(i,j)}(x)$'s is clearly referred to the circuit for syndrome calculation, which, in general, exhibits n_o input wires and r_o output wires. Extension of the scheme shown in Fig. 7.40 to the case of more than one syndrome per period will be treated in Chap. 8.

In principle, a scheme of the type described in Sect. 4.3 and here is more general than that based on MLC or MHEC equivalence (respectively Figs. 1.5 and 7.33). In fact it is valid also for a $\underline{G}_o$ sub-matrix not upper triangular, as well as for an $\underline{H}_o^T$ sub-matrix whose uppermost layer is not lower triangular. As already noted, the case of $(n_o - 1)/n_o$-rate convolutional codes is characterized by an $\underline{H}$ matrix description which is very similar to the case of $1/n_o$-rate convolutional codes described by means of a $\underline{G}$ matrix. The encoder circuits, in both cases, have only one shift register, in observer arrangement for the first, in controller arrangement for the second. Nevertheless, as long as (7.5) is exact, the computational complexity is balanced (see, for instance, Example 4.10 on the case of a 2/3-rate code). The requirement of a systematic encoding can change the situation (see Fig. 4.6). A detailed method for designing encoders with minimum number of memory elements can be found in Johannesson and Zigangirov (1999). Sometimes, to the purpose of simplifying the decoding procedure, it is possible to adopt a systematic encoder and a non-systematic model (trellis) in decoding, only the latter being in minimal form. Nevertheless this is impossible in some code concatenation schemes. A comparison comprehending also the solution based on the scheme of Fig. 7.33, where the input switch allows to have only binary branch complexity, just for a situation characterized by code rate 2/3, will be carried out in Sect. 7.10.

7.9 Tail-Biting Convolutional Codes s.s. Time-Invariant in $\underline{H}$

In Sect. 3.10 s.s. time-invariant convolutional codes in the $\underline{G}$ matrix have been treated, dealing with the possible tail-biting organization of their frames. In Sect. 4.7 this definition has been widened in order to take into account w.s. time-invariant convolutional codes still with respect to their $\underline{G}$ matrix. Here we will consider the tail-biting arrangement for s.s. time-invariant convolutional codes in their $\underline{H}$ matrix.

Definition 7.8 An s.s. time-invariant convolutional code in its $\underline{H}$ matrix, whose first $r_o m_h$ interleaved reciprocal parity check polynomials $h^*(x)$ are wrapped around over the frame length n_oZ, in such a way as to produce an $r_oZ \times n_oZ$ overall $\underline{H}$ matrix of the type shown in Fig. 7.41, assuming $r_o = 2$, $m_h = 2$, is called *tail-biting over the frame length n_oZ*.

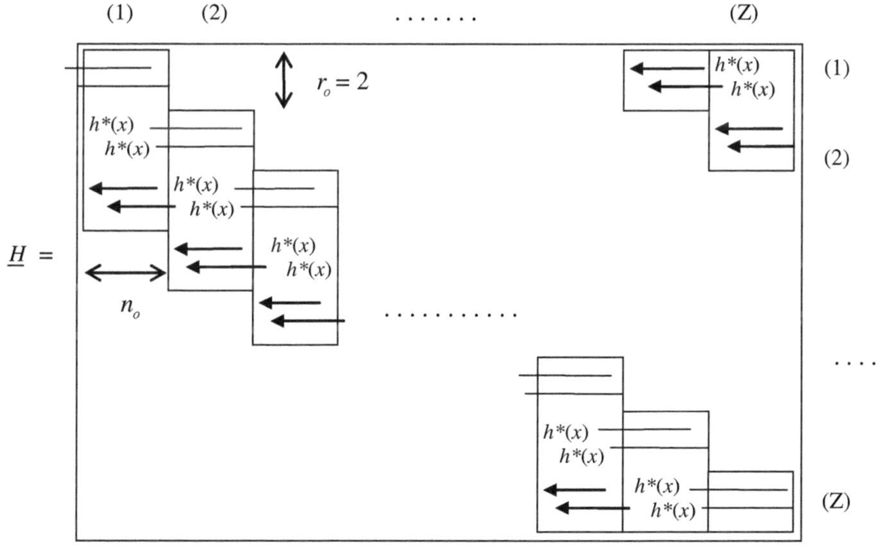

Fig. 7.41 Overall $\underline{H}$ matrix of an $(n_o - 2)/n_o$-rate tail-biting convolutional code, s.s. time-invariant in $\underline{H}$ for $m_h = 2$

In the figure, we have adopted $D_h = 3n_o - 2$, maintaining n_o as a free parameter. It is useful to compare Definition 7.8 with Definition 3.15. Here r_o replaces k_o and m_h replaces m. Furthermore the degree $D_h = \deg[h^*(x)]$ would replace the degree $D_g = \deg[g(x)] = r$, in turn related to m by the equivalence relationship (3.15) regarding MLC codes, although the last two parameters are not present in the two Definitions. A comparison between Figs. 3.22 and 7.41 clearly shows another important difference: the transformation is here made along columns, whereas in the generator matrix description it was made along rows (in Definition 7.8 the first rows are wrapped around and not the last rows, like in Definition 3.15). This difference follows from the fact that, in the present treatment, the basic element is the syndrome former sub-matrix $\underline{H}_o^T$, and not the generator sub-matrix $\underline{G}_o$. About $\underline{H}_o^T$ we can note that it has been adopted with an uppermost layer which is lower triangular, for the sake of simplicity, but this is not a restrictive choice in general. Clearly, being equal, such two polynomials cannot be here exactly superimposed.

In Definition 4.8, a tail-biting arrangement has been introduced for convolutional codes w.s. time-invariant in their $\underline{G}$ matrix. This Definition, supported by Fig. 4.8, will be useful here, although now we have a unique polynomial $h^*(x)$, as shown in Fig. 7.41. The purpose of the following considerations is to imagine which structure is that characterizing a $\underline{G}$ matrix valid for satisfying (5.4) together with the $\underline{H}$ matrix in Fig. 7.41, for whichever n_o. In order to develop a more straightforward discussion, we will assume, for the moment, that $\underline{H}_o^T$ is that in minimal form for the considered code, which is also well designed. Before wrapping around, it is possible to derive the $\underline{G}_o$ sub-matrix of such code by imposing null ex-OR sum of clusters of syndromes to a certain number of subsequent $\underline{H}_o^T$. In general, for $n_o > 3$, that is $k_o > 1$, $\underline{G}_o$ will exhibit different interleaved generator polynomials $g_{\text{int}}^{(i)}(x)$, leading to a w.s. time-invariant convolutional code. It is not restrictive to assume that this $\underline{G}_o$ sub-matrix is upper triangular, and that it is obtained with the minimum value possible for m. Starting from such $\underline{G}_o$ sub-matrix, we can obtain back the initial $\underline{H}_o^T$ by means of the column construction of the syndrome former sub-matrix (Definition 7.6), still before wrapping around. Once verified all the above conditions, $\underline{G}_o$ will be named the generator sub-matrix correspondent to $\underline{H}_o^T$.

Lemma 7.12 *Let us consider the convolutional code, s.s. time-invariant in its $\underline{H}$ matrix, with a unique reciprocal parity check polynomial $h^*(x)$, forming a proper $\underline{H}_o^T$ sub-matrix. Let $\underline{G}_o$ be the correspondent generator sub-matrix, typically made by different interleaved generator polynomials (being this code in general w.s. time-invariant in $\underline{G}$). The two overall matrices, when arranged in tail-biting form, respectively according to* Definitions 7.8 and 4.8, *satisfy the general property* (5.4) *between the generator matrix and the parity check matrix of a same code.*

Proof In the original code frame, if correctly terminated, we have a complete $\underline{G}$ matrix and a complete $\underline{H}$ matrix satisfying equality $\underline{G} \cdot \underline{H}^T = \underline{0}$. Now we start from an $\underline{H}$ matrix properly wrapped around after Definition 7.8. Application of the procedure based on null ex-OR sum of clusters of syndromes, we can observe what follows. Cyclic permutations with steps of n_o in polynomial $h^*(x)$, involving m_h

Fig. 7.42 Tail-biting generator and parity check matrix for the 1/3-rate code described in Example 7.32

$$\underline{H} = \begin{array}{ccccccccccccccc}
1&1& & & & & & & & & & &1&0&0\\
0&1&1& & & & & & & & & &0&1&0\\
1&0&0&1&1& & & & & & & & & & \\
0&1&0&0&1&1& & & & & & & & & \\
 & & &1&0&0&1&1& & & & & & & \\
 & & &0&1&0&0&1&1& & & & & & \\
 & & & & & &1&0&0&1&1& & & & \\
 & & & & & &0&1&0&0&1&1& & & \\
 & & & & & & & & &1&0&0&1&1& \\
 & & & & & & & & &0&1&0&0&1&1
\end{array}$$

$$\underline{G} = \begin{array}{ccccccccccccccc}
1&1&1&0&1&0&0&0&1& & & & & & \\
 & & &1&1&1&0&1&0&0&0&1& & & \\
 & & & & & &1&1&1&0&1&0&0&0&1\\
0&0&1& & & & & & &1&1&1&0&1&0\\
0&1&0&0&0&1& & & & & & &1&1&1
\end{array}$$

sub-matrices $\underline{H}_o^T$, imply cyclic permutations with steps of n_o in all the interleaved generator polynomials $g_{\rm int}^{(i)}(x)$, present in $\underline{G}_o$, involving all the m periods characterizing $\underline{G}_o$ itself. In all these situations, $\underline{G} \cdot \underline{H}^T = \underline{0}$ is satisfied. Conversely, starting from a $\underline{G}$ matrix properly wrapped around after Definition 4.8, the same result is obtained after application of the column construction of any type of rows in a parity check matrix, that is taking into account also the variant of this procedure, suitable for periodic polynomials, introduced in Sect. 7.6. □

Example 7.32 Let us consider a 1/3-rate convolutional code, s.s. time-invariant in its $\underline{H}$ matrix, characterized by the unique reciprocal parity check polynomial $h^*(x) = x^4 + x + 1$. Its tail-biting arrangement is shown in Fig. 7.42, over $Z = 5$ periods. The parity check matrix, reported up, is characterized by $m_h = 1$. The generator matrix, obtained by null ex-OR sum of clusters of syndromes, reported down, is characterized by $m = 2$. Either $\underline{G}$ or $\underline{H}$ have all linearly independent rows. The overall code is represented by the tern [15,5,5], since the value adopted for Z is high enough to reach the minimum distance $d = 5$, which holds also for the not tail-biting arrangement. The code rate is 1/3, as expected.

A particular situation is that of not well designed convolutional codes, not included in Lemma 7.12. As already noted in Sects. 3.10 and 4.7, the tail-biting forms of such codes show a reduced number of linearly independent rows in $\underline{G}$. Clearly this is due to the anomalous periodic rows present in their parity check matrices. Two Examples will be discussed now, both characterized by code rate equal to 2/3. The first one regards a well designed convolutional code, the second one a not well designed convolutional code.

Example 7.33 Let us consider the 2/3-rate convolutional code introduced in Example 7.28, whose $\underline{H}$ and $\underline{G}$ matrix are shown in Fig. 7.34 on the right. This code is well designed. Its tail-biting version over $Z = 6$ periods exhibits the $\underline{H}$ and

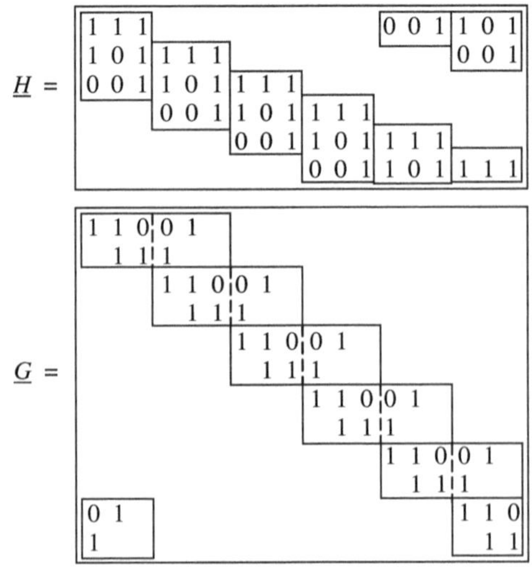

Fig. 7.43 $\underline{H}$ and $\underline{G}$ matrix for the tail-biting 2/3-rate convolutional code described in Example 7.33

$\underline{G}$ matrix shown in Fig. 7.43. The resulting code can be expressed by the tern [18,12,3] and the code rate turns out to be exactly 2/3.

As anticipated, if the considered convolutional code is not well designed, the $\underline{G}$ matrix of its tail-biting version exhibits at least one row linearly dependent on the other ones. In the $\underline{H}$ matrix the rows where a periodic polynomial is present will remain unchanged, because they already satisfy a cyclic permutation at steps of n_o. This fact contributes to maintain a correct number of information and parity check symbols.

Example 7.34 In Examples 4.4 and 4.17, we considered a 2/3-rate convolutional code characterized by $g_{\text{int}}^{(1)}(x) = 111.011$ and $g_{\text{int}}^{(2)}(x) = 010.110$. It appeared not well designed regarding its weight distribution coefficient $A(d + 1) = A(4)$.

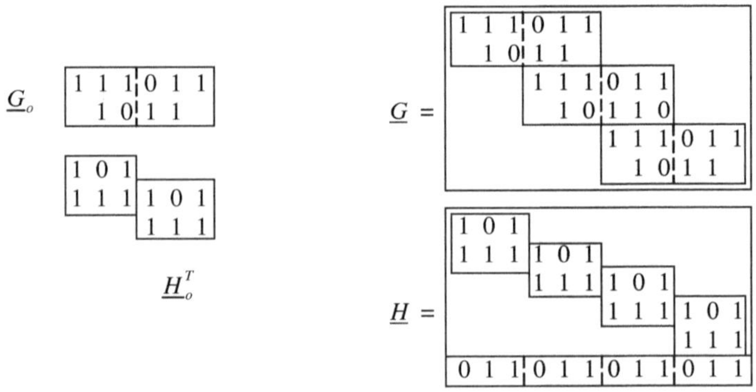

Fig. 7.44 Construction of the $\underline{H}$ matrix for the 2/3-rate not well designed convolutional code described in Example 7.34

Fig. 7.45 Tail-biting form of the code presented in Fig. 7.44

$$\underline{G} = \left[\begin{array}{cccccccccccc}
1 & 1 & 1 & 0 & 1 & 1 & & & & & & \\
 & 1 & 0 & 1 & 1 & & & & & & & \\
 & & & 1 & 1 & 1 & 0 & 1 & 1 & & & \\
 & & & & 1 & 0 & 1 & 1 & & & & \\
 & & & & & & 1 & 1 & 1 & 0 & 1 & 1 \\
 & & & & & & & 1 & 0 & 1 & 1 & \\
0 & 1 & 1 & & & & & & & 1 & 1 & 1
\end{array}\right]$$

$$\underline{H} = \left[\begin{array}{cccccccccccc}
1 & 0 & 1 & & & & & & & 1 & 1 & 1 \\
1 & 1 & 1 & 1 & 0 & 1 & & & & & & \\
 & & & 1 & 1 & 1 & 1 & 0 & 1 & & & \\
 & & & & & & 1 & 1 & 1 & 1 & 0 & 1 \\
0 & 1 & 1 & 0 & 1 & 1 & 0 & 1 & 1 & 0 & 1 & 1
\end{array}\right]$$

In Fig. 7.44 on the left, its unique non-periodic parity check polynomial is obtained, by means of the column construction procedure for calculating $\underline{H}_o^T$. Then a periodic reciprocal parity check polynomial can be found, by means of the same procedure, in the variant described in Sect. 7.6, which starts setting 1-symbols at the same time in more than one period. Such periodic reciprocal parity check polynomial turns out to be 011.011.011. … . The resulting $\underline{G}$ and $\underline{H}$ matrix for this convolutional code, over $Z' = 4$ overall periods, are reported in Fig. 7.44 on the right. Having a frame length $n = 12$, correctly terminated, we can conclude that the code so obtained is represented by the tern [12,6,3].

In Fig. 7.45 the tail-biting version of the code considered so far is constructed, over the same number of periods $Z = 4$. We can observe that the periodic parity check row remains unchanged in the $\underline{H}$ matrix, but the last row of the old $\underline{G}$ matrix, after wrapping around, is no longer present. It has been eliminated because linearly dependent on the previous rows. For this reason, the overall code becomes [12,7,3], instead of [12,8,3] (see also the evolution of the weight distribution of such tail-biting code, reported in Table 4.5). However this penalty in the code rate appears weaker and weaker with increasing Z.

As a general comment about Examples 7.33 and 7.34, it is possible to note a resemblance between the structure of a tail-biting convolutional code and that of a QC code, in its version with distributed control symbols, as expected from Theorem 4.1, but only when the code is well designed. The higher the number of anomalous periodic rows in $\underline{H}$, the larger the amount of rows in $\underline{G}$ to be eliminated after wrapping around. Just as a curiosity: the tail-biting version of a direct product code (interpreted as a not well designed convolutional code) cannot be made at all, or, better, it coincides with the original code. In fact all the rows of its $\underline{G}$ matrix after wrapping around are linearly dependent on the previous ones. Analogously, if we look at the $\underline{H}$ matrix, its non-periodic rows form $\underline{H}_o^T$ sub-matrices showing $m_h = 0$, and hence cannot be vertically subdivided into layers.

On the contrary, if the $\underline{H}$ matrix of our convolutional code is not in minimal form, a tail-biting version of it can be constructed without particular cares. Before wrapping around, such matrix showed a certain number of rows not linearly independent, the same number will remain after. This conclusion is important for LDPC convolutional codes, without 4-length short cycles of 1-symbols (Ryan and Lin 2009). It is possible to verify, in fact, that if such short cycles are not present in the original convolutional code, certainly they do not appear after wrapping around, as will be better explained in Chaps. 10–12.

7.10 Decoding Computational Complexity in the Trellis for s.s. Time-Invariant Convolutional Codes in Their $\underline{H}$ Matrix

Remaining in the framework of the soft-decision decoding procedures based on a trellis, it could be interesting to compare all the possible state diagrams associated to as many encoder circuits introduced so far, for the same convolutional code. In particular, we have the schemes in which, owing to an input switch, the branch complexity is 2 (those reported in Figs. 1.5 and in Fig. 7.33, constructed respectively on $\underline{G}$ or on $\underline{H}$), and the traditional schemes without input switch, for which the branch complexity is 2^{k_o}, independently of the choice of shift registers in controller arrangement (according to $\underline{G}$) or in observer arrangement (according to $\underline{H}$). About the last specification, it is useful to remember that, with the controller arrangement, the number of shift registers is k_o, with the observer arrangement this number is r_o.

First of all, we consider a very simple 1/2-rate convolutional code, in order to compare the trellises obtained with the encoder circuit reported respectively in Fig. 1.5 (assumed for its MLC equivalent), which is based on polynomial $g(x)$, and in Fig. 7.33 (assumed for its MHEC equivalent), which is based on polynomial $h(x)$. Traditional encoder schemes will be also considered as a reference.

Example 7.35 Assuming the 1/2-rate convolutional code characterized by interleaved generator polynomial $g(x) = x^2 + x + 1$ and by $h^*(x) = x^2 + x + 1$ (equal after Lemma 7.4), the trellis associated to an encoder circuit of the type reported in Fig. 1.5, owing to the selection in the merge states, exhibit 2^1 merge states per stage. The same number of states appears in the trellis based on the traditional encoder in controller arrangement (see Figs. 3.2 and 3.3) equivalent to it. The trellis associated to an encoder circuit, based on $h(x)$, of the type depicted in Fig. 7.33, for this code, is shown in Fig. 7.46. The first six periods have been described, with symbol times labelled from 1 to 7. In correspondence with odd symbol times we have a bifurcation in output from any state. In correspondence with even symbol times we have a transition state. After a brief transient, the trellis reaches a steady state behaviour, characterized by merging states due to the control symbols. The number of merge state is 2^2 per stage, twice the number which characterizes the two trellises based on the $\underline{G}$ matrix. Also an encoder circuit in observer arrangement, of the type described in Sect. 7.8, in non-systematic version, exhibits just one memory element, and hence the number of states per stage is 2^1. In conclusion, the adoption

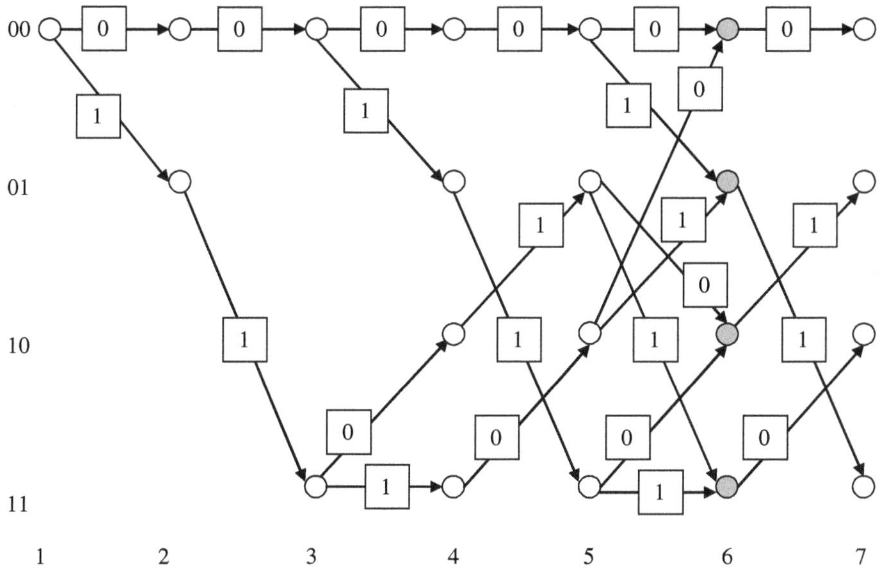

Fig. 7.46 Trellis associated to the encoder circuit based on a unique parity check polynomial (encoder circuit in Fig. 7.33) for the code treated in Example 7.35

of the trellis originated by an encoder circuit of the type depicted in Fig. 7.33 appears not convenient for what concerns overall decoding computational complexity.

Example 7.36 Let us take the 2/3-rate convolutional code characterized by $m = 1$ and $\mu = 2$, introduced in Example 4.2 and several times discussed so far. It is described by the following $\underline{G}_o$ and $\underline{H}_o^T$ sub-matrix

$$\underline{G}_o = \left[\begin{array}{ccc|ccc} 1 & 0 & 1 & 1 & 1 & 1 \\ 0 & 1 & 1 & 1 & 0 & 0 \end{array}\right], \quad \underline{H}_o^T = \begin{bmatrix} 1 & 1 & 1 \\ 0 & 0 & 1 \\ 0 & 1 & 1 \end{bmatrix}.$$

Its state diagram, associated to the $\underline{G}$ matrix, in traditional non-systematic controller arrangement, has been depicted in Fig. 4.5. It is possible to imagine how the associated trellis is made. There are 2^2 states per stage, and 4-fold multiplicity either in input or in output branches. On the other hand, the encoder circuit based on a unique parity check polynomial (see Fig. 7.33) exhibits here 7 memory elements. Therefore also in this case a decoding procedure based on a trellis constructed on $h(x)$ is not advisable. As for the encoder circuit in observer arrangement, of the type described in Sect. 7.8, it exhibits $m_h = 2$ memory elements. Therefore the number of states per stage is 2^2, exactly equal to the number we have calculated before, from the state diagram in Fig. 4.5. This compensation of performance is due, as already noted, to an equilibrium between the parameters m and m_h in (7.5).

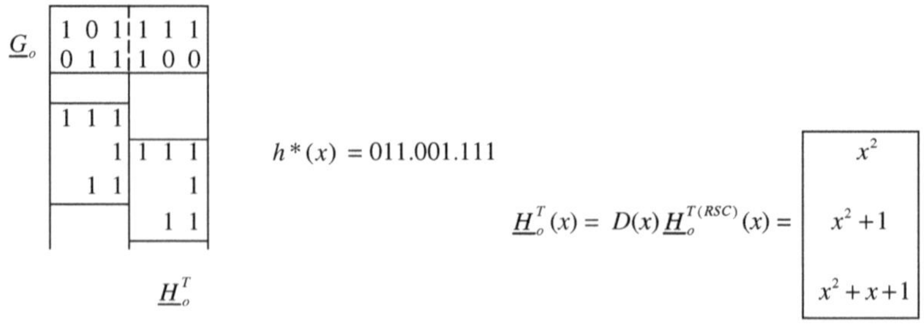

Fig. 7.47 Construction of the true $\underline{H}_o^T$ sub-matrix and of the correspondent symbolic $\underline{H}_o^T(x)$ for the convolutional code treated in Examples 4.2, 4.8, 7.30, 7.36 and 7.37

Example 7.37 About the 2/3-rate code considered in the previous Example, characterized by $m = 1, \mu = 2$, or by $m_h = 2$, the following observations can be added. In Fig. 7.47 on the left, we obtain the true $\underline{H}_o^T$ sub-matrix by means of the column construction procedure. The unique parity check polynomial $h(x)$ turns out to be 111.100.110. Therefore, after (7.13), we can write

$$h^{(3)}(x) = x^2 + x + 1, \quad h^{(2)}(x) = x^2 + 1, \quad h^{(1)}(x) = x^2,$$

leading to the symbolic polynomial sub-matrix $\underline{H}_o^T(x)$, reported in Fig. 7.47 on the right. This matrix has been denoted as $\underline{H}_o^{T\,(RSC)}(x)$ multiplied by polynomial $D(x) = x^2 + x + 1$, which, in turn, is the determinant of the 2×2 left hand sub-matrix in $\underline{G}_o(x)$, confirming that the two symbolic systematic matrices reported in Fig. 7.37 are a valid couple.

The systematic encoder circuit, in observer arrangement, associated to this code has been already shown in Fig. 4.6 down. Such code can be encoded also by means of a scheme, systematic as well, of the type introduced in Fig. 7.33. It is reported here in Fig. 7.48. The switch remains $k_o = 2$ symbol times connected to the information source and $r_o = 1$ symbol time to the feed-back, each period, and finally a proper symbol sequence is to be added, in order to assure a correct frame

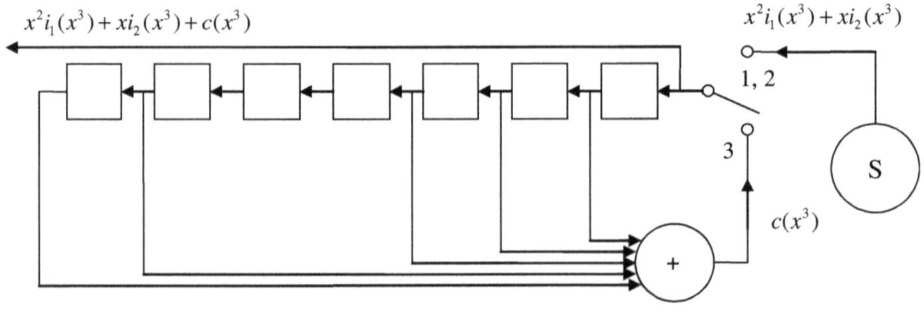

Fig. 7.48 Systematic encoder circuit based on a unique parity check polynomial for the code described in Examples 7.36 and 7.37

termination. The decoding computational complexity, adopting this alternative encoder circuit, is undoubtedly higher, so vanishing the benefit of a reduction in the branch complexity from 4 to 2.

Example 7.38 The 2/3-rate convolutional code characterized by $m = 2$ and $\mu = 3$, introduced in Example 4.1, which is w.s. time-invariant in its $\underline{G}$ matrix, exhibits the rational symbolic polynomial generator sub-matrix obtained in Example 4.9. It is reported here, together with the correspondent rational symbolic polynomial syndrome former sub-matrix,

$$\underline{G}_o^{(RSC)}(x) = \begin{bmatrix} 1 & 0 & \frac{x^3+x^2+x+1}{x^3+x^2+1} \\ 0 & 1 & \frac{x^3+1}{x^3+x^2+1} \end{bmatrix}, \quad \underline{H}_o^{T(RSC)}(x) = \begin{bmatrix} \frac{x^3+x^2+x+1}{x^3+x^2+1} \\ \frac{x^3+1}{x^3+x^2+1} \\ 1 \end{bmatrix}.$$

The latter one allows to obtain an encoder circuit in observer arrangement, of the type described in Sect. 7.8, with $m_h = 3$ memory elements. The consequent trellis exhibits 2^3 states per stage and 4 branches per state in input and in output. The same computational complexity characterizes a non-systematic encoder circuit based on the simple $\underline{G}_o(x)$. This is a case in which, in order to obtain $\underline{G}_o^{(RSC)}(x)$, it is necessary to increase the number of memory elements, considering both shift registers of the controller arrangement. Such penalty does not exist adopting $\underline{H}_o^{T(RSC)}(x)$, owing to the presence of a unique shift register for this observer arrangement.

As concluding remarks, we can summarize what follows. In the present chapter we have introduced two types of encoder circuits based on the $\underline{H}$ matrix. The first one, whose scheme is shown in Fig. 7.33, takes advantage from the fact that our code is s.s. time-invariant in its $\underline{H}$ matrix. In spite of this advantage, often the number of memory elements and consequently the number of merge states in the trellis, are too high, at the point where this solution is not advisable. This scheme has however the advantage of allowing easily scalable rate-adaptive arrangements, simply after different regulations of the switch. The second type of encoder circuit, systematic as well, which is an RSC solution in observer arrangement, of the type depicted in Fig. 4.6 down, typically shows a lower decoding computational complexity, and becomes the most advisable for high-rate convolutional codes. When the code rate is low, there are other possible situations, especially if the code is only w.s. time-invariant in its $\underline{H}$ matrix, as will be better described in Chap. 8.

7.11 A Second Bridge Between Cyclic Block Codes and s.s. Time-Invariant Convolutional Codes

In Chaps. 3 and 4 a first conceptual bridge has been established, between polynomial block and convolutional codes. It is formed by two lanes: one between cyclic block and s.s. time-invariant convolutional codes (Sect. 3.9), and the other between quasi-cyclic block and w.s. time-invariant convolutional codes (Sect. 4.9). All this first construction is based on the generator matrix. Now we have the possibility of establishing a second conceptual bridge, based on the parity check matrix, and some implications of duality. We will restrict our attention, for the moment, to a situation comprehending codes which are s.s. time-invariant in the $\underline{H}$ matrix (and generally only w.s. time-invariant in the $\underline{G}$ matrix).

The equivalence between MHEC codes and s.s. time-invariant convolutional codes in $\underline{H}$, although several times suggested, has not yet formally expressed, giving also precise correspondences between the parameters of the two models. We will try to proceed as in Sects. 3.7 and 3.8, where the $\underline{G}$ matrix was considered, stressing many analogies but also some differences with that situation. Let us begin from $(n_o - 1)/n_o$-rate codes, which are forced to be s.s. time-invariant in their $\underline{H}$ matrix.

Theorem 7.2 *The equivalence between an MHEC code, characterized by cyclic parity check polynomial $h_c(x)$, whose degree is $(N - r)$, and an s.s. time-invariant convolutional code in its $\underline{H}$ matrix, with asymptotic code rate $(n_o - 1)/n_o$, whose unique parity check polynomial $h(x)$ spans $(m_h + 1)$ periods, is supported by the following identities*

$$N - r = (m_h + 1)n_o - 1, \quad h_c(x) = h(x). \tag{7.17}$$

Proof In the upper bound (7.6) we can immediately recognize the first of (3.12), provided that parameter r is substituted to parameter D_g. This substitution is correct if the code is s.s. time-invariant in $\underline{G}$. After this premise regarding a $\underline{G}$ matrix approach, in the symbolism of cyclic codes, the degree of the parity check polynomial has been denoted $(N - r)$. Since we are treating s.s. time-invariant codes in $\underline{H}$, it is correct to take $D_h = \deg[h(x)] = \deg[h^*(x)] = N - r$, and hence the first of (7.17) coincides with the upper bound (7.7). Such equality, exactly like (3.12) in the case of a generator matrix description, refers to a situation characterized by full exploiting of the available constraint length (see Fig. 7.36). More precisely, the first of (7.17) holds for values of $(N - r)$ able to give integer values for parameter m_h, which means that $h^*(x)$ covers an integer number of periods. About the identification represented by the second of (7.17), in order to give a meaning to such symbolism, the following considerations can be developed. A unique polynomial shifts along the rows of $\underline{H}$ in both codes, with steps of n_o. In the MHEC code, this is a consequence of the periodic constant-length puncturations, leading to the suppression of $(n_o - 1)$ subsequent rows in the H-extended parity check matrix any n_o.

In the convolutional code, the parity check polynomial $h(x)$ has an interleaved nature, of the type shown, say, in expression (7.13) for $n_o = 3$, $r_o = 1$. Nevertheless, the structures of the two parity check matrices coincide, also in their external parts, where elimination of columns in the MHEC construction enables to obtain the same uniform syndrome former sub-matrix $\underline{H}_o^T$ of the convolutional code model. With equal $\underline{H}$ matrices, the two codes will be necessarily the same code. □

Comparison with Theorem 3.1 shows a sort of dual behaviour. Also here it is possible to have integer values for $(N - r)$ without corresponding integer values for m_h, but such situations are of poor interest, since the parity check constraint length is not fully exploited in such cases. Section 3.8 was devoted to s.s. time-invariant convolutional codes in their $\underline{G}$ matrix, characterized by $k_o > 1$, leading to equality (3.15), which is a sort of generalization of the first of (3.12). Here a similar discussion can regard s.s. time-invariant convolutional codes in their $\underline{H}$ matrix, characterized by $r_o > 1$. This approach gives

$$N - r = (m_h + 1)n_o - r_o, \tag{7.18}$$

which clearly results in a generalization of the first of (7.17). On the other hand, the constraint of having a unique parity check polynomial (as well as a unique generator polynomial), except for the case of $r_o = 1$ ($k_o = 1$), introduces strong limitations to the performance. The best performance, as several times noted, can be reached with w.s. time-invariant convolutional codes, which will be treated in Chap. 8, taking into account also Theorem 7.1, which allows a unitary approach. In fact, considering s.s. time-invariance as a particular case of w.s. time-invariance, the ensemble of w.s. time-invariant convolutional codes comprehends either those referred to $\underline{G}$ or those referred to $\underline{H}$.

In spite of the above similarity, between the two descriptions, there are also some differences. For instance, the $\underline{G}$ matrix of a not well designed convolutional code, at a first inspection, appears not distinguishable with respect to the one characterizing a well designed convolutional code. On the contrary, its $\underline{H}$ matrix requires the presence of some additional periodic rows. On the other hand, an $\underline{H}$ matrix not in minimal form exhibit some final rows linearly dependent on the previous ones. Such a not perfectly coincident behaviour can be related to the fact that the dual of a convolutional code exhibits an $\underline{H}$ matrix not precisely equal to the $\underline{G}$ matrix of the original one, but with some external columns eliminated, in order to allow equal protection to all the positions in the code frame.

Let us consider now a s.s. time-invariant convolutional code in its $\underline{H}$ matrix with variable code rate (see, for instance, Example 7.28, where the code rates 1/3 and 2/3 have been compared). Except for the above specification regarding some external columns of $\underline{H}$, a code of this type, interpreted as an MHEC code, can be considered as the dual code with respect to an MLC code with variable code rate as well (Sect. 3.8). With increasing parameter k_o in MLC codes, the occurrence of 2-weight code words is more and more probable. With reducing parameter r_o in MHEC codes, the occurrence of 2-weight code words is more and more probable as well, owing to a progressive reduction of 1-symbols in $\underline{H}$ acting on the same positions, up to the point where two equal columns appear, causing the correspondent 2-weight code word. Nevertheless, as long as the code rate increases, also null columns can arise, forcing to eliminate such positions, and leading to a progressive reduction of the whole frame length. This is impossible in MLC codes. On the other hand, not well designed convolutional codes cannot be obtained from MHEC codes, in which only non-periodic polynomials are present.

In traditional literature, convolutional codes are described mainly by means of their $\underline{G}$ matrix (Lin and Costello 2004). Therefore it is difficult to find a correspondence so impressive as the one shown in Table 3.10, regarding $1/n_o$-rate codes described by means of their generator matrix, also adopting here the parity check matrix. Nevertheless a similar correspondence between some very interesting $(n_o - 1)/n_o$-rate convolutional codes and certain easily recognizable cyclic block codes can be discovered, also regarding their parity check matrix description. We will consider the only case of code rate 2/3.

Example 7.39 One of the best 2/3-rate w.s. time-invariant convolutional code in its $\underline{G}$ matrix with $m = 1$ and $\mu = 2$ is that treated in Example 4.2 and several subsequent Examples. It is characterized by $g_{\text{int}}^{(1)}(x) = 101.111$ and $g_{\text{int}}^{(2)}(x) = 011.100$. The correspondent unique parity check polynomial is $h(x) = 111.100.110$ (see Example 7.36). This polynomial, which can be so decomposed $h(x) = (x+1)(x^3 + x^2 + 1)^2$, is factor of $(x^{14} + 1)$. Therefore the parent cyclic code, in an MHEC code scheme, of the 2/3 rate convolutional code here considered turns out to be a [14,7,4] self-dual cyclic code (see Example B.19).

Example 7.40 The best 2/3-rate w.s. time-invariant convolutional code in its $\underline{G}$ matrix with $m = 1$ and $\mu = 2$ (the first in Table 4.1) is characterized by $g_{\text{int}}^{(1)}(x) = 101.011$ and $g_{\text{int}}^{(2)}(x) = 011.111$. The correspondent parity check polynomial, obtained after column construction of $\underline{H}_o^T$, is $h(x) = x^7 + x^6 + x^5 + x^4 + x^2 + x + 1$. It is an irreducible primitive polynomial in the factorization of $(x^{127} + 1)$, and hence coincides with the parity check polynomial of a [127,7 64] simplex code.

Example 7.41 The best 2/3-rate w.s. time-invariant convolutional code in its $\underline{G}$ matrix with $m = 2$ and $\mu = 3$ (the second code in Table 4.1, which has also been treated in Example 4.1 and subsequent various Examples) is characterized by $g_{\text{int}}^{(1)}(x) = 101.110.000$, $g_{\text{int}}^{(2)}(x) = 011.001.101$. The correspondent unique parity check polynomial, obtained after column construction of $\underline{H}_o^T$, turns out to be $h(x) = 111.010.110.100.111$. This polynomial can be so decomposed $h(x) = (x+1)^3 (x^2 + x + 1)(x^9 + x^8 + x^4 + x + 1)$. The minimum-degree binomial, in whose factorization this parity check polynomial is present, results in $(x^{6132} + 1)$. This conclusion can be drawn observing that, among the above irreducible polynomials, there are the generator polynomial of a [3,1,3] repetition code and the generator polynomial of a [511,502,3] Hamming code, plus there is the factor $(x+1)^3$.

Remark that for all the 2/3-rate convolutional codes treated in the last three Examples, approximate prediction (7.5) is exact: for the first two with $m = 1$ and $m_h = 2$, and for the last one with $m = 2$ and $m_h = 4$. As a general consideration, we can observe that the unique parity check polynomial adopted for such codes exhibits a rather high degree D_h, and hence the resulting codes are characterized by long parity check constraint length m_h. This characteristic is more and more evident with increasing the code rate. In general, it could be a very rare event that a so long parity check polynomial is able to satisfy the condition for allowing LDPC code soft-decision decoding. This justifies the search for LDPC convolutional codes by proper direct design of the parity check matrix, or alternatively deriving them, after suitable transformations, from known LDPC block codes.

At this point, we can consider, in parallel, the correspondence between cyclic codes and s.s. convolutional codes in their $\underline{G}$ matrix, based on the MLC equivalence (Theorem 3.1), and that between cyclic codes and s.s. time-invariant convolutional codes in their $\underline{H}$ matrix, based on the MHEC equivalence (Theorem 7.2). Comprehending both these conceptual bridges, we can construct the diagram depicted in

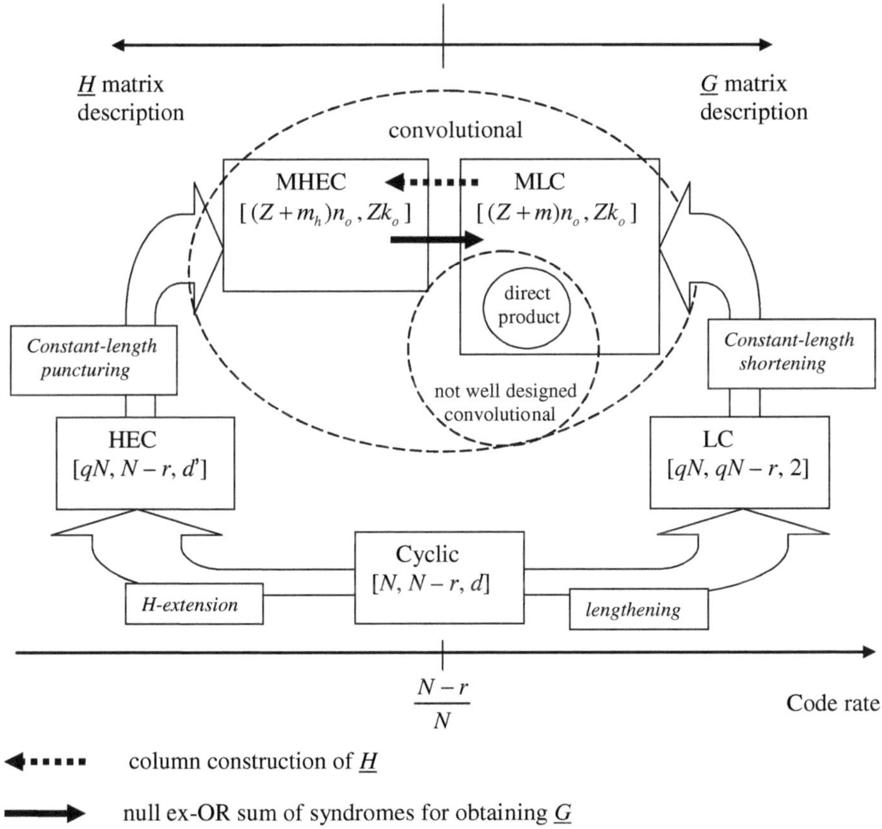

Fig. 7.49 Family tree of s.s. time-invariant codes

Fig. 7.49, representing as a sort of family tree for s.s. time-invariant codes, in which the $\underline{H}$ matrix description is reported on the left, and the $\underline{G}$ matrix description on the right. We can start from a parent cyclic code, which is characterized by parameters N, r, and hence code rate $(N - r)/N$. In order to increase the code rate, we can perform a code lengthening procedure, but this makes the code too weak (possible appearance of 2-weight code words). Therefore we can add a constant-length shortening (periodically substituting control symbols to information symbols), and reach the family of MLC codes. Starting again from the parent cyclic code, we can be interested in increasing its minimum distance. Then we perform an H-extension procedure, easily obtaining a minimum distance d' remarkably higher than d, but at the expense of a too strong reduction in the code rate. Therefore we can add a constant-length puncturation (periodically substituting information symbols to control symbols), and reach the family of MHEC codes.

Both MLC codes and MHEC codes are convolutional codes: the former ones s.s. time-invariant in their $\underline{G}$ matrix, the latter ones s.s. time-invariant in their $\underline{H}$ matrix. The two interpretations are particularly efficient for the code rates $1/n_o$ and $(n_o - 1)/n_o$ respectively. The first of (3.12) relates parameters m and r in the former case, the first of (7.17) relates parameters m_h and $(N - r)$ in the latter case.

It is possible to pass from the $\underline{H}$ matrix description to the $\underline{G}$ matrix description after application of the procedure called null ex-OR sum of clusters of syndromes, in the reverse direction after application of the procedure called column construction of $\underline{H}$. A particular subset in the set of convolutional codes is formed by those called not well designed. They show some $A(w)$ weight distribution coefficients increasing binomially with Z, for $d \leq w \leq (2d - 1)$. One can recognize immediately if a convolutional code is not well designed, looking at its generator polynomial $g(x)$, which exhibits a factor $p(x^{n_o})$, or looking at its parity check matrix, where some anomalous periodic rows must be present. In this sense, from an MHEC derivation, in which only non-periodic parity check rows are present, it is impossible to obtain a not well designed convolutional code. For this reason the subset of not well designed convolutional codes has been included only in the MLC derivation.

Inside the subset of not well designed convolutional codes we can find the family of direct product codes. They are characterized by $g(x) = p(x^{n_o})g_2(x)$, with deg $[g_2(x)]$ exactly equal to the number r_o of control symbols in a period of the considered convolutional code (Lemma 7.8).

A special case is that characterized by 1/2-rate convolutional codes. Lemma 7.4 assures that they are s.s. time-invariant either in the $\underline{G}$ matrix or in the $\underline{H}$ matrix, and that their unique parity check and generator polynomials coincide (they can be defined self-dual). Nevertheless this conclusion is true only if the considered 1/2-rate convolutional code is well designed. In fact, starting from an interleaved generator polynomial $g_{\text{int}}(x) = p(x^{n_o})g_o(x)$, and adopting $h(x) = g_{\text{int}}(x)$ we would obtain an $\underline{H}$ matrix not completely described by a syndrome former sub-matrix $\underline{H}_o^T$. More precisely, in order to construct the parity check matrix of the considered not well designed convolutional code, it is necessary to add some final periodic rows.

Conversely, when the reciprocal parity check polynomial is expressed as $h^*(x) = p(x^{n_o})h_o^*(x)$, being $h_o^*(x)$ the one related to the minimal form of $\underline{H}$, it is important to impose the null ex-OR sum of cluster of syndromes searching for the interleaved generator polynomial having minimum degree, and not simply to adopt $g_{int}(x) = h(x)$. These considerations are particularly important for the prediction of the structure characterizing the generator and the parity check matrix in the tail-biting versions of such codes, as described in Sect. 7.9.

As a final remark, we can anticipate that convolutional codes characterized by code rates $P/(2P)$, where P is a certain integer number (for instance 2/4, 3/6, …) exhibit very interesting performance with respect to simple 1/2-rate codes ($P = 1$), with analogous parameters m or m_h. We have already treated this type of codes in Example 7.10, but in the framework of an s.s. time-invariant model. Instead, the best performance is reached with different interleaved generator polynomials or parity check polynomials. The degrees of freedom gained from this constraint elimination allow the achievement of higher values for the minimum distance d. As an example, we can consider the case of 4/8-rate code obtained, by unwrapping, starting from the [24,12,8] Golay code (see Example 4.18). This topic will be treated in the next Chapter, dealing with w.s. time-invariant convolutional codes, with particular attention to their parity check matrix.

References

Jimenez-Felstrom AJ, Zigangirov KS (1999) Time-varying periodic convolutional codes with low density parity-check matrix. IEEE Trans Inf Theor 45:2181–2191

Johannesson R, Zigangirov KS (1999) Fundamentals of convolutional coding. IEEE Press, New York

Lin S, Costello DJ (2004) Error control coding. Pearson/Prentice-Hall, Upper Saddle River

MacWilliams FJ, Sloane NJA (1977) The theory of error correcting codes. North Holland, New York

Ryan WE, Lin S (2009) Channel codes: classical and modern. Cambridge University Press, Cambridge

Tanner RM (1981) A recursive approach to low complexity codes. IEEE Trans Inf Theor 27:533–547

Tanner RM, Sridhara D, Sridharan A et al (2004) LDPC block and convolutional codes based on circulant matrices. IEEE Trans Inf Theor 50:2966–2984

Chapter 8
Wide-Sense Time-Invariant Convolutional Codes in Their Parity Check Matrix

8.1 Traditional Obtainment of a Symbolic Parity Check Matrix in a w.s. Time-Invariant Convolutional Code

Let us consider a generator sub-matrix $\underline{G}_o$, or the correspondent symbolic polynomial sub-matrix $\underline{G}_o(x)$. In the most general case, the convolutional code they describe is w.s. time-invariant in both $\underline{G}$ and $\underline{H}$. As already observed, $\underline{G}_o$ and $\underline{G}_o(x)$ exhibit k_o rows. For $k_o > 1$, such rows, in general, are different. The problem of determining a syndrome former sub-matrix $\underline{H}_o^T$, or the correspondent polynomial sub-matrix $\underline{H}_o^T(x)$, compatible with the above generator matrices, can be faced as follows. First of all, Theorem 7.1 assures that, if the code is well designed, $\underline{H}_o^T$ exists and is at least w.s. time-invariant as well, furthermore there is no need of non-periodic parity check rows. Being n_o the length of the period, and having set $r_o = n_o - k_o$, as usual, $\underline{H}_o^T$ will exhibit layers of r_o rows. For $r_o > 1$, such rows are formed by as many reciprocal parity check polynomials, in general different. The symbolic sub-matrix $\underline{H}_o^T(x)$, in turn, is characterized by n_o rows and r_o columns.

We can follow a first approach, based on the procedure introduced in Sects. 4.4 and 7.8. In particular, Lemma 7.10 is completely general, and refers to a w.s. time-invariant convolutional code. This approach is what we define as traditional obtainment of the parity check matrix from the generator matrix. It exploits the symbolic rational polynomial matrices, and starts from checking the condition for assuring $\underline{G}_o(x)$ right-inversion. Sometimes this procedure turns out to be rather long. Nevertheless it is always available, under the only condition that $\underline{G}_o(x)$ is right-invertible. The traditional approach is recommended when we are interested in RSC encoder circuit schemes, in controller arrangement or in observer arrangement.

In some particular situations (with $\underline{G}_o$ upper triangular, and in order to obtain, if possible, an $\underline{H}_o^T$ characterized by a lower triangular first layer), we can follow a simpler procedure, already introduced for low-rate convolutional codes (Sect. 7.2) and for high-rate convolutional codes (Sect. 7.3), independently of the fact that they

G. Cancellieri, *Polynomial Theory of Error Correcting Codes*,
Signals and Communication Technology, DOI 10.1007/978-3-319-01727-3_8

are s.s. or w.s. time-invariant. This simpler procedure, called column construction of the parity check matrix (Definition 7.6), can be considered as a sort of inverse with respect to the procedure based on the null ex-OR sum of clusters of syndromes, by which it is possible to obtain $\underline{G}_o$ from $\underline{H}_o^T$. In Fig. 7.33 an encoder circuit, based on the parity check matrix, has been proposed, for s.s. time-invariant convolutional codes in their $\underline{H}$ matrix (or MHEC codes). In the present chapter we are going to extend the last scheme to w.s. time-invariant convolutional codes in their $\underline{H}$ matrix. Before facing this problem, in the present section we deal with the traditional approach and in the next section we will discuss some applications of the procedures of column construction and null ex-OR sum of clusters of syndromes.

An important role is played by the circuit for syndrome calculation, already introduced in Sect. 7.8 for the case with $r_o = 1$ (Definition 7.7). First of all, it is necessary to widen this Definition, in order to comprehend now the case of $r_o > 1$.

Definition 8.1 A symbolic polynomial sub-matrix $\underline{H}_o^T(x)$, with n_o rows and r_o columns, associated to a convolutional code characterized by code rate $(n_o - r_o)/n_o$, describes the *circuit for syndrome calculation*. It is shown in Fig. 8.1 in the case of $n_o = 3$ and $r_o = 2$. In this circuit, m_h represents the maximum degree of polynomials $h^{(i,j)}(x)$, $i = 1, 2, 3; j = 1, 2$.

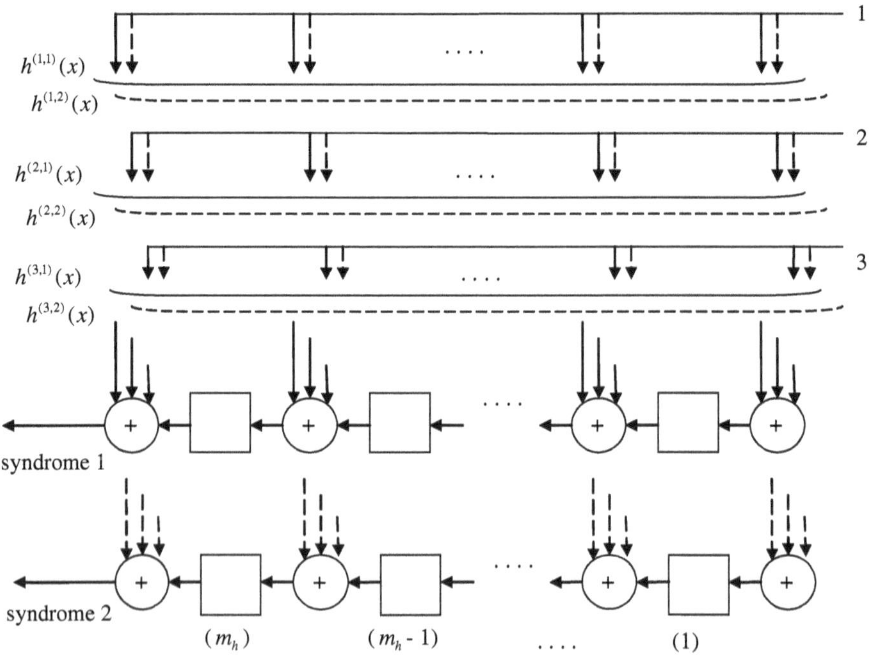

Fig. 8.1 Circuit for syndrome calculation in an 1/3-rate convolutional code

We have here 2 syndromes (in general r_o syndromes), to be calculated each period. The expression of this polynomial matrix can be obtained by an extension of (7.14) to the case characterized by, say, $n_o = 3$, $r_o = 2$, like in Fig. 8.1. Such extension is

$$\underline{H}_o^T(X) = \begin{bmatrix} h^{(1,1)}(x) & h^{(1,2)}(x) \\ h^{(2,1)}(x) & h^{(2,2)}(x) \\ h^{(3,1)}(x) & h^{(3,2)}(x) \end{bmatrix} \tag{8.1}$$

Definition 8.2 Considering a matrix $\underline{H}_o^*$, obtained after transposition and reciprocal calculation of $\underline{H}_o^T$, as well as their symbolic polynomial counterparts $\underline{H}_o^*(x)$, $\underline{H}_o^T(x)$, respectively, the polynomials in $\underline{H}_o^*(x)$ are called *input–output syndrome former polynomials* and coincide with those in (8.1), except for taking the reciprocal.

For a 1/3-rate code, the relationships between the two parity check polynomials whose reciprocals are in the rows of $\underline{H}$ and the six input–output syndrome former polynomials in (8.1) are

$$\begin{aligned} h_1^*(x) &= x^2 h^{(1,1)*}(x^3) + x h^{(2,1)*}(x^3) + h^{(3,1)*}(x^3) \\ h_2^*(x) &= x^2 h^{(1,2)*}(x^3) + x h^{(2,2)*}(x^3) + h^{(3,2)*}(x^3), \end{aligned} \tag{8.2}$$

where we can recognize a widening of expression (7.13), which holds for $n_o = 3$, $r_o = 1$. Remark that the order of the apices has been reversed because of reciprocal assumption.

In order to obtain the $\underline{G}$ matrix, starting from the knowledge of the $\underline{H}$ matrix, it is convenient to follow the same way able to allow the determination of the RSC version of the symbolic syndrome former sub-matrix (Sects. 4.3 and 4.4), with the only difference of processing the $\underline{H}$ matrix instead of the $\underline{G}$ matrix. The first step is to test if $\underline{H}_o(x)$ is right-invertible, then we find the inverse of its leftmost 2 × 2 (in general $r_o \times r_o$) square sub-matrix. Remark that this sub-matrix is the rightmost in $\underline{H}_o^*(x)$. So we obtain the RSC version of (8.1), capable for the construction of a systematic encoder, which is necessarily recursive if the original code was not systematic. This RSC version of $\underline{H}_o^T(x)$, still in the case of a 1/3-rate convolutional code, turns out to be

$$\underline{H}_o^{T(RSC)}(x) = \begin{bmatrix} \frac{h^{(1,1)\prime}(x)}{D_H(x)} & \frac{h^{(2,1)\prime}(x)}{D_H(x)} \\ 1 & 0 \\ 0 & 1 \end{bmatrix} \tag{8.3}$$

where $h^{(i,1)\prime}(x)$, $i = 1, 2$, are proper transformations of the previous correspondent polynomials, whose degrees is sometimes higher than the original ones, and $D_H(x)$ is the determinant of the lower 2 × 2 sub-matrix. We can observe a strict analogy with the structure of sub-matrix $\underline{G}_o^{(RSC)}(x)$, given by (4.9) in the case of a code rate 2/3 (Johannesson and Zigangirov 1999).

Example 8.1 Given the 1/3-rate convolutional code, whose true syndrome former sub-matrix is

$$\underline{H}_o^T = \begin{bmatrix} 1 & 1 & \\ & & 1 \\ \hline 1 & & 1 \\ & 1 & 1 \end{bmatrix},$$

we have the following reciprocal input–output syndrome former polynomials

$$h^{(1,1)*}(x) = x + 1, \quad h^{(2,1)*}(x) = 1, \quad h^{(3,1)*}(x) = x,$$
$$h^{(1,2)*}(x) = 0, \qquad h^{(2,2)*}(x) = x, \quad h^{(3,2)*}(x) = x + 1.$$

This code is characterized by parity check constraint length $m_h = 1$. Keeping in mind the convention in writing $\underline{H}_o^T(x)$, we obtain

$$\underline{H}_o^T(x) = \begin{bmatrix} (x+1) & 0 \\ 1 & x \\ x & (x+1) \end{bmatrix}$$

In order to make this symbolic polynomial sub-matrix in RSC form, that is in symbolic rational polynomial form, first of all, it is necessary to set up, in its lower part, a 2 × 2 diagonal form. Then the common factor is employed for dividing the two terms in the first row. After some manipulations, we are left with

$$\underline{H}_o^{T(RSC)}(x) = \begin{bmatrix} \frac{x^2+1}{x^2+x+1} & \frac{x+1}{x^2+x+1} \\ 1 & 0 \\ 0 & 1 \end{bmatrix}$$

It is possible to observe how, in order to obtain a 2 × 2 identity sub-matrix in the lower part of $\underline{H}_o^T(x)$, parameter m_h has been increased from 1 to 2.

In Fig. 8.2 up the circuit for syndrome calculation related to this code is shown, putting in evidence the roles of the $h^{(i,j)}(x)$ input–output polynomials. From $\underline{H}_o^{T(RSC)}(x)$ it is possible to derive the RSC encoder in observer arrangement, which is shown in Fig. 8.2 down. It exhibits one wire directly connected to the output and two shift registers, both with $D_H(x) = x^2 + x + 1$ in feed-back ex-OR sum, and with respectively $h^{(1,1)\prime}(x) = x^2 + 1$, $h^{(2,1)\prime}(x) = x + 1$, in forward ex-OR sums. This circuit has 4 memory elements on the whole. Therefore it appears not efficient in the decoding computational complexity. It could be of some interest to construct also the non-systematic encoder circuit for this code, still in observer arrangement. It exhibits 1 input wire and 3 output wires, and can be obtained multiplying $\underline{H}_o^{T(RSC)}(x)$ by

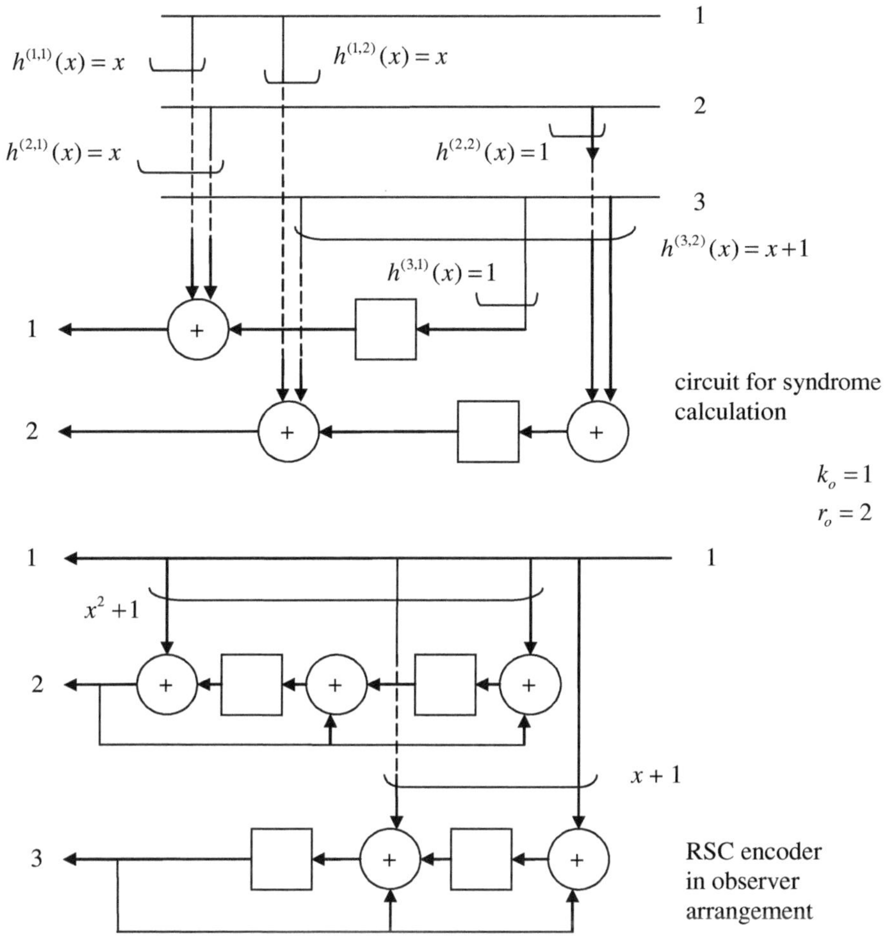

Fig. 8.2 Circuit for syndrome calculation of the 1/3-rate convolutional code introduced in Example 8.1 and its RSC encoder circuit in observer arrangement

$D_H(x) = x^2 + x + 1$. This new encoder circuit is shown in Fig. 8.3, and exhibits a number of memory elements even greater. This fact confirm that a low-rate convolutional code is more efficiently encoded employing its generator matrix.

Coming back to the problem of calculating the $\underline{G}_o(x)$ sub-matrix correspondent to (8.1), in our present case, characterized by code rate 1/3, we can expect for it an expression of the type

$$\underline{G}_o(x) = \left[g^{(1)}(x) \quad g^{(2)}(x) \quad g^{(3)}(x) \right],$$

whose RSC form can be trivially obtained as

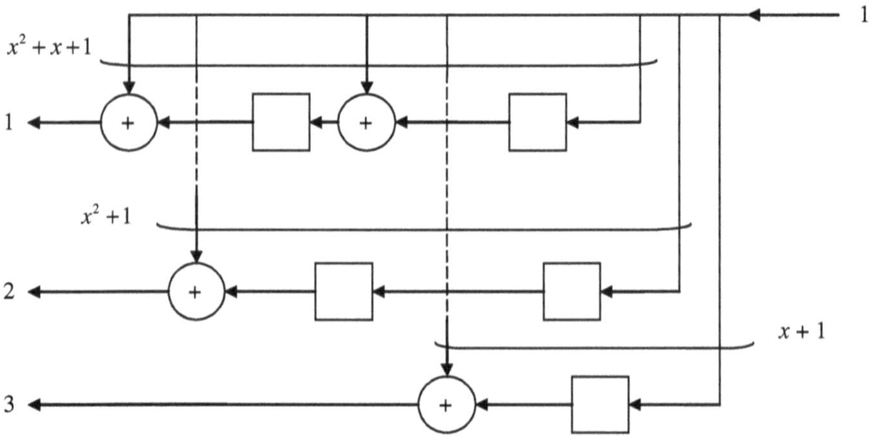

Fig. 8.3 Non-systematic encoder circuit in observer arrangement for the 1/3 rate convolutional code treated in Example 8.1

$$\underline{G}_o^{(RSC)}(x) = \begin{bmatrix} 1 & \frac{g^{(2)}(x)}{g^{(1)}(x)} & \frac{g^{(3)}(x)}{g^{(1)}(x)} \end{bmatrix},$$

and, owing to (7.8) and (7.9), the following correspondences come out

$$D_H(x) = g^{(1)}(x), \quad h^{(1,1)\prime}(x) = g^{(2)}(x), \quad h^{(2,1)\prime}(x) = g^{(3)}(x).$$

In Example 7.38 we described a 2/3-rate convolutional code, for which, given its $\underline{G}_o(x)$ sub-matrix, the correspondent $\underline{H}_o^T(x)$ sub-matrix has been calculated. Here we consider the case of a 1/3-rate convolutional code whose $\underline{H}_o^T(x)$ sub-matrix is known, and we would like to obtain the correspondent $\underline{G}_o(x)$ sub-matrix.

Example 8.2 With the 1/3-rate convolutional code already introduced in Example 8.1, from the expression of $\underline{H}_o^{T(RSC)}(x)$, we immediately obtain

$$\underline{G}_o^{(RSC)}(x) = \begin{bmatrix} 1 & \frac{x^2+1}{x^2+x+1} & \frac{x+1}{x^2+x+1} \end{bmatrix}.$$

Finally the symbolic generator sub-matrix $\underline{G}_o(x)$ and the correspondent true sub-matrix $\underline{G}_o$ will be

$$\underline{G}_o(x) = \begin{bmatrix} (x^2+x+1) & (x^2+1) & (x+1) \end{bmatrix}, \quad \underline{G}_o = [110|101|111].$$

In Fig. 8.4 up the RSC encoder in controller arrangement, which can be obtained from the above expression of $\underline{G}_o^{(RSC)}(x)$, exhibits just two memory elements. Its decoding computational complexity is imposed by parameter $m = 2$, and remains unchanged also referring to its non-systematic equivalent, characterized by just one

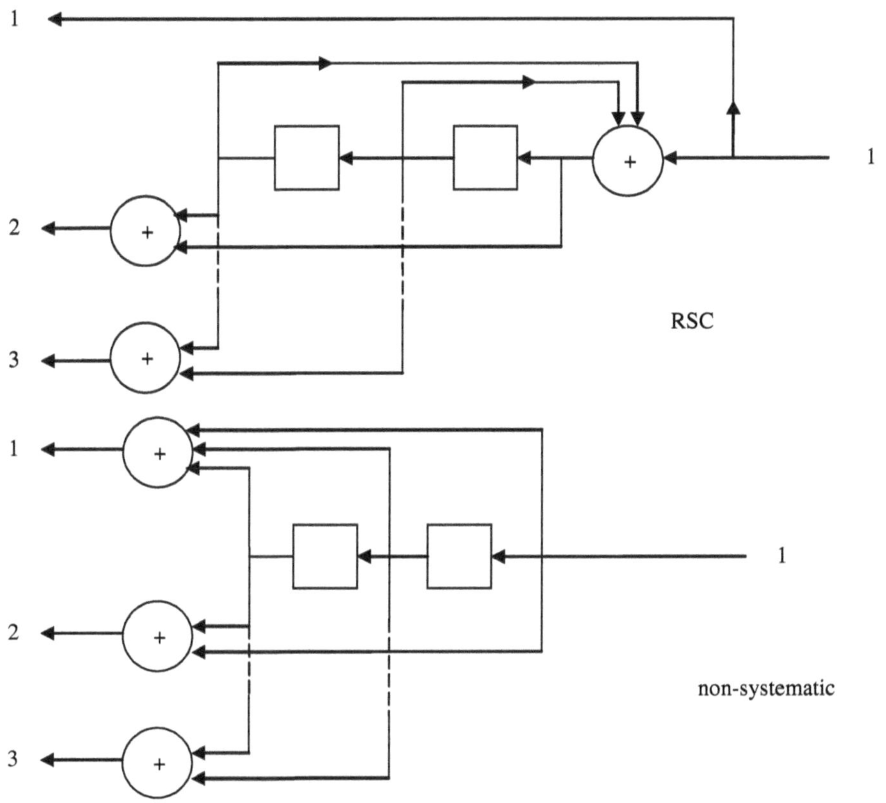

Fig. 8.4 RSC encoder in controller arrangement and non-systematic version of this encoder for the 1/3-rate convolutional code described in Examples 8.1, 8.2 and 8.3

shift register in controller arrangement and three forward ex-OR sums, as shown in Fig. 8.4 down.

Example 8.3 The code considered in Example 8.2 is not the best one, with rate 1/3 and $m = 2$ (see Table 3.4, where a code characterized by $d = 8$ is reported). Nevertheless its $\underline{H}$ matrix is interesting for application in the LDPC code context, as will be done in Chaps. 10–12, in fact it is free from 4-length short cycles of 1-symbols. The frame weight distribution coefficients of this code are as follows

$$A(7) = Z,\ A(8) = Z-1,\ A(9) = Z-2,\ A(10) = \ 2Z-5,\ A(11) = 3Z-10.$$

The situations characterized by code rate 2/3 or 1/3 are advantaged, because, in the obtainment of one of the two RSC versions of the encoder circuit (the observer arrangement in the former case, the controller arrangement in the latter case), the feed-back ex-OR sum is performed with the same wires as the feed-forward ex-OR sum of the non-systematic version. There are more general situations, in which this is not possible, and the determinant $D(x)$ or $D_H(x)$ is to be calculated.

Example 8.4 Taking the 2/4-rate convolutional code characterized by the following true generator sub-matrix and symbolic polynomial generator sub-matrix

$$\underline{G}_o = \left[\begin{array}{cccc|cccc} 1 & 1 & 0 & 1 & 1 & 0 & 1 & 0 \\ 1 & 0 & 1 & 1 & 0 & 1 & 0 & 1 \end{array}\right], \quad \underline{G}_o(x) = \begin{bmatrix} (x+1) & x & 1 & x \\ x & 1 & x & (x+1) \end{bmatrix}$$

the parameters regarding its non-systematic encoder circuit in controller arrangement are $m = 1$ and $\mu = 2$. It is possible to obtain $D(x) = D_H(x) = x^2 + x + 1$, and the following expressions (Johannesson and Zigangirov 1999) for the RSC generator and RSC syndrome-former sub-matrix

$$\underline{G}_o^{(RSC)}(x) = \left[\begin{array}{cc|cc} 1 & 0 & \frac{x^2+1}{x^2+x+1} & \frac{x^2}{x^2+x+1} \\ 0 & 1 & \frac{x^2}{x^2+x+1} & \frac{1}{x^2+x+1} \end{array}\right],$$

$$\underline{H}_o^{T(RSC)}(x) = \left[\begin{array}{cc} \frac{x^2+1}{x^2+x+1} & \frac{x^2}{x^2+x+1} \\ \frac{x^2}{x^2+x+1} & \frac{1}{x^2+x+1} \\ \hline 1 & 0 \\ 0 & 1 \end{array}\right]$$

They both show an increased decoding computational complexity, characterized by 2^4 states per stage. There is a simpler form for $\underline{H}_o^T(x)$, which leads to a non-systematic encoder circuit in observer arrangement, with only one memory element in each one of the four output shift registers. Nevertheless also here 2^4 states per stage are present in the correspondent trellis. Such non-systematic form of $\underline{H}_o^T(x)$ results in

$$\underline{H}_o^T(x) = \begin{bmatrix} 1 & x \\ x & (x+1) \\ (x+1) & x \\ x & 1 \end{bmatrix},$$

and will be justified in the next section. The simplest encoder circuit for this 2/4-rate convolutional code is then the one associated to the non-systematic $\underline{G}_o(x)$. It has a scheme in controller arrangement and with only feed-forward ex-OR sums. This non-systematic encoder circuit is shown in Fig. 8.5 and exhibits just one memory element in any of the two input shift registers. So it will be characterized by 2^2 states per stage in the correspondent trellis. In all the possible trellises considered here, the multiplicity of branches is fourfold, since 2 information symbols are contemporary generated from the information source.

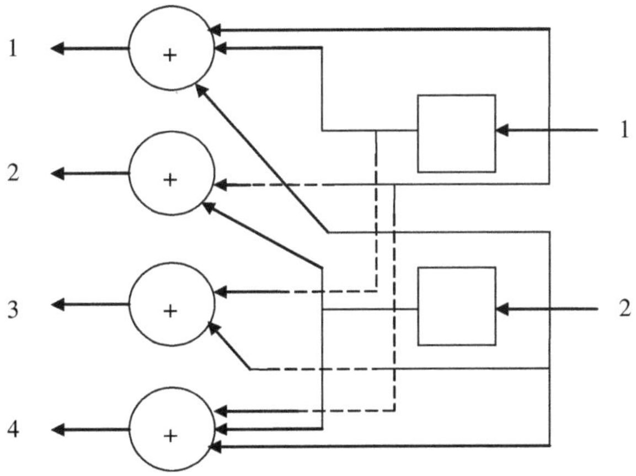

Fig. 8.5 Non-systematic encoder circuit (controller arrangement) for the 2/4 code introduced in Example 8.4

As a concluding remark, we can observe how the procedure here described is completely general, and can be applied either from a known parity check matrix in order to obtain the correspondent generator matrix or vice versa. Furthermore no constraints are to be satisfied regarding the positions occupied by information or control symbols in the period. About the decoding computational complexity, as already noted, it is possible to encode by means of a certain encoder circuit, but decode on a trellis constructed considering a simpler encoder circuit, so reducing computational complexity in decoding, although this solution introduces some limitations in possible code concatenation schemes.

8.2 Null ex-OR Sum of Clusters of Syndromes for Obtaining $\underline{G}$ and Column Construction of $\underline{H}$

In some situations, the procedures known as null ex-OR sum of clusters of syndromes (Lemma 5.11) or column construction (Definition 7.6) can allow a straightforward obtainment of $\underline{G}_o$ from the knowledge of $\underline{H}_o^T$ or vice versa, without the need of any symbolic polynomial sub-matrix. This is the case, for instance, of an upper triangular $\underline{G}_o$, as well as of a lower triangular uppermost layer in $\underline{H}_o^T$, but often, even in less restrictive situations, these tools give good results. The 2/4-rate convolutional code treated in Example 8.4 offers the opportunity for analyzing, in both directions, the most general case of w.s. time-invariant convolutional codes, either in $\underline{G}$ or in $\underline{H}$.

Example 8.5 Given the generator sub-matrix $\underline{G}_o$, describing the 2/4-rate w.s. time-invariant convolutional code introduced in Example 8.4, its $\underline{H}_o^T$ sub-matrix is directly obtained by means of the column construction procedure, leading to the result shown in Fig. 8.6 up. We find two parity check polynomials, whose expressions respectively are

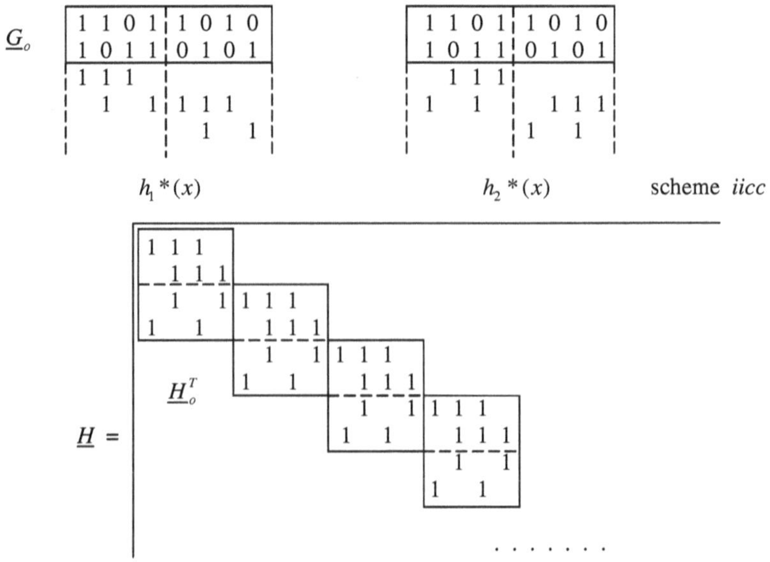

Fig. 8.6 Column construction of $\underline{H}_o^T$ and semi-infinite parity check matrix for the 2/4-rate code described in Examples 8.4 and 8.5

$$h_1(x) = x^6 + x^5 + x^4 + x^3 + x, \quad h_2(x) = x^7 + x^6 + x^5 + x^2 + 1,$$

confirming the structure of $\underline{H}_o^T(x)$ anticipated in Example 8.4. In Fig. 8.6 down four periods of the complete semi-infinite $\underline{H}$ matrix have been reported. Application of the null ex-OR sum of clusters of syndromes to a pair a subsequent $\underline{H}_o^T$ sub-matrices easily gives back $\underline{G}_o$, which is not upper triangular.

We can observe that, in the 2/4-rate w.s. time-invariant convolutional code discussed in Example 8.5, the sequence of information and control symbols in the period is of the type *iicc*. This fact can be argued from inspection of the uppermost layer in $\underline{H}_o^T$, which has been constructed intentionally in order to be lower triangular. The same conclusion can be reached, if one replaces the $\underline{G}_o$ sub-matrix with an equivalent form, made by its first row and by a second row obtained as ex-OR sum of the two present rows, so leading to an upper triangular version of $\underline{G}_o$. Also other possible types of sequence can be assumed in a code characterized by the same 2/4 code rate, for instance *icic*.

Example 8.6 Taking the 2/4-rate w.s. time-invariant convolutional code, characterized by the following generator sub-matrix

$$\underline{G}_o = \left[\begin{array}{cccc|cccc} 1 & 1 & 1 & 1 & 0 & 1 & 1 & \\ & & 1 & 0 & 1 & 1 & 1 & 1 \end{array}\right],$$

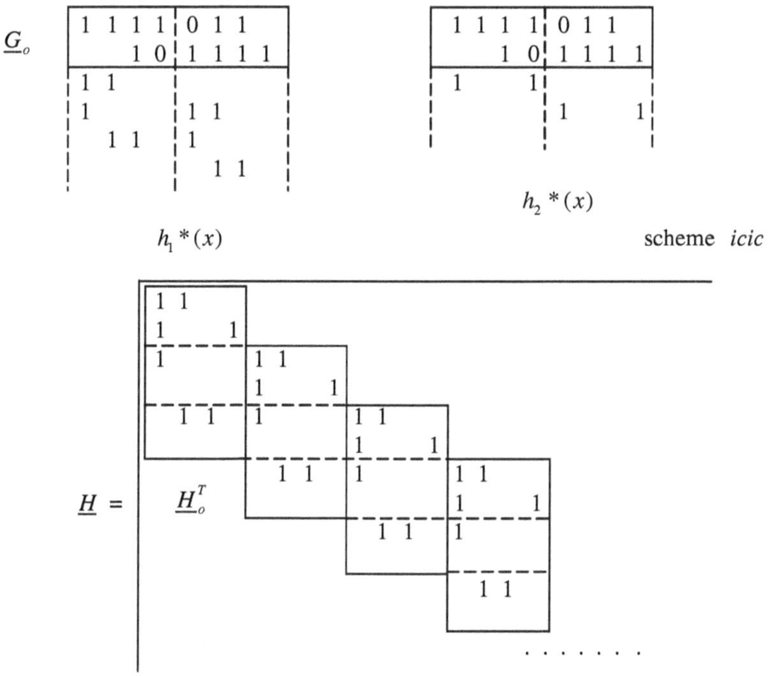

Fig. 8.7 Column construction of $\underline{H}_o^T$ and semi-infinite parity check matrix for the 2/4-rate code described in Example 8.6

it is possible to calculate the correspondent syndrome former sub-matrix, following the column construction procedure shown in Fig. 8.7 up. Here we obtain a longer parity check constraint length m_h, owing to $h_1{*}(x)$, which spans three periods. It is possible to observe that the scheme for the insertion of information and control symbols in the code frame can be written here as *icic*. The simplest encoder circuit, also for this code, is the non-systematic one, in controller arrangement, which shows 2 memory elements on the whole.

Example 8.7 The 2/4-rate w.s time-invariant convolutional codes, discussed in Examples 8.5 and 8.6 respectively, are both well designed. In spite of the fact that the second one exhibits a longer parity check constraint length m_h, its minimum distance is only $d = 3$, whereas the minimum distance of the first code is $d = 5$. The frame weight distributions characterizing such two codes, for some low weights, are formed by the following coefficients

$A(5) = 4Z - 2$, $A(6) = 5Z - 6$, $A(7) = 6Z - 14$, (code in Fig. 8.6),
$A(3) = Z - 1$, $A(4) = 2Z - 4$, $A(5) = 6Z - 12$, (code in Fig. 8.7).

As a reference in comparison with the former one, we can consider the best 1/2-rate convolutional code with $m = 2$, which is characterized by $d = 5$ as well (see Table 3.2). Such two codes have equal state complexity, but the branch complexity is smaller in the 1/2-rate code. Looking at the semi-infinite $\underline{H}$ matrix shown in Fig. 8.7 down, we can observe that it is free of 4-length short cycles of 1-symbols. Therefore this code would be suitable for a decoding as LDPC code. In this case, the computational complexity is no longer imposed by the number of states in the trellis, but by the number of 1-symbols present on average in the columns of the $\underline{H}$ matrix, as will be better described in Chaps. 10–12.

8.3 Encoder Circuits Based on Different Parity Check Polynomials

The schemes of encoder circuit shown in Figs. 2.16 and 2.18, although proposed for QC codes, can be considered also for convolutional codes which are w.s. time-invariant in their $\underline{G}$ matrix. They employ two different shift registers (in general k_o), representing as many interleaved generator polynomials, and appear as proper extensions of the encoder circuits reported respectively in Figs. 1.3 and 1.5. This scheme is valid only if the generator sub-matrix $\underline{G}_o$ has an upper triangular form. In Fig. 7.33 an encoder circuit, with just one parity check polynomial has been proposed for a convolutional code s.s. time-invariant in its $\underline{H}$ matrix (MHEC code). It derives from the encoder circuit adopted for a cyclic or shortened cyclic code, depicted in Fig. 5.6, taking only a different sequence in the switch times. An extension of such encoder scheme, where two different parity check polynomials, acting alternately, are implemented by as many shift registers, is depicted in Fig. 6.9. It has been proposed for being employed with a QC code, characterized by $q_o = 2$, but is valid also for a w.s. time-invariant convolutional code in its $\underline{H}$ matrix, provided that a proper sequence in the switch times is adopted. In particular, when we have $k_o = 1$ and $r_o = 2$, the input switches follows a 3-length period, taking one control symbol any period from each feed-back wire and one information symbol from the source. We will consider the case of this 1/3-rate convolutional code w.s. time-invariant in its $\underline{H}$ matrix (that is with two different interleaved parity check polynomials). Trivial extensions could be imagined for different code rates.

Furthermore we will take into account a lower triangular uppermost layer in the syndrome former sub-matrix $\underline{H}_o^T$, in order to assure a correct association of control symbols and times of the switches. Under such assumptions, the polynomial expression of a frame becomes

$$t(x) = x^2 i(x^3) + x c_1(x^3) + c_2(x^3), \tag{8.4}$$

where $i(x^3)$ represents the information symbol sequence, $c_1(x^3)$ the control symbols in the second position of any period, provided by polynomial $h_1(x)$, $c_2(x^3)$ the control symbols in the third position of any period, provided by polynomial $h_2(x)$. In Fig. 8.8, an encoder circuit able to give an expression of the type (8.4) is depicted. It contains two shift registers in controller arrangement and is systematic.

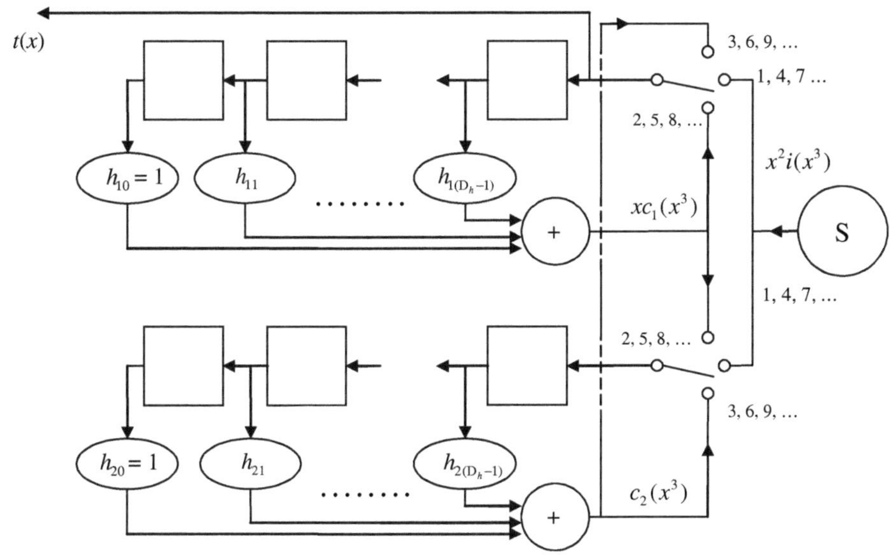

Fig. 8.8 Encoder circuit for a 1/3-rate convolutional code w.s. time-invariant in its $\underline{H}$ matrix with both parity check polynomials having degree D_h

The two polynomials have been assumed with the same degree, that is $\deg[h_1(x)] = \deg[h_2(x)] = D_h$. The highest-order coefficients, as usual, are not present. As for the parity check constraint length m_h, after (7.18), we have $D_h = 3(m_h + 1) - 1$. Clearly when the polynomials have different degrees, D_h represents the highest degree among them. As already noted for the scheme in Fig. 6.9, the same state evolution takes place in both shift registers. This allows the possibility of employing a periodically time-varying scheme, with just one shift register (Jimenez-Felstrom and Zigangirov 1999). Finally a proper symbol sequence is to be added at last, for assuring a correct frame termination.

Comparison with Fig. 6.9 shows a period length increased from $q_o = 2$ to $n_o = 3$, which means the insertion of two distributed control symbols in any period. It is interesting to note that the output frame $t(x)$ can be collected indifferently from the upper or from the lower shift register. A suitable look-up table, where proper output sequence are directly stored, can be used for allowing a correct frame termination. The number of the possible different states of a single shift register, besides the computational complexity of a soft-decision decoding procedure performed in the associated trellis, fixes also the size of this look-up table.

Example 8.8 Let us take the w.s. time-invariant convolutional code introduced in Example 8.1. It exhibits code rate 1/3, $m_h = 1$ and the following pair of interleaved parity check polynomials: $h_1(x) = x^4 + x^3 + x^2 + 1$, $h_2(x) = x^4 + x + 1$. The encoder circuit based on such polynomials, according to the general scheme reported in Fig. 8.8, turns out to be as shown in Fig. 8.9. The state complexity is 2^4, but the branch complexity only 2^1.

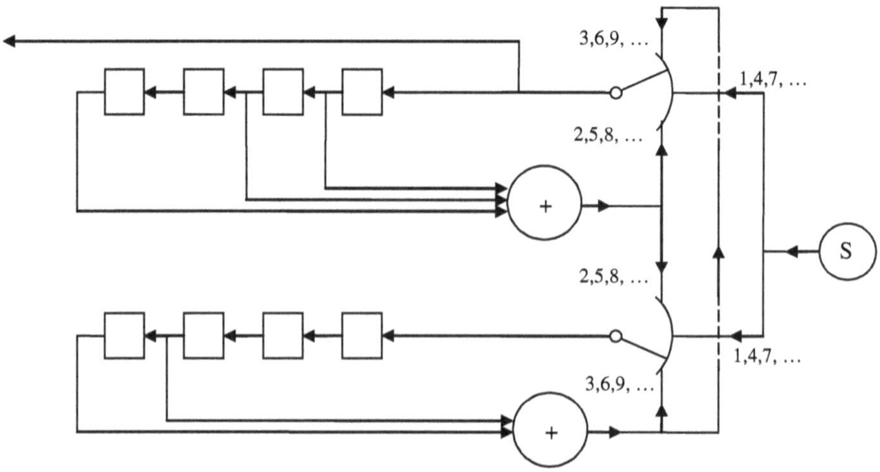

Fig. 8.9 Encoder circuit for the 1/3-rate w.s. time-invariant convolutional code described in Examples 8.1, 8.2, 8.8

8.4 Punctured Convolutional Codes and Their $\underline{H}$ Matrix

In Chap. 3 we have shown that punctured s.s. time-invariant convolutional codes in their $\underline{G}$ matrix become w.s. time-invariant in the new $\underline{G}$ matrix. Remark that we are treating now a traditional type of code puncturation, which reduces the length of the code word or frame, and not the constant-length puncturation, which is the simple elimination of a central row in the $\underline{H}$ matrix. In order to determine the parity check matrix of a punctured convolutional code, we can proceed as follows: firstly obtain the $\underline{G}_o$ sub-matrix of the punctured convolutional code under study, and then, by means of the column construction procedure, derive the correspondent $\underline{H}$ matrix. Optimization of the puncturation matrix (Definition 3.10) is easier looking at the generator matrix. In LDPC codes this observation is not valid, and typically one proceeds simply by eliminating columns in the $\underline{H}$ matrix.

Since, in general, we will reach a punctured convolutional code w.s. time-invariant in its $\underline{H}$ matrix, it is not important to distinguish between the case in which the original code was s.s. or w.s. time-invariant in $\underline{H}$. Nevertheless sometimes we can obtain also a punctured convolutional code which is s.s. time-invariant in the new $\underline{H}$ matrix. This is the case, for instance, of an $(n_o - 1)/n_o$-rate punctured convolutional code.

Example 8.9 Let us consider the best 1/2-rate convolutional code with $m = 3$ (see the third row in Table 3.2). Its interleaved generator polynomial corresponds to the sequence 11.11.01.11. Applying a puncturation, for eliminating two control symbols any three, we obtain a 3/4-rate w.s. time-invariant convolutional code in $\underline{G}$, whose $\underline{G}_o$ sub-matrix is

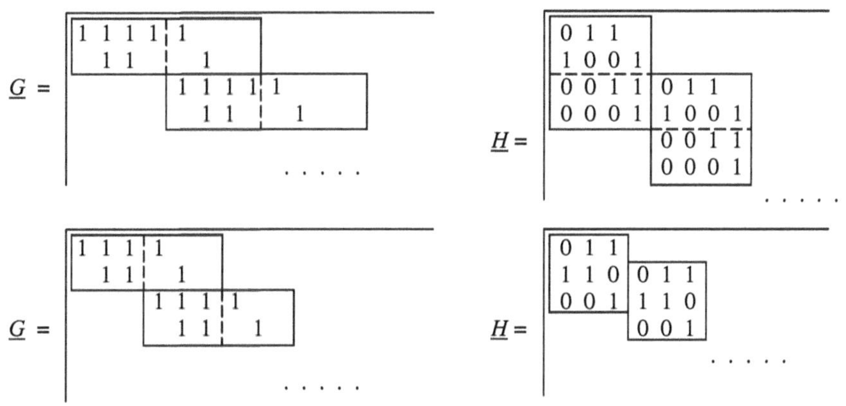

Fig. 8.10 Generator and parity check matrix for the 2/4-rate w.s. time-invariant discussed in Example 8.10 and for the 2/3-rate punctured code obtained from it

$$\underline{G}_o = \left[\begin{array}{cccc|cccc} 1 & 1 & 0 & 1 & 1 & & & \\ & 1 & 1 & 1 & 0 & 1 & & \\ & & 1 & 1 & 1 & 0 & 1 & 1 \end{array}\right].$$

After application of the column construction method, we obtain a unique parity check polynomial resulting in $h(x) = x^{13} + x^{12} + x^{10} + x^8 + x^7 + x^5 + x^2 + x + 1$. Such punctured convolutional code, which is s.s. time-invariant in $\underline{H}$, is characterized by minimum distance $d = 3$ and by weight distribution coefficients: $A(3) = 3Z - 4$, $A(4) = 14Z - 27$. This behaviour is very close to the optimum for this kind of convolutional codes (3/4-rate, $m = 1$, $\mu = 3$).

Example 8.10 Starting from the [14,10,3] QC code introduced in Examples 2.3 and 2.7, we can construct a 2/4-rate MLQC code, whose semi-infinite $\underline{G}$ matrix and $\underline{H}$ matrix are reported in Fig. 8.10 up. The latter one has been obtained from the former one by means of column construction. The sequence of information and control symbols is of the type *iicc*. This code maintains minimum distance $d = 3$, and exhibits the following frame weight distribution coefficients: $A(3) = Z$, $A(4) = 2Z - 1$, $A(5) = 4Z - 4$. Such performance is not particularly good. Nevertheless, a proper puncturation scheme can be adopted eliminating the second control symbol in any period, so obtaining a 2/3-rate punctured convolutional code (scheme *iic*), whose semi-infinite $\underline{G}$ matrix and $\underline{H}$ matrix are reported in Fig. 8.10 down. This code is s.s. time-invariant in its $\underline{H}$ matrix, in fact exhibits just one parity check polynomial. Its frame weight distribution is characterized by the following low-weight coefficients: $A(3) = 3Z - 1$, $A(4) = 7Z - 10$, $A(5) = 16Z - 43$. This is a rare example of puncturation without reducing the minimum distance, and is possible only because the original code was not optimized. The last convolutional code exhibits the same minimum distance as the best code with equal parameters (see the first row in Table 4.1), but only increased values in the normalized weight distribution coefficients.

Both codes presented in Example 8.10 are acceptable in the framework of an LDPC convolutional code model, as will be better explained in Chap. 12. It is interesting to note that, in passing from code rate 2/4 to code rate 2/3, owing to puncturation, the parity check constraint length of the code treated in Example 8.10 increases from $m_h = 1$ to $m_h = 2$. On the other hand, the degree D_h increases from 4 to 6. In Example 8.9, passing from code rate 1/2 to code rate 3/4, the value of m_h is reduced from 4 to 3. Clearly there is not a general behaviour and the result depends on the particular case.

8.5 Tail-Biting w.s. Time-Invariant Convolutional Codes and Their $\underline{H}$ Matrix

In Sect. 7.9, the tail-biting version of an s.s. time-invariant convolutional code has been treated, regarding its parity check matrix. In Fig. 7.41 the general scheme of such matrix has been reported. It is characterized by a unique reciprocal parity check polynomial $h^*(x)$. Here two different parity check polynomials $h_1^*(x)$ and $h_2^*(x)$ are considered, for a code rate which is still $(n_o - 2)/n_o$. In Fig. 8.11 this situation is depicted.

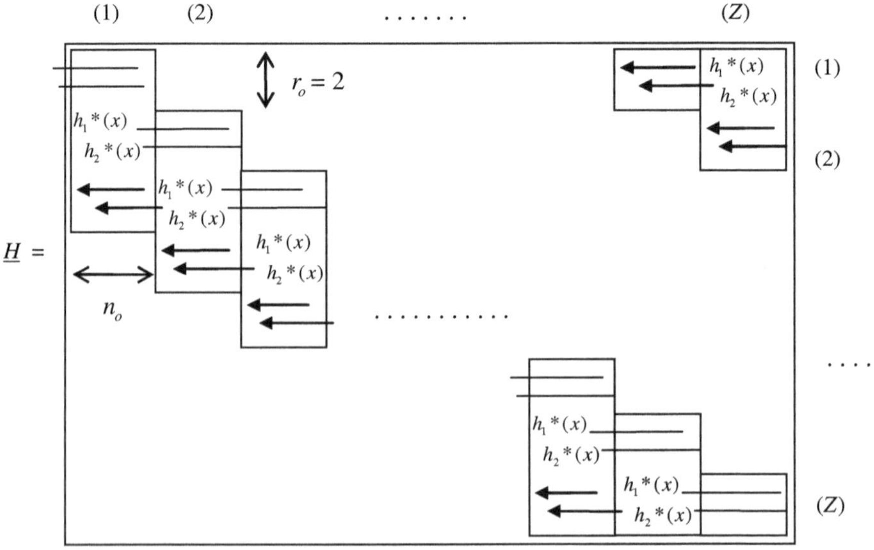

Fig. 8.11 Overall $\underline{H}$ matrix of an $(n_o - 2)/n_o$-rate tail-biting convolutional code w.s. time-invariant in $\underline{H}$ for $m_h = 2$

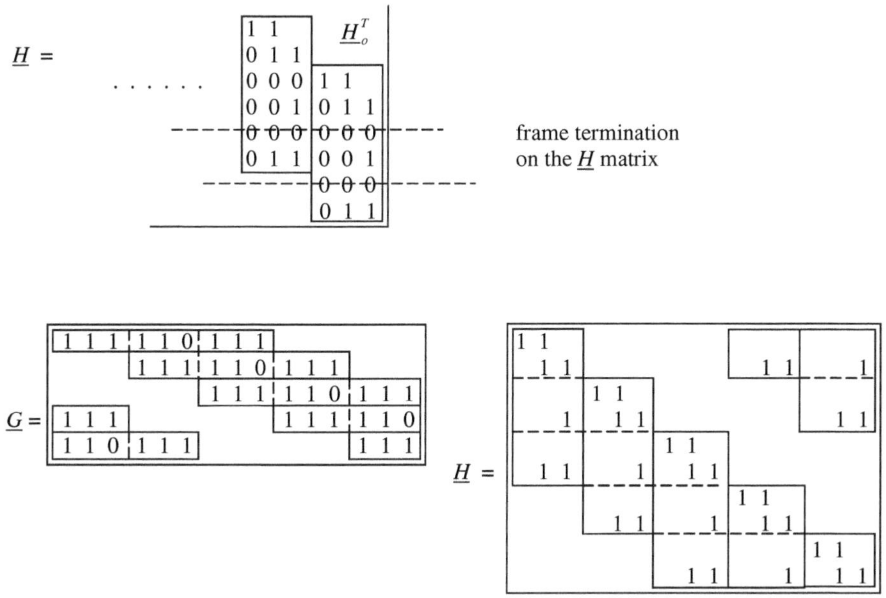

Fig. 8.12 Correct frame termination in the $\underline{H}$ matrix for the not tail-biting code and comparison between $\underline{G}$ and $\underline{H}$ matrix for the tail-biting code (Example 8.11)

Example 8.11 The best 1/3-rate convolutional code is the one characterized by an interleaved generator polynomial associated to the sequence 111.110.111 (see the second row in Table 3.4). By means of the column construction procedure, we can obtain the following reciprocal parity check polynomials: 011 and 011.001.011. Because of two empty rows in the $\underline{H}_o^T$ sub-matrix, the correct frame termination on the not tail-biting $\underline{H}$ matrix is made as shown in Fig. 8.12 up. The last two syndrome former sub-matrices appear vertically shorter, owing to elimination of such two rows. Nevertheless no columns are eliminated, so the overall number of control symbols is equal to $(2Z + 6)$. In the tail-biting arrangement, this row elimination does not occur, as it is possible to see from Fig. 8.12 down, where the generator matrix and the parity check matrix characterizing the tail-biting version of this code are reported, with $Z = 5$. It exhibits a minimum distance as small as $d = 4$. With increasing Z, either $A(4)$ or $A(6)$ vanish, up to the condition in which only $A(8)$ remains. The minimum distance is then $d = 8$, with a normalized coefficient $B(8) = 2$, like in the not tail-biting convolutional original code. This behaviour is approached for Z greater than about 15.

Example 8.12 In Example 6.22 a 2/4-rate convolutional code, w.s. time-invariant both in $\underline{G}$ and in $\underline{H}$, has been considered, as obtained from MHEQC construction. It is characterized by $m = m_h = 2$. In Fig. 6.22 its generator and parity check matrix have been reported. Here we will arrange such code in a tail-biting version, over $Z = 5$ periods. The result is shown for both matrices in Fig. 8.13. In order to reach a

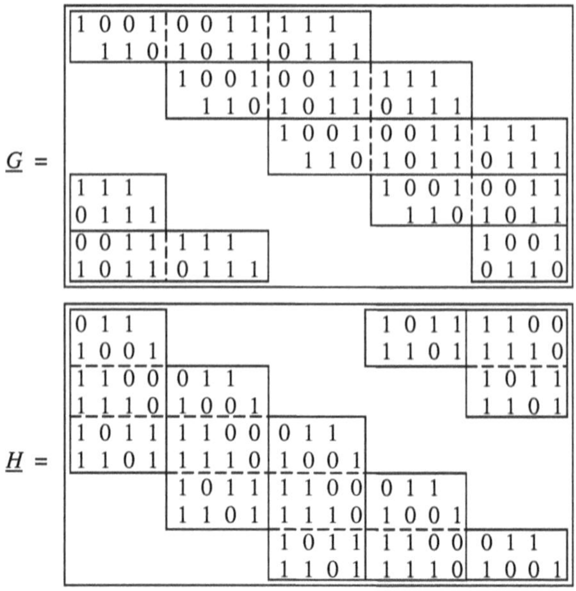

Fig. 8.13 $\underline{G}$ and $\underline{H}$ matrix for the 2/4-rate tail-biting convolutional code w.s. time-invariant in both matrices, described in Example 8.12

performance comparable with that of the original not tail-biting convolutional code ($d = 6$ with unitary normalized weight distribution coefficient), Z must increase up to about 20.

The codes considered in Examples 8.11 and 8.12 are both well designed. The existence of a minimum value for Z, needed for reaching the expected performance, is a phenomenon already noted in Chaps. 3, 4, 6. If we are dealing with a not well designed convolutional code, w.s. time-invariant in its $\underline{H}$ matrix, in such matrix there will be some periodic rows, which cannot change form after wrapping around, like in s.s. time-invariant codes (see Example 7.34 and Fig. 7.45). This anomalous behaviour is responsible of the property, already discussed in Chaps. 4 and 7, that the rank of $\underline{G}$ is reduced. Precisely this reduction is equal to the number of anomalous periodic rows in the $\underline{H}$ matrix of the original not tail-biting code.

8.6 Unwrapping QC Codes Described by Their $\underline{H}$ Matrix

In Sect. 4.8 the unwrapping procedure applied to the $\underline{G}$ matrix of a QC code has been described. Theorem 4.1 fixes the conditions for the equivalence between a tail-biting convolutional code w.s. time-invariant in $\underline{G}$ and a QC code: it is necessary, in one direction, that the convolutional code is well designed, whereas, in the other direction, that the QC code exhibits distributed control symbols and a full rank $\underline{G}$ matrix, with an invariant generator sub-matrix in its row layers.

On the $\underline{H}$ matrix description of a QC code, with a reordered parity check matrix in form of a block of truncated circulants (see Fig. 6.7), it is possible to introduce conditions similar to those imposed by Theorem 4.1 on $\underline{G}$. First of all, it is necessary that, in all the row layers of $\underline{H}$, a same parity check sub-matrix shifts. Only in this way, in fact, it is possible to put in evidence a uniform syndrome former sub-matrix $\underline{H}_o^T$. Then we have to face the condition of having full rank. In this sense, great attention is to be kept when the $\underline{H}$ matrix is not in its minimal form. For instance, a very particular manipulation of the original $\underline{H}$ matrix could be interesting as parity check matrix not in minimal form, in the framework of an LDPC model. This is the case, say, of array codes (Fan 2000).

In conclusion, there are many different possible equivalent versions of the parity check matrix in a QC code. From each of them a different reordered $\underline{H}$ matrix can be obtained. Also referring to the minimal form of $\underline{H}$, some degrees of freedom remain, due to the possibility of performing ex-OR sum of rows without varying the parity check constraint length. Therefore, in order to obtain a reliable result, it is convenient to proceed as follows. Firstly, take a QC code able to satisfy the conditions imposed on its $\underline{G}$ matrix by Theorem 4.1, then obtain, by unwrapping, the correspondent w.s. time-invariant convolutional code, and finally determine the parity check matrix of the last one.

Example 8.13 The generator matrix labelled $\underline{G}'$ in Example 4.19 refers to one of the several equivalent versions of the [8,4,4] extended Hamming code. It satisfies all the condition fixed by Theorem 4.1 and is characterized by a symbol sequence *iicciicc*. This code has been interpreted also in the framework of an MLQC model. Its unwrapped version has been studied in Example 6.21 (Fig. 6.21), where its syndrome former sub-matrix has been introduced. Now we can easily verify that it is obtainable from the $\underline{G}$ matrix after application of column construction. The overall $\underline{H}$ matrix so constructed is in minimal form. The same result could be obtained starting directly from the $\underline{H}$ matrix of the original QC code with distributed control symbols and unwrapping it. Nevertheless this situation is rather particular, since it regards a self-dual code. Not always the same result comes out, especially when we are treating $\underline{H}$ matrices not in minimal form. Once obtained, after proper unwrapping, a well designed convolutional code, over a longer frame length, it is possible to wrap around it again, so obtaining a new tail-biting version of such code. Let us adopt a frame length of only $Z = 3$ periods. We can act indifferently on the $\underline{G}$ matrix or on the $\underline{H}$ matrix, obtaining the following pair of compatible matrices

$$\underline{G}' = \left[\begin{array}{cccccccccccc}
1 & 1 & 1 & 0 & 1 & & & & & & & \\
 & 1 & 0 & 1 & 1 & 1 & & & & & & \\
\hline
 & & & & 1 & 1 & 1 & 0 & 1 & & & \\
 & & & & & 1 & 0 & 1 & 1 & 1 & & \\
\hline
1 & & & & & & & & 1 & 1 & 1 & 0 \\
1 & 1 & & & & & & & & 1 & 0 & 1
\end{array}\right]$$

$$\underline{H}' = \left[\begin{array}{cccc|cccc|cccc}
1 & 0 & 1 & & & & & & & & 1 & 1 \\
0 & 1 & 1 & 1 & & & & & & & & 1 \\
 & & 1 & 1 & 1 & 0 & 1 & & & & & \\
 & & & 1 & 0 & 1 & 1 & 1 & & & & \\
 & & & & & & 1 & 1 & 1 & 0 & 1 & \\
 & & & & & & & 1 & 0 & 1 & 1 & 1
\end{array}\right]$$

They describe a [12,6,4] tail-biting convolutional code characterized by many symmetry properties. It is possible to observe that the same couple of polynomials is present in both these matrices, confirming the property of self-duality. Only their grouping and positioning are different and is compliant with the schemes of the generator and parity check matrix in a tail-biting arrangement.

8.7 Reordered Versions of a w.s. Time-Invariant Convolutional Code

In Sect. 4.8, two types of reordering procedures have been introduced (Definitions 4.10 and 4.11), the first, starting from a tail-biting convolutional code, allows to introduce the complete circulant form of the original code, and the second, starting from a not tail-biting convolutional code, gives rise to its truncated circulant form. In both cases the generator matrices have been considered. Now we will describe the effects of such two reordering procedures on the parity check matrices. We start from a k_o/n_o-rate w.s. or s.s. convolutional code.

Corollary a of Theorem 4.1 gives the number of row layers and column layers (respectively k_o and n_o) in the reordered $\underline{G}$ matrix. Corollary b of Theorem 4.1 predicts where distributed and final control symbols are positioned, in the case of the truncated circulant form. Subsequent observations in Chap. 4 have permitted to identify in the circulants (or in the truncated circulants) the input–output sub-generator polynomials of the original convolutional code. The following proposition, treating the parity check matrix, refers to the complete circulant form only.

Lemma 8.1 *Given a k_o/n_o-rate w.s. or s.s. convolutional code, well designed so that its tail-biting form has the same code rate, the reordering procedure introduced by Definition 4.10 gives rise to a parity check matrix with $r_o = n_o - k_o$ row layers and n_o column layers. In such layers there are $Z \times Z$ circulants, whose*

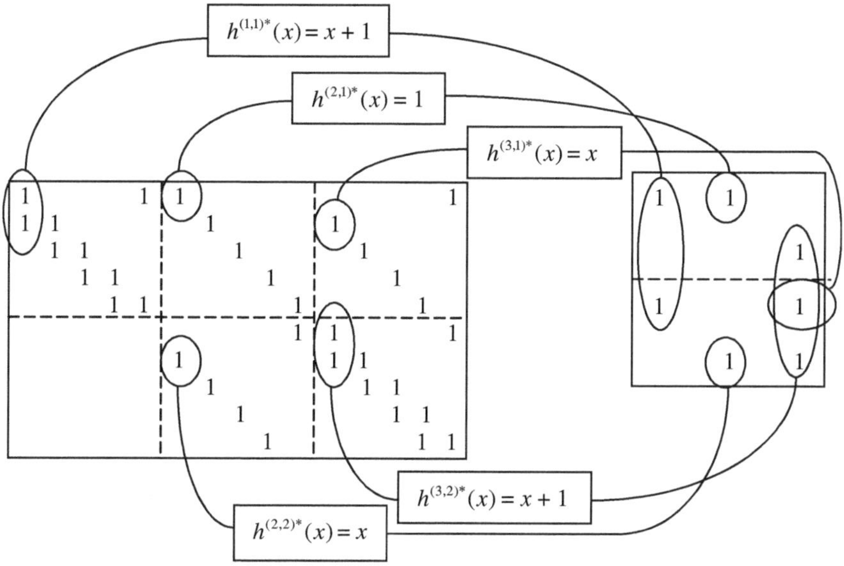

Fig. 8.14 Complete 5 × 5 circulant version of the $\underline{H}$ matrix for the 1/3-rate tail-biting convolutional code described in Example 8.14

weight and form are imposed by the reciprocal input–output syndrome former polynomials $h^{(i,j)}*(x)$, $i = 1, 2, \ldots n_o$, $j = 1, 2, \ldots r_o$.

Proof The reciprocal input–output syndrome former polynomials $h^{(i,j)}*(x)$ have been introduced in Sect. 8.1. They can be recognized in the polynomial expressions of the rows of $\underline{H}$, as stated by (8.2). Their degree is at most m_h. In the tail-biting version of this code, according to the scheme described in Fig. 8.11, such polynomials experience a cyclic permutation with steps of n_o. Then, after application of the reordering procedure introduced in Definition 4.10, with the only difference of referring to the parity check matrix and not to the generator matrix, we obtain complete circulants, whose weight and form are imposed by such $h^{(i,j)}*(x)$ polynomials. □

Example 8.14 Let us consider the 1/3-rate convolutional code introduced in Example 8.1. It is characterized by $m_h = 1$ and by two different interleaved parity check polynomials. Its reciprocal input–output syndrome former polynomials are reported here for making comparison easier,

$$h^{(1,1)*}(x) = x + 1, \quad h^{(2,1)*}(x) = 1, \quad h^{(3,1)*}(x) = x,$$
$$h^{(1,2)*}(x) = 0, \quad h^{(2,2)*}(x) = x, \quad h^{(3,2)*}(x) = x + 1.$$

In Fig. 8.14 on the left the reordered parity check matrix characterizing this code, with $Z = 5$, according to Lemma 8.1, is shown. It is important to follow the convention for understanding how the $h^{(i,j)}*(x)$ polynomials are to be interpreted.

For instance, their higher-degree power coefficients occupy lower column positions in the $\underline{H}_o^T$ sub-matrix. For comparison, in Fig. 8.14 on the right, also the true syndrome former sub-matrix is reported, and proper connection lines allow identification of any $h^{(i,j)}*(x)$ polynomial.

In the case here considered, we have no elimination of rows or columns. This leads to an exact satisfaction of the approximate relationship (7.5) between m and m_h. Such property is not always true, leading sometimes to even strong possible departures from the expected relationship between such two parameters, as already noted. It is trivial to observe that the complete circulant form of a same tail-biting convolutional code, obtained applying the reordering procedure described in Definition 4.10 to its $\underline{G}$ matrix or to its $\underline{H}$ matrix, gives rise to the same code, whose overall frame length is Zn_o. On the contrary, different results can be obtained for the truncated circulant forms.

Definition 8.3 Starting from a parity check matrix of an s.s. or w.s. time-invariant convolutional code, not in tail-biting form with Z' row layers, we take in sequence all the first, the second, … the n_oth columns, and then all the first, the second, … the r_oth rows. So we obtain a block code, with Zk_o information symbols, whereas the number of control symbols is $(Z'r_o + m_hr_o)$. This reordered form of our tail-biting convolutional code is called *truncated circulant form* of the original code *in its* $\underline{H}$ *matrix.*

In comparison with Definition 4.11, we can note two main differences: here we have vertical periods of length r_o instead of k_o, and we start from an overall number Z' of row layers in the frame, instead of the usual Z. The relationship between Z and Z' will be clarified in the following proposition.

Lemma 8.2 *Given a k_o/n_o-rate w.s. or s.s. convolutional code, not in tail-biting form, the reordering procedure introduced by Definition 8.3 gives rise to a parity check matrix with $r_o = n_o - k_o$ row layers and n_o column layers. In such layers there are $(Z' + m_h)\,xZ'$ truncated circulants, whose weight and form are imposed by the reciprocal input–output syndrome former polynomials $h^{(i,j)}*(x)$, $i = 1, 2, \ldots n_o$, $j = 1, 2, \ldots r_o$.*

Proof We can proceed exactly as in the proof Lemma 8.1. The truncated circulants are here rectangles with a vertical greater dimension. Nevertheless, in order to have the same code following the two procedures (on $\underline{G}$ and on $\underline{H}$), it is necessary to assume $Z' = Z + m$, without taking into account possible elimination of rows or columns. □

Corollary b of Theorem 4.1 states that the last m symbols in any column layer of the truncated circulant form of a convolutional code, obtained starting from its $\underline{G}$ matrix, represent as many final control symbols in the old code frame. Here we can observe that, owing to the condition $Z' = Z + m$, still such positions are occupied by these control symbols. Furthermore the parity check equations related to them are positioned in the reordered parity check matrix at the end of any row layer. Nevertheless, since in general m and m_h are different, the distribution of such parity check equations must take into account also a balance in the numbers m, m_h and Z, and even possible eliminations of rows or columns.

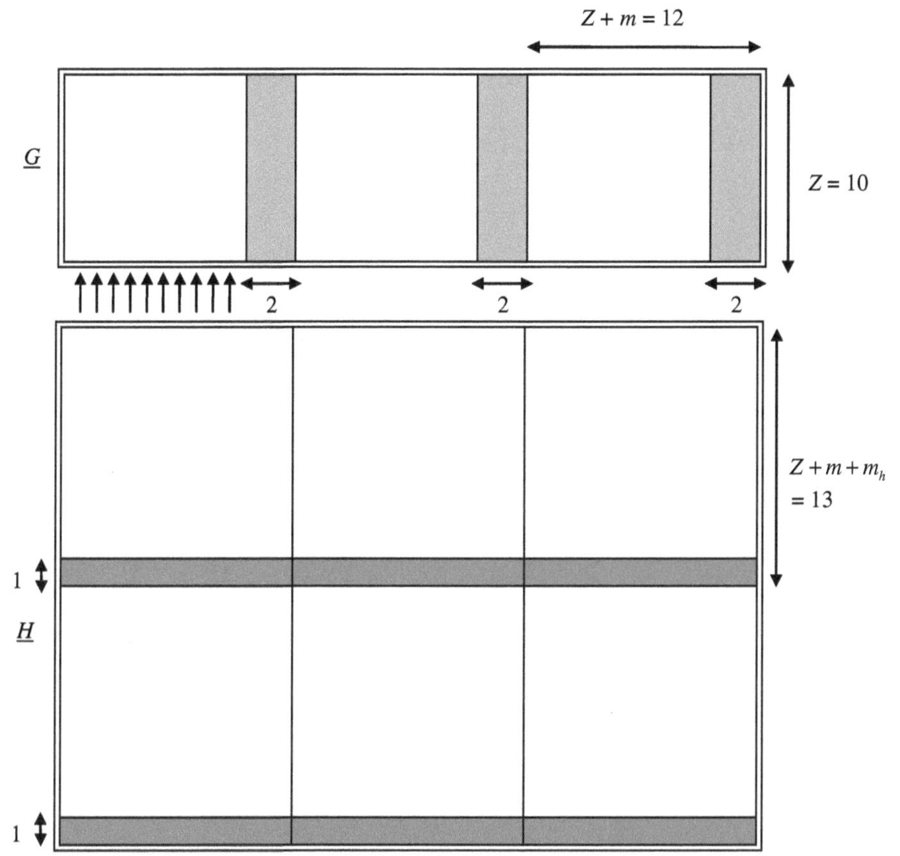

Fig. 8.15 Generator and parity check matrix for the same 1/3-rate convolutional code, after reordering to obtain truncated circulant forms ($Z = 10$, $Z' = 12$, $m = 2$, $m_h = 1$), in Example 8.15

Example 8.15 Let us consider a 1/3-rate convolutional code, with $m = 2$ and $m_h = 1$. The reordering procedure is made on $Z = 10$ periods in which information symbols are inserted. The reordered $\underline{G}$ and $\underline{H}$ matrix so obtained are qualitatively shown in Fig. 8.15, where the enlarged dimensions of the truncated circulants have been explained. The regions where 10×10 or 12×12 complete circulants are placed have been evidenced with a white colour. The $3m$ (in general $n_o m$) columns associated to as many final control symbols in the old generator matrix (now distributed) are evidenced with a bright grey colour. The $2m_h$ (in general $r_o m_h$) rows associated to as many final rows in the old parity check matrix are evidenced with a dark grey colour. Finally 10 small solid arrows denote the positions of the information symbols. The overall code exhibits $n = 36$ total symbols and $k = 10$ information symbols. In the generator matrix there are 10×12 truncated circulants. In the parity check matrix there are 13×12 truncated circulants. So the total

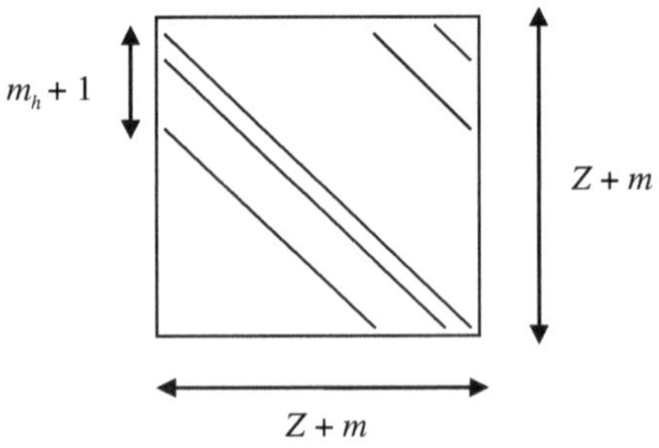

Fig. 8.16 Comparison between m_h and $(Z + m)$ in a complete square circulant

number of control symbols correctly turns out to be 26, if no rows are eliminated. Distributed control symbols in the old code frame occupy here the positions from 13 to 22 and from 25 to 34.

Obviously the m_h parity check equations representing the final control symbols in the old code frame can be positioned where one wishes in the truncated circulant $\underline{H}$ matrix (whichever reordering of the rows is always possible). Nevertheless the organization of the $\underline{H}$ matrix proposed in Fig. 8.15 down allows to exploit some advantages from what already discussed at the end of Sect. 4.8. For instance, the use of a symbolic representation for circulants and permutation circulants (see Definition 4.12 and Example 4.25), and their maximum value, fixed by an analogous of Lemma 4.5, here with m_h replacing m. All this representation is based on a schematic structure for any complete circulant in the reordered $\underline{H}$ matrix, which is of the type shown in Fig. 8.16, where the parameters m_h and $Z' = Z + m$ are compared. Clearly, in practical cases, where Z is much greater than m_h, the difference between the two dimensions of the rectangle representing the associated truncated circulant is almost negligible. Nevertheless the obtainment of such truncated circulants, after elimination of the rightmost m_h columns in the complete square circulant, is very important, from the conceptual point of view.

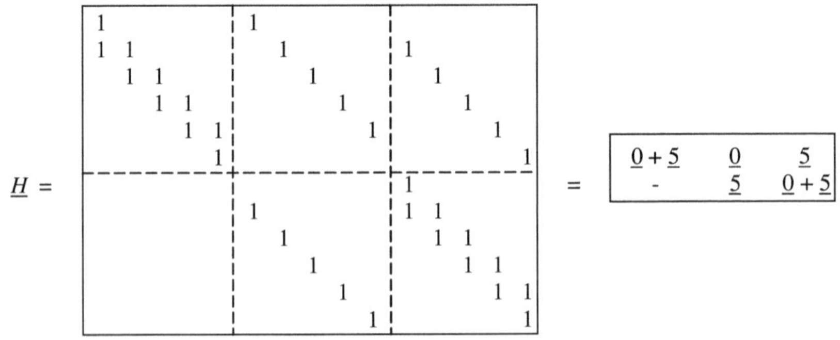

Fig. 8.17 Truncated circulant $\underline{H}$ matrix (also in symbolic form) for the 1/3-rate code treated in Example 8.16

Example 8.16 Assuming again the 1/3-rate convolutional code already treated in Example 8.14 (Fig. 8.14), with $Z = 5$, and considering that it is characterized by $m_h = 1$, we obtain the truncated circulant form for its $\underline{H}$ matrix shown in Fig. 8.17. On the right, a symbolic representation of such matrix is reported. We can observe that, adopting the convention introduced in Definition 4.12, it is necessary to label here the permutation circulants taking the complement with respect to the overall size (6 in the present case). This is due to the convention followed here for relating the $h^{(i,j)}*(x)$ polynomials to the true syndrome former sub-matrix structure.

Sometimes, in applying the procedure outlined above, some rows in the reordered $\underline{H}$ matrix can be eliminated, because useless. This is the case, for instance, of the first row in the second layer in the $\underline{H}$ matrix shown in Fig. 8.17, where a solitary 1-symbol is present. This operation is necessary also in the usual frame termination.

To summarize, Definition 4.10 regards a tail-biting convolutional code and produces a block code with complete square circulants, Definition 4.11 regards a not tail-biting convolutional code and produces a block code with horizontally truncated circulants. They both refer to transformations which can be applied to the generator matrix. Corollary b of Theorem 4.1 notes that, when Definition 4.11 is employed, the roles of distributed control symbols and final control symbols are exchanged. Here Lemma 8.1 exploits Definition 4.10, for a tail-biting convolutional code described by means of its parity check matrix, and gives rise to a block code whose parity check matrix contains complete square circulants. Definition 8.3 introduces a transformation applied to the parity check matrix of a convolutional code not in tail-biting form, leading to a block code whose $\underline{H}$ matrix contains vertically truncated circulants. Finally Lemma 8.2 notes that, when Definition 8.3 is employed, the positions of the parity check equations for distributed control symbols and for final control symbols are exchanged. In doing this, it is necessary to remember that the parity check matrix is formed by a vertical shift of the syndrome former sub-matrix $\underline{H}_o^T$ and the convention adopted in the introduction of the $h^{(i,j)}*(x)$ polynomials.

8.8 Introductory Treatment of Array Codes

Several times we have discussed about a class of QC codes, called array codes (Fan 2000), whose $\underline{G}$ matrix consists of X layers containing complete $p \times p$ circulants plus some anomalous further rows. They were firstly introduced in Sect. 2.7. Example 2.22 refers to the case characterized by $X = 2$ layers of complete circulants, subsequently we obtained a shortened version of such code in Example 2.24, a lengthened version in Example 2.28, a modified lengthened version in Example 2.33. The QC condition is characterized by a code word length $N = p^2$. All these considerations were made looking only at the $\underline{G}$ matrix. In Example 6.12, the parity check matrix of the parent QC code has been introduced, still for the same code, characterized by $p = 4$ and $X = 2$. It is formed by two layers of permutation circulants. Also the parity check matrix of its SQC version has been determined in Fig. 6.13, by simply eliminating the rightmost column layer (in order to maintain a column layer made by all $\underline{0}$ circulants on the left). In Example 6.15 the parity check matrix of a lengthened version of this code has been calculated, by adding a 5th column layer, equal to the first, and responsible of the occurrence of $p = 4$ code words having weight $w = 2$. Finally in Example 6.23 an MHEQC version of this code has been analyzed, showing that the anomalous row in the $\underline{G}$ matrix was no longer present.

Looking at the $\underline{H}$ matrix of the QC version of an array code, with only two row layers, it is possible to imagine that it belongs to a wider family of codes, whose parity check matrices exhibit more than two row layers, say Y layers, as anticipated in Sects. 6.6 and 6.7. For synthetically describing such parity check matrices, we can employ the symbolic representation of permutation circulants (Definition 4.12), in which they are associated to proper phase shifts (or simply phases). In order to take into account also the size of the circulants, we will add a subscript p to the underlined numbers adopted in this symbolism.

Definition 8.4 Given a parity check matrix made by Y row layers of square $p \times p$ permutation circulants, they are so organized: the first row layer contains all $\underline{0}_p$ circulants, the second one contains circulants whose symbolic representation increases by 1 unit at a time, the third one contains circulants whose symbolic representation increases by 2 units at a time, ... the Yth one contains circulants whose symbolic representation increases by $(Y - 1)$ units at a time. Such increases are to be calculated mod(p). The first column layer is made by all $\underline{0}_p$ circulants (but this choice is a pure convention). A code whose $\underline{H}$ matrix is so organized is defined *(Y, p) proper array code*.

About the condition on the QC code word length to be $N = p^2$, determined in the case of $Y = 2$, which has been analyzed so far, it is possible to verify that it holds also for whichever Y. As already stressed, array codes have been proposed as an interesting family of LDPC codes, whose soft-decision decoding procedure requires the avoidance of any 4-length short cycle of 1-symbols in the parity check matrix (Tanner 1981). In order to satisfy this constraint, LQC codes are to be excluded. Furthermore, when $Y > 2$, even considering code word lengths shorter than that of the QC code, it is necessary to take some cares. In particular, in order to reach a code word length as high as possible, in the case of $Y > 2$, a common factor between p and any phase increment is to be excluded.

Lemma 8.3 *The QC version of a (Y, p) proper array code, with $Y > 2$, and p prime, characterized by p column layers, and then by a code word length $N = p^2$ is free of 4-length short cycles of 1-symbols. The specification of having p prime is not necessary for $Y = 2$.*

Proof Because of the increase in the symbols representing the permutation circulants in any row layer is made with a certain step mod(p), after p column layers all the possible symbols of the alphabet have been used in all the row layers exactly one time. This is consequence of the adoption of p prime. Therefore addition of a further column layer gives rise to a 4-length short cycle. When $Y = 2$, we have only two layers, the lower one being associated to 1-unit increments in the circulant phases,

Fig. 8.18 Parity check matrices of a (Y, p) proper array code, with $p = 7$ and $Y = 3$, in QC version and in SQC version

$$\underline{H} = \begin{array}{|ccccccc|} \hline \underline{0}_7 & \underline{0}_7 & \underline{0}_7 & \underline{0}_7 & \underline{0}_7 & \underline{0}_7 & \underline{0}_7 \\ \underline{0}_7 & \underline{1}_7 & \underline{2}_7 & \underline{3}_7 & \underline{4}_7 & \underline{5}_7 & \underline{6}_7 \\ \underline{0}_7 & \underline{2}_7 & \underline{4}_7 & \underline{6}_7 & \underline{1}_7 & \underline{3}_7 & \underline{5}_7 \\ \hline \end{array}$$

$$\underline{H}' = \begin{array}{|ccccc|} \hline \underline{0}_7 & \underline{0}_7 & \underline{0}_7 & \underline{0}_7 & \underline{0}_7 \\ \underline{0}_7 & \underline{1}_7 & \underline{2}_7 & \underline{3}_7 & \underline{4}_7 \\ \underline{0}_7 & \underline{2}_7 & \underline{4}_7 & \underline{6}_7 & \underline{1}_7 \\ \hline \end{array}$$

so leading to all distinct phase differences in the p column layers, also when p is composite. Instead, when $Y \geq 3$, we can have at least three layers, the second one associated to 1-unit increments, and the third one to 2-unit increments. Taking p even, a 4-length short cycle of 1-symbols will come out, between the first and the third layer, over a length shorter than that of the QC condition. □

At this point, we can observe that the following relationship holds, relating parameters X (number of row layers in $\underline{G}$), Y (number of row layers in $\underline{H}$) and p,

$$X + Y = p$$

Clearly one can take shortened versions of the QC code considered, simply eliminating some column layers from the parent code $\underline{H}$ matrix. They typically are suppressed starting from the right hand side, so maintaining the property of having all $\underline{0}_p$ circulants in the leftmost column layer, as already noted.

Example 8.17 In Fig. 8.18 up the $\underline{H}$ matrix of the QC version of a proper array code characterized by $p = 7$ and $Y = 3$ is shown. In Fig. 8.18 down, the $\underline{H}$ matrix of its two-time shortened version is depicted. The original code is associated to the tern [49,30,6], the SQC code to the tern [35,16,6]. Such terns are justified by the fact that these $\underline{H}$ matrices show two rows linearly dependent on the others.

The following properties can be easily verified:

(i) in the $\underline{H}$ matrix of a (Y, p) proper array code there are $(Y - 1)$ rows linearly dependent on the other ones,
(ii) in the leftmost position of a certain row layer it is possible to adopt another initial permutation circulant, and then maintain the same phase increments, obtaining the same array code,
(iii) the structure of the $\underline{H}$ matrix shows a strong analogy with that of a Reed-Solomon code with $Q - 1 = p$,
(iv) the minimum distance d is always an even number.

About property (i), it is possible to eliminate one row from any row layer except one (conventionally the first). So we obtain all linear independent parity check equations. In the $\underline{G}$ matrix, correspondently, there are $R_X = Y - 1$ anomalous rows, for compensating the above exceeding rows in $\underline{H}$. Property (ii) follows from the fact that only the separation between 1-symbols in any row is important for determining the overall result. Regarding property (iii), it is possible to recognize in the $\underline{H}$ matrix discussed in Example 5.53, where a $[7,4,4]_8$ RS code is treated, a strong resemblance with the one represented in Fig. 8.18. The only difference is in employing one row layer, made by all $\underline{0}_7$ circulants, which can be interpreted as the adoption, in the uppermost position of any column layer of the permutation circulant associated of to the 8-ary symbol $\alpha^0 = 1$. Like for the rows, we can assert, in fact, that only the separation between 1-symbols in any column is important for determining the overall result. Finally, about property (iv), it can be demonstrated by induction, since it holds for $Y = 1$ (with $d = 2$), for $Y = 2$ (with $d = 4$), and, if it is valid for any Y, will be certainly valid also for $(Y + 1)$.

In spite of this simple nature, it is rather difficult to estimate the minimum distance d for $Y \geq 3$, especially if we consider some shortened versions of proper array codes (Milenkovic et al. 2008). For any further row layer which is added, augmenting Y by one unit, a binomial composition of the columns involved in the ex-OR sum of clusters of syndromes is to be taken into account. Many code words so obtained exhibit several cancellations which contribute to lower their weight. Remark that, like in Reed-Solomon codes, the $\underline{H}$ matrix described by Definition 8.4 is not in its minimal form. Nevertheless this property itself turns out to be a key feature for avoiding 4-length short cycles of 1-symbols, so making a soft-decision decoding procedure possible, for such LDPC codes (Fan 2000).

Lemma 8.4 *The exact code rate of a (Y, p) proper array code, in its QC version, is*

$$R = \frac{(p - Y)p + Y - 1}{p^2}, \quad \text{(QC)}. \tag{8.5}$$

The exact code rate of an SCQ version of the above proper array code, with code word length $n = Sp$, being $S < P$, turns out to be

$$R = \frac{(S - Y)p + Y - 1}{Sp}, \quad \text{(SQC)}. \tag{8.6}$$

Proof For the QC code, the overall word length is $N = p^2$, whereas the number of information symbols, considering also the $(Y - 1)$ rows to be eliminated in $\underline{H}$, in order to make it of full rank, is $k = Xp + (Y - 1) = (p-Y)p + Y - 1$. So the ratio k/N yields (8.5). For the SQC code, the number of $\underline{H}$ rows remains unchanged, included the $(Y - 1)$ rows to be eliminated, but the overall number of column layers decreases from p to S, so leading to (8.6). □

Typically p is much larger than Y, so that it is possible to approximate (8.5) and (8.6) respectively as

$$R \cong \frac{p - Y}{p}, \quad \text{(QC)},$$
$$R \cong \frac{S - Y}{S}, \quad \text{(SQC)}. \tag{8.7}$$

Example 8.18 Considering the two codes described in Example 8.17, they are respectively a [49,30,6] QC code and a [35,16,6] SQC code. In the latter one, in fact, we have $S = 5$. The correspondent code rates become $R = 30/49$ and $R = 16/35$. In both cases we have 19 independent parity check equations. The characteristic of maintaining invariant the number of control symbols is that shown by any type of code shortening.

Starting from Definition 8.4, we can introduce two important variants: improper array codes and generalized array codes. Improper array codes can be considered as obtained from correspondent proper array codes, after suitable constant-length puncturations. Finally generalized array codes, although maintaining the same structure, are no longer uniform along the code word length.

Definition 8.5 A code obtained from a (Y', p) proper array code, after elimination of some row layers in its $\underline{H}$ matrix, is called (Y, p) *improper array code*, with $Y < Y'$.

It is interesting to compare two array codes, the one proper and the other one improper, on the basis of equal values for parameters Y, p (and consequently code rate R). They both show uniform phase increments for the permutation circulants along any row layer, but in the proper array code such phase increments are forced to be 0, 1, … $(Y - 1)$, instead, for the improper array code, they can be chosen according to a suitable design. In this sense, such phase increments represent $(Y - 1)$ degrees of freedom, to be optimized in order to achieve the greatest minimum distance possible. Since p is prime, and the phase increase is to be calculated mod(p), the length $N = p^2$ remains that characterizing the QC condition, and also the maximum length in order to avoid 4-length short cycles. Lemma 8.4 is still valid, as well as the approximate expressions (8.7) for the code rate.

Example 8.19 Assuming still $p = 7$, and $Y = 3$, but $\Delta_2 = 2$, $\Delta_3 = 3$, instead of 1 and 2 respectively (case of the proper array code introduced in Example 8.17), we obtain a (3,7) improper array code, whose QC version is characterized by the following $\underline{H}$ matrix

$$\underline{H} = \begin{bmatrix} \underline{0}_7 & \underline{0}_7 & \underline{0}_7 & \underline{0}_7 & \underline{0}_7 & \underline{0}_7 & \underline{0}_7 \\ \underline{0}_7 & \underline{2}_7 & \underline{4}_7 & \underline{6}_7 & \underline{1}_7 & \underline{3}_7 & \underline{5}_7 \\ \underline{0}_7 & \underline{3}_7 & \underline{6}_7 & \underline{2}_7 & \underline{5}_7 & \underline{1}_7 & \underline{4}_7 \end{bmatrix}.$$

We can verify that there are no pairs of equal phase difference between any couple of row layers, so preventing the occurrence of 4-length short cycles. A possible shortened version of this code can be constructed after progressive elimination of column layers, starting from the right hand side, as usual.

All the four properties listed above are valid also for improper array codes. In fact they are consequence of the layered structure of the $\underline{H}$ matrix considered, provided that the phase increments characterizing the permutation circulants in any row are uniform. The choice of a good set of phase increments is an important challenge in order to maximize the minimum distance d, in the framework of improper array codes having comparable number of layers, code rate and overall code word length (Esmaeli et al. 2011). This choice involves the assumption of a certain minimum circulant size p.

Definition 8.6 A block code characterized by row layers of permutation circulants, like in an improper array code, but with non-uniform increasing in the phases of the permutation circulants in any row layer, is defined *generalized array code*.

When, in each row layer, all the possible permutation circulants have been used, independently of the order adopted, also this block code reaches the condition $N = p$ for being quasi-cyclic. Nevertheless, since there is no rules in its construction, it is difficult to define SQC and LQC forms. In any case, such codes are typically employed with both the number of row layers and the number of column layers much smaller than the circulant size p (Fossorier 2004). In this sense, we can interpret them as shortened (after elimination of external and also central column layers) and constant-length punctured (after elimination of row-layers), starting from a certain QC proper array code.

Some generalized array QC codes have been proposed also adopting p composite (Tanner et al. 2004). The condition for the QC behaviour remains $N = p^2$, nevertheless, as already noted, it is necessary to pay attention in order to avoid that, between two row layers, a certain difference in the circulant phases occurs more than one time. This event frustrates the minimum distance, but, more remarkably, makes soft-decision decoding as LDPC codes impossible. With strongly shortened codes, such constraint can be easily satisfied.

8.9 Generator Matrix of Improper Array Codes

In Example 2.22 we introduced the $\underline{G}$ matrix of a (2,4) proper array code ($X = 2$, $Y = 2$, $p = 4$), represented by the tern [16,9,4]. Such $\underline{G}$ matrix was characterized by $X = 2$ row layers, where the same generator sub-matrix shifts by $p = 4$ positions each time, plus an anomalous row consisting of the unitary polynomial $u_{2p}(x) = u_8(x)$. Since we have $Y = 2$, the choice of $p = 4$, as already noted, is not a problem. This [16,9,4] QC code has been demonstrated to be equivalent to the direct product code between two equal [4,3,2] parity check codes. In Example 6.12 the correspondent parity check matrix was studied, simply invoking a null ex-OR sum of clusters of syndromes, in order to verify its validity. It is characterized by $Y = 2$ row layers, formed by 4×4 permutation circulants. The second row layer shows a uniform phase increment equal to 1. In a situation so simple, also an improper array code, with a different phase increment would be equivalent.

For $Y > 2$ a different situation occurs. Clearly it is necessary to start from the $\underline{H}$ matrix and obtain the correspondent $\underline{G}$ matrix. Now we will present a procedure for doing this, with any (Y, p) improper array code. To this goal, a suitable polynomial description of both generator and parity check row layers will be adopted. Let us take a number S of subsequent sets having p positions each. Assuming $S \le p$, we can consider either SQC or QC array codes. A non-null element of the polynomial has to be placed in a precise position of a certain set. The exponent in the generic power of variable x, representing such 1-symbol, is subdivided into two parts, say x^{mp+j}. The first is multiple of p and the second plays the role of a remainder. Expression $x^{mp+j}\Big|_{R[\text{power}/p]}^{Mult[p]}$ means that this 1-symbol is placed in the set identified by integer m, and in a position identified by the remainder $R[(mp + j)/p] = R[j/p]$.

Let us consider the case of whichever number Y of row layers in $\underline{H}$, with phase increments Δ_j, $j = 2, 3, \ldots Y$, respectively in the row layers from the second to the Yth. We take $\Delta_1 = 0$ by construction, for the first row layer, as usual. The parity check polynomials (coincident with their reciprocals, being symmetric) in the Y row layers of the considered $\underline{H}$ matrix, for the QC version of such improper array code, can be symbolically written as

$$h_j(x) = u_p(x^{p+\Delta_j})\Big|_{R[\text{power}/p]}^{Mult[p]} \qquad j = 1, 2, \ldots Y, \quad (\text{QC}). \tag{8.8}$$

This symbolic polynomial, with the meaning introduced above, represents a whole row layer at a time. The unitary polynomials, to be employed in (8.8), are characterized by the trivial property of having one power in any set, so forming as many permutation circulants, all different (except for $j = 1$, since $\Delta_1 = 0$).

For the SQC version of the same code, characterized by S column layers, with $S < p$, expression (8.8) can be so modified

$$h_j(x) = u_S(x^{p+\Delta_j})\Big|_{R[\text{power}/p]}^{Mult[p]}, \quad j = 1, 2, \ldots Y, \quad (\text{SQC}), \tag{8.9}$$

in which only the weight of the unitary polynomial has been reduced from p to S.

Example 8.20 The (3,7) improper array code discussed in Example 8.19, characterized by $\Delta_2 = 2$, $\Delta_3 = 3$, exhibits the following three symbolic parity check polynomials in its QC version

$$h_1(x) = u_7(x^7)\Bigg|_{\substack{Mult[7]\\R[\text{power}/7]}}, \quad h_2(x) = u_7(x^{7+2})\Bigg|_{\substack{Mult[7]\\R[\text{power}/7]}},$$

$$h_3(x) = u_7(x^{7+3})\Bigg|_{\substack{Mult[7]\\R[\text{power}/7]}}, \quad (\text{QC}).$$

Taking the SQC version with $S = 5$, derived from it, we obtain instead

$$h_1(x) = u_5(x^7)\Bigg|_{\substack{Mult[7]\\R[\text{power}/7]}}, \quad h_2(x) = u_5(x^{7+2})\Bigg|_{\substack{Mult[7]\\R[\text{power}/7]}},$$

$$h_3(x) = u_5(x^{7+3})\Bigg|_{\substack{Mult[7]\\R[\text{power}/7]}}, \quad (\text{SQC}).$$

Theorem 8.1 *For an improper array code, characterized by parameters Y, p, Δ_j, $j = 2, 3, \ldots Y$, the $\underline{G}$ matrix is time-varying, with two types of generator polynomials. Those present in the X row layers of circulants shift with steps of p, and exhibit the following symbolic expression*

$$g(x) = \prod_{j=1}^{Y} (x^{p+\Delta_j} + 1)\Bigg|_{\substack{Mult[p]\\R[\text{power}/p]}} \tag{8.10}$$

Then there are $(Y - 1)$ anomalous rows, where the unitary generator polynomial

$$g_a(x) = u_{2p}(x) \tag{8.11}$$

shifts as well with steps of p. The SQC version of this improper array code is characterized by the same polynomials. Simply they only shifts in a lower number of row layers, since the whole code word length is reduced.

Proof Firstly we consider expression (8.10), ignoring, for the moment, the operations $Mult[p]$ and $R[\text{power}/p]$. Multiplication of whichever parity check polynomial (8.8) by (8.10) allows satisfaction of constraint (5.4). The property exploited here is analogous to the one relating the parity check polynomial and the generator polynomial in a GPC code (see Sect. 5.14). Operations $Mult[p]$ and $R[\text{power}/p]$, which introduce a p-length periodicity, do not limit the validity of the above

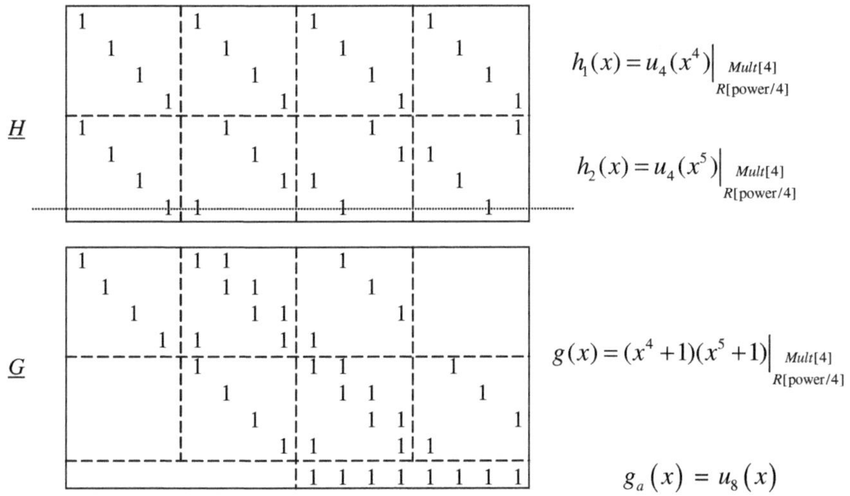

Fig. 8.19 Parity check matrix and generator matrix for the (2,4) proper array code treated in Examples 2.22, 6.12, 8.21

property, because any row of a circulant layer in $\underline{G}$ can be coupled with any row of a circulant layer in $\underline{H}$ (Mittelholzer 2002). If this is true for QC codes, it is certainly true also for SQC codes, described by (8.9) and again by (8.10). Now we consider expression (8.11). Any pair of column layers in $\underline{H}$ satisfies the null ex-OR sum of cluster of syndromes, so any possible combinations of two $u_p(x)$ polynomials (placed in as many p-length periods) can be adopted. The choice of taking two adjacent $u_p(x)$ polynomials, so obtaining expression (8.11), is made for convenience. Doing this, it is possible to easy verify that the total number of independent rows of the type $x^{jp}g_a(x)$, j integer, is $(Y - 1)$. Finally the option of adopting an exponent factor $j = 0, 1, \ldots (Y - 2)$, is advisable because, so operating, such anomalous polynomials are placed at the end of the code word, making attribution of proper positions to the resulting further information symbols easier. □

Example 8.21 The generator matrix introduced in Example 2.22 and the correspondent parity check matrix presented in Example 6.12 are compared in Fig. 8.19. On the right, also the symbolic polynomial expressions for the parity check and generator row layers have been reported. The redundant row in the $\underline{H}$ matrix has been eliminated. The anomalous additional row in the $\underline{G}$ matrix compensates the above elimination. Rigorously, this is a proper array code, nevertheless also adopting a different value for parameter Δ_1, and hence considering an improper array code, we would obtain an equivalent result. It is possible to employ also circulant symbolic representations (ignoring elimination of rows in the $\underline{H}$ matrix and addition of anomalous rows in the $\underline{G}$ matrix). For the simple case here considered, we obtain

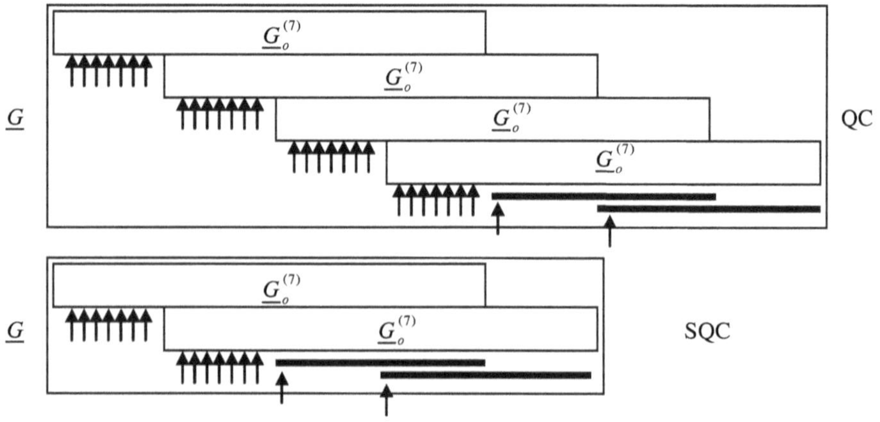

Fig. 8.20 Schematic representation of the generator matrix for the [49,30,6] QC array code and for the [35,16,6] SQC version obtained from it, as discussed in Example 8.22

$$\underline{H} = \begin{bmatrix} \underline{0}_4 & \underline{0}_4 & \underline{0}_4 & \underline{0}_4 \\ \underline{0}_4 & \underline{1}_4 & \underline{2}_4 & \underline{3}_4 \end{bmatrix}, \quad \underline{G} = \begin{bmatrix} \underline{0}_4 & (\underline{0}_4 + \underline{1}_4) & \underline{1}_4 & - \\ - & \underline{0}_4 & (\underline{0}_4 + \underline{1}_4) & \underline{1}_4 \end{bmatrix}.$$

Adopting the symbolism introduced in Chap. 2, we can label the unique sub-matrix containing $g(x)$, given by polynomial expression (8.9), as $\underline{G}_o^{(p)}$. It shifts along $\underline{G}$ with steps of p, and can be represented, by means of the symbolic expression for the circulants, with just one row, as shown above.

Example 8.22 Taking the [49,30,6] QC improper array code, with parameters $p = 7$, $Y = 3$, $\Delta_2 = 2$, $\Delta_3 = 3$, whose $\underline{H}$ matrix has been treated in Examples 8.19 and 8.20, its $\underline{G}$ matrix is characterized by the polynomial expressions

$$g(x) = (x^7 + 1)(x^{7+2} + 1)(x^{7+3} + 1)\Big|_{\substack{Mult[7] \\ R[\text{power}/7]}}, \quad g_a(x) = u_{14}(x).$$

The first expression gives rise to a generator sub-matrix which can be symbolically written as

$$\underline{G}_o^{(7)} = [\underline{0}_7 \quad (\underline{0}_7 + \underline{2}_7 + \underline{3}_7) \quad (\underline{2}_7 + \underline{3}_7 + \underline{5}_7) \quad \underline{5}_7] \ .$$

The resulting overall $\underline{G}$ matrix has a scheme of the type shown in Fig. 8.20 up, where the two anomalous rows have been placed at the end of the code word, as suggested, and represented by suitable segments. Small arrows denote where information symbols are inserted. A [35,16,6] SQC version of the same code is characterized by a generator matrix whose scheme is reported in Fig. 8.20 down. Here we have eliminated two row layers, but maintained the two anomalous rows in the same final positions. This particular type of QC code shortening procedure has been already employed in Sect. 2.7 for $Y = 2$ (see Fig. 2.20).

Except for the presence of the anomalous rows, this code can be considered w.s. time-invariant in its $\underline{G}$ matrix. In fact sub-matrix $\underline{G}_o^{(p)}$ contains p polynomials, which periodically repeat. With the general symbolism adopted in Chap. 2 about QC codes, the number of periods contained in $\underline{G}_o^{(p)}$ is $(r_p + 1)$.

Corollary a *A QC (or SQC) improper array code, characterized by parameters Y and p, exhibits a* $\underline{G}_o^{(p)}$ *sub-matrix, in which* $r_p = Y$*. Furthermore the weights of the rows in all the* $(r_p + 1)$ *periods spanned by* $\underline{G}_o^{(p)}$ *obey to a binomial law, except when cancellations occur in expression* (8.10).

Proof The equality $r_p = Y$ follows directly from the number of layers in the $\underline{H}$ matrix, equal to Y. Analogously, we can take $k_p = X$. Also in any shortened version of this array code, sub-matrix $\underline{G}_o^{(p)}$ is the same. About the non-null terms in any period of $\underline{G}_o^{(p)}$, calculation of the powers in polynomial $g(x)$, as given by (8.10), can be performed by the so called Newton's binomial power formula, well known in polynomial theory (Abramowitz and Stegun 1968). If there are no cancellations, the number of non-null terms in any period coincides with the weight contribution of such period to the overall w[$g(x)$]. □

Corollary b *The minimum distance d in an improper array code, with parameters Y and p, in QC or in SQC form, is so upper bounded*

$$d \le 2^Y. \tag{8.12}$$

Proof If no cancellation occurs, the sum of all the binomial weights in the $(Y + 1)$ periods spanned by $\underline{G}_o^{(p)}$ gives w[$g(x)$] $= 2^Y$. The minimum distance cannot be greater than this weight. □

Example 8.23 In the improper array code considered in Example 8.22, which is characterized by $Y = 3$, there are no cancellations, therefore the weights in the subsequent periods are: 1, 3, 3, 1. The sum of such weights is $2^3 = 8$.

Example 8.24 Let us consider an improper array code characterized by $Y = 4$ and $\Delta_2 = 1$, $\Delta_3 = 4$, $\Delta_4 = 5$. In this case, independently of the value of p, provided that it is higher than Δ_4, the following expression for $g(x)$ holds

$$g(x) = (x^p + 1)(x^{p+1} + 1)(x^{p+4} + 1)(x^{p+5} + 1)\Big|_{\substack{Mult[p] \\ R[\text{power}/p]}},$$

where one couple of powers vanishes in the central position $(2p + 5)$. The resulting $\underline{G}_o^{(p)}$ sub-matrix turns out to be

$$\underline{G}_o^{(p)} = \left[\underline{0}_7 \quad (\underline{0}_7 + \underline{1}_7 + \underline{4}_7 + \underline{5}_7) \quad (\underline{1}_7 + \underline{4}_7 + \underline{6}_7 + \underline{9}_7) \quad (\underline{5}_p + \underline{6}_p + \underline{9}_p + \underline{10}_p) \quad \underline{10}_p\right].$$

Since the weight of each row in such sub-matrix is w[$g(x)$] = 14, the minimum distance d cannot be greater than 14. It is interesting to note that, also in the presence of cancellations, the number of equal symbols in $\underline{G}_o^{(p)}$ is always even. This is due to a first row layer in the $\underline{H}$ matrix, where all $\underline{0}_p$ permutation circulants are present, and simplifies a rough inspection of such expressions. We can also note that property (iv), introduced in Sect. 8.8, is clearly verified.

Besides possible cancellations within a single $\underline{G}_o^{(p)}$ sub-matrix, there is the risk that the ex-OR sum of two rows of $\underline{G}$ in different layers can give a weight even lower. This makes prediction of the minimum distance in improper array codes still an open question (Baldi et al. 2012). Certainly, with increasing parameter p, the performance becomes better and better, up to a condition, after which a sort of asymptotic behaviour is reached. This property resembles the typical characteristic of a tail-biting convolutional code.

Looking at both matrices, $\underline{G}$ and $\underline{H}$, if we consider either Y or S much smaller than p, it is possible to ignore, for a moment, the anomalous rows in the former one and the eliminated rows in the latter one. So we can recognize the reordered forms of the correspondent matrices describing an $(S - Y)/S$-rate tail-biting convolutional code. This fact will be discussed in the following sections with more details, with reference to a possible inverse transformation, so leading, after unwrapping, to a convolutional version of improper array codes.

8.10 A Second Bridge Between Quasi-Cyclic Codes and w.s. Time-Invariant Convolutional Codes

In Sect. 4.9 a first conceptual bridge has been described, between quasi-cyclic block codes and w.s. time invariant convolutional codes. It was based on the generator matrix only. Now it is possible to construct a second bridge, comprehending also the parity check matrix. Let us consider the scheme depicted in Fig. 4.15, in order to reinterpret its parts, taking into account also possible effects on the parity check matrix, so enriching the meaning of some conclusions.

About the uppermost correspondence, we can observe that Theorem 4.1 still fixes the conditions for allowing a relation between QC codes and w.s. tail-biting convolutional codes. Unwrapping can be performed either in the $\underline{G}$ matrix, or in the $\underline{H}$ matrix, from a structure of the type respectively shown in Figs. 2.11 or 6.7. About the central correspondence, the considerations developed in Sect. 8.7 enable us to conclude that, not only for the generator matrix, it is possible to find the input–output polynomials in the reordered matrix describing the complete circulant form of a tail-biting convolutional code. This property holds also for the input–output syndrome former polynomials, provided that expressions of the type (8.2) are adopted for them. Figure 8.14 explains this property. The main difference with

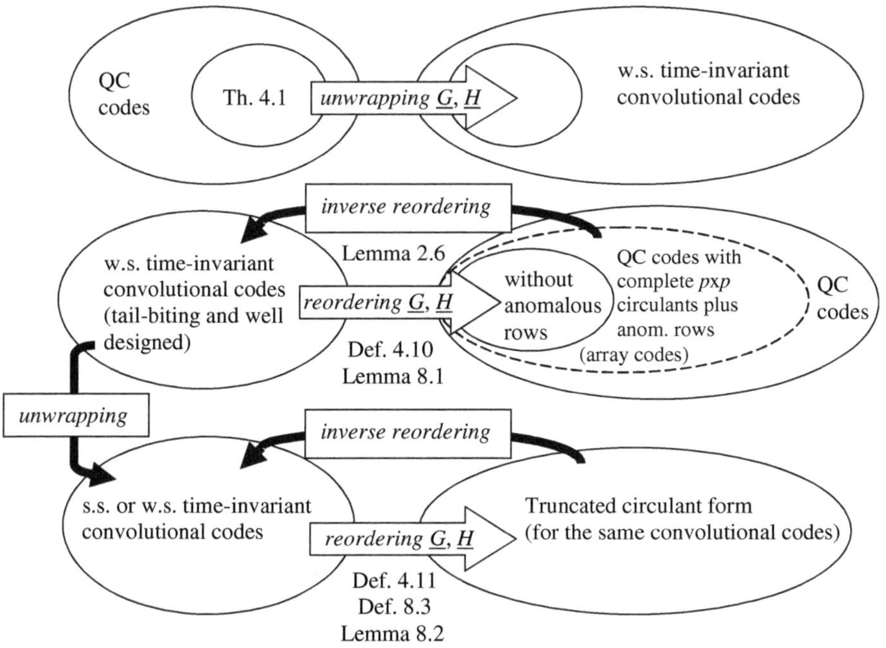

Fig. 8.21 Main ensemble correspondences in the bridge between QC codes and w.s. time-invariant convolutional codes referring to both matrices $\underline{G}$ and $\underline{H}$

respect to the $\underline{G}$ matrix transformation is in the symbolic representation of the circulants, by means of the complements with respect to $(Z + m)$.

In Fig. 8.21 the conceptual scheme presented in Fig. 4.15 has been completed, considering all the above observations. The central correspondence regards array codes. They have been proposed in the literature (Fan 2000) with reference to the only parity check matrix, as QC block codes characterized by a structural similarity with Reed-Solomon codes. The $\underline{G}$ matrix of improper array codes can be constructed according to Theorem 8.1. It is made by complete $p \times p$ circulants, collected in a generator sub-matrix $\underline{G}_o^{(p)}$, shifting in its row layers, plus $(Y - 1)$ anomalous rows. From such codes, and even from more general QC codes with similar structure, applying a reordering procedure of the type introduced in Lemma 2.6 (but substituting q_o with p), it is possible to obtain tail-biting well designed convolutional codes. Such reordering procedure can be considered as the inverse of that described in Definition 4.10 (acting on $\underline{G}$) or in Lemma 8.1 (acting on $\underline{H}$).

The lowermost correspondence deals with the possibility of applying the reordering procedure previously described directly to convolutional codes, which are not tail-biting, obtaining suitable block codes as their truncated circulant forms. In this correspondence, we will use Definition 4.11 (acting on $\underline{G}$) or Definition 8.3 (acting on $\underline{H}$). In the reverse direction, a straightforward variant of Lemma 2.6 can be employed also here. Remark that the circulants have size Z and must be

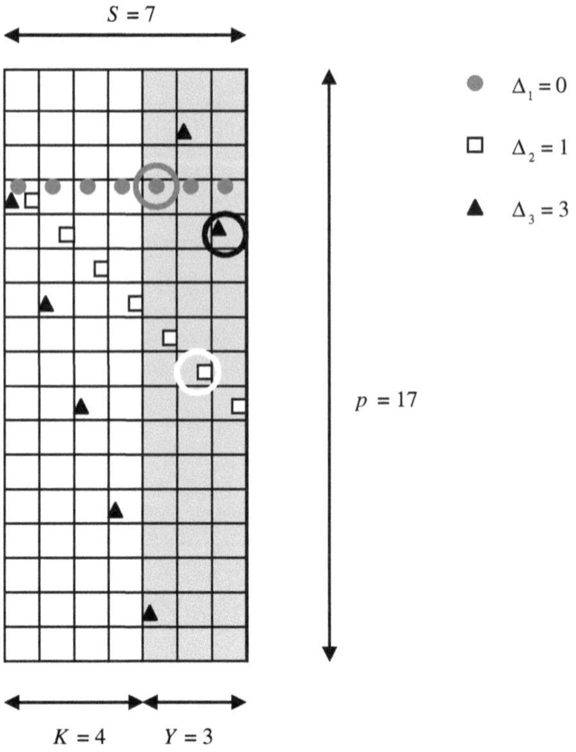

Fig. 8.22 Encoding memory for a [119,70,6] improper array code, characterized by $Y = 3$, $S = 7$, $p = 17$, $\Delta_2 = 1$, $\Delta_3 = 3$

horizontally truncated if referred to the $\underline{G}$ matrix, have size $(Z + m)$ and must be vertically truncated if referred to the $\underline{H}$ matrix. Finally unwrapping can be employed for transforming the tail-biting convolutional code equivalent to a certain QC code into its not tail-biting form, capable of being arranged over whichever frame length.

Coming back to improper array codes, they can be considered as the result of the synergic action of Y different component codes. In order to understand the mechanism responsible of the role played by each component code, in the overall parity check matrix, it is useful to employ an encoding memory. Let us consider an improper array code, characterized by parameters Y and S both remarkably smaller than p. Letting $K = S - Y$, parameter which equals the number of row layers in $\underline{G}$, we can look at a rectangle with horizontal dimension S and vertical dimension p. In Fig. 8.22 this rectangle has been depicted, taking $S = 7$ and $Y = 3$ and putting in evidence the positions influenced by rows of $\underline{H}$ belonging to different layers (different component codes). The information sequence is written by columns, from up to down, and from left to right. To fix ideas, we have chosen $p = 17$, $\Delta_2 = 1$, $\Delta_3 = 3$. Trajectories with different marks denote the actions of the three component codes, all starting from the same position, where they join. In any position of the encoding memory, there are three trajectories having a common point (representing the three

1-symbols in any $\underline{H}$ column). The overall code, taking into account also the presence of two redundant rows in $\underline{H}$, is described by the tern [119,70,6].

Although the considered code has not a parity check matrix in lower triangular form, we can assign the last three positions in any row of the encoding memory to control symbols (bright grey colour), one for each component code, as denoted by the small circles. The component codes, employing solitary parity check equations, can be considered as parity check codes, operating after a proper interleaving process, performed in this encoding memory. In a situation of this type, a reordering procedure, which represents the inverse of the one described in Definition 4.10, means simply to write by rows, instead of by columns. For instance, the first component code, instead of being a generalized parity check code with parity check polynomial $h_1(x) = u_S(x^p)$, becomes an ordered ensemble of simple parity check codes with parity check polynomial $h_1(x) = u_S(x)$. In the remaining part of the present section, we will investigate how such inverse reordering could give very interesting convolutional codes, which are also LDPC codes.

Definition 8.7 Starting from the parity check matrix of an improper array code with parameters K, Y, $S = K + Y$, p, and the circulant shift increments Δ_j, $j = 1, 2, \ldots Y$, we take in sequence the $\underline{H}$ columns occupying the 1st, the $(p + 1)$th, the $(2p + 1)$th, ... positions, then the 2nd, the $(p + 2)$th, the $(2p + 2)$th ... positions, and so on. After this, we take in sequence the $\underline{H}$ rows occupying the 1st, the $(p + 1)$th, the $(2p + 1)$th, ... positions, then the 2nd, the $(p + 2)$th, the $(2p + 2)$th ... the positions, and so on. This is made including the $(Y - 1)$ rows linearly dependent on the others. The same operation is performed on the $\underline{G}$ matrix, included its $(Y - 1)$ anomalous rows. So we obtain a code structure with distributed control symbols, having the form of a tail-biting w.s time-invariant convolutional code, plus the transformations of the anomalous rows of $\underline{G}$. This reordered form of the considered code is called *tail-biting convolutional form of an improper array code*.

The procedure described in Definition 8.7 is very similar to that adopted in Lemma 2.6. The only differences are in considering parameter p instead of parameter q_o, and in taking an initial rectangular matrix ($K \times S$ in $\underline{G}$ or $Y \times S$ in $\underline{H}$, both expressed with numbers of layers) instead of a square matrix $(pq_o) \times (pq_o)$. On the other hand, array codes do not exhibit distributed control symbols, except those related to the anomalous rows. Hence they are, by themselves, similar to QC codes in the form introduced in Sect. 2.1, and, just after this reordering procedure, they become similar to those derived in Sect. 2.3, which is the goal of the transformation here treated.

Theorem 8.2 *Under the assumption listed in Definition 8.7, the parity check polynomials in the tail-biting version of the considered SQC improper array code are*

$$h_j(x) = u_S(x^{\Delta_j S+1})\big|_{\mathrm{mod}(x^{pS}+1)}, \quad j = 1, 2, \ldots Y, \quad (\mathrm{SQC}). \tag{8.13}$$

The correspondent generator polynomials turn out to be

$$g(x) = \prod_{j=1}^{Y} (x^{\Delta_j S+1} + 1)\big|_{\mathrm{mod}(x^{pS}+1)} \quad \text{(SQC)}, \tag{8.14}$$

as transformations of the rows in the circulant layers, and

$$g_a(x) = (x + 1)u_p(x^S) \quad \text{(SQC)}, \tag{8.15}$$

as transformations of the anomalous rows.

Proof Expressions (8.13) derive from expressions (8.9) after the reordering procedure described in Definition 8.7, which can be interpreted also as produced by a change of the writing-encoding process in the rectangular memory shown in Fig. 8.22. In particular, the first parity check polynomial becomes simply $h_1(x) = u_S(x)$, in fact we have $\Delta_1 = 0$. Expressions (8.14) and (8.15) derive from expressions (8.10) and (8.11), following the same way. Alternatively we can observe that the products of all the parity check polynomials (8.13) with either (8.14) or (8.15) satisfy constraint (5.4), like in the proof of Theorem 8.1. Remark that (8.14) are present in layers having K rows, and shift with steps of S positions. □

To summarize, the tail-biting convolutional code so obtained has period length $n_o = S$, number of information symbols in any period $k_o = K$, number of control symbols in any period $r_o = Y$, number of periods $Z = p$. Rigorously, it cannot be considered a convolutional code, in fact in its $\underline{G}$ matrix there are also $(Y - 1)$ anomalous periodic rows $g_a(x)$, given by (8.15). This feature is not immediately recognizable from inspection of the reordered $\underline{H}$ matrix, although it shows $(Y - 1)$ linearly dependent rows. In this sense, such $\underline{H}$ matrix is clearly not in its minimal form, but this property is desired because it assures an LDPC behaviour.

Remark that, owing to operation mod$(x^{pS} + 1)$, when the sum of parameters Δ_j is greater than about p, expression (8.14) has a part which is self-superimposed, leading to possible cancellations (the same already mentioned in the original $\underline{G}_o^{(p)}$ sub-matrix). In spite of this fact, we can establish the entity of a constraint length for either the generator or the parity check matrix. Such considerations are useful in order to prepare a possible unwrapping operation.

Corollary a *Under the assumption of Theorem 8.2, the following relationships hold, on the generator constraint length and the parity check constraint length,*

$$m = \min\left\{ \sum_{j=2}^{Y} \Delta_j,\ (p - 1) \right\}, \tag{8.16}$$

$$m_h = \min\{(S - 1)\Delta_Y,\ (p - 1)\}, \tag{8.17}$$

where (8.16) *does not comprehend the anomalous rows* $g_a(x)$, *whereas* (8.17) *holds for the longest one among the* $h_j(x)$*'s.*

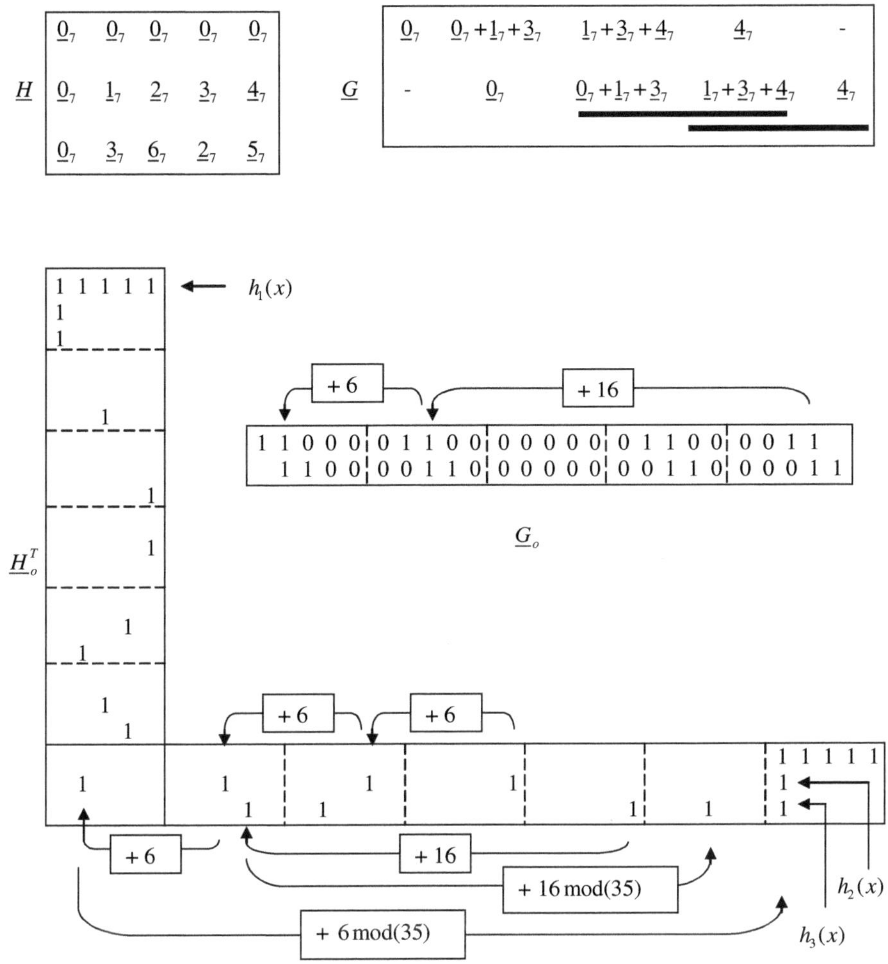

Fig. 8.23 Parity check and generator matrix before reordering, syndrome former and generator sub-matrix after reordering (Example 8.25)

Proof Ignoring operation mod(x^{pS} + 1), we introduce a polynomial $g'(x)$, and a set of polynomials $h_j'(x)$. For the first, from (8.14), it is possible to write

$$\deg[g'(x)] = \sum_{j=2}^{Y} (\Delta_j S + 1) = (Y - 1) + S \sum_{j=2}^{Y} \Delta_j,$$

so the integer part of the division of this quantity by S is $\sum_{j=2}^{Y} \Delta_j$. On the other hand, the quantity $(p - 1)$ is a sort of saturation value, reached when the whole length of the frame is occupied by the generator or by the parity check polynomials,

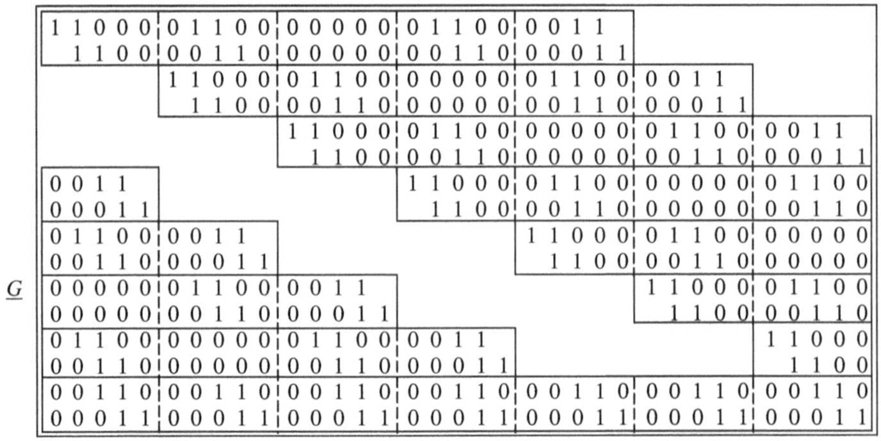

Fig. 8.24 Whole generator matrix of the tail-biting convolutional code obtained after reordering an improper array code (Example 8.25)

so taking into account operation mod($x^{pS}+1$). Therefore (8.16) is demonstrated. As for $h_j'(x)$, from (8.13), it is possible to write

$$\deg\left[h_j'(x)\right] = (S-1)\Delta_j S.$$

Since the sequence of the Δ_j's has been conventionally adopted strictly increasing, the most critical polynomial will be the Yth. So we are left with the quantity $(S-1)\Delta_Y S$, which, divided by S, yields $(S-1)\Delta_Y$. Therefore also (8.17) is demonstrated. □

In (8.16) and (8.17) the saturation value $(p-1)$ are to be considered when $g'(x)$ or $h_Y'(x)$ do not coincide with $g(x)$ or $h_Y(x)$ respectively. In all the analysis of parity check polynomials developed so far, we have ignored the term reciprocal because all these polynomials are symmetric. Furthermore, even when they are self-superimposed, cancellations cannot occur by construction. Instead the possible cancellations in $g(x)$, as previously noted, already occurred before reordering.

Example 8.25 Taking an SQC improper array code characterized by $p = 7$, $Y = 3$, $S = 5$, and hence $K = 2$, with $\Delta_2 = 1$, $\Delta_3 = 3$, we obtain the situation described in Fig. 8.23, where the parity check and generator matrix of the original code have been reported up, in symbolic form. The syndrome former sub-matrix $\underline{H}_o^T$ and the generator sub-matrix $\underline{G}_o$ of the resulting tail-biting convolutional code are constructed down. One row layer of the new $\underline{H}$ matrix has been also shown, in order to allow an easy identification of all the terms in expressions (8.13), regarding the parity check polynomials. The polynomial representations of the rows in both matrices, after reordering, according to Theorem 8.2, can be written as

$$h_1(x) = u_5(x), \quad h_2(x) = u_5(x^6)\big|_{\mathrm{mod}(x^{35}+1)}, \quad h_3(x) = u_5(x^{16})\big|_{\mathrm{mod}(x^{35}+1)},$$
$$g(x) = (x+1)(x^6+1)(x^{16}+1)\big|_{\mathrm{mod}(x^{35}+1)}, \quad g_a(x) = (x+1)u_7(x^5).$$

The two degree increments have been so calculated: $6 = S + 1$ in $h_2(x)$, $16 = 3S + 1$ in $h_3(x)$. They are taken mod(pS) = mod(35). Two black segments represent the two anomalous rows in the $\underline{G}$ matrix before reordering. About the generator sub-matrix $\underline{G}_o$, we can observe that it shows two equal rows, having weight $w = 8$ and made by couples of adjacent 1-symbols.

The whole reordered $\underline{G}$ matrix is shown in Fig. 8.24. Empty periods have been put in evidence. They are present because the generator constraint length $m = 4$, after (8.16), is smaller than $p - 1 = 6$. The two anomalous rows are characterized by a same 1-symbol pattern (8.15), whose length covers the whole code word (two lowermost rows). We can note that, except for the presence of the anomalous rows $g_a(x)$, which are periodic, the tail-biting convolutional code so obtained can be considered s.s. time-invariant in its generator matrix.

On the other hand, $\underline{H}_o^T$ exhibits a parity check constraint length $m_h = p - 1 = 6$, according (8.17), in which $(S - 1)\Delta_Y = 12$ would be greater. Therefore in this matrix there are not empty periods. Identification of $\underline{H}_o^T$ is then rather difficult in the overall parity check matrix, because of self-superimposition. On the whole, we have a [35,16,6] code. It is possible to find different groups of rows in $\underline{G}$ whose ex-OR sum give weight $w = 6$. Also the anomalous rows contribute to this result.

Instead of considering an SQC improper array code, we can describe all the reordering procedure discussed in this section, starting from the parent QC code. Definition 8.7 and Theorem 8.2 still hold, with the only difference of substituting S with p. Expressions (8.13), (8.14), (8.15) can then be replaced respectively by

$$h_j(x) = u_p(x^{\Delta_j p+1})\big|_{\mathrm{mod}(x^{p^2}+1)}, \quad j = 1, 2, \ldots Y, \quad (\mathrm{QC}). \tag{8.18}$$

$$g(x) = \prod_{j=1}^{Y} (x^{\Delta_j p+1} + 1)\big|_{\mathrm{mod}(x^{p^2}+1)}, \quad (\mathrm{QC}), \tag{8.19}$$

$$g_a(x) = (x+1)u_p(x^p), \quad (\mathrm{QC}). \tag{8.20}$$

These codes have higher code rates, but typically smaller values of the minimum distance (or at least a higher number of minimum-weight code frames), even in the presence of an optimized choice of the design parameters Δ_j. In this sense, shortened improper array codes are preferable for practical applications as very interesting rate-adaptive LDPC codes (Milenkovic et al. 2008).

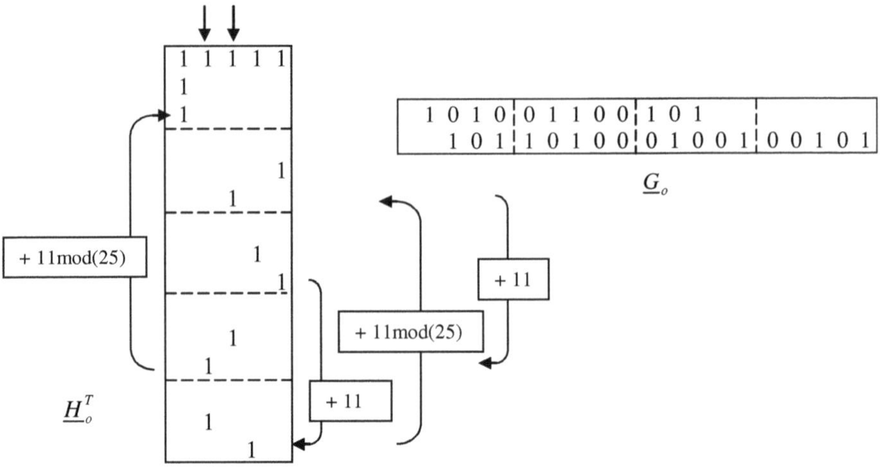

Fig. 8.25 $\underline{H}_0^T$ and $\underline{G}_0$ sub-matrix for the convolutional code obtained after unwrapping the $\underline{H}$ matrix of the QC (proper) array code described in Example 8.26

8.11 Unwrapping the Tail-Biting Convolutional Form of Improper Array Codes

Once obtained the tail-biting convolutional code equivalent to an improper array code, following Theorem 8.2, some comments have to be made about the possibility of unwrapping it. A different result is reached depending on the fact that we unwrap $\underline{G}$ or $\underline{H}$. The encoding memory, after unwrapping, can be imagined as characterized by a semi-infinite vertical dimension. This fact produces remarkable effects on the structure of the code. First of all, the anomalous rows in the $\underline{G}$ matrix vanish (since they are related to a finite length p characterizing the encoding memory), together with the rows in $\underline{H}$ which were not linearly independent. It is important to remember that parameter $n_o = S$ imposes here only the asymptotic code rate (being $k_o = K = S - Y$), and not the true code rate as is for the tail-biting code, equivalent to the parent improper array code. Since such unwrapping operation is employed for constructing convolutional LDPC codes, the $\underline{H}$ matrix, which is requested to maintain some particular properties (Tanner 1981), is unwrapped, and then the correspondent $\underline{G}$ matrix is calculated by suitable application of the procedure based on the null ex-OR sum of clusters of syndromes. We will start from the special condition of a QC code, performing unwrapping on $\underline{H}$, and then obtain the correspondent $\underline{G}$ matrix. Next we will analyze the case of an SQC code.

Example 8.26 Considering the QC array code characterized by $p = 5$, with $X = K = 2$, $Y = 3$, and phase increments $\Delta_2 = 1$, $\Delta_3 = 2$, after unwrapping its $\underline{H}$ matrix, we obtain the syndrome former sub-matrix of a convolutional code, characterized by asymptotic code rate 2/5, shown in Fig. 8.25 on the left. It is possible to note that, in the second and third row of each layer in $\underline{H}_o^T$, there is

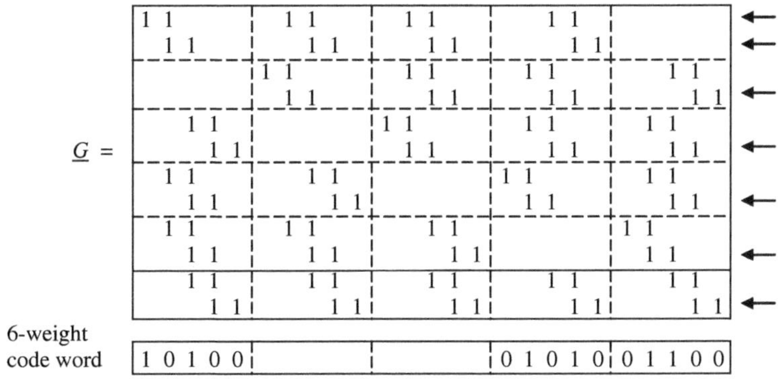

Fig. 8.26 Complete $\underline{G}$ matrix before unwrapping (Example 8.26)

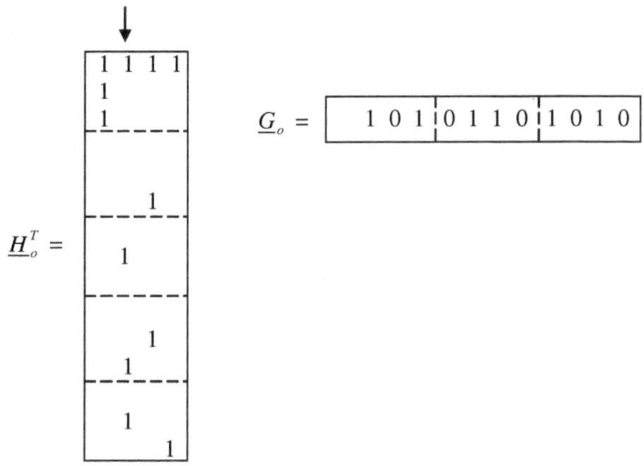

Fig. 8.27 $\underline{H}_o^T$ and $\underline{G}_o$ sub-matrix for the convolutional code obtained after unwrapping the $\underline{H}$ matrix of the SQC array code described in Example 8.27

always one and only one 1-symbol. This characteristic is peculiar of the QC versions of such reordered array codes. Only for the third rows, the vertical intervals imposed by $\Delta_3 = 2$ have been evidenced, with $11 = 2p + 1$, to be taken mod $(p^2) = \mathrm{mod}(25)$.

The $\underline{G}_o$ sub-matrix, which is compatible with the previous $\underline{H}$ matrix, is reported on the right in Fig. 8.25. We can observe that the generator constraint length turns out to be $m = 3$ and equals the sum of the Δ_j's, verifying (8.16). The positions in the period where the two information symbols are placed have been denoted with small arrows. This code is w.s. time-invariant in its $\underline{G}$ matrix. In the first row of $\underline{G}_o$ there is a 6-weight polynomial. A weight $w = 6$, which imposes also the minimum distance d, is due to the cancellation of a pair of 1-symbols. This fact can be interpreted looking at the whole generator matrix before unwrapping, which is reported in Fig. 8.26. The last two rows correspond, as usual, to the anomalous rows, in this case expressed by (8.20). The layers where $\underline{G}_o$ shifts, in such $\underline{G}$ matrix, are characterized by only one empty period. Taking the ex-OR sum of the rows whose positions are

1, 2, 4, 6, 8, 10, 12 (denoted by small arrows), we obtain the code word reported down, which, except for a cyclic permutation, coincides with the first row in the $\underline{G}_o$ sub-matrix in Fig. 8.25, confirming cancellation of a couple of 1-symbols. In this sense, the presence of the anomalous rows has contributed to reduce the minimum distance. This limitation is due to the too small value adopted for p. The first two weight distribution coefficients, characterizing this 2/5-rate w.s. time-invariant convolutional code, are $A(6) = Z$, $A(8) = 3Z - 2$. With a higher value of p, also an optimized choice of the two design parameters Δ_2, Δ_3 is possible and a minimum distance $d = 8$, with a comparable code rate, can be achieved.

The advantage of having obtained $\underline{G}_o$ by means of the null ex-OR sum of clusters of syndromes is in that we have searched for (and obtained) the solution with minimum generator constraint length. Adoption of a first polynomial with the lowest degree possible will give an help in better understanding the case of a shortened code. Anyway, in each period, we have still an even number of 1-symbols. They form a couple of adjacent 1-symbols as a consequence of the assumption $\Delta_1 = 0$.

Example 8.27 Let us take now a one-time shortened version of the QC array code assumed in Example 8.26, so we have an asymptotic code rate 1/4. After unwrapping the $\underline{H}$ matrix, and obtainment of the correspondent $\underline{G}$ matrix by imposing the null ex-OR sum of clusters of syndromes, the resulting $\underline{H}_o^T$ and $\underline{G}_o$ sub-matrix appear as shown in Fig. 8.27. In this case, $\underline{G}_o$ exhibits only one row. Unfortunately the 6-weight row remains. It resembles the former one in the $\underline{G}_o$ sub-matrix of the 2/5-rate QC code (its structure is unchanged, except for a scaling factor). In the $\underline{H}_o^T$ sub-matrix we can note here the presence of several empty rows (besides those associated to the first parity check polynomial). Such number of empty rows increases proportionally to $(p - S)$, that is to the number of single shortening operations performed on the original QC code. The first two coefficients of the weight distribution, characterizing this 1/4 rate s.s. time-invariant convolutional code, are $A(6) = Z$, $A(8) = Z - 1$. The position in the period, where the unique information symbol is placed, has been denoted with a small arrow.

Clearly, with increasing parameter p, maintaining Y constant, it is possible to perform a stronger and stronger code shortening. This choice, together with an accurate design of parameters Δ_j, can allow the achievement of rather high values of the minimum distance d, not too far from bound (8.12) (Baldi et al. 2014), as will be better described in Chap. 12.

As a final remark, we would like to stress that the unwrapping procedure of array codes here described is not related to a possible MHEQC construction of convolutional codes, starting from its parent QC code. The last procedure, already described in Sect. 6.10, discloses other possible opportunities for designing good convolutional codes after proper derivation from known QC array codes, and will be further detailed in Chap. 12 too.

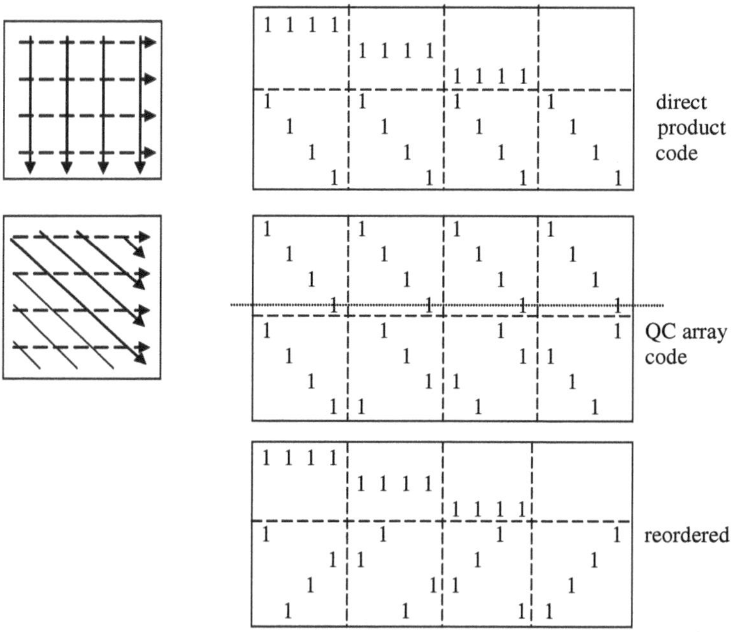

Fig. 8.28 Encoding memories and parity check matrices for a direct product code, a QC array code, and a reordered version of the last one

8.12 Further Considerations on Direct Product Codes Related to QC Array Codes

In Sect. 5.17 two types of code concatenation were considered in a rectangular encoding memory: direct concatenation of generalized parity check codes (or briefly DC-GPC code) and direct product of generalized parity check codes (or briefly DP-GPC code). The latter family of codes (see Definition 5.19) is very similar to QC array codes with $Y = 2$. The only difference is in the trajectories of the component codes in the encoding memory, which, in the array code, are subjected to a sort of reflection at the boundaries. This difference entails the use of permutation circulants, instead of identity matrices with different sizes. In Lemma 5.43 the equivalence between a DP-GPC code and the direct product of two parity check codes was demonstrated. Such two parity check codes necessarily have different lengths, so that the encoding memory is rectangular. Now we will treat a possible equivalence between a direct product code and an array code, with $Y = 2$ component codes.

Let us start from the complete QC version of this array code, which can be set in correspondence with a product code between two equal parity check codes having length p. The encoding memory for the direct product code and that for the present array code, both square and characterized by the same size p, are compared. In the former one we have two ordered ensembles of parity check codes, operating on row

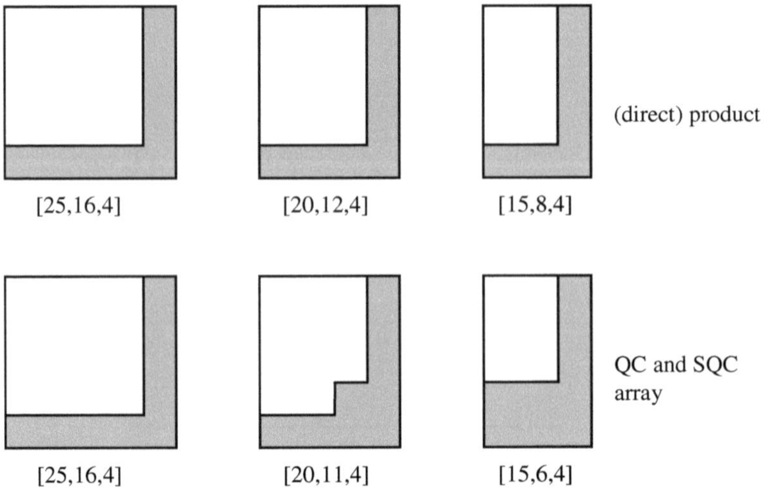

Fig. 8.29 Encoding memories for a family of direct product codes and a family of array codes

trajectories and on column trajectories. In the latter encoding memory, we have two ordered ensembles of parity check codes, operating on row trajectories and on tilted trajectories (with slope established by parameter Δ_2).

Theorem 8.3 *A QC array code with $Y = 2$ and whichever p is equivalent to a direct product code between two $[p,p - 1,2]$ parity check codes. This result is independent of the choice of parameter Δ_2.*

Proof The two encoding memories described above are depicted in Fig. 8.28. For fixing ideas, we have taken $p = 4$, $\Delta_2 = 1$. The correspondent $\underline{H}$ matrices are reported aside. Proceeding like for the proof of Lemma 5.43, it is possible to show that such two codes are equivalent. This result is valid for proper array codes ($\Delta_2 = 1$) and also for improper array codes ($\Delta_2 > 1$, provided that Δ_2 is co-prime with p). The last conclusion can be reached imagining tilted trajectories with steeper and steeper slopes, reflected by the lower and upper boundary in the encoding memory. □

The lowermost $\underline{H}$ matrix in Fig. 8.28 has been obtained, after elimination of the fourth row, linearly dependent on the others, in the following way. We have performed a reordering procedure only on the columns, according to Definition 8.7, but maintaining unchanged the rows. It is possible to observe that the first layer in the new $\underline{H}$ matrix, made by three only rows, coincides with the upper part in the parity check matrix of the direct product code. Then we have a second layer, made by 4 permutation matrices which can be interpreted as proper rotation of as many permutation circulants. It is possible to conclude that many other forms of parity check matrices can be obtained, after proper transformations, anyway leading to all equivalent codes associated to the tern $[p^2,(p - 1)^2,4]$. On the other hand,

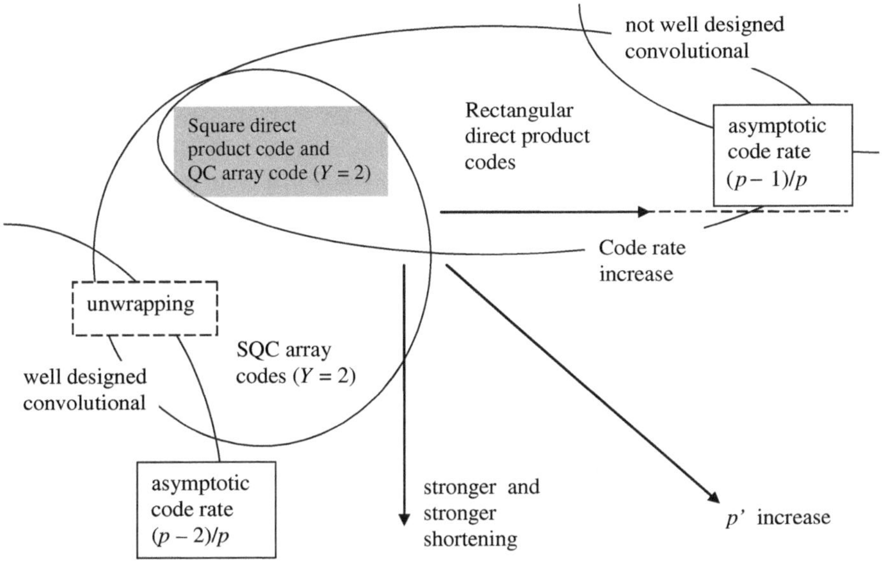

Fig. 8.30 Different behaviours of unwrapped array codes and direct product codes between two parity check codes

equivalence is no longer true between rectangular direct product codes and SQC array codes, both with two component codes.

Corollary a *An SQC array code, with $Y = 2$ and whichever values for parameters p and S, is equivalent to a subcode of a direct product code, constructed taking a $[p,p-1,2]$ parity check code and an $[S,S-1,2]$ parity check code.*

Proof After a one-time shortening, the number of control symbols of the array code remains equal to $Yp - (Y-1) = 2p - 1$. On the contrary, reducing the horizontal dimension of the encoding memory in the direct product code by one unit, the number of control symbols is reduced by one unit. This concept can be repeated for further single shortening operations. □

Example 8.28 Taking a [25,16,4] direct product code and its equivalent QC array code, after a progressive reduction in the horizontal dimension of the two rectangular encoding memories, we obtain the situations shown in Fig. 8.29, where the positions occupied by control symbols have been evidenced. In the case of the SQC array codes, the number of such control symbols remains 9, and they have been progressively accumulated in the lower rows. Instead, in the case of the direct product codes, the number of control symbols progressively reduces (like the number of rows in the first layer of their parity check matrix). Remark that such direct product code remains equivalent to other product codes, employing proper permutations in an interleaver, provided that couples of trajectories in the encoding memory joint at most one time.

These observations can help in attempting to give an answer to the following question: which mechanism enables an array code with $Y = 2$ to appear as a well designed convolutional code, after unwrapping? In fact, as several times stressed, direct product codes exhibit the behaviour of not well designed convolutional codes. First of all, the QC condition, supported by a square $p \times p$ encoding memory, is not capable of being interpreted within a convolutional code model, because, in the last model, one dimension should be much longer than the other one. Let us take a new variable $p' > p$ and form a rectangular encoder memory $p \times p'$. The asymptotic code rate, after a progressive increase of p' (which represents the number of periods in the frame of the SQC equivalent convolutional code, after unwrapping), remains $(p - 2)/p$. This means that the array code is more and more strongly shortened with respect to the dimension p'. Such behaviour is schematically shown on the left in Fig. 8.30. The direct product code, after increasing of one dimension in the rectangular encoding memory, originates a family of codes which can be interpreted as a convolutional code too, but the code rate becomes here asymptotically $(p - 1)/p$, for very high values of p'. In this condition, its behaviour is that of a not well designed convolutional code, as shown in Fig. 8.30 on the right. In conclusion, the array code, by scaling the whole parameter p', with circulants having higher and higher size, originates a family of SQC codes, which, even without unwrapping, will be characterized by low-weight coefficients $A(w)$ growing linearly with p'.

A link between these two classes of codes (array codes and direct product of parity check codes) is no longer possible for $Y > 2$. If one consider, say, 3 parity check component codes, the minimum distance of their direct product is anyway $d = 8$, and this performance is paid by a reduced code rate. The correspondent improper array code can exhibit $d = 8$ only when strongly shortened, otherwise its minimum distance is $d = 6$, even in the form of a well designed convolutional code obtained from it, after unwrapping. In other terms, the direct product code exploits a third spatial dimension for assuring iteration of orthogonality relationships among its component codes, whereas the array code remains in a bi-dimensional encoding memory. This observation confirms how the behaviours of these two families of codes have a point of correspondence only for $Y = 2$.

References

Abramowitz M, Stegun IA (1968) Handbook of mathematical functions. Dover, New York

Baldi M, Bianchi M, Cancellieri G et al (2012) On the generator matrix of array codes. In: Proceedings of Softcom 2012, Split (Croatia)

Baldi M, Cancellieri G, Chiaraluce F et al (2014) Array convolutional low-density parity-check codes. IEEE Comm Lett 18:336–339

Esmaeli M, Tadayon MH, Gulliver TA (2011) More on the stopping and minimum distances of array codes. IEEE Trans Comm 59:750–757

Fan JL (2000) Array codes as low-density parity check codes. In: Proceedings of 2nd international symposium on Turbo Codes, Brest, France, pp 543–546

Fossorier MPC (2004) Quasi-cyclic low-density parity check codes from circulant permutation matrices. IEEE Trans Inf Theor 50:1788–1793

Jimenez-Felstrom AJ, Zigangirov KS (1999) Time-varying periodic convolutional codes with low density parity-check matrix. IEEE Trans Inf Theor 45:2181–2191

Johannesson R, Zigangirov KS (1999) Fundamentals of convolutional coding. IEEE Press, New York

Milenkovic O, Kashyap N, Leyba D (2008) Shortened array codes of large girth. IEEE Trans Inf Theor 52:3707–3722

Mittelholtzer T (2002) Efficient encoding and minimum distance bound for Reed-Solomon-type array codes. In: Proceedings of ISIT 2002, Lousanne, Switzerland, p 282

Tanner RM (1981) A recursive approach to low complexity codes. IEEE Trans Inf Theor 27:533–547

Tanner RM, Sridhara D, Sridharan A et al (2004) LDPC block and convolutional codes based on circulant matrices. IEEE Trans Inf Theor 50:2966–2984

Part III
Modern Coding

Chapter 9
Turbo Codes

9.1 The Basic Idea of Turbo Codes

Turbo codes were firstly presented in 1993 (Berrou et al. 1993a), although the basic idea for a similar application can be found in the literature much before, speaking about general forms of concatenated codes (Forney 1966). Two component codes, in parallel concatenation with interleaver, are decoded by a soft-decision algorithm, after a converging iterative procedure. Two convolutional component codes were proposed, in order to exploit the small number of low-weight code frames characterizing them. The objective to be reached is as follows: the low-weight code frames in one component code have to become high-weight code frames in the other, and vice versa. This goal was expected on a large overall code size, but surprisingly it appeared possible also with code word lengths n as small as few thousands. In this sense, the use of RSC codes (see Sect. 3.6) was determinant, as will be better explained in the following. The component codes were chosen among the best 1/2-rate convolutional codes, with m varying from 3 to 5, in order to maintain decoding computational complexity acceptable.

The overall code is a block code, where a sliding window decoding procedure (Definition 3.6) is not possible. The number of iterations necessary for reaching convergence typically varies from few tenths to one hundred, leading to some time latency drawback (see Appendix D), and forcing to not increase the block length n too much, especially for isochronous applications (Proakis 1983). In spite of these two constraints, turbo codes have been highly appreciated, since their performance, for medium values of the error probability p (from 10^{-5} to 10^{-3}), appeared to be less than 1 dB away from Shannon limit for soft-decision decoding (see Figs. D.7 and D.8 in Appendix D). For larger values of the *SNR*, an abrupt change of slope, in the error probability curve, gives rise to an asymptotic coding gain G_{∞} remarkably smaller, but, at the moment of their first presentation, such codes were proposed not for high-*SNR* applications. Subsequent results have shown that the change of slope can be controlled, by means of a particular design, in such a way as to extend the

G. Cancellieri, *Polynomial Theory of Error Correcting Codes*,
Signals and Communication Technology, DOI 10.1007/978-3-319-01727-3_9

best operation conditions up to values of p as small and 10^{-7}, end even lower (Berrou and Glavieux 1996).

Nevertheless some problems appeared immediately evident. They regarded how to perform a correct termination of both code frames adding less than the overall number of final control symbols required and how to increase the code rate, whose value, after what discussed in Sect. 1.6 about parallel code concatenation, would become $R = k/(k + r_1 + r_2)$, being k the number of information symbols and r_1, r_2 the overall control symbols introduced by the two component codes respectively.

About the first problem, taking two equal 1/2-rate component codes with constraint length m, the total number of final control symbols should be $4m$. Various solutions have been proposed for introducing proper constraints to the interleaver, in order to reduce this number to $2m$. This goal can be achieved when both RSC encoders reach the end of the frame in the same state (Barbulescu and Pierobon 1995). Nevertheless it was noted that, taking the second component code without a frame termination at all, the performance obtained is only slightly worse than in the presence of both correct terminations. This solution has been frequently adopted in practice, in order to maintain the true code rate not too low.

Yet the main problem remained the asymptotic code rate, which, assuming the value 1/2 for each component code, would fall at 1/3. This limitation was solved by properly puncturing the resulting block code (Ryan and Acikel 1999) or the original component codes (Kim 1997). Surprisingly, an elimination of one control symbol any two, so leading asymptotically to a 1/2-rate overall block code, produces a modest degradation of performance. Even a stronger code puncturation, if made according to an optimized criterion, does not make the overall code too weak, allowing a great variety of possible applications where a scalable code rate is required (Benedetto et al. 1998a).

Two decoding algorithms were proposed, already available since several years at the moment of turbo code presentation: BCJR algorithm, from the names of the four inventors (Bahl et al. 1974) and Soft-Output Viterbi Algorithm (SOVA) (Hagenauer and Hoeher 1989). They exhibit analogous performance, although the former one has been more widely employed. In Sect. 9.5, a brief description of such algorithms will be given, with the purpose of evaluating their computational complexity. After the above brief review on the history of turbo codes, which, together with LDPC codes, represent the best up-to-date solutions for channel coding, we will proceed with their structural analysis.

Definition 9.1 A parallel-concatenation of two RSC 1/2-rate convolutional codes, with an interleaver I acting before the input of the second encoder circuit, as shown in Fig. 9.1, is called *turbo code*. From the two encoders, just the control symbols are taken.

Although in the firstly proposed schemes the two component codes were different, subsequent studies have shown that the interleaver is, by itself, able to assure a complementary contribution from them. So typically two equal component codes are adopted, but the interleaver is to be synthesized taking into account their structure (Benedetto et al. 1998a).

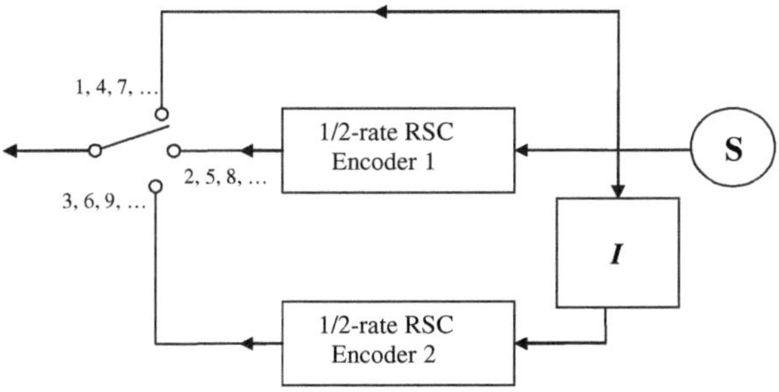

Fig. 9.1 Scheme of the encoder circuit for a 1/3-rate turbo code

Lemma 9.1 *In a turbo code, with two equal component codes whose encoder circuits are characterized by m memory elements each, after a correct frame termination of both component codes, the true code rate is*

$$R = \frac{1}{3 + \frac{4m}{Z}}, \tag{9.1}$$

in which Z is the number of periods in the frame where information symbols are inserted.

Proof To the frame length $L = 2(Z + m)$ of the first component code, further $(Z + 2m)$ control symbols must be added from the second component code. So we have an overall code word length $n = 3Z + 4m$, whereas the number of information symbols is $k = Z$. The interleaver, which is represented by a proper permutation matrix, does not change these numbers. From the ratio k/n, we are left with expression (9.1). □

Parameter Z represents also the size of interleaver I. As anticipated, it is chosen as a trade-off between the need of having a good error probability curve and the requirement of limiting time latency. Typically with m of the order of 4, it is enough to choose Z of some hundreds for having R smaller than 1/3 by an approximate quantity $4m/Z$, whose value is few percent. Remark that, if one accepts a non-terminated frame in the second component code, such code rate penalty is reduced to about $2m/Z$. Size and design of the interleaver I are responsible for the negative slope in the error probability curve, which can be more or less steep, and is usually named water fall in the region of low *SNR* values (see Appendix D).

Definition 9.2 Once obtained an error probability curve of the type reported in Fig. D.7, the maximum coding gain assured by the use of a certain interleaver I, in a turbo code scheme, is called *interleaver coding gain*, and denoted G_I.

In Fig. 9.2, a qualitative increase in the interleaver coding gain G_I with increasing interleaver size Z is presented. Two values for Z have been considered: 1,000 and 5,000. On the right the evolutions of the coding gain $G(p)$, in the two cases, are reproduced. Nevertheless also an optimized interleaver design is important, after having fixed its size (Garello et al. 2001a), to the point where

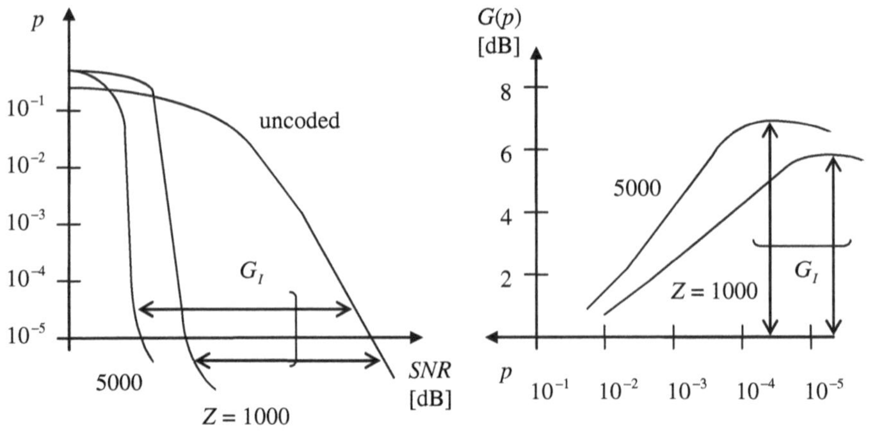

Fig. 9.2 Qualitative behaviour of the interleaver coding gain depending on interleaver size Z

length-adaptive turbo codes are still difficult to be synthesized, without changing the interleaver at any new operation condition.

A particular question is that regarding code puncturation, in order to reach asymptotic code rate at least equal to 1/2, like in each one of the two component codes taken separately. We have to observe that the entire turbo code can be interpreted as a block code with periodically distributed control symbols. So we can adopt for it a periodic puncturation matrix of the type introduced in Definition 3.10, which is typically employed with convolutional codes. The only care is to start from a 1/3-rate code.

Lemma 9.2 *In a scheme of the type reported in* Fig. 9.1, *insertion of a pair of puncturation devices after each one of the two encoder circuits, or insertion of a unique puncturation device after the output switch, is equivalent, provided that the following puncturation matrices are adopted in the two cases*

$$\underline{P}' = \begin{bmatrix} 1 & 1 \\ 1 & 0 \end{bmatrix}, \quad \underline{P}'' = \begin{bmatrix} 1 & 1 \\ 1 & 0 \\ 0 & 1 \end{bmatrix}.$$

Proof Matrix $\underline{P}'$ is to be applied separately to both the component codes, in order to increase the code rate from 1/2 to 2/3 (see Example 3.16). Matrix $\underline{P}''$, instead, is to be applied to the resulting turbo code, in order to increase the code rate from 1/3 to 1/2. The two 1-symbols in the first row of $\underline{P}''$ mean that both information symbols in two adjacent periods are taken. Then, in the former period the control symbol from the second encoder is eliminated, and, in the latter period, the control symbols from the first encoder is eliminated. This cycle repeats up to the frame termination. Such behaviour is exactly what matrix $\underline{P}'$ produces acting separately after each one of the two encoder circuits. □

Turbo codes offer one of the most significant example of anomalous behaviour in the error probability curve p'' after correction (see Fig. D.7), with a maximum in the coding gain for central values of p. The interleaver coding gain G_I is typically some dB higher than the asymptotic coding gain G_∞. Puncturation reduces the value of G_∞, but, often, does not change appreciably the water fall and hence the value of G_I. This is due also to the increases in the *SNR* by a factor of 3/2, consequence of the increased code rate.

Definition 9.3 The word weight distribution of a turbo code, with respect to that of each component code, shows a translation, on average, towards higher values. This phenomenon is called *spectral thinning* (Lin and Costello 2004). Nevertheless the presence of few residual low-weight code words originates a phenomenon called *BER floor* (BER being the acronym of Bit Error Ratio).

A qualitative example of spectral thinning is shown in Fig. 9.3, where up we have the behaviour of the 1/3-rate turbo code and down that of the 1/2-rate punctured turbo code. For both the frame weight distribution of the component codes is taken as a comparison (grey segments), being d_a their minimum distance, d the minimum distance of the turbo code without puncturation, finally d' that of the turbo code with puncturation. Such two minimum distances, not much greater than d_a, have negligible effects on the water fall region, imposed by the average weight values.

Definition 9.4 Given a certain turbo code, characterize by an interleaver coding gain G_I, and by a code rate R (considering also the possible puncturation), the following parameter

$$d_{eff} = \frac{G_I}{R} \tag{9.2}$$

is called its *effective minimum distance*.

Remark that the effective minimum distance is typically a non-integer value. In Fig. 9.3 the effective minimum distances of the two turbo codes considered have been qualitatively evidenced in dashed-dotted lines (d_{eff} at rate 1/3, $d_{eff}{}'$ at rate 1/2 after puncturation). Such rough statistical parameter is always obtainable from (9.2) after having plotted the p'' curve, resulting from a computer simulation. It can be justified taking into account a sort of averaging process (Perez et al. 1996) in an expression like (D.7), but with a number of terms in the summation much greater than $(2d - 1)$. As anticipated, the reduction of $d_{eff}{}'$ with respect to d_{eff} is not so large, although the phenomenon of spectral thinning is much more evident without the puncturation. The change in the code rate R has an important role in this compensation. In conclusion, code puncturation appears as a very good solution, to the point where also the elimination of a greater fraction of control symbols is sometimes suggested (reaching code rates of 3/4, 4/5 and even 5/6, rarely more). Unfortunately puncturation does not allow an easy prediction of the performance of

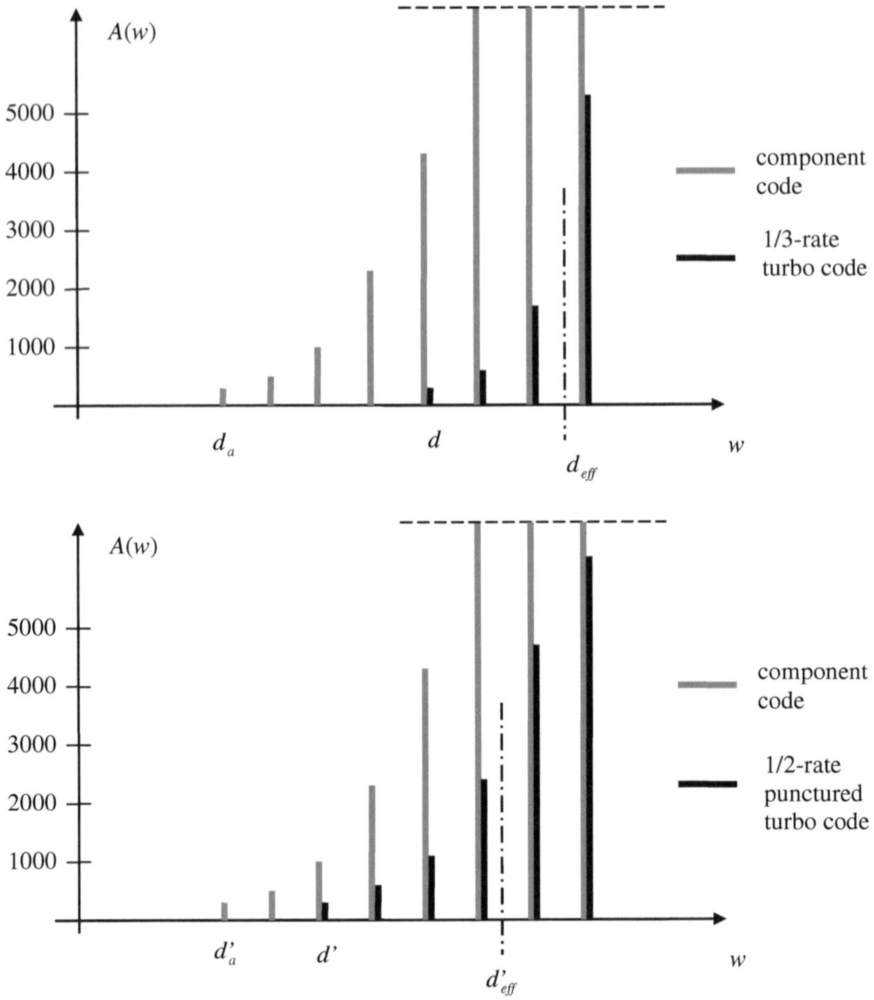

Fig. 9.3 Example of spectral thinning and estimate of the effective distance

the resulting code, therefore often it is necessary to proceed by attempts in order to reach the optimized combined choice of interleaver and puncturation matrix.

The BER floor, in turn, represents the change of slope in the p'' curve for higher values of the *SNR*, imposing the entity of the asymptotic coding gain G_∞. Coefficient $A(d)$, where d is the true minimum distance, can be as small as few units. In this case the BER floor appears, but for extremely small values of the error probability, not easy to be revealed by computer simulations. A correct frame termination also in the second component code, leading to an increase of d, can mitigate somewhat the BER floor, but typically does not give any benefit to the water fall.

9.2 Some Particular Aspects of RSC Codes

Lemma 9.3 *A one-weight information symbol* $i(x) = x^p$*, given in input to an RSC code (or to its equivalent MLC code), produces an output sequence with an infinite number of* 1*-symbols, provided that the code is well designed.*

Proof The division of x^p by $g^{(1)}(x^2)$, or by the whole $g_{int}(x)$, yields their respective periodic co-factors. They represent the state evolution in the two cases, but in both the state $\underline{0}$ cannot be reached again, except at the end of the frame. The condition that the code is well designed is necessary, because in a not well designed convolutional code there are closed paths in the state diagram associated to all 0-symbols in output (see Sect. 7.4). □

The same conclusion can be reached also observing that a circuit like the one depicted in Fig. C.8 up, which is of the type able to encode an RSC code, represents a digital filter having an infinite impulse response (Oppenheim and Schafer 1974).

Example 9.1 Taking the 1/2-rate RSC code characterized by $g^{(1)}(x) = x^2 + x + 1$ and $g^{(2)}(x) = x^2 + 1$, already treated in Example 3.20, its interleaved generator polynomial is 11.10.11, to be taken as $g(x)$ for its equivalent MLC version. In Fig. 9.4 up the polynomial divisions leading to the respective current remainders (originating the sequences of distributed control symbols in the two cases) are reported. For the RSC version, we divide by $g^{(1)}(x^2)$ and multiply the quotient Q by $g^{(2)}(x^2) = x^4 + 1$. For the MLC version, we simply divide by $g_{int}(x)$ and insert the current remainder. So the semi-infinite systematic generator matrices of this RSC code or of its equivalent MLC version will appear as shown in Fig. 9.4 down. They

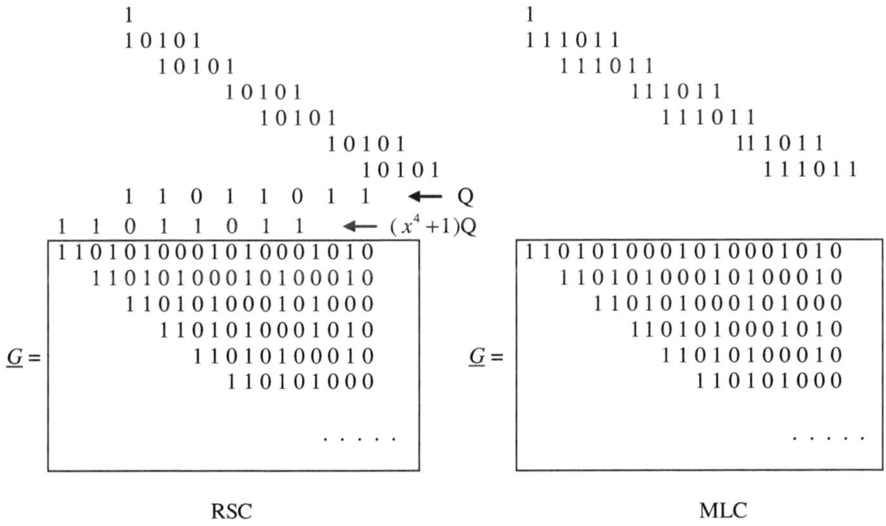

Fig. 9.4 Polynomial divisions and $\underline{G}$ matrices for the codes introduced in Example 9.1

are obtained repeating the first row after a shift of $n_o = 2$ positions towards the right. As expected, they are coincident and show columns with just one 1-symbol in correspondence with the positions where information symbols are to be placed. On the right hand side of these solitary 1-symbols, proper sequences of control symbols are inserted, without occupying the positions reserved to other information symbols.

In conclusion, information sequences with just one 1-symbol are not points of weakness for turbo codes, since they collect a very high number of 1-symbols as control symbols from both RSC encoders. It can be interesting to compare Fig. 3.15, where the behaviour of the RSC version of a not well designed convolutional code is reported, with Fig. 9.4. This comparison confirms how a not well designed convolutional code is to be avoided as a component code in a turbo code scheme.

On the contrary, 2-weight information sequences may represent a severe point of weakness, because of the following aspect. Division by $g^{(1)}(x)$ yields null remainder in the case of a binomial $(x^N + 1)$, $(x^{2N} + 1)$, ..., among whose factors $g^{(1)}(x)$ is present. So the first RSC code can contribute to the overall weight of the code frame with the control symbols produced by such pair of information 1-symbols. If they are too close each other, the resulting weight is not high. In this case, the second RSC code should produce a higher weight, by means of its control symbols, to be added, provided that such code does not work in the same condition (2-weight information sequences separated by a particularly unlucky interval). This problem, obviously, is present also in the MLC model of the RSC code behaviour. Among the characteristics that the interleaver should exhibit, the possibility of avoiding the occurrence of such situation is to be carefully imposed.

Remark that, often, in the total output weight given by an RSC (MLC) code, the contribution of the control symbols are only a fraction. Just when a pair of information 1-symbols are remarkably separated, it is possible to predict that the total output weight is at least about half their separation. In fact this situation corresponds to a route in the state diagram whose length, before coming back to state $\underline{0}$, has the same entity. When two information 1-symbols are less separated, the output weight can be remarkably smaller, owing to a very poor contribution from control symbols.

Example 9.2 Taking again the 1/2-rate RSC code characterized by $g^{(1)}(x) = x^2 + x + 1$ and $g^{(2)}(x) = x^2 + 1$, we have critical binomials of the type $(x^3 + 1)$, $(x^6 + 1)$, ... Assuming the first one of these binomials (whose exponent has to be doubled because of the convolutional code period), the first encoder produces the overall 6-weight output sequence

$$\begin{aligned} t'(x) &= x(x^6 + 1) + [(x^6 + 1)/(x^4 + x^2 + 1)](x^4 + 1) \\ &= x^7 + x^6 + x^4 + x^2 + x + 1. \end{aligned}$$

In it, two 1-symbols are information symbols, and the remaining 4 symbols are control symbols. Denoting with c_2, in their proper positions, the control symbols attached by the second encoder, the overall 1/3-rate code frame will appear as follows

$$t(x) = 11\ c_2.01\ c_2.01\ c_2.11\ c_2.00\ c_2.00\ c_2.\ldots$$

Assuming the same RSC encoder also for the second component code, the risk of collecting only 4 further 1-symbols in the positions labelled c_2 exists. In this case, the resulting minimum distance will be not higher than 10. Instead, if they are remarkably separated, say as fixed by the second binomial listed above, we easily obtain w$[c_2]$ = 8, and the value of the minimum distance, due to this bound, will increase up to 14.

A turbo code, on the whole, is a linear code. Therefore, if there are low-weight sequences of information symbols collecting a high number of control 1-symbols, it is unavoidable that some high-weight sequences of information symbols collect a low number of control 1-symbols. The purpose of the interleaver is to find an optimum balance between these two tendencies, but it is to be constructed considering the particular RSC component codes adopted.

Definition 9.5 An interleaver, after which two 1-symbols in the input sequence, separated by less than S positions, will be spread over a higher number of positions in the output sequence, is called an *S-spread interleaver.*

Clearly parameter S can grow only if the overall size of the interleaver grows as well. Deterministic interleavers (Divsalar and Pollara 1995) can give S values of the order of the square root of their overall size. On the other hand, an S-spread interleaver will be efficiently employed with a pair of component codes whose $g^{(1)}(x)$ is factor of $(x^N + 1)$, with $N < S$. About the importance of further binomials $(x^{2N} + 1)$, …, we can observe that the quotients they produce will have twice, … the weight of the quotient produced by the fundamental binomial $(x^N + 1)$. So they are expected to collect a progressively higher number of 1-symbols as control symbols, from the action of $g^{(2)}(x)$, and hence their importance is rapidly decreasing. In this sense, an S-spread interleaver, if sufficiently long, is expected to avoid the effect of 2-weight information sequences.

A deterministic approach, as that based on the S-spread interleaver can help in the design of a particular interleaver, in relation to a precise couple of RSC encoders, but is less efficient in the estimate of the behaviour of a whole class of turbo codes, taken on average. In doing this, it is possible to develop a statistical model for the interleaver, as will be outlined in the next section.

Finally we can observe that the performance of the turbo code described so far is strongly influenced by the fact that code puncturation is performed on control symbols only and that RSC encoders are employed. Nevertheless good results have been obtained also removing one of the two above assumptions or even both (Banerjee et al. 2005), although in the framework of a rather different approach. In common we have, however, the principle that the parallel concatenation of convolutional codes, enforced by the action of an interleaver, is able to produce the phenomenon of spectral thinning, which is rather resistant against frame non-termination and code puncturation.

9.3 Statistical Prediction of Turbo Code Performance

Definition 9.6 We define *statistical uniform interleaver* of length Z the model of a probabilistic device (Benedetto and Montorsi 1996) which, on average, maps a given input sequence with weight w_i into all the $\binom{Z}{w_i}$ different permutations, with equal probability given by

$$P(w_i, Z) = \frac{1}{\binom{Z}{w_i}}. \tag{9.3}$$

Let us write an approximate expression for the average binary symbol error probability, after the soft-decision decoding algorithm here applied (see Lemma D.5 in Appendix D). It results in

$$p'' \cong \sum_w \frac{w}{n} A(Z,w) \frac{1}{2} erfc\left(\sqrt{w \frac{Z}{n} SNR}\right), \tag{9.4}$$

where w is the frame output weight, n is the overall frame output length, Z is the total number of information symbols, $A(Z,w)$ is the number of frames having output weight w. About the lower and upper bounds of this summation, we can consider either the true minimum distance and few further terms, in order to estimate the BER floor, or several terms around the expected effective minimum distance d_{eff}, if we attempt to obtain a prediction about the water fall behaviour. The latter purpose is more difficult to be reached, since typically an expression of the type (9.4) is only roughly approximated for small values of the *SNR*. The output weight results in

$$w = w_i + c_1 + c_2, \tag{9.5}$$

where w_i is the weight of the information sequence, c_1, c_2 are the numbers of non-null control symbols introduced by the first and second RSC encoder respectively.

Coefficient $A(Z,w)$, in turn, can be expressed as

$$A(Z,w) = \sum_{\substack{w_i, c_1, c_2 \\ (w = w_i + c_1 + c_2)}} \frac{M_1(Z, c_1|w_i) M_2(Z, c_2|w_i)}{\binom{Z}{w_i}}, \tag{9.6}$$

where $M_1(Z, c_1|w_i)$ is the number of frames from the first RSC code which, considering also a possible puncturation, are associated to input sequences having weight w_i, over a length Z, and $M_2(Z, c_2|w_i)$ is the same parameter for the second RSC code. The summation is made in an ensemble conditioned by (9.5).

The numerator in each term of summation (9.6) is the product of two rather high numbers, but the denominator can be even higher. In the previous section, we have concluded that, for $w_i = 1$, w turns out to be very large, and hence the contribution of the correspondent coefficients $M_1(Z, c_1|1)$, $M_2(Z, c_2|1)$ have a role just for high-order terms in (9.4). On the other hand, for $w_i = 2$, w becomes moderately high, but the product between $M_1(Z, c_1|2)$ and $M_2(Z, c_2|2)$ remains rather small, so that even a not extremely high denominator $\binom{Z}{w_i}$ is enough for confining the effects of such term only to the BER floor behaviour. With increasing the input weight w_i, more and more terms exhibit a sort of equilibrium between the numerator and the denominator in (9.6), both showing very large entities. This equilibrium justifies the phenomenon of spectral thinning. Often also 3-weight information symbol sequences produce low-weight frames, responsible of some performance limitations.

As already stressed, the statistical treatment described so far is able to give average values over an ensemble of interleavers having similar properties. About the variance in the expected minimum distance distribution, which expresses how the choice of a particular interleaver can give possible better performance, an exhaustive investigation with small-size randomly generated interleavers has been developed. On 1/3-rate turbo codes (without puncturation) with only one of the two component codes correctly terminated, adopting equal RSC component codes with $m = 3$, the following results have been obtained (Garello et al. 2001b). For an interleaver size Z of the order of 100, the average minimum distance $\langle d \rangle$ is about 12, whereas the best result is $d_{best} = 15$. For Z of the order 1,000, it is possible to obtain $\langle d \rangle$ of about 14, and $d_{best} = 22$. Adopting a pair of RSC component codes with $m = 4$, the above average values $\langle d \rangle$ increase by about 20 %, and the best values, especially for an interleaver size Z of the order of 1,000, show even stronger possible improvements.

9.4 $\underline{G}$ Matrix of a Turbo Code and Correct Frame Termination

An RSC code, with $m = 2$, like the one considered in Example 9.1, exhibits a $\underline{G}$ matrix which can be constructed as shown in Fig. 9.5, where we have inserted dots between subsequent periods in order to make interpretation easier. In this representation, a_j, $j = 1, 2, \ldots Z$, are distributed control symbols, whereas $r_i(k)$, $k = 1, 2, \ldots Z$, and $i = 1, 2, 3, 4$ (in general up to $2m$), are control symbols needed for the correct frame termination.

$$\underline{G} = \begin{array}{|l|}\hline 1\ a_1\ .\ 0\ a_2\ .\ 0\ a_3\ .\ 0\ a_4\ .\ \ldots\ldots\quad .\ 0\ a_Z\ .\ r_1(1)\ r_2(1)\ .\ r_3(1)\ r_4(1) \\ \qquad 1\ a_1\ .\ 0\ a_2\ .\ 0\ a_3\ .\ \ldots\ldots\quad .\ 0\ a_{Z-1}\ .\ r_1(2)\ r_2(2)\ .\ r_3(2)\ r_4(2) \\ \qquad\qquad 1\ a_1\ .\ 0\ a_2\ .\ \ldots\ldots\quad .\ 0\ a_{Z-2}\ .\ r_1(3)\ r_2(3)\ .\ r_3(3)\ r_4(3) \\ \qquad\qquad\qquad 1\ a_1\ .\ \ldots\ldots\quad .\ 0\ a_{Z-3}\ .\ r_1(4)\ r_2(4)\ .\ r_3(4)\ r_4(4) \\ \qquad\qquad\qquad\qquad \ldots\ldots\ldots\ldots\ldots\ldots \\ \hline \end{array}$$

Fig. 9.5 Structure of the $\underline{G}$ matrix of a 1/2-rate RSC code with $m = 2$

$$\begin{array}{|cccc|}\hline 0 & 1 & 0 & 0 \\ 0 & 0 & 1 & 0 \\ 1 & 0 & 0 & 0 \\ 0 & 0 & 0 & 1 \\ \hline \end{array} \qquad\qquad \begin{array}{|c|}\hline i_3\ i_1\ i_2\ i_4 \\ \hline \end{array}$$

$$\underline{G} = \begin{array}{|l|}\hline 1\ 0\ a_1^{(2)}\ .\ 0\ 0\ a_2^{(2)}\ .\ 0\ a_1^{(1)}\ a_3^{(2)}\ .\ 0\ a_2^{(1)}\ a_4^{(2)}\ .\ r_1^{(1)}(3)\ r_2^{(1)}(3)\ r_3^{(1)}(3)\ r_4^{(1)}(3)\ .\ r_1^{(2)}(1)\ r_2^{(2)}(1)\ r_3^{(2)}(1)\ r_4^{(2)}(1) \\ 0\ a_1^{(1)}\ 0\ .\ 1\ a_2^{(1)}\ a_1^{(2)}\ .\ 0\ a_3^{(1)}\ a_2^{(2)}\ .\ 0\ a_4^{(1)}\ a_3^{(2)}\ .\ r_1^{(1)}(1)\ r_2^{(1)}(1)\ .\ r_3^{(1)}(1)\ r_4^{(1)}(1)\ .\ r_1^{(2)}(2)\ r_2^{(2)}(2)\ r_3^{(2)}(2)\ r_4^{(2)}(2) \\ 0\ 0\ 0\ .\ 0\ a_1^{(1)}\ 0\ .\ 1\ a_2^{(1)}\ a_1^{(2)}\ .\ 0\ a_3^{(1)}\ a_2^{(2)}\ .\ r_1^{(1)}(2)\ r_2^{(1)}(2)\ r_3^{(1)}(2)\ r_4^{(1)}(2)\ .\ r_1^{(2)}(3)\ r_2^{(2)}(3)\ r_3^{(2)}(3)\ r_4^{(2)}(3) \\ 0\ 0\ 0\ .\ 0\ 0\ 0\ .\ 0\ 0\ 0\ .\ 1\ a_1^{(1)}\ a_1^{(2)}\ .\ r_1^{(1)}(4)\ r_2^{(1)}(4)\ r_3^{(1)}(4)\ r_4^{(1)}(4)\ .\ r_1^{(2)}(4)\ r_2^{(2)}(4)\ r_3^{(2)}(4)\ r_4^{(2)}(4) \\ \hline \end{array}$$

Fig. 9.6 Interleaver effect and overall $\underline{G}$ matrix for the 1/3-rate turbo code introduced in Example 9.3

Let us consider now a $Z \times Z$ permutation matrix (see Definition 1.2) describing the interleaver. In the second RSC code, assumed identical to the first one, we have a permuted version of the input sequence, and hence a permuted order in the rows of the above $\underline{G}$ matrix. The only care, for constructing the overall generator matrix of our turbo code, is to eliminate the information symbols in the output of the second RSC code and associate the control symbols to the rows where the correspondent information symbols were placed, in the original $\underline{G}$.

Example 9.3 Given a sequence characterized by $Z = 4$ information symbols, and an interleaver described by the 4×4 permutation matrix shown up-left in Fig. 9.6, we obtain the interleaved sequence shown up-right. Assuming a pair of equal RSC component codes, with $m = 2$, like the one introduced in Example 9.1, we obtain an overall $\underline{G}$ matrix of the type shown in Fig. 9.6 down. Superscripts denote the first or the second RSC code. All the 0-symbols have been represented in order to make understanding easier. The final control symbols, needed for both frame terminations, are $4 + 4 = 8$ on the whole, and have been reported at the end (firstly those regarding the first RSC code, and then those regarding the second RSC code). No puncturation has been made. The resulting code word is characterized by $n = 20$, $k = Z = 4$. Clearly, as long as Z increases, the code rate approaches 1/3. It is possible to note that the generator matrix of the turbo code, so constructed, is systematic only in wide sense (Definition 1.5). However this matrix is trivially invertible.

Example 9.4 Assuming the 1/3-rate turbo code, with two equal RSC component codes, with $m = 2$, and described in Example 9.1, 9.2, we can particularize the complete generator matrix $\underline{G}$ shown in Fig. 9.6. Taking $Z = 4$ periods in which information symbols are inserted, and the interleaver adopted in Example 9.3, we obtain a certain $\underline{G}$ matrix. In it, for instance, the last row shows a structure which derives from the fact that we have chosen an interleaver which maintains the same fourth position in both RSC encoders. This choice is not good, especially when this position is close to the frame end. On the other hand, $Z = 4$ is a frame length too small for allowing an optimization in the interleaver structure. The resulting turbo code is a [20,4,7] block code. Its word weight distribution turns out to be

$$A(0) = 1,\ A(7) = 1,\ A(8) = 1,\ A(9) = 2,\ A(10) = 2,$$
$$A(11) = 5,\ A(12) = 3,\ A(16) = 1.$$

With a better choice of the interleaver, we can reach $d = 8$, but the code so obtained is strongly influenced by puncturation. With increasing Z, it is possible to reach $d = 10$. Coming back to the very simple code considered above, if we proceed with a code puncturation, by eliminating one control symbol any two, only in the periods where information symbols are inserted, we remain with a [16,4,5] block code, whose word weight distribution is

$$A(0) = 1,\ A(5) = 1,\ A(7) = 3,\ A(8) = 5,\ A(9) = 3,\ A(10) = 2,\ A(13) = 1.$$

With increasing Z, the code rate will approach 1/2. The 5-weight code words remain few, and are responsible for a BER floor which will be influent only at very small values of the error probability. This property is almost general and can be so summarized. Some very-low-weight code frames have a number not proportional to the frame length Z, but they represent few solitary 1-symbol patterns (not shifting along the frame length as occurs in convolutional codes). Clearly, their effect is more and more negligible with increasing the size of the interleaver. On the other hand, it is difficult to consider a scalable interleaver, based on a certain structure, which can be enlarged as one wishes, simply increasing Z.

In order to produce a growth in the true code rate, as anticipated, especially for values of Z not too high, which are strongly frustrated by the addition of $4m$ final control symbols, one of the two component codes can be taken with its frame not terminated. Surprisingly the performance of the new code does not become remarkably worse. In fact this procedure acts only on those low-weight code words whose multiplicity is independent of Z.

Example 9.5 In the 1/3-rate turbo code treated in Example 9.4, we can eliminate the last four symbols, due to the correct frame termination in the second RSC code. So we obtain a [16,4,6] code, whose word weight distribution turns out to be

$$A(0) = 1,\ A(6) = 1,\ A(7) = 2,\ A(8) = 5,\ A(9) = 5,\ A(10) = 1,\ A(13) = 1.$$

If we adopt this solution on the punctured code, we obtain a [12,4,4] code, whose 1/2 asymptotic code rate is more rapidly approached, and whose word weight distribution becomes

$$A(0) = 1,\ A(4) = 1,\ A(5) = 5,\ A(6) = 4,\ A(7) = 2,\ A(8) = 2,\ A(10) = 1.$$

For understanding the role played by the interleaver, we will consider now other possible interleavers, all applied to the same pair of equal RSC component codes, and all with $Z = 4$. Often a very good interleaver with the complete 1/3-rate code word suffers some weakness when puncturation or frame non-termination in the second component code, or both, are applied.

Example 9.6 Taken the same pair of RSC codes as in Examples 9.4, 9.5, still with $Z = 4$, we will describe the effects of two alternative interleavers. For both no 1-symbols are on the main diagonal of the permutation matrix describing them. This means that all the positions are changed. The results obtained are collected in the tables reported in Fig. 9.7. The first solution is characterized by just even weights and a very pronounced spectral thinning for the original 1/3-rate code, but it suffers either puncturation or frame non-termination in one of the two component

0	0	1	0
1	0	0	0
0	0	0	1
0	1	0	0

$i_2\ i_4\ i_1\ i_3$

	A(0)	A(3)	A(5)	A(6)	A(7)	A(8)	A(9)	A(10)	A(11)	A(12)	A(13)	A(16)
complete [20,4,8]	1					2		7		5		1
not terminated [16,4,6]	1			1	3	3	4	3			1	
punctured [16,4,5]	1		1	1	2	3	3	4			1	
not ter. and punct. [12,4,3]	1	1	3	4	4	1	1	1				

0	1	0	0
0	0	1	0
0	0	0	1
1	0	0	0

$i_4\ i_1\ i_2\ i_3$

	A(0)	A(4)	A(5)	A(6)	A(7)	A(8)	A(9)	A(10)	A(11)	A(12)	A(13)	A(16)
complete [20,4,7]	1				1		2	2	7	2		1
not terminated [16,4,7]	1				4	5	4	1			1	
punctured [16,4,7]	1				4	4	3	3			1	
not ter. and punct. [12,4,4]	1	4	5	1	3	1		1				

Fig. 9.7 Results obtained with two other possible interleavers (Example 9.5)

codes. On the contrary, the second solution, which is associated to an interleaver described by a permutation circulant, shows better performance after puncturation, frame non-termination and even both.

In order to evaluate the performance of a [20,4,8] block code, like that described in Fig. 9.7 up, we can compare it with a direct product code: assuming an [8,4,4] extended Hamming code and a [2,1,2] parity check (or repetition) code, we obtain a [16,4,8] direct product code, which is clearly better. Nevertheless this direct product code exhibits $A(8) = 14$, whereas in the turbo code described above, we have only $A(8) = 2$.

Example 9.7 With the purpose of understanding the role of a large interleaver, to which the model of the statistical uniform interleaver can be related, let us consider the two only code words having output weight $w = 8$ in the complete form of the code described in Fig. 9.7 up. In this [20,4,8] 1/3-rate turbo code, such two code words are both originated by 3-weight information sequences. One of them is characterized by $w_i = 3$, $c_1 = 2$, $c_2 = 3$, and the other by $w_i = 3$, $c_1 = 3$, $c_2 = 2$. So the former one will contribute to coefficients $M_1(4,2|3)$, $M_2(4,3|3)$, the latter one to coefficients $M_1(4,3|3)$, $M_2(4,2|3)$. At the denominator in expression (9.6), the term $\binom{4}{3} = 4$ is not interesting by itself, but is useful for attempting an extrapolation: with increasing Z, we can expect that the above coefficients M_1 and M_2, still related to the minimum distance $w = d = 8$, grows both linearly with Z, so at the numerator we have Z^2, but at the denominator, owing to $w_i = 3$, we have Z^3.

As a general comment, after the results reported in Examples from 9.4 to 9.7, and taking into account also the properties of a statistical uniform interleaver discussed in the previous section, we can observe what follows. There are three main types of code words:

(i) those whose number $A_o(w)$ is substantially independent of Z,
(ii) those whose number can be written $A_1(w) = B_1(w)Z - \delta_1(w)$,
(iii) those whose number $A_2(w)$ increases more than linearly with Z.

For fixing ideas, we can take the following binomial growth, for the last type of coefficients, $A_2(w) = C_2(w)\binom{Z}{2} - \delta_2(w)$. Remark that the delays δ_1 and δ_2 depend on the interleaver, and hence on Z again, but they can be strongly increased optimizing the structure of the interleaver at any new design. In expression (9.4), where Z is to be set at the denominator, we will obtain three types of effects: the first one due to $A_o(w)$ practically negligible in the water fall region, and useful only for BER floor prediction, the second one, due to $A_1(w)$, which, after a possible long delay δ_1, appears independent of Z, and the third one, which, after a possible very long delay δ_2, makes any further increase in the interleaver size no longer efficient. Clearly the second effect is dominant in the water fall region, and hence in fixing the interleaver coding gain G_I.

Recently an enhanced scheme of turbo codes has been proposed, based on sending more than one symbol at a time to each encoder. Taking, for instance, a pair of equal $(n_o-1)/n_o$-rate RSC convolutional codes, generally in RSC encoder observer arrangement, and passing to the interleaver (n_o-1)-tuples of information symbols, it is possible to obtain a so called multiple-binary turbo code (Douillard and Berrou 2005). The overall code rate turns out to be $(n_o-2)/n_o$, without punc-turation. In this situation, the characteristics of the interleaver and those of the two component codes can be efficiently optimized without the risk of a possible dete-rioration due to puncturation.

9.5 Outline of the Decoding Algorithm

Modern soft-decision decoding algorithms are based on iterative processes, during which convergence to a maximum likelihood estimate of the code word or code frame is reached, after a more or less long series of iterations. At an intermediate step of the decoding procedure, they need, besides a certain symbol identity in a given position, also a measure of its reliability.

Definition 9.7 A procedure able to give the prediction of the identity of a certain symbol and a measure of its reliability is called *Soft-In Soft-Output decoding algorithm* (*SISO*). Soft-in means that the input signal is a continuous variable, soft-output that a measure of the expectation reliability is given. A possible method for expressing this measure is in analyzing the probability of a decoded symbol (0 or 1 in a binary alphabet) in a certain position, given the observed value of the input signal. It is known as *A Posteriori Probability* (*APP*).

In order to give an outline of the principle on which such algorithms are based, we will briefly describe the procedure called Soft Output Viterbi Algorithm (SOVA) (Hagenauer and Hoeher 1989; Berrou et al. 1993b). It operates on a traditional binary generator trellis (see Definition 1.40), where, as usual, a com-parison between the metrics associated to two paths merging in the same node is performed, allowing the choice of the most likely one of them. In SOVA, at this point of the procedure, a further output is given, regarding the separation between such two metrics. Then this information can be cumulated and processed in various ways. In any case, a sort of weight is associated to any position in the decoded word or frame. Such weight, representing a higher or lower level of reliability of the symbol identity, is updated at each step of the decoding procedure. Clearly, employed alone, this algorithm does not give any advantage, except an evaluation of the quality of the information received. Nevertheless, when we have two encoding processes, acting on the same information sequence, the two decoders can operate taking into account also a mutual help deriving from crossed estimates and weights (extrinsic information).

In the original paper where turbo codes were firstly presented (Berrou et al. 1993a), the BCJR algorithm was suggested to be employed. It was known since

1974 (Bahl et al. 1974), for the soft-decision decoding of convolutional codes with long constraint lengths. A logarithmic ratio between the two *APP*'s is considered.

Definition 9.8 Referring to the prediction of a certain symbol s_j, the natural logarithm of the ratio between the two *APP*'s is defined Logarithm of Likelihood Ratio (*LLR*). It is possible to write

$$LLR(s_j) = \text{Log}\left[\frac{P\{s_j = 1|\text{obs. value}\}}{P\{s_j = 0|\text{obs. value}\}}\right], \tag{9.7}$$

where $P\{s_j = x \mid \text{obs.value}\}$, $x = 0, 1$, represents the conditional probability that s_j is 0 or 1, given a certain observed value of the input signal.

The *LLR* can be positive (when $s_j = 1$ is the most probable result) or negative (when $s_j = 0$ is the most probable result). The higher the absolute value of the *LLR*, the more reliable the result. It is possible to set a minimum absolute value for this parameter, in such a way as to stop the calculations when a prefixed reliability has been reached for all the code frame. A more efficient method for stopping the decoding procedure is as follows: at any iteration, the parity check test is made in order to verify if there are residual errors. When this test for error detection gives a negative result the procedure can end. Nevertheless such a test is possible only if the parity check matrix of the considered turbo code is available.

For application in a turbo coding context, the procedure described so far has been properly adapted considering two separate modified BCJR algorithms (one for each component code), with a crossed extrinsic information passing from one to the other. A conceptual scheme of this principle is shown in Fig. 9.8. It is necessary to distinguish if symbol s_j represents information (i_j), control from the first RSC encoder (c_{1j}) or control from the second RSC encoder (c_{2j}). All these symbols, at the first step of the process, are affected by noise. Then the two modified BCJR decoders calculate a pair of estimates $\hat{s}_{1j}$, $\hat{s}_{2j}$, and the correspondent logarithm of likelihood ratios $LLR_1(s_j)$, $LLR_2(s_j)$.

Considered the form adopted for $LLR(s_j)$ in (9.7), for a binary channel it is enough to send forward the only *LLR*, in fact the estimate, at the step considered each time, is included in its sign. Furthermore, with a binary channel affected by additive Gaussian noise (Definition D.2) the expression of the *LLR* contains the noise variance. Therefore, in order to make initialization more efficient, it should be

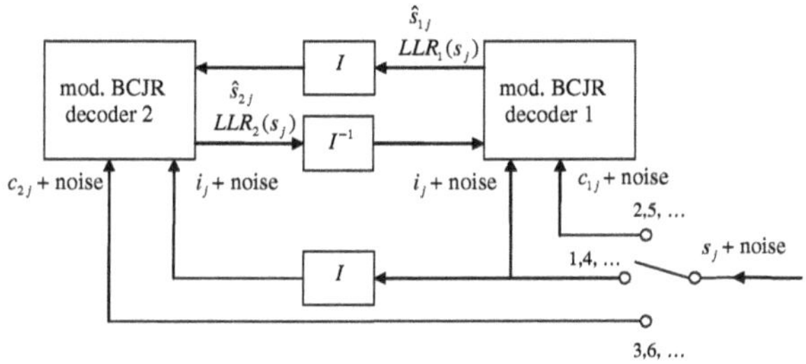

Fig. 9.8 Conceptual scheme of turbo code decoding

advisable to have an approximate knowledge also of such noise variance (Lin and Costello 2004).

In subsequent steps, progressive refinements of the *LLR* are passed forward and backward between the two decoders. In all these passages, the interleaver I and the de-interleaver I^{-1} are involved, in order to give the right sequence order for the next decoding step. Roughly speaking, one decoder exploits an help from the extrinsic information provided by the other decoder, which previously operated as well. The name "turbo" derives from the analogy with a similar well known arrangement in engine mechanics (Berrou et al. 1993a).

The procedure here summarized can be applied similarly, with a pair of SOVA decoders (Hagenauer and Papke 1994), instead of modified BCJR decoders. This solution was firstly proposed for a direct serial concatenation of two convolutional codes (Berrou et al. 1993b). One of the main difference is that SOVA does not require a knowledge of the noise variance (Ryan and Lin 2009), as is necessary, instead, for the initial evaluation of the *LLR* with BCJR. Finally we can note that SOVA is optimized in order to minimize the word error probability, instead BCJR to minimize the binary symbol error probability. In conclusion, SOVA exhibits generally a less computational complexity, but the best results are typically obtained with modified BCJR, although a wide variety of situations may occur, depending also on the particular interleaver employed.

Such two decoding algorithms are examples of more general iterative decoding procedures called message passing algorithms (Ryan and Lin 2009), in which two or more processors produce complementary decoding information and transfer their instantaneous result to each other, in form of proper messages. Such message transfer is supported by a suitable geometrical structure. In the case of turbo codes, the geometrical structure is offered by the two code trellises and by the interleaver (or de-interleaver). A decoding algorithm of this type can be used also for a soft-decision decoding of product codes (direct or with interleaver), which will be better described in Sect. 9.7. Probably the most interesting application of the message passing principle is however in LDPC codes, to which Chaps. 10–12 are devoted.

The size of the code words, in all the above applications, is so great that the soft-decision decoding algorithms proposed for them necessarily give sub-optimal results. Nevertheless they can reach error probability curves very close to that achievable with an optimum decoding algorithm (a small fraction of dB apart, depending also on the number of quantization levels adopted in signal processing). On the other hand, except rare situations in which a parallel computation can be performed, the decoding procedure is carried out on the whole code word, sometimes leading to a rather heavy time latency, especially at low symbol rates.

Regarding the convergence of the iterative message passing procedures which can be adopted with a proper concatenation of two component codes, it is convenient to distinguish the following three main situations (Hagenauer et al. 1996):

(i) two convolutional codes,
(ii) one convolutional code and one block code,
(iii) two block codes.

In situation (*i*) the role of the interleaver is fundamental, either in fixing the performance or in allowing the possibility of a rapid converge. In situation (*ii*) and even more in situation (*iii*) the structure of the block codes plays the most important role. Sometimes, in order to simplify decoding, it is convenient to develop the message passing algorithm on a support offered by the parity check matrix, especially when high-rate block codes are adopted or in the framework of an LDPC code model.

In order to estimate the computational complexity of the whole iterative decoding procedure employed with turbo codes, we have to take into account the trellises of the two component codes, but also the size and structure of the interleaver. Clearly, the higher the size, the greater the decoding computational complexity, like occurs in any block code. Nevertheless also the structure of the interleaver has an important role, as confirmed by the number of iterations requested for reaching convergence. It is useful to distinguish between: causal interleavers (with a very small memory, periodically filled and emptied), and non-causal interleavers (with a memory as large as the whole code frame, or of the same order of magnitude). Turbo code applications show better performance with non-causal interleavers, but, also in the framework of this type of devices, we can have strong differences.

In particular, it is possible to define a state number profile as a parameter able to synthesize the behaviour of an interleaver (Garello et al. 2001a). In causal interleavers this parameter is practically uniform and equals the memory size, in non-causal interleavers it is typically variable and achieves a maximum value, related to the convergence feature of the decoding algorithm. A good interleaver design, besides the constraints regarding the relation between coding gain and size, should take into account also the above considerations about decoding computational complexity, with direct impact on circuit costs and time latency.

9.6 Turbo Codes in Serial Concatenation

Traditionally, turbo codes are only those codes characterized by a scheme based on a parallel concatenation, of the type depicted in Fig. 9.1. Nevertheless, in an enlarged sense, also a modified version of that scheme, in which a serial code concatenation is employed, can be referred to turbo codes as well. In Fig. 9.9 the scheme of a turbo code in serial concatenation, employing two 1/2-rate convolutional codes, is depicted (Benedetto et al. 1998b). A first strong difference with

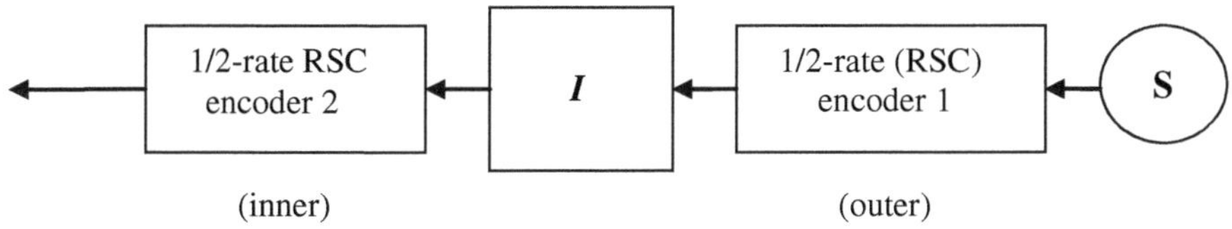

Fig. 9.9 Scheme of the encoder circuit for a 1/4-rate serially concatenated turbo code

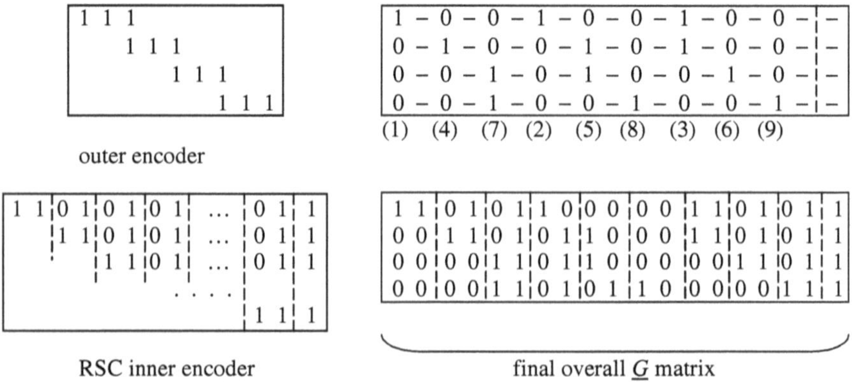

Fig. 9.10 Generator matrix construction for the 1/4-rate serially concatenated turbo code

respect to parallel concatenation is in the fact that here, before possible punctur-ations, the asymptotic code rate is only 1/4. Furthermore we have different roles played by the two component codes. As in any serial concatenation (see Fig. 1.9), we can distinguish an inner code (RSC encoder 2) and an outer code (encoder 1). The latter one can or cannot be in RSC form. The interleaver is useful for spreading its 1-symbols patterns, whereas the RSC inner encoder has the role of adding 1-symbols to the low-weight code patterns outgoing from the outer encoder.

Example 9.8 Let us take the pair of equal 1/2-rate convolutional codes having $g_{\text{int}}(x) = x^2 + x + 1$. Adopting $Z = 4$ periods for the frame of the inner code, since we have $m = 1$, but the last symbol (always zero) can be eliminated, the interleaver size is only 9. We assume that it is characterized by the output symbol sequence $[s_1\, s_4\, s_7\, s_2\, s_5\, s_8\, s_3\, s_6\, s_9]$. This is a rather regular permutation, but it is enough for comprehending the mechanism involved in the process. In Fig. 9.10 up-left we have the generator matrix of the non-systematic outer encoder, then the effect of the interleaver is to apply a permutation in the columns of such generator matrix, as shown in Fig. 9.10 up-right, where the positions devoted to the control symbol in any period and to the final control symbol due to the inner encoder have been already displayed in the overall frame (denoting them by small segments).

The inner encoder, RSC version of the one considered before, is characterized by the $\underline{G}$ matrix reported down-left in Fig. 9.10. This matrix is applied to any row of the preceding matrix. The final overall $\underline{G}$ matrix is then as described down-right. The resulting code is a [19,4,6] 1/4-rate serially concatenated turbo code, whose word weight distribution turns out to be

$$A(0) = 1,\ A(6) = 3,\ A(8) = 3,\ A(10) = 6,\ A(12) = 3.$$

It is possible to observe that the outer code in Example 9.8 is a [9,4,3] terminated 1/2-rate non-systematic convolutional code, whereas the inner code is a [19,9,3] terminated 1/2-rate RSC code.

Also for serially concatenated turbo codes, it is advisable to perform a code puncturation. For instance, taking one control symbol any two in the inner component code, we obtain a 1/3-rate punctured code, typically not too worse than the original one. On the contrary, some cares can be taken in adopting non-terminated frames, since the minimum distance could be strongly sacrificed. If the code frame is long enough, the code rate penalty due to both frame terminations is generally acceptable.

Example 9.9 To the code introduced in Example 9.8, let us apply a puncturation, taking two control symbols any three on average. Adopting a suitable puncturation matrix, the resulting 1/3-rate serially concatenated punctured turbo code is a [15,4,5] code, with the following word weight distribution

$$A(0) = 1,\ A(5) = 3,\ A(6) = 1,\ A(7) = 3,\ A(8) = 4,\ A(9) = 1,\ A(10) = 2,\ A(11) = 1.$$

In general, with respect to a traditional parallel concatenation having comparable parameters, serially concatenated turbo codes show a smaller interleaver coding gain, but also a less evident BER floor. Nevertheless many variants (Maunder and Hanzo 2010) on such schemes have been proposed, to the point where very different design goals and code concatenation schemes have been related, in last years, to the wide and fruitful family of turbo codes.

About the decoding procedures, it is generally preferable to employ ad hoc algorithms, based on a proper post-processing, in order to generate reliability data after the decision of the outer decoder (Seshadri and Hoeher 1993). The non-reciprocal role played by the two component codes, following this way, can be more efficiently exploited. However such message passing procedures are very similar, in their principle of operation, to SOVA or to BCJR algorithm.

9.7 Turbo-Product Codes

In a direct product code, we have an ordered set of code words (Definition 1.34) organized as a sequence of rows obtained from the first component code. Then the second component code is applied to such code words operating along the columns of a rectangular encoding memory. The role of the two component codes are interchangeable. Nevertheless, differently from the case of a parallel concatenation, the code rate is more strongly reduced, owing to the presence of controls on controls.

Now we are going to widen this concept, taking a pair of convolutional codes, instead of block codes. The only difference is in the substitution of an ensemble of terminated code frames to an ensemble of code words. The scheme we can adopt for the rectangular encoding memory is that reported in Fig. 9.11 (Gazi and Yilmaz

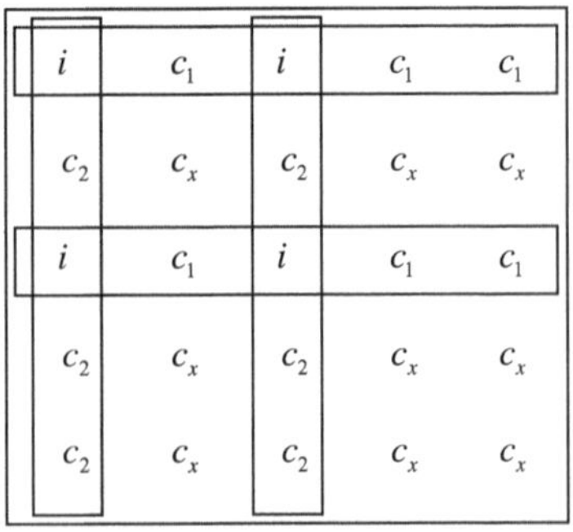

Fig. 9.11 Encoding memory for a turbo-product code without interleaver with $k = 4$

2006), where different types of symbols have been evidenced: i means information symbols, c_1 control symbols of the first component code on information symbols, c_2 control symbols of the second component code on information symbols, finally c_x control symbols on control symbols, or briefly crossed control symbols. We have considered the case of $Z = 2$ (which means a total number of information symbols $k = 4$), with a pair of equal component codes of the type employed in Example 9.8, that is with $m = 1$, and only one final control symbol for a correct frame termination. The resulting turbo-product code is a [25,4,9] code, and we have on the whole: 4 symbols of type i, 6 symbols of type c_1, 6 symbols of type c_2, 9 symbols of type c_x. The name turbo-product follows clearly from the particular soft-decision decoding algorithm which can be adopted, typically derived from application of the message passing principle, in the framework of SISO algorithms (Pyndiah 1998), initially proposed only for a pair of block codes. In the following, we will restrict our attention to the case of two convolutional codes (to which Fig. 9.11 is referred, although with very short frames).

A first remarkable difference of this scheme with respect to parallel or serial concatenation schemes is that here a frame of one component code does not act on all the information symbols, but only on a subset of them. On the other hand, many code frames are contemporary treated. Typically this leads to very long overall code words.

Lemma 9.4 *Given a pair a 1/2-rate convolutional codes, characterized by generator constraint length m_1 and m_2 respectively, fully exploited, the construction of a turbo-product code taking Z_1 periods from the first and Z_2 periods from the second, without any puncturation and with both frames correctly terminated, yields an [n, k, d] block code with the following parameters*

$$n = 4Z_1Z_2 + 4(m_1Z_2 + m_2Z_1) + 4m_1m_2, \quad k = Z_1Z_2, \quad d = d_1d_2. \tag{9.8}$$

where d_1 and d_2 are the minimum distances of the two component codes.

Proof The first expression of (9.8) follows from the product between the two frame lengths $n_1 = 2(Z_1 + m_1)$ and $n_2 = 2(Z_2 + m_2)$. The third expression of (9.8) is due to Corollary a of Theorem 1.3, about direct product codes in general. □

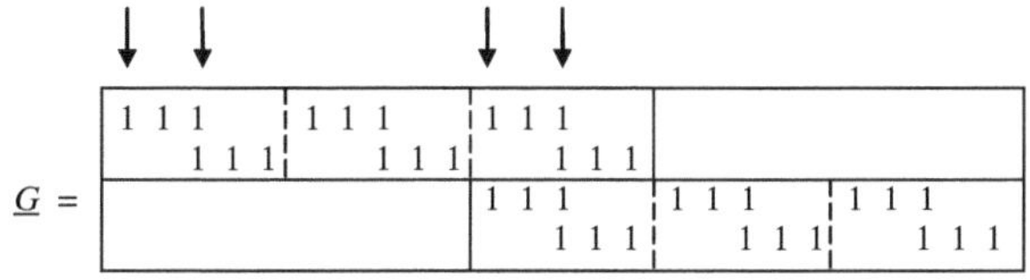

Fig. 9.12 Kronecker generator matrix of the turbo-product code described in Example 9.10

Remark that the first expression of (9.8) collects three terms having different orders of magnitude. For instance, taking $m_1 = m_2 = 3$, $Z_1 = Z_2 \cong 30$, the first term is of the order of 3,600, the second term of the order of 700, the third term is 36. The code rate is low, with the same asymptotic value as in a serial concatenation, but the obtainment of this asymptotic value is much slower. Clearly code puncturation or frame non-termination, or both, are recommended. Here even both component codes can be non-systematic. The advantage of starting from a direct product code scheme is in the predictability of the resulting code behaviour, say from its generator matrix in Kronecker product form (see Theorem 1.4).

Example 9.10 Assuming the two equal codes considered in Example 9.8, and characterized by $g_{int}(x) = x^2 + x + 1$, over a number of periods $Z = 2$ for both, we obtain the scheme already reported in Fig. 9.11. Considering both codes in non-systematic form, that is with a generator matrix of the type

$$\underline{G}_1 = \underline{G}_2 = \begin{bmatrix} 1 & 1 & 1 & 0 & 0 \\ 0 & 0 & 1 & 1 & 1 \end{bmatrix},$$

we are left with the Kronecker matrix for the overall turbo-product code shown in Fig. 9.12. Small arrows denote the positions where information symbols are inserted.

This is a [25,4,9] code, with the following word weight distribution

$$A(0) = 1,\ A(9) = 4,\ A(12) = 4,\ A(16) = 3,\ A(17) = 2,\ A(18) = 2,$$

which can be compared with the weight distributions of the two component codes, consisting of $A(0) = 1$, $A(3) = 2$, $A(4) = 1$. Remark that, when the two frame lengths increase, an asymptotic code rate 1/4 is approached. For instance, adopting $Z = 10$, we obtain a [484,100,9] overall code. We can imagine the structure of the above $\underline{G}$ matrix as long as Z increases in both component codes. There will be a growth either in the row number within any layer or in the overall number of layers. This behaviour, already noted in Sects. 4.9 and 6.10 (with a pair of different codes, and one of the two in tail-biting version), has been named that of a doubly convolutional code. In the present case, it is due to both component codes acting in the same way on orthogonal trajectories, without any tail-biting arrangement.

Example 9.11 Starting from the complete [25,4,9] turbo-product code presented in Example 9.10 (Fig. 9.12), we can apply a code puncturation eliminating the fourth symbol in a period, and the second symbol in the subsequent period. So we remain with a punctured code whose asymptotic code rate is 1/3. In the case under study it is characterized by the tern [21,4,7]. Adopting a non-terminated frame for one of the two component codes, without puncturation, we obtain a [20,4,6] code. Superimposing both solutions we are left with a [16,4,5] code.

Frequently the name turbo-product codes is employed, with reference to the soft-decision decoding procedure adopted, also for the simple product of two block codes. In this case, they are often separated by an interleaver. The possibility of one block code and one convolutional code is also frequently considered. For instance, the pair of component codes adopted in Example 9.10 are equivalent to a pair of [5,2,3] two-time shortened Hamming codes, which are block codes. Nevertheless, with increasing parameter Z, this equivalence obviously is no longer true. A cyclic block code, which is strongly shortened, exhibits an evolution of the weight distribution resembling that of a convolutional code, as already noted. This property makes the two schemes capable of obtaining comparable results. Many solutions have been proposed (Gracie and Hamon 2007; Sidorenko et al. 2012), in particular showing the role of the interleaver. Most attention is however devoted to the possibility of maintaining good performance also after a strong puncturation. Finally, when a high minimum distance is to be obtained, the drawback of a too long code word can arise in a direct product turbo-code, if both component codes are convolutional.

Example 9.12 Starting from a pair of convolutional codes, the first with $m = 3$ and the second with $m = 4$, respectively with $d_1 = 6$ and $d_2 = 7$ (see Table 3.2 for adopting the best 1/2-rate convolutional codes with such parameters), we can obtain an overall 1/4-rate turbo-product code with minimum distance $d = 42$. Nevertheless, in order to reach an asymptotic coding gain of the order of 10 dB, it is necessary to adopt Z_1 and Z_2 at least of about 100, which means a block length somewhat greater than 40,000. A code of this size is close to the maximum limit for admitting a SISO decoding algorithm with acceptable computational complexity. Furthermore, when the minimum distance is so high, the number of code words having $w = d$ is very large, so that the error probability curve exhibits a break-even point for values not too small of the *SNR*.

A completely different situation is that occurring when the product is made including in the overall encoder an interleaver. The scheme of encoder circuit reported in Fig. 1.19 (direct product of two block codes stressing their equal role) is here to be modified for better understanding the case of two convolutional encoders separated by an interleaver. As shown in Fig. 9.13, the memory of the interleaver is rectangular, with one dimension equal to the frame length of the first convolutional code. Let us consider a pair of 1/2-rate convolutional codes. Taking $n_1 = 2(Z_1 + m_1)$ such frame length and choosing Z_2 periods for the second convolutional code, the memory will be a $Z_2 \times n_1 = Z_2 \times 2(Z_1 + m_1)$ rectangular structure.

Fig. 9.13 Scheme of a turbo-product code, employing two 1/2-rate convolutional codes, with interleaver

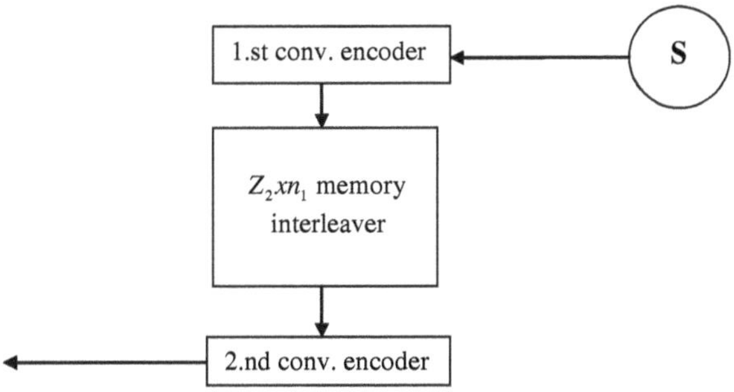

It is important to distinguish two types of interleavers (Gazi and Yilmaz 2006):

(i) the so called column interleaver, in which any column is interleaved independently, without elements passing from one column to another,
(ii) the so called full interleaver, in which elements can pass from one column to another.

Clearly only adopting a column interleaver the property of reaching minimum distance $d = d_1 d_2$ is maintained. The product codes so arranged show a behaviour not remarkably different from the one of direct product codes, and the improvement they can produce is in a break-even point slightly shifted towards the left hand side, but often accompanied by a less steep water fall.

Nevertheless, adopting a full interleaver, although the above constraint on the minimum distance is no longer satisfied, and consequently the BER floor could be more pronounced, it is possible to reach very high values of interleaver coding gain. The phenomenon of spectral thinning appears more smoothed, the minimum distance becomes smaller, but the number of low-weight code words are very few. With the data of Example 9.12, accepting a minimum distance of the order of 15, but with very small values of all the weight distribution coefficients $A(w)$ up to about 45, the break-even point in the curve of the error probability can shift remarkably to the left hand side.

An interesting comparison can be made between serially concatenated turbo codes and turbo product codes. In fact, assuming two 1/2-rate codes, in both schemes we have an asymptotic code rate 1/4, without puncturation. Nevertheless the interleaver employed in turbo product codes must contain several code frames, whereas the one needed in serially concatenated turbo codes contains just one frame. More precisely, letting Z the number of periods in the frames of any component code of the turbo product code and Z_s the number of periods in the unique frame of the outer component code in the serial concatenation scheme, with a comparable size of the interleaver, it is possible to adopt $Z \cong \sqrt{Z_s}$. Assuming Z not smaller than 50, and hence Z_s not smaller than 2,500, we obtain two behaviours not remarkably different, especially if a full interleaver is employed in the turbo product scheme. For smaller sizes of the interleaver, a serial concatenation is typically preferable. An advantage to be considered for the product code finally is that a parallel decoding procedure can be partially set up, so the time latency can be reduced to be proportional to Z (Gazi and Yilmaz 2006). Nevertheless the computational complexity, in its part related to the interleaver, remains proportional to its overall size, that is to $2Z^2$.

9.8 Parity Check Matrix of Turbo Codes

Although not greatly investigated in the literature, the structure assumed by the parity check matrix of a turbo code can have some interest. We will treat only two cases: the traditional parallel concatenation and the turbo-product. This analysis is useful for assuring a parity check test on the decoded frames in traditional turbo code schemes, and also in the perspective of a comparison with LDPC codes (Esmaeli 2006).

First of all, let us consider the parity check matrix of an RSC encoder circuit. Proceeding like in Example 9.1, it is possible to obtain a recursive systematic version of $\underline{H}$, simply knowing the interleaved generator polynomial of our 1/2-rate code. In this matrix, only one 1-symbol is present in the column having odd positions (those devoted to information symbols). The columns with even positions, instead, collect many control symbols. All the rows are non-periodic, but they are unlimited on the left hand side. It is possible to describe the behaviour of such parity check matrix as follows. The last 1-symbol in any row represents a control symbol acting on the unique information symbol in the position before it, but also on many control symbols associated to previous periods. This matrix, which is not in minimal form, is however equivalent to the traditional $\underline{H}$ matrix, where the syndrome former sub-matrix $\underline{H}_o^T$ vertically shifts, generally designed in its minimal form.

Example 9.13 Assuming the same RSC code taken also in Example 9.1, that is the one associated $g_{int}(x)$ corresponding to the sequence 11.10.11, with $Z = 4$, the expression for its RSC parity check matrix can be calculated as shown in Fig. 9.14 up. It has been constructed from that in minimal form after proper ex-OR sums of rows. This matrix is characterized by semi-infinite syndrome former sub-matrices, whose columns in even positions, after the first two 1-symbols show an alternation of 1- and 0-symbols. It is interesting to compare this recursive systematic form of the $\underline{H}$ matrix with the correspondent form of the $\underline{G}$ matrix, which can be obtained from Example 9.1, with the addition of a proper frame termination. This generator matrix is reported down in Fig. 9.14.

Example 9.14 The turbo code considered in Example 9.4, which exploits an interleaver of the type shown in Fig. 9.6 up, that is $[i_3\ i_1\ i_2\ i_4]$, can be analyzed as follows. In Fig. 9.15 up the $\underline{H}$ matrix is reported in the only parts provided by the second encoder. The positions to be occupied by the control symbols of the first encoder, taking into account also the effect of the interleaver, have been already displayed (with small segments). Bold characters refer to the information symbol positions. The conventional sequence of information and control symbols is reported in the Figure, with the following notation: i is, as usual, an information

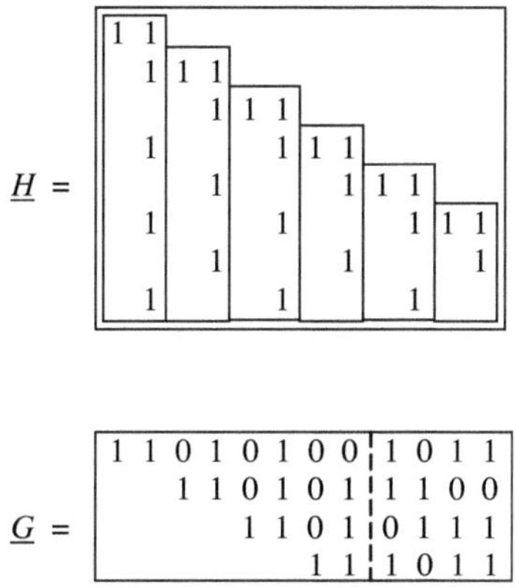

Fig. 9.14 Parity check matrix and generator matrix for the RSC code described in Example 9.13

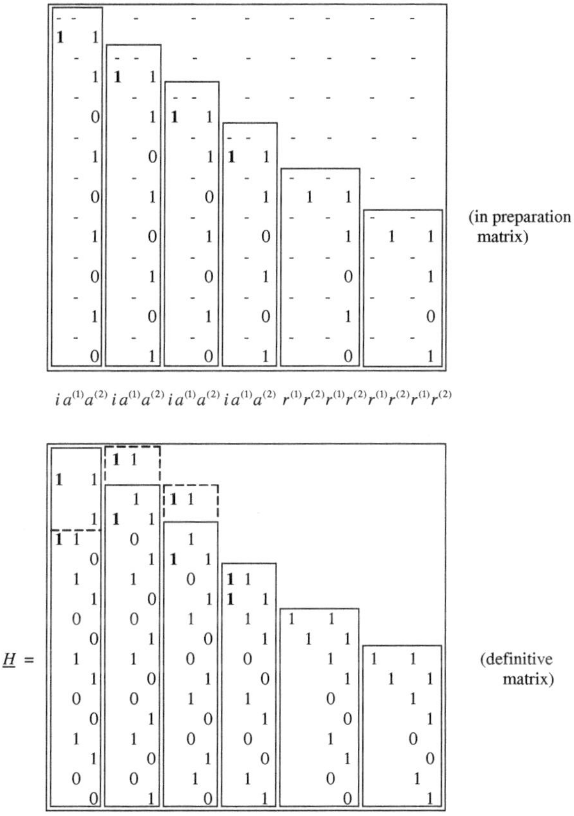

Fig. 9.15 Construction of the parity check matrix for the 1/3-rate convolutional code treated in Example 9.14

symbol, $a^{(1)}$ and $a^{(2)}$ represent distributed control symbols of the two encoders, $r^{(1)}$ and $r^{(2)}$ the respective final control symbols. Taking into account the permutation introduced by the interleaver, the final result is that shown in Fig. 9.15 down. It is

Fig. 9.16 Parity check matrix of the 1/4-rate turbo-product code without interleaver discussed in Examples 9.10 and 9.15

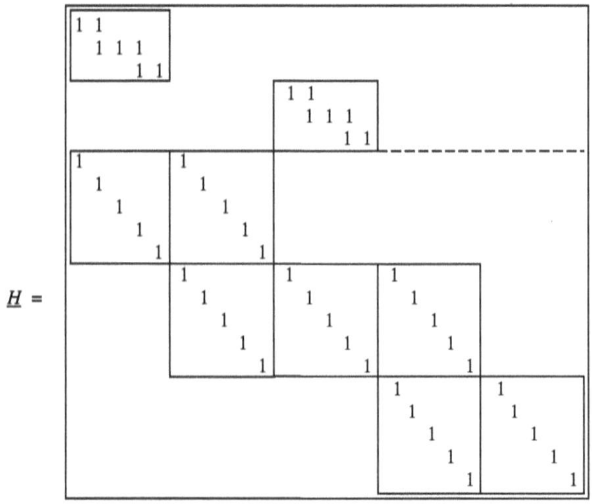

possible to see the structure of the overall syndrome former sub-matrix, as it would be without the action of the interleaver, looking at the fourth period. The effects of the interleaver are evidenced with dashed borders. Looking at Figs. 9.6 and 9.15, it is possible to observe, in spite of the very simple interleaver adopted, how strong is the time-varying characteristic exhibited by the $\underline{G}$ and $\underline{H}$ matrix in a turbo code of this type.

Example 9.15 The same direct product code already considered in Example 9.10, with two equal non-systematic component codes, characterized by $g_{int}(x) = x^2 + x + 1$, over a number of periods $Z = 2$ for both, is taken here for the determination of its parity check matrix. Following Lemma 5.39, it can be constructed as shown in Fig. 9.16, considering that the parity check matrix of each component code results in

$$\underline{H} = \begin{bmatrix} 1 & 1 & & & \\ & 1 & 1 & 1 & \\ & & & 1 & 1 \end{bmatrix}.$$

In fact the last column and the last row are to be eliminated. As already noted, the overall result is a [25,4,9] code, therefore its $\underline{H}$ matrix exhibits 21 rows. In Fig. 9.12 the correspondent generator matrix was reported. Comparison allows to justify the structure of the upper part of such $\underline{H}$ matrix. Clearly, on longer frame lengths, the result appears not so extreme, and the asymptotic code rate can be rather rapidly approached. The introduction of a column interleaver is expected to produce a substitution of the identity matrices $\underline{I}_5$, in the lower part of the overall $\underline{H}$ matrix, with suitable permutation matrices. The variation in the weight distribution is modest, even for values of Z remarkably higher. Only a full interleaver, which completely transforms the lower part of the parity check matrix, is able to originate a less abrupt spectral thinning, improving the coding gain for central values of the error probability. Remark that the $\underline{H}$ matrix in Fig. 9.16 is free of 4-length short cycles, and hence this turbo-product code (even with a column interleaver) is also capable of been decoded as an LDPC code. This property is no longer true, in general, when a full interleaver is employed.

References

Bahl LR, Cocke J, Jelinek F, Raviv J (1974) Optimal decoding of linear codes for minimizing symbol error rate. IEEE Trans Inf Theory 20:248–287

Banerjee A, Vatta F, Scanavino B, Costello DJ (2005) Non-systematic turbo codes. IEEE Trans Commun 53:1841–1849

Barbulescu AS, Pierobon SS (1995) Terminating the trellis of turbo-codes in the same state. Electron Lett 31:22–23

Benedetto S, Montorsi G (1996) Unveiling turbo codes: some results on parallel concatenated coding schemes. IEEE Trans Inf Theory 42:409–428

Benedetto S, Divsalar D, Montorsi G (1998a) A search for good convolutional codes to be used in the construction of turbo codes. IEEE Trans Commun 46:1101–1105

Benedetto S, Divsalar D, Montorsi G, Pollara F (1998b) Serially concatenation of interleaved codes: design and performance analysis. IEEE Trans Inf Theory 44:909–926

Berrou C, Glavieux A, Thitimajshima P (1993a) Near Shannon limit error-correcting coding and decoding: turbo codes. In: Proceedings of IEEE international conference communications (ICC93), Geneva:1064–1070

Berrou C, Adde P, Angui E, Faudeil S (1993b) A low complexity soft-output Viterbi decoder architecture. In: Proceedings of IEEE international conference communications (ICC93), Geneva:737–740

Berrou C, Glavieux A (1996) Near optimum error correcting coding and decoding: turbo codes. IEEE Trans Commun 44:1261–1271

Divsalar D, Pollara F (1995) Weight distributions for turbo codes using random and non-random permutations. JPL, TDA Prog Rep, pp 42–122

Douillard C, Berrou C (2005) Turbo codes with rate-m/(m + 1) constituent convolutional codes. IEEE Trans Commun 53:1630–1638

Esmaeli M (2006) The minimal product parity check matrix and its application. In: Proceedings of IEEE ICC 2006, Instanbul, pp 1113–1118

Forney GD (1966) Concatenated codes. MIT Press, Cambridge

Garello R, Montorsi G, Benedetto S, Cancellieri G (2001a) Interleaver properties and their applications to trellis complexity analysis of turbo codes. IEEE Trans Commun 49:793–807

Garello R, Chiaraluce F, Pierleoni P et al (2001b) On error floor and free distance of turbo codes. In: Proceedings of IEEE international conference communications 1, Helsinki, pp 45–49

Gazi O, Yilmaz AO (2006) Turbo product codes based on convolutional codes. ETRI J 28:453–460

Gracie K, Hamon MH (2007) Turbo and turbo-like codes: principles and applications in telecommunications. Proc IEEE 95:1228–1254

Hagenauer J, Hoeher P (1989) A Viterbi algorithm with soft-decision outputs and its applications. Ibn: Proceedings IEEE global telecommunications conference, Dallas, pp 1680–1686

Hagenauer J, Papke L (1994) Decoding turbo codes with the soft-output Viterbi algorithm. In: Proceedings of international Symposium information theory, Trondheim (Norway):164

Hagenauer J, Offer E, Papke L (1996) Iterative decoding of binary block and convolutional codes. IEEE Trans Inf Theory 42:429–445

Kim MG (1997) On systematic punctured convolutional codes. IEEE Trans Inf Theory 45:133–139

Lin S, Costello DJ (2004) Error control coding. Pearson/Prentice-Hall, Upper Saddle River

Maunder RG, Hanzo L (2010) Iterative decoding convergence of serially concatenated codes. IEEE Trans Vehic Tech 59:216–224

Oppenheim AV, Schafer RW (1974) Digital signal processing. Prentice-Hall, Englewood Cliffs

Perez L, Seghers J, Costello DJ (1996) A distance spectrum interpretation of turbo codes. IEEE Trans Inf Theory 42:1698–1709

Proakis JC (1983) Digital communications. McGraw Hill, Tokyo

Pyndiah R (1998) Near-optimum decoding of product codes: block turbo codes. IEEE Trans Commun 46:1003–1010

Ryan WE, Acikel OF (1999) Punctured turbo-codes for BPSK/QPSK channels. IEEE Trans Commun 47:1315–1323

Ryan WE, Lin S (2009) Channel codes: classical and modern. Cambridge University Press, Cambridge

Seshandri N, Hoeher P (1993) On post-decision symbol-reliability generation. In: Proceedings of IEEE international conference communications (ICC93), Geneva, pp 741–745

Sidorenko V, Bossert M, Vatta F (2012) Properties and encoding aspects of direct product convolutional codes. In: Proceedings of IEEE international symposium informations theory, Cambridge (MA), pp 2351–2355

Chapter 10
Low Density Parity Check Codes

10.1 Tanner Graph and Message Passing Decoding Algorithms Constructed on It

Low density parity check codes are characterized by a parity check matrix where 1-symbols are much less frequent than 0-symbols. Their possible good performance was early suggested in Gallager (1962). Only several years later (Tanner 1981), an efficient soft-decision decoding algorithm was proposed for them. Only after some further years, a behaviour very close to Shannon limit was predicted for codes of this type properly decoded (MacKay and Neal 1997), in this being comparable with or even better than turbo codes. The geometric structure on which message passing between two processors in the decoding algorithm takes place is here offered by a particular graph representation of the parity check matrix itself.

Definition 10.1 The positions of 1-symbols in the $\underline{H}$ matrix of a code, with n total symbols and r control symbols in its code words, can be represented by a graph (called *Tanner graph*). In this structure, all the positions (columns of $\underline{H}$) are collected on the one side, and the parity check equations (rows of $\underline{H}$) are collected on the other side. Edges represent the 1-symbols of the considered row. The n nodes on the one side are called *variable nodes*. The r nodes on the other side are called *control nodes*.

Example 10.1 The [7,3,4] simplex code, whose reciprocal parity check polynomial is $h^*(x) = x^3 + x + 1$, can be represented by the Tanner graph shown in Fig. 10.1.

It is possible to note that the number of edges touching a variable node equals the number of 1-symbols in the correspondent $\underline{H}$ column. Instead, the number of edges touching a control node equals the number of 1 symbols in the row represented by such node. The fact that some variable nodes are touched by a smaller number of edges does not mean, in general, that the correspondent positions in the code word are less protected. As already explained in the introduction of the $\underline{H}$ matrix (see Chap. 5), there is a system of linear equations, represented by as many rows of $\underline{H}$, which interest,

G. Cancellieri, *Polynomial Theory of Error Correcting Codes*,
Signals and Communication Technology, DOI 10.1007/978-3-319-01727-3_10

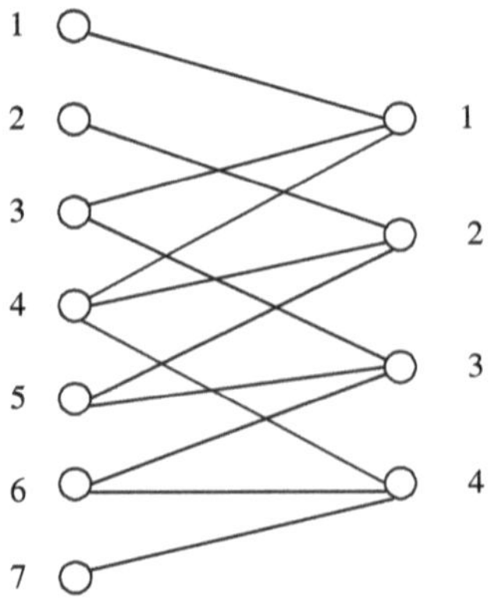

$$\underline{H} = \begin{array}{c} \begin{matrix} 1 & 2 & 3 & 4 & 5 & 6 & 7 \end{matrix} \\ \left[\begin{matrix} 1 & 0 & 1 & 1 & & & \\ & 1 & 0 & 1 & 1 & & \\ & & 1 & 0 & 1 & 1 & \\ & & & 1 & 0 & 1 & 1 \end{matrix}\right] \end{array} \begin{matrix} 1 \\ 2 \\ 3 \\ 4 \end{matrix}$$

Fig. 10.1 Tanner graph for the [7,3,4] simplex code described in Example 10.1

in the same way, all the positions of the code word. This is not for few particular codes, for instance LC codes, but the less protected external positions in their code words are not immediately evidenced in a graph of this type, because $h(x)$ would be periodic. The Tanner graph is able to support a message passing soft-decision decoding algorithm, in which estimates of symbol identities and proper reliability functions are transferred. Clearly this procedure, independently of how is realized, will become feasible if the number of 1-symbols in $\underline{H}$ is not too high.

Definition 10.2 A code is defined *low density parity check* (*LDPC*) when its $\underline{H}$ matrix shows a low number of 1-symbols with respect to the total number nr of its elements (of the order of 5 % or less).

Definition 10.3 An LDPC code described by a parity check matrix $\underline{H}$ whose columns have all the same weight w_c is called *column-regular*. When the rows of such matrix have all the same weight w_r, it is called *row-regular*.

Example 10.2 The code described in Example 10.1 is not column-regular, but is row-regular. Typically column-regularity is considered more important than row-regularity, in order to evaluate decoding computational complexity and even the expected performance.

Definition 10.4 Let us consider an LDPC code (not necessarily regular), whose $\underline{H}$ matrix collects a same module periodically repeated varying some parts in it at any replica. A code of this type is called *structured LDPC code*.

Example 10.3 Array codes (see Sects. 8.8 and 8.9) are both column-regular and row-regular LDPC codes. This property can be argued directly from inspection of their $\underline{H}$ matrix, provided that it is taken with the linear dependent rows included. The possible presence of parity check rows not linearly independent does not influence the decoding procedure constructed on the Tanner graph. Array codes are also structured LDPC codes. The module, in the $\underline{H}$ matrix, which is periodically repeated is represented by a permutation circulant.

Lemma 10.1 *Given a regular LDPC code, either in the column weights or in the row weights, with parameters n, r,* w_r*,* w_c*, the density* ρ *of* 1*-symbols in its* $\underline{H}$ *matrix can be calculated as*

$$\rho = \frac{w_c}{r} = \frac{w_r}{n}. \tag{10.1}$$

Proof The distribution of the edges among both control nodes and variable nodes is uniform. All the control nodes are touched by w_r edges, and all the variable nodes are touched by w_c edges. The overall number of positions in $\underline{H}$ is nr. The number of 1-symbols in such positions can be calculated as $nw_c = rw_r$. □

An LDPC code which is at least column-regular generally can be easily encoded, and its design is typically rather scalable. The same is for a code which is at least w. s. time-invariant in its parity check matrix. On the contrary, irregular LDPC codes, especially when strongly time-varying, resemble turbo codes, in which the design has to be typically repeated for any new situation. Sometimes they can show also an encoding computational complexity, comparable with the decoding computational complexity. A structured LDPC code, finally, often exhibits a straightforward encoding procedure. Its scalability is measured by the size of the $\underline{H}$ module, which is periodically repeated, and by the type of variants to be introduced at any period.

Example 10.4 DC-GPC codes (see Definition 5.18) are structured but not column-regular LDPC codes (in the up-right part of their $\underline{H}$ matrix there are null sub-matrices). DP-GPC codes (Definition 5.19), instead, are structured, column-regular but not row-regular. The module, in both, can be assumed as an identity matrix, whose size represents the variant.

Example 10.5 Array codes are regular and structured LDPC codes, as already noted. The module is a permutation circulant, whose phase represents the variant. Adopting the symbolism introduced at the end of Chap. 8, about these codes, we obtain the following parameter correspondences

$$Y = w_c, \quad X = w_r, \quad p^2 = n, \quad pY = r, \quad (QC),$$
$$Y = w_c, \quad S = w_r, \quad Sp = n, \quad pY = r, \quad (SQC).$$

Remark that, in these equalities, the presence of parity check rows linearly dependent has been ignored.

About the similitude between LDPC codes and turbo codes in their constructive properties (and even in the possible soft-decision decoding procedure to be employed), we can look at the scheme shown in Fig. 10.2 (Wang and Fossorier 2009). The central part of the Tanner graph can be interpreted as an interleaver, through which the ports on the right hand side (control nodes) and the ports on left hand side (variable nodes) are properly interconnected. Regular LDPC codes have a uniform organization of the ports on both sides. Structured LDPC codes show an interleaver whose geometry is modular. Finally irregular LDPC codes exhibit an

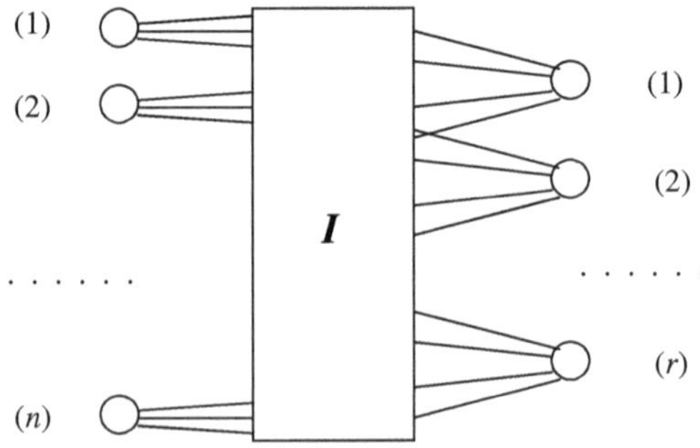

Fig. 10.2 Conceptual scheme of the Tanner graph

interleaver with non-uniform nature. Attempting to interpret LDPC codes as the result of the actions of many component codes, we can imagine that any parity check equation is expressed by a single binary parity check component code, properly interleaved.

We can have either block LDPC codes or convolutional LDPC codes. Any convolutional code typically shows, by itself, a low average density of 1-symbols in its $\underline{H}$ matrix (if the code frame is long enough), therefore it could be always classified as an LDPC code. Furthermore, since it is at least w.s. time-invariant, can be considered also structured, taking the syndrome former sub-matrix $\underline{H}_o^T$ as the module, and its periodic vertical shift as the variant.

In the first paper where LDPC block codes have been originally proposed, three hard-decision decoding algorithms have been suggested (Gallager 1962), precisely one called bit-flipping, and two resembling majority logic procedures (Lin and Costello 2004), but already oriented to message passing. The latter one of this pair (which is a refinement of the former one, including a correction made taking advantage from an estimate of the channel state) can be considered, in some way, as a hint for all the subsequently proposed soft-decision decoding algorithms. At present, the procedure most frequently adopted is the so called sum-product algorithm. It can be considered strictly related, although independently invented, to an algorithm, developed originally in Bayesian networks, and called belief propagation algorithm (Pearl 1988). Such two denominations can be considered substantially equivalent. In Sect. 10.8 we will describe more in depth the sum-product principle and some simplified algorithms derived from it.

10.2 Short Cycles and the Row-Column Constraint

Independently of the particular message passing algorithm employed for soft-decision decoding applied on an LDPC code, in the Tanner graph there are closed paths, whose length can strongly influence the convergence of the algorithm itself. The shortest closed paths possible have length 4. They correspond to short rectangular cycles connecting 1-symbols in the $\underline{H}$ matrix, and have been already discussed several times in previous chapters.

Definition 10.5 A closed path in the Tanner graph, correspondent to a short cycle in $\underline{H}$ connecting 1-symbols, of length λ_g (always an even number not lower than 4), will be called *λ_gth-order girth.*

Definition 10.6 A parity check matrix for which, in the correspondent Tanner graph, 4th-order girths are avoided, is said to satisfy the so called *row-column constraint* on LDPC codes.

LDPC codes for practical uses are requested to satisfy the row-column constraint, in order to guarantee the convergence of the decoding algorithm. Furthermore, the presence of 6th-order girths, 8th-order girths, and so forth, on the one side, contributes to lower the minimum distance, and, on the other side, makes decoding algorithm convergence a slower process.

Example 10.6 The [7,3,4] simplex code considered in Example 10.1 has a Tanner graph free of 4th-order girths. There are 6th-order girths, like the one shown in Fig. 10.3, together with the correspondent short cycle of 1-symbols in the $\underline{H}$ matrix. The minimum distance d is influenced by the presence of low-order girths. It is intuitive that the lower the length λ_g, the smaller the d, but a precise relationship depends also on the remaining 1-symbols along the columns involved in the short cycle.

QC and SQC array codes (see Sect. 8.8 and subsequent sections) satisfy the row-column constraint provided that the code word length n does not exceed p^2. Also a DC-GPC code or a DP-GPC code can satisfy the row-column constraint, provided that the code word length does not exceed a certain upper bound (see Lemma 5.42). This topic will be treated more in depth in Chap. 11. In QC and SQC array codes, reordering rows and column in the $\underline{H}$ matrix yields tail-biting w.s. time-invariant convolutional codes. The following proposition regards 4th-order girths in the two forms of such codes.

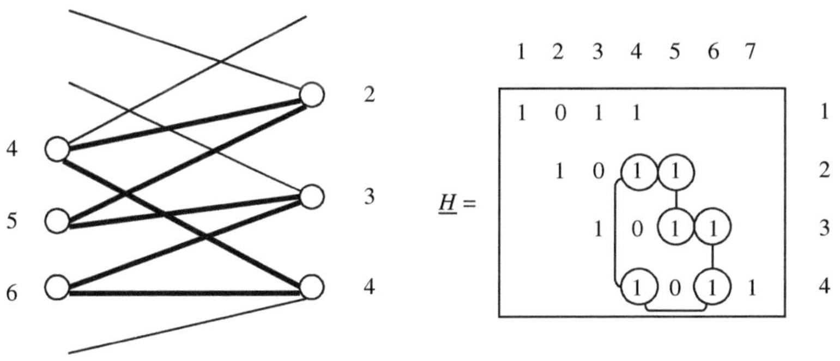

Fig. 10.3 Tanner graph and parity check matrix with a 6th-order girth (Example 10.6)

Lemma 10.2 *After application of a column and a row reordering, of the type described in* Definition 8.7, *on the* $\underline{H}$ *matrix of an LDPC code satisfying the row-column constraint, this constraint turns out to be still satisfied.*

Proof Acting firstly on the columns, the row-column constraint remains satisfied. Acting then on the rows as well. □

A situations in which the row-column constraint may be no longer satisfied, maintaining code equivalence, is when we take the ex-OR sum of rows. Inversely, when an $\underline{H}$ matrix in minimal form does not satisfy the row-column constraint, the same code with an $\underline{H}$ matrix not in minimal form sometimes satisfies it.

Example 10.7 A 1/2-rate convolutional code, characterized by the interleaved generator polynomial $g_{\text{int}}(x)$ associated to the sequence 11.11.01, exhibits a parity check matrix in minimal form which can be constructed taking the reciprocal parity check polynomial $h^*(x)$ associated to the sequence 10.11.11 (after Lemma 7.4). It shows a parity check constraint length $m_h = 2$. In Fig. 10.4, where this situation is depicted, it is possible to note the presence of several 4-length short cycles. Nevertheless, multiplying $h^*(x)$ by $(x^2 + 1)$, we obtain a new reciprocal parity check polynomial, associated to the sequence 10.01.00.11. The correspondent parity check matrix, not in minimal form, is reported in Fig. 10.4 down, showing how 4-length short cycles are no longer present. The new parity check constraint length is $m_h = 3$. At the end of the frame, two rows and as many columns can be eliminated, leading to an equal frame length. The new convolutional code is the same as before, as we can immediately verify applying the procedure based on the null ex-OR sum of clusters of syndromes, in order to obtain back polynomial $g_{\text{int}}(x)$.

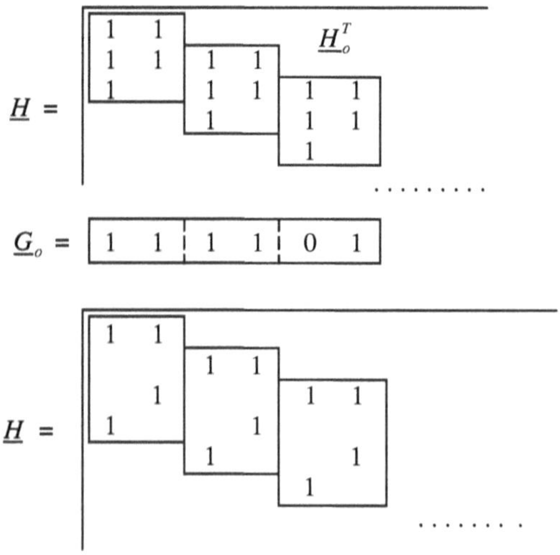

Fig. 10.4 Construction of a parity check matrix without 4-length short cycles, by ex-OR sums of rows taken from an $\underline{H}$ matrix in minimal form with 4-length short cycles (Example 10.7)

Suitable computer-based procedures can provide the design of $\underline{H}$ matrices with a prefixed minimum order for the girths present in the correspondent Tanner graphs. The most popular among such procedures exploits a principle called PEG (Progressive Edge Growth), in which further edges are added to an initial structure maximizing short cycle lengths and progressively involving all the variable nodes (Hu et al. 2001).

Lemma 10.3 *The direct product code between a pair of codes, whose* $\underline{H}$ *matrices satisfy the row-column constraint, is characterized by an overall* $\underline{H}$ *matrix satisfying the row-column constraint too.*

Proof The upper part of the $\underline{H}$ matrix in a direct product code (see Fig. 5.30) is made by several orthogonal parity check matrices of one component code. Here 4-length short cycles cannot be present, if $\underline{H}_2$ satisfies the row-column constraint. The lower part of the $\underline{H}$ matrix in a direct product code is made by identity matrices, whose size equals the length n_2, forming a macroscopic structure which resembles that of the parity check matrix $\underline{H}_1$ of the other component code. In this lower part, 4-length short cycles cannot be present, if $\underline{H}_1$ satisfies the row-column constraint. Between the two parts there are no 4-length short cycles, because up the horizontal separation between two 1-symbols is at most $(n_2 - 1)$, down it is an integer multiple of n_2. □

Example 10.8 Taking the direct product between a [7,3,4] simplex code and a [3,2,2] parity check code, we obtain a [21,6,8] direct product code, whose $\underline{H}$ matrix and encoding memory can be constructed as shown in Fig. 10.5. The simplex code, already treated also in Examples 10.1 and 10.6, satisfies the row-column constraint. The parity check code exhibits the same property, although in a sort of degenerate way, since its $\underline{H}$ matrix has a unique row. It is possible to verify that also the overall parity check matrix of the resulting direct product code does not contain any 4-length short cycle.

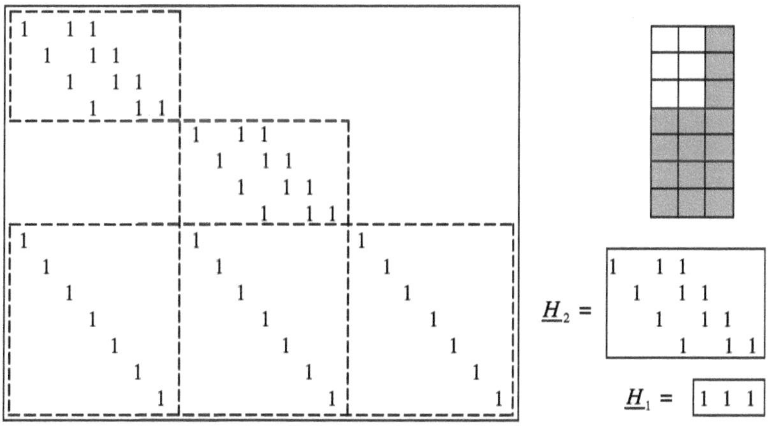

Fig. 10.5 $\underline{H}$ matrix and encoding memory of the [21,6,8] direct product code (Example 10.8)

In Example 9.15 a turbo product code has been presented, taking two equal 1/2-rate convolutional codes as component codes. Since they separately satisfy the row-column constraint, it is possible to verify that also the resulting code exhibits the same property (see Fig. 9.16).

Coming back to the scheme depicted in Fig. 10.2, on the left hand side, the variable nodes are directly interfaced with the channel, forming a sort of inner code in a serial concatenation scheme. On the whole, they can interpreted as an ordered ensemble of repetition codes (Ryan and Lin 2009). On the right hand side, we have, instead, a sort of outer code, which can be interpreted as an ordered ensemble of parity check codes, one for each control node. The interleaver scrambles the order of the first ones with respect to the second ones. In this view, message passing in the Tanner graph is very similar to an iterative decoding algorithm employed for turbo codes.

In particular, message passing in LDPC codes occurs between the ensemble of variable nodes and the ensemble of control nodes. It interests, during the first iterations, the variable nodes and the next neighbouring control nodes. Then it propagates along the Tanner graph progressively to other variable nodes superimposing the effects of localized decoding procedures. For this reason, 4th-order girths strongly frustrate the convergence of such iterative algorithm. LDPC codes whose Tanner graphs show longer girths perform better, and exhibit also a reduced time latency in the overall convergence behaviour, because the whole graph is rapidly involved in a unique process. Finally we have to note that irregular LDPC codes are characterized by nodes of different importance (depending on the number of edges touching them). Typically a subset of control nodes are of this type. They can support also isolated variable nodes (those touched by just one edge). Propagation of messages ends in correspondence of such variable nodes. In order to have an idea of how convergence can occur in message passing algorithms between two decoding procedures (that on variable nodes and that on control nodes), the following situation is examined, dealing with erasure correction (see Definition D.35).

Example 10.9 Let us take the direct product code discussed in Example 10.8, with a received signal showing a large amount of erasures, but no errors. Depending on where they are distributed, we can have a more or less efficient decoding. After Lemma D.8, since the considered code has $d = 8$, up to seven single erasures can always be corrected. We will consider a particular case characterized by 13 single erasures, but whose positioning is particularly favourable, and hence the iterative procedure which can be followed for their correction will be advantaged. In Fig. 10.6 on the left, the initial distribution of the erasures is shown. After a first iteration (decoding by columns), we obtain a situation of the type denoted (1). After a second iteration (decoding by rows) situation (2) is reached. The third iteration (decoding again by columns) allows to achieve the final result, without residual erasures at all. Remark that the [7,3,4] simplex code is able to correct up to three single erasures, and the [3,2,2] parity check code up to one single erasure. The principle adopted by the decoder is as follows: correct the single erasures in each component code word, maintaining unchanged the code words where the single erasure number equals or exceeds d.

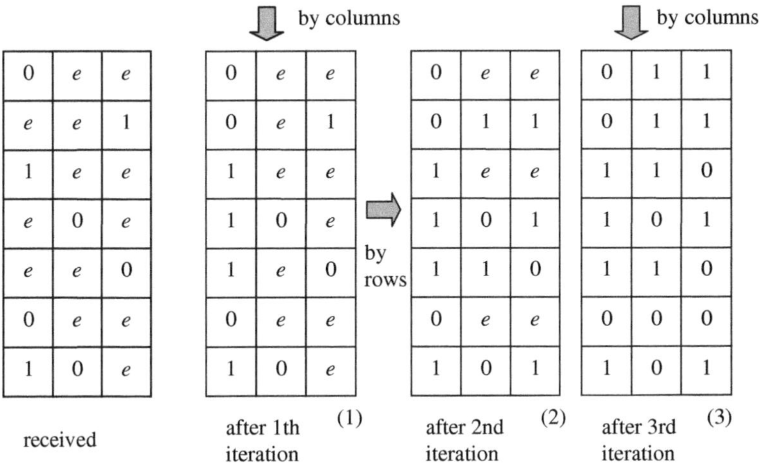

Fig. 10.6 Erasure correction process by a message passing algorithm in a [21,6,8] direct product code (Example 10.9)

A so good performance like the one described in Example 10.9 is possible because we have completely excluded the presence of errors. With errors and erasures contemporary present, the approach should be strongly modified. Nevertheless the situation here treated shows how it is possible to work well beyond the correction capability of the considered code, if it is constructed exploiting the complementary action of several component codes and the decoding procedure is made according to a synergic strategy. Coming back to LDPC codes and to a soft-decision message passing decoding algorithm, we can consider to have as many component codes as the single parity check codes related to the control nodes. At the second iteration, a new estimate of the symbols related to all the variable nodes can be performed. Then the procedure repeats. The total LLR (see Definition 9.8) is updated at any iteration. Sometimes the performance of an LDPC code (especially when irregular), although decoded by means of a sub-optimum algorithm, can appear better than what its minimum distance could allow to predict, owing to a mechanism of the type described in Example 10.9.

There are several families of LDPC codes (in the following sections we will try to give a sort of classification, based on their design principle), but typically they show rather steep curves of error probability after decoding p'', not far from Shannon limit, over an extended range of variability in this parameter. A possible BER floor, when present, occurs typically for very low values of p''. Such codes can work at high code rates (regular and even structured codes typically better), as well as at low code rates. Here irregular codes seem to be advantaged (Richardson and Urbanke 2001). So they compete, showing at least comparable merit, with all the variants of turbo codes.

10.3 Main Families of LDPC Codes

There are systematic and non-systematic LDPC codes (MacKay 1999). This distinction is not restricted to possible encoding problems (which however are also to be taken into account), but regards the structural properties of the codes themselves. A first version of systematic LDPC codes was proposed by Gallager in his original work (Gallager 1962). They are based on a layered $\underline{H}$ matrix. All the L layers exhibit the same number μ of rows. The generic layer is a μ x($\mu\gamma$) sub-matrix with all columns having weight 1. The resulting $\underline{H}$ matrix is a (μL)x($\mu\gamma$) matrix. Nevertheless, owing to the possible presence of some linearly dependent rows, the code rate can be only at most equal to $(\gamma - L)/\gamma$. The first layer is constructed with μ rows where as many unitary polynomials $u_\gamma(x)$ shift with steps of γ. The subsequent layers are made by column permutations of the first layer. A code of this type is regular, with $w_c = L$ and $w_r = \gamma$. Owing to the random column permutation, we cannot define such codes also as structured LDPC codes. The row-column constraint can be easily imposed.

MacKay and Neal, in 1995, proposed a less regular form of LDPC codes (MacKay and Neal 1995; MacKay 1999), in general not systematic. Although available in several variants, their design can be summarized as follows. Firstly it is necessary to dispose a random sequence of different columns, all with weight w_c. In doing this, we have to avoid 4th-order girths, and maintain the weights of the rows as close as possible to a prefixed value γ. In order to satisfy the row-column constraint, it is necessary that two columns exhibit coincident 1-symbols just in one position. Then, optionally, it is possible to partition $\underline{H}$ into two sub-matrices, the one on the right being square and easily invertible. The last condition is clearly added in order to simplify the encoding procedure, allowing to obtain a systematic equivalent of the considered code.

Lemma 10.4 *In a column-regular LDPC code, characterized by all the $\underline{H}$ matrix columns having weight $w_c = 2$, the presence of a λ_gth-order girth is a necessary and sufficient condition for the existence of a code word t(x) with* $\mathrm{w}[t(\mathrm{x})] = \lambda_g/2$.

Proof We remember that, as observed in Definition 10.5, λ_g is always an even number, because it represents the perimeter of a geometric figure in the $\underline{H}$ matrix with parallel vertical and horizontal segments. In Fig. 10.7 up the form of the short cycle associated to a generic girth is depicted. Since both the 1-symbols of any column are employed as end points of a vertical segments, there are no other 1-symbols in such columns. This is enough for allowing verification of the null ex-OR sum of the correspondent syndromes. So the considered code word $t(x)$ exhibits 1-symbols in the positions occupied by the above syndromes and its weight turns out to be $\lambda_g/2$. □

Remark that an $\underline{H}$ matrix, which is subdivided into row layers, with one 1-symbol each, like occurs in array codes, DC-GPC codes, DP-GPC codes and also in the Gallager codes previously described, represents a very particular case. The situation

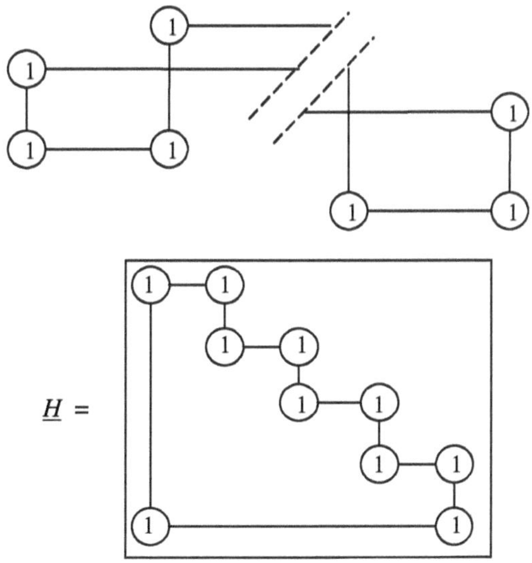

Fig. 10.7 Generic short cycle in an $\underline{H}$ matrix with $w_c = 2$ in any column, and LDPC version of the $\underline{H}$ matrix of a [5,1,5] repetition code

described by Lemma 10.4 (see Fig. 10.7 up) is more general. For instance, in Fig. 10.7 down, the parity check matrix of a [5,1,5] repetition code is reported. This $\underline{H}$ matrix cannot be partitioned into two row layers. In it, the lowermost row can be eliminated, because it is linearly dependent on the other ones (so allowing the presence of just one information symbol), but the 10th-order girth is anyway valid in the correspondent Tanner graph, together with what predicted by Lemma 10.4.

Both Gallager codes and MacKay-Neal codes have been demonstrated to approach Shannon limit for a block length n tending to infinity (MacKay and Neal 1997; MacKay 1999). This conclusion is based on the use of an ideal optimum decoding algorithm. Nevertheless even employing a sub-optimum iterative decoding algorithm, like the so called sum-product algorithm (which will briefly described in Sect. 10.8), they both appear very attractive. Gallager codes are typically superior at high code rates, MacKay-Neal codes at low code rates. The best turbo codes, having comparable block length, can perform slightly better than the latter ones, but only at very low code rates and in a region characterized by rather high values of the error probability.

After the above description of the early proposed types of LDPC codes, we can list the following main families of codes, which, in the last ten years, have been investigated, starting from their construction principle:

(i) codes based on protographs,
(ii) codes constructed employing repeat and accumulation devices,
(iii) codes derived from the decomposition of finite geometries (Euclidean, affine, projective),
(iv) codes obtained starting from MDS codes and in particular Reed-Solomon codes.

Protograph-based LDPC codes were proposed in Thorpe (2003) as structured LDPC codes, whose Tanner graph is built starting from a very simple prototype. This module is properly scaled with the addition of a transformation, consisting, for instance, in the substitution of single edges with suitable permutations. In this case, permutation size becomes the scale factor. In other situations, a repeated copy of the protograph is firstly operated. In protograph-based LDPC codes it is possible to adopt a partially parallel architecture for the decoding procedure. Better results, however, are obtained with proper variants of this construction method, leading to less regular codes. A first variant (weak) consists of assuming some permutation matrices interesting a group of edges instead of a single edge. This design technique is sometimes known as multi-edge procedure (Ryan and Lin 2009). A second variant (strong) is that in which a procedure briefly called "copy and permute" is applied (Costello et al. 2007). In this case, after a multiple copy of the same prototype, a permutation is applied to the whole graph.

A repeat devices is the encoder circuit of a repetition code. An accumulation device is a circuit which calculates the quotient of the polynomial division by $(x + 1)$. Combination of such devices can yield very good LDPC codes (typically at low code rates), whose encoding procedure is extremely simple. With the addition of an interleaver, also the structured form intrinsically present in this scheme can be broken, leading to irregular LDPC codes, and even moderately high code rates can be reached by proper code puncturation. The simplest code of the family obtained employing repeat and accumulation devices is called RA code (Repeat-Accumulate). The repetition code adopted can have whichever length n, leading to a code whose overall code rate remains equal to $1/n$ (Divsalar et al. 1998). The role of the accumulator is to insert 0-symbols in long sequences of 1-symbols and vice versa. Puncturation finally allows to reach the desired code rate. A first very simple variant of RA codes consists of inserting an interleaver. Its presence has given motivation for the name IRA codes (Irregular Repeat-Accumulate) (Jin et al. 2000). The use of the interleaver implies that more than one repetition code word are contemporary treated. Sometimes a more involved structure is exploited, where, instead of a simple puncturation, a proper superimposition of control symbols is made. A further variant, able to improve the performance of RA and IRA codes, is in the addition of a pre-processing, consisting of a further accumulator. So we obtain an enhanced scheme called ARA (Accumulate-Repeat-Accumulate) (Abbasfar et al. 2007).

Proper subspaces of finite geometries can be efficiently adopted for constructing traditional linear block codes (Rudolph 1967). They admit a polynomial description in both their $\underline{G}$ and $\underline{H}$ matrix. The minimum distance d is often known and even some bounds on the coefficient $A(d)$ can be calculated. To such traditional codes, a majority-logic hard-decision decoding procedure has been proposed to be applied. LDPC codes, based on finite geometries (Euclidean, affine and projective), have been recently presented with the same features, but with the advantage of a soft-decision decoding, assured by message passing algorithms. Typically they show a regular structure, but a rather high value of column weight w_c. LDPC codes based on lines in an Euclidean geometry have been proposed in Johnson and Weller (2004). They can be either cyclic or quasi-cyclic. Having chosen a finite field with Q elements, in the case of QC-codes, we can adopt a sequence of $(Q^m - 1) \times (Q^m - 1)$ circulants, where m is an integer (Ryan and Lin 2009).

An affine geometry offers some more degrees of freedom in the code design. Points and lines in a finite affine plane have been used for constructing a family of LDPC cyclic codes (Kou et al. 2001), characterized by total code word length $n = 2^{2s} - 1$ and number of control symbols $r = 3^s - 1$, where s is a proper integer parameter. For instance, taking $s = 5$, we obtain a code with $n = 1{,}023$ and $r = 242$. Typically high code rates can be reached, with minimum distance d exactly predictable, as well as rather tight bounds on coefficient $A(d)$ (Lin and Costello 2004). Quasi-cyclic codes can show either just one layer or more than one layers, each made by proper circulants. In particular, it is possible to consider a high-rate code, which can be interpreted as the dual of a $1/a$-rate code, whose generator matrix has just one row layer. Its structural properties can be employed for predicting the performance of such family of LDPC codes (Kamiya 2007). In Chap. 11, such codes will be better investigated. In a projective geometry, an interesting case is that of a family of LDPC codes characterized by $n = 2^{2s} + 2^s + 1,\ r = 3^s + 1$, and minimum distance lower bound $d \geq 2^s + 2$. Adopting $s = 5$, as before, it is possible to obtain a code with $n = 1{,}057$, r = 244, $d \geq 34$, which are very interesting parameters (Lin and Costello 2004), although the number of minimum-distance code words is rather high.

As an example of the use of a finite geometry, in the analysis of the weight distribution of a code constructed according to a certain rule, we can consider the encoding memory of this code. The rows of $\underline{H}$ can be viewed as trajectories in this finite plane. When two trajectories intersect in two points, a 4th-order girths occur. When three trajectories intersect in as many points we have a 6th-order girth, and so on. This analysis and the consequent code behaviour prediction will be carried out, for a certain family of codes in an Euclidean geometry, either block or convolutional, in Chap. 11.

MDS codes (see Sect. 1.22) are codes attaining Singleton bound (Theorem D.3 in Appendix D and Fig. 1.31) on their minimum distance, in a Q-ary alphabet. Among them, Reed-Solomon codes show fine scalability, in either the code rate or the overall size, owing also to the possibility of multiple extension or shortening. Here a one-to-one correspondence is taken between $p \times p$ permutation circulants and the $p = Q - 1$ non-null elements of the field. The minimum distance comes out from the presence of short cycles, across the different layers in the parity check matrix. All such codes are quasi-cyclic, or some shortened versions of a parent QC code. Clearly array codes (already treated in Chaps. 6 and in 8) belong to this family of codes, and their $\underline{H}$ matrix can be synthesized taking subsequent powers of the primitive element α. For all such codes, the following proposition, already outlined, holds.

Lemma 10.5 *In a QC code whose $\underline{H}$ matrix is constructed by means of permutation circulants, necessary and sufficient condition for satisfaction of the row-column constraint is that two column layers does not show, between the same positions, two equal differences in their circulant phases.*

Proof In the vertices of a rectangle let us imagine to have the permutation circulants $\underline{a}_p$, $\underline{b}_p$ on one column layer and $\underline{c}_p$, $\underline{d}_p$ on the other column layer. The vertical separation between 1-symbols in the former circulant pair is $(a - b)$, or $(p + a - b)$ plus a certain multiple of p. The same occurs in the latter circulant pair, with the vertical separation $(c - d)$, or $(p + c - d)$ plus the same multiple of p. If the equality $a - b = c - d$ holds, a 4-length short cycle occurs. Since $a - b < p$, as well as $c - d < p$, if the above equality does not hold, this 4-length short cycle cannot be formed. □

About the LDPC codes still derived from RS codes, but not in the framework of array codes, we can mention, say, the following construction (Ryan and Lin 2009). Let us take $Q = 2^b$, b integer, and $p = 2^b - 1$ be a composite integer (for instance $p = p_1p_2$). In this case it is possible to construct an $\underline{H}$ matrix with layers of $p \times p$ circulants, taken from separate sets, whose number of elements equals the factors of p (for instance p_1 sets of p_2 elements or p_2 sets of p_1 elements). The largest $\underline{H}$ matrix so obtainable has one of the factors as the number of row layers and the other one as the number of column layers (for instances p_1 row layers and p_2 column layers). From this parent code, code shortening and code puncturation, always acting on entire layers, allow to find the desired code rate and overall block length. Since the component sub-matrices employed in each layer are different, a proper sequence of them will prevent 4.order girths, following Lemma 10.5. In this arrangement, there is also a null $p \times p$ sub-matrix, in just one row layer.

Example 10.10 Taking $p = 255$, and $p_1 = 5$, $p_2 = 51$, it is possible to arrange medium-size codes with excellent characteristics, whose code rate can vary between about 1/2 and 3/4. The unique null sub-matrix present in this construction allows to obtain a moderate level of irregularity.

Example 10.11 A different construction is that based on a finite field for which Q is integer power of a prime number. An RS code, with only two information symbols (say [26,2,25], for which $Q = 3^3 = 27$), is characterized by the complete square parity check matrix, in symbolic form, shown in Fig. 10.8 up. From it we can derive the $\underline{H}$ matrix reported down, where there are null sub-matrices along the diagonal, denoted with small segment, and sub-matrices $\underline{P}_j$ are proper $p \times p$ permutation circulants (with $p = Q - 1$ as usual) associated to as many expressions of the primitive elements α. We can adopt the following correspondence: $\underline{P}_j$ is represented by $(\alpha^j - 1)$. Such parity check matrix characterizes an interesting family of quasi-cyclic LDPC codes (Djurdjevic et al. 2003). Elimination of row layers in this complete $\underline{H}$ matrix allows to obtain the best trade-off between code rate and minimum distance.

After this brief overview on structured and often regular LDPC codes, it is important to observe that a completely different approach characterizes irregular codes. Especially when they are thought to be irregular directly from their initial design (and are not obtained by suitable modifications of some regular structures), a computer optimization, devoted to avoid girths up to a prefixed length (if possible), is generally employed. Examples of such design techniques are PEG (already

$$\underline{H} = \begin{bmatrix} 0 & (\alpha^1 - 1) & (\alpha^2 - 1) & \cdots & (\alpha^{Q-2} - 1) \\ (\alpha^{Q-2} - 1) & 0 & (\alpha^1 - 1) & \cdots & (\alpha^{Q-3} - 1) \\ & & \cdots\cdots & & \\ (\alpha^1 - 1) & (\alpha^2 - 1) & (\alpha^3 - 1) & \cdots & 0 \end{bmatrix}$$

$$\underline{H} = \begin{bmatrix} - & \underline{P}_1 & \underline{P}_2 & \cdots & \underline{P}_{Q-2} \\ \underline{P}_{Q-2} & - & \underline{P}_1 & \cdots & \underline{P}_{Q-3} \\ & & \cdots\cdots & & \\ \underline{P}_1 & \underline{P}_2 & \underline{P}_3 & \cdots & - \end{bmatrix}$$

Fig. 10.8 LDPC code construction from a [26,2,25] RS code (Example 10.11)

mentioned) and ACE (Approximate Cyclic Extrinsic message degree) (Tian et al. 2004). They will be briefly described in Sect. 10.7. In the next section, we will present, instead, some procedures for increasing irregularity in structured LDPC codes.

10.4 Masking and Row or Column Splitting

When an LDPC code is structured, with modules not too large (see Definition 10.4), like in protograph-based codes, RA codes, array codes, finite geometry codes, it is possible to eliminate some of these modules, according to a certain rule, more or less uniform.

Definition 10.7 Elimination of some modules in a structured LDPC code is called *masking*. If the parity check matrix where masking is to be performed exhibits n_o column layers and r_o row layers, where its modules are placed, a *masking matrix* $\underline{M}$ can be defined as a proper $r_o \times n_o$ binary matrix with 1-symbols in the positions where the modules are to be eliminated, and 0-symbols in all the other positions.

It is useful to describe the topology of the original code by means of a binary $r_o \times n_o$ parity check matrix $\underline{H}_s$ with 1-symbols in the positions where its modules are initially present. For instance, in LDPC array codes, $\underline{H}_s$ is made by all 1-symbols. After masking, performed according to a masking matrix $\underline{M}$, the new topologic parity check matrix $\underline{H}'_s$ will be $\underline{H}'_s = \underline{H}_s + \underline{M}$. The effects of a moderate masking can be summarized as follows. Typically the minimum distance d decreases with respect to the original code, but a more pronounced effect of spectral thinning can be reached, to the point where, for medium values of the error probability, the performance may be better. Furthermore the decoding algorithm generally converges more rapidly and is able to correct particular errors even beyond the limit imposed by d.

$$\underline{M} = \begin{array}{|cccccccc|}\hline 0&0&0&0&0&0&0&1\\ 0&1&0&0&0&0&0&0\\ 0&0&0&1&0&0&0&0\\ 0&0&0&0&0&1&0&0\\ \hline\end{array} \qquad \begin{array}{|cccccccc|}\hline \underline{0}_p & \underline{0}_p & \underline{0}_p & \underline{0}_p & \underline{0}_p & \underline{0}_p & \underline{0}_p & - \\ \underline{0}_p & - & \underline{2\Delta_2}_p & \underline{3\Delta_2}_p & \underline{4\Delta_2}_p & \underline{5\Delta_2}_p & \underline{6\Delta_2}_p & \underline{7\Delta_2}_p \\ \underline{0}_p & \underline{\Delta_3}_p & \underline{2\Delta_3}_p & - & \underline{4\Delta_3}_p & \underline{5\Delta_3}_p & \underline{6\Delta_3}_p & \underline{7\Delta_3}_p \\ \underline{0}_p & \underline{\Delta_4}_p & \underline{2\Delta_4}_p & \underline{3\Delta_4}_p & \underline{4\Delta_4}_p & - & \underline{6\Delta_4}_p & \underline{7\Delta_4}_p \\ \hline\end{array}$$

Fig. 10.9 Application of a uniform masking to an improper array LDPC code

Example 10.12 Given an improper array code, with $r_o = Y = 4$ and $n_o = S = 8$, with parameters Δ_j, $j = 2, 3, 4$, to be optimized, with a scale factor p high enough, let us adopt a masking matrix $\underline{M}$ as shown in Fig. 10.9 on the left. The modified $\underline{H}$ matrix becomes as shown on the right. The usual symbolic form has been employed for representing circulants. Small segments denote null matrices. The masking procedure here adopted is strongly uniform and is obtained with one null matrix any two column layers, in different row layers. The result can lead to performance improvement if one chooses the parameters Δ_j, and also the value of p, together with $\underline{M}$. On the other hand, a less uniform masking can provide even better results, without a particular optimization in the choice of the Δ_j's, but requires typically a higher value for parameter p. In any case, a column-regular code typically becomes column-irregular.

Let us consider a selective code shortening of an improper array code, that is the elimination of column layers not only progressively from the right hand side of its $\underline{H}$ matrix, as usual in traditional shortening. This solution represents an intermediate step towards the introduction of generalized array codes (Definition 8.6). The combined use of selective shortening and masking allows to overcome a constraint in the maximum length λ_g of the girths present in any type of array codes (included generalized ones), on which we will come back in Chap. 11 (Fossorier 2004; Milenkovic et al. 2008). This can lead, in principle, to a remarkable increase in the minimum distance d, especially at low code rates.

Clearly, when the topologic matrix $\underline{H}_s$, by itself, is characterized by a rather small number of 1-symbols, masking is not efficient. On the other hand, when this number is high, as in the case of array codes, masking can improve by far the performance. It does not vary the length of the code, the number of its control symbols, and consequently the code rate, but simply reduces the number of 1-symbol in columns and rows, adopting a more or less random choice. We can observe that, if selective shortening (elimination of columns), as well as constant-length puncturation (elimination of rows) or masking, is performed on a parity check matrix able to satisfy the row-column constraint, this property remains valid.

Another technique able to reduce regularity in structured LDPC codes consists of splitting the rows or the columns of $\underline{H}$ into two or more than two parts (Lin and Costello 2004). In principle, a procedure of this type, which changes the code parameters (number of control symbols or overall length), can be applied to whichever type of LDPC code, in order to enrich the number of possible variants derived from it.

Definition 10.8 Given an LDPC code, one row in its $\underline{H}$ matrix can be split into V new rows, subdividing its 1-symbols among such new rows in the old positions. This procedure is called *V-fold row splitting*.

After row splitting the number of control symbols increases, whereas the overall word length remains unchanged. It is necessary that at least two 1-symbols are left in each of the new rows, in order to have a valid parity check equation. Furthermore it is necessary to avoid that any new row is linearly dependent on the other rows of the parity check matrix, in order to still have an $\underline{H}$ matrix with the maximum number of independent rows. Finally row splitting produces a variation in the row weight but not in the column weight, and reduces the code rate.

Example 10.13 Taking the block code, whose $\underline{H}$ matrix is shown in Fig. 10.10 on the left, we will apply a 2-fold row splitting to the second and to the fourth row. In both cases, the first pair of 1-symbols are collected in the upper row, the second pair of 1-symbols in the lower row. The old and the new parity check matrix are compared. Instead of a uniform row weight $w_r = 4$, we have now four rows with weight 2 and four rows with weight 4, leading to an average weight $\langle w_r \rangle = 3$ on the whole. The new $\underline{H}$ matrix remains with one row linearly dependent on the other ones, therefore the code so constructed is a [12,5,3] code, but with a strongly reduced number of 3-weight code words. Initially the code was a [12,7,3] regular code, free of 4-length short cycles. Its $\underline{H}$ matrix was made by circulants (of weights 0, 1, 2). Therefore it was a QC code, but such property is no longer true in the new code.

Definition 10.9 Given an LDPC code, a column in its $\underline{H}$ matrix can be split into V new columns, subdividing its 1-symbols among such new columns in the old positions. This procedure is called *V-fold column splitting*.

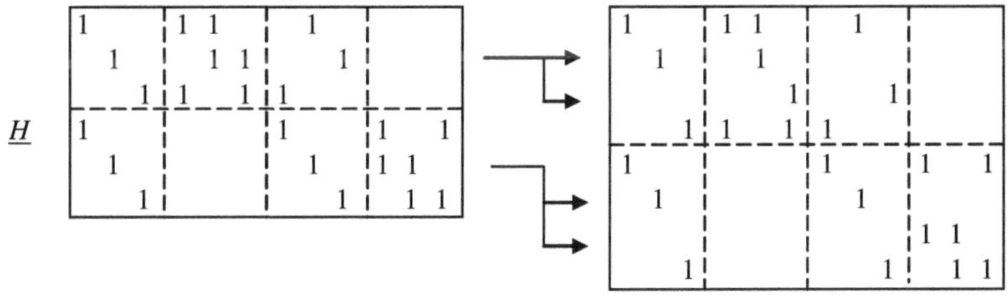

Fig. 10.10 Example of application of row splitting

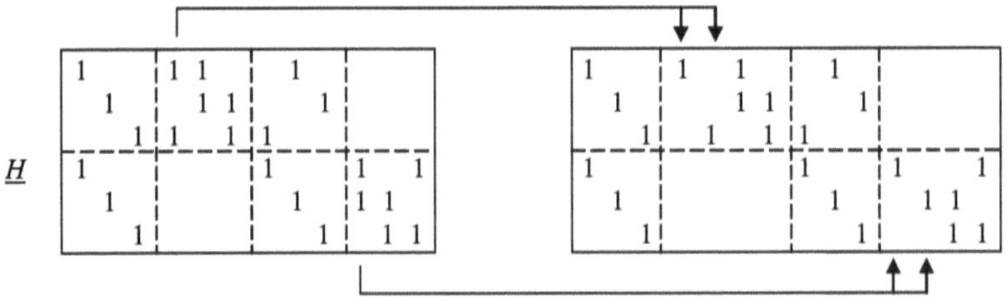

Fig. 10.11 Example of application of column splitting

After column splitting the number of control symbols remains unchanged, whereas the overall word length grows, and hence the number of information symbols increases. New columns with just one 1-symbol are acceptable. Column splitting produces a variation in the column weight but not in the row weight, and increases the code rate.

Example 10.14 Taking again the block code, introduced in Example 10.13, we will apply a 2-fold column splitting to the fourth and to the tenth row of its $\underline{H}$ matrix. This choice is suggested by the fact that such columns belong to the 2-weight circulants having no rows in common. The minimum distance of the new code is maintained $d = 3$. The old and the new parity check matrix are compared in Fig. 10.11. The new $\underline{H}$ matrix has full rank, therefore the code so constructed is characterized by the tern [14,8,3]. It shows a uniform row weight $w_r = 4$ and an average column weight $\langle w_c \rangle \cong 1.7$. Also here the property of being QC is no longer true in the new code.

Lemma 10.6 *If the row-column constraint is satisfied in the original LDPC code, it will be satisfied as well in the new code, obtained after row or column splitting.*

Proof Row splitting contributes to increases the number of control nodes in the Tanner graph of the considered code. Column splitting has the same effect on the number of variable nodes. So a girth of a certain length, if present, can maintain the same length or increase it. If there are no 4th-order girths, this property will be satisfied also in the new code. □

Row and column splitting are often employed for constructing LDPC codes, by progressive elimination of low-order girths (sometimes even of 4th order). Nevertheless, row splitting typically enforces the code, because introduces new control symbols. On the contrary, column splitting makes the code weaker, because here the number of positions in the code word increases, without varying the number of control symbols. Typically such two procedures are employed in more involved techniques for computer based design of LDPC codes, especially irregular.

10.5 LDPC Codes Obtained from Superimposition

Superimposition is a very general method for obtaining a composite LDPC code, more or less structured, depending on the constituent parts of parity check matrix adopted. We start from an $r \times n$ binary base matrix $\underline{B}$, which satisfies the row-column constraint. Then a set of $\underline{A}_j$ matrices, $j = 1, 2, \ldots, q$, is assumed, all bxc, and characterized by the following properties (Xu et al. 2005): each of them satisfies the row-column constraint by itself, and, in a same row or column, any pair of them also satisfies this constraint (pairwise row-column constraint). Generally the $\underline{A}_j$'s, which are called constituent matrices, have homogeneous characteristics (for instance the same row weight or column weight). Finally the $\underline{H}$ matrix of the composite code is obtained after replacement of any 1-symbol in the base matrix $\underline{B}$ with one member of the $\underline{A}_j$ set, with the further constraint of having all different $\underline{A}_j$'s in each row and in each column.

Clearly the resulting *rbxnc* overall $\underline{H}$ matrix, under the above assumptions, cannot show 4-length short cycles. Furthermore, it collects the properties of both the base matrix and the constituent matrices. For instance, if they are regular with $w_c^{(B)}$, $w_c^{(A)}$ column weights and $w_r^{(B)}$, $w_r^{(A)}$ row weights, $\underline{H}$ is regular with $w_c^{(B)}w_c^{(A)}$ column weight and $w_r^{(B)}w_r^{(A)}$ row weight. Typically the minimum distance is remarkably increased with respect to the one associated to a code with $\underline{B} = \underline{H}$. If the column weight obtained after superimposition is too high, it is possible to employ masking procedures. In this case the $\underline{H}_s$ matrix describing the topology will coincide with the base matrix $\underline{B}$. Remark that only when the $\underline{A}_j$'s are $b \times b$ square matrices, the code rate remains equal to $(n - r)/r$, that is to the code rate due to the base matrix.

A particular situation is that characterized by circulants (of weight 1 or greater than 1) as the $\underline{A}_j$ matrices to be employed. From Lemma 8.1, any code, whose $\underline{H}$ matrix can be subdivided into square sub-matrices, null or made by proper circulants, is a QC code. Therefore the superimposition procedure can be considered also one of the methods for an efficient design of quasi-cyclic LDPC codes. If we adopt only permutation circulants, the constraint of having all different $\underline{A}_j$'s in each row and in each column is no longer necessary, in order to assure satisfaction of the row-column constraint, after the choice of the base matrix $\underline{B}$ free of 4-length short cycles. On the contrary, in the presence of circulants with weights greater than 1, the row-column constraint is to be imposed in each constituent matrix and in any pairwise combination.

Example 10.15 Taking as $\underline{B}$ the parity check matrix of a [7,3,4] simplex code (like in Example 10.1), and assuming $b \times b$ permutation circulants as constituent matrices, it is possible to obtain a QC expanded version of the original code. It is certainly free of 4th-order girths. If b is high enough (of the order some tenths), also the density ρ of 1-symbols becomes suitable for a true LDPC code. Unfortunately some 4-weight code words remain, because of the presence in $\underline{B}$ of two external columns having weight 1. A different situation occurs, say, if we take the following set of constituent matrices: $\underline{A}_0^{(1)} = \underline{0}_b$, $\underline{A}_1^{(1)} = \underline{1}_b$, $\underline{A}_2^{(1)} = \underline{2}_b$, $\underline{A}_{01}^{(2)} = \underline{0}_b + \underline{1}_b$, where the symbolic representation of $b \times b$ permutation circulants has been employed. The last constituent matrix is a 2-weight circulant. It is possible to verify that this matrix, together anyone of the remaining three $\underline{A}_j^{(1)}$ matrices, satisfies also the pairwise row-column constraint. Then, adopting the following overall matrix

$$\underline{H} = \begin{bmatrix} \underline{A}_{01}^{(2)} & - & \underline{A}_1^{(1)} & \underline{A}_2^{(1)} & - & - & - \\ - & \underline{A}_{01}^{(2)} & - & \underline{A}_1^{(1)} & \underline{A}_2^{(1)} & - & - \\ - & - & \underline{A}_0^{(1)} & - & \underline{A}_1^{(1)} & \underline{A}_2^{(1)} & - \\ - & - & - & \underline{A}_0^{(1)} & - & \underline{A}_1^{(1)} & \underline{A}_{01}^{(2)} \end{bmatrix},$$

where external column layers have now weight $w = 2$, any 4-weight code word will be eliminated. The distribution of the column weights is here as follows: six columns layers with $w = 2$ and just one column layer with $w = 3$. Also the weights of the row layers are changed, in fact we have now one central row layer with $w = 3$, and three external row layers with $w = 4$. The resulting code is a [7b,3b,5] QC code, provided that $b \geq 5$. It is finally interesting to note that odd-weight code words are present here, owing to different column weights.

Example 10.16 To the code introduced in Example 10.15, obtained after superimposition starting from the [7,3,4] simplex code, it is possible to apply a masking matrix $\underline{M}$ with just one 1-symbol, placed in correspondence with position (2,4). So the central column layer of the new parity check matrix will exhibit a weight $w = 2$, like all the other column layers. Such new column-regular matrix then results in

$$\underline{H}' = \begin{bmatrix} \underline{A}_{01}^{(2)} & - & \underline{A}_1^{(1)} & \underline{A}_2^{(1)} & - & - & - \\ - & \underline{A}_{01}^{(2)} & - & - & \underline{A}_2^{(1)} & - & - \\ - & - & \underline{A}_0^{(1)} & - & \underline{A}_1^{(1)} & \underline{A}_2^{(1)} & - \\ - & - & - & \underline{A}_0^{(1)} & - & \underline{A}_1^{(1)} & \underline{A}_{01}^{(2)} \end{bmatrix}.$$

This code is characterized by $d = 4$, instead of $d = 5$, confirming that, when the column weight is rather small, masking can reduce the minimum distance.

Example 10.17 Let us consider the same code as in Example 10.15, but with a difference, regarding the use of truncated circulants, instead of complete square circulants. In $\underline{H}$ matrices, truncated circulants are rectangles whose vertical dimension is greater than the horizontal dimension (see Lemma 8.2). They can be efficiently employed when the size of the circulants are much greater than the phases characterizing them, and play an important role in the description of convolutional codes which could be obtained by means of unwrapping, starting from the reordered form of a QC code directly in its $\underline{H}$ matrix (see Sect. 8.9). In Fig. 10.12, we consider the case of 6×4 truncated circulants with only phases 0, 1, 2. Various rows can be eliminated (those denoted by small arrows), since they have weight 1 or even a null

$$
\underline{H} = \left[\begin{array}{cccccccccccccccccccccccccccc|c}
0&0&0&0&&&&&0&0&0&0&1&0&0&0&&&&&&&&&&&&&\leftarrow \\
1&0&0&0&&&&&1&0&0&0&0&1&0&0&&&&&&&&&&&&& \\
1&1&0&0&&&&&0&1&0&0&0&0&1&0&&&&&&&&&&&&& \\
0&1&1&0&&&&&0&0&1&0&0&0&0&1&&&&&&&&&&&&& \\
0&0&1&1&&&&&0&0&0&1&0&0&0&0&&&&&&&&&&&&& \\
0&0&0&1&&&&&0&0&0&0&0&0&0&0&&&&&&&&&&&&&\leftarrow \\
&&&&0&0&0&0&&&&&0&0&0&0&1&0&0&0&&&&&&&&&\leftarrow \\
&&&&1&0&0&0&&&&&1&0&0&0&0&1&0&0&&&&&&&&& \\
&&&&1&1&0&0&&&&&0&1&0&0&0&0&1&0&&&&&&&&& \\
&&&&0&1&1&0&&&&&0&0&1&0&0&0&0&1&&&&&&&&& \\
&&&&0&0&1&1&&&&&0&0&0&1&0&0&0&0&&&&&&&&& \\
&&&&0&0&0&1&&&&&0&0&0&0&0&0&0&0&&&&&&&&&\leftarrow \\
&&&&&&&&0&0&0&0&&&&&0&0&0&0&1&0&0&0&&&&&\leftarrow \\
&&&&&&&&0&0&0&0&&&&&1&0&0&0&0&1&0&0&&&&& \\
&&&&&&&&1&0&0&0&&&&&0&1&0&0&0&0&1&0&&&&& \\
&&&&&&&&0&1&0&0&&&&&0&0&1&0&0&0&0&1&&&&& \\
&&&&&&&&0&0&1&0&&&&&0&0&0&1&0&0&0&0&&&&& \\
&&&&&&&&0&0&0&1&&&&&0&0&0&0&0&0&0&0&&&&&\leftarrow \\
&&&&&&&&&&&&0&0&0&0&&&&&0&0&0&0&0&0&0&0&\leftarrow \\
&&&&&&&&&&&&0&0&0&0&&&&&1&0&0&0&1&0&0&0& \\
&&&&&&&&&&&&1&0&0&0&&&&&0&1&0&0&1&1&0&0& \\
&&&&&&&&&&&&0&1&0&0&&&&&0&0&1&0&0&1&1&0& \\
&&&&&&&&&&&&0&0&1&0&&&&&0&0&0&1&0&0&1&1& \\
&&&&&&&&&&&&0&0&0&1&&&&&0&0&0&0&0&0&0&1&
\end{array}\right]
$$

Fig. 10.12 Parity check matrix obtained by the use of truncated circulants (Example 10.17)

weight at all. The value adopted for b is here very small, but the number of eliminated rows does not increase with b. We have preferred to maintain such rows in order to simplify understanding of the origin of an $\underline{H}$ matrix so constructed. It can be interpreted as the reordered version of the parity check matrix describing a 3/7-rate convolutional code, not in tail-biting form.

10.6 Procedures for Obtaining LDPC Codes from Known Non-LDPC Codes

Many traditional textbook codes (either block or convolutional) are characterized by good distance properties, but often show an $\underline{H}$ matrix whose Tanner graph is frustrated by 4th-order girths. In other situations, 4th-order girths are not present, but the density ρ of 1-symbols is too high. In this section we will present some possible procedures for reaching the above two goals and obtaining good LDPC codes. In some situations, it is possible to achieve an increased minimum distance too.

Taking a known code, as a starting point for this LDPC code construction, we can mention the following three ways:

(i) choice of some linear combinations of parity check rows,
(ii) row and/or column splitting, as described in Sect. 10.4,
(iii) substitution of 1-symbols with proper permutation matrices.

Procedure *i* clearly gives a parity check matrix not in minimal form. It has been adopted for the 1/2-rate convolutional code described in Example 10.7. Assuming a BCH cyclic code, one can start writing the complete square $\underline{H}$ matrix, where $(n - r)$ rows are linearly dependent on the other ones. Next, after proper linear combinations and positioning, the number of 4-length short cycles can be minimized. This procedure is somehow advantaged, if one refers to shortened versions of the original cyclic code, in fact after code shortening we can eliminate the rows in the resulting $\underline{H}$ matrix with higher weights. Following this way it is possible to discover many interesting well known codes without 4th-order girths at all in their Tanner graph, or with a residual small number of such girths (Halford et al. 2006). An efficient tool for the selection of rows, when they are taken from an s.s. time-invariant code in its parity check matrix, consists of calculating the periodic autocorrelation function. In correspondence with the shifts generating the maximum of autocorrelation function, it is possible to choose the pair of rows whose ex-OR sum yields a new row with minimum weight (Baldi and Chiaraluce 2008). Good results have been obtained with several BCH codes, although with values of n rather small and consequently density ρ of 1-symbols not low enough, for allowing very impressive performance.

As regards procedure ii, in Lin and Costello (2004) the [23,12,7] Golay code has been taken as a starting point. In its parity check matrix row and column splitting have been applied in combined fashion, up to the obtainment of a new code characterized by $n = 46$ and $k = 24$. A remarkable effect of spectral thinning has been reached. The new code is characterized by a minimum distance $d = 5$, but low-weight coefficients $A(5) = 1$, $A(6) = 4$, $A(7) = 18$. The effective minimum distance can be estimated to be as high as about 10. Typically row splitting is the most straightforward method for breaking 4th-order girths. The cost of this operation is in a code rate reduction, therefore column splitting can be added in order to balance this trend.

Procedure iii, instead, maintains the code rate unchanged, but has a cost in code word length (however useful in order to reduce the density ρ). In this sense, it can be particularly attractive for convolutional codes, entailing only a longer frame termination. In fact the decoding algorithm does not suffer a parity check constraint length increase, if not extremely strong. If properly organized, this procedure can break a great number of 4-length short cycles in $\underline{H}$. We can consider the present approach distinct from superimposition, which starts from a base matrix $\underline{B}$ already free of 4-length short cycles. If the $\underline{H}$ matrix of the original code was column- or row-regular, or both, such properties will be maintained also in the new matrix. In spite of this constraint, often the minimum distance increases. Nevertheless the degrees of freedom in the choices of permutation matrices of whichever type are too many. It is then convenient to restrict ourselves only to a particular class of permutation matrices, in order to make code design more practical, and a good choice appears that of permutation circulants.

Definition 10.10 Let us consider a procedure, performed on a parity check matrix $\underline{H}$ (satisfying or not the row-column constraint), in which an optimized substitution

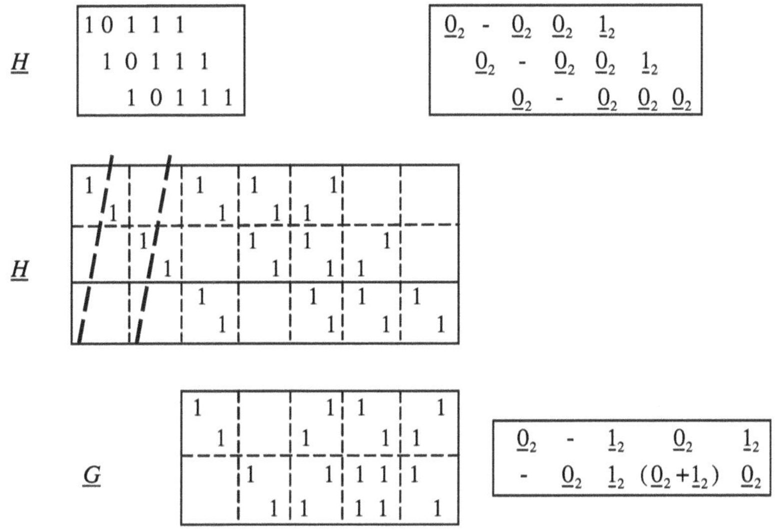

Fig. 10.13 Parity check matrices for a code on which circulant expansion is performed, and $\underline{G}$ matrix of the [10,4,4] final shortened code (Example 10.18)

of all the 1-symbols with permutation circulants is made. It is called *circulant expansion of the original* $\underline{H}$ *matrix.*

Circulant expansion is particularly efficient when the 1-symbol patterns involved in a 4-length short cycle are few. A situation of this type can be expected to occur for SC block codes and even more frequently for convolutional codes. The density ρ of 1-symbols turns out to be reduced by a factor equal to the size p of the permutation circulants. Often an increase in the minimum distance can be achieved, provided that p is high enough and the choice of the circulant phases is optimized.

Example 10.18 Assuming a [7,4,3] Hamming code, we have a parity check matrix where three 4-length short cycles are present, as shown in Fig. 10.13 up-left. In order to break the associated girths, by means of circulant expansion, it is enough to choose permutation matrices with a size as small as $p = 2$, and operate the substitution described in Fig. 10.13 up-right. This symbolic $\underline{H}$ matrix corresponds to the detailed matrix depicted down. Just two permutation circulants are $\underline{1}_2$, instead of $\underline{0}_2$. The row-column constraint is satisfied, but the minimum distance remains $d = 3$, as in the original code. Considering the shortened version of such code, characterized by 5 periods instead of 7, after elimination of the leftmost two column layers, we obtain a [10,4,4] code, whose $\underline{G}$ matrix is reported down (in both detailed and symbolic form). The $\underline{H}$ matrix is made by only permutation circulants (or null square sub-matrices), whereas in the $\underline{G}$ matrix also higher-weight circulants can be present. In order to allow LDPC decoding, it is however necessary to increase the size p of the circulants, at least up to some tenths.

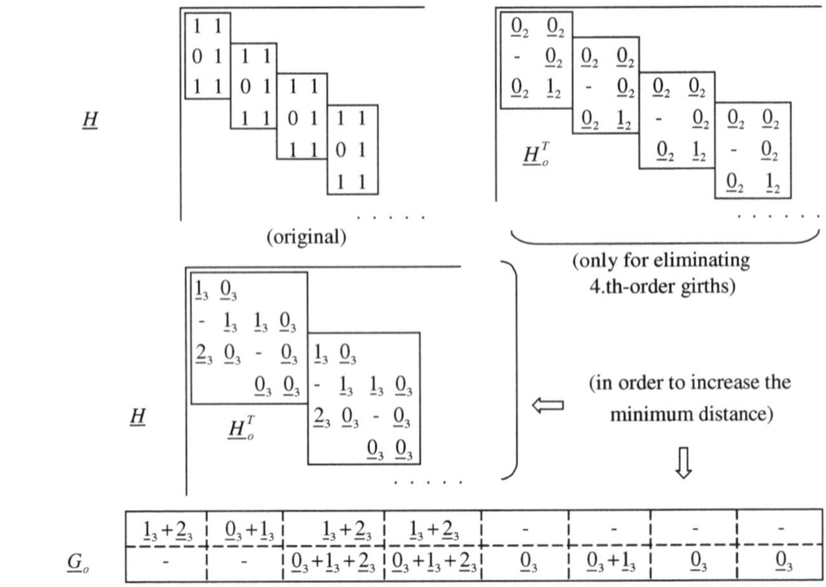

Fig. 10.14 Parity check matrices of the original convolutional code, of the code without 4th-order girths and of the code with increased minimum distance (Example 10.19)

Example 10.19 The best s.s. time-invariant 1/2-rate convolutional code is the one characterized by interleaved generator polynomial 11.10.11 (see the second row in Table 3.2). Its semi-infinite parity check matrix, over the first 4 periods is shown in Fig. 10.14 up-left. There are two types of 4-length short cycles: the one within a same $\underline{H}_o^T$ sub-matrix, and the other across two adjacent $\underline{H}_o^T$ sub-matrices. In order to eliminate them, by means of circulant expansion, it is enough to adopt $p = 2$, and insert a $\underline{1}_2$ circulant in the positions common to both types of 4-length short cycles. The new parity check matrix so obtained, free of 4-length short cycles, is reported in Fig. 10.14 up-right. Nevertheless, in order to increase the minimum distance, it is possible to assume $p = 3$, and the structure for the parity check matrix shown in the middle of Fig. 10.14. This is a 6/12-rate convolutional code, in fact the new syndrome former sub-matrix collects two old sub-matrices. The correspondent generator sub-matrix $\underline{G}_o$ is reported down. The new code exhibits $d = 8$, confirming a remarkable increase with respect to the value $d = 5$ of the original code. The information and control symbol sequence is *iiiccciiiccc*. An LDPC decoding is possible, even without enlarging further parameter p, since the 1-symbol density is rather low, especially on long frames. This fact represents an advantage of convolutional codes over block codes.

A result so good, reached employing a value of the circulant size p as low as 3, is possible only because the original convolutional code had normalized coefficients $B(5)$, $B(6)$ and $B(7)$ rather small. In other terms, circulant expansion is expected to be able to originate a remarkable minimum distance increase in low-rate convolutional codes. When the number of low-weigh 1-symbol patterns is too high, as typically occurs for high-rate convolutional codes, the goal of increasing d, even by few units, is much more difficult.

Let us consider now the following trivial $\underline{H}$ matrix, made by only 1-symbols in each position,

$$\underline{H} = \begin{bmatrix} 1 & 1 & 1 & 1 & & \ldots \\ 1 & 1 & 1 & 1 & & \\ 1 & 1 & 1 & 1 & & \\ & & & & \ldots & \\ 1 & 1 & 1 & 1 & & \ldots \end{bmatrix},$$

where an extraordinary high number of 4-length short cycles is present. In spite of this fact, after a proper choice of the permutation circulants to replace 1-symbols, all these 4-length short cycles can be eliminated. This is the case of an array code. Therefore array codes (proper, improper and even generalized) can be imagined as obtained from the above $\underline{H}$ matrix, after a suitable procedure of circulant expansion.

For improper array codes, the choice of the design parameters Δ_j and of the circulant size p, which is a sort of scale factor for the whole code word length, is of fundamental importance. The evaluation of the presence of girths with various orders λ_g, in the associated Tanner graph, can give some helps in this optimization. The following lower and upper bounds on d can be calculated for generalized array codes having $Y = 3$ (Milenkovic et al. 2008), related to the length of the correspondent minimum-order girths, independently of p,

$$\begin{aligned} \lambda_g &= 8, & 6 \le d \le 24, \\ \lambda_g &= 10, & 10 \le d \le 24, \\ \lambda_g &= 12, & 14 \le d \le 24. \end{aligned} \tag{10.2}$$

They hold for whichever length of the whole code word (hence also for SQC codes). This is due to the fact that such types of girths take place over a very small number of subsequent column layers. The separation between the lower and upper bound depends on the type of girth, and on how it can combine with other girths.

For proper array codes, on the full length of the QC code, adopting an approach not based on the girths but on algebraic constructions, the following upper bounds have been obtained: $d \le 6$, 12, 20, respectively for $Y = 3$, 4, 5 (Mittelholzer 2002). Improper array codes, as already noted, satisfy the upper bound (8.12), derived from the general form of their $\underline{G}$ matrix (Baldi et al. 2012). A more detailed analysis of such codes will be carried out in Chap. 11. In that occasion, we will return also on generalized array codes (Definition 8.6), for which we can anticipate here the upper bound on the minimum distance $d \le (Y + 1)!$ (MacKay and Davey 1999), already present in (10.2) for the only case of $Y = 3$.

Once performed a circulant expansion, it is possible to proceed with a moderate masking. As usual, in this second operation, the needs of breaking girths of orders progressively higher and of not reducing too much the average column weight $\langle w_c \rangle$ are to be balanced. In this framework, we can construct a variant of improper array codes, in a form resembling the structure of a serial concatenation of Y component codes.

Example 10.20 Let us consider $Y = 3$ row layers and a total number of column layers equal to 6. The last two column layers exhibit weight 2 and 1 respectively, whereas the preceding four column layers are characterized all by weight 3. The sequence of the circulants along each row layer can be chosen according to the rule of an improper array code, with parameters Δ_2, Δ_3, so obtaining

$$\underline{H} = \begin{bmatrix} \underline{0}_p & \underline{0}_p & \underline{0}_p & \underline{0}_p & & \\ \underline{-4\Delta_2}_p & \underline{-3\Delta_2}_p & \underline{-2\Delta_2}_p & \underline{-\Delta_2}_p & \underline{0}_p & \\ \underline{-5\Delta_3}_p & \underline{-4\Delta_3}_p & \underline{-3\Delta_3}_p & \underline{-2\Delta_3}_p & \underline{-\Delta_3}_p & \underline{0}_p \end{bmatrix},$$

where negative phases have been employed with obvious meaning, in order to construct a lower triangular parity check matrix. Such a structure is similar to that of a DC-GPC code (see Definition 5.18). This sort of concatenated form of an improper array code can be of some interest when the code rate is low, in order to introduce a variation in the column layer weights, so favouring the convergence of the decoding algorithm. The average column weight in the above $\underline{H}$ matrix turns out to be $\langle w_c \rangle = 2.5$.

Stronger actions in order to randomize an originally structured $\underline{H}$ matrix can be performer only varying the main properties of the code itself, say its word length and number of control symbols, as occurs in column or row splitting. Such irregular codes can be designed manually, or by means of a moderate use of computer optimizations. In the next section, instead, a brief review of the main computer based design procedures will be presented. They can lead to strongly irregular and unstructured LDPC codes.

10.7 Computer Based Design of Irregular LDPC Codes

In order to put in evidence a girth, evaluate its length, and attempt to remove it, we can construct a trellis, repeating several times the Tanner graph. So two different paths connecting a pair of subsequent different nodes represent a girth. The number of stages measures its length. Elimination of an edge, or edge subdivision by means of row or column splitting can help in increasing the orders of the girths.

Example 10.21 The Tanner graph shown in Fig. 10.15 on the left exhibits three girths (one of length 4 and two of length 6), as confirmed by the $\underline{H}$ matrix reported down. The trellis, which represents the periodic repetition of this Tanner graph, is

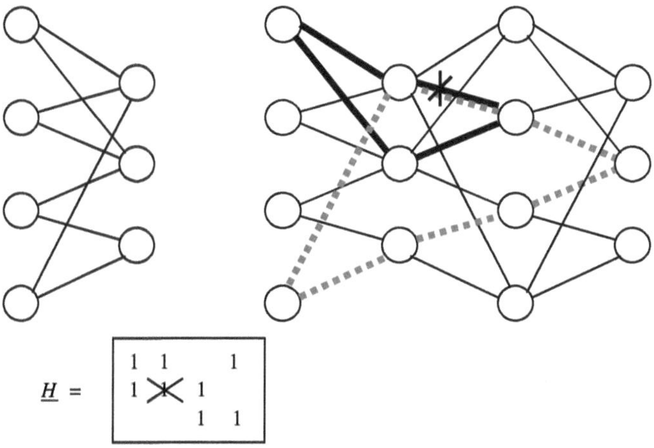

Fig. 10.15 Trellis derived from a Tanner graph and girth elimination (Example 10.21)

shown on the right. The first repetition (from a first stage of control nodes to a second stage of variable nodes) is formed as a mirror image of the Tanner graph itself. Then the period so constructed can be repeated as many times as one wishes. It is necessary to ignore the connections from a node to its replicas. The remaining non-trivial connections between different nodes are to be interpreted as follows. A λ_gth-order girth occurs when two distinct paths connect a pair a different nodes, after $\lambda_g/2$ single branches each. For instance, in the case here considered, the 4th-order girth, evidenced across the first two stages, and one of the two 6th-order girths, evidenced across all the three stages, can be both eliminated by dropping out a unique edge. It corresponds to the 1-symbol crossed in the $\underline{H}$ matrix. Remark that elimination also of the remaining 6th-order girth, in a proper position (leftmost 1-symbol in the first row) would lead to an equivalent of the [4,1,4] repetition code.

As already noted, instead of pure elimination, it is possible to perform splitting of a variable node (column splitting) or of a control node (row splitting). An edge touching a certain control node, after doubling such node because of row splitting, will be attributed to only one of the new nodes. Following this way, we can introduce a moderate degree of irregularity in an already existing structured graph.

A completely different approach, by which a strong irregular graph can be constructed, is that called PEG (already introduced in Sect. 10.2). It starts from a graph without any edge at all, or with a thin backbone of edges, and adds edges progressively, satisfying prefixed constraints on the maximum length of the girths (Hu et al. 2005). A variant of this tool consists of adding also some control nodes, realizing a sort of extended code (Jacobsen and Soni 2007). This procedure (similar to row splitting) entails code rate lowering, and could be balanced with the use of column splitting too. So it is possible to design very efficient irregular LDPC codes, whose scalability, in code rate and word length, becomes very fine, although the design is to be made any time ad hoc.

A statistical treatment of mutual information, called ExIT chart (Extrinsic Information Transfer), developed when an iterative procedure is carried out by two component processors in proper co-operation, can provide a guide for optimizing PEG procedures (Richardson and Urbanke 2008). ExIT charts are also employed for the design of irregular interleavers in turbo codes (Ten Brink 2001). In Sect. 10.9, we will describe an ExIT chart application for studying the convergence of LDPC decoding algorithm and a threshold in the *SNR* value, beyond which convergence can occur.

Variable nodes and control nodes have different roles in the decoding process. Typically control nodes are touched by a higher number of edges. Therefore the decoding algorithm extends its action, in subsequent iterations, mainly through control nodes. We can look at the Tanner graph as a tree connecting progressively all the variable nodes. In this picture, we have first neighbours of a certain variable node, second neighbours, and so forth. Not only short cycles in the Tanner graph are responsible of possible convergence problems. There are also other figures which can make convergence slower and sometimes even frustrate performance.

Definition 10.11 A set of variable nodes in which all the nodes are connected to their neighbours with at least two paths (crossing different control nodes) is called a *stopping set of variable nodes* (Ryan and Lin 2009).

The denomination follows from the fact that, in an erasure channel (like that described in Example 10.9) in correspondence with a stopping set, the decoding algorithm stops. The smaller the size of the stopping sets, the slower the process by which all the variable nodes of the Tanner graph are involved in the decoding process, along the trees starting from few nodes (Richter 2006). Typically a stopping set results in a combination of various girths. There is a relation between stopping set of variables nodes and the code words, but it depends widely on the particular structure of the considered code.

A computer-based tool for designing irregular LDPC codes in which the analysis of the stopping sets is performed, together with a maximization of the girth lengths is called ACE (already mentioned at the end of Sect. 10.3) (Tian et al. 2004). It allows to obtain very low BER floors. This procedure can be also employed for optimizing the choice of the circulants in a QC code, and in particular of the phase increments Δ_j in improper array codes (Esmaeli et al. 2011).

10.8 Outline of the Sum-Product Algorithm, Its Computational Complexity and Points of Weakness

As already anticipated, Gallager, in its early proposal of LDPC codes (Gallager 1962), suggested three types of hard-decision decoding algorithms. In the first two of them (called algorithm A and algorithm B), the principle of message passing was outlined, although with a Hamming metric. The third algorithm is called bit-flipping and is based on treating the parity check equations in a proper sequence, turning the identities (flipping) of some symbols in proper positions at any step.

After introduction of Tanner graph, to support an efficient message passing procedure, and owing to the availability of high-speed electronic technology, a soft-decision decoding algorithm able to process contemporary a suitable amount of data has been arranged (Kschischang et al. 2001). The sum-product algorithm was so called because of the elementary operations performed by the processors in the nodes, although this characteristic is no longer evident, after the logarithm calculation of the likelihood ratio.

We can write the *LLR* as expressed by (9.7), exactly as for a BCJR algorithm employed in turbo decoding (see Sect. 9.5). The sum-product algorithm starts in all variable nodes, with a proper initialization, which is helped by knowing an estimate of noise variance and signal amplitude. Then it proceeds, involving all the control nodes, again all the variable nodes, and so on. At any step, the total *LLR*, sum of all the $LLR(s_j)$, $j = 1, 2, \ldots, n$, is updated. On this parameter proper stopping criteria can be fixed. Alternatively a test on the parity check equations can be performed, in order to verify if the estimated symbol sequence is a code word. Other more involved stopping procedures can be adopted, depending also on the particular structure of the LDPC code considered (Ryan and Lin 2009).

Looking at a single initial variable node, we can imagine the formation of a tree, over which the decoding operations propagate. It has to be extended progressively to the whole Tanner graph. An ideal sum-product algorithm should give an exact *LLR*. This condition is achievable only when all the messages passed can be considered statistically independent. The presence of finite-order girths in the Tanner graph originates a statistical dependence, stronger and stronger with increasing their number. So this algorithm turns out to be sub-optimum. Nevertheless, if the girths are long and the stopping sets collect many nodes, it can allow error correction even beyond the limit imposed by the minimum distance d, exploiting wide interstices in the code word spatial distribution.

The computational complexity of the sum-product algorithm grows with the average column weight $\langle w_c \rangle$ and with the overall word length n. More precisely, we can take an evaluation of the computational complexity linear with both (Lin and Costello 2004). Nevertheless the structure of the code (in particular the actual distribution of row weights and column weights) can simplify by far the process. When the LDPC code is highly structured, some parallel processing can be arranged, leading to a remarkable reduction in time latency. On the other hand, a strongly unstructured and irregular code, as already stressed, is preferable from the point of view of decoding algorithm convergence, although its encoding operation could be paradoxically even more difficult (with computational complexity proportional to n^2). In this respect, a family of irregular codes, characterized by low density generator matrix (briefly called LDGM codes), have received remarkable attention, owing to their reduced computational complexity in encoding (Oenning and Moon 2001).

With respect to turbo codes, LDPC codes are characterized by a fundamental difference: each iteration in the sum-product algorithm contributes to increase the number of elementary trees involved (each starting from a different variable node), whose superimposition becomes larger and larger. It is possible to develop a statistical analysis of message transferring along the tree, starting from an initial set of variable nodes all referred to the same control node. This statistical analysis will be described in the next section, in order to give theoretic predictions about the possibility of approaching Shannon limit for a soft-decision decoding (see Theorem D.7 and Fig. D.8).

Sometimes a BER floor may appear, even when the code minimum distance is very high, because of the possible presence of clusters of variable nodes where the algorithm experiences a sort of stand point. They form sub-graphs which, in particular error configurations, force the algorithm to run inside, without possibility of escaping. They are called trapping sets of variable nodes (Richardson 2003; Zhang and Ryan 2009). Their presence can be revealed and the effects they produce can be controlled, by proper modifications of the sum-product algorithm (Han and Ryan 2009).

Another cause for non-ideal behaviour of the this algorithm, depending also on the particular code considered, is the so called decoding threshold. It can be defined as the value of an operation variable (typically the *SNR*), above which the fraction of incorrect messages approaches zero when the number of iterations tends to infinity (Richardson et al. 2001). In more general terms, the decoding threshold is a set of values for certain code parameters (code rate, word length, column weights, …) and operation condition parameters (*SNR* in an AWGN channel, expected error probability values before and after coding, …) which subdivides the space where a family of codes can be positioned into two regions: the one where a reliable information transfer is possible, and the other one where this is not (Ryan and Lin 2009). Considering very long LDPC codes, substantially free of girths in the Tanner graph, it is possible to develop a statistical treatment of their behaviour. However it is necessary to maintain separate two classes of codes: the regular ones (for which some parts of the treatment is deterministic) and the irregular ones (for which average results are to be calculated, in a way similar to that described in Sect. 9.3, for the analysis of a statistical interleaver in turbo codes). This treatment will be outlined in the next section.

Several variants of the sum-product algorithm, typically for reducing its computational complexity, have been proposed. Among them, one is widely diffused and is called min-sum algorithm (Chen et al. 2005). It takes into account only few dominant terms in the summations for control node update, where a hyperbolic tangent inverse function is to be approximated. In the initialization of this algorithm, the knowledge of the noise variance is not necessary (showing a similar advantage as that offered by SOVA in comparison with BCJR, in turbo decoding). Min-sum, or some other algorithms, all derived from sum-product, exhibit performance often not too worse, especially when designed for the particular code under study.

10.9 Statistical Analysis of the Asymptotic Behaviour of Regular and Irregular LDPC Codes

The principle on which the sum-product decoding algorithm is based, outlined in the previous section, allows now to justify a statistical treatment of its behaviour, in order to predict asymptotic performance (for code word length n tending to infinity) in regular and irregular LDPC codes. Also the number of iterations will be assumed to be infinite. This procedure is known as density evolution (Richardson and Urbanke 2001), meaning the evolution of the probability density function of the messages being passed in the tree, which starts from a certain initial set of variable nodes, with increasing the number of iterations. In particular, the code rate and the *SNR* are compared, having Shannon limit for soft-decision decoding as a reference. Firstly we will consider regular LDPC codes, for which the following proposition allows a simplification in the treatment.

Lemma 10.7 *In an LDPC code, regular in both columns and rows, with w_c and w_r respectively as column and row weight, the code rate R can be written as*

$$R = 1 - \frac{w_c}{w_r}. \tag{10.3}$$

Proof From (10.1), where the overall 1-symbol density ρ is a constant, we can write $nw_c = rw_r$, and hence the code rate $R = (n - r)/n$ turns out to be as expressed by (10.3). □

The structure of the tree, in which density evolution takes place, is organized as follows. From a control node to one of the variable nodes on which it acts, there is an edge. From this variable node to the other control nodes there are $(w_c - 1)$ edges. From each one of such control nodes to the other variable nodes there are $(w_r - 1)$ edges. In any situation, the original node is not to be accounted for. In Fig. 10.16 such a structure is depicted, assuming $w_c = 3$ and $w_r = 6$.

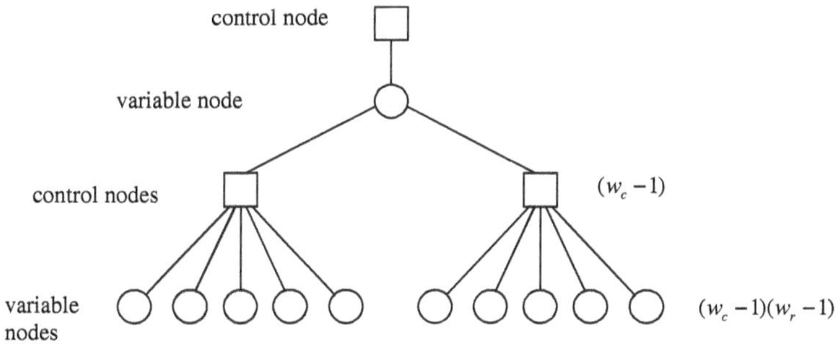

Fig. 10.16 Tree in the density evolution model for $w_c = 3$ and $w_r = 6$

The message produced in the first variable node can be written as the sum of a message m_0 from the uppermost control node plus $(w_c - 1)$ messages $m_j, j = 1, 2, \ldots (w_c - 1)$, from as many control nodes. This elementary structure repeats. If the code word length is very long, as already noted, the effect of possible cycles is negligible. Under this assumption, the messages become statistically independent. With a regular LDPC code, we can adopt all equal random variables for describing these messages. So the convolution of their density probability functions is to be taken into account (Papoulis 1965).

The message m_j outgoing from the control nodes, in turn, can be expressed as the product of $(w_r - 1)$ factors containing the *LLR*'s, in the form of a sign function multiplied by a reliability constant. Adopting a Gaussian approximation, for which all the variables are specified by just two parameters (mean value and variance), it is possible to reach an iterative solution, able to relate a threshold value for the *SNR* to a given couple (w_c, w_r), and hence to the code rate, owing to (10.3). This enables us to estimate the separation of the behaviour characterizing the considered regular LDPC code from Shannon limit (Richardson and Urbanke 2001). With $w_c = 3$, $w_r = 6$, that is for a 1/2-rate code, the threshold in the *SNR* turns out to be of about 1.1 dB, and the separation from Shannon limit is smaller than 1 dB. For codes characterized by a lower code rate, this separation increases.

Considering now irregular LDPC codes, it is necessary to introduce an averaging process over all the possible types of variable and control nodes (taking into account the number of edges touching them). An external tool is then able to optimize the distribution of column weights, fixed its maximum value $w_{\max} = \text{Max}\{w_c\}$, the code rate, and some working conditions (type of channel, value of the error probability before encoding, …). Once obtained the best distribution of column weights, it is easy to calculate its mean value $\langle w \rangle$. Taking regular codes with $w_c = \langle w \rangle$ as a reference, such irregular LDPC codes perform typically better and are able to approach Shannon limit with reduced asymptotic separation, especially at low code rates (Richardson et al. 2001).

Some simulations have confirmed this prediction (obtained under the assumption of n tending to infinity and hence of a statistical independence of the messages). With a code word size of the order of 200 thousands and code rate 1/2, a threshold *SNR* value of about 0.25 dB (break-even point in the curve of the error probability), only 0.006 dB away from Shannon limit, is achievable with an irregular LDPC code. A binary symbol error probability after decoding of the order 10^{-5} is reached at about 0.65 dB (Ryan and Lin 2009). This result has been obtained with a distribution of the column weights characterized by non-null values from $w = 2$ to $w_{\max} = 30$.

ExIT chart, as already anticipated, is a general method for the statistical analysis of the convergence of an iterative procedure, when two processors co-operate. In the case of LDPC codes, there are two sets of processors: the variable nodes and the control nodes. It is possible to report on a graph the evolution of the mutual information going into and coming out from a control node, on average. This graph shows the progressive increment in this mutual information, from 0 up to 1. A similar graph can be made for the variable nodes, with analogous result, on average.

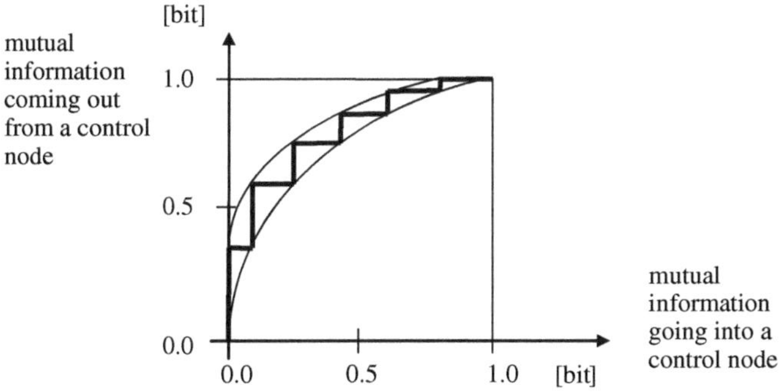

Fig. 10.17 Example of ExIT chart applied to an LDPC decoding procedure

If one bit of mutual information is finally achieved, no uncorrected errors remain in the decoded code word. This graph, which appears as shown qualitatively in Fig. 10.17, is formed by a sequence of vertical and horizontal segments, starting from point (0,0), up to reach point (1,1), if decoding is successful. It enables one to determine the *SNR* threshold, intended as the break-even point in the error probability curve.

The two continuous curves touching the external points of the vertical and horizontal segments can be predicted, by means of a statistical analysis based on an ensemble of codes having similar properties (Ashikhmin et al. 2004). The wider the separation between such two curves, the lower the number of iterations required. When the two continuous curves intersect, the iterative procedure turns out to be no longer convergent and decoding fails. By means of this technique, for either regular or irregular LDPC codes, the *SNR* threshold values can be determined, leading to a confirmation of what predicted by density evolution. In particular, an application to repeat-accumulate LDPC codes has given the possibility of designing codes for which the decoding algorithm is able to satisfy some desired convergence properties (Ten Brink and Kramer 2003).

An ExIT chart procedure, applied to a turbo code with a BCJR decoding algorithm, shows the trajectories of the two RSC decoders. They follow traces which depart, on the left or on the right, from the quadrant diagonal in a diagram of the type shown in Fig. 10.17 (Ryan and Lin 2009). The smaller the *SNR* the thinner the separation between these two trajectories, up to a situation where they intersect, making convergence impossible.

A statistical analysis of the type described so far, employing a different approach, was performed also in some original works of MacKay and Neal (MacKay and Neal 1997, MacKay 1999), describing the expected performance of the firstly proposed LDPC codes. The basic idea of sum-product decoding algorithm was introduced, and investigated in comparison with an optimum decoding algorithm. In particular, the possibility of reaching an error probability value as low as one wishes, like in Shannon limit, has been analyzed. Including in the model, besides an unconstrained overall block length n, also an unconstrained maximum column weight, the answer was that this is possible, so encouraging the use of LDPC codes also in practical situations. Curves of code rate R as a function of the power noise spectral density, with a uniform column weight w_c as a parameter, have shown that, for $w_c \geq 6$, especially at code rates of the order of 1/2, Shannon limit can be closely approached.

Let us consider a family of long block codes, in an AWGN channel, characterized by the following property: their minimum distance d increases proportionally to the word length n. A property of this type assures that a finite code rate is maintained with an error probability tending to zero, so preventing the presence of any BER floor. Regular LDPC codes with $w_c = 3$, $w_r = 6$ are characterized by d growing as about $0.023 \cdot n$, although their *SNR* threshold is rather high (not lower than 1.1 dB, as already noted). Irregular LDPC codes, with the same code rate, can reach $0.11 \cdot n$, accompanied by a lower *SNR* threshold. An asymptotic weight distribution for protograph-based LDPC codes has been calculated, showing that such codes can approach the limit $0.015 \cdot n$ at code rate 1/2 and maintain proportionality between d and n, although with a factor which is smaller and smaller, even at higher code rates (Divsalar et al. 2005).

10.10 A First Approach to LDPC Convolutional Codes

There are two main ways for designing LDPC convolutional codes: to start from a good LDPC block code, operating some manipulations on its $\underline{H}$ matrix, or to perform a direct design exploiting the convolutional nature of the code since the beginning. The search for linear combinations of rows in order to satisfy the row-column constraint, described in Example 10.7, can be included in the second type of procedure. The same we can say about application of circulant expansion to the best traditional convolutional codes, as described in Example 10.19. Both such approaches, in fact, begin from an initial code already in convolutional form. On the other hand, in LDPC convolutional codes (not tail-biting and well designed), a sliding window decoding algorithm can be applied, like in traditional convolutional codes (Definition 3.6). The length of such window turns out to be proportional to the parity check constraint length m_h (Jimenez-Felstrom and Zigangirov 1999).

In many situations, especially when good irregular LDPC block codes are taken as starting points, the first way listed above has been preferred. In this section we will treat only this type of codes, mainly for interpreting which properties of the original LDPC block code are expected to be transferred to its convolutional transformation. Chapter 12 will be devoted to an exhaustive analysis of both types of possible design of LDPC convolutional codes.

In order to fix ideas, let us consider ARA or protograph-based LDPC block codes, for which the following procedure was firstly proposed to be applied, in order to derive their convolutional counterparts (Costello et al. 2007). It is possible to adopt a suitable block module of length n_o, with r_o parity check rows (both multiples of a same integer y). This module is cut by means of a staircase sequence of segments, so obtaining y stairs, according to an asymptotic code rate, already

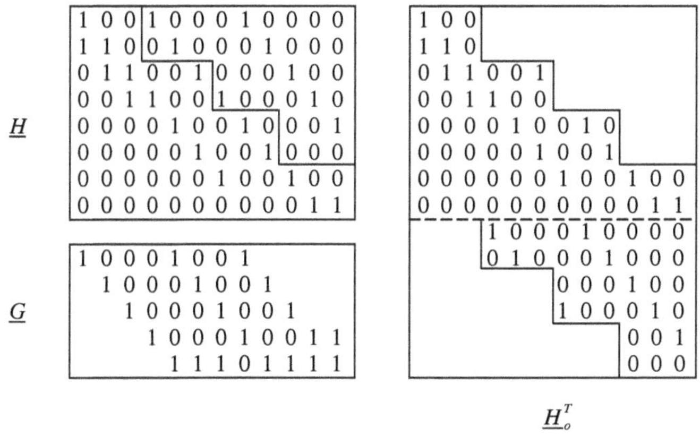

Fig. 10.18 Original parity check matrix (and correspondent $\underline{G}$ matrix), effect of the cutting procedure, and resulting $\underline{H}_o^T$ sub-matrix of the convolutional code described in Example 10.22

imposed by the above choice of parameters n_o and r_o. Then the upper part is eliminated and reported down. This composite module (which turns out to be the $\underline{H}_o^T$ sub-matrix of the convolutional code under construction) is repeated a number of times as high as one wishes, forming the desired semi-infinite parity check matrix. If the original $\underline{H}$ matrix satisfies the row-column constraint, this property will characterize also the semi-infinite parity check matrix so constructed (Jimenez-Felstrom and Zigangirov 1999).

Example 10.22 In Fig. 10.18 the cutting procedure above described is shown taking $n_o = 12$, $r_o = 8$, and hence $y = 4$. The original $\underline{H}$ matrix, column-regular, satisfies the row-column constraint. All the columns have weight $w_c = 2$, whereas the weights of the rows vary from 2 to 4. The girth of minimum length has $\lambda_g = 6$, so leading to $d = 3$ (according to Lemma 10.4). Since we have uniformly $w_c = 2$, there is a row in such $\underline{H}$ matrix linearly dependent on the other rows, but this is not a problem for this type of convolutional code construction. The generator matrix $\underline{G}$ which can be associated to the original $\underline{H}$ matrix is reported down. This is a [12,5,3] code, time-varying both in $\underline{H}$ and in $\underline{G}$. The resulting convolutional code has asymptotic code rate apparently 4/12 (but we will verify that its true value is 4/11). Such convolutional code could be considered w.s. time-invariant in its $\underline{H}$ matrix. The syndrome former sub-matrix $\underline{H}_o^T$, shown on the right, vertically shifts by 8 rows each period, in order to form the semi-infinite parity check matrix. The number of 6th-order girths is reduced and they now occupy two adjacent periods. The minimum distance remains $d = 3$. Clearly, with a more irregular initial module (for instance adopting a variable column weight), better results could be achieved.

Looking at Fig. 10.18, it is difficult to establish where information symbols and control symbols are located. This is consequence of the fact that the uppermost layer in the $\underline{H}_o^T$ sub-matrix is not lower triangular. They are certainly distributed with

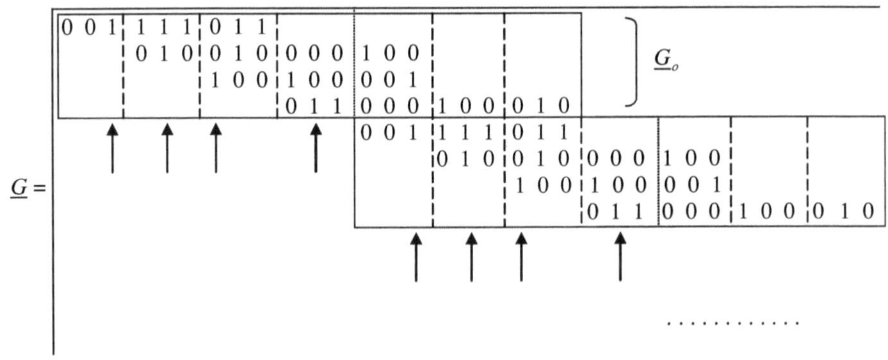

Fig. 10.19 Semi-infinite generator matrix compatible with the $\underline{H}$ matrix constructed by the cutting procedure shown in Fig. 10.18

1 information symbol and 2 control symbols per sub-period (the overall period is formed by $y = 4$ sub-periods). It is easy to verify that a valid generator matrix for the new code is of the type shown in Fig. 10.19, where two subsequent $\underline{G}_o$ sub-matrices have been reported. This generator matrix has been obtained after null ex-OR sum of clusters of syndromes, applied to the semi-infinite parity check matrix, imposing the presence of one control symbol in any sub-period. Small arrows denote the positions where information symbols are expected to be placed (leftmost 1-symbol in any row). It is possible to observe how the second position in any period is useless (it shows only 0-symbols). All the correspondent columns in $\underline{G}$, as well as in $\underline{H}$, can then be eliminated, and the correct asymptotic code rate becomes 4/11.

The column weight distribution in the parity check matrix does not change, after the cutting procedure here described. For the particular case considered in Example 10.22, it remains uniform with $w_c = 2$. This constraint is too strong for giving the possibility of reaching higher values for the minimum distance d, considering also the residual presence of two 6th-order girths (producing as many 3-weight rows in $\underline{G}_o$).

Comparison of performance between the original block code and the convolutional code obtained by the above construction procedure, typically shows better results associated to the convolutional code. In order to assume comparable situations, the original block code and a proper tail-biting form of the convolutional code are to be examined (Mitchell et al. 2008). The latter one has been taken over a sufficiently high number Z of periods, in order to be sure that the transient, during which anomalous low-weight code frames could be present, is ended.

Remark that the cutting procedure discussed so far can be interpreted as a sort of unwrapping, like the one already introduced for QC codes in Chap. 8, regarding their $\underline{H}$ matrix. The only difference is that, here, the original small block code is, in general, completely time-varying in its $\underline{H}$ matrix. For this reason, the convolutional codes so derived are usually called time-varying. Looking at the final $\underline{H}_o^T$ sub-matrix as a unique part, instead, we can continue in defining them w.s. time-invariant.

On the other hand, by construction, with reference to this $\underline{H}_o^T$ sub-matrix, we have a parity check constraint length m_h necessarily equal to 1. In this sense, although $\underline{H}_o^T$ spans a rather long period, the symbols belonging to subsequent periods can be considered weakly correlated, and the length of the sliding-window decoding procedure which can be applied is not too long.

Another possibility for obtaining good LDPC convolutional codes from well known LDPC block codes is that regarding QC codes, whose $\underline{H}$ matrix is characterized by circulants (of weight 0, 1 or greater than 1). This $\underline{H}$ matrix can be reordered according to Lemma 2.6, leading to a tail-biting convolutional form, which is unwrapped. So a convolutional LDPC code, whose frame length can be as high as one wishes, is obtained. A particular case of this situation is that regarding array codes (in which the circulants are only permutation circulants and there are no null sub-matrices). Theorem 8.2 gives closed form expressions for the rows of the parity check and the generator matrix characterizing improper array LDPC convolutional codes, preferably starting from a suitable SQC version, which imposes the code rate value. In Examples 8.25 and 8.26, this procedure has been already discussed.

Nevertheless the convolutional codes so obtained are too regular and structured, and often it is necessary to introduce some not uniform elements, although of deterministic nature. Instead of proper or improper array codes, where the sequence of phases is linear, a power law can be adopted. Furthermore some null square sub-matrices can be introduced, after application of a moderate masking (Tanner et al. 2004).

Example 10.23 Let us take a size $p = 31$ for the permutation circulants, and the following $\underline{H}$ matrix for the original quasi-cyclic array code

$$\underline{H} = \begin{bmatrix} \underline{0}_{31} & \underline{1}_{31} & \underline{3}_{31} & \underline{7}_{31} & \underline{15}_{31} \\ \underline{0}_{31} & \underline{5}_{31} & \underline{15}_{31} & \underline{4}_{31} & \underline{13}_{31} \\ \underline{0}_{31} & \underline{25}_{31} & \underline{13}_{31} & \underline{20}_{31} & \underline{3}_{31} \end{bmatrix},$$

where in the first row the sequence of phase increments is associated to $1 \cdot 2^j$, in the second row to $5 \cdot 2^j$, in the third row to $25 \cdot 2^j$, $j = 1, 2, \ldots$, all these integer numbers being calculated mod(31). In the selection of the multiplicative constants, it is necessary to avoid a situation like that discussed in Lemma 10.5 (equal aligned phase difference in two columns). This is not difficult if the number of column layers is noticeably smaller than p (the code is strongly shortened with respect to its natural QC length p^2). Remark that the above construction represents a sort of first step in a conceptual transition from an improper array code towards a generalized array code, where the permutation circulants are chosen completely without rules (see Definitions 8.5 and 8.6). A similar passage is that based on the selective code shortening, described in Sect. 10.4. In order to give a further random contribution to the above $\underline{H}$ matrix, a suitable masking can be employed, inserting few null 31×31 sub-matrices. This passage allows to obtain an irregular LDPC block code. Only after such operation, we can proceed with reordering and unwrapping. For what discussed in Chaps. 7 and 8, the phases of the circulants in the block code $\underline{H}$ matrix are properly related to the coefficients of the symbolic polynomial syndrome former sub-matrix $\underline{H}_o^T(x)$. So the parity check matrix of the LDPC convolutional code, obtained after reordering and unwrapping, can be easily designed (see the concluding remarks at the end of Sect. 8.7).

In Tanner et al. (2004) many codes of this type have been proposed. The more the randomization, the better the behaviour. The original SQC code, with increasing the circulant size p, asymptotically tends to approach the convolutional code performance. With three component codes, a minimum distance d of the order of 20 can be achieved, with a code rate of about 1/2. The upper bound $d \leq (Y+1)!$ (which holds for generalized array block codes, as anticipated in Sect. 10.6) is expected to be achieved. In principle, it could be also overcome, owing to masking. Unfortunately higher code rates show results not so good. Another drawback is in the very long parity check constraint length m_h, which characterizes such convolutional codes. In Chap. 12 we will return on a construction of this type, but starting from improper array codes, and optimizing the code word length over which unwrapping is made, so obtaining a shorter constraint length m_h (Baldi et al. 2014). In this case, good results at rather high code rates can be achieved too.

A further possibility for obtaining LDPC convolutional codes, starting from cyclic or QC block codes, is that based on an MHEC code or MHEQC code construction. In this case, we can start from a code which either satisfies or does not satisfy the row-column constraint. In the first situation, the following proposition is of interest.

Lemma 10.8 *If an HEC code or HEQC code is characterized by an $\underline{H}$ matrix where the row-column constraint is satisfied, this property holds also for any MHEC o MHEQC code obtained from it.*

Proof An HEC code exhibits a certain number of control symbols, and as many parity check equations collected in its $\underline{H}$ matrix. An MHEC code, which is obtained from the previous HEC code by eliminating some central rows in the $\underline{H}$ matrix (after constant-length puncturation), cannot experience the occurrence of 4th-order girths if they are not already present in the Tanner graph of the old $\underline{H}$. In fact, this graph is transformed only by the elimination of all the edges touching the control nodes corresponding to such parity check equations. The same can be said about HEQC codes and MHEQC codes. □

Unfortunately, the case of an H-extended code directly satisfying the row-column constraint is not frequent. For instance, the HEC code obtained from a [7,3,4] simplex code, treated in Example 10.1 would be of this type (see Fig. 10.1). Nevertheless, even when the $\underline{H}$ matrix of the original H-extended code does not satisfy the row-column constraint, it is possible that a periodic elimination of parity check rows enables satisfaction of such constraint. The asymptotic code rate of the convolutional code, after this type of design, plays an important role, in the sense

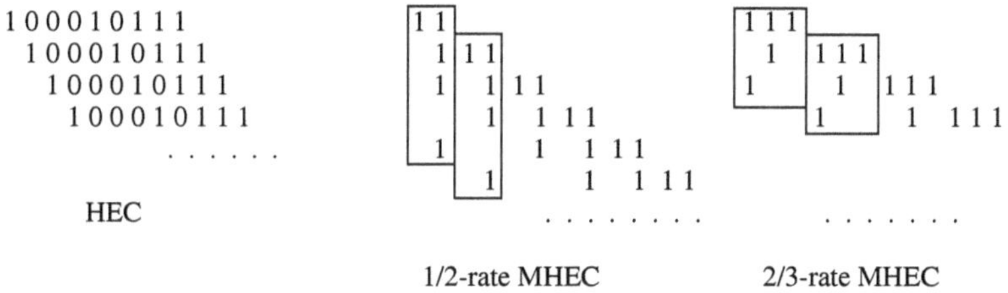

Fig. 10.20 Parity check rows in three situations regarding row-column constraint satisfaction in MHEC codes (Example 10.24)

that higher code rates can be advantaged, but at the expense of a minimum distance reduction.

Example 10.24 Starting from the reciprocal parity check polynomial associated to the binary sequence 100010111, which describes the dual code of a [15,7,5] BCH code, we can form different MHEC codes. The full H-extended version of this code is characterized by several 4-length short cycles. The 1/2-rate MHEC version exhibits the sequence 01.00.01.01.11 and is characterized by just two residual 4-length short cycles per period. The 2/3-rate MHEC version can be associated to the sequence 100.010.111, and has no 4-length short cycles. Further increase in the code rate gives rise to positions in the frame not protected by the code. The situations described above are shown in Fig. 10.20, where two consecutive syndrome-former sub-matrices have been evidenced in each one of the two convolutional codes considered. The 2/3-rate code obtained here, whose minimum distance is $d = 3$, belongs to a family which has been already treated in Example 7.8 (Fig. 7.9) and Example 7.13. Finally it is possible to observe that the 1/2-rate code can be interpreted also as the MLC code version of the [15,7,5] BCH code.

After the present brief introduction, it is possible to anticipate that the convolutional versions of known LDPC block codes often outperform their parent codes. This fact is justified by the possibility of breaking a large number of low-order girths, offered by the pseudo-diagonal form of the $\underline{H}$ matrix in a convolutional code. Nevertheless, just as a consequence of this property, one can feel that the direct design of LDPC convolutional codes could give even better results.

References

Abbasfar A, Divsalar D, Yao K (2007) Accumulate-repeat-accumulate codes. IEEE Trans Commun 55:692–702

Ashikhmin A, Kramer G, Ten Brink S (2004) Extrinsic information transfer functions: model and erasure channel properties. IEEE Trans Inf Theory 50:2657–2673

Baldi M, Chiaraluce F (2008) A simple scheme for belief propagation decoding of BCH and RS codes in multimedia transmissions. Int J Dig Multim Broadcas 8:ID957846

Baldi M, Bianchi M, Cancellieri G et al (2012) On the generator matrix of array codes. In: Proceedings of Softcom 2012, Split (Croatia)

Baldi M, Cancellieri G, Chiaraluce F (2014) Array convolutional low-density parity-check codes. IEEE Commun Lett 18:336–339

Chen J, Tanner RM, Jones C, Li Y (2005) Improved min-sum decoding algorithms for irregular LDPC codes. In: Proceedings of international symposium on information theory, 2005, Adelaide (Australia), pp 449–453

Costello DJ, Pusane AE, Jones CR, Divsalar D (2007) A comparison of ARA- and protograph-based LDPC block and convolutional codes. In: Information theory and applications workshop, San Diego (CA)

Divsalar D, Jin H, McEliece R (1998) Coding theorems for turbo-like codes. In: Proceedings of 36th annual allerton conference on communication, Monticello (IL), pp 201–210

Divsalar D, Jones C, Dolinar S, Thorpe J (2005) Protograph based LDPC codes with minimum distance linearly growing with block size. In: Proceedings of IEEE global telecommunications conference, St. Louis (MI), pp 1152–1156

Djurdjevic I, Xu J, Abdel-Ghaffar K, Lin S (2003) A class of low-density parity-check codes constructed based on Reed-Solomon codes with two information symbols. IEEE Commun Lett 7:317–319

Esmaeli M, Tadayon MH, Gulliver TA (2011) More on the stopping and minimum distances of array codes. IEEE Trans Commun 59:750–757

Fossorier MPC (2004) Quasi-cyclic low-density parity check codes from circulant permutation matrices. IEEE Trans Inf Theory 50:1788–1793

Gallager RG (1962) Low-density parity-check codes. IRE Trans Inf Theory 8:21–28

Halford TR, Grant AJ, Chugg KM (2006) Which codes have 4-cycle-free Tanner graphs? IEEE Trans Inf Theory 52:4219–4223

Han Y, Ryan WE (2009) Low-floor decoders for LDPC codes. IEEE Trans Commun 57:1663–1773

Hu XY, Eleftheriou E, Arnold DM (2001) Progressive edge-growth Tanner graphs. In: Proceedings on 2001 IEEE global telecommunications conference, San Antonio (TX), pp 995–1001

Hu XY, Eleftheriou E, Arnold DM (2005) Regular and irregular progressive edge-growth Tanner graphs. IEEE Trans Inf Theory 51:386–398

Jacobsen N, Soni R (2007) Design of rate compatible irregular LDPC codes based on edge growth and parity splitting. In: Proceedings IEEE 66th vehicular technology conference, Baltimore (MD), pp 1052–1056

Jimenez-Felstrom AJ, Zigangirov KS (1999) Time-varying periodic convolutional codes with low density parity-check matrix. IEEE Trans Inf Theory 45:2181–2191

Jin H, Khandekar A, McEliece R (2000) Irregular repeat-accumulate codes. In: Proceedings of 2nd international conference on turbo codes, Brest (France), pp 1–8

Johnson SJ, Weller SR (2004) Codes for iterative decoding from partial geometries. IEEE Trans Commun 52:236–243

Kamiya N (2007) High-rate quasi-cyclic low-density parity-check codes derived from finite affine planes. IEEE Trans Inf Theory 53:1444–1459

Kou Y, Lin S, Fossorier MPC (2001) Low-density parity-check codes based on finite geometries: a rediscovery and new results. IEEE Trans Inf Theory 47:2711–2736

Kschischang F, Frey B, Loeliger HA (2001) Factor graphs and the sum-product algorithm. IEEE Trans Inf Theory 47:485–519

Lin S, Costello DJ (2004) Error control coding. Pearson/Prentice-Hall, Upper Saddle River

MacKay DJC (1999) Good error-correcting codes based on very sparse matrices. IEEE Trans Inf Theory 45:399–431

MacKay DJC, Davey M (1999) Evaluation of Gallager codes for short block length and high rate applications. In: Proceedings of IMA Workshop codes systems and graphical models, Minneapolis (MN), pp 113–130

MacKay DJC, Neal RM (1995) Good codes based on very sparse matrices. In: Boyd C (ed) Proceedings of cryptography and coding 5th IMA conference. Lecture notes computer science. Springer, Berlin, pp 110–111
MacKay DJC, Neal RM (1997) Near Shannon limit performance of low density parity check codes. Electron Lett 33:457–458
Milenkovic O, Kashyap N, Leyba D (2008) Shortened array codes of large girth. IEEE Trans Inf Theory 52:3707–3722
Mitchell DGM, Pusane AE, Zigangirov KS, Costello DJ (2008) Asymptotically good LDPC convolutional codes based on protographs. Proc ISIT, Toronto (Canada), pp 1030–1034
Mittelholzer T (2002) Efficient encoding and minimum distance bound for Reed-Solomon-type array codes. In: Proceedings of ISIT 2002, Lousanne (Switzerland), p 282
Oenning TR, Moon J (2001) A low-density generator matrix interpretation of parallel concatenated single bit parity check codes. IEEE Trans Magn Rec 37:737–741
Papoulis A (1965) Probability, random variables, and stochastic processes. McGraw Hill, New York
Pearl J (1988) Probabilistic reasoning in intelligent systems: networks of plausible inference. Morgan Kauffman, San Mateo (CA)
Richardson T (2003) Error floors of LDPC codes. In: Proceedings of 41th annual, allerton conference on communications, Monticello (IL), pp 1426–1435
Richardson T, Urbanke R (2001) The capacity of low-density parity-check codes under message-passing decoding. IEEE Trans Inf Theory 47:599–618
Richardson T, Urbanke R (2008) Modern coding theory. Cambridge University Press, Cambridge
Richardson J, Shokrollahi A, Urbanke R (2001) Design of capacity-approaching irregular low-density parity-check codes. IEEE Trans Inf Theory 47:2001
Richter G (2006) Finding small stopping sets in the Tanner graphs of LDPC codes. In: Proceedings of 4th international symposium turbo codes, Munich (Germany)
Rudolph LD (1967) A class of majority logic decodable codes. IEEE Trans Inf Theory 13:305–307
Ryan WE, Lin S (2009) Channel codes: classical and modern. Cambridge University Press, Cambridge
Tanner RM (1981) A recursive approach to low complexity codes. IEEE Trans Inf Theory 27:533–547
Tanner RM, Sridhara D, Sridharan A et al (2004) LDPC block and convolutional codes based on circulant matrices. IEEE Trans Inf Theory 50:2966–2984
Ten Brink S (2001) Convergence behaviour of iteratively decoded parallel concatenated codes. IEEE Trans Commun 49:1727–1737
Ten Brink S, Kramer G (2003) Design of repeat-accumulate codes for iterative detection and decoding. IEEE Trans Sig Proc 51:2764–2772
Thorpe J (2003) Low-density parity-check (LDPC) codes constructed from protographs. JPL INP Prog Rep 42(154):42–154
Tian T, Jones G, Villasenor JD, Wesel RD (2004) Selective avoidance of cycles in irregular LDPC code construction. IEEE Trans Commun 52:1242–1247
Wang CL, Fossorier MPC (2009) On asymptotic ensemble weight enumerators of LDPC-like codes. IEEE J Sel Areas Commun 27:899–907
Xu J, Chen L, Zeng LQ et al (2005) Construction of low-density parity-check codes by superimposition. IEEE Trans Commun 53:243–251
Zhang Y, Ryan WE (2009) Toward low LDPC-code floor: a case study. IEEE Trans Commun 57:1566–1573

Chapter 11
Binomial Product Generator LDPC Block Codes

11.1 Cyclic Version of a Composite Code Based on GPC Codes

In Sect. 5.17 we have introduced composite codes based on GPC component codes. We considered firstly a direct concatenation, or DC-GPC (see Definition 5.18, valid for whichever number M of component codes) and then a direct product concatenation or DP-GPC (see Definition 5.19, up to now, just for the case of $M = 2$ component codes). In Sect. 10.2 we observed that they are interesting structured LDPC codes (the row-column constraint is satisfied in both, provided that the code word length is properly upper bounded). A DP-GPC code is also column-regular. Lemmas 5.43 and 5.44 finally show an equivalence between such codes and suitable direct product codes (valid only in the case of $M = 2$, we are treating now).

We can observe a resemblance between the parity check matrix of such codes, made by layers of identity matrices of different sizes, and the parity check matrix of improper array codes, made by circulants with different phases, but all with equal size. Restricting our attention, for the moment, to the case of only two layers ($M = 2$ and $Y = 2$), it is possible to exploit this resemblance for predicting some similar features. In Fig. 11.1, we summarize analogies and differences of such two families of codes, by comparing their $\underline{H}$ matrices in symbolic form (see Figs. 5.36 and 8.18, in which $\underline{I}_p$ and $\underline{0}_p$ have the same meaning). In improper array codes we take $p = 5$, $\Delta_2 = 2$, whereas in the codes based on GPC component codes we take $r_1 = 5$, $r_2 = 7$. Finally, in order to make comparison more evident, we have shortened also GPC codes with steps of r_1 (starting from the right hand side in the DP-GPC case, from the left hand side in the direct concatenation). These choices originate some truncated identity matrices labelled $|\underline{I}_7$ or $\underline{I}_7|$, respectively for a truncation on the left or on the right hand side.

All the codes in Fig. 11.1 have minimum distance $d = 4$. In spite of the similitude described so far, we obtain rather different terns for representing such codes, but this fact is due to the smallness of parameter $p = r_1 = 5$. For much larger values of

G. Cancellieri, *Polynomial Theory of Error Correcting Codes*,
Signals and Communication Technology, DOI 10.1007/978-3-319-01727-3_11

$$\underline{H}: \begin{bmatrix} \underline{0}_5 & \underline{0}_5 & \underline{0}_5 & \underline{0}_5 & \underline{0}_5 \\ \underline{0}_5 & \underline{2}_5 & \underline{4}_5 & \underline{1}_5 & \underline{3}_5 \end{bmatrix}$$

[25,16,4] QC improper array

$$\begin{bmatrix} \underline{0}_5 & \underline{0}_5 & \underline{0}_5 & \underline{0}_5 \\ \underline{0}_5 & \underline{2}_5 & \underline{4}_5 & \underline{1}_5 \end{bmatrix}$$

[20,11,4] SQC improper array

$$\underline{H}: \begin{bmatrix} \underline{I}_5 & \underline{I}_5 & \underline{I}_5 & \underline{I}_5 & \underline{I}_5 & \underline{I}_5 & \underline{I}_5 \\ \underline{I}_7 & \underline{I}_7 & \underline{I}_7 & \underline{I}_7 & \underline{I}_7 & & \end{bmatrix}$$

[35,24,4] DP-GPC

$$\begin{bmatrix} \underline{I}_5 & \underline{I}_5 & \underline{I}_5 & \underline{I}_5 & \underline{I}_5 & \underline{I}_5 \\ \underline{I}_7 & \underline{I}_7 & \underline{I}_7 & \underline{I}_7 & \underline{I}_7 & \end{bmatrix}$$

[30,19,4] shortened DP-GPC

$$\underline{H}: \begin{bmatrix} \underline{I}_5 & \underline{I}_5 & \underline{I}_5 & \underline{I}_5 & \underline{I}_5 & \underline{I}_5 & \underline{I}_5 \\ \underline{I}_7 & \underline{I}_7 & \underline{I}_7 & \underline{I}_7 & \underline{I}_7 & \underline{I}_7 & \end{bmatrix}$$

[42,30,4] DC-GPC

$$\begin{bmatrix} \underline{I}_5 & \underline{I}_5 & \underline{I}_5 & \underline{I}_5 & \underline{I}_5 & \underline{I}_5 \\ \underline{I}_7 & \underline{I}_7 & \underline{I}_7 & \underline{I}_7 & \underline{I}_7 & \underline{I}_7 \end{bmatrix}$$

[37,25,4] shortened DC-GPC

Fig. 11.1 Comparison of improper array codes, DP-GPC codes and DC-GPC codes ($M = Y = 2$)

$p = r_1$, this separation is less evident. In Example 10.20 a variant of improper array codes has been presented, adopting an $\underline{H}$ matrix with a scheme which is typical in a direct concatenation, so showing that such arrangement is possible also in this case. We will attempt here to treat such families of codes from a unitary point of view.

First of all we are interested in studying what happens, in DP-GPC codes, when the number M of layers increases. Since they will be no longer equivalent to a direct product code of two parity check codes, it is better to introduce a new denomination.

Definition 11.1 Considering M component GPC codes, characterized by parameters r_j, $j = 1, 2, \ldots M$, all increasing and co-prime, and assuming an overall code word length $n = r_1 r_2$, we can arrange an $\underline{H}$ matrix with M row layers, where the jth one of them is made by a sequence of $\underline{I}_{r_j}$ identity sub-matrices. This code is called *parallelepiped concatenation of GPC codes*, or briefly *P-GPC* code.

Clearly, in the case of $M = 2$, DP-GPC codes and P-GPC codes coincide. Remembering property *i* in Sect. 8.8 about array codes, we can state what follows about the total number of control symbols: also here, some parity check rows (whose number is equal to that of the row layers minus one) are not linearly independent. A shortened version of the code so introduced can be obtained reducing parameter n with respect to the above condition $n = r_1 r_2$ (by column elimination as usual). So a P-GPC code, with parameters described according to Definition 11.1, but whichever $n \leq r_1 r_2$, is characterized by the following total number of information symbols

$$k = n - \sum_{j=1}^{M} r_j + (M - 1). \tag{11.1}$$

Also in this situation, the minimum distance d is always an even number. In principle, the structure described by Definition 11.1 could be defined also for longer code word lengths, continuing addition of columns. Nevertheless beyond the upper bound $n = r_1 r_2$ it is not possible to work, because the row-column constraint is no longer satisfied. A small advantage, from the code rate point of view, can be offered by the direct concatenation. Lemma 5.42, in fact, fixes the maximum length n in order to satisfy the row-column constraint in $n = r_1 r_2 + \sum_{j=2}^{M} r_j$.

Example 11.1 In Fig. 11.2 symbolic and detailed expressions for the $\underline{H}$ matrices associated to the codes characterized by $M = 3$, $r_1 = 3$, $r_2 = 4$, $r_3 = 5$, are depicted. We can observe that there are no 4-length short cycles. After addition of one more identity sub-matrix in the first layer, three 4-length short cycles would appear between the first two layers. The P-GPC code is represented by the tern [12,2,6], where the two information symbols are due to the elimination of as many linearly dependent rows. The DC-GPC code, instead, is a [21,9,6] code. Here the $\underline{H}$ matrix has full rank. The strong difference between such two codes is due to the fact that the r_j parameters are extraordinarily small. In order to have true LDPC codes (with 1-symbol density ρ low enough), they must be of the order of some tenths, so leading to codes very similar. A difference remains however in their structure,

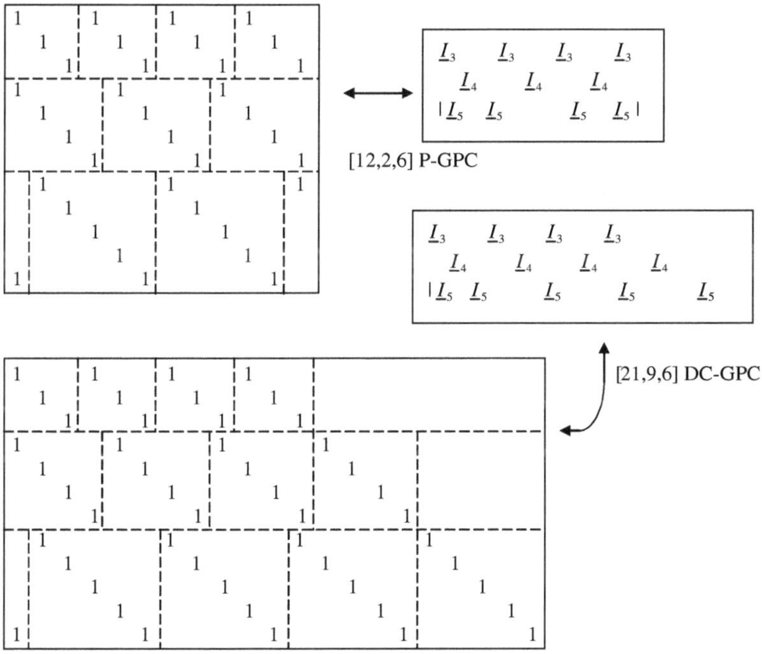

Fig. 11.2 Parity check matrices for the [12,2,6] P-GPC code and for the [21,9,6] DC-GPC code (Example 11.1)

which is column-regular in the first case and column-irregular in the second. A more detailed discussion about the minimum distance will be presented in Sect. 11.3. It is possible to verify that, in the P-GPC version, equivalent codes are obtained after any possible cyclic permutation of the layers. We have chosen a layer alignment easily comparable with that characterizing the $\underline{H}$ matrix of the DC-GPC version, whose rightmost part is lower triangular.

We can observe that, for array codes (proper, improper and even generalized), the maximum length in order to avoid any 4th-order girth coincides with the QC code period length $N = p^2$. Nevertheless, also in this case, the possible direct concatenation version (see Example 10.20) leads to some advantages in the code rate.

Lemma 11.1 *For an improper array code, whose* $\underline{H}$ *matrix is characterized by* Y *row layers, with parameters* p *and* Δ_j, $j = 1, 2, \ldots Y$, *in direct concatenation arrangement, the maximum length for satisfying the row-column constraint is* $n = p(p + 1)$.

Proof The row-column constraint is to be verified between any pair of layers. The $\underline{H}$ matrix of the considered code exhibits a first row layer with $(p - Y + 2)$ non-null subsequent circulants, the second row layer with $(p - Y + 3)$, … the Yth row layer with $(p + 1)$. In any case, the most critical pair is that between the last two row layers. □

The concatenation model stresses the characteristic, present in either P-GPC or in array codes, to collect the result of single subsequent actions, due to as many component codes. Now, we would like to interpret such two families of composite codes as deriving from a unitary integrated action. As known, array codes are quasi-cyclic over the length $N = p^2$, independently of the number Y of component codes involved. It is evident that DP-GPC codes ($M = 2$) are cyclic over the length $N = r_1 r_2$, in fact, the 2-weight code word appearing after addition of one column, for $n = r_1 r_2 + 1$, is clearly the effect of a code lengthening procedure. One could expect that P-GPC codes (M > 2) are cyclic over the same code word length. This is a wrong deduction. In fact the following proposition holds.

Lemma 11.2 *A P-GPC code, with parameters* r_j, $j = 1, 2, \ldots M$, *increasing and all co-prime, is a cyclic code characterized by length* $N = \prod_{j=1}^{M} r_j$.

Proof Let us consider the case of $M = 3$, and then generalize it. In order to find the minimum value for N able to satisfy the desired requirement, we can refer to the co-factor of all the parity check rows. Taking the GPC code in each one of the three layers as a reference, the following multiple equality holds

$$x^N + 1 = (x^{r_1} + 1)u_{w_1}(x^{r_1}) = (x^{r_2} + 1)u_{w_2}(x^{r_2}) = (x^{r_3} + 1)u_{w_3}(x^{r_3}).$$

Owing to the assumption that all the r_j's are co-prime, the solution turns out to be $w_1 = w_2w_3$, $w_2 = w_1w_3$, $w_3 = w_1w_2$ and $N = r_1r_2r_3$. A straightforward generalizations yields the following weights for the unitary polynomials

$$w_i = \prod_{\substack{j=1 \\ j \neq i}}^{M} r_j, \qquad j = 1,\ 2,\ \ldots M, \tag{11.2}$$

and the overall length N turns out to be $N = \prod_{j=1}^{M} r_j$. □

In conclusion, a P-GPC code, for $M \geq 3$, is a cyclic code over a very long word length, but its application as a good LDPC code (satisfying the row-column constraint and hence allowing convergence of the decoding algorithm) is limited to strongly shortened versions of the above cyclic code, whose block length can be at most $n = r_1r_2$.

11.2 Generator Polynomial for a Parallelepiped Concatenation of GPC Codes

A DP-GPC code ($M = 2$), is equivalent to the direct product between two parity check codes (see Lemma 5.43). This property is no longer valid for a P-GPC code with $M \geq 3$. In this, it is possible to find an analogy with array codes with $Y \geq 3$.

Example 11.2 Let us consider a P-GPC code, with $M = 3$ component codes, and parameters $r_1 = 5$, $r_2 = 7$, $r_3 = 11$. The LDPC code with maximum length obtained from it is a [35,14,6] code. It can be interpreted looking at a rectangular encoding memory of the type shown in Fig. 11.3. Taking 5 control symbols from the first component code (c_1), 6 control symbols from the second component code (c_2) and 10 control symbols from the third component code (c_3), such control symbols can be placed firstly on the lowest border row and on the rightmost border column (like in a direct product code). Then, those regarding the third component code can be inserted by progressively filling the remaining positions from the bottom. In this sense, a code with a third component code turns out to be a subcode of the one with two component codes only. This organization of the encoding memory resembles the one adopted for array codes (see Example 8.28, Fig. 8.29 down).

Theorem 11.1 *A P-GPC code, with parameters r_j, $j = 1, 2, \ldots M$, increasing and all co-prime, is an s.s. time-invariant code in its generator matrix. The unique generator polynomial able to describe such $\underline{G}$ matrix can be written as*

Fig. 11.3 Encoding memory for the [35,14,6] P-GPC code described in Example 11.2

i	i	i	i	c_2
i	i	i	i	c_2
i	i	i	i	c_2
i	i	c_3	c_3	c_2
c_3	c_3	c_3	c_3	c_2
c_3	c_3	c_3	c_3	c_2
c_1	c_1	c_1	c_1	c_1

i information symbols

c_1 control symbols of the 1.th component code

c_2 control symbols of the 2.nd component code

c_3 control symbols of the 3.rd component code

$$g(x) = (x+1)\prod_{j=1}^{M} u_{r_j}(x). \tag{11.3}$$

Proof We will proceed with the case of $M = 3$, further generalizations are trivial. After Lemma 11.2, we have a length of the parent cyclic code which results in $N = r_1r_2r_3$. Expression (B.18) (see Appendix B) yields the factorization of a binomial, whose power exponent is the product of two co-prime numbers, a, b. It can be extended to the present case, in order to comprehend the situation of three co-prime numbers, r_j, $j = 1, 2, 3$. Such extension turns out to be

$$x^{abc} + 1 = (x+1)u_a(x)u_b(x)u_c(x)Q_{a,b,c}(x),$$

where $Q_{a,b,c}(x)$ is the co-factor of what precedes. It depends on the particular choice of the numbers a, b, c. The above expression allows one to write

$$x^{r_1r_2r_3} + 1 = (x+1)u_{r_1}(x)u_{r_2}(x)u_{r_3}(x)Q_{r_1,r_2,r_3}(x). \tag{11.4}$$

The degree of $g(x)$ is equal to the total number of control symbols, that is $r = r_1 + r_2 + r_3 - 2$. We have to search for a polynomial, with such degree, among the factors here present. The three types of parity check rows in the $\underline{H}$ matrix can be written in the following polynomial from

$$h_1(x) = u_{r_2r_3}(x^{r_1}), \quad h_2(x) = u_{r_1r_3}(x^{r_2}), \quad h_3(x) = u_{r_1r_2}(x^{r_3}). \tag{11.5}$$

In order to allow satisfaction of (5.8) with $N = r_1r_2r_3$, for all these parity check rows, the desired generator polynomial is then

$$g(x) = (x+1)u_{r_1}(x)u_{r_2}(x)u_{r_3}(x). \tag{11.6}$$

The case characterized by a greater value of M can be treated similarly, leading to the general expression (11.3). □

We can note that factor $(x + 1)$ could be joined with whichever of the remaining unitary polynomials, leading to a binomial. Nevertheless expression (11.3) is preferable, because it stresses the fact that all the component codes play an equal role. The unique difference is in their magnitude, which imposes the maximum length $n = r_1r_2$, for assuring row-column constraint satisfaction, in an LDPC model. Clearly, the same generator polynomial holds also for any (further) shortened version with respect to the parent LDPC code having length $n = r_1r_2$. This new code is constructed eliminating some columns in the original $\underline{H}$ matrix. In doing this, it is not necessary to eliminate entire identity sub-matrices at a time, but we can suppress single columns. The code rate scalability so obtained is at the level of single symbols (Baldi et al. 2008). Anyway the information symbols can be associated to the first positions in the code words. Their number, in general, is given by (11.1), which, in the particular case of the maximum length $n = r_1r_2$, becomes

$$k = r_1r_2 - \sum_{j=1}^{M} r_j + (M - 1). \tag{11.7}$$

Example 11.3 Let us take a P-GPC code, with $M = 3$, and parameters $r_1 = 5$, $r_2 = 6$, $r_3 = 7$. Its longest version is a [30,14,6] code. Eliminating the rightmost 6 columns in its $\underline{H}$ matrix, we obtain a [24,8,8] shortened code. In all the codes of this family we have the unique generator polynomial $g(x) = (x+1)u_5(x)u_6(x)u_7(x)$. It corresponds to the binary sequence 10101100000110101. Then the generator matrix of this [24,8,8] code can be written as

$$\underline{G} = \begin{bmatrix} 101011000001101010000000 \\ 010101100000110101000000 \\ \\ \ldots \\ \\ 000000010101100000110101 \end{bmatrix}.$$

In code shortening, the minimum distance is increased from 6 to 8. The word weight distribution of this code, up to $w = 10$, turns out to be: $A(0) = 1$, $A(8) = 29$, $A(10) = 44$. With $n = 25$, the minimum distance reduces to $d = 6$.

Corollary a *There is an equivalent form for the $\underline{H}$ matrix of a P-GPC code, which is in minimal form. It shows that this code can be assumed to be s.s. time-invariant also in its $\underline{H}$ matrix. The unique parity check polynomial h(x), whose reciprocal is expected to shift in the rows of this parity check matrix, turns out to be*

$$h(x) = Q_{r_1, r_2, \ldots r_M}(x). \tag{11.8}$$

Proof Taking again the only case of $M = 3$, the parity check polynomial we are considering is the co-factor of (11.6) with respect to the binomial (11.4). Generalization is straightforward. □

It could be interesting to find the factors which must be assumed, in order to obtain the parity check rows, of the type (11.5), in the Y row layers of the original $\underline{H}$ matrix, starting from (11.8). After long but straightforward calculations, it is possible to obtain

$$h_j(x) = h(x) \prod_{\substack{i=1 \\ i \neq j}}^{M} u_i(x), \quad j = 1,\ 2,\ \ldots M. \tag{11.9}$$

Unfortunately the minimal form of $\underline{H}$ characterized by (11.8) is useless, because it does not satisfy the row-column constraint. One could introduce the following variant, acting on the generator polynomial: replace $g(x)$ with $g'(x) = (x+1)^{M-1} g(x)$. The new code, whose generator polynomial is the product of all the involved binomials, is still cyclic, although on a longer cyclic period, but it is impossible, also here, to arrange a parity check matrix for it, able to satisfy the row-column constraint. For a code of this type a hard-decision decoding algorithm, similar to bit flipping, has been proposed (Cancellieri et al. 2004).

We can note how, in P-GPC codes, either $g(x)$ or $h(x)$ are symmetric polynomials. About the minimum distance of such codes (topic which will be investigated more in detail in the next section), it is possible to observe that, like in improper array codes with (8.12), we have an upper bound of the type

$$d \leq 2^M. \tag{11.10}$$

In fact, especially if the r_j's are not too small and k is remarkably greater than $(M - 1)$, a very critical code word is that associated to the information sequence $u(x) = (x+1)^{M-1}$. It leads to the code word

$$t(x) = \prod_{j=1}^{M} (x^{r_j} + 1), \tag{11.11}$$

whose maximum weight (if no cancellations occur) is 2^M.

Remark the analogy between (11.11) and (8.10), which holds for improper array codes. Comparison, except for the operations *Mult*[p] and *R*[position/p], yields the following correspondence relationships

$$M : Y, \quad r_1 : p, \quad r_j : p + \Delta_j, \quad j = 1, 2, \ldots M. \tag{11.12}$$

A P-GPC code admits $(M - 1)$ information symbols more, all multiplying the same unique generator polynomial. On the other hand, an improper array code admits $(Y - 1)$ additional anomalous rows in its $\underline{G}$ matrix, of the type $u_{2p}(x)$. They

Table 11.1 Comparison showing the main differences between the two families of LDPC codes considered

	Improper array codes	P-GPC codes
Row-column constraint	$n \leq p^2$	$n \leq r_1 r_2$
Time-invariance in $\underline{G}$	w.s. + anom. rows	s.s
Period length (QC or cyclic)	$N = p^2$	$N = r_1 r_2 r_3 \ldots r_M$
Equival. with tail-bit. convolutional	Yes	No

both have a similar origin, in the not full rank of $\underline{H}$. If r_1 or p are sufficiently high, the minimum distance can be evaluated from a code word model based on (11.11) in the one case, or ignoring anomalous rows in the other case.

Before concluding the present section, it is useful to remark that the family of codes called P-GPC codes and the family of improper array codes show some differences. They are summarized in Table 11.1, stressing the properties which are important for an interpretation as very good LDPC codes. For a QC improper array code, its equivalence with a tail-biting convolutional code can be efficiently exploited for obtaining a convolutional version of this code, after unwrapping its $\underline{H}$ matrix, as discussed in Sect. 8.11. This property does not hold for a P-GPC code, although there are other possible ways for constructing a convolutional code starting from it. The arrangement in a concatenated version, in order to gain some further length increment and consequent code rate increase, is possible for both, although with different solutions. These arrangements would have also the advantage of breaking the column weight regularity (more and more strongly as the code rate decreases). Finally the application of a proper masking (see Definition 10.7), acting on complete square sub-matrices, is able to reach the desired degree of irregularity, so favouring, at any code rate, decoding algorithm convergence.

In spite of the above differences, in order to treat such two families of codes according to a unitary approach, based on the analogy in their generator matrix description, we will introduce the following denomination, able to collect the main characteristics they show in common.

Definition 11.2 Improper array codes and P-GPC codes, including also their concatenated or masked arrangements, will be named in the following *binomial product generator codes* (or briefly *BPG codes*).

11.3 Evaluation of the Minimum Distance in BPG Block Codes

The presence of short cycles favours occurrence of low-weight code words. Combinations of the correspondent low-order girths, leading to possible stopping sets of variable nodes (see Definition 10.11), are responsible of minimum distance reduction, but their analysis is not easy. On the other hand, we can write equalities between linear combinations of parameters r_j, in P-GPC codes, or of parameters $(p + \Delta_j)$, in improper array codes, which allows minimum distance evaluation, at least for numbers M or Y of the component codes not too high (at most of the order of 5). This model will be carried out here for the only case of a P-GPC code. Some particular aspects of the results we are going to obtain can be transferred also to improper array codes, owing to the similitude between the generator matrices of such two families of LDPC codes. Nevertheless the presence of *Mult*[*p*] and *R*[power/*p*] operations, introduced in Theorem 8.1, will require some further considerations, which will be developed in subsequent sections.

First of all, it is possible to observe that, in the codes here considered, a sort of spectral thinning, of the type already seen in turbo codes (see Definition 9.3), can occur, especially when the r_j's are large numbers. The effective minimum distance d_{eff} can be considered of the order of the upper bound (11.10). This conclusion follows from the fact that a code word having this binomial product form

$$t(x) = \prod_{j=1}^{M} (x^{a_j r_j} + 1), \tag{11.13}$$

in which a_j can be whichever integer number, is always valid, provided that the length of the code word is long enough. Expression (11.13) can be considered as a sort of generalization of (11.11). Its weight is $w = 2^M$, if there are no coincidences in its resulting terms (and hence no cancellation pairs). The most part of such code words have full weight, when r_j parameters are large, so justifying code word crowding around $w = 2^M$. The information sequence able to originate (11.13) is

$$u(x) = (x + 1)^{M-1} \prod_{j=1}^{M} u_{a_j}(x^{r_j}).$$

When parameters r_j's are very small, also the true generator polynomial (11.3) may have a small weight. For instance, in Example 11.3, with $M = 3$, $r_1 = 5$, $r_2 = 6$, $r_3 = 7$, we have $w[g(x)] = 2^3 = 8$ as well. Nevertheless, with increasing the r_j's, $w[g(x)]$ becomes greater and greater. This makes the analysis based on the binomial product code words (11.13) more and more exhaustive.

Example 11.4 In Table 11.2 some P-GPC codes having $M = 3$ are analyzed, on the basis of the weight of their true generator polynomials (11.3). The terns at $n = r_1r_2$, length beyond which the row-column constraint is no longer satisfied, are also reported.

Definition 11.3 In a P-GPC code, let us consider linear combinations of r_j parameters, $j = 1, 2 \ldots M$, which can be written as $\sum_j a_j r_j$, where a_j are positive integer numbers. A linear combination of this type is called *zth-order linear*

Table 11.2 Weights of some the generator polynomials of some P-GPC codes with $M = 3$ (Example 11.4)

r_1	r_2	r_3	Tern at $n = r_1 r_2$	w[$g(x)$]
4	5	7	[20,6,6]	6
5	6	7	[30,14,6]	8
5	7	8	[35,17,6]	10
7	8	9	[56,34,6]	10
7	9	10	[63,39,6]	12

combination in the ensemble of the r_j's, when it involves z different r_j parameters. The case of a single parameter is permitted, leading to 1st-order linear combinations.

Definition 11.4 Under the same assumption as in Definition 11.3, an equality between two linear combinations, involving different sets of parameters, is called *Zth-order equality in the ensemble of the r_j's*, when the overall number of different parameters involved is Z. The equality order Z turns out to be the sum of the orders of the two linear combinations which are equated.

Lemma 11.3 *In a P-GPC code, characterized by M component codes, the existence of a 2nd-order equality is a sufficient condition for the occurrence of a 2^{M-1}-weight code word. This code word takes place over a length of twice the integer value of the equality plus the sum of all the r_j parameters not involved in the equality plus one.*

Proof Assuming

$$L = a_i r_i = a_k r_k, \tag{11.14}$$

a code word of the type (11.13), with all the parameters r_j different from r_i, r_k, with unitary coefficients, can be written as

$$t(x) = (x^{2L} + 1) \prod_{\substack{j=1 \\ j \neq i,k}}^{M} (x^{r_j} + 1). \tag{11.15}$$

It shows a weight which is halved with respect to the maximum 2^M fixed by (11.10), because of the cancellation of a pair of central terms in the product of the two binomials associated to r_i, r_k. Such code word is the shortest one of the type considered. From (11.15) its length is $2L$ plus the sum of the remaining r_j coefficients plus one. □

We can observe that the above low-weight code word is symmetric. There are also asymmetric low-weight code words. They are always present in the form of couples of reciprocal code words. As long as the code word length n increases, the number of low-weight code words increases too, and the role of asymmetric code words becomes more and more important. Typically they are originated by combinations of more than one equality of the type introduced in Definition 11.4, supported by asymmetric trinomials.

Coming back to Lemma 11.3, we can observe that equality (11.14) clearly implies the presence of a 4th-order girth. On the other hand, after assumption that all coefficients r_j are co-prime numbers, it is possible to conclude that such equality can be verified only for $a_i = r_k$, $a_k = r_i$, or multiple of this quantities. Therefore the contribution of this proposition is poor, since it regards lengths greater than the one over which the row-column constraint is satisfied. A more useful proposition is in what follows.

Lemma 11.4 *In a P-GPC code, characterized by M component codes, the existence of a 3rd-order equality is a sufficient condition for the occurrence of a* $3 \cdot 2^{M-2}$*-weight code word over a length of twice the integer value of the equality plus the sum of all the* r_j *parameters not involved in the equality plus one.*

Proof The 3rd-order equality here considered can be written

$$L = a_i r_i = a_k r_k + a_h r_h. \tag{11.16}$$

If the code is made by just three component codes, this equality gives rise to a 6-weight code word of the type

$$\begin{aligned} t(x) &= x^{a_i r_i + a_k r_k + a_h r_h} + x^{a_i r_i + a_h r_h} + x^{a_i r_i + a_k r_k} + x^{a_h r_h} + x^{a_k r_k} + 1 \\ &= x^{2L} + x^{a_i r_i + a_h r_h} + x^{a_i r_i + a_k r_k} + x^{a_h r_h} + x^{a_k r_k} + 1. \end{aligned} \tag{11.17}$$

In fact, starting from eight terms, two central terms form a coincident pair, owing to (11.16). Considering a more general situation, with $M > 3$, the remaining binomials can be taken all with unitary coefficients, in order to obtain the shortest code word possible, leading to

$$t(x) = \left(x^{2L} + x^{a_i r_i + a_h r_h} + x^{a_i r_i + a_k r_k} + x^{a_h r_h} + x^{a_k r_k} + 1\right) \prod_{\substack{j=1 \\ j \neq i,\, k,\, h}}^{M} \left(x^{r_j} + 1\right).$$

This code word is characterized by the weight $w[t(x)] = 6 \cdot 2^{M-3} = 3 \cdot 2^{M-2}$, and its length turns out to be $2L$ plus the sum of the remaining r_j coefficients plus one. □

The 6-weight polynomial considered in the above demonstration is symmetric, because we can write also $x^{a_i r_i + a_h r_h} = x^{2L - a_k r_k}$ and $x^{a_i r_i + a_k r_k} = x^{2L - a_h r_h}$, and the resulting code word $t(x)$ is symmetric too, after property *i* in Lemma B.13 of Appendix B. All the situations considered in Lemmas 11.3 and 11.4 are based on the assumption that there are no other equalities besides the one taken into account. If more than one equalities are contemporary present, they can combine, leading either to symmetric code words or to pairs of reciprocal asymmetric code words with lower and lower weights.

When an equality of the type (11.16) is valid with all unitary a_j coefficients, polynomial (11.11) itself shows a reduced weight. Precisely it is characterized by $w[t(x)] = 3 \cdot 2^{M-2}$ instead of 2^M, and this fact occurs after only $(M - 1)$ information symbols. Such choice is clearly a bad choice. The best choice would be, instead, to adopt a set of r_j parameters, able to produce a series of equalities with a_j coefficients as high as possible.

About equalities of order greater than 3, the following proposition generalizes Lemma 11.4, but it is to be taken with some cares, paying attention to the fact that the equality considered is assumed to be alone. It remains only a sufficient condition for the occurrence of a low-weight code word, but nothing is established regarding the fact that it can be the shortest one with the considered weight.

Lemma 11.5 *In a P-GPC code, characterized by M component codes, the existence of a Zth-order equality is a sufficient condition for the occurrence of a* $(2^{Z-1} - 1) \cdot 2^{M-Z+1}$*-weight code word over a length of twice the integer value of the equality plus the sum of all the* r_j *parameters not involved in the equality plus one.*

Proof It is possible to iterate the discussion developed in the proof of Lemma 11.4, still subdividing the r_j parameters into two sets: those involved in the equality and the remaining ones. As a check, we can verify that the statements of Lemmas 11.3 and 11.4 are already included in the above expression. □

Example 11.5 Taking the P-GPC code characterized by $M = 5$ and r_j parameters 17, 18, 25, 29, 31, the following 4th-order equality holds

$$65 = 2 \cdot 18 + 1 \cdot 29 = 2 \cdot 17 + 1 \cdot 31,$$

leading to a 28-weight code word, whose length is at least $n = 2 \cdot 65 + 25 + 1 = 156$. Remark that the sum of control symbols is 120, therefore the occurrence of this 28-weight code word can frustrate the performance of a code whose code rate is higher than about 1/4.

In order to map the predictions collected in Lemma 11.5 on actual situations, it is useful to summarize all the possibilities they disclose, for M up to 6. This result is shown in Table 11.3, where also the degree of the polynomial representing the shortest code word with the considered weight has been reported. In this representation, we have employed an average value $\langle r \rangle$ for the r_j's just as a reference value.

As a first conclusion, we can suggest a choice for the r_j's able to guarantee that the differences between any pair of them are the most separate possible, maintaining their values as small as possible. Clearly such two requirements are in contrast. Furthermore it is necessary to pay great attention to the possible combinations of equalities, whose effects are less predictable. It is useful, however, to reverse the point of view and introduce the following necessary condition.

Table 11.3 Possible situations originated by Lemma 11.5

M	Maximum weight	Order Z of the equality	Weight produced	Deg[$t(x)$]
3	8	3	6	$2L$
4	16	3	12	$2L + \langle r \rangle$
		4	14	$2L$
5	32	3	24	$2L + 2\langle r \rangle$
		4	28	$2L + \langle r \rangle$
		5	30	$2L$
6	64	3	48	$2L + 3\langle r \rangle$
		4	56	$2L + 2\langle r \rangle$
		5	60	$2L + \langle r \rangle$
		6	62	$2L$

Theorem 11.2 *In a P-GPC code, characterized by M component codes, a necessary condition for assuring the propertie* $d = 2^M$, *when the* r_j*'s are high enough to avoid effects directly due to* (11.3), *is that n is so small to prevent the existence of any equality of the type introduced in* Definition 11.4.

Proof When the low-weight code words are due to expressions of the type (11.13), if we exclude the presence of any possible equality, like those discussed in Lemma 11.5, for Z up to M, there are no conditions for reducing the minimum distance under the value $d = 2^M$. □

11.4 Effects of Combined Equalities and of Independent Equalities

That described so far is not the only mechanism after which low-weight code words can appear in a P-GPC code, especially when high values of M are considered. In fact, if more than one equality are contemporary present, they can combine or remain independent. In order to understand the consequences of this double possibility, it is useful to investigate firstly how several short cycles in $\underline{H}$, related to the same equality, can produce a low-weight code word. Then we will take into account the contemporary presence of more than one equality.

Lemma 11.6 *In a P-GPC code, characterized by* $M = 3$, *the presence of a 6-length short cycle in* $\underline{H}$ *implies a 6-weight code word, whose length is twice the horizontal displacement of the considered short cycle plus one.*

Proof Without loss of generality, we can take as a reference the situation depicted in Fig. 11.4, where $A_1 = a_1 r_1$, $A_2 = a_2 r_2$, $A_3 = a_3 r_3$. The 3rd-order equality supporting the considered short cycle is $A_1 + A_2 = A_3$. From Lemma 11.4, taking $L = A_3$, the statement is demonstrated. □

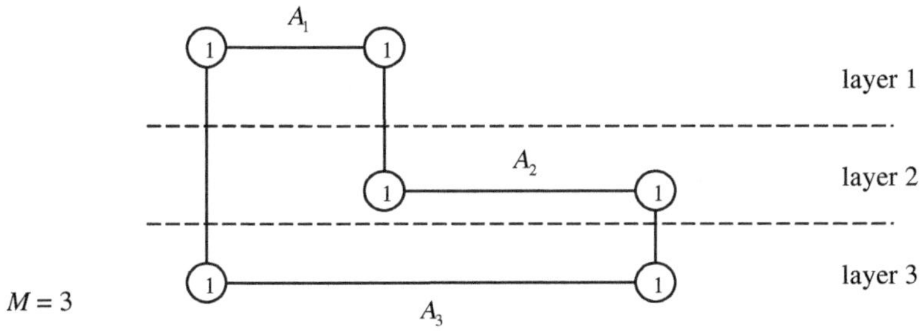

Fig. 11.4 A 6-length short cycle in the $\underline{H}$ matrix of a P-GPC code with $M = 3$

To be more precise, there are other short cycles, of length 6, 8, 12, supported by the same equality, involving the third 1-symbol in any column of the considered short cycle and other three columns. They, all together, contribute to form the 6-weight symmetric code word under study. Ex-OR sum of six syndromes is related to the superimposition of all the above mentioned short cycles. We can expect the existence of a further 6-length short cycle on one side of the short cycle depicted in Fig. 11.4. Then the central position will drop out, and a code word of the type $(0, A_1, A_2, 2A_3 - A_2, 2A_3 - A_1, 2A_3)$ finally will come out. This long description is needed for a unique equality acting alone.

Taking now a code with $M = 4$, we can construct a particularly unfortunate situation of the type reported in Fig. 11.5. It is originated by the contemporary presence of the following two 3rd-order equalities: $A_1 + A_2 = A_3$, $A_1 + A_4 = A_2$. It is easy to verify that a further 3rd-equality holds, $2A_2 = A_3 + A_4$. In this situation, an

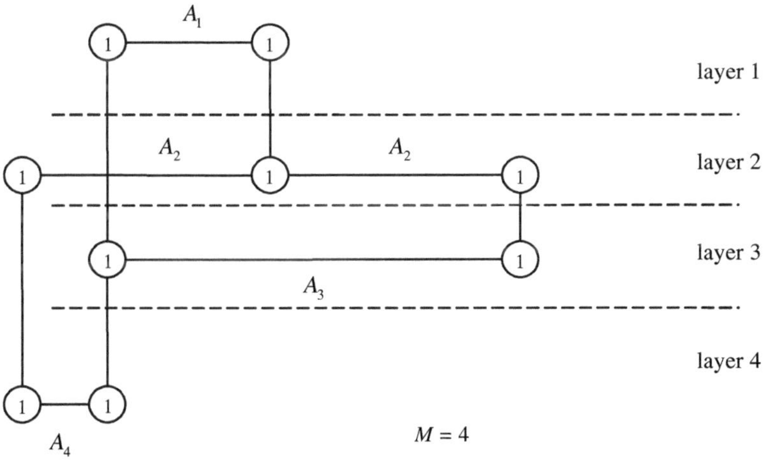

Fig. 11.5 Combination of two 6-length short cycles in the $\underline{H}$ matrix of a P-GPC codes with $M = 4$

8-weight code word occurs, representing a strong reduction with respect to the original maximum value $d = 16$.

As already anticipated, a different situation is that regarding pairs of reciprocal asymmetric low-weight code words. They are characterized by longer lengths with respect to their symmetric counterpart, but only with $M = 3$. Adopting $M \geq 3$, and in the contemporary presence of more than one equality, the reciprocal asymmetric code words can have also shorter lengths.

Example 11.6 Assuming $M = 3$, and $r_1 = b$, $r_2 = b + 1$, $r_3 = b + 2$, the 3rd-order equality $2r_2 = r_1 + r_3$ always holds, independently of b, which is a sort of scale factor. It produces a symmetric 6-weight code word whose polynomial expression is $(x^{4r_2} + x^{4r_2 - r_1} + x^{4r_2 - r_3} + x^{r_3} + x^{r_1} + 1)$, and is obtained simply multiplying (11.11) by $(x^{r_2} + 1)$. The above equality originates also a pair of asymmetric reciprocal 6-weight code words. Precisely they come out from a further multiplication of the above 6-weight symmetric code word by the asymmetric trinomial $(x^{2r_2} + x^{r_1} + 1)$ or by its reciprocal. The resulting code word is

$$t(x) = x^{2(r_1+r_2+r_3)} + x^{2r_1+2r_2+r_3} + x^{2(r_2+r_3)} + x^{2r_3} + x^{r_1} + 1, \qquad (11.18)$$

and also $t^*(x)$ will appear for the same value of n. In Table 11.4 the situation characterized by $b = 31$ is described. Approximate values of the code rate R at the appearance of any low-weight code word are reported too.

As a reference, we can consider that the value of $r_1 r_2$, up to which the decoding procedure is possible, is of the order of one thousand, corresponding to a code rate R of about 9/10. Nevertheless this code, at a code rate so high, is affected by an unacceptably large number of 6-weight code words. On the other hand, being M as small as 3, it is impossible to reach the value $d = 2^{M-1} = 4$ (like occurs in Fig. 11.5 with $M = 4$). The first row in the Table represents the 8-weight binomial product (11.11) which appears with $k = 3$. Finally w[$g(x)$] is of the order of some tenths, and hence completely not influent.

The pair of asymmetric code words obtained in Example 11.6 are about 3/2 longer than the unique symmetric code word from which they derive. This conclusion is valid also for more general situations, but always in the presence of just one equality. If there are other equalities, as already noted, particular combinations

Table 11.4 Some low-weight code words for the code characterized by $M = 3$, $r_1 = 31$, $r_2 = 32$, $r_3 = 33$ (Example 11.6)

t(x) from expression	Non-null power coefficients	w[$t(x)$]	Minimum length n	Approx. code rate R
(11.11)	(0, 31, 32, 33, 63, 64, 65, 96)	8	97	1/50
(11.17)	(0, 31, 33, 95, 97, 128)	6	129	1/3
(11.18)	(0, 31, 66, 130, 159, 192)	6	193	1/2
Rec. of (11.18)	(0, 33, 62, 126, 161, 192)	6	193	1/2

of them may give also asymmetric code words whose length is remarkably shorter. So, when more than one equality combine, there is the risk that a sort of collapse in the minimum distance occurs.

On the other hand, when we have independent equalities the minimum distance collapse can take place in an even stronger way. Assuming, for instance, $M = 6$, with two separate sets of three parameters each, related by as many independent equalities, the minimum distance d turns out to be reduced from 64 to 48 at the occurrence of the first and from 48 to 36 at the occurrence of the second (involving the remaining three parameters). In conclusion, a value of M greater than 5 is not advisable, except for low values of the code rate. Perhaps, in order to increase the minimum distance, it is preferable to adopt a direct product of two P-GPC codes, or a product with interleaver, as will be discussed in the next section.

As long as the code word length increases and new equalities appear, they tend to be combined in more and more involved fashion. It is possible to distinguish two types of equalities: those depending on the particular values of the r_j's and those depending only on the difference between any pair of them, so that a common scale factor (like b in Example 11.6) drops out.

Definition 11.5. The equalities depending on the particular values of the r_j's are called *special equalities*. Those independent of a common scale factor are called *systematic equalities*.

For instance, the third-order equality described in Example 11.5 is special. In order to develop the design of a good P-GPC code, it is convenient to fix firstly the difference between pairs of r_j's, in such a way as to prevent the occurrence of systematic equalities for values of n too small, and then choose the scale factor, which imposes also the number of control symbols and consequently the code rate R.

Example 11.7 Assuming $M = 4$, a good sequence of differences can be 3, 5, 2. In order to maintain the total number of control symbols not too high, we can adopt a scale factor of the order of some tenths. A very good trade-off is the following set of r_j parameters: 31, 34, 39, 41. There are two prime numbers and two composite numbers consisting of twice a prime number and three times a prime number respectively. With these parameters, the first equalities are as follows

$$109 = 2 \cdot 39 + 31 = 2 \cdot 34 + 41 \quad \text{(4th order, systematic)}, \quad n \geq 219,$$
$$155 = 5 \cdot 31 = 34 + 39 + 2 \cdot 41 \quad \text{(4th order, special)}, \quad n \geq 311.$$

Such two equalities produce the following 3rd-order special new equality

$$279 = 9 \cdot 31 = 4 \cdot 39 + 3 \cdot 41 \quad \text{(3rd order, special)}, \quad n \geq 558.$$

On the other hand, there are two further 3rd-order systematic equalities

$$272 = 8 \cdot 34 = 5 \cdot 31 + 3 \cdot 39 \quad \text{(3rd order, systematic)}, \quad n \geq 585.$$
$$273 = 7 \cdot 39 = 5 \cdot 41 + 2 \cdot 34 \quad \text{(3rd order, systematic)}, \quad n \geq 577.$$

The last equality can be combined with the first one, since both have in common the term $2 \cdot 34$. It is important to observe that the overall number of control symbols is 142. Therefore we can conclude that a minimum distance $d = 16$ is guaranteed up to a code rate R of about 1/3. Next it is possible to predict progressively the presence of 14-weight, 12-weight, and even lower-weight code words, in number larger and larger with increasing parameter n. Beyond a code rate of about 3/4, any prediction is complicated by the presence of combined equalities and asymmetric low-weight code words with length rather short.

11.5 Main Results on DC-GPC Codes and Direct Products of DC-GPC Codes

The long discussion about equalities, outlined in the previous section, is useful also for designing good DC-GPC codes. Direct concatenation is preferable with respect to the parallelepiped concatenation owing to the following two main aspects: it gives LDPC codes which can be very easily encoded, and the column weights in the $\underline{H}$ matrix are not uniform. The last property, especially at low code rates, favours the decoding algorithm. On the other hand, when we are looking for a code rate as high as possible, especially employing small r_j values, a certain advantage can be gained with a direct concatenation. About the last observation, from expression (11.1), taken for $n = r_1 r_2$, and from Lemma 5.42, we can fix the upper bound for the code rate R, in the two types of codes, as follows

$$\begin{aligned} R &\leq 1 - \frac{\sum_{j=1}^{M} r_j - (M-1)}{r_1 r_2}, \quad &\text{(P-GPC codes)}, \\ R &\leq 1 - \frac{\sum_{j=1}^{M} r_j}{r_1 r_2 + \sum_{j=2}^{M} r_j}, \quad &\text{(DC-GPC codes)}. \end{aligned} \tag{11.19}$$

Example 11.8 Assuming $M = 3$ and $r_1 = 11$, $r_2 = 13$, $r_3 = 14$, at most we obtain $R = 107/143 \cong 0.61$ with the parallelepiped concatenation and $R = 122/160 \cong 0.76$ with the direct concatenation. Minimum distance $d = 6$ is the same in both and coefficients $A(6)$ in the word weight distribution are comparable. Sometimes few 4-weight code words can appear, because of the last columns in the $\underline{H}$ matrix. This fact have negligible consequence on the performance, except for a possible BER floor.

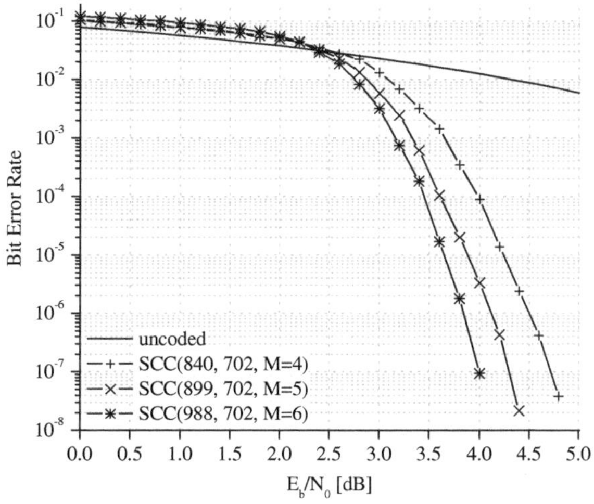

Fig. 11.6 Simulation results obtained on an AWGN channel for small-size LDPC codes constructed as the direct concatenation of GPC codes (from Baldi et al. (2010) copyright by authors)

An efficient encoder circuit, suitable for being applied also for the parallelepiped concatenation of GPC codes (provided that the last code word positions are eliminated), has been proposed in Baldi et al. (2009a). It is based on *M* pipelines and a pair of rotating switches, one in input and the other one in output.

In Fig. 11.6 some computer simulation results are reported, on the binary symbol error probability for an AWGN channel (Baldi et al. 2010). They regard small-size DC-GPC codes (SCC means serially concatenated codes). In this notation parameter E_b/N_o replaces *SNR*. The sequence of r_j parameters is 29, 31, 35, 43, 59, 89. All the three codes considered are characterized by $k = 702$ information symbols. The length and code rate vary, adopting $M = 4, 5, 6$ (and correspondently $R = 0.84, 0.78, 0.71$), with progressive addition of the last component code. This family of codes represents an example of rate-adaptive codes, in which the encoder circuit is the same. They all show an overlapping between the first two layers in their $\underline{H}$ matrix always remarkably smaller than the product $r_1r_2 = 899$, so reducing the occurrence of equalities of the type described in the previous section.

Medium-size codes (*n* of the order 5 or 10 thousands) and large-size codes (*n* of the order 20 thousands or more) have shown to be competitive with LDPC codes of comparable size, constructed following other design techniques, only at very high code rates (of about 0.9 or more). In this case, the choice of $M = 5$ is considered as an optimum trade-off.

Lemma 10.3 assures that a direct product code performed between codes separately satisfying the row-column constraint is able to satisfy the row-column constraint as well. A good solution for medium-size and large-size codes is in the possibility of arranging a pair of very small DC-GPC codes and performing the direct product between them. Letting M_a and M_b the numbers of GPC codes in each of the two component codes, the resulting column weights of the overall code varies from 1 to $(M_a + M_b)$, so assuring a relatively high degree of irregularity. In Fig. 11.7 two direct product codes so constructed (PC means product codes) are compared

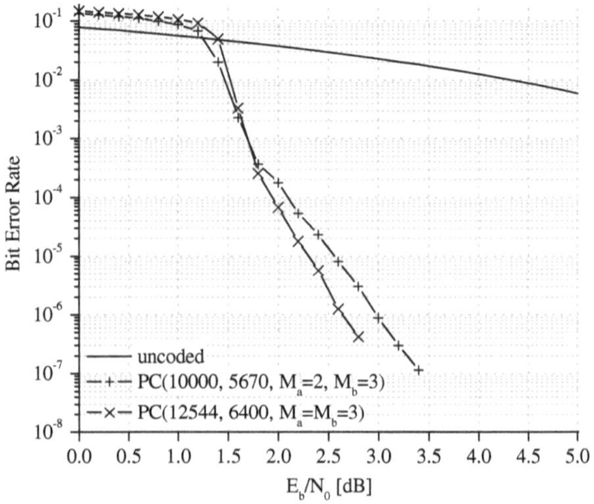

Fig. 11.7 Simulation results obtained on an AWGN channel for the direct product of two DC-GPC codes (from Baldi et al. (2010) copyright by authors)

(Baldi et al. 2010), the first with $M_a = 2$ and $M_b = 3$ (hence $d = 24$), the second with $M_a = 3$ and $M_b = 3$ (hence $d = 36$). They both exhibit a code rate R a little higher than 1/2. We can observe a change in the slope of the curves, which can be attributed to a remarkable effect of spectral thinning.

These codes show rather large block lengths. Smaller block lengths can be obtained selecting for instance $M_a = 4$, $M_b = 1$, that is a second component code consisting of the equivalent of a single parity check code. This choice, which has a very small cost in terms of code rate, is able to give minimum distance doubling with respect to what obtained from the first component code (typically 10 or 12).

An interleaver can also be inserted between the two component codes, with the purpose of modelling the error probability curve in such a way as to match the requirements of the considered channel (Baldi et al. 2012a), adopting the most suited trade-off between water fall and BER floor. It is possible to adopt an interleaver able to maintain the property of having a minimum distance equal to $d_a d_b$, being d_a and d_b those of the two component codes. This constraint can be satisfied employing the so called column interleaver introduced in Sect. 9.7. Such type of interleaver, in the present context, has the further advantage of maintaining the property of row-column constraint satisfaction. On the contrary, with a full interleaver, this property has to be imposed again. A PEG algorithm (see Sect. 10.7), after proper adaptation, has been employed in order to select the permutation matrices to be chosen as column interleavers, with the goal of maximizing the order of the girths. Looking at the typical $\underline{H}$ matrix of a direct product code, characterized by a lower part where large identity matrices are distributed according to disposition of 1-symbols in the parity check matrix of the second component code (see Sect. 5.16), after introduction of a column interleaver, such identity matrices will be substituted with suitable permutation matrices. In Fig. 11.8 the benefit in the introduction of an interleaver is shown, starting from a condition characterized by $M_a = M_b = 2$, with two types of column interleavers (CP means circulant permutations, RP more

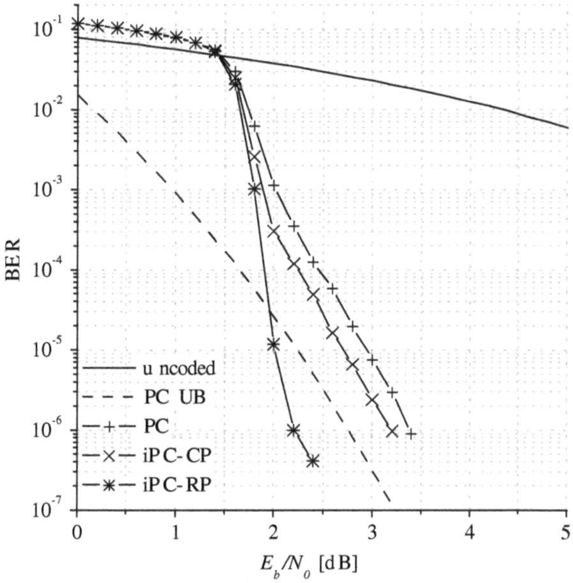

Fig. 11.8 Simulation results obtained on an AWGN channel for a 2/3-rate direct product code, without and with interleaver (reprinted by permission from (Baldi et al. 2012a))

general permutations). The total length is of ten thousand, and the overall code rate about 2/3. In a circulant permutation, the degrees of freedom are not enough for assuring the desired spectral thinning and hence improved coding gain. Finally a dashed curve shows qualitatively an asymptotic performance for the direct product code, based on the number of code words having minimum distance $d = 16$. UB means Union Bound (Proakis 1983).

The performance of such LDPC direct product codes or product codes with interleaver can be compared with those offered by a different class of concatenated codes, constructed as the direct product of various single parity check codes. For them a SISO iterative procedure allowing soft-decision decoding is available (Rankin and Gulliver 2001). We briefly mentioned this type of code in Sect. 5.17. The encoding memory shows here as many dimensions as the number of component codes. The maximum value for this number, in order to maintain the overall block length not too large and the code rate not too low, is five. In general, taking M single parity check component codes, all of the type $[n_a, n_a - 1, 2]$, the overall code length and number of information symbols turn out to be $n = n_a^M$, $k = (n_a - 1)^M$, so that the code rate becomes

$$R = \left(\frac{n_a - 1}{n_a}\right)^M.$$

Computer simulation results have shown decoded curves of the error probability very close to the prediction which can be obtained from the minimum distance d and the number $A(d)$ of code words having weight d, easily calculated, after Corollary a of Theorem 1.3, as

$$d = 2^M, \qquad A(d) = \binom{n_a}{2}^M.$$

Since parameter $A(d)$ is here very high, insertion of an interleaver offers the possibility of modifying the error probability curve: a pronounced water fall can be obtained at the expense of a more and more evident BER floor, especially if one accepts a smaller value for d, adopting a full interleaver.

In conclusion, the multi-dimensional direct product of single parity check codes typically gives superior performance, but its design parameters are less flexible. In fact the code rate and the number M of component codes, as well as the overall block length, are related by the above equations. On the contrary, the LDPC construction based on a pair of short DC-GPC codes appears less constrained in the code rate, overall size and number of GPC codes to be separated into the two orthogonal sets of component codes (so better controlling the expected performance).

One of the most interesting aspects of LDPC codes constructed as BPG codes (DC-GPC, P-GPC and even improper array codes), in order to reach very flexible performance, is in the adoption of a proper masking procedure. It gives the benefit of making the parity check matrix less regular, so advantaging decoding algorithm convergence. A whole identity matrix or permutation circulant matrix is here eliminated, in positions which can be chosen in more or less random way. The numbers of edges interesting control and variable nodes can be so accurately designed, and some benefits in the performance have been evidenced by computer simulations (Baldi et al. 2009b). In particular, the number of 6th-order girths is strongly reduced, leading to a consequent reduction in the number of $3 \cdot 2^{M-2}$-weight code words (see Lemma 11.4). Such code words are the main responsible of the form characterizing the error probability curve, especially for high values of the *SNR*. In fact the short cycles in the $\underline{H}$ matrix originating them may combine, leading to a minimum distance collapse. On the other hand, a too strong masking makes the code too weak, therefore the percentage of eliminated identity matrices (or permutation circulant matrices) is to be accurately chosen. With rather high code rates (of the order of 5/6 or more), the benefit deriving from masking in these codes practically vanishes, owing to the too large number of residual equality combinations. On the contrary, for small code rates (typically 1/2 or 2/3), it can give an improvement of up to 0.5 dB in the required *SNR* for error probability values ranging from 10^{-7} to 10^{-4}.

Other forms of product codes, not necessarily based on parity check codes as component codes, have been presented in Fengfan et al. (2007). Starting from a pair of irregular LDGM codes (see Sect. 10.8), taken in order to simplify the encoding procedure, medium-size and large-size codes have been obtained, with a code rate varying from 2/5 to 2/3. Interleaving is not necessary, owing to the irregular nature of the component codes. The decoding algorithm is able to reach convergence after a small number of iterations. In Esmaeli (2006) the parity check matrix with minimum 1-symbol density is studied still for LDPC product codes, direct or with interleaver. The cases of up to three orthogonal component codes have been analyzed, with particular attention to the possibility of improving the performance, owing to elimination of low-order girths. Such product codes can be interpreted also as LDPC block codes obtained from superimposition (see Sect. 10.5).

11.6 Improper Array LDPC Codes

Equalities of the type described in Sect. 11.4 for P-GPC codes can be written also for array codes, with the only care of the modification regarding mod(p) operations. We can note how (8.10) represents a sort of cyclic permutation of (11.11). In similar way, anomalous rows (8.11) become dominant in the determination of d only for very small values of p, exactly as the true generator polynomial (11.3) for very small values of r_1. The greater the p, the more adaptable the results of Lemma 11.5 also to the case of improper array codes, provided that we consider equalities including in them also the effects of mod(p) operations.

Before treating improper array codes in detail, let us give few results about proper array codes (see Definition 8.4). In Mittelholzer (2002), considering proper array codes over the full length of the QC code ($n = p^2$), the following upper bounds on the minimum distance has been calculated, assuming a number of component codes Y up to 6,

$$\begin{aligned} &d \leq 6, && (Y = 3), \\ &d \leq 12, && (Y = 4), \\ &d \leq 20, && (Y = 5), \\ &d \leq 32, && (Y = 6). \end{aligned} \tag{11.20}$$

The first of them is tight and also the number of the 6-weight code word has been determined in this case, giving $A(6) = p\binom{p}{3}$. The above results holds independently of the values assumed for parameter p.

Some particular cases, still for proper array codes on the QC code full length, have been treated in Yang and Helleseth (2003), with small values of p and $Y \leq 4$. These choices can be interesting for designing small LDPC codes, to be employed also in concatenated schemes as component codes. For instance, a lower bound $d \geq 10$, for $Y = 4$ can be reached provided that we assume $p > 7$ and prime. As already noted, adoption of p prime is always advisable, although, if its value is very high and the code under study is strongly shortened, it could be not absolutely necessary. Better results can be obtained with improper array codes (see Definition

8.5), for which an optimized design of Δ_j parameters is available, considering the value of p adopted and efficiently exploiting the possibility of code shortening.

Lemma 11.7 *For a proper or improper array code, also in SQC form, the minimum distance d is upper bounded by the more restrictive between such two conditions* $d \le 2^Y$, $d \le 2p$.

Proof The first upper bound is (8.12) and is imposed by the maximum weight which $g(x)$ can reach, when no cancellations of any pair of terms occur in its expression (8.10). The second upper bound is imposed by the presence in the $\underline{G}$ matrix of some anomalous rows, whose polynomial expression is given by (8.11). Both such bounds are to be considered, and the more restrictive becomes dominant. □

Clearly just for very small values of p and rather high values of Y, $d \le 2p$ is dominant. Nevertheless possible ex-OR sums of the two types of rows can limit the overall behaviour, contributing to produce values of d slightly smaller than 2^Y. On the QC code full length, an optimized choice of Δ_j parameters is analyzed in Esmaeli et al. (2011), considering stopping sets (see Definition 10.11) and their relationship with code words. The following design goals can be reached: $d = 10$ for $Y = 4$, $p = 7$, and $d = 12$ for $Y = 5$, $p = 11$. In comparison with the maximum possible values (respectively $2^4 = 16$ and $2^5 = 32$), the first one appears as an acceptable reduction, whereas the second one undoubtedly represents a too strong penalty. The occurrence of combined equalities (strictly related to stopping sets) makes such codes not recommended, taking into account also that the number Y of component codes equals the column weights of $\underline{H}$ and hence influences directly the decoding computational complexity.

An important overcoming of this situation is offered by adopting a large value for p and a strong code shortening. In Milenkovic et al. (2008) a very general treatment of SQC improper array codes is carried out, taking into account the length λ_g of the girths. The main results obtained for improper shortened array codes and even for generalized array codes are collected in the lower and upper bounds (10.2), which holds for $Y = 3$. As already noted, they are independent of p and of the code word length. We can observe how the lower bound depends on λ_g, whereas the upper bound is anyway coincident with $(Y + 1)!$ as already anticipated in Sect. 10.6 and on which we will return during the next section.

The advantage of an improper array code, with respect to a generalized array code, is in its uniform behaviour, which allows easy extrapolations, since the generator sub-matrix is w.s. time-invariant, except for the $(Y - 1)$ anomalous rows. In Baldi et al. (2012b), sequences of sets of values which is possible to take as the Δ_j's, having minimum entities, in order to reach bound (8.12) within a row in the $\underline{G}_o^{(p)}$ sub-matrix, have been studied. For instance, the following choices are possible

$$\begin{array}{llllll} Y = 3, & \Delta_2 = 1, & \Delta_3 = 2, & & & p \geq 3, \\ Y = 4, & \Delta_2 = 1, & \Delta_3 = 2, & \Delta_4 = 4, & & p \geq 7, \\ Y = 5, & \Delta_2 = 1, & \Delta_3 = 2, & \Delta_4 = 4, & \Delta_5 = 7, & p \geq 11. \end{array}$$

Such sets of values are progressively included in, one another, and can be interesting for a modular code construction, with the goal of easy adapting to different channel conditions. Nevertheless all these p values are too small for satisfying the other upper bound $d \leq 2p$. So, with the same Δ_j values, it is necessary to adopt $p \geq 5$, 11, 17, respectively for $Y = 3, 4, 5$, in order to have the possibility of really reaching the goal $d = 2^Y$. On the other hand, different code words in the same generator sub-matrix $\underline{G}_o^{(p)}$, after ex-OR sum, may give a new code word with lower weight.

Example 11.9 Let us consider the case of $Y = 3$, $S = 4$, with the above choice ($\Delta_2 = 1$, $\Delta_3 = 2$) with exactly $p = 5$, which guarantees that the anomalous rows, taken alone, do not produce the stronger constraint on the minimum distance, since $2p = 10$. It is important to remember that the $\underline{G}$ matrix of an SQC improper array code is characterized by $(S - Y)$ row layers, where $\underline{G}_o^{(p)}$ shifts with steps of p, over S column layers. In turn, $\underline{G}_o^{(p)}$ spans $(Y + 1)$ consecutive p-length periods. So, in the case considered, we have just one sub-matrix. On the other hand, the two anomalous code words produce all the $\binom{S}{2} = 6$ possible pairs of $u_p(x)$ polynomials, occupying as many pairs of p-length periods. The binary sequence associated to the generator polynomial characterized by the above Δ_j parameters turns out to be 10000.11100.01110.00010. It is possible to easily verify that the ex-OR sum of this code word with the anomalous row 00000.11111.11111.00000 yields a 6-weight code word, so reducing the minimum distance d from 8 to 6. Therefore it is necessary to take at least $p = 7$, even for a code whose code rate is as low as about 1/4, in order to have $d = 8$.

Furthermore, as S increases, different code words, belonging the more than one $\underline{G}_o^{(p)}$ sub-matrix, will contribute to give new possibilities for a reduction in the value of d. For this reason, p has to be further increased. The last solution is important also for having a true LDPC code. In this sense, the above discussion about the minimum values for p is useful only from a theoretic point of view. Values of p of the order of some tenths enable us to design good SQC improper array codes, with code rate from 1/2 to about 3/4. To this goal, we can adopt optimized choices for the Δ_j parameters, based on considerations similar to those developed in Sect. 11.3, 11.4 and 11.5, treating P-GPC codes and their r_j parameters.

11.7 Generalized Array LDPC Codes

In any type of array code (even generalized, according to Definition 8.6), there are always 12th-order girths (Milenkovic et al. 2008; Fossorier 2004). The short cycle in the $\underline{H}$ matrix correspondent to a girth of this type involves only three column layers. Such situation occurs starting from $S = 3$, $Y = 2$, and can be extended to higher and higher values of the couple (S,Y) always for $Y = S - 1$, involving progressively other 12th-order girths. The growth of the upper bound in the minimum distance for generalized array codes, according to a factorial law, already anticipated, is consequence of this property. It has been demonstrated in MacKay and Davey (1999), adopting the calculation of a matrix determinant (see Definition A.16) as a reference. In conclusion we obtain

$$d \leq (Y+1)!. \tag{11.21}$$

Let us focus our attention to the case characterized by $Y = 2$ and $S = 3$. The resulting 6-weight code word can be calculated in polynomial form. Nevertheless, as long as S increases, the code becomes time-varying in its $\underline{G}$ matrix, so different generator sub-matrices have to be taken into account. They all extend over three p-length periods. In general, like in improper array codes, any one of these time-varying generator sub-matrices spans $(Y + 1)$ consecutive p-length periods.

Example 11.10 Let us consider a generalized array code, with $Y = 2$, $S = 3$, and characterized by the following parity check matrix

$$\underline{H} = \begin{bmatrix} \underline{0}_p & \underline{0}_p & \underline{0}_p \\ \underline{0}_p & \underline{1}_p & \underline{3}_p \end{bmatrix},$$

where we have employed a symbolism for circulant representation like that adopted in Example 8.19. This $\underline{H}$ matrix resembles that of an improper array code (with all 0-phase circulants conventionally in the first column layer and in the first row layer). Only in the second row layer we can note a non-uniform phase sequence. The following generator matrix, coincident with a unique generator sub-matrix $\underline{G}_o^{(p)}$, can be easily associated to the above parity check matrix

$$\underline{G}_o^{(p)} = \left[(\underline{0}_p + \underline{-2}_p) \quad (\underline{1}_p + \underline{-2}_p) \quad (\underline{0}_p + \underline{1}_p) \right].$$

Example 11.11 Considering the generalized array code already treated in Example 11.10, if we add a fourth column layer, it is necessary to employ, for the column

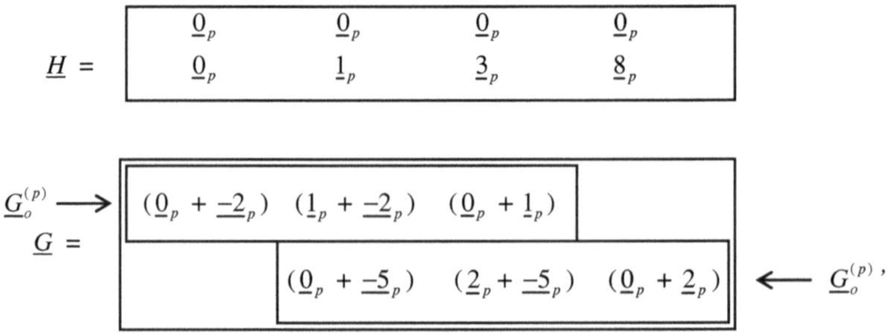

Fig. 11.9 Parity check and generator matrix for the generalized array code having $S = 4$ and $Y = 2$ (Example 11.11)

layers from the 2nd to the 4th, a different generator sub-matrix $\underline{G}_o^{(p)'}$. This situation is shown in Fig. 11.9. Remark that the property of having pairs of equal circulants either in $\underline{G}_o^{(p)}$ or in $\underline{G}_o^{(p)'}$, already noted for improper array codes, is due to the first layer conventionally made by only $\underline{0}_p$ permutation circulants.

In this simplified situation [which however can be demonstrated to be completely general (Fossorier 2004)], also a polynomial expression, in the form of a product, can be provided. In order to do this with easy possibility of further extensions, it is convenient to write the $\underline{H}$ matrix with negative phase symbols (representing complementary circulants with respect to p). Adopting this parity check matrix

$$\underline{H} = \begin{bmatrix} \underline{0} & \underline{0} & \underline{0} \\ \underline{-i_1} & \underline{-i_2} & \underline{-i_3} \end{bmatrix},$$

we simply obtain

$$g(x) = [(x^{2p}+1)x^{i_2} + [(x^p+1)(x^p x^{i_3} + x^{i_1})]\Big|_{\substack{Mult[p] \\ R[\text{position}/p]}} .$$

This polynomial, if there are no cancellations, has weight $w = 6 = 3!$ and is so constructed. One contribution is multiplied by $(x^{2p} + 1)$, choosing the exponent associated to the unique column excluded by this binomial. Similarly two contributions are multiplied by $(x^p + 1)$. Their exponents are associated to the columns excluded by this binomial, in the two possible positions it can occupy.

Extension to the case of $Y = 3$ is straightforward. Assuming here

$$\underline{H} = \begin{bmatrix} \underline{0} & \underline{0} & \underline{0} & \underline{0} \\ \underline{-i_1} & \underline{-i_2} & \underline{-i_3} & \underline{-i_4} \\ \underline{-j_1} & \underline{-j_2} & \underline{-j_3} & \underline{-j_4} \end{bmatrix},$$

it is possible to obtain this symbolic polynomial expression

$$\begin{aligned} g(x) = \{&(x^{3p}+1)(x^{i_3+j_2} + x^{i_2+j_3})\} \\ &+ (x^{2p}+1)[x^p(x^{i_4+j_2} + x^{i_2+j_4}) + (x^{i_1+j_3} + x^{i_3+j_1})] \\ &+ (x^p+1)[x^{2p}(x^{i_4+j_3} + x^{i_3+j_4}) + x^p(x^{i_4+j_1} + x^{i_1+j_4}) \\ &+ (x^{i_2+j_1} + x^{i_1+j_2})]\}\Big|_{\substack{Mult[p] \\ R[\text{position}/p]}} . \end{aligned}$$

We have put in evidence three terms: the first, with two contributions, is multiplied by $(x^{3p} + 1)$, the second, with four contributions, is multiplied by $(x^{2p} + 1)$, the third, with six contributions, is multiplied by $(x^p + 1)$. The overall weight, if there are no cancellations, is $w = 24 = 4!$. The above polynomial construction justifies the factorial law (11.21), after a mechanism which resembles a matrix determinant calculation. Remark that such an upper bound on the minimum distance d can be approached only for very low code rates, even when, in the above generator polynomial (which can be assumed for representing symbolically any row in the considered layer), any possible cancellation has been excluded. More precisely, the generator matrix here considered refers to a subcode of the array code. In fact a common factor can be always eliminated from all the binomials. Nevertheless the weight of the rows of this true $\underline{G}$ matrix will be remarkably higher.

The drawback of a code which is completely time-varying is in that its behaviour cannot be extrapolated. A situation which can be considered still time-varying, but based on a well predictable rule in the sequence of the phase increments characterizing the row layers, has been proposed in Tanner et al. (2004), with the purpose of transforming the original block code into a convolutional code. It exploits a geometric instead of an arithmetic phase progression, and has been already mentioned in Sect. 10.10 (Example 10.23). Before concluding this section it is possible to note that, in spite of the fact that bound (11.21) allows values of minimum distance d very high, such bound could be even overcome if we employ masking. Nevertheless, as already stressed, although the presence of 12th-order girth can be avoided in this case, elimination of 1-symbols in general makes the code weaker. So the possibility of reaching a value of d close to a bound so ambitious appears by itself a very good result.

11.8 Some Variants of Improper Array Codes

In Example 10.20 a concatenated form of an array code has been presented. A lower triangular matrix is to be assumed. A code so arranged becomes moderately irregular (with variable column weights) and an increase in the code rate can be gained, if parameter p is not too high. Lemma 11.1 fixes the maximum length of this directly concatenated array code. Correspondingly the maximum code rate turns out to be

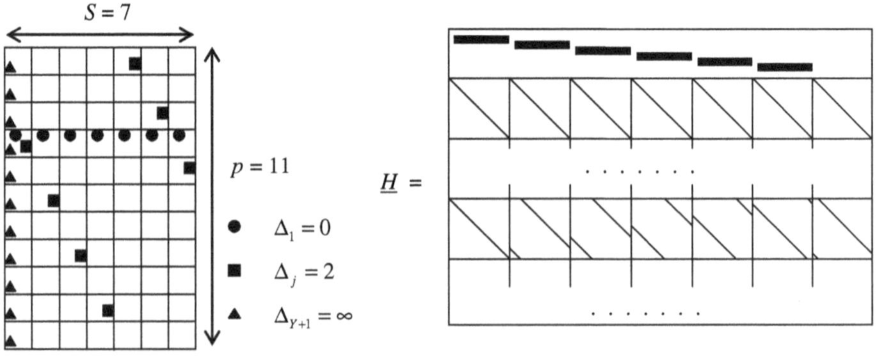

Fig. 11.10 Encoding memory and parity check matrix for an SQC array code with addition of a further orthogonal component code (Example 11.12)

$$R = \frac{p+1-Y}{p+1}, \qquad (11.22)$$

which is not approximate, in fact the parity check matrix, in this case, has full rank. When this condition holds, in the first $(Y-2)$ row layers of the $\underline{H}$ matrix, there is a number of subsequent permutation circulants smaller than p, in the last layer they are exactly $(p + 1)$.

Another possible variant of array codes consists in the addition of an orthogonal component code. Let us consider the rectangular encoding memory where an array code can be constructed. There is an horizontal trajectory, related to the component code in the first row layer of $\underline{H}$ (associated to $\Delta_1 = 0$) and then a number $(Y-1)$ of tilted trajectories, related to the remaining component codes. We can add now a vertical trajectory, and hence a further component code.

Example 11.12 In Fig. 11.10 on the left, a rectangular encoding memory, characterized by $p = 11$ and $S = 7$, is depicted. It refers to an SQC improper array code. In this encoding memory, after having written by columns, we encode on a horizontal trajectory (associated to $\Delta_1 = 0$) and then on $(Y-1)$ tilted trajectories (associated to the remaining Δ_j parameters), only one of which has been represented for simplicity ($\Delta_j = 2$). At this point, it is possible to add a vertical trajectory. It will be associated to a $(Y + 1)$th component code, which is different from the other ones. It results in a parity check code with word length p. The number of control symbols so added is S, equal to the number of the columns in the encoding memory (where as many vertical trajectories take place). Only on the full length of the parent QC code such number would be equal to that of all the other component codes, that is p. The new component code has been represented by a degenerate circulant phase shift $\Delta_{Y+1} = \infty$.

The overall parity check matrix of the code so constructed is reported on the right in Fig. 11.10. The original $\underline{H}$ matrix has been modified by addition of $S - 1 = 6$ rows made by all 1-symbols. They can be represented by as many unitary polynomials $u_p(x) = u_{11}(x)$. The Sth row would be linearly dependent on the other parity check rows, so it has been eliminated. Remark that here $(Y-1)$ rows in $\underline{H}$ are still linearly dependent on the other ones, nevertheless anomalous rows in the $\underline{G}$ matrix, according to the polynomial expression (8.11), are absent, as one can immediately verify (remember that p is an odd prime number). Addition of the further orthogonal component code finally does not introduce any 4-length short cycle, if no 4-length short cycles were present in the original $\underline{H}$ matrix.

Lemma 11.8 *Given an array code (proper or improper), with parameters Y, Δ_j, $j = 1, 2, \ldots Y$, (and S if shortened), the generator matrix of the code obtained from it after addition of an orthogonal component code can be obtained from the generator polynomial.*

$$g(x) = \left[(x+1)\prod_{j=1}^{Y}(x^{p+\Delta_j}+1)\right]\Big|_{\substack{Mult[p] \\ R[\text{position}/p]}} . \qquad (11.23)$$

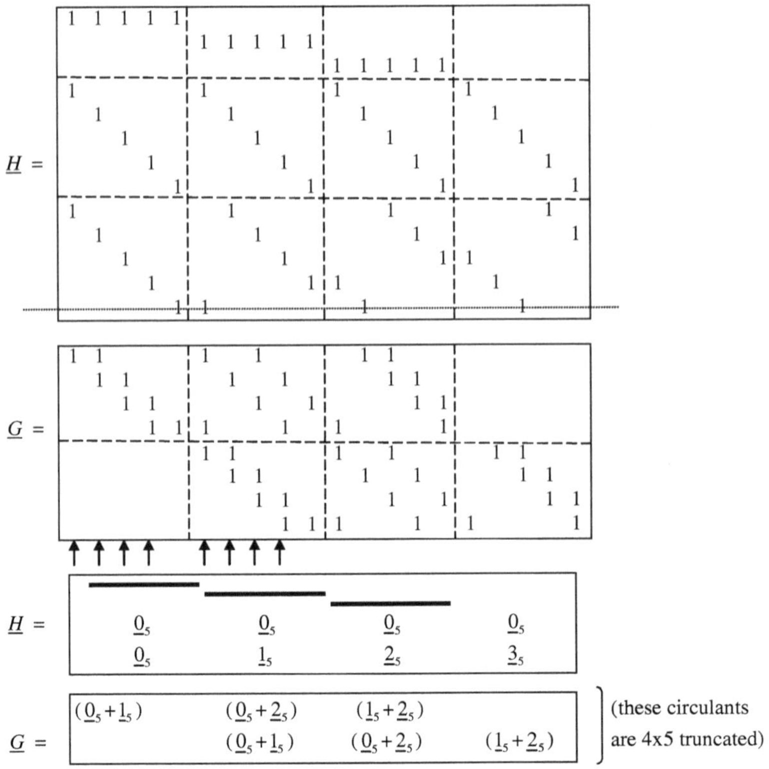

Fig. 11.11 Parity check matrix and generator matrix for the SQC array code plus orthogonal component code described in Example 11.13

There are no anomalous rows. Any row layer of the $\underline{G}$ *matrix is made by the same generator sub-matrix, shifting by p positions at a time. This generator sub-matrix collects* $(Y + 1)$ *subsequent truncated circulants, each made by* $(p - 1)$ *rows and p columns.*

Proof With respect to Theorem 8.1, we have only to add the constraint of an even number of 1-symbols per period. This constraint can be satisfied by taking two consecutive code words, each given by (8.10). So (11.23) is demonstrated. The pth position in any p-length period is occupied by one distributed control symbol. This entails the adoption of $(p - 1) \times p$ truncated circulants. Then the last Y periods are still made by all control symbols. Finally anomalous rows are not needed in this $\underline{G}$ matrix, because the total number k of information symbols, given by $(S - Y)(p - 1)$, is exactly equal to the number of positions devoted to information symbols in the first $(S - Y)$ periods. □

Example 11.13 Assuming the SQC code characterized by $p = 5$, $S = 4$, $Y = 2$, $\Delta_2 = 1$, the addition of the orthogonal component code gives rise to a [20,8,6] code, whose parity check matrix and generator matrix are shown in Fig. 11.11 (either in detailed form or in symbolic form). The generator polynomial (11.23), producing all the rows of $\underline{G}$, turns out to be

$$g(x) = [(x+1)\prod_{j=1}^{Y}(x^5+1)(x^{5+1}+1)]|_{\substack{Mult[5]\\ R[\text{position}/5]}} .$$

A rapid inspection of the generator matrix enables us to verify that there are 4 information symbols in each one of the first two 5-length periods. Their overall number is $k = 8$. Small arrows denote their positions, as usual.

With respect to traditional array codes, we have here a shorter length in the generator polynomial. In other terms, the $\underline{G}_o^{(p)}$ sub-matrix spans the same number $(Y+1)$ of p-length periods, but with one more component code. In order to develop the following final consideration, we can imagine the square encoding memory of the QC code plus the orthogonal additional component code. It is a square of size p with $(Y+1)$ different types of trajectories: one vertical, one horizontal, and the remaining ones properly tilted (according to the Δ_j's, $j = 2, 3, \ldots Y$). It is possible to exchange the roles between the two component codes associated to $\Delta_1 = 0$ and to $\Delta_{Y+1} = \infty$ simply adopting, for the remaining Δ_j parameters, the substitution $\Delta_j' = p - \Delta_j$.

The addition of an orthogonal component code is possible also in a P-GPC code. In this case, looking at the parity check matrix, the rows with unitary polynomials to be added must have length equal to the smallest parameter r_1. Otherwise the row-column constraint is not satisfied. It is easy to verify that the maximum number of parity check independent equations due to this orthogonal component code is $(r_2 - 1)$, and is reached when $n = r_1 r_2$. All these considerations confirm how improper array codes and P-GPC codes are very similar, provided that the comparison is carried out taking parameter p and parameter r_1 as the basic scale factor in the two cases. Nevertheless the following difference is immediately evident: in array codes, the trajectories jump at the boundaries, so filling all the available positions, instead, in P-GPC codes, they interrupt at the first boundary, leading to row weights lower and lower with increasing the entity of the considered r_j parameter. In spite of this difference, we can consider such two families of codes within the comprehensive class of BPG codes, owing to the very similar polynomial expression of their low-weight code words, say (8.10) and (11.11). In this sense, an expression similar to (11.23) could be assumed now, eliminating in it the operations *Mult*[p] and *R*[position/p], in order to obtain independent rows in the generator matrix of such variant of P-GPC codes. Nevertheless the structures of these rows are deformed, because of the need of having always two 1-symbols in any r_1-length period, and considering that such periods are not aligned with the remaining row layers of $\underline{H}$. So the resulting $\underline{G}$ matrix becomes time-varying.

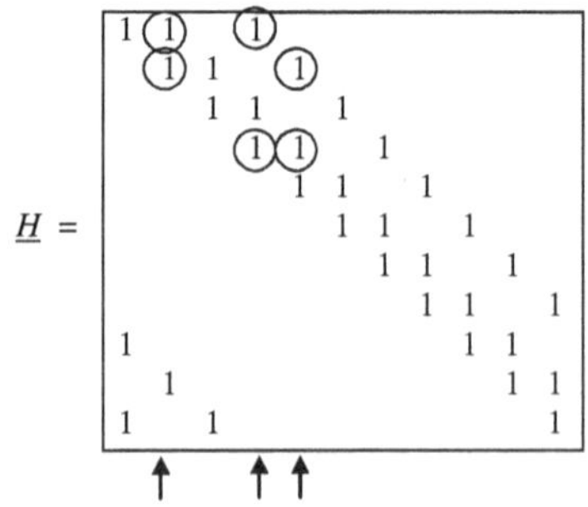

Fig. 11.12 Occurrence of a 6-length short cycle, always present within a single 3-weight circulant

11.9 Single-Layer Parity Check Matrix QC Codes

Before concluding the present chapter, it is interesting to treat a particular class of quasi-cyclic LDPC block codes. They cannot be considered strictly as BPG codes, since their generator matrix does not show generator polynomials expressed as the product of binomials. Nevertheless some procedures adopted in BPG codes will turn out to be useful also in this new description. Such codes can be interpreted as derived by proper planes in an affine geometry and have been already mentioned in Sect. 10.3. They are characterized by an $\underline{H}$ matrix made by a unique row layer, where circulants with weights greater than one take place. Such QC codes can be considered as the dual of LDPC codes having the generator matrix with only one $\underline{G}_o^{(p)}$ sub-matrix (Kamiya 2007). Their parity check matrix can be written as

$$\underline{H} = [\underline{H}_1|\underline{H}_2|\cdots|\underline{H}_a], \qquad (11.24)$$

where $\underline{H}_j$, $j = 1, 2, \ldots a$, are circulants. The weights of the columns depend on the weights of such circulants. In general they are not with uniform weight, therefore the code is not column-regular, although the weights of the rows remain uniform. Clearly the row-column constraint has to be imposed by a careful choice of the vertical separations between 1-symbols in the circulants, which must be all different.

The code rate turns out to be $R = (a - 1)/a$. Furthermore, looking at their $\underline{G}$ matrix, it is possible to obtain some versions of it having low weight, so leading to LDGM codes (see Sect. 10.8), which in turn can be easily encoded. To this purpose, in Baldi et al. (2011) some conditions on the circulant matrices $\underline{H}_j$ to be employed, in order to assure their inversion capability, have been presented, with an estimate of the encoding computational complexity of such codes.

Within a single circulant of weight not lower than 3, there is always a 6-length short cycle, as shown in Fig. 11.12, where the case characterized by $p = 11$, and phases 0, 1, 3, has been considered. It is easy to verify that the columns involved in this short cycle are to be chosen taking the reciprocal of the phase sequence in the three permutation circulants forming the considered 3-weight overall circulant, except for possible operations mod(p). The number of such 6-length short cycles is

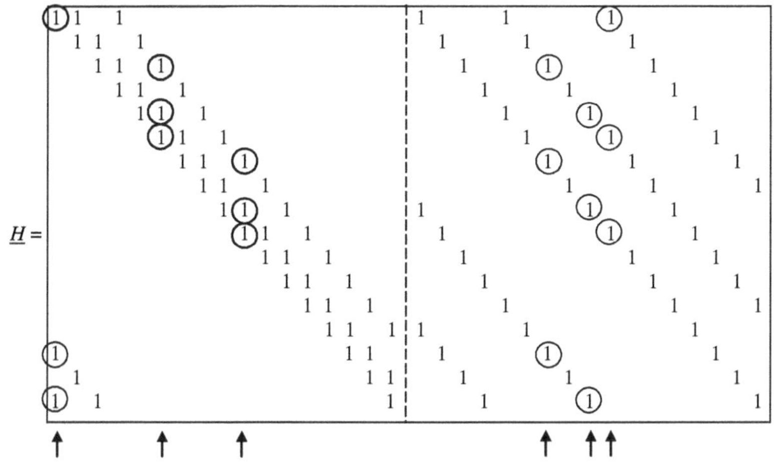

Fig. 11.13 Origin of a 6-weight code word in a pair of sub-matrices, taking $p = 17$, and phases 0, 1, 3 in the former sub-matrix, 0, 4, 9 in the latter (Example 11.14)

$p = 11$, on the whole. Considering now two circulants, of the same weight, not lower than 3, it is possible to find two 6-length short cycles (one in the former sub-matrix and the other in the latter) matching together. So they give rise to an overall 6-weight code word.

Example 11.14 Let us consider the case of $p = 17$, with phases 0, 1, 3 in the former circulant, 0, 4, 9 in the latter circulant, as shown in Fig. 11.13. We obtain a unique 1-symbol pattern having weight 6. It is possible to observe that here the columns involved in this code word, are the reciprocal of the two phase sequences, properly shifted and exchanged. The symbolic expressions for such parity check matrix and generator matrix respectively are

$$\underline{H} = [(\underline{0}_{17} + \underline{1}_{17} + \underline{3}_{17}) \qquad (\underline{0}_{17} + \underline{4}_{17} + \underline{9}_{17})],$$
$$\underline{G} = [(\underline{0}_{17} + \underline{5}_{17} + \underline{9}_{17}) \qquad (\underline{6}_{17} + \underline{8}_{17} + \underline{9}_{17})].$$

This LDPC (and also LDGM) code turns out to be described by the tern [34,17,6], and is characterized by $A(6) = 17$, since the above 1-symbol pattern is the unique, with weight $w = 6$. The general rule we can draw from Example 11.14, about the relationship between the $\underline{G}$ and the $\underline{H}$ matrix in a 1/2-rate code of this type, with both circulants having weight 3, is

$$\begin{aligned} \underline{H} &= \left[(\underline{0} + \underline{i}_1 + \underline{i}_2) \qquad (\underline{0} + \underline{j}_1 + \underline{j}_2)\right], \\ \underline{G} &= \left[(\underline{0} + \underline{j_2 - j_1} + \underline{j}_2) \qquad (\underline{j_2 - i_2} + \underline{j_2 - i_1} + \underline{j}_2)\right]. \end{aligned} \tag{11.25}$$

With higher values of the code rate, that is with $a > 2$, the circulant size p is to be enlarged, in order to allow availability of a greater number of phases, whose differences have to be all distinct. In this situation the number of code words having weight coincident with the minimum distance grows binomially with parameter a. As for the construction of the $\underline{G}$ matrix associated to a single-layer parity check matrix of the type (11.24), with $a > 2$, it is possible to proceed as in the case of a generalized array code. We can subdivide such parity check matrix into $(a - 1)$ overlapped sub-matrices, made by as many couples of subsequent circulants. For any of such sub-matrices, a proper generator sub-matrix can be calculated. It is made by two circulants as well, constructed adopting the rule discussed so far. For instance, in the case of $a = 3$, we have the following situation

$$\underline{H} = [\underline{H}_1|\underline{H}_2|\underline{H}_3], \quad \underline{G} = \left[\begin{array}{c|c|c} \multicolumn{2}{c|}{\underline{G}_{12}} & \\ \hline & \multicolumn{2}{|c}{\underline{G}_{23}} \end{array}\right],$$

where sub-matrix $\underline{G}_{12}$ depends only on the couple $\underline{H}_1, \underline{H}_2$, and sub-matrix $\underline{G}_{23}$ only on the couple $\underline{H}_2, \underline{H}_3$. Clearly this code is time-varying in its $\underline{G}$ matrix. Also in this case, the generator matrix so obtained can refer to a subcode of the considered code. Nevertheless, here too, the weight of the rows in the true $\underline{G}$ matrix will be remarkably higher.

When all the circulants have the same weight w_c, it is possible to demonstrate the following upper bound on the minimum distance (Kamiya 2007)

$$d \leq 2w_c, \quad (w_c \geq 3), \tag{11.26}$$

which can be justified from the considerations here developed.

If w_c is even, the overall $\underline{H}$ matrix has not full rank, and there is an anomalous code word represented by the unitary polynomial $u_p(x)$, as we can easily verify after application of the null ex-OR sum of clusters of syndromes. Anyway interesting results can be obtained only when p is at least of some tenths, so making negligible the presence of such anomalous code word.

The model treated so far can be generalized to the case of an irregular code, for which $\underline{H}$, in (11.24), is made by circulants $\underline{H}_j$ of different weights (Baldi et al. 2011). In this case, calling w_j, $j = 1, 2, \ldots a$, the weights of the circulants, the following upper bound on d exists

$$d \leq \min\{w_i + w_j\}, \quad i \neq j, \quad w_i, w_j \geq 3. \tag{11.27}$$

A particular situation is that characterized by $\min\{w_i + w_j\} = 3 + 2$, where (11.27) is still valid, leading to $d = 5$, but, owing also to the presence of a unitary code word $u_p(x)$ in the p-length period corresponding to the 2-weight sub-matrix $\underline{H}_j$, there are several possible 1-symbol patterns of weight 5. Their number and form depend also on the particular choice adopted for the phases of the circulants.

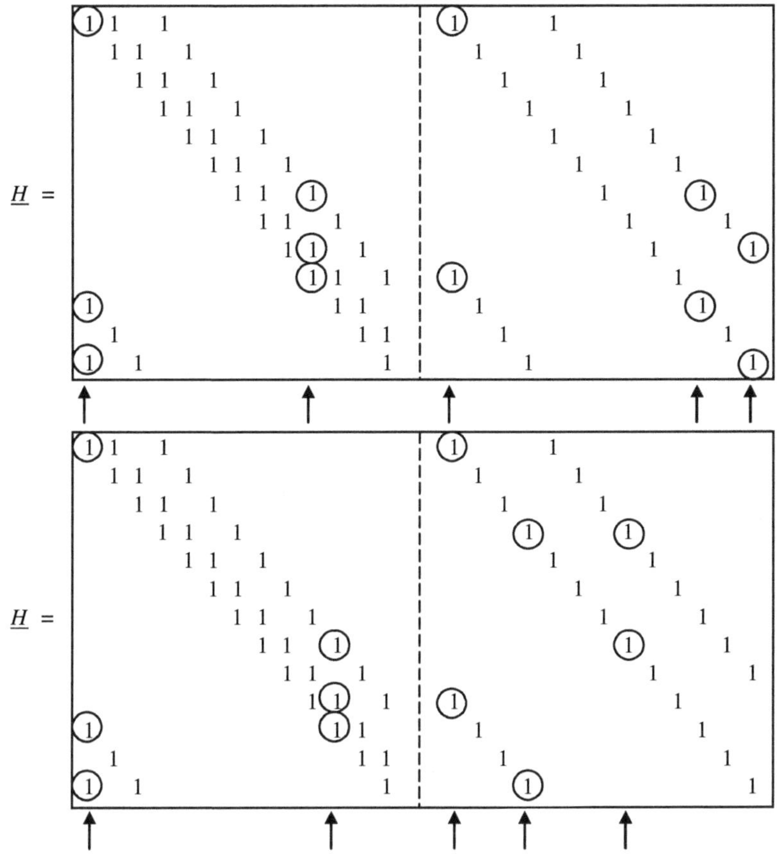

Fig. 11.14 Two 5-weight code words in a pair of sub-matrices of the same $\underline{H}$ matrix, taking $p = 13$, and phases 0, 1, 3 in the former sub-matrix, 0, 4, in the latter (Example 11.15)

Example 11.15 Let us consider the case of $p = 13$, with phases 0, 1, 3 in the former circulant, 0, 4 in the latter circulant, as shown in Fig. 11.14. We obtain two 1-symbol patterns having weight 5. It is possible to observe that the first is similar to the one described by (11.25), considering also mod(p) operations. The second derives from a special equality (like those introduced in Sect. 11.4 about P-GPC codes). Here also an anomalous row is present in the generator matrix, of the type $u_{13}(x)$, in correspondence with the latter 13-length period. So the tern representing this code becomes [26,14,5]. If we choose 0 and 5 as phases in the latter circulant, instead of 0 and 4, the second 1-symbol pattern of weight 5 does not occur, so that coefficient $A(5)$ is reduced from 26 to 13.

The $\underline{H}$ matrix of the codes described in the present section (regular and irregular) can be reordered, in its columns only, and unwrapped, leading to an interesting family of LDPC convolutional codes, which are s.s. time invariant in their parity check matrix (Baldi et al. 2012c). They will be treated more in detail in Chap. 12. About property (11.25), we can note, after the establishment of this correspondence, that it is related to Lemma 7.4, regarding 1/2-rate convolutional codes, for which the rows in $\underline{G}$ are reciprocal of the rows in $\underline{H}$.

References

Baldi M, Cancellieri G, Chiaraluce F, Carassai A (2008) Easily encodable LDPC codes based on polynomial codes. In: Proc Softcom 2008, Split, Dubrovnik (Croatia)

Baldi M, Cancellieri G, Carassai A, Chiaraluce G (2009a) LDPC codes based on serially concatenated multiple parity-check codes. IEEE Comm Lett 13:142–144

Baldi M, Cancellieri G, Chiaraluce F, De Amicis A (2009b) Regular and irregular multiple serially concatenated multiple-parity-check codes for wireless applications. J Comm Soft Syst 5: 140–148

Baldi M, Cancellieri G, Chiaraluce F (2010) Exploiting concatenation in the design of low-density parity-check codes. Int J Adv Telecom 3:28–38

Baldi M, Bambozzi F, Chiaraluce F (2011) On a family of circulant matrices for quasi-cyclic low-density generator matrix codes. IEEE Trans Inf Th 57:6052–6067

Baldi M, Cancellieri G, Chiaraluce F (2012a) Interleaved product LDPC codes. IEEE Trans Comm 60: 895–901

Baldi M, Bianchi M, Cancellieri G et al (2012b) On the generator matrix of array codes. Proc Softcom 2012, Split (Croatia)

Baldi M, Bianchi M, Cancellieri G, Chiaraluce F (2012c) Progressive differences convolutional low-density parity check codes, IEEE Comm Lett 16:1848–1851

Cancellieri G, Chiaraluce F, Garello G (2004) Theory and simulation of binomial product generator codes. Proc ISITA 2004, Parma (Italy): 970–975

Esmaeli M (2006) The minimal product parity check matrix and its application. Proc Int Conf Comm, Istanbul (Turkey): 1113–1118

Esmaeli M, Tadayon MH, Gulliver TA (2011) More on the stopping and minimum distances of array codes. IEEE Trans Comm 59:750–757

Fengfan Y, Ming Y, Lin L (2007) High-performance simple-encoding generator-based systematic irregular LDPC codes and resulted product codes. J Electron (China) 24:613–621

Fossorier MPC (2004) Quasi-cyclic low-density parity check codes from circulant permutation matrices. IEEE Trans Inf Th 50:1788–1793

Kamiya N (2007) High-rate quasi-cyclic low-density parity-check codes derived from finite affine planes. IEEE Trans Inf Th 53:1444–1459

MacKay DJC, Davey M (1999) Evaluation of Gallager codes for short block length and high rate applications. Proc IMA Workshop Codes Syst Graph Mod, Minneapolis (MN): 113–130

Milenkovic O, Kashyap N, Leyba D (2008) Shortened array codes of large girth. IEEE Trans Inf Th 52:3707–3722

Mittelholzer T (2002) Efficient encoding and minimum distance bound for Reed-Solomon-type array codes. Proc ISIT 2002, Lousanne (Switzerland): 282

Proakis JC (1983) Digital Communications. MacGraw Hill, Tokyo

Rankin DM, Gulliver TA (2001) Single parity check product codes. IEEE Trans Comm 49:1354–1362

Tanner RM, Sridhara D, Sridharan A et al (2004) LDPC block and convolutional codes based on circulant matrices. IEEE Trans Inf Th 50:2966–2984

Yang K, Helleseth T (2003) On the minimum distance of array codes as LDPC codes. IEEE Trans Inf Th 49:3268–3271

Chapter 12
LDPC Convolutional Codes

12.1 General Considerations

LDPC convolutional codes have been already introduced in several occasions, typically in relation with LDPC block codes or with known convolutional codes, not able to satisfy the row-column constraint. They can be summarized as follows.

In Examples 8.25, 8.26, 8.27 as many convolutional codes have been obtained by unwrapping the reordered form of the correspondent array codes. Precisely, Example 8.25 treats an SQC improper array code with code rate 2/5, and $p = 7$. Example 8.26 takes into account, instead, a QC proper array code still with code rate 2/5, but $p = 5$. Finally Example 8.27 analyzes the one-time shortened version of the previous QC code, leading to a 1/4-rate convolutional code. An analysis of these three situations, allows to conclude that the $\underline{H}_o^T$ sub-matrix, for a code so derived, shows the following main characteristics:

i. it exhibits row layers having Y rows each (the number of component codes in the original array block code),
ii. in the first row of the uppermost layer the unitary polynomial $u_{n_o}(x)$ is present, being $n_o = S$ in the SQC case and $n_o = p$ in the QC case,
iii. in any row, except the first one, of the remaining layers, there is exactly one 1-symbol in the QC case, there is at most one 1-symbol in the SQC case,
iv. the upper bound on the minimum distance remains as fixed by (8.12), that is $d \leq 2^Y$.

The most interesting situation is that associated to the choice of an improper SQC array code, since it offers the highest number of degrees of freedom to be optimized (Baldi et al. 2014). The goal of this design will be to reach a minimum distance as close as possible to the above upper bound. The discussion on such codes and on other possible codes which can be considered as variants of them, will be continued in Sect. 12.3.

G. Cancellieri, *Polynomial Theory of Error Correcting Codes*,
Signals and Communication Technology, DOI 10.1007/978-3-319-01727-3_12

In Example 10.19 an application of circulant expansion has been presented, starting from a well known 1/2-rate convolutional code with $m = 2$ not satisfying the row-column constraint. Proper choice of the permutation circulants, with size as small as $p = 3$, besides breaking all the 4-length short cycles, has allowed to reach an increase in the minimum distance from $d = 5$ to $d = 8$.

In Example 10.22 the cutting procedure suggested in Jimenez-Felstrom and Zigangirov (1999) has been applied, starting from a protograph-based LDPC block code, with code rate 1/3, so achieving a 4/11 final LDPC convolutional code, characterized by a uniform column weigh $w_c = 2$. Example 10.23 describes a convolutional code obtained from a variant of an improper array code, taken as original block code, with the possible introduction of some random parts (Tanner et al. 2004).

Another type of derivation is that presented in Sect. 5.15 for MHEC codes and in Sect. 6.10 for MHEQC codes. More precisely, Example 5.41 treats a 2/3-rate convolutional code in whose $\underline{H}$ matrix the row-column constraint is verified. The same is for the MHEC code described in Example 10.24 (Fig. 10.20). In Example 6.23, an MHEQC code solution is presented, starting from an array code with just two component codes, but whose principle can be easily generalized. This procedure will be analyzed more in detail in the next section.

Example 10.7 describes how, for a convolutional code whose $\underline{H}$ matrix in minimal form does not satisfy the row-column constraint, an equivalent $\underline{H}$ matrix, not in minimal form, can be adopted, which, instead, is able to satisfy such constraint. On the other hand, the 1/2-rate MLC code obtained from the [7,4,3] Hamming code with interleaved generator polynomial 11.01 satisfies, as is, the row-column constraint. Starting from the [15,7,5] BCH code whose parity check polynomial is 11.01.00.01, the 1/2-rate MHEC obtained from it also satisfies the row-column constraint as is. A [31,26,3] Hamming code, with generator polynomial 11.11.01, originates a 1/2-rate MLC code which is that mentioned above (Example 10.7). For it, taking the new parity check polynomial, not in minimal form, 10.01.00.11, the row-column constraint becomes satisfied. Another [31,26,3] Hamming code, associated to the generator polynomial 11.10.11, gives rise to the 1/2-rate MLC version treated in Example 10.19, already mentioned regarding a suitable use of circulant expansion.

In this chapter we will show analogies in the behaviour of some of the above listed situations, attempting to draw some hints for the possibility of directly designing good LDPC convolutional codes, without the need of any parent block code or known traditional convolutional code. This goal is expected to give the opportunity of fully exploiting the main features of convolutional codes, like the small number of low-weight code frames or a sliding window decoding procedure. About the last possibility, the length of the window clearly increases as the parity check constraint length m_h. More precisely, the following property holds.

Lemma 12.1 *In LDPC convolutional codes, with period length n_o, number of control symbols per period r_o, and parity check constraint length m_h, the number of rows in the syndrome former sub-matrix $\underline{H}_o^T$ is confined between $(m_h r_o + 1)$ and*

$(m_h + 1)r_o$. A sliding-window decoding algorithm can give good results only if the length of the window is at least $(m_h + 1)n_o$.

Proof The number of row layers over which $\underline{H}_o^T$ extends is $(m_h + 1)$. Each layer has r_o rows, but the last layer can have some final rows empty. The number of rows between the first and the last non-null rows in $\underline{H}_o^T$ is therefore confined between the extreme values $(m_h r_o + 1)$ and $(m_h + 1)r_o$. The sequence of the $\underline{H}_o^T$ sub-matrices forms a pseudo-diagonal in the semi-infinite overall $\underline{H}$ matrix. The horizontal width of this pseudo-diagonal then becomes $(m_h + 1)n_o$. The code frames having low weights, important for decoding, could be even longer. In this sense, the above length can be considered only as a lower bound. □

For an LDPC convolutional code characterized by low-weight code frames whose length is greater than $(m_h + 1)n_o$ (typically much less than twice this length), a decoding algorithm with proper window is to be adopted. This behaviour resembles that regarding Viterbi algorithm, employed in a generator trellis for traditional convolutional codes, with the only difference of taking here the parity check constraint length m_h, instead of the generator constraint length m.

After these considerations, the decoding computational complexity can be estimated to be proportional to the average value $\langle w_c \rangle$ of the column weights in the syndrome former sub-matrix and to the product $n_o m_h$. The goal of a good design is therefore to minimize both $\langle w_c \rangle$ (as in LDPC block codes) and the product $n_o m_h$. Instead, parameter r_o is to be considered independent, and is employed in the code design for fixing the code rate. On the other hand, it is useful to remember that, when n_o and r_o have a common factor, it is possible to improve the performance with respect to the case of an equal code rate obtained with n_o and r_o co-prime.

In Jimenez-Felstrom and Zigangirov (1999), after the cutting procedure applied to a small $\underline{H}$ module, the value of m_h is constrained to be equal to 1, but the period n_o is typically very long (it coincides with the length of the original module). Such codes are often classified as time-varying over this long period. On the contrary, in Tanner et al. (2004), after unwrapping applied on a variant of an improper array code, with possible introduction of some randomization, the constraint length m_h is very long. By contrast n_o is typically rather short. In order to maintain acceptable the width of the window in decoding, both such types of codes appear not suited for high code rates.

An interesting question regards the correct frame termination in the $\underline{H}$ matrix of an LDPC convolutional code when the quantity $(m_h + 1)r_o$ is rather high. In this case, inclusion in the $\underline{H}$ matrix of the whole last syndrome former sub-matrices (although some rows can be eliminated as useless) can produce a strong sacrifice in the true code rate. In Bates et al. (2006) this problem has been faced considering different possible solutions, especially from the point of view of circuit construction. The tail-biting termination is always possible, without any cost in code rate lowering, but, as already noted, this solution does not admit a sliding window decoding. Furthermore, if m_h is not remarkably smaller than the number of periods Z in the frame, there is the risk of a possible minimum distance reduction.

A particular frame termination of an LDPC convolutional code can be adopted, taking a solution suggested also for irregular block codes designed with PEG algorithm. It is based on a zigzag arrangement for the edges in the final part of the Tanner graph (Hu et al. 2001). This solution yields a lower triangular form for the right hand part of $\underline{H}$, but can be adopted only when a unitary vertical separation is not present in the parity check rows to be terminated. In fact it produces, in any column, a pair of adjacent 1-symbols, shifting by one position at a time (so forming a sort of staircase). Minimum distance preservation is typically guaranteed.

Example 12.1 Let us apply a zigzag frame termination to a 4/5-rate LDPC convolutional code with $w_c = 2$ and $m_h = 9$. Four only regular periods have been considered, leading to the overall $\underline{H}$ matrix shown in Fig. 12.1. One of the rows of the new matrix $\underline{H}$ is linearly dependent on the others, therefore the last row has been

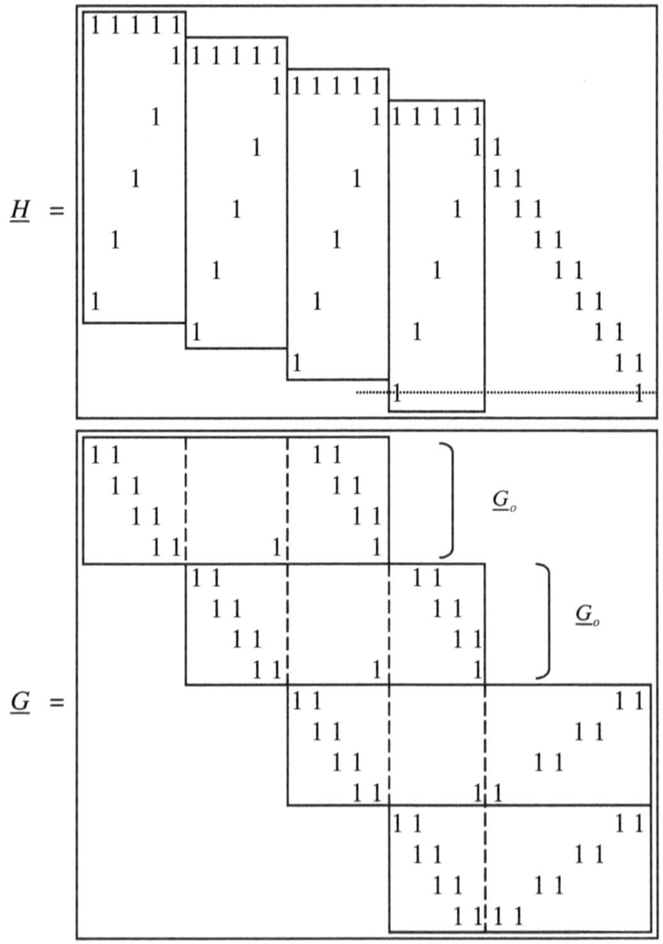

Fig. 12.1 Zigzag termination of a short frame for a 4/5-rate LDPC convolutional code

eliminated. Starting from a [20,8,4] initial code, we have obtained a [28,16,4] final code. The generator matrix of this code is reported down. It is possible to observe the 8 additional rows in the lowermost part of this matrix, which appear remarkably changed with respect to the first rows. Nevertheless the minimum distance remains $d = 4$. Two regular generator sub-matrices $\underline{G}_o$ have been evidenced, showing that $m = 2$. The third sub-matrix is modified by the presence of the zigzag frame termination, and then there are four rows else, which span twelve final positions on the whole. Besides the distributed control symbols, there are 9 final subsequent control symbols.

There are two possible uses of LDPC convolutional codes: as solitary codes, or as component codes in a composite encoding scheme. In the latter case, the component codes must be rather small. For instance it is possible to arrange the direct serial concatenation of a convolutional code with a block code, both LDPC, simply adding the parity check rows of the second to those of the first. In order to match the two parity check matrices, leading to an overall lower triangular form, the zigzag frame termination can be adopted in the convolutional code, but with great cares in order to avoid 4-length short cycles between the two. For traditional SISO decoding algorithms, the computational complexity in a code concatenation of this type is rather high. With LDPC codes, an overall single block code can be considered, to which a unique sum-product decoding algorithm is applied.

In the following two sections, we will treat LDPC convolutional codes derived from BPG block codes (see Definitions 11.2). This derivation can be performed after a sort of modified H-extension procedure, leading to codes with long periods n_o but rather short parity check constraint length m_h, or after a reordering and unwrapping procedure, typically with opposite characteristics.

12.2 Modified H-Extended Codes from BPG Block Codes

Let us consider a DP-GPC code (Definition 5.19). In order to simplify this analysis, we will assume the case of $M = 2$, that is two only layers in the $\underline{H}$ matrix, but the treatment can be easily extended also to higher values of M (P-GPC code according to Definition 11.1). As usual, parameters r_1 and r_2 are co-prime and satisfy the relationship $r_2 > r_1$. Taking two row layers of finite length, the one with $r_1 \times r_1$ identity matrices, and the other with $r_2 \times r_2$ identity matrices, it is possible to alternate them, as shown qualitatively in Fig. 12.2. Two syndrome former sub-matrices $\underline{H}_o^T$ are evidenced. This code remains w.s. time-invariant in $\underline{H}$, but the information and control symbol sequence in the super-period so obtained is of the type *iii…ccc…iii…ccc*. Such super-period must be shorter than the length over which a 4-length short cycle can occur, so satisfying the row-column constraint. This requirement introduces a limitation in the maximum code rate which can be reached, maintaining parameters r_j not too high.

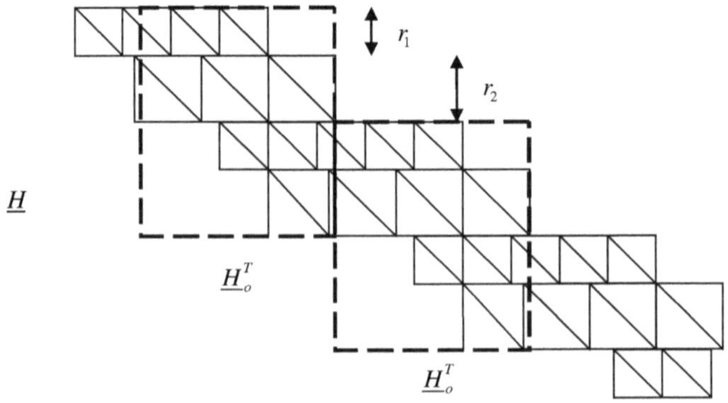

Fig. 12.2 Qualitative form of the parity check matrix in the w.s. time-invariant convolutional code obtained starting from a DP-GPC code

It is possible to realize also an overlap between two layers having equal r_j, leading to 3-weight columns in certain positions, so constructing a code which is column-irregular. The maximum overlap possible, for guaranteeing satisfaction of the row-column constraint, is characterized by $(r_1 + r_2)$ columns with weight 3. This is the case depicted in the figure. The procedure described so far cannot be classified as an MHEC code derivation of the convolutional code, because we have not a unique parity check polynomial. Nevertheless the same principle is adopted if one takes the whole DP-GPC code as a unitary element.

Example 12.2 Taking $r_1 = 3$ and $r_2 = 4$, the maximum asymptotic code rate which can be obtained, in order to satisfy the row-column constraint, is $R = 5/12$. The up-left corner of the parity check matrix, with two syndrome former sub-matrices $\underline{H}_o^T$ (evidenced with dashed borders) is shown in Fig. 12.3. The two alternated layers are organized in such a way as to assure that the first five position in the period are devoted to information symbols. Such code, which exhibits a convolutional period n_o exactly equal to $r_1 r_2 = 12$, is characterized by $d = 6$ (remark that the minimum distance of the original DP-GPC code was only 4). This good performance is due to the overlap. It is possible to verify that the number of subsequent 3×3 identity matrices in the one layer is $r_2 + 1 = 5$, and the number of subsequent 4×4 identity matrices in the other layer is $r_1 + 1 = 4$. The LDPC code so constructed is column-irregular with $\langle w_c \rangle \cong 2.6$. Clearly with larger values of r_1 and r_2, it is possible to obtain longer periods and higher code rates.

The above design procedure can be easily generalized, increasing the number M of the component codes (assuming a P-GPC code as the unitary element). Nevertheless the different sizes of the identity matrices to be employed makes somehow involved the principle to be followed. It is preferable to develop this model with reference to array codes, in an MHEQC code framework. In Example 6.23, a procedure of this type was already proposed, but simply placing side by side the replicas of the same parity check module (see Figs. 6.23 and 6.24). On the contrary, what is suggested here consists of continuing each layer along adjacent periods. As in the

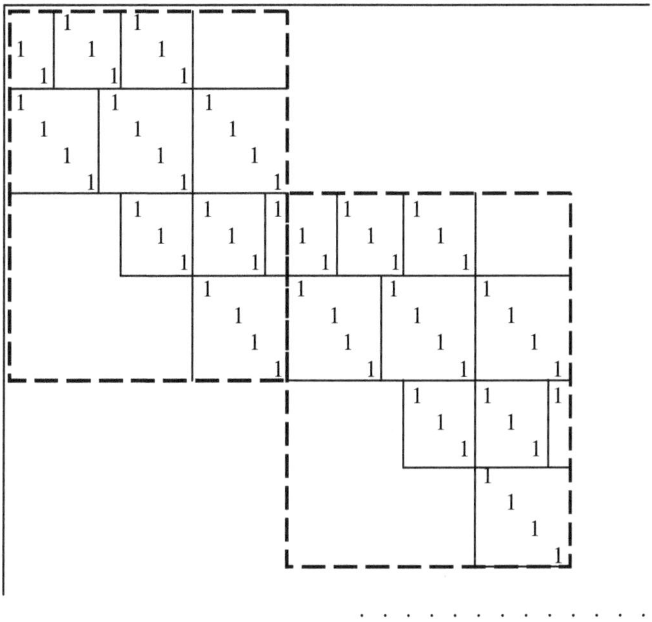

Fig. 12.3 Up-left corner of the semi-infinite $\underline{H}$ matrix characterizing the convolutional code with $r_1 = 3$ and $r_2 = 4$ described in Example 12.2

previous analysis, we will start from two only layers ($Y = 2$), characterized by circulants having size p, with parameters $\Delta_1 = 0$ and whichever $\Delta_2 < p$. Before introducing the possibility of overlapping for the same type of row layer (at most in only one column layer), it is convenient to analyze the non-overlapped situation, which is schematically shown in Fig. 12.4. A code rate $4p/6p$ has been considered up, and a code rate $6p/8p$ down. The initial part of a second syndrome former sub-matrix is evidenced in both cases on the right. Any symbol represents the phase of

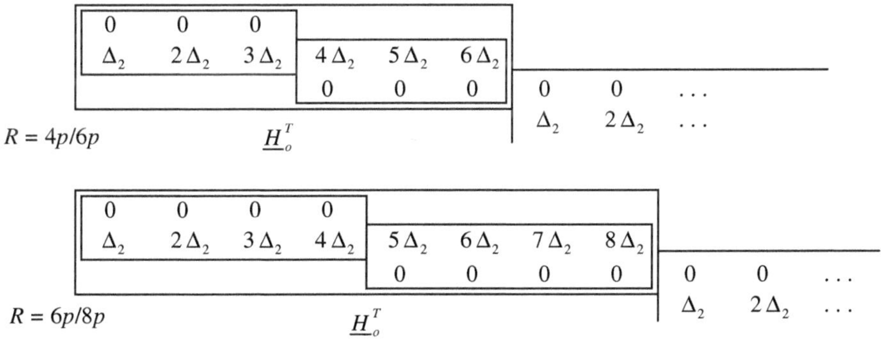

Fig. 12.4 Parity check matrices for the convolutional codes obtained from an array code with two layers without overlapping (code rate $4p/6p$ and $6p/8p$)

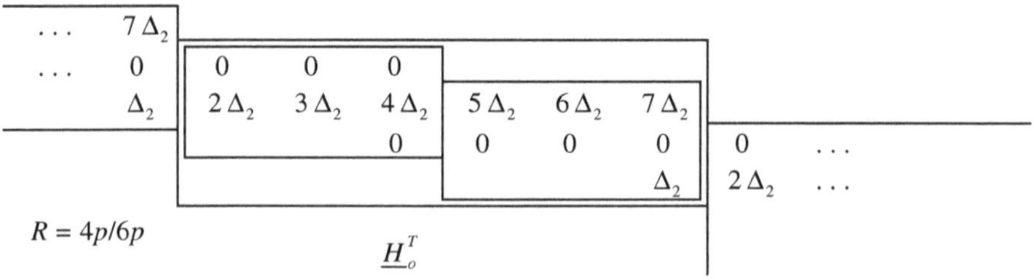

Fig. 12.5 Parity check matrix for the convolutional code obtained from an array code with two layers in the presence of overlapping (code rate $4p/6p$)

the correspondent permutation circulant, and the sequence is constructed mod(p) as usual.

Let us take into account now the possibility of overlapping. In Fig. 12.5 the new situation is depicted with reference to the only case of a code rate $4p/6p$. One further circulant is added continuing each circulant sequences in the considered row layer. So the asymptotic code rate does not change, but the column weight is no longer uniform (there are also 3-weight columns).

The code rate can be varied as one wishes, independently of the choice of an overlapped or a non-overlapped solution, and also of the number Y of component codes, provided that p is high enough. In Fig. 12.6 the case of $Y = 3$ is described, for a code rate $R = 3p/6p$ (with and without overlapping). A third parameter Δ_3 for the third component code has been adopted, whereas parameter p is necessarily greater than 7.

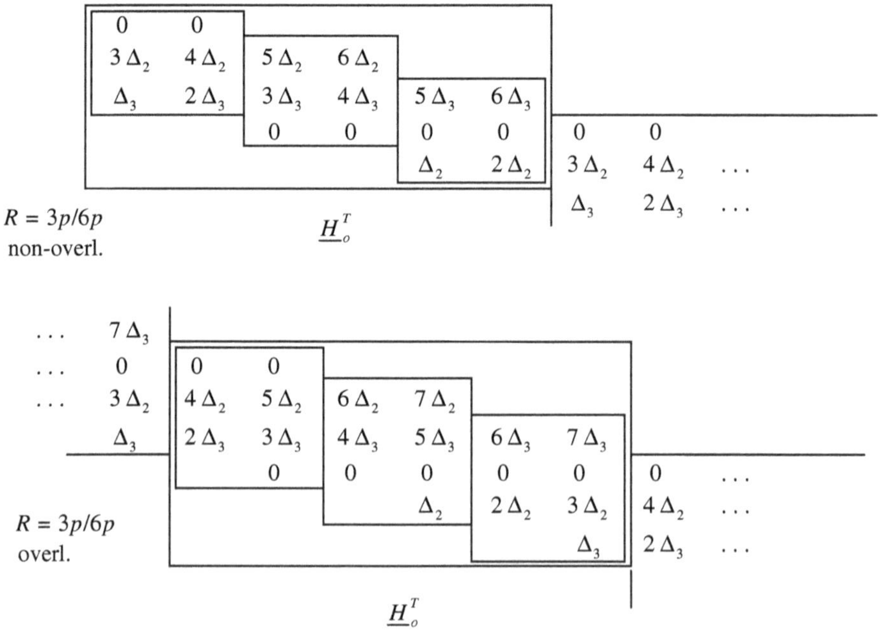

Fig. 12.6 Parity check matrices for the convolutional codes obtained from an array code with three layers ($R = 3p/6p$)

The length of convolutional code super-period is $n_o = p_o Yp$, where p_o is the number of columns forming each element in the overall $\underline{H}_o^T$ sub-matrix. The number of elements, in turn, equals Y. In Fig. 12.4 we have $p_o = 3$ up and $p_o = 4$ down, in Fig. 12.5 $p_o = 3$, finally in Fig. 12.6 $p_o = 2$. In any super-period, $r_o = Yp$ control symbols are inserted, like in the original parent QC code. The asymptotic code rate can then be calculated as follows

$$R = \frac{(p_o - 1)Yp}{p_o Yp} = \frac{p_o - 1}{p_o} \tag{12.1}$$

Example 12.3 Assuming $p = 5$, we consider the case of two component codes with $\Delta_2 = 1$, without overlapping and with overlapping, both with a code rate 10/20 (that is adopting $p_o = 2$). In Fig. 12.7 these two cases are described. The parity check matrices and also the correspondent generator matrices are shown in the usual symbolic form. From inspection of the two $\underline{G}$ matrices, it is possible to note how addition of overlap yields, also here, a noticeable increment in the weight of the

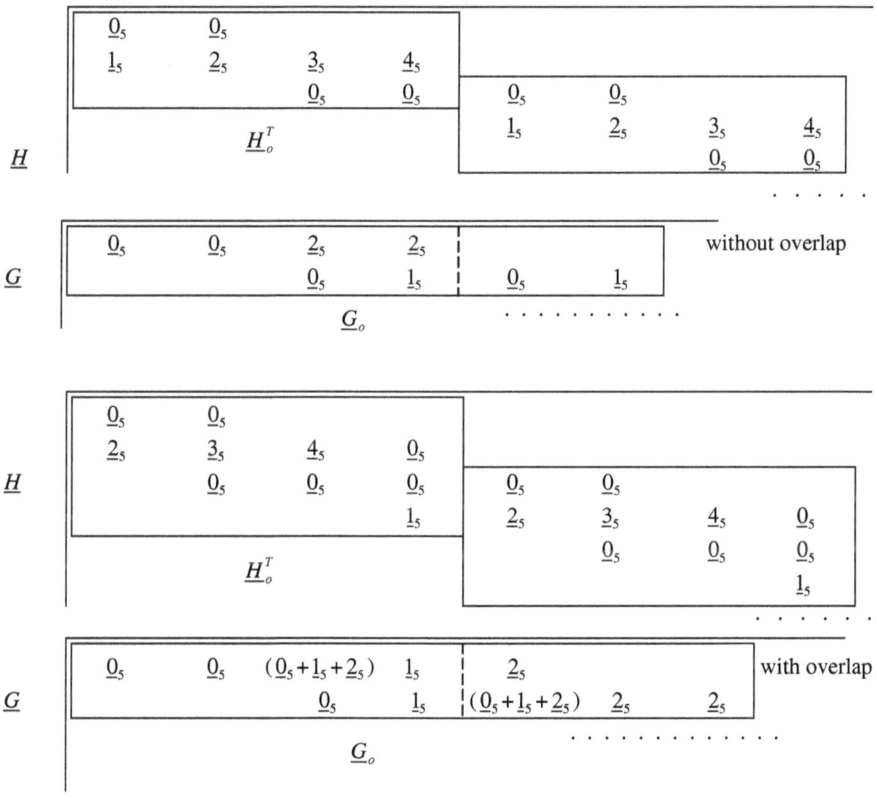

Fig. 12.7 Parity check and generator matrices of the convolutional codes having code rate 10/20 discussed in Example 12.3

generator polynomials. The minimum distance of the first code is $d = 4$, that of the second code, instead, becomes $d = 7$. In both cases we have very short constraint lengths ($m_h = m = 1$, with the second period not completely exploited).

With the choices adopted, the number of permutation circulants in a row layer of the overall $\underline{H}$ matrix is $L = Yp_o$ without overlapping and $L = Yp_o + 1$ with overlapping. This parameter is important for establishing the window length in a sliding window decoding procedure. It is useful also for giving a lower bound on p, once fixed a certain code rate, controlled by parameter p_o.

Lemma 12.2 *In the MHEQC construction of an LDPC convolutional code, starting from an improper array code, with Y component codes, assuming a certain value for parameter* p_o, *and hence for the code rate, in order to satisfy the row-column constraint, the size p of the circulants is so lower bounded*

$$\begin{aligned} &p \geq (Y-1)p_o, && \text{(non.overl.)}, \\ &p \geq (Y-1)p_o + 1, && \text{(overl.)} \end{aligned} \qquad (12.2)$$

Proof The number of subsequent permutation circulants on which two adjacent row layers in $\underline{H}$ overlap (which is the most critical situation) turns out to be $(L - p_o)$. A 4-length short cycle occurs when the number of subsequent overlapped circulants exceeds p, provided that p is prime, as usual. From the above two calculations for parameter L, the lower bounds (12.2) can be finally derived, in the two cases. □

It is however advisable to adopt a value for p remarkably higher than the lower bounds (12.2). In fact, if p is high enough, there is the possibility of optimizing the design parameters Δ_j, $j = 2, 3, \ldots, Y$, like in the improper array QC codes, from which such convolutional codes derive. Nevertheless a value of p of the order of some tenths can be considered as the maximum possible, because the length of the window in a sliding-window decoding procedure increases proportionally to p_o, Y and p itself, becoming rapidly unacceptable, especially for rather high code rates. As a final remark, we can note that no attention has been paid in fixing precise initial phases, or in obtaining lower triangular upper layers in the syndrome former sub-matrices. This simplification is due to the observation that only phase differences are important, also in the obtainment of the $\underline{G}$ matrix, for improper array codes and even for generalized array codes, as deduced in Chap. 11.

12.3 Convolutional LDPC Codes Obtained by Reordering and Unwrapping BPG Block Codes

A different approach, although still starting from BPG codes, will be developed here, following a design procedure based on position reordering in the code word. The model already adopted for array codes will be extended to P-GPC codes (see Definition 11.1) confirming how they can be interpreted as members of a unique family (that of BPG codes). Let us consider an improper array code with Y component codes, in its SQC or QC version. After reordering its $\underline{H}$ matrix according to Definition 8.7, we obtain the tail-biting convolutional form of such code. This double point of view can be explained looking at an encoding memory consisting of a rectangle of the type shown in Fig. 8.22. With reference to an SQC code, this rectangle has horizontal dimension S and vertical dimension p, whereas the trajectories expressing the parity check equations of the Y component codes are imposed by their Δ_j parameters.

The code rate of the tail-biting convolutional code obtained from reordering remains $R = (S - Y)/S$, as that characterizing the original SQC code. Coming back to the rectangular encoding memory reported in Fig. 8.22, it is possible to note that the third trajectory (that imposed by $\Delta_3 = 3$) is characterized by a jump with two points reported up, due to mod(p) operation. This picture regards a tern of parity check equations in the middle of the code frame. There are positions before this one, for which a lower number of points are subjected to the jump in the third trajectory. Clearly any point represents a 1-symbol in the $\underline{H}$ matrix.

In Sect. 8.11 unwrapping was made over the length of the parent QC code. This leads to a value for the parity check constraint length resulting in $m_h = p - 1$, also when we consider an asymptotic code rate progressively reduced (see Examples 8.26 and 8.27). In conclusion, the two parameters S and p act independently, the first fixing the code rate, and the second the parity check constraint length. As already observed, the higher the p with respect to $S = n_o$, the lower the average density of 1-symbols in the $\underline{H}_o^T$ sub-matrix.

Let us consider now the encoding memory shown in Fig. 12.8 on the left. It describes the only tilted trajectories associated to a second component code characterized by $\Delta_2 = 1$, with $S = 4$ and $p = 8$ (a value of p so small is employed just for simplifying the treatment). Such two parameters coincide with the dimensions of the rectangle. First of all, we can observe that shorter tilted trajectories are present in the initial and final part (in the uppermost and lowermost rows of the memory). The first trajectory touches the positions 3, 8, the second the positions 2, 7, 12, the

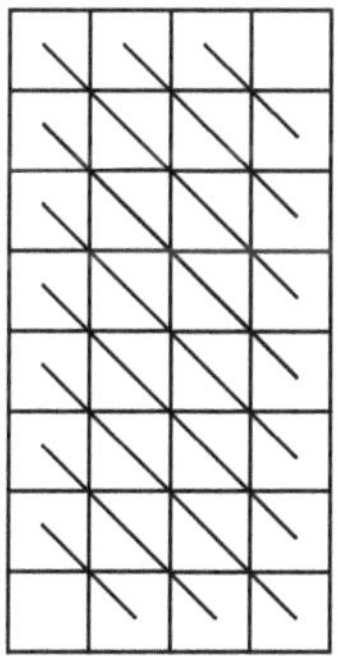

1	5	9	13	17	21	25	29
0 0 1 0	0 0 0 1						
0 1 0 0	0 0 1 0	0 0 0 1					
1 0 0 0	0 1 0 0	0 0 1 0	0 0 0 1				
	1 0 0 0	0 1 0 0	0 0 1 0	0 0 0 1			
		1 0 0 0	0 1 0 0	0 0 1 0	0 0 0 1		
			1 0 0 0	0 1 0 0	0 0 1 0	0 0 0 1	
				1 0 0 0	0 1 0 0	0 0 1 0	0 0 0 1
					1 0 0 0	0 1 0 0	0 0 1 0
						1 0 0 0	0 1 0 0

Fig. 12.8 Encoding memory, trajectories and correspondent rows in the parity check matrix due to the second component code, for an improper SQC array code with $S = 4$, $p = 8, \Delta_2 = 1$

third the positions 1, 6, 11, 16, and so on. Finally the 7th trajectory touches the positions 17, 22, 27, 32, the 8th the positions 21, 26, 31, the 9th the positions 25, 30. Positions 4 and 29 are not protected by this component code (the first component code, characterized by $\Delta_1 = 0$, will provide such protections). On the right in Fig. 12.8 the rows of the parity check matrix associated to the considered component code are represented. They will be properly interleaved with the rows of the other component codes.

Assuming a value of p intentionally much higher than the maximum phase increment Δ_Y, we can imagine that all the trajectories do not experience any jump. The unwrapping procedure will be performed after satisfaction of this condition. If one represents in polynomial form the rows shifting in the $\underline{H}$ matrix of the convolutional code so obtained, in the general case of Y component codes, the following expressions come out

$$h_j(x) = u_s(x^{\Delta_j S+1}), \quad j = 1, 2, \ldots Y \tag{12.3}$$

They are the unwrapped form of expressions (8.13), in fact coincide with them, except for mod $(x^{pS} + 1)$ operation, confirming our construction (also for the first component code, characterized by $\Delta_1 = 0$).

On the other hand, the parity check rows in the M layers of the $\underline{H}$ matrix in a P-GPC code can be written as $u_{w_j}(x^{r_j})$, $j = 1, 2\ldots M$ (see Sect. 11.1). Remembering the transformation occurred from (8.9) and (8.13), which describes the effect of reordering, we can conclude that expressions (12.3) represent the transformation of the rows $u_{w_j}(x^{r_j})$, once adopted the following correspondences: $w_j = S$, $r_j = p + \Delta_j$, $M = Y$. This enables us to interpret the convolutional code described by (12.3) as the reordered version of a P-GPC block code. Remark that condition $w_j = S$, with $j > 1$, holds for a high number of central parity check rows. Some anomalous unitary polynomials with lower weights are present in the initial and final part of the frame (whose true length is very long but finite), as usual in convolutional codes.

Directly unwrapping (8.14), or by imposing the null ex-OR sum of clusters of syndromes in the $\underline{H}$ matrix, the expression of the unique interleaved generator polynomial which can be coupled with (12.3) turns out to be

$$g(x) = \prod_{j=1}^{Y} (x^{\Delta_j S+1} + 1) \tag{12.4}$$

This expression appears as derived from (11.11) and not from (11.3), as a consequence of the elimination of the $(Y - 1)$ linearly dependent rows in the original $\underline{H}$ matrix. Remark that parameter $p = r_1$ is no longer present in (12.3) and (12.4), in fact it had the role of allowing the development of this model, but tending to an infinitely large value. In the following, we will treat the convolutional versions of P-GPC codes, so obtained, although some symbols will be still taken from the improper array code model.

Lemma 12.3 *Given the convolutional version of a P-GPC code, with M component codes, period length n_o = S, number of control symbols per period r_o = M, parameters Δ_j, j = 2, … M, all different and strictly increasing (Δ_1 = 0 as usual), the generator constraint length and parity check constraint length respectively result in*

$$m = \sum_{j=2}^{M} \Delta_j \tag{12.5}$$

$$m_h = (S-1)\Delta_M \tag{12.6}$$

Proof Equality (12.5) follows from (12.4), considering that $n_o = S$. Equality (12.6) can be easily derived from direct inspection of the encoding memory, noting that Δ_M measures the steeper trajectory slope. □

Remark the analogy with Corollary a of Theorem 8.2. More precisely, comparison between (8.16) and (12.5) shows that the former implies the latter when the sum of the Δ_j's is not greater than $(p-1)$, as expected. Comparison between (8.17) and (12.6) yields a similar conclusion.

Example 12.4 Taking $Y = M = 3$, $S = 5$, $\Delta_2 = 1$, $\Delta_3 = 3$, exactly as in Example 8.25, the following expressions can be adopted for the convolutional version of this P-GPC code

$$h_1(x) = u_5(x), \quad h_2(x) = u_5(x^6), \quad h_3(x) = u_5(x^{16}),$$
$$g(x) = (x+1)(x^6+1)(x^{16}+1)$$

The last interleaved generator polynomial is present in two rows any five positions. In Fig. 12.9 the up-left corner of the $\underline{H}$ matrix characterizing this 2/5-rate convolutional code is depicted. Since the $\underline{H}_o^T$ sub-matrix exhibits a very high number of rows, some lower rows have not been represented. The remaining 1-symbols of the third component code will be placed continuing the dashed traces with steeper slope (imposed by $\Delta_3 = 3$). The correspondent generator sub-matrix $\underline{G}_o$ is reported down.

It is interesting to note that, in situations of the type treated in Example 12.4, after a proper reordering of columns and rows in the $\underline{H}$ matrix, it is however necessary to perform an unwrapping procedure, which turns out to be similar to the cutting method employed in Jimenez-Felstrom and Zigangirov (1999) (see Example 10.22). In order to better explain this observation, it is useful to consider a simpler situation, with only two component codes.

Example 12.5 Taking a P-GPC code with $M = 2$ component codes, characterized by $r_1 = 4$ and $r_2 = 5$, hence $\Delta_2 = 1$, we consider its maximum length, corresponding to the tern [20,12,4]. Column reordering is made taking one column any four. So we obtain the new $\underline{H}'$ matrix shown up in Fig. 12.10. Since in the array code model, we

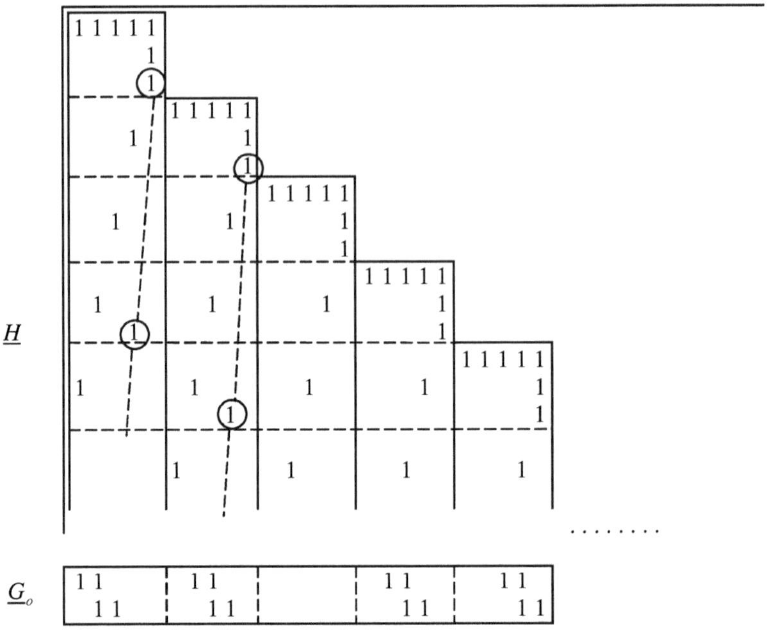

Fig. 12.9 Parity check matrix and generator sub-matrix of the 2/5-rate LDPC convolutional code described in Example 12.4

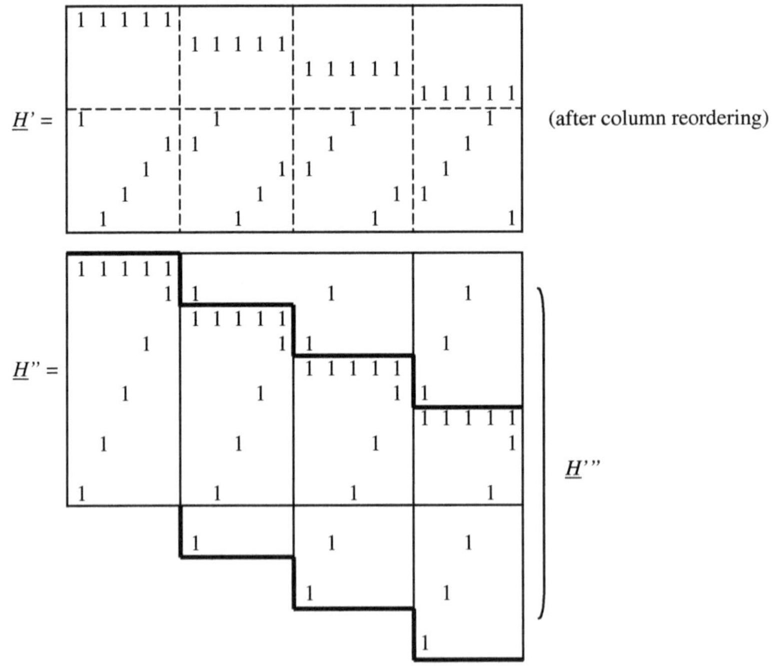

Fig. 12.10 Reordering of columns and rows for obtaining the syndrome former sub-matrix in the convolutional code discussed in Example 12.5

would have $S = 5$, now we take one row any five, and obtain the $\underline{H}''$ matrix reported down, where we have alternate effects from the two component codes. One empty row has been introduced for balancing the parameters. The code associated to such new matrix can be interpreted as a 3/5-rate tail-biting convolutional code, to be unwrapped. Such operation can be made by means of the staircase cut depicted in the Figure. Thus the final version is that called $\underline{H}'''$, which collects four consecutive syndrome former sub-matrices $\underline{H}_o^T$ (all equal) and can be prolonged over whichever frame length. With respect to the traditional cutting procedure (see Fig. 10.18), we can observe a strong resemblance. Only the last stair exhibits a double height. The value of parameter m_h, according to (12.6), becomes $m_h = 4$, but it is calculated on a single small-scale $\underline{H}_o^T$ sub-matrix. In Fig. 10.18 it is only $m_h = 1$, but on a composite syndrome former sub-matrix, made by different small-scale sub-matrices.

In conclusion, the LDPC convolutional code obtained from an array code, following the construction procedure described in Sects. 8.10 and 8.11, exhibits trajectories in the encoding memory which show jumps of vertical length equal to p. After unwrapping, there is a module collecting the trajectories, characterized by a precise value of p, which shifts along the new encoding memory (that of the convolutional code, whose vertical dimension is much longer). This module maintains horizontal size S, but has a length p along the vertical direction (Baldi et al. 2014). On the contrary, the convolutional code described by (12.3) and (12.4), which can be interpreted as derived from a P-GPC block code, exhibits trajectories in the encoding memory without jumps. After unwrapping, in this case, there is a module where parameter p is absent, and whose vertical length is imposed only by Δ_Y. Such module shifts along the encoding memory of the convolutional code, which maintains the horizontal size of the original P-GPC code, measured by the number of circulants in the first layer of its parity check matrix. Both types of modules here introduced in the encoding memory have the form of the syndrome former sub-matrix $\underline{H}_o^T$ to be adopted.

The choices of the design parameters Δ_j, $j = 2, 3, \ldots Y$, are very important in order to guarantee a close approach to the bound $d \leq 2^Y$ (Corollary b of Theorem 8.1) or to the bound $d \leq 2^M$, as given by (11.10). With P-GPC codes, owing to the trajectories without jumps which characterize the module vertically shifting along their encoding memory, such close approach appears favoured. Nevertheless this advantage is reached at the expense of an increase in the parity check constraint length m_h.

12.4 The Best Performance of LDPC Convolutional Codes up to Now

Improper array codes, owing to the linear increase which characterizes the phases of the permutation circulants in the layers of their $\underline{H}$ matrix, are too regular for giving very good performance in the convolutional version obtained from them (after reordering and unwrapping). In Tanner et al. (2004) a power law in the phase increments of the sequence of permutation circulants in any layer of an array scheme is suggested. By proper choice of the bases of such geometrical progressions, a remarkable improvement in the performance can be reached, especially for code rates not too high. In Example 10.23 a convolutional code so constructed has been presented. The length of the sliding window in the decoding algorithm coming out from this approach varies from few hundreds (for code rates of the order of 1/2) to several thousands (for code rates up to 7/8).

The cutting procedure described in Sect. 10.10 (see Fig. 10.18) is not so different, but starts from an even less regular block code, like the one obtained from ARA or protograph-based LDPC block codes. A super-period, very similar to that described in Sect. 12.2, is present here, whose length equals the one of the whole parent block code. Such codes are often called time-varying convolutional (Jimenez-Felstrom and Zigangirov 1999), but they are only w.s. time-invariant on a very long super-period, and typically insert packets of information and control symbols alternately within this super-period. A code of this type has been described in Example 10.22. On the other hand, as already noted, this family of codes exhibits $m_h = 1$, so that the very long super-period is balanced by such a small value for the parity check constraint length, leading to typical sliding window lengths shorter than those of the convolutional codes of the type previously described.

Such two design methods for obtaining the convolutional version of an LDPC block code, although suggested some years ago, are still at the state of the art. The first one takes advantage from the wide literature on array codes, fixing bounds on their minimum distance and good practices in designing the permutation circulant phases. In principle, it can be applied also to any type of quasi-cyclic LDPC block code (Pusane et al. 2011), therefore also to those QC block codes characterized by circulants with weight higher than one (Smarandache and Vontobel 2012). In all these applications, the input-output syndrome former polynomials (Definition 8.2) can be directly obtained by the symbolic expressions of the original circulants. In this sense, if we start from an array code, these expressions consist of only monomials, instead when circulants having weight greater than one are employed, there are also binomials, trinomials and so on. The second design method, based on the cutting procedure, has the advantage of being more general, although it always starts from small block codes, already representing as many LDPC behaviours by themselves. The code rate of such small block codes is the same as the code rate of the convolutional code to be designed. In a variant of this procedure, the cut trajectory can be subjected to random variations (Pusane et al. 2011), but anyway on average it follows a staircase diagonal in the rectangle of the $\underline{H}$ matrix characterizing the original small block code.

Reordering and unwrapping procedures or cutting procedures generally give rise to convolutional codes showing better performance with respect to the original LDPC block code, as already said. Furthermore column-irregular codes are able to reach an improved behaviour exploiting more efficiently the decoding algorithm, as expected. For instance, in Fig. 12.11, a comparison between the performance of the convolutional version of a block code obtained after cutting, in the case of protograph-based column-irregular original codes, and the performance of these original

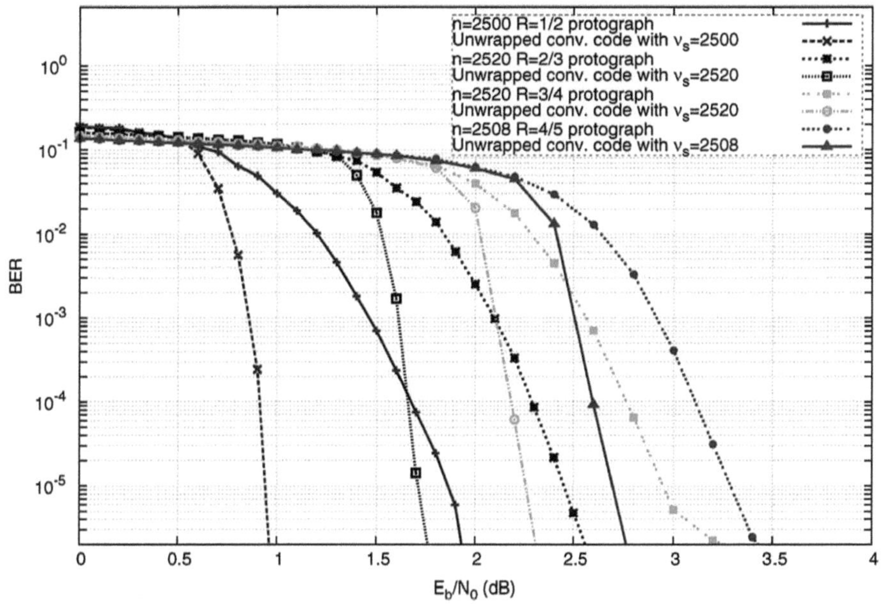

Fig. 12.11 Comparison between the performance of the convolutional version of some LDPC codes and the original LDPC block codes [from (Pusane et al. 2011), reprinted by permission]

parent codes is reported. These computer simulations give the binary symbol error probability for an AWGN channel, and parameter v_s represents the width of the pseudo-diagonal in the $\underline{H}$ matrix $(m_h + 1)n_o$ (Pusane et al. 2011). In this notation parameter E_b/N_o replaces *SNR*. For code rates of 1/2 or 2/3, a typical increase in the coding gain of about 0.7 dB can be reached. For larger code rates this advantage is less evident. Remark that the values of parameter v_s here adopted are very large for this type of construction, especially with a low code rate.

Although the LDPC convolutional codes here briefly reviewed offer the best performance up to now, the fact that they derive from LDPC block codes gives a motivation for supposing that even better results could be obtained from a direct design, able to exploit the intrinsic nature of a convolutional code. In the following sections we will present considerations useful in order to design good LDPC convolutional codes without deriving them from suitable families of LDPC block codes, like made in this presentation so far. We will start from 1/2-rate codes, which, as known, are s.s. time-invariant both in $\underline{H}$ and in $\underline{G}$. Then we will extend the treatment to $(n_o - 1)/n_o$-rate codes (s.s. time-invariant only in $\underline{H}$), and finally to $1/n_o$-rate codes (s.s. time-invariant only in $\underline{G}$), maintaining separated the cases of column-regular and column-irregular LDPC codes. In some situations, circulant expansion (see Definition 10.10) will be added.

12.5 1/2-Rate LDPC Regular Convolutional Codes Directly Designed

With reference to a 1/2-rate convolutional code, let us consider the parameters shown in Fig. 12.12 on the left, for describing its syndrome former sub-matrix $\underline{H}_o^T$. Such scheme is completely general for the case of a column-regular code, characterized by $w_c = 2$ (and $w_r = 4$ except in the outermost parts of the code frame). The choice of two 1-symbols in the first row of $\underline{H}_o^T$ allows to have an information symbol in the first position and a control symbol in the second position of any period. Clearly for $a_2 = a_1$ the row-column constraint would be not satisfied. From the semi-infinite $\underline{H}$ matrix depicted in Fig. 12.12 on the right, it is possible to argue that no 4-length short cycles can arise if a_2 and a_1 are different. The choice $a_1 > a_2$ is still able to describe a completely general situation, in fact if $a_2 > a_1$ the same results will be achieved with reciprocal polynomials.

This type of $\underline{H}$ matrix is s.s. time-invariant, and is characterized by a reciprocal parity check polynomial which can be written as

$$h^*(x) = x(x^{2a_1} + 1) + (x^{2a_2} + 1), \quad a_1 > a_2 \tag{12.7}$$

leading to a parity check constraint length $m_h = a_1$. The width of the pseudo-diagonal characterizing the $\underline{H}$ matrix turns out to be $(m_h + 1)n_o = (a_1 + 1)2$.

Let us consider now a column-regular 1/2-rate convolutional code, with column weight $w_c > 2$ and row weight $w_r = 2w_c$. In order to use a symbolism not too heavy, we will analyze the case of $w_c = 3$. In the situation characterized by the $\underline{H}_o^T$ sub-matrix depicted in Fig. 12.13, where $b_j > a_j$, we can write the following reciprocal parity check polynomial

$$h^*(x) = x(x^{2b_1} + x^{2a_1} + 1) + (x^{2b_2} + x^{2a_2} + 1), \quad b_1 > b_2 \tag{12.8}$$

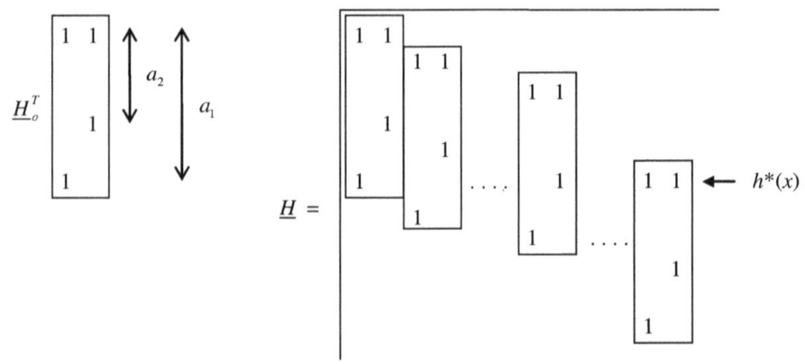

Fig. 12.12 Syndrome former sub-matrix and semi-infinite overall matrix for 1/2-rate LDPC column-regular convolutional codes characterized by $w_c = 2$

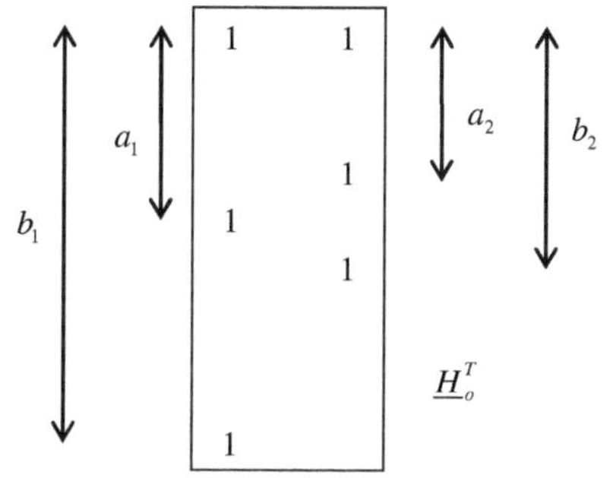

Fig. 12.13 Symbolism adopted for describing the case of $w_c = 3$, for column-regular 1/2-rate LDPC convolutional codes

Comparison with (12.7) shows that here a pair of trinomials substitutes a correspondent pair of binomials. Here we clearly obtain $m_h = b_1$.

The two situations analyzed so far allows easy obtainment of general results. Some of them regard only the 1/2-rate code family. Some others can be extended also to the whole $(n_o - 1)/n_o$-rate code family, and will be treated, in part, in the next section. About the minimum distance we can conclude what follows. For $w_c = 2$, in the case characterized by $a_1 = 2a_2$, we have $d = 3$, in fact a 6-length short cycle comes out, and Lemma 10.4 holds. This situation is, for instance, that treated in Fig. 7.14 on the right, for $a_2 = 1$. In all the other cases, still with $w_c = 2$, we have $d = 4$. More precisely, if a_1 and a_2 are both odd or both even, there are only even-weight code frames. If one of the two is even and the other one is odd, there are also odd-weight code frames, like in the case treated in Example 10.7, after having eliminated 4-length short cycles by adoption of an $\underline{H}$ matrix not in minimal form. For $w_c \geq 3$, it is possible to expect the upper bound $d \leq 2w_c$. In fact a code frame of the type $t(x) = h(x)$ will be always present. Remark that this conclusion, supported by Lemma 7.4, is valid also when $\underline{H}$ is not in minimal form, simply in this case the generator polynomial will have degree lower than deg$[h(x)]$. About the last property, it is useful to introduce the following proposition.

Lemma 12.4 *Whichever column-regular 1/2-rate LDPC convolutional code, characterized by* $w_c = 2$, *has an* $\underline{H}$ *matrix not in minimal form.*

Proof In both the terms of expression (12.7) a factor $(x^2 + 1)$ is present. When a polynomial in the variable x^2 can be extracted from the original $h(x)$, it is possible to eliminate it and obtain the same code. The residual factor represents the polynomial $h_c(x)$, whose reciprocal shifts in the rows of the parity check matrix in minimal form, which is equivalent to the original one. □

After Lemma 7.4, about the self-dual property of the $\underline{G}$ and the $\underline{H}$ matrix in a 1/2-rate convolutional code, we obtain obtaining

$$g^*(x) = h_c^*(x) = xu_{a_1}(x^2) + u_{a_2}(x^2), \quad a_1 > a_2 \tag{12.9}$$

where the last expression comes out from the division of (12.7) by $(x^2 + 1)$, taking into account the equality $x^{2a_j} + 1 = (x^2 + 1)u_{a_j}(x^2)$, where $u_{a_j}(x^2)$ is an enlarged unitary polynomial of weight a_j.

Remark that, besides the reciprocal code frame having minimum length, expressed by (12.9), the following reciprocal code frame is also present

$$t^*(x) = h^*(x) = x(x^{2a_1} + 1) + (x^{2a_2} + 1), \quad a_1 > a_2 \tag{12.10}$$

so confirming what anticipated in the above discussion on the minimum distance achievable for $w_c = 2$. Typically (12.9), which is more compact, exhibits a weight not lower than the weight of (12.10), which is always $w[t(x)] = 4$. Only for the case in which $a_1 = 2a_2$, we have $w[g(x)] = 3$ in (12.9) and hence $d = 3$, as previously noted. Finally in (12.9) there is a justification of what anticipated about the frame weight distribution, related to even or odd values for the a_j's.

Example 12.6 Let us consider the case with $a_1 = 3$, $a_2 = 1$. We obtain the interleaved generator polynomial which, from (12.9), considering also elimination of the reciprocal, results in

$$g(x) = xu_1(x^2)x^4 + u_3(x^2) = x^5 + x^4 + x^2 + 1,$$

and the code frame deriving from (12.10) turns out to be

$$t(x) = (x^2 + 1)g(x) = x^7 + x^6 + x^5 + 1,$$

with weight equal to 4 as well.

The family of codes here treated for $w_c = 2$ then has a generator constraint length $m = a_1 - 1 = m_h - 1$. The case analyzed in Example 10.7 is characterized by $a_1 = 3$, $a_2 = 2$, and hence $m_h = 3$, $m = 2$, as before. This code exhibits $w[g(x)] = 5$, but the weight of (12.10) remains $w[t(x)] = 4$, therefore the minimum distance is again $d = 4$.

Clearly the best codes with $w_c = 2$ are those characterized by $d = 4$ and the smallest values of parameters a_j, in order to maintain short constraint lengths. This goal is reached for both the codes characterized by $m_h = 3$, and parameters $a_1 = 3$, $a_2 = 1$, or $a_1 = 3$, $a_2 = 2$ (respectively treated in Examples 12.6 and 10.7). Their weight distribution coefficients, up to $w = 7$, are listed in Table 12.1. The sequence of binary symbols associated to expressions (12.9) and (12.10) are reported too. Z represents, as usual, the number of periods where information symbols are inserted. Being for both codes $m = 2$, the overall number of periods in the correctly terminated frame will be $(Z + 2)$.

It is possible to note a resemblance between (12.9), (12.10) here and (11.3), (11. 11), which hold for P-GPC codes. In both cases, the lowest-weight code words are related not to the generator polynomial itself, but to a superimposition of properly

Table 12.1 Performance of the best codes in the family of LDPC regular convolutional codes with code rate 1/2 and $w_c = 2$

a_1	a_2	$g(x)$	$t(x)$	$A(4)$	$A(5)$	$A(6)$	$A(7)$
3	1	11.01.01	11.10.00.01	2Z − 1		5Z − 14	
3	2	11.11.01	11.00.10.01	Z − 1	2Z − 2	2Z − 7	5Z − 20

shifted replicas of $g(x)$. This similitude is valid also for greater values of w_c, but only if they are even numbers. In fact the expression of $h(x)$, assuming w_c odd, does not contain, in general, a polynomial in the variable x^2, as a factor. See, for instance, expression (12.8), which holds for $w_c = 3$. In other terms, there will be very good 1/2-rate convolutional codes, characterized by w_c odd, with $\underline{H}$ matrix in minimal form.

Let us consider now two important propositions, valid also in a widened context: all the $(n_o-1)/n_o$-rate column-regular codes with w_c even on the one side, and with w_c odd on the other side. They strongly contribute to maintain separate such two families of regular LDPC convolutional codes.

Lemma 12.5 *Whichever regular* $(n_o - 1)/n_o$*-rate LDPC convolutional code, characterized by* w_c *even, has an* $\underline{H}$ *matrix not in minimal form.*

Proof The ex-OR sum of all the rows of $\underline{H}$ gives a null result. So it has not full rank. There is at least one row linearly dependent on the others. Such row, generally eliminated at the end of the frame, means that the $\underline{H}$ matrix is not in its minimal form. □

Lemma 12.6 *Column-regular LDPC codes with* w_c *odd have code words with only even weights. This property holds for convolutional codes having any code rate and even for block codes.*

Proof The ex-OR sum of all the rows of $\underline{H}$ leads to a row made by all 1-symbols, linearly dependent on the other ones, but to be satisfied as well by any word of the considered code. A code word so constrained, as occurs in a binary parity check code, must have even weight. □

In the final part of this section we will present some codes with $w_c = 3$, so verifying such property. After Lemma 12.5, for $w_c > 2$ and even number, with any $n_o \geq 2$, it is possible to proceed adopting a generalization of expression (12.7), from which a common factor, function of x^{n_o}, can be extracted.

A generalization of (12.10), able to comprehend the case of any $n_o \geq 2$, still with $w_c = 2$, can be written as

$$t^*(x) = x^{(i-j)}(x^{n_o a_i} + 1) + (x^{n_o a_j} + 1), \quad i > j, \tag{12.11}$$

confirming that $d = 4$ is maintained, as in the case of $n_o = 2$. Taking into account also the enumeration of all the possible couples i, j, we can estimate a normalized weight distribution coefficient $B(4) = \binom{n_o}{2}$. Expression (12.11) holds independently of which parameter, between a_i and a_j, is greater, but under the condition that there are no pairs of parameters for which it is possible to write $a_i = 2a_j$ or a tern of parameters in a same equality.

The upper bound $d \leq 2w_c$, already justified, does not hold for code rates of the type $P/(2P)$, where P is an integer common factor (like, for instance, 2/4, case treated in Example 7.10). They will be analyzed separately, in Sect. 12.9. After this important premise, the following proposition can give operative support to the code design (Baldi et al. 2012).

Lemma 12.7 *In a convolutional code, regular or irregular, with code rate $(n_o - 1)/n_o$, in order to satisfy the row-column constraint, the columns of the syndrome former sub-matrix have to show all different vertical separations between pairs of 1-symbols.*

Proof If in $\underline{H}_o^T$ there are two equal vertical separations they can be aligned or not. If they are aligned, a 4-length short cycle is formed directly in the same $\underline{H}_o^T$ sub-matrix. If they are not aligned, and vertically separated by b positions, a 4-length short cycle will occur between two $\underline{H}_o^T$ sub-matrices horizontally separated by b periods. The situations here considered are the only two able to produce 4-length short cycles. □

Remark that Lemma 12.7 does not hold when the vertical shift of the $\underline{H}_o^T$ sub-matrix consists of more than one row at a time, like in $1/n_o$-rate and also in $P/(2P)$-rate codes. In fact, for such codes, two equal vertical separations, in turn separated by b positions, provided that b is not an exact multiple of n_o, are not aligned.

In a 1/2-rate regular code, characterized by whichever w_c, able to reach the upper bound $d = 2w_c$, minimization of m_h is then achieved searching for the most compact set of the possible vertical separations between pairs of 1-symbols, in the two columns of the $\underline{H}_o^T$ sub-matrix. This combinatorial problem can be easily faced by means of computer search. The following lower bound on m_h exists

$$m_h \geq 2\binom{w_c}{2}, \tag{12.12}$$

in which equality holds only when all the integer numbers from 1 to $2\binom{w_c}{2}$ are employed. The last number, whose value is the highest one, in this case, coincides with m_h.

Example 12.7 Taking $w_c = 3$, we have three regular good codes, all characterized by $d = 6$ and $m_h = 7$, whereas $m_h = 6$ is impossible. The three vertical separations in each column are listed in Table 12.2. They all employ six integer numbers taken among the set {1, 2, 3, 4, 5, 6, 7}. Only one integer number (respectively 6, 4, 2) is excluded. For all these three codes, the normalized weight distribution coefficients

Table 12.2 Best 1/2-rate LDPC convolutional codes with $w_c = 3$ (all characterized by $m_h = 7$)

a_1, a_2, b_1, b_2	Vertical separations		Integer number not employed	$g(x) = h(x)$
2,1	2	1		
7,4	5	3	6	11.10.01.00.10.00.00.01
	7	4		
2,1	2	1		
5,7	3	6	4	11.10.01.00.00.01.00.10
	5	7		
3,1	3	1		
7,6	4	5	2	11.10.00.01.00.00.10.01
	7	6		

result in $B(6) = 1$, $B(8) = 0$, and $B(10)$ ranging from 6 to 10. It is possible to confirm what predicted by Lemma 12.6, in fact only even-weight code frames are present. Furthermore a null $B(8)$ contributes to give a good performance.

12.6 High-Rate LDPC Regular Convolutional Codes Directly Designed

The reciprocal parity check polynomial, shifting in the rows of the $\underline{H}$ matrix which describes a 1/2-rate convolutional code, is

$$h^*(x) = xh^{(1)*}(x^2) + h^{(2)*}(x^2) \tag{12.13}$$

where $h^{(1)}(x)$ and $h^{(2)}(x)$ are the two input-output syndrome former polynomials (Definition 8.2). With the symbolism adopted so far, the regular 1/2–rate convolutional codes characterized by parameters $a_j, b_j, c_j, \ldots, j = 1, 2$, exhibit input-output syndrome former reciprocal polynomials which can be written as

$$h^{(j)*}(x) = \cdots + x^{c_j} + x^{b_j} + x^{a_j} + 1, \quad j = 1, 2 \tag{12.14}$$

Generalization of (12.13) to the case of a $(n_o - 1)/n_o$-rate convolutional code, for whichever $n_o \geq 2$, is then straightforward and yields

$$h^*(x) = x^{n_o-1}h^{(1)*}(x^{n_o}) + x^{n_o-2}h^{(2)*}(x^{n_o}) + \cdots + h^{(n_o)*}(x^{n_o}), \tag{12.15}$$

where all the $h^{(j)}(x)$'s have the same weight w_c, because of the assumption that the code is column-regular with column weight w_c. Any input-output syndrome former polynomial acts on the same position in the period. Taking only two positions in subsequent periods, it is possible to form a $2w_c$-weight code frame, like in a 1/2-rate convolutional code. Such code frames are typically longer than the lengths of the polynomials in the $\underline{G}_o$ sub-matrix, even when w_c is odd, but their presence is responsible for the upper bound on the minimum distance, which remains

$$d \leq 2w_c, \quad (n_o \geq 2, w_c \geq 3) \tag{12.16}$$

The last condition on w_c is due to the possible presence of a 6-length short cycle when $w_c = 2$, leading to $d = 3$. On the other hand, for $n_o > 2$, a higher number of couples of columns have now to be examined applying Lemma 12.7, in order to satisfy the row-column constraint. More precisely, instead of (12.12), the following lower bound on the parity check constraint length holds here

$$m_h \geq n_o \binom{w_c}{2} \tag{12.17}$$

Furthermore, as already noted for $w_c = 2$ regarding coefficient $B(4)$, in a regular code able to satisfy the upper bound $d = 2w_c = 4$, the number of d-weight 1-symbol patterns increases as the binomial $\binom{n_o}{2}$. Such increase is clearly a strongly limiting feature for the performance of the considered high-rate convolutional codes.

In the following, except for some general considerations, we will treat only the cases of $w_c = 2$ and $w_c = 3$, with the purpose of listing some conditions for achieving the goal $d = 4$ or $d = 6$ respectively. In order to have information symbols associated to the first $(n_o - 1)$ positions in the period, and hence the last position occupied by the unique control symbols, we will restrict our attention to only $\underline{H}_o^T$ sub-matrices characterized by a first row with all 1-symbols.

Example 12.8 Taking the 2/3-rate regular convolutional code, with $w_c = 2$, and parameters $a_1 = 6$, $a_2 = 4$, $a_3 = 1$, we have the situation shown in Fig. 12.14. The semi-infinite parity check matrix is reported up. In the middle we have the correspondent generator sub-matrix, in which the most compact pair of interleaved generator polynomials are present. Finally, down, three low-weight code frames of the type discussed above (employing only the same two positions in subsequent periods), whose weight is $w = 4$, are reported. They represent situations of the type expressed by (12.11). This code exhibits $m_h = 6$, $m = 3$, $d = 4$, and coefficients $A(4) = 6Z - 12$, $A(5) = 8Z - 16$. The long delay in the first of such two coefficients is due to the long 1-symbol patterns here considered.

The condition $a_i = 2a_j$, for $w_c = 2$, is to be avoided in order to exclude that 3-weight code words occur, like for the 1/2-rate codes. Nevertheless the above condition is no longer the only one for preventing occurrence of 3-weight codes frames here. In fact they can be formed also by 6-length short cycles, consequence of an equality involving three different a_j parameters.

Example 12.9 Taking the 2/3-rate regular convolutional code, with $w_c = 2$, and parameters $a_1 = 4$, $a_2 = 3$, $a_3 = 1$, we have the situation shown in Fig. 12.15. Here this equality holds $a_1 = a_2 + a_3$, which originates two 6-length short cycles and as many 3-weight 1-symbol patterns (one in the first row in the generator sub-matrix).

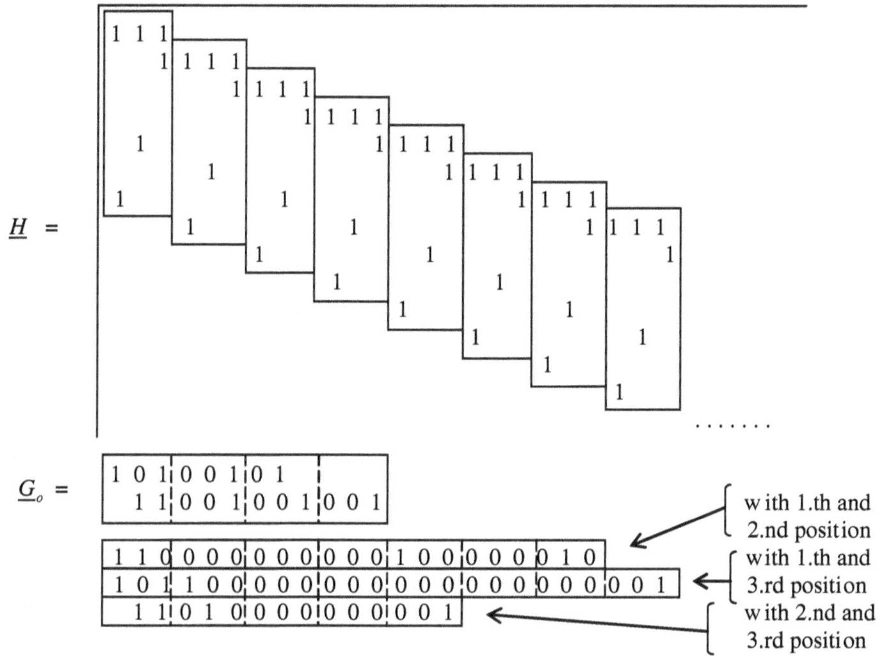

Fig. 12.14 Semi-infinite parity check matrix, generator sub-matrix and three long 1-symbol patterns having weight $w = 4$, for the 2/3-rate code in Example 12.8

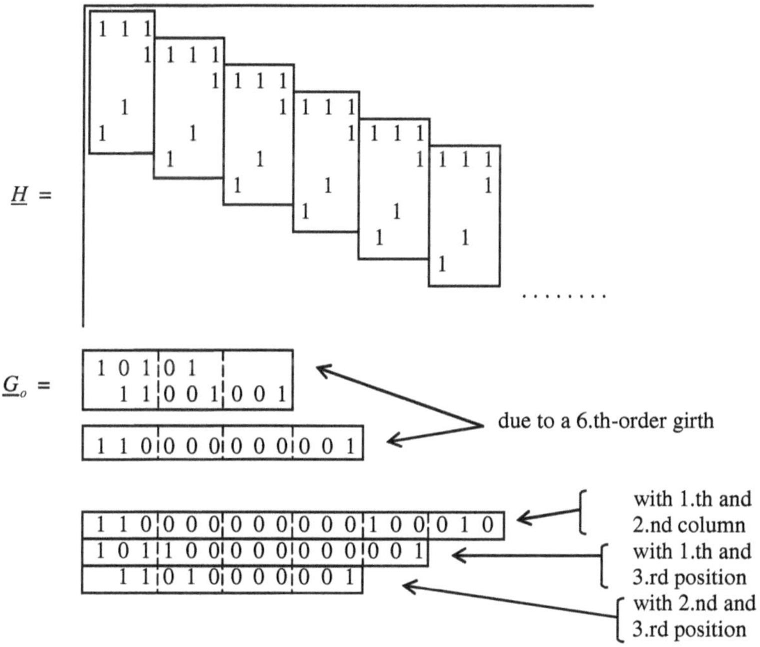

Fig. 12.15 Semi-infinite parity check matrix, generator sub-matrix and three code frames having weight $w = 4$, for the 2/3-rate code described in Example 12.9

Such code is characterized by $m_h = 4$, $m = 2$, $d = 3$, and coefficients $A(3) = 2Z - 1$, $A(4) = 4Z - 6$. Three long 4-weight 1-symbol patterns, of the type expressed by (12.11), have been reported also here.

Let us consider now the case of $w_c = 3$. In order to have $d = 2w_c = 6$, it is necessary to find n_o sequences of three integer numbers assuring all different vertical separations, in order to satisfy the row-column constraint, according to Lemma 12.7. Next, the closer the m_h to bound (12.17), the better the code design. Nevertheless also possible equalities involving three vertical separations belonging to different columns, so leading to a 6-length short cycles, are to be considered. Fortunately, they represent here a less important problem, with respect to what occurs for $w_c = 2$, because any column has a third 1-symbol to be properly coupled.

We remember that, for $n_o = 2$, it is impossible to employ all the 6 first integer numbers, and hence satisfy (12.12) with equality sign, but we are forced to adopt at least $m_h = 7$, so excluding one number in the set $\{1,2,3,4,5,6,7\}$, as already discussed in Example 12.7 (Table 12.2). A similar situation occurs for $n_o = 3, 4$, whereas for $n_o = 5$ there is the possibility of employing all the available integer numbers from 1 to $5\binom{3}{2} = 15$.

Example 12.10 Let us consider an LDPC regular convolutional code with $w_c = 3$, for which we adopt a symbolism of the type assumed in expression (12.14), with the only difference of considering a higher number of parameters a_j, b_j (that is with j up to 3, 4, 5). Taking $n_o = 3$, it is possible to adopt $m_h = 10$, one unit more than the condition of equality in (12.17). For instance, in the first row of Table 12.3, a possible choice is reported, where the only number 3 has been excluded. With $n_o = 4$, we can assume $m_h = 13$, as before (see for instance the second row of the Table where the only number 11 has been excluded). Finally, with reference to $n_o = 5$, (12.17) can be satisfied, instead, with equality sign, for instance, after the choice reported in the third row of the Table, so leading to $m_h = 15$. The first two

Table 12.3 Examples of choices for the best parameters a_j, b_j, in a convolutional code with $w_c = 3$ and code rate $(n_o - 1)/n_o$

n_o	a_1 b_1	a_2 b_2	a_3 b_3	a_4 b_4	a_5 b_5	Vertical separations					Integer numbers not employed
3	1	8	4			1	8	4			
	7	10	9			6	2	5			3
						7	10	9			
4	1	2	3	5		1	2	3	5		
	9	6	13	12		8	4	10	7		11
						9	6	13	12		
5	1	2	3	4	5	1	2	3	4	5	
	13	8	10	15	14	12	6	7	11	9	–
						13	8	10	15	14	

convolutional codes exhibit $d = 6$. On the other hand, as expected, coefficient $B(6)$ strongly increases from $n_o = 3$ to $n_o = 4$. The third convolutional code exhibits $d = 4$ with $B(4) = 1$.

Clearly the above choices are those associated to the shortest parity check constraint length possible, but the procedure for their obtainment neither can be generalized for higher values of w_c, nor is naturally predisposed for a rate-adaptive approach. This goal can be achieved, instead, with a particular family of LDPC regular convolutional codes, with code rate $(n_o - 1)/n_o$, proposed in Baldi et al. (2012). They have been named uniform Progressive Difference Convolutional (PDC) LDPC codes, paying attention to the property that the vertical separations between 1-symbols in the columns follow a rule, in which the differences progressively grow by a uniform quantity with increasing w_c and n_o.

In order to design their syndrome former sub-matrix $\underline{H}_o^T$, it is useful to assume a doubly ordered set of parameters $\delta_{i,s}, i = 1, 2, \ldots n_o,\ s = 1, 2, \ldots (w_c - 1)$, where i denotes the column position in the period, whereas s expresses subsequent intervals in the same column, finally $\delta_{i,s}$ is the differential vertical separation. Uniform PDC codes show trajectories connecting 1-symbols characterized by zigzag behaviours. The slopes of the segments of such trajectories are imposed by parameters $\delta_{i,s}$, and have to be treated separately from up-right to down-left and from up-left to down-right. In Fig. 12.16 such integer numbers have been reported in the case of $n_o = 4$ and $n_o = 5$, for w_c up to 4 (s up to 3). Remark that the first row of $\underline{H}_o^T$ is made by all 1-symbols, so that $\delta_{i,1}$ represent the separations of the points in the first segment of the zigzag trajectory from such row, $\delta_{i,2}$ the separations of the points in the second segment from the correspondent in the previous segment, and so on. Also a rough picture of the resulting zigzag trajectories is reported in the Figure. It is possible to observe that their slope is independent of n_o, but grows very rapidly as w_c increases.

The parity check constraint length can be calculated as

$$\begin{aligned} m_h &= \sum_{s=1}^{w_c-1} \delta_{1,s} \quad (w_c \text{ even}), \\ m_h &= \sum_{s=1}^{w_c-1} \delta_{n_o,s} \quad (w_c \text{ odd}). \end{aligned} \tag{12.18}$$

Also this very important parameter is reported in the Figure. Comparison with (12.17) can be considered for appreciating how large is the departure from such lower bound. Clearly we have an exponential growth in m_h with increasing w_c, which makes this solution acceptable only for values of w_c not higher than 5. A particular case is that characterized by $w_c = 2$, for which parameters $\delta_{i,1}$ turn out to be all the subsequent odd integer numbers, and it is possible to verify that the first of (12.18) tends to become about twice the value of equality in (12.17).

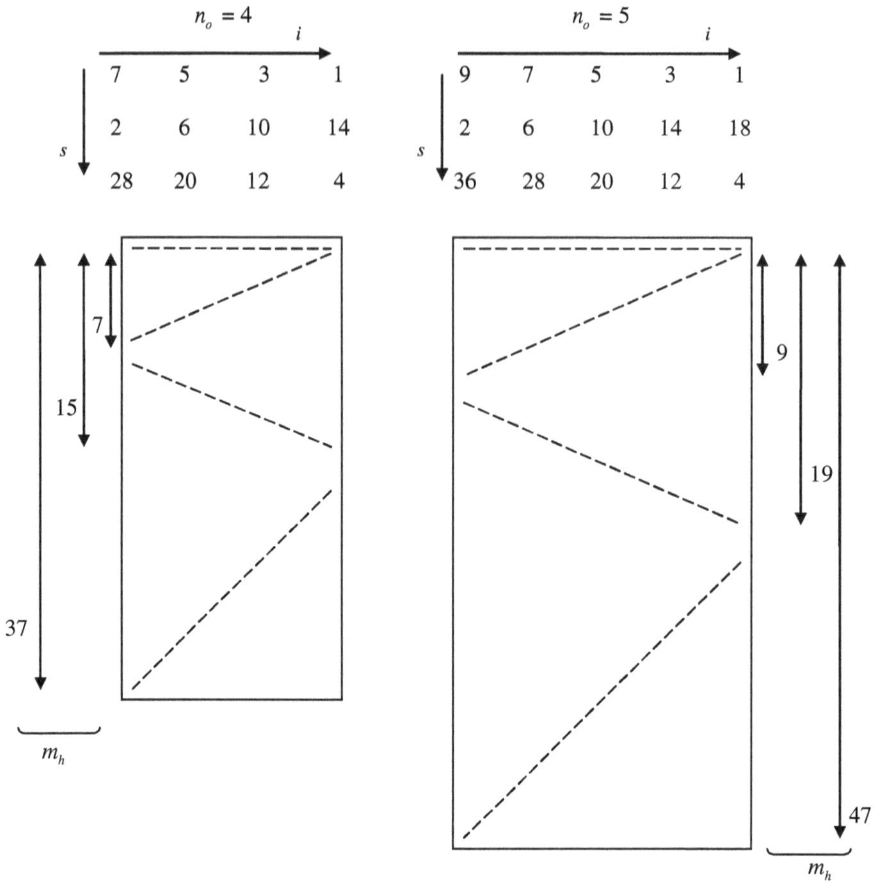

Fig. 12.16 Parameters $\delta_{i,s}$ for uniform PDC codes, with w_c up to 4, and code rate 3/4 or 4/5

On the other hand, the upper bound on the minimum distance imposed to these column-regular codes remains a limiting constraint. Not better results can be obtained with column-irregular solutions, as long as we maintain the code rate equal to $(n_o - 1)/n_o$, as justified by the following observation. The tail-biting versions of all codes treated in this and in the previous section can be interpreted also as reordered versions of QC codes characterized by single-layer parity check matrices, of the type introduced in Sect. 11.9, assuming $a = n_o$ (Kamiya 2007; Baldi et al. 2011). In particular the upper bound (11.26) obtained for that family of codes coincides with the upper bound characterizing the present regular convolutional codes. Following this equivalence, (11.27) can be adopted as the upper bound on d, in the case of irregular convolutional codes, still with code rate $(n_o - 1)/n_o$, which will be analyzed in the next section. The possibility of direct obtainment of the phases characterizing the circulants in the block code version from the input-output syndrome-former polynomials in the convolutional code version makes the passage extremely straightforward in both directions. This observation confirms how QC block codes and tail-biting convolutional codes are two aspects of the same phenomenon. In Bocharova et al. (2009) a unitary analysis of both types of LDPC codes is presented, under the condition of being column-regular and with $w_c = 2$. It is based on a graph theory employed for determining the possible presence of girths, and shows how, for low-rate codes, instead, rather large values of the minimum distance can be achieved.

An interesting opportunity for increasing the minimum distance is offered by the choice of a code rate $(Pk_o)/(Pn_o)$, that is by a code in which the period length and the number of information symbols per period are not co-prime. We have already mentioned the case of $k_o = 1$, $n_o = 2$. In particular, with $P = p$, adopting p-size permutation circulants to replace 1-symbols, starting from very simple base $\underline{H}$ matrices, it is possible to perform a circulant expansion procedure, of the type introduced in Sect. 10.6. An alternative approach consist of taking a high-rate LDPC code (say a uniform PDC code with w_c up to 3 and n_o not greater than 4), as component code, in a code concatenation scheme. It is possible to assume, as a second code, either a block code, or another convolutional code. In the latter case, for instance adopting a turbo-product arrangement, we obtain a scheme, already classified as doubly convolutional code. In this sense, the fine scalability of uniform PDC codes, assured by the modular structure of their $\underline{H}_o^T$ sub-matrix, remains the best feature of such family of codes.

12.7 High-Rate Irregular LDPC Convolutional Codes Directly Designed

Let us consider now briefly the case of column-irregular codes, firstly with code rate 1/2, and then with code rate $(n_o - 1)/n_o$. Being such codes s.s. time-invariant in $\underline{H}$, they will be still row-regular with $w_r = \mathrm{w}[h(x)]$, although column-irregular. With code rate 1/2, expression (12.13) is still valid for the unique parity check polynomial $h(x)$, but here with two input-output syndrome-former polynomials $h^{(1)}(x), h^{(2)}(x)$, having different weights. The minimum distance in a code of this type is so upper bounded

$$d \leq \mathrm{w}[h^{(1)}(x)] + \mathrm{w}[h^{(2)}(x)], \tag{12.19}$$

provided that one of these two weight is not lower than 2 and the other not lower than 3. The above statement is clear consequence of (11.27) and of the equivalence already described at the end of the previous section. It is possible to observe that the above expression includes also the case of a regular code, for $\mathrm{w}[h^{(1)}(x)] = \mathrm{w}[h^{(2)}(x)] = w_c$.

Example 12.11 Let us take the syndrome former sub-matrix shown on the left in Fig. 12.17. It corresponds to the following pair of reciprocal input-output syndrome former polynomials

$$\underline{H}_o^T = \begin{bmatrix} 1 & 1 \\ 1 & 0 \\ 0 & 0 \\ 1 & 0 \\ 0 & 1 \end{bmatrix} \qquad \underline{G}_o = \begin{bmatrix} 1 & 1 & | & 0 & 1 & | & 0 & 0 & | & 0 & 1 & | & 1 & 0 \end{bmatrix}$$

Fig. 12.17 Syndrome former sub-matrix and generator sub-matrix for the 1/2-rate irregular LDPC convolutional code described in Example 12.11

$$h^{(1)*}(x) = x^3 + x + 1, \qquad h^{(2)*}(x) = x^4 + 1$$

They have no factors in common, therefore this $\underline{H}$ matrix is in its minimal form. The correspondent $\underline{G}_o$ sub-matrix is then made by the unique row $g(x) = h(x)$, and turns out to be as shown in Fig. 12.17 on the right. This 1/2-rate irregular code exhibits $m = m_h = 4$, $d = 5$, and the following expressions for the first five weight distribution coefficients

$$A(5) = Z, \quad A(6) = A(7) = 0, \quad A(8) = 5Z{-}14, \quad A(9) = 8Z{-}38.$$

As a general comment, a minimum distance $d = 5$ with $\langle w_c \rangle = 2.5$, like obtained in Example 12.11, can be considered as a good result. This goal can be assumed as a reference also for irregular codes with higher code rates. The 1/2-rate QC block code equivalent to the one described in Example 12.11, is characterized by two only circulants, whose phases correspond to the coefficients of the two syndrome-former polynomials describing the original 1/2-rate convolutional code. The symbolic parity check matrix of this QC block code, obtained after inverse column reordering, results in

$$\underline{H} = [(\underline{0}_p + -\underline{1}_p + -\underline{3}_p)\,(\underline{0}_p + -\underline{4}_p)],$$

where the circulant size p is the number of periods of the tail-biting convolutional codes before inverse reordering.

We can observe how the number of d-weight code frames in the 1/2-rate irregular convolutional code described in Example 12.11 is due to a normalized coefficient $B(d) = 1$. Then we have several null higher-weight normalized coefficients. This property, which follows from the sparseness of $\underline{H}$, suggests the following variant. Taking a longer period (exactly $2p$ long), and substituting proper permutation circulants of size p to 1-symbols, like in the procedure of circulant expansion (see Sect. 10.6), it is possible to increase by far the minimum distance d.

In irregular 1/2-rate convolutional codes, which typically have an $\underline{H}$ matrix in minimal form, the lowest-weight code frame, resulting from the equality $t(x) = h(x)$, can be originated by the unique 1-symbol pattern of weight $w = d$. For higher code rates this property is typically not so frequent, first of all because the code frames based on an equality of this type are more than one, and secondly because the probability of having cancellations in the ex-OR sum of low-weight code frames increases with the code rate. A good trade-off can be obtained adopting two weights (for instance just one column having weight 2 and all the remaining weight 3), and aiming at satisfying an upper bound of the type (12.19) with equality sign. This upper bound has anyway to be properly modified, in order to take into account the presence of more than two input-output syndrome former polynomials. From (11.27) and (12.11), we can write here

$$d \leq \min\{w[h^{(i)}(x)] + w[h^{(j)}(x)]\}, \ i \neq j, \tag{12.20}$$

which holds when one of the two polynomials has weight at least 2 and the other one at least 3. Remark that (12.19), which refers to 1/2-rate codes, represents a particular case of (12.20).

Before concluding this treatment, it is useful to remember that the $(n_o - 1)/n_o$-rate convolutional codes (regular or irregular), have a tail-biting form, which, after inverse column reordering, become a single-layer parity check matrix. The symbolic form of the parity check matrix, for the QC block code equivalent to the 1/2-rate code presented in Example 12.11, has been already calculated. The general relationship between the coefficients $h_i^{(j)}$ of the input-output syndrome former polynomials $h^{(j)}(x)$ and the phases of the permutation circulants, present in all the circulants of the QC code $\underline{H}$ matrix, can be obtained as follows. The symbolic expression for the single-layer parity check matrix describing such QC code is

$$\underline{H} = \left[\left(\underline{0}_p + \underline{-\Sigma(1)}_p\right)\left(\underline{0}_p + \underline{-\Sigma(2)}_p\right)\cdots\left(\underline{0}_p + \underline{-\Sigma(n_o)}_p\right)\right], \tag{12.21}$$

where $\underline{-\Sigma(j)}_p, j = 1, 2, \ldots n_o$, collects all the symbols i, corresponding to non-null $h_i^{(j)}$ coefficients, except $h_{m_h}^{(j)} = 1$, with minus sign. The presence of $\underline{0}_p$ in all the circulants is consequence of the adoption of a first row in $\underline{H}_o^T$ with all 1-symbols (which means the above convention on coefficient $h_{m_h}^{(j)}$). Negative signs take into account that in the $\underline{H}$ matrix the reciprocal of the parity check polynomial is present. It is finally useful to note that (12.15) is related to the symbolic polynomial syndrome former sub-matrix and the to associated RSC sub-matrix (see Sect. 7.8) as follows

$$\underline{H}_o^T(x) = \begin{bmatrix} h^{(1)}(x) \\ h^{(2)}(x) \\ \cdots \\ h^{(n_o)}(x) \end{bmatrix}, \quad \underline{H}_o^{T(RSC)}(x) = \begin{bmatrix} \frac{h^{(1)}(x)}{h^{(n_o)}(x)} \\ \frac{h^{(2)}(x)}{h^{(n_o)}(x)} \\ \cdots \\ 1 \end{bmatrix}$$

Such expressions lead to the possibility of constructing a very simple encoder circuit, in observer arrangement, with just one shift register, also in systematic form.

12.8 Low-Rate LDPC Convolutional Codes Directly Designed

In this section we are interested in codes having rate $1/n_o$, which are s.s. time-invariant in their $\underline{G}$ matrix. Excluding the case, already treated, of a code rate 1/2, it is possible to say that here more than one parity check polynomials will be present. The syndrome former sub-matrix $\underline{H}_o^T$ vertically shifts, in the overall $\underline{H}$ matrix, by $(n_o - 1)$ rows any period. Furthermore a code which is column-regular can or cannot be also row-regular.

In order to fix ideas, it is convenient to treat in detail the case of 1/3-rate convolutional codes, and then draw some general conclusions. There are two possible convenient arrangements for the first layer in the $\underline{H}_o^T$ sub-matrix (made by two rows and three columns): the first row with all 1-symbols, or the whole layer in a lower triangular form. The former choice typically requires a longer constraint length m_h for a comparable behaviour (in terms of minimum distance and number of lowest-weight code frames). Nevertheless, it allows an easy comparison with 1/2-rate and high-rate codes, described so far. The latter choice is able to assign precise positions, in the final part of the period, to the $n_o - 1 = 2$ control symbols. So it is capable of giving a description of the type introduced in Chap. 8, with two interleaved parity check polynomials leading to as many control symbols in different positions. Both the choices, anyway, assure that the first position in the period is occupied by an information symbol, as one can easily verify after application of the null ex-OR sum of one cluster of syndromes.

In spite of the fact that we are treating a very simple situation, there are many variables to be managed in the code design. First of all, here Lemma 12.7 is no longer valid for allowing satisfaction of the row-column constraint. In fact we can subdivide the couples of $\underline{H}$ rows (regarding their vertical separation) into four types: from even to even (ee), from odd to odd (oo), from even to odd (eo), from odd to even (oe). Even and odd are referred to the positions of the single rows of $\underline{H}$ in the vertical period whose length is 2. A 4-length short cycle comes out only if the equality in two vertical separations occurs between two equal types of row couples. Also the vertical separations, in their turn, are to be subdivided into: even and odd. The same vertical separation, therefore, can be used at most two times. If it is even, one time in ee row couples and the other time in oo row couples. If it is odd, one time in eo row couples and the other time in oe row couples.

Example 12.12 In Fig. 12.18 up, two semi-infinite parity check matrices are shown, for as many column-irregular 1/3-rate convolutional codes. The one on the left adopts the first row with all 1-symbols, and the one on the right is characterized by a first layer in lower triangular form. They both satisfy the row-column constraint. As regards the vertical separations, we have two times 1, 4, in the first, and two times 1, 2, 3, in the second, according to what previously discussed. Both codes are characterized by $m_h = 2$. Remark that we have two columns with weight 3 and one column with weight 2. In Fig. 12.18 down, the generator polynomials of the two

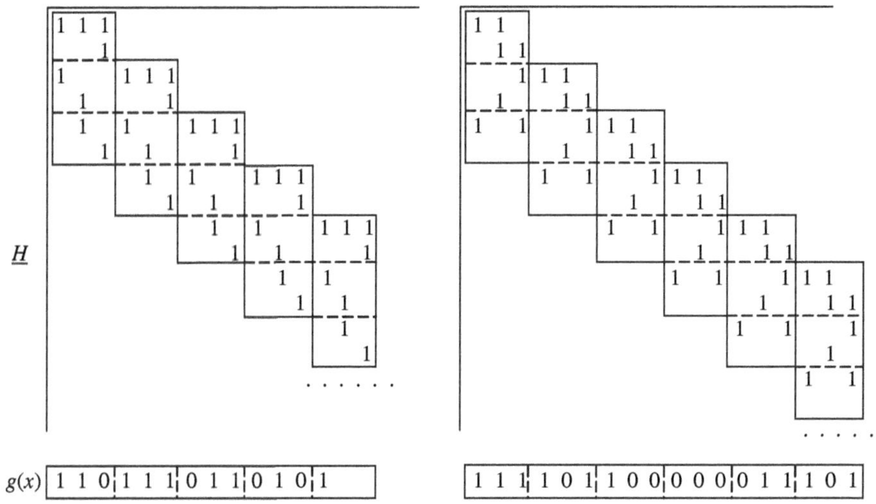

$A(8) = Z - 1$, $A(9) = Z$, $A(10) = Z - 6$ $A(10) = Z$, $A(12) = 2Z - 5$, $A(14) = 2Z - 9$

Fig. 12.18 Parity check matrices for two column-irregular 1/3-rate convolutional codes with $m_h = 2$ (Example 12.12) and correspondent generator polynomials

codes are reported. The generator sub-matrices are made by just one generator polynomial, as expected. These codes exhibit the following performance: $d = 8$ and also odd-weight code words the first, $d = 10$ and only even-weight code words the second. The latter result is obtained with a longer generator matrix constraint length ($m = 5$ instead of $m = 4$). The values of d here obtained, taking into account that both such irregular codes are characterized by $\langle w_c \rangle = 2.66$, are interesting, but not excellent.

We attempt now to develop a more detailed analysis for regular codes having $w_c = 2$, able to predict also their minimum distance, which obviously can increase with respect to the case of $n_o = 2$. Columns able to couple even and odd positions are preferable in order to reach this goal with a not too high value of m_h.

Example 12.13 In Fig. 12.19 two semi-infinite parity check matrices, describing as many 1/3-rate column-regular codes, with $w_c = 2$, are reported, maintaining the usual difference in the first layer of their $\underline{H}_o^T$ sub-matrix. They both exhibit $d = 6$, but for the code on the left $m_h = 3$ is needed, whereas for the code on the right only $m_h = 2$ is enough. For both we have $m = 2$. Some weight distribution coefficients $A(w)$ characterizing the two codes are finally reported. Both codes are also row-regular with $w_r = 3$, and both employ only odd vertical separations (able to couple even with odd positions, as suggested).

With m_h as small as 1, still for $w_c = 2$, it is impossible for column-regular 1/3-rate codes to achieve $d = 6$, and we remain at most with $d = 4$. On the other hand, increasing m_h beyond 3, a minimum distance even greater than 6 can be obtained. A possible choice able to give $d = 8$, with $m_h = 5$, is to adopt the structure on the

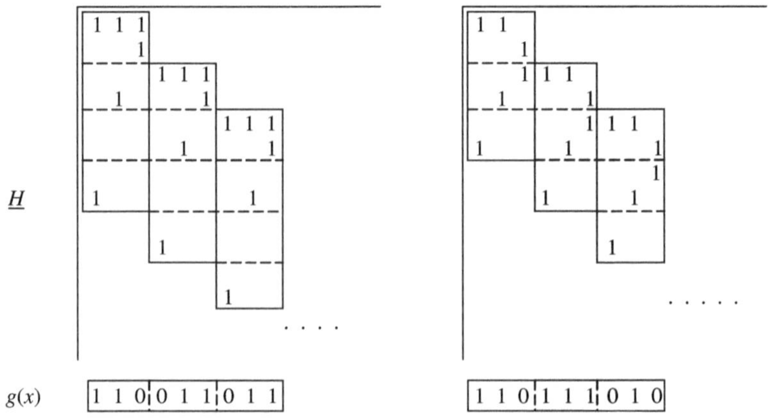

A(6) = 2Z − 1 , A(8) = Z − 2   A(6) = 2Z − 1 , A(8) = Z − 1

Fig. 12.19 Parity check matrices for two column-regular 1/3-rate convolutional codes with $w_c = 2$, $d = 6$ (Example 12.13) and correspondent generator polynomials

right in Fig. 12.19 for the first layer of sub-matrix $\underline{H}_o^T$, and the following set of vertical separations: $a_1 = 7$, $a_2 = 10$, $a_3 = 2$.

Example 12.14 With $w_c = 2$ and $m_h = 2$, there is also the possibility of having odd-weight code words, in particular $d = 5$, which can be considered a sub-optimum result, but accompanied by some subsequent frame weight coefficients null. This is the case of the code characterized by the following pair of $\underline{H}_o^T$, $\underline{G}_o$ sub-matrices

$$\underline{H}_o^T = \begin{bmatrix} 1 & 1 & \\ & & 1 \\ \hline & & \\ & & \\ & & 1 \\ \hline & 1 & \\ 1 & & \end{bmatrix}, \quad \underline{G}_o = [1 \quad 1 \quad 0 | 0 \quad 0 \quad 0 | 1 \quad 0 \quad 1 | 0 \quad 0 \quad 1]$$

The code so obtained is also row-regular, and exhibits the following frame weight distribution coefficients, up to $w = 8$: $A(5) = Z$, $A(6) = A(7) = 0$, $A(8) = 2Z - 3$. This code is capable of assuring a remarkable minimum distances increase if a circulant expansion procedure is applied, also with not too high values of the scale factor p.

On the other hand, 1/3-rate column-irregular codes, characterized by an average column weight $\langle w_c \rangle = 2$, with m_h as small as 1, are able to reach $d = 5$. For irregular codes an analysis based on the girths id not efficient, especially if we have also columns with weight 1.

Example 12.15 In Fig. 12.20, two irregular 1/3-rate convolutional codes are described, both characterized by $m_h = 1$ and six 1-symbols in their $\underline{H}_o^T$ sub-matrices, leading to $\langle w_c \rangle = 2$. The column weight distribution is: one 3-weight column,

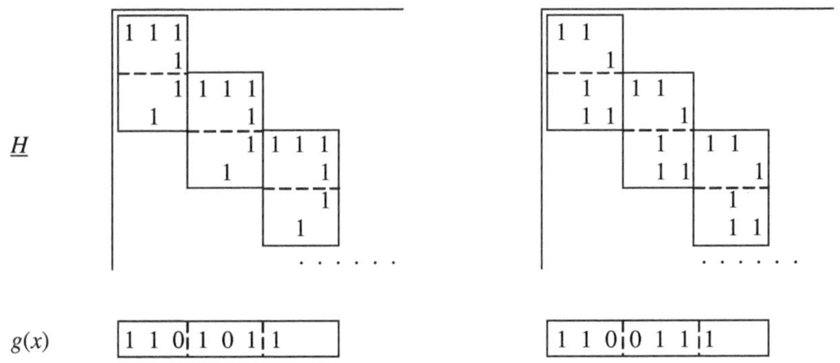

A(5) = Z , A(6) = Z − 1 , A(7) = 0 , A(8) = Z − 2   A(5) = Z , A(6) = A(7) = 0 , A(8) = Z − 3

Fig. 12.20 Semi-infinite parity check matrices and interleaved generator polynomials for the 1/3-rate convolutional code treated in Example 12.15

one 2-weight column, and one 1-weight column in both. We can observe several vertical interval reused. Both codes show $d = 5$. Remark that the latter code, although column-irregular, is row-regular, having even and odd rows both with weight $w_r = 3$. It is characterized by $B(5)$ as small as 1 and $B(6) = B(7) = 0$, like the code described in Example 12.14, but with a reduced value for m_h and m.

Example 12.16 The convolutional codes introduced in Example 12.15, in their tail-biting version over a number p of periods, can be transformed into equivalent QC block codes, according to the procedure described at the end of the previous section. An expression like (12.21) can be employed for the symbolic form of their $\underline{H}$ matrix, but with two row layers. In this case we have column reordering and also row reordering. Precisely we obtain the following symbolic parity check matrices

$$\underline{H} = \begin{bmatrix} \underline{0}_p & \underline{0}_p & (\underline{0}_p + \underline{-1}_p) \\ - & \underline{-1}_p & \underline{0}_p \end{bmatrix},$$

$$\underline{H} = \begin{bmatrix} \underline{0}_p & (\underline{0}_p + \underline{-1}_p) & - \\ - & \underline{-1}_p & (\underline{0}_p + \underline{-1}_p) \end{bmatrix}.$$

In this representation the total weight of a column of circulants coincide with those of the correspondent columns of the syndrome former sub-matrix. So regular convolutional codes will be characterized by column of circulants having all the same weights.

With $w_c = 3$, the minimum distance which can be achieved, even with regular codes, is much higher. For instance, with an arrangement derived, after row and column reordering and unwrapping, from a generalized array code, there is the bound $d \leq 24$, given by (11.21). This goal appears achievable with code rates of 1/4, 2/5 and 3/6, all characterized by $Y = w_c = 3$, provided that p is high enough.

Even higher upper bounds can be established removing the constraint of having only permutation circulants (Smarandache and Vontobel 2012). Irregular codes are not constrained by these upper bounds, and, in principle, can show even a better behaviour, in this case with $\langle w_c \rangle = 3$. The following brief notes give an outline about the possible performance of convolutional codes characterized by a code rate lower than 1/3, considering either m_h minimization (permitted by the reuse of vertical separations) or d maximization.

Lemma 12.8 *In a $1/n_o$-rate convolutional code, the same vertical separation between 1-symbols can be used at most $(n_o - 1)$ times, in order to satisfy the row-column constraint.*

Proof We can proceed as in the proof of Lemma 12.7, repeating the enumeration with all the $(n_o - 1)$ rows in each row layer of $\underline{H}$. For two rows $(n_o = 3)$, subdividing the rows in even or odd, this conclusion was already drawn, and confirmed by Example 12.12. □

This proposition assures the possibility of having very short constraint lengths m_h. About the minimum distance, even in very simple situations, like that of regular codes with $w_c = 2$, an exhaustive analysis is rather complicate (Bocharova et al. 2009). Anyway irregular codes typically perform better, either in the achievable values for d, or in the decoding algorithm behaviour and hence in the overall performance. In Sect. 10.9, we observed that regular LDPC codes are not recommended, especially at low code rates, because of threshold problems in the sum-product algorithm, evidenced by means of a density evolution analysis. In conclusion, the most efficient exploitation of the general principle expressed by Lemma 12.8 and of the high values available for d is with irregular low-rate convolutional codes.

Table 12.4 Parameters of some 1/4-rate LDPC convolutional codes with $w_c = 2$ or $\langle w_c \rangle = 2$

Type	$h_j^*(x)$	$g(x)$	m_h	m	d
Reg.	0000.1100	1100.1111	1	1	6
	0101.0010				
	1010.0001				
Reg.	1000.0000.1100	1100.0001.0110.0011	2	3	7
	0000.0011.0010				
	0000.0100.0001				
Reg.	0010.0000.1100	1100.0000.0011.0001.1001.0001	2	5	8
	0100.0000.0010				
	1000.0001.0001				
Irreg.	1000.1100	1100.1011.0100.0001	1	3	7
	1010.0001				
	0100.0001				

Example 12.17 In Table 12.4 four codes, having code rate 1/4 and $w_c = 2$ or $\langle w_c \rangle = 2$, are described, giving the three reciprocal parity check polynomials, the generator polynomial, the constraint lengths m_h, m, and the minimum distance d. The one reported in the first row has a minimum distance only one unit less than the best 1/4-rate convolutional code with $m = 1$ (see Table 3.5). With increasing m_h, d grows, as expected, together with m. The code in the last row is irregular, and exhibits a value for the minimum distance d, which is one unit greater than the correspondent value obtained with the regular code having equal m_h (first row). This result is possible because m is greater. Finally the maximum value for d which is achievable with a regular code characterized by $m_h = 2$ is $d = 8$, as obtained by the third code. It is shows a generator constraint length m as high as 5.

12.9 LDPC Convolutional Codes with Period Length and Number of Information Symbols Per Period not Co-prime

As already noted at the end of Sect. 10.7, a code whose period length n_o and number of information symbols per period k_o have a factor P in common typically exhibits better performance with respect to a code with the same code rate k_o/n_o, but with n_o and k_o co-prime. We will discuss here briefly the case of 2/4-rate LDPC convolutional codes, in comparison with 1/2-rate LDPC convolutional codes with similar values of the sliding window length and column weights.

The 2/4-rate convolutional code discussed in Example 8.6 satisfies the row-column constraint (see Fig. 8.7). Therefore it could be adopted as an LDPC code. The scheme of the information and control symbol sequence in the period is *icic*. The constraint lengths respectively turn out to be $m_h = 2$ and $m = 1$. This code is column-irregular, with two columns of $\underline{H}_o^T$ on four having weight 1. The minimum distance turns out to be rather sacrificed ($d = 3$), not only because of an average column weight as small as $\langle w_c \rangle = 1.75$, but mainly as a consequence of a bad choice in the column weight distribution. It is possible to obtain better results in a comparable situation.

Example 12.18 In Fig. 12.21 another 2/4-rate LDPC convolutional code is described. It is characterized by the scheme *iicc*, exhibits $m_h = m = 2$, and is column-irregular with average column weight $\langle w_c \rangle = 1.75$ (as in the previous case). Its minimum distance is $d = 4$, exactly as the best 1/2-rate column-regular convolutional code with $w_c = 2$.

Example 12.19 In Fig. 12.22 a column-regular 2/4-rate LDPC convolutional code is described. It is characterized by the scheme *iicc*, exhibits $m_h = m = 3$, and column weight $w_c = 2$. Its minimum distance is $d = 5$, one unit more with respect to the best 1/2-rate column-regular convolutional code with $w_c = 2$.

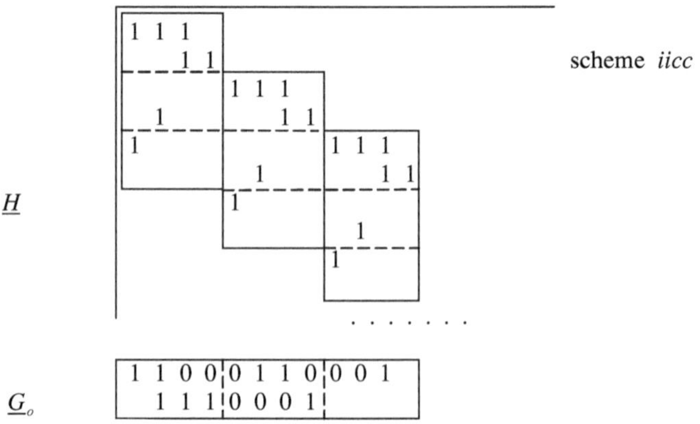

$$A(4) = Z \ , A(5) = 5Z - 6 \ , A(6) = 6Z - 16$$

Fig. 12.21 Semi-infinite parity check matrix and interleaved generator polynomials for the 2/4-rate irregular convolutional code treated in Example 12.18

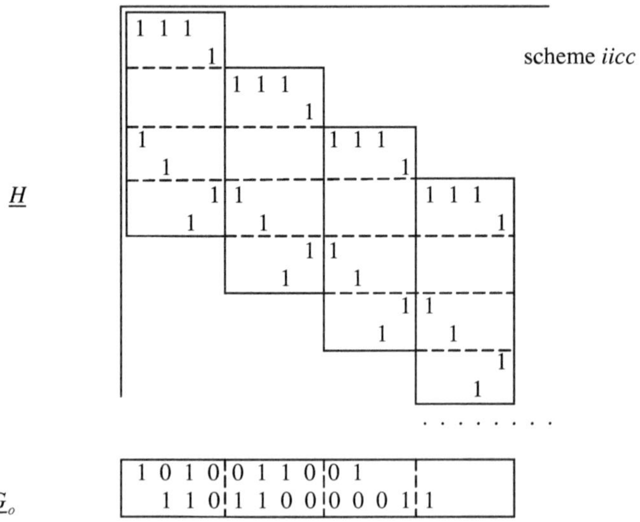

$$A(5) = 2Z - 1 \quad , \quad A(6) = 7Z - 14 \quad , \quad A(7) = 8Z - 23 \quad ,$$

Fig. 12.22 Semi-infinite parity check matrix and interleaved generator polynomials for the 2/4-rate regular convolutional code treated in Example 12.19

Example 12.20 The symbolic parity check matrices of the QC block codes obtained from the convolutional codes described in Examples 12.18 and 12.19, in tail-biting arrangement over p periods, after inverse reordering, are respectively as follows

$$\underline{H} = \begin{bmatrix} (\underline{0}_p + \underline{-2}_p) & \underline{0}_p & \underline{0}_p & - \\ - & \underline{-1}_p & \underline{0}_p & \underline{0}_p \end{bmatrix},$$

$$\underline{H} = \begin{bmatrix} (\underline{0}_p + \underline{-2}_p) & \underline{0}_p & \underline{0}_p & \underline{-3}_p \\ - & \underline{-2}_p & \underline{-3}_p & \underline{0}_p \end{bmatrix}.$$

12.10 Final Comments

The last Examples have shown that, although independently designed, the relation between QC block LDPC codes and LDPC convolutional codes, in tail-biting forms, enables us to consider them as two different organizations for the same geometrical structure. It is even possible also to consider directly in the block code a modified version of its $\underline{H}$ matrix with truncated circulants instead of square circulants, which becomes equivalent to the not tail-biting arrangement of its convolutional counterpart (see Sect. 8.7). QC codes whose $\underline{H}$ matrix is made by just one layer of circulants (see Sect. 11.9) have been demonstrated to be equivalent to tail-biting high-rate convolutional LDPC codes, of the type treated here in Sects. 12.6 and 12.7. Also for them, it is possible to employ a truncated circulant version, corresponding to not tail-biting convolutional codes.

In Sect. 10.5, in the framework of superimposition method, truncated circulants have been adopted in a block code (see Fig. 10.12). Clearly this code is equivalent to a convolutional LDPC code, not in tail-biting arrangement. In that case the row-column constraint satisfaction and the choice of the code rate have been already imposed by the original block code (a simplex code). Nevertheless sometimes it is possible to start from a base code, which does not satisfy the row-column constraint. Circulant expansion has been exploited to this purpose in several cases, often starting directly from a convolutional code (see, for instance, the best 1/2-rate convolutional code with $m = 2$, described in Fig. 10.14). The minimum distance of a convolutional code, after the permutation circulant substitution, can increase, especially if the low-weight normalized coefficient $B(d)$, $B(d+1)$, … have small values. This fact is similar to what occurs in a code concatenation scheme, with the introduction of an interleaver.

Example 12.21 There is a very simple 1/2-rate convolutional code, whose $\underline{H}_o^T$ sub-matrix, with $w_c = 2$, is not in minimal form. It has been already treated in Example 7.13 and subsequent considerations, and is characterized by $d = 3$, $B(3) = B(4) = 1$. More precisely it is described by $h^*(x) = x^5 + x^2 + x + 1$, but is equivalent to an irregular code having $h^*(x) = x^3 + x^2 + 1$ (see Fig. 7.9). They both satisfy the row-column constraint, but the former one is better suited for circulant expansion application. We can substitute permutation circulants to its 1-symbols as described in the following. Adopting $p \geq 2$, it is possible to eliminate the unique 3-weight 1-symbol pattern with the $p/(2p)$-rate convolutional code described in Fig. 12.23 on

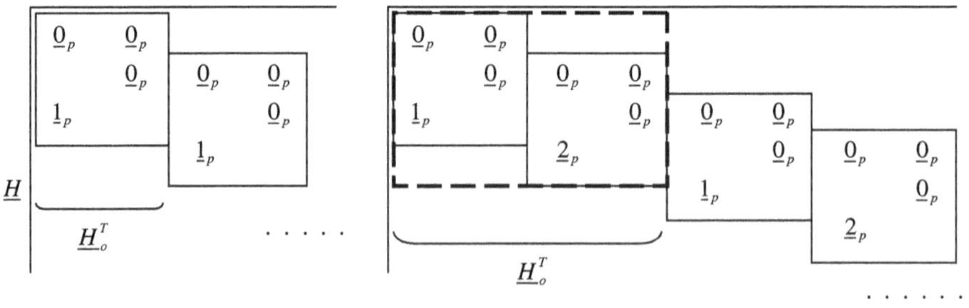

Fig. 12.23 Parity check matrices and low-weight normalized coefficients for the two convolutional codes obtained after application of circulant expansion

the left. Adopting $p \geq 3$, it is possible to eliminate also the unique 4-weight 1-symbol pattern with the $2p/(4p)$-rate convolutional code described in Fig. 12.23 on the right. Taking $p = 3$, we obtain a scheme of the type *iiiccc* in the first and of the type *iiicccciiiccc* in the second. The normalized weight distributions respectively are $B(4) = p$ and $B(5) = p$ (the last result, that is a unique 5-weight 1-symbol pattern, only for $p \geq 4$). These outcomes show that, in a situations so simple, it is useless to take values of p greater than few units, but this is enough for increasing the minimum distance from $d = 3$ respectively to $d = 4$ and $d = 5$. With larger circulant sizes, and an optimized choice of the circulant phases, even higher minimum distance increases can be obtained.

Clearly the cases treated in Example 12.21 are extremely simple and circulant expansion gives a benefit with p higher than a very low value. As long as the initial structure becomes more and more involved, this value grows, and the design of the resulting desired code is progressively more complicated. Nevertheless, once adopted an optimized design, a threshold remains, above which a further increase in p becomes no longer necessary. This phenomenon is well known in convolutional codes, about the length of the frame, and is justified by the observation that such expanded codes are only a transformed form of the tail-biting version of suitable convolutional codes (seen from another point of view). Nevertheless the benefit with respect to the original convolutional code (before circulant expansion) is here only due to the presence of an interleaver. In fact, any circulant can be interpreted as produced by a suitable interleaver. In other terms, circulant expansion does not lower the code rate.

A different situation is that characterizing the codes named doubly convolutional (see Sect. 4.9 in the framework of an MLQC model, Sect. 6.10 treating MHEQC codes, and Sect. 12.2 for further possible applications deriving from BPG codes). Here two distinct processes for introducing distributed control symbols are employed, in a way very similar to what occurs in turbo-product codes, constructed with two convolutional codes (see Examples 9.10 and 9.15). Adopting this solution, the resulting code rate is typically the product of those characterizing the two original convolutional codes. In such types of codes, we have two degrees of freedom for fixing the code rate, which can be jointly optimized. Nevertheless the external structure of the $\underline{H}$ matrix remains that of a convolutional code, with an $\underline{H}_o^T$ sub-matrix (typically rather large), shifting along a pseudo-diagonal. This property is evident in MHEQC codes or in the code treated in Sect. 12.2. About turbo-product codes, for which the row-column constraint, if verified in the two component codes, is certainly verified also in the resulting code (after Lemma 10.3), we can look at Fig. 9.16. Like in any direct product code, we have an $\underline{H}$ matrix made by two parts, but proper row reordering can lead to an interleaved version of such two parts, in which a unique $\underline{H}_o^T$ sub-matrix comes out. In all these solutions, one of the two convolutional codes maintains the property of having a frame length whose number of periods can be as high as one wishes. Nevertheless the possible sliding window decoding procedure will be characterized by a window size proportional to the parity check constraint length m_h of such convolutional code, but also to the overall frame of the other convolutional code. In this sense, the encoding memory, where these doubly convolutional codes can be constructed, must have one dimension much smaller than the other.

Modern codes (mainly related to turbo codes and LDPC codes) are progressively substituting more traditional codes in standard recommendations. For instance, in satellite digital video broadcasting (ETSI EN 301 790 2009) and in UMTS or LTE radio mobile systems (ETSI TS 136 212 2013), medium-size 1/3-rate turbo codes are suggested. In analogous way, in wireless communications (IEEE 802.16e 2012) and in 10GBASE-T Ethernet connections (IEEE 802.3an 2012) medium- and high-rate LDPC codes are being progressively adopted. They all are block codes. On the contrary, LDPC convolutional codes, although proposed in several possible situations, see for example (Kishigami et al. 2009), are not yet accepted in standards, because of their difficult compliance with traditional convolutional codes, characterized by a generator constraint length much more shorter. This observation should advise to develop a particular study of short-constraint-length LDPC convolutional codes, directly designed. To this purpose, the brief analysis presented in this chapter could give a starting contribution.

References

Baldi M, Bambozzi F, Chiaraluce F (2011) On a family of circulant matrices for quasi-cyclic low-density generator matrix codes. IEEE Trans Inf Th 57:6052–6067

Baldi M, Bianchi M, Cancellieri G, Chiaraluce F (2012) Progressive differences convolutional low-density parity check codes. IEEE Comm Lett 16:1848–1851

Baldi M, Cancellieri G, Chiaraluce F (2014) Array convolutional low-density parity-check codes. IEEE Comm Lett 18:336–339

Bates S, Elliot DC, Swamy R (2006) Termination sequence generation circuits for low-density parity-check convolutional codes. IEEE Trans Circ Syst 53:1909

Bocharova IE, Kudryashov BD, Satyukov RV (2009) Graph-based convolutional and block LDPC codes. Prob Inf Transm 45:357–377

(ETSI EN 301 790) Digital Video Broadcasting (DVB) Interaction Channel for Satellite Distribution Systems, ETSI EN 301 790, 2009

(ETSI TS 136 212) Evolved Universal Terrestrial Radio Access (E-UTRA), Multiplexing and Channel Coding, 2013

Hu XY, Eleftheriou E, Arnold DM (2001) Progressive edge-growth tanner graphs. Proc IEEE Globecom 01, San Antonio (TX): 995–1001

(IEEE 802.16e) IEEE Standard for Air Interface for Broadband Wireless Access Systems, 802.16e, 2012

(IEEE 802.3an) IEEE Standard for Ethernet, 802.3an, 2012

Jimenez-Felstrom AJ, Zigangirov KS (1999) Time-varying periodic convolutional codes with low density parity-check matrix. IEEE Trans Inf Th 45:2181–2191

Kamiya N (2007) High-rate quasi-cyclic low-density parity-check codes derived from finite affine planes. IEEE Trans Inf Th 53:1444–1459

Kishigami T, Murakami Y, Yoshii I (2009) LDPC-convolutional codes for E-MBS FEC, proposal to IEEE 802.16 broadband wireless access working group, for 802.16 m discussion and adoption

Pusane AE, Smarandache R, Vontobel O, Costello DJ (2011) Deriving good LDPC convolutional codes from LDPC block codes. IEEE Trans Inf Th 57:835–857

Smarandache R, Vontobel PO (2012) Quasi-cyclic LDPC codes: influence of proto- and Tanner graph structure on minimum Hamming distance upper bounds. IEEE Trans Inf Theory 58:585–607

Tanner RM, Sridhara D, Sridharan A et al (2004) LDPC block and convolutional codes based on circulant matrices. IEEE Trans Inf Th 50:2966–2984

Appendix A
Matrix Algebra in a Binary Finite Field

The Binary Finite Field

Definition A.1 Given a binary alphabet, with symbols 0, 1, the operation called *ex-OR sum* (or *addition*) is made according to the table shown in Fig. A.1 on the left.

Definition A.2 Given a binary alphabet, with symbols 0, 1, the operation called *product* (or *multiplication*) is made according to the table shown in Fig. A.1 on the right.

Addition is expressed by sign +, and multiplication by simply writing the two symbols (called the factors) one close to the other.

It is straightforward to verify the following properties (Hankerson et al. 2004):

1. the ensemble of symbols is closed for both operations,
2. 0 is the null element in the ex-OR sum, in the sense that $a + 0 = a$, for any symbol a,
3. 1 is the unitary element in the multiplication, that is $a1 = a$, for any symbol a,
4. the addition inverse element—a, also called the opposite, satisfies the equation $a + (-a) = 0$,
5. the multiplication inverse a^{-1}, also called the reciprocal, satisfies the equation $a^{-1}a = 1$, for any non zero symbol a,
6. both operations show commutative properties, that is $a + b = b + a$, $ab = ba$, for any pair of symbols a, b,
7. addition is associative, in the sense $a + (b + c) = (a + b) + c$, for any tern of symbols a, b, c,
8. multiplication is associative, in the sense $a(bc) = (ab)c$, for any tern of symbols a, b, c,

Fig. A.1 Addition and multiplication in a binary finite field

	0	1
0	0	1
1	1	0

addition

	0	1
0	0	0
1	0	1

multiplication

G. Cancellieri, *Polynomial Theory of Error Correcting Codes*,
Signals and Communication Technology, DOI 10.1007/978-3-319-01727-3

9. addition is distributive with respect to multiplication, that is $(a + b)c = ac + bc$, for any tern of symbols a, b, c.

Definition A.3 Also in a non-binary alphabet, when the above nine properties are satisfied, the symbols and the two operations considered are named to be a *field*. If the symbols (called also elements of the field) have a finite number, we deal with a *finite field*.

The binary case, described by the tables in Fig. A.1, is a particular case of a finite field. In it, symbol 1 is the unique non-null symbol. It coincides with its opposite and with its reciprocal. For Q-ary finite fields, proper generalizations of addition and multiplication are to be made (see also Appendix D). Not for all integer numbers a finite field exists. It can be constructed for prime numbers and integer powers of prime numbers. Rational numbers form a field. Also real numbers form a field. They both have infinite elements. Natural numbers do not form a field, because in their ensemble there are no opposites and no reciprocals. Integer numbers do not form a field, because, although each of them has an opposite, there are no reciprocals.

In both the fields of rational numbers and real numbers (and even in the ensembles of natural or integer numbers), a relationship is present, after which one element is said to be greater than another. This relationship comes out from the fact that an ordering can be adopted in a sequence of such elements, with reference to their magnitude. In finite fields, an ordering of this type is not admitted, therefore we cannot say that one element is greater than another.

Vectors and Matrices

Definition A.4 An ordered ensemble of elements (represented by as many consecutive symbols) is called a *vector*. It can be written either as a row or as a column. Conventionally we will write $\underline{u}$ for a *row vector* and $\underline{u}^T$ for the correspondent *column vector*. The transformation of the one into the other is called *vector transposition*. In a row vector, the order is conventionally from the left hand side to the right hand side. In a column vector the order is conventionally from up to down.

When the above definition is applied to symbols taken from a binary alphabet, we are considering a so called binary vector.

Example A.1 Given the binary row vector $\underline{u} = [0 \quad 1 \quad 1 \quad 0 \quad 0 \quad 1]$ the correspondent column vector, obtained from vector transposition,

$$\underline{u}^T = \begin{bmatrix} 0 \\ 1 \\ 1 \\ 0 \\ 0 \\ 1 \end{bmatrix}$$

We can adopt symbol u_j for denoting the j-th element in a vector $\underline{u}$. Subscript j expresses the position of the considered element in the vector. A vector having n elements (hence with j varying from 1 to n) is denoted as an n-vector. Transposition applied to a column vector gives back the original row vector.

Definition A.5 A doubly-ordered ensemble of elements is called a *matrix*. Symbolically the elements of a matrix $\underline{A}$ can be written A_{ij}, where i denotes the row and j the column where the considered element is placed. A matrix of this type is represented by a rectangle having two dimensions, along which ordering is performed from up to down and from left to right.

The definition of matrix can be extended also to the case of more than two dimensions. A matrix having s dimensions collects elements which can be represented by s subscripts. In the following, except when the contrary is explicitly said, we will consider only two-dimension matrices. A matrix having m rows and n columns (hence with i varying from 1 to m and j varying from 1 to n) is called an $m \times n$ matrix. When $m = n$, instead of a rectangular matrix, we have a square matrix. Finally, when the above definitions are applied to symbols taken from a binary alphabet, we can specify that the considered matrix is a binary matrix.

Definition A.6 Given an $m \times n$ matrix $\underline{A}$, whose elements are A_{ij}, another matrix $\underline{A}^T$ can be obtained from it, whose elements are A_{ji}. The latter matrix is called the *transpose* of the former one, and is an $n \times m$ matrix. All the vectors in the new matrix are proper transpositions of the vectors of the original matrix.

Example A.2 Given the 2 × 3 binary matrix $\underline{A}$ reported below on the left, its transpose $\underline{A}^T$ turns out to be as shown on the right,

$$\underline{A} = \begin{bmatrix} 0 & 1 & 1 \\ 1 & 1 & 0 \end{bmatrix}, \quad \underline{A}^T = \begin{bmatrix} 0 & 1 \\ 1 & 1 \\ 1 & 0 \end{bmatrix}$$

A vector can be considered as a particular matrix, with just one row (or one column). In the following we will introduce two operations between matrices, addition and multiplication. Precise assumptions on their dimensions are to be preliminary fixed. The same is, as particular cases, for such two operations performed on vectors.

Definition A.7 Given two binary matrices, $\underline{A}$ and $\underline{B}$, with the same numbers of rows and columns, the *matrix addition* between them is a new matrix $\underline{C} = \underline{A} + \underline{B}$, whose elements are $C_{ij} = A_{ij} + B_{ij}$, where A_{ij} are the elements of $\underline{A}$ and B_{ij} the elements of $\underline{B}$ in the same positions. Addition + is to be made according to the analogous operation adopted for the symbols representing the elements A_{ij} and B_{ij}. For a binary field, addition turns out to be an ex-OR sum.

Example A.3 An ex-OR sum matrix is shown below, assuming $m = 3$ rows and $n = 5$ columns,

$$\begin{bmatrix} 1 & 1 & 0 & 1 & 0 \\ 0 & 1 & 1 & 0 & 0 \\ 1 & 0 & 0 & 0 & 1 \end{bmatrix} + \begin{bmatrix} 1 & 0 & 0 & 1 & 1 \\ 1 & 1 & 1 & 1 & 1 \\ 0 & 0 & 0 & 1 & 1 \end{bmatrix} = \begin{bmatrix} 0 & 1 & 0 & 0 & 1 \\ 1 & 0 & 0 & 1 & 1 \\ 1 & 0 & 0 & 1 & 0 \end{bmatrix}$$

Matrix Product Row-by-Column

Definition A.8 Given an $m \times n$ matrix $\underline{A}$, whose elements are A_{ij}, and an $n \times q$ matrix $\underline{B}$, whose elements are B_{hk}, the *matrix product row-by-column* between such matrices is an $m \times q$ matrix $\underline{C} = \underline{A} \cdot \underline{B}$, whose elements can be written as

$$C_{jk} = \sum_{j=1}^{n} A_{ij} B_{jk} \tag{A.1}$$

Addition and multiplication are defined in the field of the symbols representing the elements A_{ij} and B_{hk}. For a binary filed, we have ex-OR sum and multiplication as shown in Fig. A.1.

Clearly this product is not commutative. Even when the two matrices are square (necessary condition for allowing calculation of either $\underline{A} \cdot \underline{B}$ or $\underline{B} \cdot \underline{A}$), in general, the two products are different.

Example A.4 A matrix product row-by-column is reported below, with $n = 5$, $m = 2$ and $q = 3$, with reference to a binary field,

$$\begin{bmatrix} 0 & 1 & 1 & 0 & 0 \\ 1 & 1 & 1 & 0 & 1 \end{bmatrix} \cdot \begin{bmatrix} 1 & 1 & 1 \\ 1 & 0 & 1 \\ 1 & 1 & 0 \\ 0 & 0 & 1 \\ 0 & 1 & 1 \end{bmatrix} = \begin{bmatrix} 0 & 1 & 1 \\ 1 & 1 & 1 \end{bmatrix}$$

The result is a 2×3 binary matrix as expected.

Example A.5 Two products are shown below, between the 2×2 binary square matrices $\underline{A}$ and $\underline{B}$, allowing to verify that, in general, $\underline{A} \cdot \underline{B} \neq \underline{B} \cdot \underline{A}$,

$$\underline{A} \cdot \underline{B} = \begin{bmatrix} 1 & 1 \\ 0 & 1 \end{bmatrix} \cdot \begin{bmatrix} 1 & 0 \\ 1 & 0 \end{bmatrix} = \begin{bmatrix} 0 & 0 \\ 1 & 0 \end{bmatrix},$$

$$\underline{B} \cdot \underline{A} = \begin{bmatrix} 1 & 0 \\ 1 & 0 \end{bmatrix} \cdot \begin{bmatrix} 1 & 1 \\ 0 & 1 \end{bmatrix} = \begin{bmatrix} 1 & 1 \\ 1 & 1 \end{bmatrix}$$

Lemma A.1 *This equality holds about the transpose of a product*

$$(\underline{A} \cdot \underline{B})^T = \underline{B}^T \cdot \underline{A}^T \tag{A.2}$$

Proof This result follows from Definition A.6 and (A.1). □

Example A.6 The product of a 2 × 3 matrix $\underline{A}$ by a column 3-vector $\underline{b}^T$ and the product of the row 3-vector $\underline{b}$ by $\underline{A}^T$ are reported below,

$$\underline{A} \cdot \underline{b}^T = \begin{bmatrix} 1 & 0 & 1 \\ 1 & 1 & 0 \end{bmatrix} \cdot \begin{bmatrix} 0 \\ 1 \\ 1 \end{bmatrix} = \begin{bmatrix} 1 \\ 1 \end{bmatrix},$$

$$\underline{b} \cdot \underline{A}^T = \begin{bmatrix} 0 & 1 & 1 \end{bmatrix} \cdot \begin{bmatrix} 1 & 1 \\ 0 & 1 \\ 1 & 0 \end{bmatrix} = \begin{bmatrix} 1 & 1 \end{bmatrix}$$

The latter is the transpose of the former, so confirming (A.2).

Definition A.9 Given a set of vectors with the same number of elements, they are called *linearly independent* if none of them can be obtained as a *linear superimposition* of the other ones. Linear superimposition (or *linear combination*) consists of an addition after multiplication by constants. In a binary field this operation is reduced to a simple ex-OR sum.

Definition A.10 An $n \times n$ square matrix with all 0 elements, except those on its diagonal, equal to 1, is called *$n \times n$ identity matrix*. The symbol here adopted for such matrix is $\underline{I}_n$.

Lemma A.2 *Given a square $n \times n$ matrix $\underline{A}$, the following double equality holds*

$$\underline{I}_n \cdot \underline{A} = \underline{A} \cdot \underline{I}_n = \underline{A} \tag{A.3}$$

In this sense, $\underline{I}_n$ is the unitary element in the product row-by-column between square $n \times n$ matrices.

Proof This result follows directly from (A.1) and Definition A.10. □

Definition A.11 Vectors and matrices are proper collections of symbols or variables (the elements). In order to distinguish them from single elements, say single symbols or variables, the latter ones are called *scalar constants*.

Systems of Linear Equations

In traditional linear algebra, matrices are employed in the theory of the systems of linear equations. Given N scalar unknowns, the system of as many linear equations on them can be written as follows

$$\underline{A} \cdot \underline{y}^T = \underline{b}^T, \tag{A.4}$$

where $\underline{A}$ is an $N \times N$ matrix collecting the equation coefficients, $\underline{y}^T$ is the column vector of the scalar unknowns, finally $\underline{b}^T$ is the column vector of some scalar known constants.

Definition A.12 A system of linear equations, of the type described by (A.4), contains all *linearly independent equations* if matrix $\underline{A}$ has just linearly independent rows, without any constraint on $\underline{b}^T$. Linear independent rows means that the row vectors forming our matrix are all linearly independent (see Definition A.9).

With reference to square matrices only, it is possible to introduce an inverse element in the product row-by-column performed on them. In doing this we refer to the unitary element $\underline{I}_n$ introduced in (A.3).

Definition A.13 Matrix $\underline{A}^{-1}$ represents the *inverse* of matrix $\underline{A}$. The relationship between $\underline{A}^{-1}$ and $\underline{A}$, both $n \times n$ square matrices, is

$$\underline{I}_n = \underline{A} \cdot \underline{A}^{-1} = \underline{A}^{-1} \cdot \underline{A} \tag{A.5}$$

Nevertheless, not all the square matrices can be inverted. For example, where the rows or columns of $\underline{A}$ are not all linearly independent, $\underline{A}^{-1}$ does not exist.

Theorem A.1 (Leibniz-Cramer) *If the N equations in a linear system are all linearly independent, and* $\underline{b}^T$ *is a non-null vector,* (A.4) *can be solved* (Cramer 1750), *leading to a set of solutions* $\underline{y}^T$, *so formally expressed*

$$\underline{y}^T = \underline{A}^{-1} \cdot \underline{b}^T \tag{A.6}$$

Proof Multiplying (A.4) on its left hand side by $\underline{A}^{-1}$ yields

$$\underline{A}^{-1} \cdot \underline{A} \cdot \underline{y}^T = \underline{A}^{-1} \cdot \underline{b}^T$$

After (A.5) and (A.3) we are left with (A.6). □

Now our attention will be focused on the calculation of the inverse of a square matrix $\underline{A}$.

Definition A.14 A matrix $\underline{A}$ which admits inversion will be named *non-singular*.

Definition A.15 Given an element A_{ij} of a square matrix $\underline{A}$, the sub-matrix obtained eliminating in $\underline{A}$ the *i*-th row and the *j*-th column is called *(i, j) minor* of A_{ij} in $\underline{A}$, and denoted $Mi(A_{ij})$.

Definition A.16 Given a non-singular matrix $\underline{A}$, its *determinant* $\det(\underline{A})$ is the scalar quantity which can be obtained by the following iterative procedure

$$\det(\underline{A}) = \sum_{j=1}^{N} A_{ij}(-1)^{i+j}\det\left[Mi\left(A_{ij}\right)\right] \tag{A.7}$$

The same result can be obtained for whichever row i. The same result can be obtained also substituting i to j, and hence summing along a column and not along a row.

Example A.7 Assuming a 3 × 3 matrix $\underline{A}$ of the type reported below on the left, with symbols taken in the ensemble of integer numbers (positive and negative), the calculation of its determinant is made as described on the right,

$$\underline{A} = \begin{bmatrix} 1 & 2 & -3 \\ 0 & -1 & 3 \\ 2 & -2 & 2 \end{bmatrix}, \det(\underline{A}) = 1 \cdot (-2+6) - 2 \cdot (0-6) - 3 \cdot (0+2) = 10.$$

The elements of the first row and their minors have been used, according to (A.7).

Now we will give some important properties without demonstrating them. For a proof of these results, see, for instance, (Brown 1991).

Theorem A.2 *The inverse* $\underline{A}^{-1}$*of an* $n \times n$ *matrix* $\underline{A}$*, if it is non-singular, is a square matrix, still* $n \times n$*, whose generic element turns out to be*

$$A^{-1}{}_{ij} = (-1)^{i+j} \frac{\det[Mi(A_{ji})]}{\det(\underline{A})} \tag{A.8}$$

Example A.8 The inverse of matrix $\underline{A}$ introduced in Example A.7 results in

$$\underline{A}^{-1} = \frac{1}{10} \begin{bmatrix} 4 & 2 & 3 \\ 6 & 8 & -3 \\ 2 & 6 & -1 \end{bmatrix}$$

Remark that, in order to perform this calculation, the ensemble of elements has been extended to the field of rational numbers.

From Theorem A.1, expression (A.8) allows to calculate the solution of a system of N linear equations in as many unknowns of the type (A.4). As anticipated, and also from direct inspection of expression (A.8) where $\det(\underline{A})$ is at the denominator, such N linear equations must be all linearly independent.

Finally we can observe that, in a binary field, any multiplication by $(-1)^{i+j}$ in (A.7) and (A.8) can be ignored. Furthermore the determinant of a binary matrix can be only 0 or 1. Therefore the division where it is placed at the denominator is not influent on the result, provided that it is not 0.

Example A.9 The following binary matrix has non-null determinant

$$\det(\underline{A}) = \det\left(\begin{bmatrix} 1 & 1 & 1 \\ 1 & 1 & 0 \\ 0 & 1 & 1 \end{bmatrix}\right) = 1 \cdot 1 + 1 \cdot 1 + 1 \cdot 1 = 1$$

Its inverse turns out to be

$$\underline{A}^{-1} = \begin{bmatrix} 1 & 0 & 1 \\ 1 & 1 & 1 \\ 1 & 1 & 0 \end{bmatrix}$$

Lemma A.3 *Necessary and sufficient condition for having det($\underline{A}$) = 0 is that there are rows in $\underline{A}$ linearly dependent on other rows.*

The above property holds also if one substitutes columns to rows, although in the following, since we start from the form (A.4) for representing a system of linear equations, we will refer typically to rows only.

Lemma A.4 *In a square matrix $\underline{A}$, with det($\underline{A}$) = 0, the numbers of linearly independent rows equals the number of linearly independent columns.*

Example A.10 Let us consider the following 3 × 3 binary matrix $\underline{A}$, characterized by det($\underline{A}$) = 0,

$$\underline{A} = \begin{bmatrix} 1 & 0 & 1 \\ 1 & 1 & 0 \\ 0 & 1 & 1 \end{bmatrix}$$

In it, the ex-OR sum of the first two rows gives the third row. Analogously the ex-OR sum of the first two columns gives the third column.

Lemma A.5 *The maximum number of vectors having n elements, all linearly independent, is n.*

Proof Let us consider n row vectors, having n elements, with just one 1-symbol each in as many different positions. Constructing an $n \times n$ square matrix from them, in proper sequence, gives rise to the $n \times n$ identity matrix $\underline{I}_n$. Any possible further row turns out to be linearly dependent on the above n rows. □

Uniqueness and Multiplicity for the Solutions of a System of Linear Equations

Lemma A.6 *Given the linear system expressed by* (A.4), *where $\underline{A}$ is an $N \times N$ matrix, and $\underline{b}^T \neq \underline{0}^T$ if it admits solution, such solution is unique.*

Proof After Theorem A.1, this solution is given by (A.6). On the other hand, (A.7) and (A.8) show that $\underline{A}^{-1}$ is unique. Therefore solution (A.6) is unique. □

Let us consider now N unknowns, but a different number M of equations. It is necessary to distinguish two situations: $M > N$ and $M < N$. For both we will still assume $\underline{b}^T \neq \underline{0}^T$.

Definition A.17 An $m \times n$ matrix $\underline{A}$, in general with $n \neq m$, is said to have *rank p*, if p is the maximum number of rows or columns in it, which are linearly independent. It is denoted as rank($\underline{A}$).

After Lemma A.5, the following inequality holds

$$p \leq \min\{n, m\} \tag{A.9}$$

A generalization of equality (A.4) regards the situation in which matrix $\underline{A}$ is an $M \times N$ matrix, characterized by M rows and still N columns. In this situation the unknown vector $\underline{y}^T$ remains a column N-vector, whereas vector $\underline{b}^T$, containing known constants, becomes a column M-vector.

Definition A.18 In an equation of the type (A.4), $\underline{A}$ is called *coefficient matrix* of the systems. A new matrix can be obtained adding a last column where $\underline{b}^T$ is inserted. This $M \times (N + 1)$ matrix is called *extended coefficient matrix* of the system.

Theorem A.3 (Frobenius-Capelli) *A system of M linear equations in N unknowns, with* $\underline{b}^T \neq \underline{0}^T$, *admits solutions only when the rank of its coefficient matrix coincides with the rank of its extended matrix* (Capelli 1886; Frobenius 1911).

Proof For $M \leq N$ this condition is always verified after (A.9). For $M > N$, if this condition is not verified, there are incompatible equations, and hence they cannot be contemporary satisfied. □

Lemma A.7 *In a system of M linear equations with N unknowns, characterized by* $M > N$, *once verified the condition expressed by* Theorem A.3, *in order to have a unique solution, it is necessary that rank*($\underline{A}$) $= N$.

Proof From (A.9), rank($\underline{A}$) cannot be greater than N. If rank($\underline{A}$) $= N$, there are $(M - N)$ equations which depend linearly on the remaining N. They can be eliminated, and we are left with a system with N independent equations in as many unknowns, already treated, and able to admit a unique solution. □

Theorem A.4 (Rouché-Fontené) *In a system of M linear equations in N unknowns, with non-null* $\underline{b}^T$, *and characterized by rank*($\underline{A}$) $< N$, *the solutions are more than one* (Rouché 1875; Fonténé 1892).

Proof We can treat this situation as follows. Selecting the first rank($\underline{A}$) unknowns, a solution of this sub-system is unique. A number $[N - \text{rank}(\underline{A})]$ of unknowns remain undetermined. Therefore to the above solution, obtained for the first rank($\underline{A}$) unknowns, can be attributed a multiplicity as high as the total number of the different possible sequences of $[N - \text{rank}(\underline{A})]$ elements belonging to the considered field. □

Corollary a *In the situation described by* Theorem A.4, *for a finite field characterized by Q elements, the multiplicity of the solution is* $Q^{N-rank(\underline{A})}$.

Proof This is the number of all the possible sequences of $[N - \text{rank}(\underline{A})]$ elements taken from a Q-ary alphabet. They play the role of as many independent parameters in the unique solution regarding the first $\text{rank}(\underline{A})$ unknowns. □

So if we are working in a binary field the multiplicity of the solution is $2^{N-rank(\underline{A})}$. Instead, if the considered field has infinite elements, like for rational or real numbers, the multiplicity becomes $\infty^{N-rank(\underline{A})}$.

Example A.11 In a binary field, let us consider the following situation, with $M = 2$ and $N = 3$,

$$\underline{A} = \begin{bmatrix} 1 & 0 & 1 \\ 1 & 1 & 1 \end{bmatrix} \qquad \underline{b}^T = \begin{bmatrix} 1 \\ 0 \end{bmatrix}$$

The rank of $\underline{A}$ is 2. The two linear equations of this system can be written as

$$y_1 + y_3 = 1,$$
$$y_1 + y_2 + y_3 = 0.$$

The third unknown can be transferred to the right hand side of both equations, leading to

$$y_1 = 1 + y_3,$$
$$y_1 + y_2 = y_3.$$

In this new system, with 2 equations in 2 unknowns $(y_1 \text{ and } y_2), y_3$ can be considered as an independent parameter. Since the rank of the coefficient matrix is 2, Theorem A.1 can be applied, leading to the solution

$$\begin{bmatrix} y_1 \\ y_2 \end{bmatrix} = \begin{bmatrix} (1 + y_3) \\ 1 \end{bmatrix},$$

which holds either with $y_3 = 0\,(\text{and } y_1 = 1, y_2 = 1)$ or with $y_3 = 1\,(\text{and } y_1 = 0, y_2 = 1)$.

Example A.12 In the field of rational numbers, let us consider the following situation, with $M = 3$ and $N = 3$,

$$\underline{A} = \begin{bmatrix} 1 & -1 & 0 \\ 2 & 0 & -3 \\ 1 & 1 & -3 \end{bmatrix}, \underline{b}^T = \begin{bmatrix} 2 \\ 0 \\ -2 \end{bmatrix}.$$

Here $\text{rank}(\underline{A})$ is 2, in fact the third row of $\underline{A}$ is the difference of the second minus the first. Therefore one equation can be eliminated. Remark that even the 2×4 extended matrix formed by addition of one more column containing $\underline{b}^T$ has still rank 2, so Theorem A.3 is satisfied. After elimination of the third equation, we are left with a situation similar to the one treated in Example A.11, that is

$$y_1 - y_2 = 2,$$
$$2y_1 - 3y_3 = 0,$$

leading to the solution

$$y_1 = \frac{3}{2}y_3, \quad y_2 = \frac{3}{2}y_3 - 2.$$

This solution has multiplicity ∞^1, owing to all the infinite possible values assumed by y_3. We can consider this result as a sort of parametric solution, being y_3 the parameter.

Homogeneous Linear Systems

A quite different situation occurs when $\underline{b}^T = \underline{0}^T$, that is when all the known constants are null.

Definition A.19 A system of linear equations expressed by (A.4), with $\underline{b}^T = \underline{0}^T$, that is

$$\underline{A} \cdot \underline{y}^T = \underline{0}^T, \tag{A.10}$$

is called a *homogeneous system*. It always admits a null solution $\underline{y}^T = \underline{0}^T$, called the *trivial solution*.

Lemma A.8 *If system* (A.10) *admits a solution different from the trivial one, any other solution obtained multiplying it by a non-null scalar constant is valid.*

Proof Once obtained a solution $\underline{y}^T = \underline{p}^T$, (A.10) can be substituted by $\underline{A} \cdot \underline{p}^T = \underline{0}^T$. Therefore multiplying both sides of such identity by whichever non-null scalar constant λ, we obtain $\underline{A} \cdot (\lambda \underline{p}^T) = \underline{0}^T$, which means that also $\lambda \underline{p}^T$ is solution of (A.10). This conclusion is independent of the number of elements in the considered field.□

The following proposition gives a condition for the presence of such a non-trivial solution (which could be also derived from Theorem A.3) and an expression for it.

Lemma A.9 *Given the homogeneous system* (A.10), *where* $\underline{A}$ *is an* $N \times N$ *square matrix, a necessary and sufficient condition in order to have solutions, besides the trivial one, is det*($\underline{A}$) = 0. *If we have rank*($\underline{A}$) = $N - 1$, *the non-trivial solution can be calculated as*

$$y_j = \lambda(-1)^{N+j} \det[Mi(N, j)], \quad j = 1, 2, \ldots, N, \tag{A.11}$$

where $Mi(N, j)$ is the minor obtained suppressing the jth column and the Nth row in the $\underline{A}$ matrix, whereas λ is the scalar constant introduced by Lemma A.8, which can be properly expressed as a function of one unknown, say y_N.

Proof Transferring the Nth unknown to the right hand side of the equations, like in Examples A.11 and A.12, gives rise to a new system, characterized by $(N-1)$ equations in the $(N-1)$ unknowns from y_1 to y_{N-1}. It admits a unique solution, of the type (A.6), in which $\underline{b}^T$ is due to y_N. The equality det($\underline{A}$) = 0 justifies expression (A.11), where we have assumed $\lambda = y_N/\{(-1)^{N+N}\det[Mi(N,N)]\} = y_N/\det[Mi(N,N)]$, provided that $\det[Mi(N,N)] \neq 0$. If necessary, a permutation in the order of the rows can be made with the purpose of having $\det[Mi(N,N)] \neq 0$, leading to a different unknown to be taken as a parameter. □

Example A.13 Let us consider the following homogeneous system, characterized by $N = 3$, defined in the field of rational numbers,

$$\begin{aligned} y_1 - y_2 - 2y_3 &= 0 \\ 2y_2 + y_3 &= 0. \\ 2y_1 - 3y_3 &= 0 \end{aligned}$$

First of all, we can observe that det($\underline{A}$) = 0, since the third equation is obtained from the sum of the second plus the first multiplied by 2. Next we obtain

$$\underline{A} = \begin{bmatrix} 1 & -1 & -2 \\ 0 & 2 & 1 \\ 2 & 0 & -3 \end{bmatrix}, \ \det[Mi(3,3)] = 2, \det[Mi(3,1)] = 3, \det[Mi(3,2)] = -1.$$

So the non-trivial solution is here represented by the two relationships

$$y_1 = \frac{3}{2}y_3, \quad y_2 = -\frac{1}{2}y_3,$$

where y_3 can assume whichever rational value. Remark that also the trivial solution is included in this expression, after the choice $y_3 = 0$.

Example A.14 Assuming the binary homogeneous system, with $N = 3$,

$$\begin{aligned} y_1 + y_3 &= 0 \\ y_2 + y_3 &= 0\,, \\ y_1 + y_2 &= 0 \end{aligned}$$

it is possible to write

$$\underline{A} = \begin{bmatrix} 1 & 0 & 1 \\ 0 & 1 & 1 \\ 1 & 1 & 0 \end{bmatrix}, \quad \det[Mi(3,3)] = 1, \quad \det[Mi(3,1)] = 1, \quad \det[Mi(3,2)] = 1.$$

Also here det($\underline{A}$) = 0, in fact the third row of $\underline{A}$ is the sum of the first two rows. The trivial solution is $y_1 = 0, y_2 = 0, y_3 = 0$,whereas the non-trivial solution can be represented by $y_1 = y_3, y_2 = y_3$, that is $y_1 = 1, y_2 = 1, y_3 = 1$.

Remark that in a Q-ary finite field the non-trivial solution has multiplicity $(Q-1)$, related to all the non-null elements of the field. So in a binary field, it is characterized by multiplicity 1, exactly as the trivial solution. This situation is similar to that of a non-homogeneous system with $M = N - 1$, and a coefficient matrix with full rank.

In general, it is possible to recognize a resemblance between the case of non-homogeneous systems with rank($\underline{A}$) < N and the case of a homogeneous system treated here. Remark that the condition of having rank($\underline{A}$) = $N - 1$ is of basic importance for the validity of the treatment developed so far. A more general situation, characterized by rank($\underline{A}$) < $N - 1$, can be faced by increasing the multiplicity of the non-trivial solution. For instance, in a Q-ary finite field, the number of non-trivial solutions of a homogeneous system is $(Q^{N-rank(\underline{A})} - 1)$. In other terms, they are $Q^{N-rank(\underline{A})}$ included the trivial one.

As a final remark, we can observe that often the solution of a system of linear equations can be obtained by subsequent substitutions, and this procedure may be simpler than that described so far. In this sense, it is always possible to take two equations, perform a linear superimposition of them, and then replace one of the two with such linear superimposition. This is equivalent to operate subsequent substitutions, if the number of unknowns in any row progressively decreases. Solutions based on subsequent substitutions, although sometimes simpler, suffer the limitation of being suitable only for particular situations. Nevertheless, the solutions (A.6) for a non-homogeneous system with rank($\underline{A}$) = N and (A.11) for a homogeneous system with rank($\underline{A}$) = $N - 1$ are completely general.

Matrices and Sub-matrices

Definition A.20 In an $m \times n$ matrix $\underline{A}$, it is possible to identify some *sub-matrices*, separating it into parts by proper horizontal or vertical straight boundaries.

Example A.15 Given a 3×5 matrix $\underline{A}$ of this type

$$\underline{A} = \left[\begin{array}{ccc|cc} 1 & 0 & 0 & 1 & 1 \\ 0 & 1 & 0 & 0 & 1 \\ 0 & 0 & 1 & 1 & 0 \end{array}\right],$$

it can be decomposed as follows

$$\underline{A} = [\underline{I}_3 \mid \underline{B}],$$

where $\underline{I}_3$ is the 3×3 identity matrix and $\underline{B}$ is the remaining 3×2 matrix.

Inversion or transposition can be transferred to the sub-matrices, employing some useful equalities. Among them, the following ones have straightforward demonstration. For a non-singular square matrix $\underline{A}$, we have

$$\det(\underline{A}^T) = \det(\underline{A}), \quad (\underline{A}^T)^{-1} = (\underline{A}^{-1})^T. \tag{A.12}$$

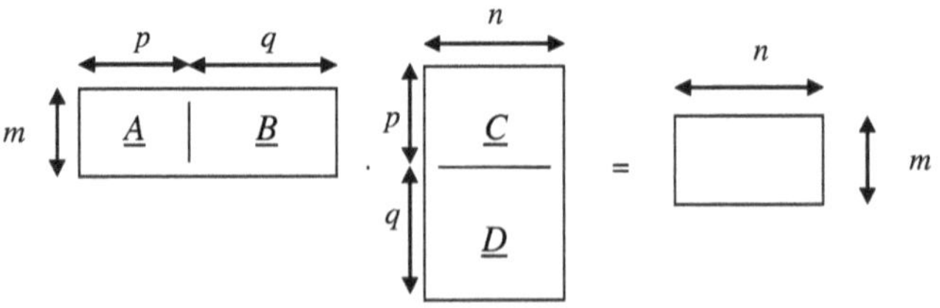

Fig. A.2 Product of two partitioned matrices giving a not partitioned matrix as result

For any rectangular matrix $A = [\underline{a}_1 | \underline{a}_2]$, it is possible to write

$$\underline{A}^T = \begin{bmatrix} \underline{a}_1^T \\ \hline \underline{a}_2^T \end{bmatrix}. \tag{A.13}$$

Assuming the two partitioned matrices shown in Fig. A.2 on the left, where p and q are suitable integer numbers, their product row-by-column gives rise to the matrix depicted on the right, characterized by the same number m of rows as $\underline{A}$ or $\underline{B}$, and the same number n of columns as $\underline{C}$ or $\underline{D}$. Although $\underline{A} \cdot \underline{C}$ and $\underline{B} \cdot \underline{D}$ could be calculated, the overall product does not exhibit such partial results in any form. In fact multiplication row by column requires employment of a whole row and of a whole column.

A completely different situation is that depicted in Fig. A.3, where the resulting matrix can be partitioned into four sub-matrices, all capable of being expressed by proper products of pairs of sub-matrices. Sub-matrices $\underline{A} \cdot \underline{C}$, $\underline{A} \cdot \underline{D}$, $\underline{B} \cdot \underline{C}$, $\underline{B} \cdot \underline{D}$ are evidenced, showing a sort of distributive property in the final rectangular matrix.

Sometimes a matrix is expressed as a function of a scalar parameter. This means that all its elements are scalar functions of such parameter. In some particular situations, the function depends on another set of matrices. This is, for instance, the case treated below.

Definition A.21 Given a pair of binary matrices $\underline{A}$ and $\underline{B}$, *Kronecker product* of the first by the second is the new matrix, where any 1-symbol of $\underline{A}$ is replaced by $\underline{B}$, whereas any 0-symbol is substituted by a proper $\underline{0}$ matrix. The new matrix is denoted $\underline{A} \otimes \underline{B}$.

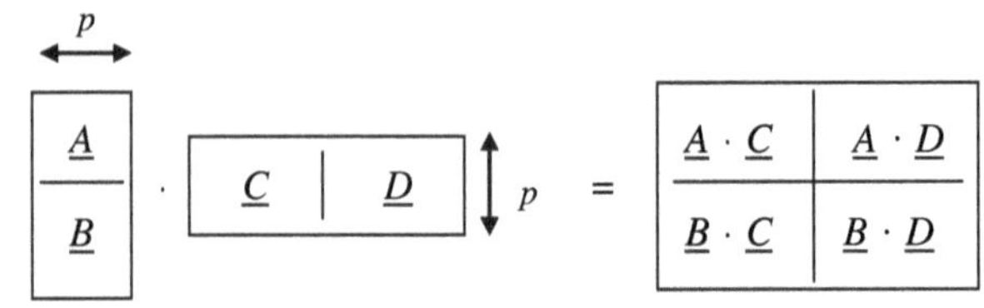

Fig. A.3 Product of two partitioned matrices giving a partitioned matrix also as result

$$\underline{A} \otimes \underline{B} = \left[\begin{array}{cc|cc|cc|cc} 1 & 0 & 1 & 0 & 0 & 0 & 1 & 0 \\ 1 & 1 & 1 & 1 & 0 & 0 & 1 & 1 \\ \hline 0 & 0 & 1 & 0 & 1 & 0 & 1 & 0 \\ 0 & 0 & 1 & 1 & 1 & 1 & 1 & 1 \\ \hline 0 & 0 & 1 & 0 & 0 & 0 & 1 & 0 \\ 0 & 0 & 1 & 1 & 0 & 0 & 1 & 1 \end{array}\right] \qquad \underline{B} \otimes \underline{A} = \left[\begin{array}{cccc|cccc} 1 & 1 & 0 & 1 & 0 & 0 & 0 & 0 \\ 0 & 1 & 1 & 1 & 0 & 0 & 0 & 0 \\ 0 & 1 & 0 & 1 & 0 & 0 & 0 & 0 \\ \hline 1 & 1 & 0 & 1 & 1 & 1 & 0 & 1 \\ 0 & 1 & 1 & 1 & 0 & 1 & 1 & 1 \\ 0 & 1 & 0 & 1 & 0 & 1 & 0 & 1 \end{array}\right]$$

Fig. A.4 Kronecker products $\underline{A} \otimes \underline{B}$ and $\underline{B} \otimes \underline{A}$ with matrix $\underline{A}$ and $\underline{B}$ given in Example A.16

Example A.16 Let us take the binary matrices

$$\underline{A} = \begin{bmatrix} 1 & 1 & 0 & 1 \\ 0 & 1 & 1 & 1 \\ 0 & 1 & 0 & 1 \end{bmatrix}, \quad \underline{B} = \begin{bmatrix} 1 & 0 \\ 1 & 1 \end{bmatrix}.$$

Their Kronecker product $\underline{A} \otimes \underline{B}$ turns out to be as shown in Fig. A.4 on the left. Their Kronecker product $\underline{B} \otimes \underline{A}$ is, instead, as shown in Fig. A.4 on the right. Clearly Kronecker product is not commutative.

It is possible to observe that the number of rows in $\underline{A} \otimes \underline{B}$ is the product of the number of rows in $\underline{A}$ by the number of rows in $\underline{B}$. The same is for the number of columns. Clearly analogous properties hold also for $\underline{B} \otimes \underline{A}$, leading, in this case, to the same result.

Lemma A.10 *In the situation described by* Definition A.21, *if rank*($\underline{A}$) = *a and rank*($\underline{B}$) = *b, we obtain*

$$\text{rank}(\underline{A} \otimes \underline{B}) = \text{rank}(\underline{B} \otimes \underline{A}) = ab. \tag{A.14}$$

Proof The number of linearly independent rows in one the two original matrices, reproduced in any row layer of the resulting Kronecker matrix (imposed by the other original matrix), gives the total number of linearly independent rows in it. The same is for the columns. □

Linear Transformations of Vectors

Definition A.22 Equation (A.4) can be interpreted as a *linear transformation* of vector $\underline{b}^T$ into vector $\underline{y}^T$. In this sense, the solution of our system of linear equations, formally expressed by (A.6), can be seen as an *inverse linear transformation*. Its existence requires that $\underline{A}$ is non-singular.

Traditionally, in error correcting code theory, it is more frequent to consider row vectors instead of column vectors, therefore we will write

$$\underline{w} = \underline{v} \cdot \underline{A}, \tag{A.15}$$

where $\underline{v}$ and $\underline{w}$ are both row n-vectors and $\underline{A}$ is an $n \times n$ non-singular square matrix. This expression replaces (A.4). The inverse transformation is therefore

$$\underline{v} = \underline{w} \cdot \underline{A}^{-1}, \tag{A.16}$$

which replaces (A.6).

It is possible to observe that vector $\underline{v} = \underline{0}$ always corresponds to vector $\underline{w} = \underline{0}$, and vice versa. This is consequence of the fact that the transformation is linear, but we can interpret also such result as due to the presence of the trivial solution in any homogeneous system. In this case, the trivial solution is common, although unknowns and known constants play exchanged roles.

Sometimes the new vector $\underline{w}$ has a different number of elements with respect to the original vector $\underline{v}$. In this case, an m-vector $\underline{v}$ is transformed into an n-vector $\underline{w}$. For $m < n$, the number of all possible $\underline{w}$ vectors is greater than the number of distinct $\underline{v}$ vectors. So there will be vectors $\underline{w}$ without correspondent $\underline{v}$. Nevertheless the linear transformation considered remains a one-to-one transformation, like in the case described before, so assuring reversibility. On the other hand, $\underline{A}$ is now an $m \times n$ matrix, and the above reverse transformation will be possible only if $\text{rank}(\underline{A}) = m$. In this situation, after Lemma A.3 and property (A.9), in $\underline{A}$ there are m linearly independent rows. In order to describe the inverse transformation, it is necessary to modify the concept of matrix inverse (restricted to square matrices only).

Definition A.23 Given a rectangular $m \times n$ matrix $\underline{A}$, with $m < n$, provided that $\text{rank}(\underline{A}) = m$, it is possible to introduce an inverse transformation with respect to (A.15), by means of an $n \times m$ matrix $\underline{A}^{-1}_{right}$, called the *right-inverse* of $\underline{A}$. The following relationships hold

$$\underline{A} \cdot \underline{A}^{-1}_{right} = \underline{I}_m, \; \underline{v} = \underline{w} \cdot \underline{A}^{-1}_{right}, \tag{A.17}$$

but there is not a unique solution for $\underline{A}^{-1}_{right}$.

Lemma A.11 *A possible choice for* $\underline{A}^{-1}_{right}$ *is*

$$\underline{A}^{-1}_{right} = \underline{A}^T \cdot (\underline{A} \cdot \underline{A}^T)^{-1}. \tag{A.18}$$

Proof $\underline{A}^T$ is an $n \times m$ matrix, $\underline{A} \cdot \underline{A}^T$ and $(\underline{A} \cdot \underline{A}^T)^{-1}$ are both $m \times m$ square matrices, so that $\underline{A}^{-1}_{right}$ is an $n \times m$ matrix too. From (A.15) and (A.18), considering also (A.5), equalities (A.17) are satisfied. □

Example A.17 Given this 2×3 binary matrix

$$\underline{A} = \begin{bmatrix} 1 & 1 & 0 \\ 0 & 1 & 1 \end{bmatrix},$$

it is possible to verify that the first of (A.17) is satisfied with both the 3×2 matrices

$$\underline{A}_{right}^{-1} = \begin{bmatrix} 1 & 1 \\ 0 & 1 \\ 0 & 0 \end{bmatrix}, \quad \underline{A}_{right}^{-1} = \begin{bmatrix} 0 & 0 \\ 1 & 0 \\ 1 & 1 \end{bmatrix}.$$

Definition A.24 With reference to column vectors, it is possible to introduce an inverse transformation by means of a matrix $\underline{A}_{left}^{-1}$, called the *left-inverse* of $\underline{A}$. The following relationship holds, between $\underline{A}$ and $\underline{A}_{left}^{-1}$,

$$\underline{A}_{left}^{-1} \cdot \underline{A} = \underline{I}_m. \tag{A.19}$$

Also in this case, there is not a unique solution for $\underline{A}_{left}^{-1}$.

Lemma A.12 *A possible choice for* $\underline{A}_{left}^{-1}$ *is*

$$\underline{A}_{left}^{-1} = (\underline{A}^T \cdot \underline{A})^{-1} \cdot \underline{A}^T \tag{A.20}$$

Proof We can proceed as for the proof of Lemma A.11. □

Owing to the traditionally preferred choice of using row vectors in error correcting code theory, when $m = n$, we can take (A.16) as the inverse of (A.15), whereas, when $m < n$, we can limit our attention to the concept of right-inverse and related expressions (A.17), (A.18). The following matrix arrangement allows a very general procedure for the obtainment of a unique right-inverse matrix, which can be taken as a reference. It is particularly useful when the considered matrix depends on a parameter.

Definition A.25 Given a rectangular $m \times n$ matrix $\underline{G}$, with $m < n$ and rank($\underline{G}$) = m, it is always possible to write this equality

$$\underline{G} = \underline{A} \cdot \underline{\Gamma} \cdot \underline{B}, \tag{A.21}$$

where $\underline{A}$ is an $m \times m$ matrix, $\underline{B}$ is an $n \times n$ matrix and $\underline{\Gamma}$ is an $m \times n$ matrix. Expression (A.21) is called *Smith form* of matrix $\underline{G}$ (Smith 1861). If $\underline{G}$ is right-invertible, $\underline{\Gamma}$ can be arranged in such a way as to show non non-null elements just in its leftmost $m \times m$ square sub-matrix. Furthermore such sub-matrix will exhibit non-null elements only on its diagonal, as shown in Fig. A.5, for $m = 3$.

After this construction, the right-inverse of $\underline{G}$ can be calculated as

$$\underline{G}_{right}^{-1} = \underline{B}^{-1} \cdot \underline{\Gamma}_{right}^{-1} \cdot \underline{A}^{-1}, \tag{A.22}$$

where $\underline{\Gamma}_{right}^{-1}$ is the right-inverse of $\underline{\Gamma}$. The situation described by (A.22) is shown in Fig. A.6. $\underline{\Gamma}_{right}^{-1}$ has non-null elements just in its uppermost $m \times m$ square sub-matrix, which is diagonal as before. This diagonal sub-matrix is the inverse of the previous one. With obvious symbolism, we can then write

$$\text{diag}\{\alpha_1, \alpha_2, \alpha_3\} = [\text{diag}\{\gamma_1, \gamma_2, \gamma_3\}]^{-1}.$$

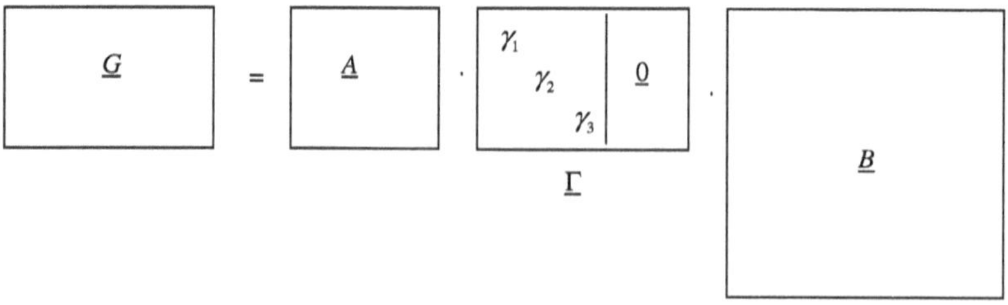

Fig. A.5 Situation described by (A.21) for $m = 3$

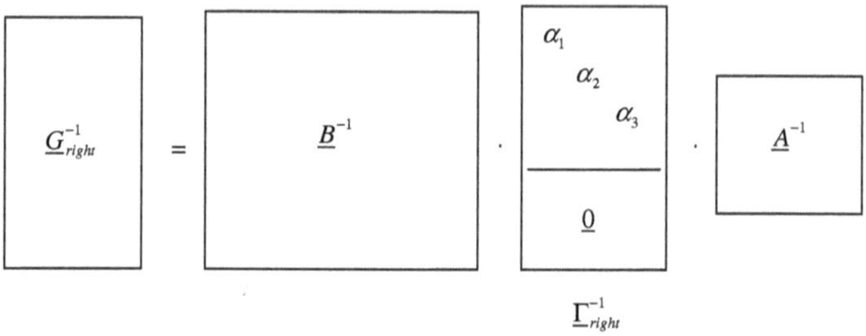

Fig. A.6 Situation described by (A.22) related to the one shown in Fig. A.5

The procedure here summarized, based on the Smith form of a certain rectangular matrix, is more general than the one suggested by Lemma A.11, and enables us to adopt a reference choice for the right-inverse matrix to be analyzed. In error correcting code theory, the generator matrix $\underline{G}$ is that employed for encoding (that is transforming the information symbol sequence into a longer symbol sequence imposed by the code). Its right-inverse $\underline{G}^{-1}_{right}$ is necessary for decoding, even without exploiting the correction capability of the code, but simply for obtaining back the information symbol sequence from the received longer symbol sequence.

Vector Spaces

Definition A.26 A *vector space* having dimension m can be defined by means of a matrix product of the type (A.15). It is associated to the $m \times n$ matrix $\underline{A}$, provided that $\text{rank}(\underline{A}) = m$. The rows of $\underline{A}$ are called *base row vectors* of the considered vector space $\underline{w}$. The dimension m expresses the number of *degrees of freedom* present in vectors $\underline{v}$, which control as many degrees of freedom in the correspondent vectors $\underline{w}$.

Definition A.27 If a rectangular $m \times n$ matrix $\underline{A}$, with $m < n$, has $\text{rank}(A) = m$, that is shows all linearly independent rows, it is said to have a *full rank*.

Equation (A.15), when $\underline{A}$ has full rank, can be interpreted as a bi-univocal correspondence between vectors $\underline{v}$ and vectors $\underline{w}$. Although vectors $\underline{w}$ are characterized by n elements, their vector space has still dimension m, since the number of degrees of freedom available is m. In a finite field with Q elements, there are Q^m vectors $\underline{w}$, corresponding to as many vectors $\underline{v}$. Moreover, it will be Q^{n-m} possible further vectors $\underline{w}$, without correspondence to any vector $\underline{v}$.

Definition A.28 The substitution of any row of $\underline{A}$ with a linear superimposition of its original rows gives an *equivalent description of the same vector space*, with the same dimension m. This operation can be interpreted as a change in the base of our vector space (Brown 1991).

Definition A.29 Given a vector space associated to a rectangular $m \times n$ matrix $\underline{A}$, with full rank, elimination of one or more rows in it gives rise to a new vector space, called a proper *sub-space* of the vector space considered. The dimension of such sub-space is equal to m minus the number of eliminated rows.

Definition A.30 Given a vector space associated to a rectangular $m \times n$ matrix $\underline{A}$, with full rank, it is possible to introduce a *null space* with respect to the original vector space, as that associated to a rectangular $(n - m) \times n$ matrix $\underline{A}_\perp$, after the following equality

$$\underline{A} \cdot \underline{A}_\perp^T = \underline{0}, \tag{A.23}$$

where $\underline{0}$ is an $m \times (n - m)$ null matrix.

This null space, whose matrix product is $w' = v' \cdot \underline{A}_\perp$, has dimension $(n - m) \times n$. Obtainment of $\underline{A}_\perp$, from (A.23), does not give a unique solution. Nevertheless all the possible results correspond to the same vector space. Matrix $\underline{A}_\perp$ must have itself full rank. Two possible equivalent matrices $\underline{A}_\perp$ so obtained will differ only for the substitution of some rows with proper linear superimpositions of them.

In the geometric physical space, the null space is related to the concept of orthogonal sub-spaces. For instance, a straight line (one-dimension sub-space) is orthogonal to a certain plane (2-dimension sub-space), in an overall 3-dimension geometric physical space.

Example A.18 Given this 2×5 binary matrix

$$\underline{A} = \begin{bmatrix} 1 & 1 & 0 & 1 & 0 \\ 0 & 1 & 1 & 0 & 1 \end{bmatrix},$$

it is possible to verify that (A.23) is satisfied with both the following two 3×5 matrices

$$\underline{A}_\perp = \begin{bmatrix} 1 & 0 & 1 & 1 & 1 \\ 0 & 1 & 0 & 1 & 1 \\ 0 & 0 & 1 & 0 & 1 \end{bmatrix}, \quad \underline{A}_\perp = \begin{bmatrix} 1 & 0 & 1 & 1 & 1 \\ 0 & 1 & 0 & 1 & 1 \\ 1 & 1 & 0 & 0 & 1 \end{bmatrix}.$$

They differ only in the third row, which has been obtained in the latter one as an ex-OR sum of all the three rows in the former one. This confirms that these are two equivalent representations for the same vector space. Simply they express two different bases for its representation.

Lemma A.13 *After* (A.23), Eq. (A.15) *can be rewritten as*

$$\underline{w} \cdot \underline{A}_{\perp}^{T} = \underline{0}, \tag{A.24}$$

where $\underline{0}$ *is a null row vector having* $(n - m)$ *elements.*

Proof Left-multiplication of (A.23) by $\underline{v}$, considering also (A.15), allows to obtain (A.24). □

Equality (A.24), in a binary field, can be interpreted as a linear homogeneous system, whose number of possible solutions is 2^m. They represent as many n-vectors, included vector $\underline{0}$. In error correcting code theory, the parity check matrix $\underline{H}$ describes a certain code by means of its null vector space.

References

Brown WA (1991) Matrices and vector spaces. Dekker, New York

Capelli A (1886) Corso di analisi algebrica (Course on algebraic analysis). Sacchetto, Rome

Cramer G (1750) Introduction a l'analyse del courbes algebraiques (Introduction to algebraic curve analysis). Freres Cramer & Cl. Filbert, Geneva

Fonténé GF (1892) L'hyperspace a n − 1 dimensions (The hyperspace at n − 1 dimensions). Gautier-Villars, Paris

Frobenius G (1911) Uber den rang einer matrix (On the rank of a matrix). Reiner, Berlin

Hankerson D, Menezes A, Vanstone S (2004) Guide to elliptic curve cryptography, Chap. 2. Springer, New York

Rouché E (1875) Sur la discussion des equations du premier degree (About the discussion of first order equations). Comptes Rendus, Acad. des Sciences 81, Paris

Smith HJS (1861) On systems of linear indeterminate equations and congruences. Phil Trans R Soc London 151:293–326

Appendix B
Polynomial Representation of Binary Sequences

Polynomial ex-OR Sum

Definition B.1 A *binary polynomial* is an ensemble of powers of a dummy variable x, representing a sequence of binary symbols 0, 1. The exponents of the powers denote the positions in the sequence. Assuming that the ith-order power is multiplied by p_i, where p_i is a binary symbol, a polynomial $p(x)$ can then be written as $p(x) = \sum_i p_i x^i$.

Conventionally higher powers refer to preceding symbols. Since multiplication by 0-symbol makes the product null, the only powers surviving are those multiplied by 1-symbols.

Example B.1 The sequence 100101, expressed from the first (leftmost) to the last (rightmost) symbol, is represented by the binary polynomial $x^5 + x^2 + 1$.

The convention here adopted in representing symbol sequences by polynomials is the same as in (Peterson and Weldon 1972; Oppenheim and Schafer 1974; MacWilliams and Sloane 1977; Anderson and Mohan 1991), and opposite to that adopted, for instance, in (Johannesson and Zigangirov 1999; Lin and Costello 2004; Ryan and Lin 2009).

Definition B.2 The *sum* (or *addition*) of two polynomials, $p(x)$ and $q(x)$, is made considering binary ex-OR sums between the two symbols related to the same power in the two polynomials.

Example B.2 Taking $p(x) = x^5 + x^2 + 1$ and $q(x) = x^7 + x^3 + x^2 + x$, we obtain

$$p(x) + q(x) = x^7 + x^5 + x^3 + x + 1.$$

Since ex-OR sum exhibits the property that one of the two addends can replace the result and vice versa (see Fig. A.1 in Appendix A), in binary polynomial sums, the following property holds: the equality $p(x) + q(x) = z(x)$ is equivalent to the equalities $p(x) + z(x) = q(x)$ or $q(x) + z(x) = p(x)$. This is due to the fact that, in a binary field, the addition opposite of a certain element coincides with the element itself.

G. Cancellieri, *Polynomial Theory of Error Correcting Codes*,
Signals and Communication Technology, DOI 10.1007/978-3-319-01727-3

Polynomial Multiplication

Definition B.3 The *multiplication* (or *product*) of a polynomial $p(x)$ by x^j gives rise to a new polynomial, in which any power is increased by j. The correspondent binary sequence exhibits j additional 0-symbols on its right hand side.

Example B.3 Taking $p(x) = x^5 + x^2 + 1$ and multiplying it by x^3, we obtain

$$x^3 p(x) = x^8 + x^5 + x^3.$$

The correspondent binary sequence is changed from 100101 to 100101000. We can interpret this phenomenon as a time shift in advance.

Definition B.4 The *multiplication* (or *product*) of two polynomials, $p(x)$ and $q(x)$, is made taking any power of $q(x)$ and multiplying it by $p(x)$, as described by Definition B.3.

Example B.4 Taking $p(x) = x^5 + x^2 + 1$, and $q(x) = x^3 + x$, we obtain

$$p(x)q(x) = x^8 + x^5 + x^3 + x^6 + x^3 + x = x^8 + x^6 + x^5 + x.$$

It is useful to remark that multiplication between single coefficients in the two polynomials is made according to the rules defined in the binary field (see Fig. A.1 in Appendix A and subsequent properties).

Definition B.5 The highest-order non-null power of a polynomial $p(x)$ is called its *degree*, and denoted $\deg[p(x)]$.

Lemma B.1 *In a multiplication of two polynomials,* $p(x)$ *and* $q(x)$, *we have*

$$\deg[p(x)q(x)] = \deg[p(x)] + \deg[q(x)]$$

Proof The product of the two highest powers in the two polynomials cannot be cancelled, because of ex-OR sum, by any other product. So it is always present and exhibits an exponent which is the sum of the two exponents. □

Definition B.6 The number of 1-symbols in a polynomial $p(x)$ is called its *weight*, and denoted $w[p(x)]$.

Example B.5 Taking $p(x) = x^5 + x^2 + 1$, we have $w[p(x)] = 3$.

In the following, except when the contrary is explicitly said, we will consider only polynomials without 0 symbols on the right, that is with the rightmost non-null coefficient coincident with the 0th-order coefficient. About the weight characterizing the product of two polynomials, we have the next two propositions.

Lemma B.2 *The weight of the product of* $p(x)$ *and* $q(x)$ *satisfies the condition*

$$w[p(x)q(x)] = w[p(x)]w[q(x)] - 2C \qquad \text{(B.1)}$$

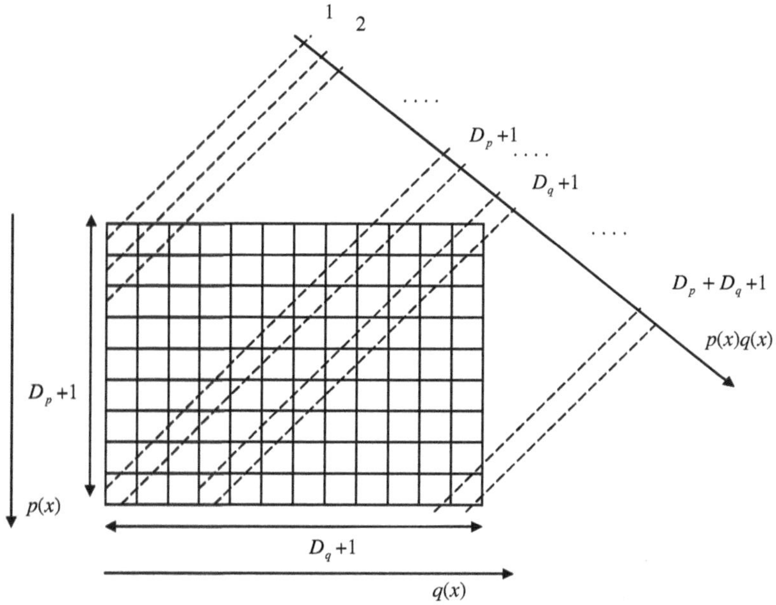

Fig. B.1 Rectangular matrix for evaluating the weight of $p(x)q(x)$

where C is the number of coincidences between two sums of couples of exponents, being the exponents of any couple taken the one from p(x) and the other from q(x).

Proof In Fig. B.1 a rectangular matrix has been depicted, with the purpose of helping in understanding how the product $p(x)q(x)$ comes out. We have set $D_p = \deg[p(x)]$ and $D_q = \deg[q(x)]$. Let us imagine that polynomial $p(x)$ is written along a vertical line, and that polynomial $q(x)$ is written along a horizontal line, according to the arrows. The up-right oblique line collects the result, in $(D_p + D_q + 1)$ possible positions, as stated by Lemma B.1. Along any anti-diagonal (from down-left to up-right), an ex-OR sum is to be made. Each anti-diagonal joints all the matrix elements whose co-ordinate sum identifies a precise position on the oblique line collecting the result. In the matrix, an element is equal to 1 only if both polynomials exhibit a 1-symbol in correspondence of the row and column identifying such element (because of the multiplication). C expresses the number of pairs of 1-symbols on the same anti-diagonal, so eliminating each other (because of the ex-OR sum). □

Lemma B.3 *The weight of the product of two polynomials, p(x) and q(x), is so bounded*

$$2 \le \mathrm{w}[p(x)q(x)] \le \deg[p(x)] + \deg[q(x)] + 1 \tag{B.2}$$

Proof As for the lower bound, we can observe that, in Fig. B.1, on the up-right oblique line, we have just two positions without possibility of a cancellation, since they correspond to as many anti-diagonals with only one element in the matrix: the first one and the last one. They are respectively associated to the lowest-order power and to the highest-order power of $p(x)q(x)$. The upper bound is due to all the possible positions in the result, without any cancellation. □

The lower bound in (B.2) can be justified also keeping in mind what stated by Lemma B.1, which deals with the only highest non-null power. Both the lowest and the highest power cannot be cancelled. Anyway, we can observe that the lower bound in (B.2) is reached when the maximum number of possible cancellations occurs.

Theorem B.1 *The product between two polynomials expresses the temporal convolution of the two binary sequences that they represent.*

Proof Let us denote with $\tilde{p}(t)$ and $\tilde{q}(t)$ the time-signal sequences respectively associated to $p(x)$ and $q(x)$. Their temporal convolution (Proakis 1983) is expressed by the integral

$$\tilde{C}(t) = \int_{-\infty}^{+\infty} \tilde{p}(\tau)\tilde{q}(t-\tau)d\tau \tag{B.3}$$

It is a discrete function of time containing the result of elementary ex-OR sums collected in the same position (output symbol time t). This resulting position is the sum of the positions characterizing the two addends, exactly as in the correspondent polynomial product the resulting power is the sum of the powers in the two polynomial factors. □

An intuitive view for understanding the convolution is based on expressing the result in such a way as to form a parallelogram, like that depicted in Fig. B.2, where we have assumed $D_p < D_q$, so the parallelogram exhibits a zone where the two horizontal bases (whose lengths are D_q) are superimposed. This representation can be adopted also for the correspondent polynomial product.

A completely equivalent construction, although less intuitive, is that depicted in Fig. B.3, where the roles of the two polynomials are exchanged. We have still $D_p < D_q$, but here the two bases (whose lengths are D_p) are not superimposed.

Fig. B.2 Polynomial product represented as a parallelogram

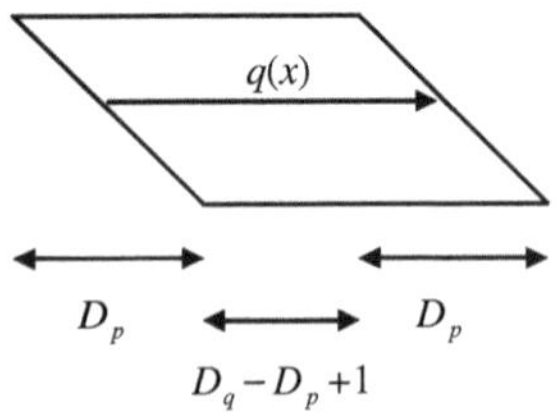

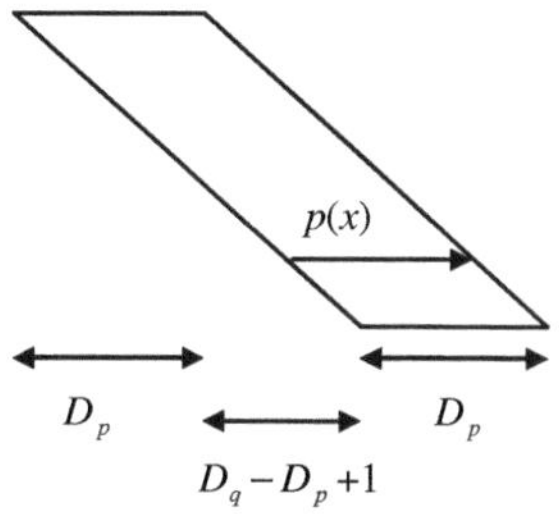

Fig. B.3 Parallelogram construction of a polynomial product, equivalent to the one in Fig. B.2

Polynomial Division

In evaluating $p(x)/x^j = x^{-j}p(x)$, it would be necessary to consider negative powers. This fact is not acceptable, if a reference position has to be maintained in correspondence with the 0th-order power. So the procedure for obtaining the division by x^j will stop, leading to a possible remainder. Now, instead of a simple monomial x^j, we will take a polynomial $z(x)$ as divisor. Linearity of the ex-OR sum operation assures that the above considerations can be repeated as well for all the powers of $z(x)$. Two situations must be considered, depending on the degree $\deg[z(x)]$ with respect to that of the polynomial $p(x)$ to be divided.

Definition B.7 Let us consider a *polynomial division* of $p(x)$ (the dividend) by $z(x)$ (the divisor). If $\deg[p(x)] < \deg[z(x)]$, we just have

$$\frac{p(x)}{z(x)} = \frac{1}{z(x)} R\left[\frac{p(x)}{z(x)}\right],$$

where $R[\ ]$ is called the *remainder* of our division, and in this case it coincides with the dividend. On the contrary, if $\deg[p(x)] \geq \deg[z(x)]$, it is possible to write

$$\frac{p(x)}{z(x)} = Q\left[\frac{p(x)}{z(x)}\right] + \frac{1}{z(x)} R\left[\frac{p(x)}{z(x)}\right] \tag{B.4}$$

where $Q[\ ]$ is called the *quotient* of our division. The procedure for calculating Q consists of repeated ex-OR sums of $z(x)$, in order to eliminate 1-symbols, up to reach a residual degree lower than $\deg[z(x)]$.

Example B.6 Taking $p(x) = x^5 + x^2 + 1$ and $z(x) = x^3 + x + 1$, we can perform the division of the first by the second as depicted in Fig. B.4. The result given by this operation comprehends the quotient and the remainder, respectively as

$$Q\left[\frac{p(x)}{z(x)}\right] = x^2 + 1, \quad R\left[\frac{p(x)}{z(x)}\right] = x$$

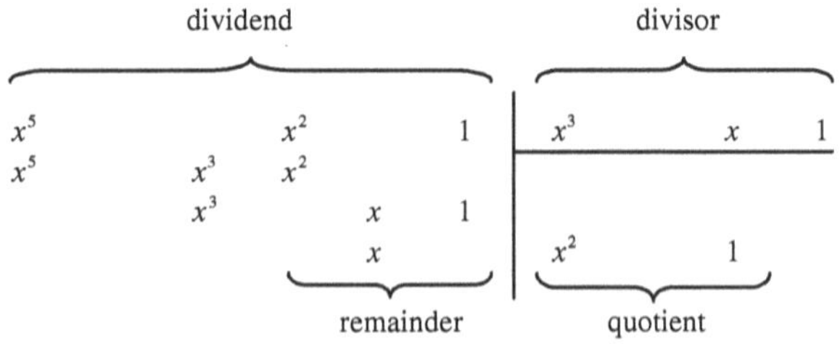

Fig. B.4 Procedure for calculating quotient and remainder of a polynomial division

Lemma B.4 *In a division of p(x) by z(x), provided that* $\deg[p(x)] \geq \deg[z(x)]$, *we have*

$$\deg\left[Q\left[\frac{p(x)}{z(x)}\right]\right] = \deg[p(x)] - \deg[z(x)]$$

Proof The first step of the procedure depicted in Fig. B.4, always present, is devoted to obtain the highest-order power in the quotient, whose exponent results in the difference between deg[$p(x)$] and deg[$z(x)$]. □

Lemma B.5 *The remainder of a division has degree always smaller than that of the divisor.*

Proof This result follows directly from Definition B.7. □

Definition B.8 An *exact quotient* is that associated to a division with null remainder.

Lemma B.6 *An exact quotient multiplied by the divisor yields the dividend.*

Proof For whichever couple of polynomials $p(x)$ and $z(x)$, provided that deg[$p(x)$] ≥ deg[$z(x)$], after multiplication of (B.4) by $z(x)$, it is possible to write

$$p(x) = Q\left[\frac{p(x)}{z(x)}\right]z(x) + R\left[\frac{p(x)}{z(x)}\right] \tag{B.5}$$

So, if the quotient is exact, we are left with just the first term on the right hand side. □

We can observe how Lemma B.6 states that an exact quotient is the inverse operation of a multiplication. Furthermore, any couple of polynomials can be multiplied, giving rise to a new polynomial. On the contrary, also any couple of polynomials can be divided, with the only care that dividend degree is not smaller than divisor degree, but only sometimes we obtain an exact quotient. In a multiplication of some polynomials, they are called the factors with respect to the result. In order to test if, in a certain polynomial $p(x)$, another polynomial $z(x)$ of lower degree is present as a factor, we have to divide $p(x)$ by $z(x)$, and, if there is a null remainder, the answer is yes.

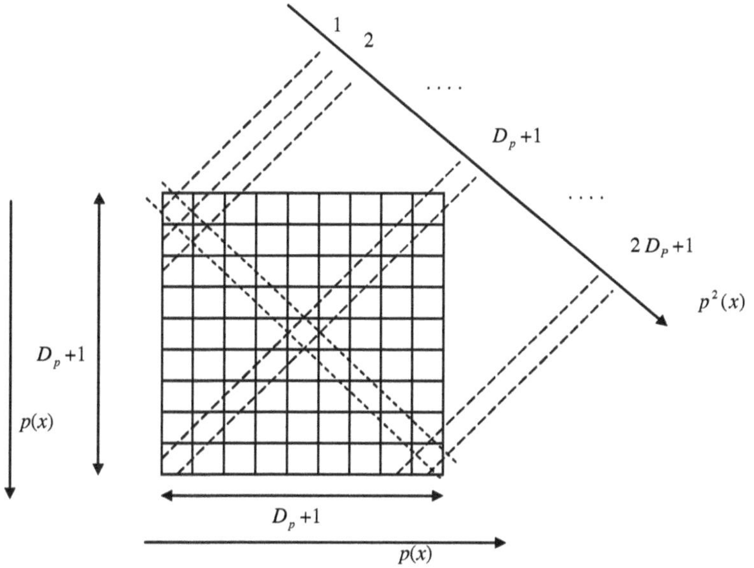

Fig. B.5 Square matrix for evaluating the weight of $p(x^2)$

Square and Square Root

Definition B.9 A polynomial $p(x)$ multiplied by itself gives rise to its *square*, denoted $p^2(x)$.

Example B.7 Assuming the binary polynomial $p(x) = x^5 + x^2 + 1$, we obtain

$$p^2(x) = x^{10} + x^4 + 1$$

Theorem B.2 *The square of a binary polynomial p(x) exhibits the following property*

$$p^2(x) = p(x^2) \tag{B.6}$$

Proof Modification of the matrix depicted in Fig. B.5, in such a way as to describe the case of a square, yields a square matrix with the same polynomial $p(x)$ on both dimensions, as shown in Fig. B.5. In this square matrix, we call diagonals those from up-left to down-right, and anti-diagonals those from down-left to up-right. If there is the sum $(a + b)$, where a is an exponent taken from polynomial $p(x)$ on the vertical axis, and b is an exponent taken from polynomial $p(x)$ on the horizontal axis, certainly the sum $(b + a)$ will be present too, provided that $a \neq b$. These two 1-symbols belong to the same anti-diagonal, and hence they vanish. Just the sums of the type $a + a = 2a$ will be present alone. So they contribute to as many 1-symbols in the square $p^2(x)$. In conclusion, any exponent of the original

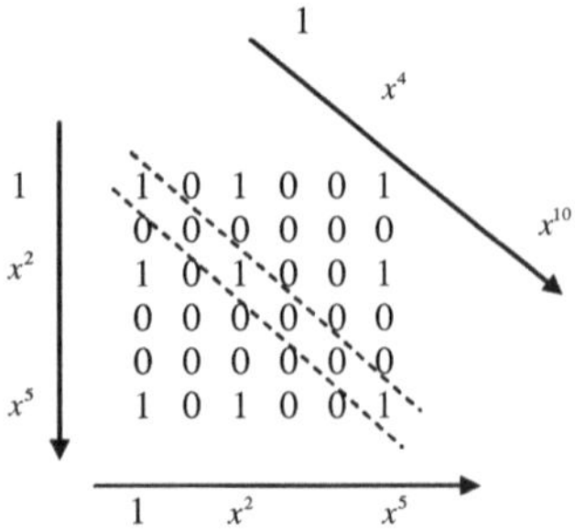

Fig. B.6 Square matrix for representing the square of the polynomial described in Example B.8

$p(x)$ turns out to be doubled in $p^2(x)$, and these powers are the only non-null powers of $p^2(x)$. □

Example B.8 Taking the polynomial $p(x) = x^5 + x^2 + 1$, the square matrix depicted in Fig. B.5 for calculating its square $p^2(x) = x^{10} + x^4 + 1$, is arranged as shown in Fig. B.6, where dotted parallel lines evidence also here the longest diagonal. Along it three uncoupled 1-symbols are aligned.

Remark that (B.6) is a strict consequence of the use of a binary alphabet. It holds also for any finite field having a number of elements equal to an integer power of 2.

Corollary a *Assuming a binary polynomial p(x), its 2^L-th-order power, where L is an integer, exhibits the following property*

$$p^{2^L}(x) = p(x^{2^L}) \tag{B.7}$$

Proof Iterating L times (B.6), it is possible to obtain (B.7). □

Corollary b *The weight of the square of any binary polynomial coincides with the weight of the original polynomial, that is*

$$w[p^2(x)] = w[p(x)] \tag{B.8}$$

Proof Polynomials $p(x)$ and $p(x^2)$ are characterized by an equal number of non-null powers. □

Coming back to the general expression (B.1) about the weight of a product, we can observe that, in the case of a square, the number of coincidences C results in

$$C = \binom{w[p(x)]}{2} \tag{B.9}$$

This is in perfect agreement with the discussion made on the square matrix in Fig. B.5, since such a binomial coefficient represents the total number of couples which can be formed in an ensemble of $w[p(x)]$ elements.

Definition B.10 When a polynomial is a square, it is possible to obtain its *square root*, that is the original polynomial, before squaring. The square root of polynomial $P(x)$ is denoted as $\sqrt{P(x)}$.

So polynomial squaring produces a one-to-one correspondence between a certain polynomial and its square. This correspondence can be inverted, leading to an unique possible result. We can recognize immediately if a binary polynomial $P(x)$ is the square of another polynomial, and determine the latter one.

Lemma B.7 *A binary polynomial $P(x)$ is a square if and only if it exhibits just powers with even (or null) exponents. Its square root is calculated by halving the exponent of any non-null power present in it.*

Proof This property and the procedure for obtaining $\sqrt{P(x)}$ is a direct consequence of Theorem B.2. □

With reference to Example B.7, assuming $P(x) = x^{10} + x^4 + 1$, we can obtain back $p(x) = \sqrt{P(x)} = x^5 + x^2 + 1$.

Further Properties of Polynomial Division

Lemma B.8 *Quotient and remainder of a polynomial division are linear functions of the dividend.*

Proof For any couple of polynomials $p_1(x)$, $p_2(x)$, and any divisor $z(x)$, we can write

$$\begin{aligned}\frac{p_1(x)+p_2(x)}{z(x)} &= \frac{p_1(x)}{z(x)} + \frac{p_2(x)}{z(x)} \\ &= Q\left[\frac{p_1(x)}{z(x)}\right] + Q\left[\frac{p_2(x)}{z(x)}\right] + \frac{1}{z(x)}R\left[\frac{p_1(x)}{z(x)}\right] + \frac{1}{z(x)}R\left[\frac{p_2(x)}{z(x)}\right]\end{aligned}$$

so that

$$\begin{aligned}Q\left[\frac{p_1(x)+p_2(x)}{z(x)}\right] &= Q\left[\frac{p_1(x)}{z(x)}\right] + Q\left[\frac{p_2(x)}{z(x)}\right] \\ R\left[\frac{p_1(x)+p_2(x)}{z(x)}\right] &= R\left[\frac{p_1(x)}{z(x)}\right] + R\left[\frac{p_2(x)}{z(x)}\right]\end{aligned} \tag{B.10}$$

□

Let us now consider a non-null power, whose exponent can vary. We will denote it as x^L. Dividing this non-null power by $z(x)$, being $\deg[z(x)] < L$, different situations may occur. Owing to Lemma B.2 (lower bound), it is impossible to have a null remainder. We can imagine that parameter L progressively increases. Division can be performed in an iterative way, forming a current remainder and a current quotient. The degree of the first is constrained to be smaller than that of the divisor, owing to Lemma B.5. The degree of the second, instead, increases following the growth of L, as stated by Lemma B.4. As long as the process above described is going on, we can observe two only possibilities, described as follows.

Lemma B.9 *In the polynomial division of the dividend x^L, after an increase of L by one unit, the current remainder can experience one of the two only following transformations: it simply shifts (maintaining form but just increasing the degree) or it changes, because of an ex-OR sum with $z(x)$.*

Proof Assuming $z(x) = \sum_{i=0}^{D} z_i x^i$, in order to have $D = \deg[z(x)]$, coefficient z_D must be 1. After the first ex-OR sum during the division, $z_D = 1$ vanishes, and there are D positions on the right occupied by the new remainder. This means a change in the form of the current remainder at this first step. Then, if z_{D-1} and the coefficient of the previous remainder in the same position are different, the next step will be a further change. Otherwise (if they are both 0 or both 1), the old remainder simply shifts, without changing form. This behaviour continues up to reach the final remainder. □

Lemma B.10 *The current remainder, at a certain step of the division procedure applied to x^L, completely determines the remaining part of the division itself.*

Proof Since the dividend is x^L, there are no other polynomial coefficients to be taken into account after x^L itself, eliminated by the first ex-OR sum with $z(x)$. Then the current remainder completely determines the division procedure. □

Theorem B.3 *Division of x^L by $z(x)$, provided that $w[z(x)] > 1$, with L increasing, exhibits a current remainder and a current quotient both periodic.*

Proof Assuming $D = \deg[z(x)]$, at most we can have 2^D possible different remainders, either simple shifts or changed forms. After this evolution, since a null remainder is impossible, as already noted, the current remainder is forced to repeat.

L	current quotient	current remainder	
3	1	$x+1$	
4	x	x^2+x	conclusion
5	x^2+1	x^2+x+1	of the 1.st period
6	x^3+x+1	x^2+1	
7	x^4+x^2+x+1	1	
8	$x^5+x^3+x^2+x$	x	
9	$x^6+x^4+x^3+x^2$	x^2	
10	$x^7+x^5+x^4+x^3+1$	$x+1$	
11	$x^8+x^6+x^5+x^4+x$	x^2+x	2.nd period
12	$x^9+x^7+x^6+x^5+x^2+1$	x^2+x+1	
13	$x^{10}+x^8+x^7+x^6+x^3+x+1$	x^2+1	
14	$x^{11}+x^9+x^8+x^7+x^4+x^2+x+1$	1	

(Under the quotient of $L = 14$: "1.st period" braces $x^{11}+x^9+x^8+x^7$; "2.nd period" braces x^4+x^2+x+1.)

Fig. B.7 Current quotient and current remainder in the division of x^L by $z(x) = x^3 + x + 1$

From Lemma B.10, the entire process repeats. So both the current remainder and the current quotient are periodic. □

Example B.9 Let us consider $z(x) = x^3 + x + 1$. The table shown in Fig. B.7 collects current quotient and current remainder in the division of x^L by $z(x)$. It is possible to recognize the conclusion of a first period at $L = 7$, and the complete formation of a second period from $L = 8$ to $L = 14$. Finally we can observe that, when a coefficient 1 is added to the current quotient, the expression of the current remainder changes, otherwise it simply shifts by one position, according to Lemma B.9.

Corollary a *The maximum length of the period introduced in* Theorem B.3 *is* $(2^D - 1)$.

Proof The number of different combinations of D coefficients (forming all the possible current remainders) is 2^D, but the one consisting of all 0 coefficients is not to be considered, since it would represent an exact quotient. So their number will be not greater than $(2^D - 1)$. After at most such a number of steps, the division procedure repeats. □

We can observe how the case described in Example B.9 regards a situation in which the period is maximum, in fact we have $D = 3$, and hence $2^D - 1 = 7$.

Corollary b *Division of any polynomial $x^L p(x)$ by $z(x)$, with L increasing, exhibits a current remainder and a current quotient both periodic, except when the division leads to an exact quotient.*

Proof Taking into account Lemma B.8 about linearity, for any power in $p(x)$ it is possible to apply Theorem B.3. Here the transient for eliminating all the 1-symbols present in $p(x)$ is more involved, but always of finite length, leading to a current remainder and a current quotient both periodic, with period length at most equal to $(2^D - 1)$, also in this case. □

Theorem B.4 *An exponent N always exists, such that any polynomial $z(x)$, whose degree is finite, is factor of the binomial $(x^N + 1)$.*

Proof Taking $L = N$ as the period in the division of x^L by $z(x)$, with L increasing, treated so far, after an entire period we can write

$$R\left[\frac{x^N}{z(x)}\right] = 1 \tag{B.11}$$

On the other hand, the following sequence of equalities holds

$$R\left[\frac{x^N}{z(x)}\right] = R\left[\frac{x^N + 1 + 1}{z(x)}\right] = R\left[\frac{x^N + 1}{z(x)}\right] + R\left[\frac{1}{z(x)}\right] = R\left[\frac{x^N + 1}{z(x)}\right] + 1.$$

Comparison with (B.11) finally yields $R\left[\frac{x^N+1}{z(x)}\right] = 0$, which means that $z(x)$ is factor of $(x^N + 1)$. □

Example B.10 Polynomial $z(x) = x^3 + x + 1$, already taken in Example B.9, is factor of $(x^7 + 1)$. Looking at Fig. B.7, it is possible to observe how $N = 7$ turns out to be the length of the period shown by the current remainder and by the current quotient in the division mentioned in Theorem B.3.

Definition B.11 Given a polynomial $z(x)$, factor of $(x^N + 1)$, where N is the smallest integer for which this property holds, the polynomial

$$\bar{z}(x) = \frac{x^N + 1}{z(x)} \tag{B.12}$$

is called the *co-factor* of $z(x)$.

It is interesting to note that factor and co-factor satisfy, with equality sign, the lower bound expressed by (B.2), about the weight of a product between two polynomials.

Example B.11 Polynomial $z(x) = x^3 + x + 1$, already taken in Examples B.9 and B.10, is factor of $(x^7 + 1)$, and is characterized by a co-factor expressed by polynomial $\bar{z}(x) = x^4 + x^2 + x + 1$.

Example B.12 Polynomial $z(x) = x^4 + x^3 + x + 1$ is factor of $(x^6 + 1)$. Its co-factor results in $\bar{z}(x) = x^2 + x + 1$.

Symmetry Properties

Definition B.12 Given a binary polynomial $p(x) = \sum_{i=0}^{D} p_i x^i$, with non-null 0th-order coefficient, its *reciprocal* is defined as

$$p^*(x) = \sum_{i=0}^{D} p_{D-i} x^i \tag{B.13}$$

The reciprocal exhibits the same degree and weight of the original polynomial, but the order of its non-null power coefficients are reversed (Peterson and Weldon 1972). Clearly we have $p^{**}(x) = p(x)$, that is a double reciprocal coincides with the original polynomial.

Example B.13 Given $p(x) = x^5 + x^3 + x + 1$, representing the sequence 101011, its reciprocal is $p^*(x) = x^5 + x^4 + x^2 + 1$, representing the sequence 110101.

For a symmetric polynomial $p(x)$, taking the reciprocal, the equality $p^*(x) = p(x)$ holds. An asymmetric polynomial is, instead, a polynomial for which the above equality is not valid. The following three Examples enable us to introduce important relationships between weight and degree in symmetric polynomials.

Example B.14 The binary polynomial $p(x) = x^5 + x^3 + x^2 + 1$, associated to the sequence 101101, is symmetric. Its reciprocal is again $p^*(x) = x^5 + x^3 + x^2 + 1$.

This polynomial is the product between the two polynomials $(x^3 + 1)$ and $(x^2 + 1)$, both symmetric as well.

Example B.15 The binary polynomial $p(x) = x^6 + x^4 + x^2 + 1$, associated to the sequence 1010101, is symmetric. Precisely it can be interpreted as the square of polynomial $(x^3 + x^2 + x + 1)$, showing that, when a polynomial is symmetric, also its square is symmetric.

Example B.16 The binary polynomial $p(x) = x^6 + x^4 + x^3 + x^2 + 1$, associated to the sequence 1011101 is symmetric. Precisely it is the product between the two polynomials $(x^4 + x^3 + x^2 + x + 1)$ and $(x^2 + x + 1)$.

Lemma B.11 *A symmetric polynomial $p(x)$, whose degree is odd, must have even weight.*

Proof If $\deg[p(x)]$ is odd, such polynomial does not exhibit a central position. Its symmetry axis is placed between two adjacent positions, and divides the polynomial overall length into two reciprocal parts, associated to the two sides with respect to it. So the overall weight must be even. □

Lemma B.12 *A symmetric polynomial $p(x)$, whose degree is even, can have either even weight or odd weight. In the latter case in its central position a 1-symbol must be present.*

Proof Differently from the case treated before, here a central position is always present. For the 1-symbols out of this central position, what discussed in the proof of the Lemma B.11 still holds. Their total number of 1-symbols is even. The remaining amount for reaching the overall weight of $p(x)$ is due to the central position. □

We can note how Example B.14 represents a situation related to Lemma B.11. Examples B.15 and B.16 regard, instead, Lemma B.12, respectively with a 0-symbol and with a 1-symbol in their central position. In any polynomial $p(x)$, besides the two outermost 1-symbols (the one associated to coefficient p_D and the other associated to coefficient p_0), which contribute to its symmetric part, there are typically a symmetric part and an asymmetric part. The symmetric part is made by pairs of coupled 1-symbols or by the possible solitary 1-symbol in the central position (only if D is even), whereas the asymmetric part is made by uncoupled 1-symbols. The following statement and subsequent observations offer an accurate analysis of possible situations regarding symmetry properties in a polynomial product.

Lemma B.13 *Given some symmetric polynomials $s_j(x)$, $s_k(x)$, and some asymmetric polynomials $a_h(x)$, $a_l(x)$, we have the following possibilities for their product:*

i. *$s_j(x)s_k(x)$, which is always symmetric,*
ii. *$s_j(x)a_h(x)$, which is always asymmetric,*
iii. *$a_h(x)a_l(x)$, which can be either symmetric or asymmetric.*

Proof Situation *i* is characterized by a symmetric distribution of the powers due to $s_k(x)$ around any power of $s_j(x)$. Hence the product exhibits a new symmetry axis, but remains symmetric. Situation *ii* is characterized by an asymmetric distribution of the powers due to $a_h(x)$ around any power of $s_j(x)$, so leading to an overall asymmetric result. Situation *iii* generally gives an asymmetric polynomial. Two exceptions are of interest. The first one deals with the factor and the co-factor in the decomposition of a binomial $(x^N + 1)$. The second one regards the product between an asymmetric polynomial and its reciprocal, and will be better investigated below. □

About the degree $\deg[p(x)p^*(x)]$, for either a symmetric or an asymmetric $p(x)$, it is always $2\deg[p(x)]$, like in a square. Finally, coming back to situation *iii*, we will demonstrate now the property regarding the product between two reciprocal polynomials, anticipated above.

Theorem B.5 *Given an asymmetric polynomial* $a(x)$, *the product* $a(x)a^*(x)$ *is symmetric.*

Proof Assuming a_k and a_{D-h} as non-null coefficients in $a(x)$ placed in not symmetric positions, and setting $D = \deg[a(x)]$, the coefficients $a_{D-k}{}^*$ and $a_h{}^*$ will be non-null coefficients in $a^*(x)$. From their products, we obtain two couples of non-null coefficients in $a(x)a^*(x)$: a'_{k+h}, a'_{2D-k-h}, clearly in symmetric positions, and a'_{D-k+h}, a'_{D-h+k} in symmetric positions too. The case of $h = k$, regards the central position, which does not influence symmetry. In conclusion, the whole product turns out to be symmetric. □

Example B.17 Taking $p(x) = x^4 + x^3 + 1$, that is the sequence 11001, which is asymmetric, we obtain $p(x)p^*(x) = x^8 + x^7 + x^5 + x^4 + x^3 + x + 1$, that is 110111011, which is symmetric.

Remark that $p(x)p^*(x)$ has always even degree, as already noted, so there is a central positions in it, associated to the power D, where a 1-symbol or a 0-symbol is present, respectively if $w[p(x)]$ is odd or even. The case treated in Example B.17 refers to the first situation. If one takes, instead, $p(x) = x^4 + x^3 + x^2 + 1$, the result is 110000011, which is related to the second situation.

Factor and co-factor in the decomposition of a binomial $(x^N + 1)$ are often both asymmetric. This represents, as discussed in the proof of Lemma B.13, the other case in which two asymmetric polynomials exhibit a symmetric product. In this case all the asymmetric coefficients are eliminated, and the only outermost two remain.

Example B.18 Assuming $z(x) = x^3 + x + 1$ and $\overline{z}(x) = x^4 + x^2 + x + 1$, their product gives $(x^7 + 1)$. They both are asymmetric, in fact correspond to the sequences 1011 and 10111. Nevertheless not always factor and co-factor are asymmetric. They can be also both symmetric (see, for instance, Example B.12).

As for the weight of the product $a(x)a^*(x)$ of an asymmetric polynomial with its reciprocal, it is possible to fix only the following bounds

$$2 \leq \mathrm{w}[a(x)a^*(x)] \leq 2D+1 \tag{B.14}$$

both assured by Lemma B.3, for whichever couple of polynomials. Nevertheless here the lower bound means that the two polynomials must be, at the same time, reciprocal and a factor/co-factor couple. The upper bound, in its turn, refers to the situation in which the product exhibits all 1-symbols. The spread in the possible results bounded by (B.14) is consequence of a smaller or larger amount of cancellations C in expression (B.1) about the weight of a generic product.

Example B.19 Taking $p(x) = x^7 + x^6 + x^5 + x^4 + x + 1$, that is 11110011, which is asymmetric, we obtain $p(x)p^*(x) = x^{14} + 1$, satisfying with equality sign the lower bound in (B.14). In this case, the reciprocal $p^*(x)$ coincides with the co-factor $\overline{p}(x)$ with respect to the binomial $(x^{14} + 1)$, so matching the lower bound in (B.14).

Example B.20 Taking $p(x) = x^3 + x + 1$, that is 1011, already assumed in Example B.9, we obtain $p(x)p^*(x) = x^6 + x^5 + x^4 + x^3 + x^2 + x + 1$, that is 1111111, which means $\mathrm{w}[p(x)p^*(x)] = 7$, satisfying with equality sign the upper bound in (B.14). The present case can be classified also as an even-degree symmetric polynomial with a 1-symbol in its central position (and hence odd weight), according to Lemma B.12. There are also many irreducible polynomials satisfying this characteristic as well (see for instance those associated to the sequences 111 or 11111).

Factorization of the Binomial $(x^N + 1)$

Definition B.13 Assuming a polynomial characterized by w subsequent 1-symbols and no other symbols, it will be called *unitary polynomial* having weight w, and labelled $u_w(x)$, so that

$$u_w(x) = \sum_{j=0}^{w-1} x^j \tag{B.15}$$

Lemma B.14 *The following factorization is always possible*

$$x^N + 1 = (x+1)u_N(x) \tag{B.16}$$

Proof In this case all the central positions exhibit coincidences, so (B.16) easily comes out, independently of the particular considered value of N. □

In (B.16) the number of cancellations C, assuming expression (B.1) for the weight of a product, turns out to be $C = N - 1$. Unitary polynomials $u_N(x)$ sometimes can be factorized further, sometimes they are irreducible (see also Example B.20).

Example B.21 Restricting our attention to the prime numbers up to 23, the following results hold. With $N = 3, 5, 11, 13, 19$, the unitary polynomial $u_N(x)$ is irreducible. With $N = 17$ it is decomposable into two symmetric polynomials. Finally, with $N = 7, 23$, it is decomposable into a pair of reciprocal polynomials.

Lemma B.15 *The multiplication of* $(x + 1)$ *by whichever polynomial* $q(x)$ *gives a result whose weight* $w[(x + 1)q(x)]$ *is even.*

Proof In an expression of the type (B.1) both terms on the right hand side are even numbers. So the result turns out to be even as well. □

Reversely, Lemma B.15 can be interpreted also as follows: from an even-weight polynomial, factor $(x + 1)$ can be always extracted. But also a stronger statement is true: from an odd-weight polynomial factor $(x + 1)$ cannot be extracted. In order to justify this proposition, we can imagine the division of this odd-weight polynomial by $(x + 1)$. It easy to verify that the remainder will be always equal to 1.

Lemma B.16 *When N is odd, in the decomposition of* $(x^N + 1)$ *there is just one factor* $(x + 1)$.

Proof In decomposition (B.16), $u_N(x)$ has an odd weight N. Since $w[u_N(x)] = N$ is odd, this unitary polynomial cannot contain another factor $(x + 1)$, owing to the considerations developed after Lemma B.15. □

With N even, the unitary polynomials $u_N(x)$ can be always factorized, and, among the factors so obtained, at least another factor $(x + 1)$ is present.

Example B.22 Assuming $N = 14$, after extraction of factor $(x + 1)$ in an expression of the type (B.16), we remain with $u_{14}(x)$, which can be rewritten as

$$u_{14}(x) = (x^7 + 1)u_7(x) = (x + 1)u_7(x^2),$$

so that a further factor $(x + 1)$ has been shown to be present. In this case, the remaining factor $u_7(x^2)$ does not contain any other binomial $(x + 1)$.

From another point of view, we can assert that $(x + 1)$ is present in the factorization of any binary binomial, according to Lemma B.14, at least one time (this is the case of N odd). Instead, if N is even, different situations may occur.

Lemma B.17 *When the degree N of the binomial is N = 2b, with b odd,* $(x + 1)$ *is present in it exactly two times.*

Proof In this case we can write

$$x^N + 1 = (x^b + 1)(x^b + 1) = (x + 1)u_b(x)(x + 1)u_b(x) \quad \text{(B.17)}$$

and $u_b(x)$ cannot contain other factors $(x + 1)$, owing to Lemma B.16. □

This result explains what anticipated in Example B.22, where the last three factors in (B.17) have been collected in $u_{2b}(x) = u_{14}(x)$. The following statement generalizes what expressed by Lemma B.17.

Lemma B.18 *When the degree N of the binomial is even, it is always possible to write* $N = b2^L$, *with b odd (included the case of* $b = 1$*), and L a certain integer. In this situation, factor* $(x + 1)$ *is present exactly* 2^L *times.*

Proof Here our binomial is the 2^L-th power of $(x^b + 1)$. In any of such binomials, $(x + 1)$ is present just one time, therefore in $(x^N + 1)$ it will be present 2^L times on the whole. □

It is possible to conclude that, when N is even, the problem of factorizing $(x^N + 1)$ can be reduced to the factorization of a binomial having a lower odd degree. Hence we are left with the case characterized by N odd, for which it is useful to distinguish three situations:

i. N is a prime number,
ii. N is the product of two odd co-prime numbers,
iii. N is the power of a prime number.

In a more general context, including also non-binary alphabets, the decomposition into irreducible factors of a binomial $(x^N + 1)$, with N odd, is traditionally faced by means of a procedure called search for the Nth root of unity (MacWilliams and Sloane 1977), which will be outlined in the following. Nevertheless, it is convenient to derive here some preliminary properties characterizing the factors of $(x^N + 1)$, with N odd, simply taking advantage from weight and symmetry aspects.

Lemma B.19 *In the factorization of* $(x^N + 1)$, *with N odd, if there is an asymmetric irreducible polynomial* $a(x)$, *there is certainly also its reciprocal* $a^*(x)$.

Proof In (B.16) the unitary polynomial $u_N(x)$, being N odd, cannot exhibit further $(x + 1)$ factors. Moreover it is symmetric. Theorem B.5 states that the product of an asymmetric polynomial by its reciprocal gives rise to a symmetric polynomial. On the other hand, in case *ii* of Lemma B.13 the possibility that the product between a symmetric polynomial and an asymmetric polynomial is symmetric is excluded. Then the only remaining possibility is as follows: if there is an irreducible polynomial $a(x)$, there will be also $a^*(x)$. Finally the requirement that $a(x)$ is irreducible is due to the following observation. If one multiplies $a(x)$ by $(x + 1)$, the resulting polynomial is still asymmetric, but its reciprocal $a^*(x)(x + 1)$ cannot be found in the original binomial $(x^N + 1)$, since there is just one factor $(x + 1)$ on the whole. □

Yet, not always the couple of reciprocal polynomials is present. Sometimes the factorization up to irreducible factors leads to only symmetric polynomials. For instance, as anticipated in Example B.21, the case of $N = 17$ is characterized by the decomposition of $u_N(x)$ into two irreducible polynomials both symmetric.

Example B.23 The decomposition of $u_{17}(x)$ into irreducible factors yields

$$u_{17}(x) = (x^8 + x^5 + x^4 + x^3 + 1)(x^8 + x^7 + x^6 + x^4 + x^2 + x + 1).$$

These two symmetric polynomials correspond to the binary sequences 100111001 and 111010111, clearly both symmetric.

Lemma B.20 *When N is odd, in the factorization of $u_N(x)$, there are only irreducible factors characterized by odd weight.*

Proof After Lemma B.15 and subsequent observations, since there are no further $(x + 1)$ factors in $u_N(x)$, the factorization of $u_N(x)$ can give only odd-weight irreducible factors. □

In case *ii* of the above classification, we can let $N = ab$, but it is necessary to distinguish, more in detail, if the two odd co-prime integers a, b, are both prime or not. Let us consider firstly $N = ab$, with a and b both prime and odd. The following important statement holds.

Lemma B.21 *When $N = ab$, with a and b both prime and odd, the decomposition of binomial $(x^N + 1)$ gives*

$$x^{ab} + 1 = (x + 1)u_a(x)u_b(x)Q_{a,b}(x) \tag{B.18}$$

where polynomial

$$Q_{a,b}(x) = \frac{u_{ab}(x)}{u_a(x)u_b(x)}$$

has degree $deg[Q_{a,b}(x)] = ab - a - b + 1$, and is symmetric.

Proof After Lemma B.16, polynomial $(x + 1)$ in (B.18) is unique. Elimination of this factor in $(x^{ab} + 1)$ enables us to write the above expression for $Q_{a,b}(x)$. Furthermore, being $x^{ab} + 1 = u_a(x^b)(x^b + 1) = u_b(x^a)(x^a + 1)$, the following double equality holds

$$u_{ab}(x) = u_a(x^b)u_b(x) = u_b(x^a)u_a(x) \tag{B.19}$$

leading to these two further expressions for $Q_{a,b}(x)$

$$Q_{a,b}(x) = \frac{u_a(x^b)}{u_a(x)} = \frac{u_b(x^a)}{u_b(x)}. \tag{B.20}$$

A unitary polynomial is symmetric. Its enlarged version, obtained after substituting a power of x to the simple x, will be symmetric too. From property *i* of Lemma B.13, the division of a symmetric polynomial by a symmetric polynomial is symmetric itself. Thus $Q_{a,b}(x)$ is symmetric. About its degree, keeping in mind that $\deg[u_w(x)] = w - 1$, we can write $\deg[u_a(x^b)] = ab - b$, as well as $\deg[u_b(x^a)] = ab - a$. Therefore from both the equalities (B.20), we are left with $\deg[Q_{a,b}(x)] = ab - a - b + 1$. The same result can be also obtained directly from its definition in B(18). □

Lemma B.22 *Under the assumptions of* Lemma B.21, *polynomial $Q_{a,b}(x)$ has even degree, odd weight and exhibits, in its central position, a non-null coefficient.*

Proof Parameters a and b are odd, then product ab is odd too. Subtraction of $(a + b - 1)$, which is odd, makes $(ab - a - b + 1)$ even. Lemma B.12 states that the weight of a symmetric polynomial, whose degree is even, can be either even or odd, depending on its central position. Nevertheless, from (B.20) we obtain the equality $u_a(x^b) = Q_{a,b}(x)u_a(x)$, in which both $u_a(x^b)$ and $u_a(x)$ have odd weight. From (B.1) the weight of $Q_{a,b}(x)$ must then be odd as well. Finally Lemma B.12 assures that its central position exhibits a non-null coefficient. □

Example B.24 Let us consider $(x^{21} + 1)$, which is a binomial of the type (B.18) with $a = 3$ and $b = 7$. It can be factorized as follows

$$x^{21} + 1 = (x + 1)\left(x^2 + x + 1\right)\left(x^6 + x^5 + x^4 + x^3 + x^2 + x + 1\right)Q_{3,7}(x),$$

where $Q_{3,7}(x) = x^{12} + x^{11} + x^9 + x^8 + x^6 + x^4 + x^3 + x + 1$, representing the binary sequence 1101101011011.

Lemma B.23 *Under the assumptions of* Lemma B.21, *polynomial* $Q_{a,b}(x)$, *if decomposable, can be factorized into irreducible factors, all characterized by odd weight.*

Proof Being $w[Q_{a,b}(x)]$ odd, in polynomial $Q_{a,b}(x)$ irreducible factors whose weight is even cannot be present. In fact the product between an even-weight polynomial and an odd-weight polynomial would give an even-weight resulting polynomial. □

Like for $u_N(x)$ when N is prime, sometimes polynomial $Q_{a,b}(x)$ can be further factorized, sometimes it is irreducible. In the first case, $Q_{a,b}(x)$ can be factorized into either a pair of reciprocal polynomials or two symmetric polynomials.

Example B.25 Polynomial $Q_{3,7}(x)$ obtained in Example B.24 can be decomposed as

$$\begin{aligned} &x^{12} + x^{11} + x^9 + x^8 + x^6 + x^4 + x^3 + x + 1 \\ &\quad = \left(x^6 + x^5 + x^4 + x^2 + 1\right)\left(x^6 + x^4 + x^2 + x + 1\right) \end{aligned}$$

Such two reciprocal polynomials (both irreducible) represent the binary sequences 1110101 and 1010111 respectively. Therefore the complete decomposition of binomial $(x^{21} + 1)$ originates the following irreducible polynomials and correspondent binary sequences

$$\begin{array}{ll} x + 1 & 11, \\ x^2 + x + 1 & 111, \\ x^3 + x^2 + 1 & 1101, \\ x^3 + x + 1 & 1011, \\ x^6 + x^5 + x^4 + x^2 + 1 & 1110101, \\ x^6 + x^4 + x^2 + x + 1 & 1010111, \end{array}$$

where the third and the fourth derive from the decomposition of $u_7(x)$, that is 1111111.

Example B.26 Assuming $(x^{33} + 1)$, which is a binomial of the type (B.18) with $a = 3$ and $b = 11$, it can be factorized as follows

$$x^{33} + 1 = (x + 1)u_3(x)u_{11}(x)Q_{3,11}(x),$$

where we have $Q_{3,11}(x) = \left(x^{10} + x^7 + x^5 + x^3 + 1\right)\left(x^{10} + x^9 + x^5 + x + 1\right)$. Such two polynomials, corresponding to the sequences 10010101001 and 11000100011, are both symmetric.

What discussed here about the decomposition of $Q_{a,b}(x)$, when $N = ab$, with a and b both prime and odd, can be easily extended to the case in which one of these two parameters, in turn, is a product of two prime numbers, and so on. We are then left with case *iii* of the above classification, in which N is the power of a prime odd number.

Lemma B.24 *For* $N = a^2$, *with a prime and odd, the following equality holds*

$$x^N + 1 = (x + 1)u_a(x)u_a(x^a) \tag{B.21}$$

Proof Since $(x + 1)u_a(x) = x^a + 1$, expression (B.21) can be rewritten as $(x^a + 1)u_a(x^a)$, which easily yields $(x^{a^2} + 1)$. □

One could attempt to extend the definition of polynomial $Q_{a,b}(x)$ to the present case, where $a = b$, after (B.19) or (B.20). Nevertheless comparison with (B.21) would give $Q_{a,a}(x) = u_a(x^a)/u_a(x)$, but this division is not an exact quotient (as long as a is prime). Therefore this approach is not possible: the situation characterized by N coincident with a square of a prime must be treated differently.

Example B.27 Assuming $a = 3$, and $N = 3^2 = 9$, we have

$$x^9 + 1 = (x + 1)u_3(x)u_3\left(x^3\right) = (x + 1)\left(x^2 + x + 1\right)\left(x^6 + x^3 + 1\right),$$

where both the last two factors are irreducible.

Sometimes polynomial $u_a(x)$ can be decomposed. If this occurs, $u_a(x^a)$ can be decomposed too. In the cases of $a = 3$ and $a = 5$, $u_a(x)$ is irreducible, but this is not for $a = 7$, situation in which $u_a(x)$ shows two irreducible factors, one the reciprocal of the other. This property is useful in the decomposition of, say, $(x^{49} + 1)$. The following statement generalizes Lemma B.24.

Lemma B.25 *For* $N = a^L$, *with a odd prime and L an integer, we have*

$$x^N + 1 = (x + 1)u_a(x)u_a(x^a)\ldots u_a\left(x^{a^{L-1}}\right) \tag{B.22}$$

Proof Iteration L times what stated by Lemma B.24 makes it possible to demonstrate (B.22). □

Example B.28 Assuming $L = 3$, and still $a = 3$, that is $N = 27$, we obtain

$$\begin{aligned} x^{27} + 1 &= (x+1)u_3(x)u_3(x^3)u_3(x^9) \\ &= (x+1)(x^2+x+1)(x^6+x^3+1)(x^{18}+x^9+1) \end{aligned}$$

It is possible to conclude that, when $u_a(x)$ is irreducible, $u_a(x^a)$, $u_a\left(x^{a^2}\right)$, … are also irreducible. This property holds for whichever odd integer a.

Decomposition of (x^N + 1), N Odd, for a Binary Field, by the Search for the *N*th Root of Unity

In this treatment, several statements will not be demonstrated. Our attention is restricted, for the moment, only to a binary alphabet, whose elements are 0, 1. Such elements, together with the operation of ex-OR sum and multiplication, form a binary finite field (see also Appendix A). The particular case characterized by $N = 2^m - 1$, being m an integer, will be considered with more details, since it exhibits some particular properties. Let us start from the formal equation

$$x^N = 1 \tag{B.23}$$

Definition B.14 In a proper extension of the considered finite field, there are N roots for Eq. (B.23). Such roots can be imagined to form a circle, like complex numbers in Gauss plane. If α_j, $j = 0, 1, \ldots, (N-1)$, are these roots (each being called an *Nth root of unity*), it is possible to write

$$x^N + 1 = \prod_{j=0}^{N-1} (x + \alpha_j) \tag{B.24}$$

Since factor $(x + 1)$ is always present, conventionally we adopt α_0 associated to this factor.

It is convenient to substitute the single roots a_j with proper sets of roots, which are collected in as many irreducible polynomials. In each set, any element is obtained as a suitable power of a certain element, in a sequence of subsequent powers (MacWilliams and Sloane 1977). This procedure stops because of periodicity, in turn due to the finite value of N. If we have, say, M sets, it is then possible to replace (B.24) with

$$x^N + 1 = (x+1) \prod_{j_1=1}^{n_1} \left(x + \alpha_2^{j_1}\right) \cdot \prod_{j_2=1}^{n_2} \left(x + \alpha_2^{j_2}\right) \cdot \ldots \cdot \prod_{j_M=1}^{n_M} \left(x + \alpha_M^{j_M}\right), \tag{B.25}$$

$$\sum_{i=1}^{M} n_i = N - 1$$

If s is an element of the ensemble containing the first $(N - 1)$ non-null integers, other elements can be obtained from it, after multiplication by subsequent powers of a basis 2, so forming sets of the type

$$C_s = \{s, s2, s2^2, \ldots, s2^{n_s-1}\}, \quad (s2^{n_s}) \bmod N = s \tag{B.26}$$

The union of all the sets C_s, which can collect different numbers of elements, contains all the elements of the original ensemble. Construction is made in such a way as to assure that the same element cannot be present in more than one set.

Example B.29 Assuming $N = 9$, we obtain the following three sets

$$C_0 = 0,$$
$$C_1 = \{1, 2, 4, 8, 7, 5\},$$
$$C_3 = \{3, 6\}.$$

In this partitioning we have considered that (16)mod9 = 7, (32)mod9 = 5, and C_2 has not been taken into account because element 2 of our finite field was already present in C_1.

Definition B.15 The sets C_s are called *cyclotomic cosets* in N with basis 2. The corresponding factorization of the binary binomial $(x^N + 1)$ is characterized by a set of roots for any cyclotomic coset. The number of elements in each coset coincides with the degree D_s of the irreducible polynomial, factor of $(x^N + 1)$, associated to it.

Definition B.16 The D_s roots collected in the same irreducible polynomial, as described in Definition B.15, are called *conjugate roots*.

Example B.30 Factorization of $(x^9 + 1)$ starts from the subdivision into cyclotomic cosets of the elements of a binary field characterized by $N = 9$, shown in Example B.29. C_0 corresponds to $(x + 1)$, always present. C_3 corresponds to $(x^2 + x + 1)$. Finally C_1 corresponds to $(x^6 + x^3 + 1)$. In this situation, after the first trivial identification, the remaining ones should be constructed by attempts.

When, in a binary alphabet, we assume $N = 2^m - 1$, it is possible to have a help. In fact, known a particular irreducible polynomial, all the other ones can be obtained from it. Furthermore, the following property holds, whose demonstration can be found in MacWilliams and Sloane (1977).

Theorem B.6 *In the case of $N = 2^m - 1$, a particular irreducible polynomial $p(x)$, able to generate all the other irreducible polynomials, always exists, and its degree is m.*

Definition B.17 Polynomial $p(x)$, introduced in Theorem B.6, is called the *primitive polynomial* in the factorization of $(x^N + 1)$, being $N = 2^m - 1$.

Definition B.18 In a finite field with 2^m elements, associated to as many sequences of m binary symbols, an irreducible polynomial $p(\alpha)$, characterized by $\deg[p(\alpha)] = m$, can form all the non-null elements of the field, by proper subsequent powers of the

Power-like	Polynomial form	Sequence of 3 symbols	
0	0	0 0 0	(null element)
α^0	1	0 0 1	(non-null elements)
α^1	α	0 1 0	
α^2	α^2	1 0 0	
α^3	α^2+1	1 0 1	
α^4	$\alpha^2+\alpha+1$	1 1 1	
α^5	$\alpha+1$	0 1 1	
α^6	$\alpha^2+\alpha$	1 1 0	

Fig. B.8 Construction of the 7 non-null elements of the finite field with $m = 3$ (Example B.31)

formal variable α. This is made taking such powers mod[$p(\alpha)$], and α is called the *primitive element* of such 2^m-ary field.

Although, in principle, such irreducible polynomial and the primitive polynomial to be exploited for the factorization of $(x^{2^m-1}+1)$ could be different, it is convenient to take the same polynomial. This fact justifies the use of an identical notation.

Example B.31 For $m = 3$, we have 7 non-null elements. As the irreducible polynomial, it is possible to take $\rho(\alpha) = \alpha^3 + \alpha^2 + 1$. In Fig. B.8, the eight elements of such field are listed, in power-like form, in polynomial form, and in the form of a sequence of 3 binary symbols. The polynomial form is obtained by expressing the power-like form mod$(\alpha^3 + \alpha^2 + 1)$, whereas the sequence of binary symbols is related to the polynomial form by the usual convention.

Remark that it is possible to write also $\alpha^7 = 1$, so this construction turns out to be periodic with period 7. This means that α, besides to be a 3rd root of the considered primitive polynomial, is also a 7th root of the binomial $(\alpha^7 + 1)$. In general, α, besides to be an mth root of the primitive polynomial $p(\alpha)$, is also a $(2^m - 1)$th root of the binomial $(\alpha^{2^m-1}+1)$, from which we started the present discussion. In conclusion, when $N = 2^m - 1$, the m roots collected in the primitive polynomial $p(x)$, are also as many roots of $(x^N + 1)$.

Example B.32 Factorization of (x^7+1) can be performed as follows. After assumption of $p_1(x) = x^3 + x^2 + 1$ as the primitive polynomial, it can be associated to the cyclotomic coset $C_1 = \{1,2,4\}$. There is another non-trivial cyclotomic coset, $C_3 = \{3,6,5\}$, in fact (12)mod7 = 5. It is associated to the polynomial $p_2(x) = x^3 + x + 1$. More precisely, $p_1(x) = x^3 + x^2 + 1$ collects the three roots α, α^2, α^4, whereas $p_2(x) = x^3 + x + 1$ collects the three roots α^3, α^6, $\alpha^{12} = \alpha^5$. The last conclusion is justified by the following long but straightforward computations

$$
\begin{aligned}
(x+\alpha)(x+\alpha^2)(x+\alpha^4) &= x^3 + (\alpha^4+\alpha^2+\alpha)x^2 + (\alpha^6+\alpha^5+\alpha^3)x + 1 \\
&= x^3 + x^2 + 1 \\
(x+\alpha^3)(x+\alpha^6)(x+\alpha^5) &= x^3 + (\alpha^6+\alpha^5+\alpha^3)x^2 + (\alpha^4+\alpha^2+\alpha)x + 1 \\
&= x^3 + x + 1
\end{aligned}
$$

where we have taken various identities from the table in Fig. B.8, together with $\alpha^7 = 1$. In conclusion, the factorization of $(x^7 + 1)$ into irreducible polynomials, following the method here described, yields

$$x^7 + 1 = (x + 1)(x^3 + x^2 + 1)(x^3 + x + 1)$$

Example B.33 With $N = 9$ (case already considered in Examples B.27, B.29, B.30), clearly not in the form $N = 2^m - 1$, we have three cyclotomic cosets, and correspondently three sets of conjugated roots. Irreducible polynomials and such sets are so related

$$\begin{array}{ll} C_0 = \{0\} & (x+1) \\ C_1 = \{1, 2, 4, 8, 7, 5\} & (x^6 + x^3 + 1) \\ C_3 = \{3, 6\} & (x^2 + x + 1) \end{array}$$

in fact, as already noted, $(16)\mathrm{mod}9 = 7$ and $(32)\mathrm{mod}9 = 5$. Therefore the factorization of $(x^9 + 1)$ into irreducible polynomials becomes

$$x^9 + 1 = (x + 1)(x^6 + x^3 + 1)(x^2 + x + 1).$$

confirming what obtained in Examples B.27 and B.29, for a situation where N is the square of a prime odd number. Nevertheless, since N is not of the type $N = 2^m - 1$, it is not possible to identify, between the two non-trivial irreducible polynomials listed above, any primitive polynomial, able to allow the construction of the other irreducible factor of $(x^9 + 1)$.

When N is odd prime and not equal to $(2^m - 1)$, a primitive polynomial exists, but it has to be searched in a finite field with a higher number of elements (Wicker 1995). How this search and the consequent construction can be made is beyond the objectives of this brief review. Another important property regarding cyclotomic cosets and the correspondent factorization of $(x^N + 1)$ can be enunciated by the following statement, whose demonstration can be found in Peterson and Weldon (1972).

Theorem B.7 *If N is a prime odd number, all the cosets, except C_0, have the same number of elements, therefore all the irreducible factors of ($x^N + 1$), except $(x + 1)$, have the same degree.*

The justification of this result is in the process for reaching the condition $(s2^{n_s}) \bmod N = s$, which is characterized by an equal number of steps for all the cyclotomic cosets. Finally it is possible to observe that, when $N = 2^m - 1$, number always odd, independently if prime or product of odd factors, the number of different irreducible polynomials, having degree m, which can be assumed as primitive polynomials, increases with increasing parameter m. The following two propositions regard whichever N odd, not necessarily equal to $(2^m - 1)$.

Lemma B.26 *When N is product of two odd factors, say N = ab, a primitive polynomial must be searched among the factors of polynomial $Q_{a,b}(x)$ introduced in* Lemma B.21.

Proof In all the other possible groups of factors, we have just factors of $u_a(x)$ or $u_b(x)$, and hence primitive polynomials in the decomposition of $(x^a + 1)$ or of $(x^b + 1)$, but not in the decomposition of $(x^{ab} + 1)$. □

Lemma B.27 *In the factorization of $(x^N + 1)$, with N odd, if a certain asymmetric primitive polynomial exists, also its reciprocal will be a primitive polynomial.*

Proof In a pair of reciprocal polynomials, $p(x)$, $p^*(x)$, there is no preference regarding symmetry, since $(x^N + 1)$ is symmetric. Therefore, if $p(x)$ is a primitive polynomial, also $p^*(x)$ exhibits the same property. □

Example B.34 Taking $N = 15$, that is $N = 2^m - 1$, with $m = 4$, it is possible to verify that polynomial $p(\alpha) = \alpha^4 + \alpha^3 + 1$ is irreducible. So, we could arrange a table similar to that reported in Fig. B.8, with 15 rows, associated to as many subsequent powers of the primitive element α, besides the null element of such finite field. Cyclotomic cosets and corresponding irreducible factors then will be as follows

$$\begin{array}{ll} C_0 = \{0\}, & x + 1, \\ C_1 = \{1, 2, 4, 8\}, & x^4 + x^3 + 1, \\ C_3 = \{3, 6, 9, 12\}, & x^4 + x^3 + x^2 + x + 1, \\ C_5 = \{5, 10\}, & x^2 + x + 1, \\ C_7 = \{7, 11, 13, 14\}, & x^4 + x + 1. \end{array}$$

Remember that the last elements of any cyclotomic coset are obtained mod(15). The complete factorization into irreducible polynomials of $(x^{15} + 1)$ finally results in

$$x^{15} + 1 = (x + 1)(x^4 + x^3 + 1)(x^4 + x + 1)(x^4 + x^3 + x^2 + x + 1)(x^2 + x + 1).$$

As a polynomial $Q_{a,b}(x)$, we have here $Q_{3,5}(x) = (x^4 + x^3 + 1)(x^4 + x + 1)$, where one of these two factors can be assumed as a primitive polynomial.

Example B.35 Taking $m = 5$, that is $N = 31$, prime number, we have 30/5 = 6, according to Theorem B.7. Cyclotomic cosets and correspondent irreducible polynomials, written just as sequences of binary symbols for simplicity, can be listed as follows

$$\begin{array}{ll} C_0 = \{0\}, & 11, \\ C_1 = \{1, 2, 4, 8, 16\}, & 100101, \\ C_3 = \{3, 6, 12, 17, 24\} & 111101, \\ C_5 = \{5, 9, 10, 18, 20\}, & 110111, \\ C_7 = \{7, 14, 19, 25, 28\}, & 101111, \\ C_{11} = \{11, 22, 13, 21, 26\}, & 111011, \\ C_{15} = \{15, 23, 27, 29, 30\}, & 101001. \end{array}$$

Table B.1 Factorization of $(x^N + 1)$, for N odd, up to 63

N	Irreducible polynomials
3	11, 111
5	11, 11111
7	11, 1101, 1011
9	11, 111, 1001001
11	11, 11111111111
13	11, 1111111111111
15	11, 111, 11111, 11001, 10011
17	11, 100111001, 111010111
19	11, 1111111111111111111
21	11, 111, 1101, 1011, 1010111, 1110101
23	11, 101011100011, 110001110101
25	11, 11111, 100001000010000100001
27	11, 111, 1001001, 1000000001000000001
29	11, 11111111111111111111111111111
31	11, 101001, 100101, 101111, 111101, 110111, 111011
33	11, 111, 11111111111, 10010101001, 11000100011
35	11, 111, 1101, 1011, 1011110010111, 1110100111101
37	11, 1111111111111111111111111111111111111
39	11, 111, 1111111111111, 1111000111101, 1011110001111
41	11, 101111100111001111101, 110110100111001011011
43	11, 100111111111001, 101010010010101, 110100010001011
45	11, 111, 1001001, 11111, 10011, 11001, 1000000001001, 1001000000001
47	11, 100011000111011101111, 111101110110111000110001
49	11, 1101, 1011, 1000000100000000000001, 1000000000000010000001
51	11, 111, 100111001, 111010111, 100011011, 110110001, 110011111, 111110011
53	11, 11111111111111111111111111111111111111111111111111111
55	11, 11111, 11111111111, 101101101001011100111, 111001110100101101101
57	11, 111, 1111111111111111111, 1011100001000011101, 1111011101011101111
59	11, 11111111111111111111111111111111111111111111111111111111111
61	11, 1111111111111111111111111111111111111111111111111111111111111
63	11, 111, 1001001, 1101, 1011, 1010111, 1110101, 1100001, 1000011, 1110011, 1100111, 1101101, 1011011

Also here the last elements of any cyclotomic cosets are calculated mod(31). Each one of the above irreducible polynomials can be assumed as a primitive polynomial. This choice induces a permutation in the order of the remaining ones.

Before concluding the present brief review, we can summarize all the results of the factorization of $(x^N + 1)$, with N odd, into irreducible polynomials, up to $N = 63$, as shown in Table B.1. The following short comments can be added to all what observed so far.

For N = 3, 5, 11, 13, 19, 29, 37, 53, 59, 61, we have the only trivial decomposition $(x+1)u_N(x)$. For N = 7, 23, 47, we are in a situation characterized by a prime number for which $u_N(x)$ factorizes into a pair of reciprocal irreducible polynomials. For N = 17, 41, prime number as well, $u_N(x)$ factorizes into a pair of symmetric irreducible polynomials. For N = 43, prime number too, $u_N(x)$ factorizes into a tern of symmetric irreducible polynomials. Finally for N = 31, prime number, but also of the type $N = 2^5 - 1$, there are 30/5 = 6 irreducible polynomials, according to Theorem B.6 and Theorem B.7. They form three couples of reciprocal polynomials (see also Example B.35).

Let us discuss now values of the type $N = ab$, with a, b prime. For N = 15, 21, 35, 39, 55, besides the non-trivial factors of $(x^a + 1)$ and $(x^b + 1)$, there is a polynomial $Q_{a,b}(x)$ which factorizes into a pair of reciprocal irreducible polynomials. For N = 33, 57, $Q_{a,b}(x)$ is, instead, the product of two symmetric irreducible polynomials. For N = 51, $Q_{a,b}(x)$ is the product of four irreducible polynomials, grouped into two pairs, reciprocal each other.

Considering the square of a prime number, that is N = 9, 25, 49, we can observe how the non-trivial factors of the square root are simply scaled to form the additional factors. The cube N = 27 exhibits a repetition of this rule one time more (according to Lemma B.25).

We are left finally with situations characterized by $N = ab$, where a is prime, but b is itself the product of two primes, or the square of a prime. Here not only the non-trivial factors of the three binomials must to be present, but also those of the three possible products between two of them. The cases n = 45, 63 are of these type, but just with b = 9 being a square. In the first case, also the non-trivial factors of $(x^{15} + 1)$ are present. In the second case, also the non-trivial factors of $(x^{21} + 1)$ are present. For this reason, among the 8 irreducible asymmetric polynomials, grouped into pairs where the one is reciprocal of the other, only the last 6 are primitive (according to Lemma B.26). The case of N = 63 is noticeable also because it can be written as $N = 2^6 - 1$.

It is interesting to compare some situations characterized by $N = 2^m - 1$, with m increasing. Table B.2 contains the numbers of different primitive polynomials, and their weights, for m = 3, 4, 5, 6, 7. The number of different primitive polynomials grows with increasing m, as already said, and such growth appears rather rapid.

Table B.2 Number and weights of the primitive polynomials when $N = 2^m - 1$

m	N	Number of primitive polynomials	Weights
3	7	2	3, 3
4	15	2	3, 3
5	31	6	3, 3, 5, 5, 5, 5
6	63	6	3, 3, 5, 5, 5, 5
7	127	18	3, 3, 3, 3, 5, 5, 5, 5, 5 , 5, 5, 5, 5, 7, 7, 7, 7

Remark that $N = 127$ is prime, so we have $126/7 = 18$, as expected, according to Theorem B.7. These 18 irreducible (and primitive) polynomials are listed below, grouping them on the basis of their weight w,

$$11000001, 10000011, 10010001, 10001001 \quad w = 3$$

$$\begin{array}{l} 11110001, 10001111, 10111001, 10011101 \\ 11100101, 10100111, 11010101, 10101011 \\ 10110101, 10101101 \qquad\qquad\qquad\qquad w = 5 \end{array}$$

$$11111101, 10111111, 11110111, 11101111 \quad w = 7$$

Also for larger values of m, there will be polynomials characterized by weight 3, and by all the odd weights between 3 and the maximum value w_{max}, which is so expressed

$$w_{max} = \begin{array}{ll} m, & (m\ odd) \\ m-1, & (m\ even) \end{array}. \qquad \text{(B.27)}$$

Polynomial Interleaved Multiplication

Polynomial interleaved multiplication is a generalization of traditional polynomial multiplication. It is based on a periodic behaviour characterized by different positions where two ordered sets of polynomials (instead of simply two polynomials) are involved. They represent the two factors of such a multiplication. In defining polynomial interleaved multiplication, it is necessary to keep well in mind the need of assuring, in the associated interleaved division, that an exact quotient, multiplied by the divisor, gives back the dividend. This concept is related to the existence of the inverse (or right-inverse) in a coefficient matrix expressing a vector transformation (see Appendix A). In polynomial code theory, one of the two ordered sets of polynomials represents the information symbol sequence properly partitioned. The other one describes the action of the encoder circuit (see also Appendix C).

In order to satisfy the above condition, first of all we have to adopt the constraint of devoting different positions to the coefficients of the polynomials in the set describing the encoder circuit. Furthermore it is necessary that such polynomials are properly aligned on their left hand side, and hence have the same degree. When they originally exhibit different degrees, it is necessary to shift their highest-order coefficient towards higher powers, after multiplication by a proper shift factor, in such a way as to place all the new highest-order coefficients in the same period.

Let us consider the two sets of polynomials $u_i(x)$, $i = 1, 2, \ldots, n_o$, and $p_i(x)$, $i = 1, 2, \ldots, n_o$. The polynomials of the first set, which express the information sequence, is to be taken just with powers multiple of a certain period n_o, so forming a sort of raster structure, expressed by $u_i(x^{n_o})$. Remark that here symbol $u_i()$ does not express a unitary polynomial, but has been adopted following the traditional symbolism in polynomial code theory.

Definition B.19 The *polynomial interleaved product* (or *multiplication*) between the set of polynomials $u_i(x^{n_o})$ and the set of polynomials $p_i(x)$, $i = 1, 2, \ldots, n_o$, can be written as

$$m(x) = x^{n_o-1}u_1(x^{n_o})p_1(x) + x^{n_o-2}u_2(x^{n_o})p_2(x) + \cdots + u_{n_o}(x^{n_o})p_{n_o}(x). \qquad \text{(B.28)}$$

In order to obtain all the highest-order coefficients in different positions of the same period, it is necessary to select the polynomial $p_j(x)$ having maximum degree, and apply a proper shift factor $x^{\Delta i}$ to any $p_i(x)$, with $\Delta_i = \deg[p_j(x)] - \deg[p_i(x)]$. This operation is not necessary if all the polynomials $p_i(x)$ have the same degree.

Example B.36 Taking $p_1(x) = x^2 + 1$, $p_2(x) = x^5 + x^4 + 1$, their interleaved product by two binary sequences $u_1(x^2)$ and $u_2(x^2)$ is

$$\begin{aligned} m(x) &= xu_1(x^2)x^3p_1(x) + u_2(x^2)p_2(x) \\ &= xu_1(x^2)(x^5 + x^3) + u_2(x^2)(x^5 + x^4 + 1), \end{aligned}$$

where we have assumed $\Delta_1 = 3$ in order to set the highest-order coefficients of the two polynomials in two different positions of the same period.

Polynomial Interleaved Division

Interleaved division is the inverse operation with respect to interleaved multiplication. It gives rise to an interleaved quotient and to a unique final remainder. After adoption of possible shift factors, we can consider that all the polynomials in the divisor have the same degree.

Definition B.20 Given a certain polynomial $p(x)$ (the dividend), its *polynomial interleaved division* by an ordered set of polynomials $z_j(x)$, $j = 1, 2, \ldots, n_0$ (the interleaved divisors), provided that $\deg[p(x)]$ is high enough, yields quotient and remainder as follows

$$\begin{aligned} &\frac{p(x)}{x^{n_0-1}z_1(x) + x^{n_0-2}z_2(x) + \cdots + z_{n_0}(x)} \\ &= Q\left[\frac{p(x)}{x^{n_0-1}z_1(x) + x^{n_0-2}z_2(x) + \cdots + z_{n_0}(x)}\right] \\ &\quad + \frac{1}{x^{n_0-1}z_1(x) + x^{n_0-2}z_2(x) + \cdots + z_{n_0}(x)} R\left[\frac{p(x)}{x^{n_0-1}z_1(x) + x^{n_0-2}z_2(x) + \cdots + z_{n_0}(x)}\right] \end{aligned} \qquad \text{(B.29)}$$

Polynomials $z_j(x)$ are associated to as many positions in a period of length n_0. The operation described in Definition B.20, like traditional polynomial division, is performed in subsequent steps. Firstly the highest power of the dividend, and then the highest power of the current remainder are divided by the proper divisor $z_j(x)$. The concept of current remainder is the same as that introduced in Lemma B.9 and subsequent considerations, for a traditional polynomial division. Proper means that the divisor $z_j(x)$ has to be selected as the one with its highest power in the position where the highest power of the original polynomial (or of the current remainder) is placed.

Lemma B.28 *In order to have a univocal result, in an interleaved division, it is necessary that the ordered set of interleaved divisors $z_j(x)$, $j = 1, 2, \ldots, n_0$, exhibit all the same degree, independently of the presence or not of the zeroth order power.*

Proof Only under this condition, for any position in the period it will be a proper $z_j(x)$. This constraint allows to align the highest-order coefficient of any $z_j(x)$ in subsequent adjacent positions of the same period on their left hand side (as already imposed to interleaved multiplication). Naming D the degree of all the $z_j(x)$'s, after the above alignment, the final remainder will exhibit a degree smaller than D, and the interleaved division will end. □

In conclusion, the dividend and the current remainder are unique, although n_0 current ordered quotients come out, one for each ordered divisor (each position in the period).

Lemma B.29 *The result of* (B.29), *under the assumption of* Lemma B.28, *can be written as*

$$\frac{p(x)}{x^{n_o-1}z_1(x)+x^{n_o-2}z_2(x)+\cdots z_{n_o}(x)}$$
$$= x^{n_o-1}q_1(x^{n_o})+x^{n_o-2}q_2(x^{n_o})+\cdots+q_{n_o}(x^{n_o})$$
$$+\frac{1}{x^{n_o-1}z_1(x)+x^{n_o-2}z_2(x)+\cdots+z_{n_o}(x)}R\left[\frac{p(x)}{x^{n_o-1}z_1(x)+x^{n_o-2}z_2(x)+\cdots z_{n_o}(x)}\right] \tag{B.30}$$

Proof The various $z_j(x)$ act, by means of their highest-order coefficients, in distinct positions of the period n_o, so leading to the correspondent ordered quotients $q_j(x^{n_o})$, $j = 1, 2, \ldots, n_o$. The remainder, instead, remains unique. □

Definition B.21 On the right hand side of expression (B.30), there are two parts. The sum forming the first part is called *interleaved quotient*, whereas the ratio forming the unique term in the second part has a numerator $R[\;]$, simply called the final remainder, as in a traditional polynomial division.

In order to obtain back to the original dividend, we have to perform, inversely, an interleaved product between the ordered set of polynomials $q_i(x^{n_o})$and the ordered set of polynomials $z_i(x)$. Finally we have to add the unique remainder if present.

Fig. B.9 Interleaved division introduced in Example B.37

	1101	$z_1(x)$
100100010	1110	$z_2(x)$
1101		
1110		
1101	10100	$q_1(x^2)$
1110	10001	$q_2(x^2)$
$R(x)$: 0100		

Example B.37 Taking $p(x) = x^8 + x^5 + x$, and the following pair of ordered divisors $z_1(x) = x^3 + x^2 + 1$, $z_2(x) = x^3 + x^2 + x$, the procedure described above evolves as shown in Fig. B.9. The ordered quotients turn out to be $q_1(x^2) = x^4 + x^2$, $q_2(x^2) = x^4 + 1$ and the final remainder is $R(x) = x^2$. It is possible to verify that, inversely, we have

$$x^8 + x^5 + x = x(x^4 + x^2)(x^3 + x^2 + 1) + (x^4 + 1)(x^3 + x^2 + x) + x^2,$$

where the first two terms are due the interleaved product, whereas the last term is the remainder.

Lemma B.3 states that the product of two polynomials has at least weight 2, and this is the case of the binomial $(x^N + 1)$. Such property implies that polynomial division of x^L for whichever polynomial $z(x)$, with $\deg[z(x)] < L$, has remainder 1. This property is no longer true for an interleaved division.

Example B.38 Assuming x^6, and the pair of ordered divisors $z_1(x) = x^3 + x^2 + x$, $z_2(x) = x^3 + x^2 + 1$, we obtain a null remainder. In fact the following equality holds

$$x^6 = x(x^2 + 1)(x^3 + x^2 + x) + x^2(x^3 + x + 1),$$

which means $q_1(x^2) = (x^2 + 1)$, $q_2(x^2) = x^2$, forming an exact quotient.

It is possible to show some properties regarding interleaved division and the periodic behaviour (on a large scale) of its interleaved quotient or of its remainder, but only when the above anomalous situation is excluded. Such properties are similar to those already seen for traditional polynomial division, although some particular features have to be remarked.

Lemma B.30 *In an interleaved division of the dividend x^L, after an increase of L by one unit, the current remainder simply shifts (maintaining form but just increasing the degree) or changes, because of the ex-OR sum with a proper interleaved divisor $z_j(x)$.*

Proof We can proceed like for Lemma B.9, with the only difference of considering an interleaved division. In correspondence with the highest-order coefficient of the current remainder the highest-order coefficient of a proper interleaved divisor $z_j(x)$ will be assumed. This highest-order coefficient vanishes, the current remainder changes form, and the procedure goes on. □

Lemma B.31 *The current remainder, at a certain step of the of an interleaved division procedure applied to* x^L, *completely determines the remaining part of the division itself.*

Proof Like in Lemma B.10, also here, since the dividend is x^L, there are no other polynomial coefficients to be taken into account after x^L itself, eliminated by the first ex-OR sum. Then the current remainder completely determines the procedure of an interleaved division. □

Theorem B.8 *The interleaved division of* x^L *by an ordered set of interleaved divisors* $z_j(x)$, *with L increasing, provided that it is not exact, exhibits a current remainder and a current interleaved quotient, both periodic on large scale.*

Proof We can proceed like for Theorem B.3, with the only difference of considering here an interleaved division. First of all, the possibility of having null final remainder, that is an exact interleaved division, is to be taken into account and must be excluded. Otherwise, a remainder 1 will appear in the same position of the period where x^L was initially placed. In fact the current remainder completely controls the process after Lemma B.31, so, assuming $D = \deg[z_j(x)]$, at most we can have $2^{n_o D}$ different remainders, either simple shifts or changed forms. Still from Lemma B.31, the new 1-symbol, in the same position where x^L was initially placed, gives rise to the repetition of the same evolution. Therefore the current interleaved quotient and the current remainder will be both periodic on large scale (this specification is needed in order to avoid confusion with the small-scale period n_o). □

Corollary a *The interleaved division of any polynomial* $x^L\ p(x)$ *by an ordered set of interleaved divisors, with L increasing, exhibits a current remainder and a current interleaved quotient, both periodic on large scale, provided that the division is not exact.*

Proof We can proceed as in the proof of Corollary b of Theorem B.3. This property is no longer valid if the remainder, at a certain step of the procedure, becomes null, leading to an exact interleaved division. □

Example B.39 Assuming x^{15} and the pair of interleaved divisors $z_1(x) = x^4 + x^3 + 1$, $z_2(x) = x^4 + x^3 + x$, we obtain a remainder with a 1-symbol in the same position (the former one). In fact, the following equality holds

$$x^{15} + x = x(x^{10} + x^4 + x^2 + 1)(x^4 + x^3 + 1) + (x^{10} + x^8 + x^6 + x^2)(x^4 + x^2 + x).$$

Let us adopt now a binomial as the dividend, in an interleaved division by the two ordered divisor $z_1(x)$, $z_2(x)$, chosen in such a way as to exclude the possibility that a single power of our binomial gives an exact quotient. After Theorem B.8, we will obtain an exact interleaved quotient (a null final remainder). The original binomial can be written also here $(x^N + 1)$. In the case considered in Example B.39, we have $N = 14$. The interleaved division of this binomial taken as the dividend, is shown in Fig. B.10. Such interleaved division is exact. This assert is equivalent to

```
                                    | 1 1 0 0 1              z1(x)
 1 0.0 0.0 0.0 0.0 0.0 0.0 0.0 0.10 |  1 0 1 1 0             z2(x)
 1 1 0 0 1                          |________________
   1 0 1 1 0                        |
       1 0 1 1 0                    | 1 0.0 0.0 0.1 0.1 0.1 0   q1(x^2)
           1 0 1 1 0                | 0 1.0 1.0 1.0 0.0 1.0 0  q2(x^2)
             1 1 0 0 1
                 1 1 0 0 1
                   1 0 1 1 0
                     1 1 0 0 1
                         -
```

Fig. B.10 Exact interleaved division introduced in Example B.39

say that the interleaved division of x^{15}, taken alone, exhibits interleaved quotient and remainder both periodic with period $N = 14$.

It is important to stress that a remainder with just one 1-symbol can be obtained also before the reaching of the whole large-scale period length N, if this 1-symbol occupies a different position in the small-scale period, with respect to that where x^L is placed. Therefore, in order to establish the length of the large-scale period, it is necessary to have x^L and the final remainder consisting of one 1-symbol in the same position of the small-scale period n_o. From all the above considerations, it is possible to conclude that, in order to have a periodic behaviour, the large-scale period N must be an integer multiple of the small-scale period n_o.

Outline of the Search for the *N*th Root of Unity in a Non-binary Finite Field

In a Q-ary finite field we have to consider suitable operations of addition and multiplication, described by as many tables (see Appendix D). Remark that the minus sign is here, in general, needed (the opposite of an element does not coincide with it, except when Q is 2 or an integer power of 2). The concept of cyclotomic coset can be widened in order to consider a Q-ary finite field as follows. Instead of (B.26), it is possible to write

$$C_s = \{s, sQ, sQ^2, \ldots, sQ^{n_s-1}\}, \quad (sQ^{n_s}) \bmod N = s \tag{B.31}$$

The formal decomposition (B.24) is replaced by

$$x^N - 1 = \prod_{j=0}^{N-1} (x - \alpha_j) \tag{B.32}$$

Here the factor always present is $(x - 1)$. To this factor, conventionally the root α_0 is associated.

Lemma B.32 *The following expression is valid in general, instead of* (B.16),

$$x^N - 1 = (x-1)(x^{N-1} + \cdots + x + 1) \tag{B.33}$$

Proof In the result, couples of adjacent terms cancel together. Only the two external terms remain. □

The second factor on the left hand side of (B.33) can be or cannot be further decomposable. An expression analogous to (B.25), with sets of grouped roots, appears as follows

$$x^N - 1 = (x-1)\prod_{j_1=1}^{n_1}(x-\alpha_1^{j_1}).\prod_{j_2=1}^{n_2}(x-\alpha_2^{j_2})\ldots\prod_{j_M=1}^{n_M}(x-\alpha_M^{j_M}) \tag{B.34}$$

$$\sum_{i=1}^{M} n_i = N - 1.$$

A particular situation is that characterized by $N = Q - 1$.

Lemma B.33 *For* $N = Q - 1$, *expression* (B.34) *becomes*

$$x^N - 1 = (x-1)(x-\alpha_1)(x-\alpha_2)\ldots(x-\alpha_{N-1}) \tag{B.35}$$

Proof The number of non-null elements of the finite field equals the degree of the binomial to be factorized into irreducible polynomials, so leading to sets made by just one factor each. □

Assuming α as the primitive element of our Q-ary field, (B.35) can be also formally written as

$$x^N - 1 = (x-\alpha^0)(x-\alpha^1)(x-\alpha^2)\ldots(x-\alpha^{Q-2}), \quad N = Q - 1 \tag{B.36}$$

in which we have set $\alpha^0 = 1$. In (B.35) or (B.36) there is the product of $(Q - 1)$ binomials, each having degree 1 and 0th-order term equal to one different non-null element of the field.

References

Anderson JB, Mohan S (1991) Source and channel coding. Kluwer, Boston

Johannesson R, Zigangirov KS (1999) Fundamentals of convolutional coding. IEEE Press, New York

Lin S, Costello DJ (2004) Error control coding. Pearson/Prentice-Hall, Upper Saddle River

MacWilliams FJ, Sloane NJA (1977) The theory of error correcting codes. North Holland, New York

Oppenheim AV, Schafer RW (1974) Digital Signal Processing. Prentice-Hall, Englewood Cliffs
Peterson WW, Weldon EJ (1972) Error-correcting codes. MIT Press, Boston
Proakis JC (1983) Digital Communications. McGraw Hill, Tokyo
Ryan WE, Lin S (2009) Channel codes: classical and modern. Cambridge University Press, Cambridge
Wicker SB (1995) Error control systems for digital communication and storage. Prentice-Hall, Englewood Cliffs

Appendix C
Electronic Circuits for Multiplication or Division in Polynomial Representation of Binary Sequences

Circuits for Performing Polynomial Multiplication

Definition C.1 A *shift register* is an electronic circuit arranged in such a way as to form a chain of memory elements, in which binary sequences can go forth according to a first-input first-output rule.

There is a clock, whose time reference controls binary symbols going forth along the shift register. Any elementary operation in the circuit constructed employing a shift register is made in precise symbol times, imposed by this clock.

Definition C.2 We can have two types of shift registers: those in which the same binary sequence $p(x)$ crosses all the memory elements (*controller arrangement*), and those in which the memory elements are separated by ex-OR sums (*observer arrangement*).

In Fig. C.1, the structures of the two arrangements are shown, for a shift register having 3 memory elements. The controller arrangement is better suited for extracting a sequence from a certain stage of the shift register. The observer arrangement is better suited for adding a sequence to that travelling along the shift register, by means of an ex-OR sum. In many applications of linear digital signal processing, it is necessary to perform a multiplication or division of the polynomial representing an input binary sequence by a second polynomial, whose coefficients are wired in the circuit.

Lemma C.1 *The polynomial product of an input binary sequence $u(x)$ by a wired polynomial $p(x)$, whose degree is* $\deg[p(x)] = D$, *can be performed by means of a shift register in controller arrangement, having D memory elements and whose scheme is depicted in* Fig. C.2.

Proof Let us suppose that $u(x) = 1$. The output sequence will be $p(x) = \sum_{i=0}^{D} p_i x^i$. Then we can invoke the distributive property of the polynomial multiplication with respect to the ex-OR sum, and the superimposition of effects, property certainly verified for the circuit in Fig. C.2, owing to its linearity. □

Lemma C.2 *The polynomial product of an input binary sequence $u(x)$ by a wired polynomial $p(x)$, whose degree is* $\deg[p(x)] = D$, *can be performed by means of a shift register in observer arrangement, having D memory elements and whose scheme is depicted in* Fig. C.3.

G. Cancellieri, *Polynomial Theory of Error Correcting Codes*,
Signals and Communication Technology, DOI 10.1007/978-3-319-01727-3

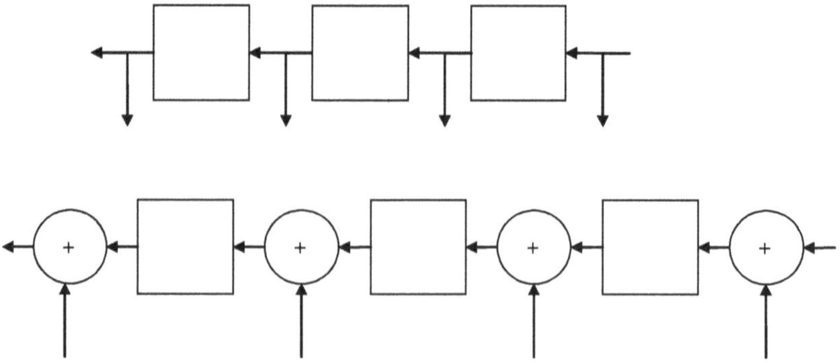

Fig. C.1 Controller arrangement (*up*) and observer arrangement (*down*) of a shift register

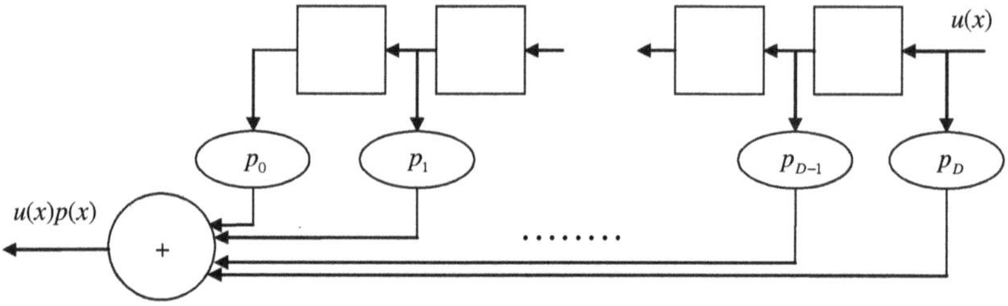

Fig. C.2 Shift register able to calculate a polynomial product in controller arrangement

Proof We can proceed as in the proof of Lemma C.1, with the only difference that here the overall ex-OR sum is distributed. □

Example C.1 Taking $p(x) = x^3 + x + 1$, that is $p_0 = p_1 = p_3 = 1$, and $p_2 = 0$, we obtain the two shift register arrangements shown in Fig. C.4.

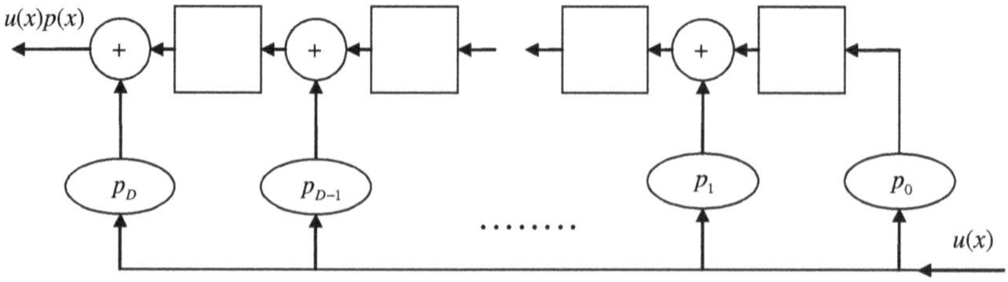

Fig. C.3 Shift register able to calculate a polynomial product in observer arrangement

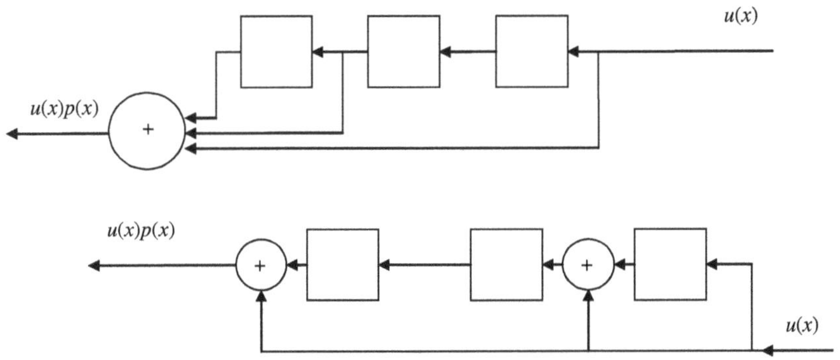

Fig. C.4 Controller and observer arrangements of a shift register able to perform multiplication by $p(x) = x^3 + x + 1$

It is possible to remark that the number of memory elements is equal in the two cases, and results in $D = 3$. The order in the presence of wires (corresponding to non-null coefficients in the wired polynomial), instead, is reversed. About this order, it is important to stress that the convention here adopted, coherently with what made in Appendix B, is the same as in (Peterson and Weldon 1972; Oppenheim and Schafer 1974; MacWilliams and Sloane 1977; Anderson and Mohan 1991) and opposite to that adopted, for instance (Johannesson and Zigangirov 1999; Lin and Costello 2004; Ryan and Lin 2009).

Circuits for Performing Polynomial Division

The current remainder and the current quotient in the division of a polynomial $x^L i(x)$ by a polynomial $z(x)$, as long as L increases, are strictly related (see Lemma B.9). In particular, the current quotient shifts towards higher and higher powers, and sometimes a coefficient 1 is added on its right hand side. When this occurs, the current remainder changes, because of an ex-OR sum with $z(x)$. We will show here two possible realizations for circuits able to calculate polynomial division, so implementing the above procedure. The first one is in controller arrangement, and the second one in observer arrangement.

Lemma C.3 *The current quotient of the division of an input sequence* $x^L i(x)$ *by a polynomial* $z(x)$, *whose degree is* $\deg[z(x)] = D$, *can be calculated by means of the shift register in controller arrangement, whose scheme is depicted in* Fig. C.5. *It gives also the current remainder* $R\left[\frac{x^L i(x)}{z(x)}\right]$.

Proof Let us suppose that $i(x) = 1$. The output sequence will be $Q\left[\frac{x^L}{z(x)}\right]$. Then we can consider Lemma B.8, in its part regarding the linearity property of a quotient

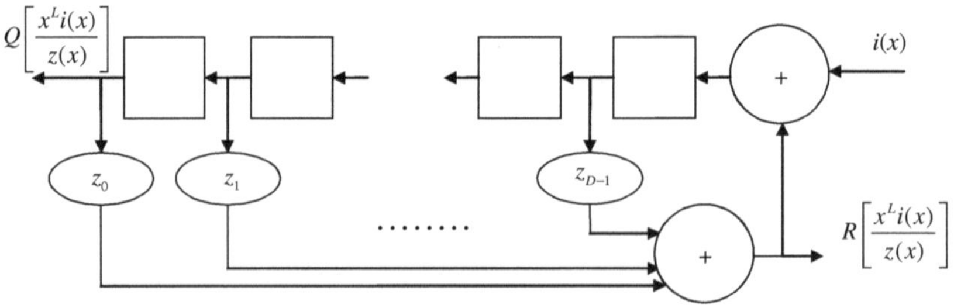

Fig. C.5 Shift register able to calculate the current quotient and the current remainder in controller arrangement

with respect to the dividend, and the superimposition of effects, certainly valid for the circuit depicted in Fig. C.5, owing to its linearity. During the last D steps, when calculation of $Q\left[\frac{x^L}{z(x)}\right]$, or in general $Q\left[\frac{x^L i(x)}{z(x)}\right]$, is going to end, at the output of the feed-back summation the final remainder (if any) can be found. In previous symbol times, the same output gives the current remainder. □

Lemma C.4 *The current quotient and the current remainder of the division of an input sequence* $x^L i(x)$ *by a polynomial* $z(x)$*, whose degree is* $\deg[z(x)] = D$*, can be calculated by means of the shift register in observer arrangement, whose scheme is depicted in* Fig. C.6.

Proof Let us suppose that $i(x) = 1$. The output sequence will be $Q\left[\frac{x^L}{z(x)}\right]$. As noted before, when a coefficient 1 is added on the right hand side of the current quotient, the current remainder changes, because of an ex-OR sum with $z(x)$. This property allows the appearance of the current remainder $R\left[\frac{x^L}{z(x)}\right]$ in the D memory elements of the shift register. Then we can proceed as in the proof of Lemma C.3. At the end of the operation, the final reminder (if any) is left in the D memory elements of the shift register. □

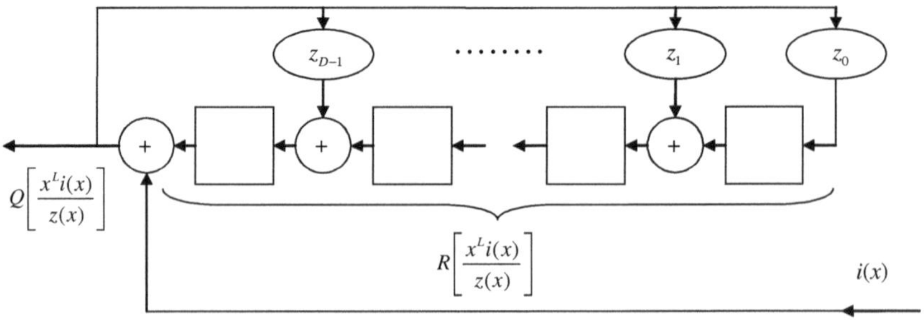

Fig. C.6 Shift register able to calculate the current quotient and the current remainder in observer arrangement

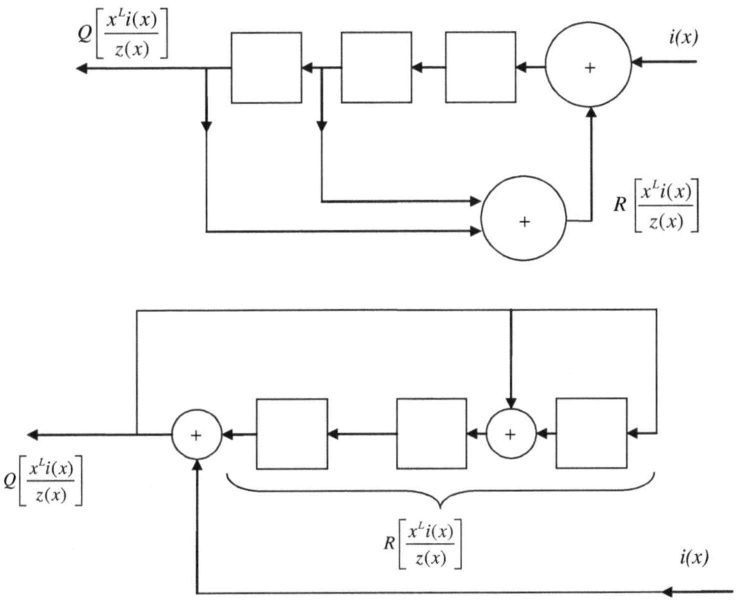

Fig. C.7 Controller and observer arrangements of a shift register able to perform division by $z(x) = x^3 + x + 1$

It is interesting to remark that, for multiplication, the shift registers of the schemes in Figs. C.2 and C.3 exhibit feed-forward summations, instead, for division, feed-back summations are needed. Furthermore, among the wires enabling summation, that corresponding to coefficient z_D is absent (although it must be non-null by assumption). As a consequence, the divisor polynomial $z(x)$ must have at least weight 2.

Example C.2 Taking $z(x) = x^3 + x + 1$, that is $z_0 = z_1 = z_3 = 1$, and $z_2 = 0$, we obtain the two shift register arrangements shown in Fig. C.7.

Here too, the number of memory elements is equal in the two cases, and results in $D = 3$. The order in the presence of wires, instead, is reversed, exactly as in the case of a multiplication.

It is possible to perform also a multiplication and a division at the same time, employing a unique shift register. In Fig. C.8, the circuits able to do this are described, respectively in a controller and in an observer arrangement. The performed operation is

$$\begin{aligned} f(x) &= \frac{p(x)i(x)}{z(x)} \\ &= \frac{(p_D x^D + p_{D-1}x^{D-1} + \cdots + p_1 x + p_0)i(x)}{x^D + z_{D-1}x^{D-1} + \cdots + z_1 x + z_0}. \end{aligned} \tag{C.1}$$

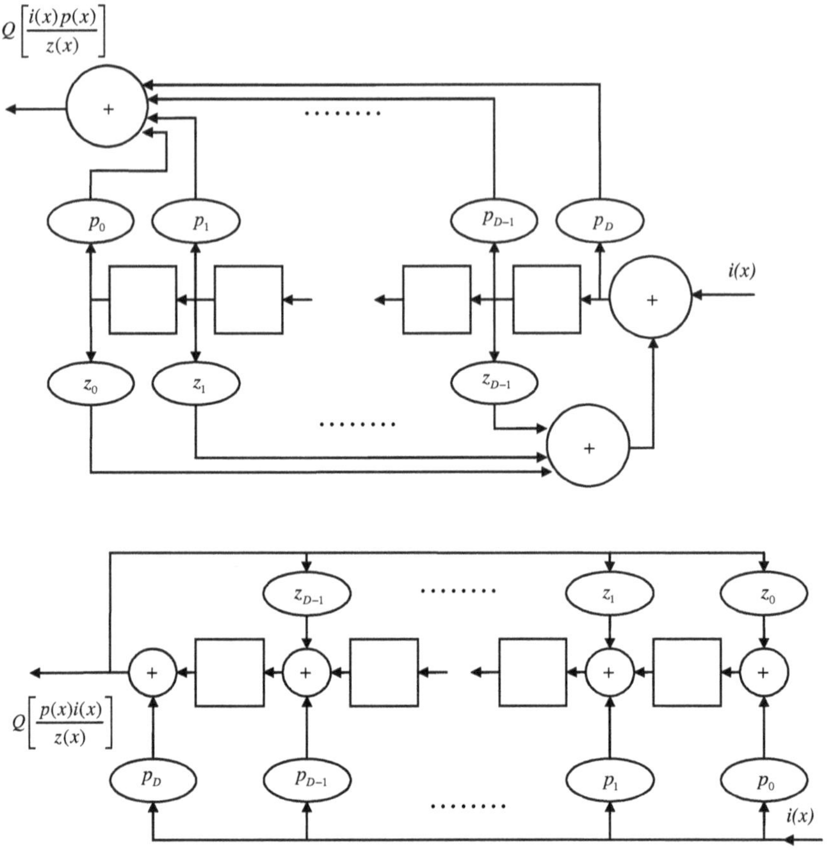

Fig. C.8 Circuits able to perform a multiplication and a division at the same time (the *upper* in controller and the *lower* in observer arrangement)

The electronic circuits shown in Fig. C.8 are widely employed in digital filtering (Oppenheim and Schafer 1974), for practical implementation of FIR filters (Finite Impulse Response), with the only feed-forward sum, or IIR filters (Infinite Impulse Response), with both feed-forward and feed-back sum.

Example C.3 Assuming $p(x) = x^2 + x$ and $z(x) = x^3 + x + 1$, we obtain the two circuits shown in Fig. C.9.

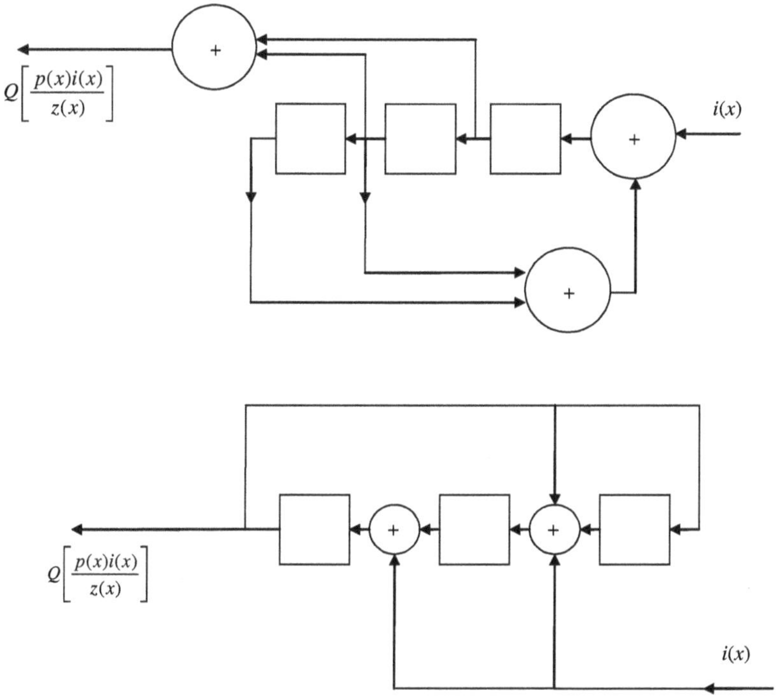

Fig. C.9 Controller and observer arrangements of a shift register able to perform multiplication by $p(x) = x^2 + x$ and division by $z(x) = x^3 + x + 1$

Circuits for Performing Polynomial Interleaved Multiplication and Division

In Appendix B a polynomial interleaved multiplication is introduced (Definition B.19). The effects of the single polynomial multiplications are collected in a unique ex-OR sum, as given by expression (B.28).

Lemma C.5 *The polynomial interleaved productof an input binary sequence* $[xu(x^2) + u_2(x^2)]$ *by a couple of wired polynomials* $p_1(x)$, $p_2(x)$, *whose degree is* $\deg[p_1(x)] = \deg[p_2(x)] = D$, *also owing to a possible shift factor* x^{Δ_i}, *can be performed by means of a couple of shift registers, in controller arrangement, properly interconnected, according to the scheme depicted in* Fig. C.10.

Proof The flux of the input binary sequence is subdivided by the switch, in such a way as to assure that the symbols in odd times are multiplied by $p_1(x)$, whereas the symbols in even times are multiplied by $p_2(x)$. The final ex-OR sum provides the superimposition of the two output sequences, according to (B.28). □

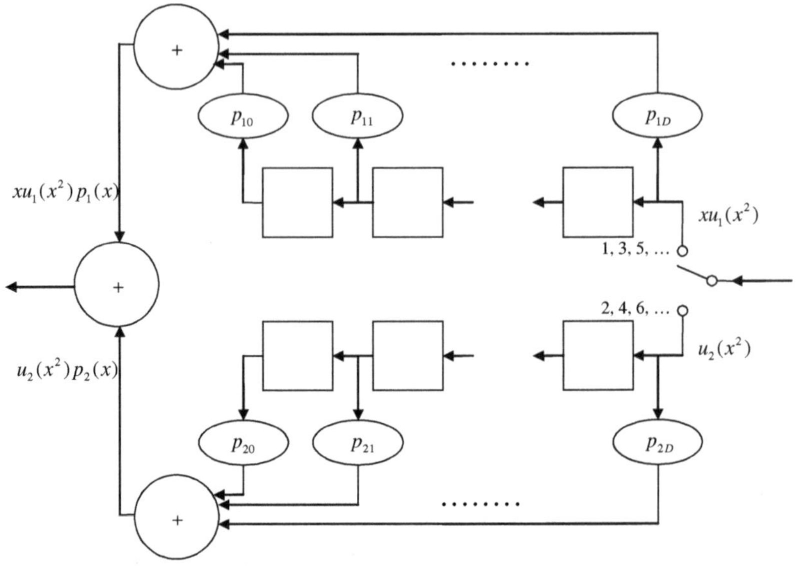

Fig. C.10 Circuit able to perform an interleaved multiplication in controller arrangement

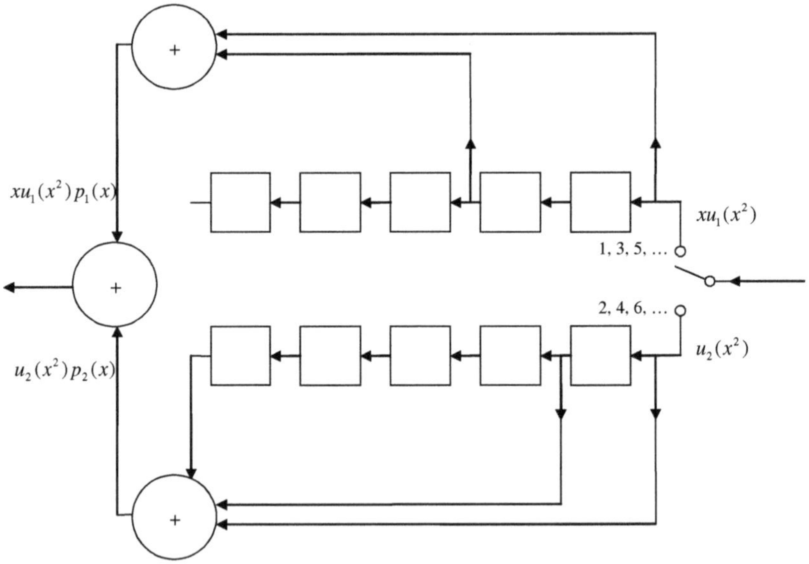

Fig. C.11 Circuit able to perform the interleaved multiplication of an input sequence by $p_1(x) = x^5 + x^3$ and $p_2(x) = x^5 + x^4 + 1$

Example C.4 Taking the situation described in Example B.36, that is $p_1(x) = x^2 + 1$ and $p_2(x) = x^5 + x^4 + 1$, we obtain the circuit shown in Fig. C.11, where we have adopted $\Delta_1 = 3$, in order to have the same polynomial degree, equal to 5.

It is possible to note that the first memory element of one shift register, in the symbol times where the input symbols are sent to the other shift register, is filled by a 0-symbol. This 0-symbol then progressively pass through all the memory elements in the considered shift register. A circuit similar to that described so far could be designed also in observer arrangement, but the above property would be no longer true, owing to the distributed ex-OR sums. Since evolution of the contents in the memory elements of the shift registers represents an important characteristic for the description of circuit operations, in the following we will restrict our attention to shift registers in controller arrangement only. Finally an extension of such description to the case of whichever number of wired polynomials is long but straightforward.

In Appendix B a polynomial interleaved division is introduced (Definition B.20). In this operation, the current remainder is unique, whereas there are n_o current (ordered) quotients, n_o being the number of polynomial divisors.

Lemma C.6 *The polynomial interleaved division of an input binary sequence* $i(x)$ *by a couple of wired polynomials* $z_1(x)$, $z_2(x)$, *whose degree is* $\deg[z_1(x)] = \deg[z_2(x)] = D_z$, *also owing to a possible* shift factor x^{Δ_i}, *can be performed by means of a couple of shift registers, in controller arrangement, properly interconnected,* according to the scheme depicted in Fig. C.12.

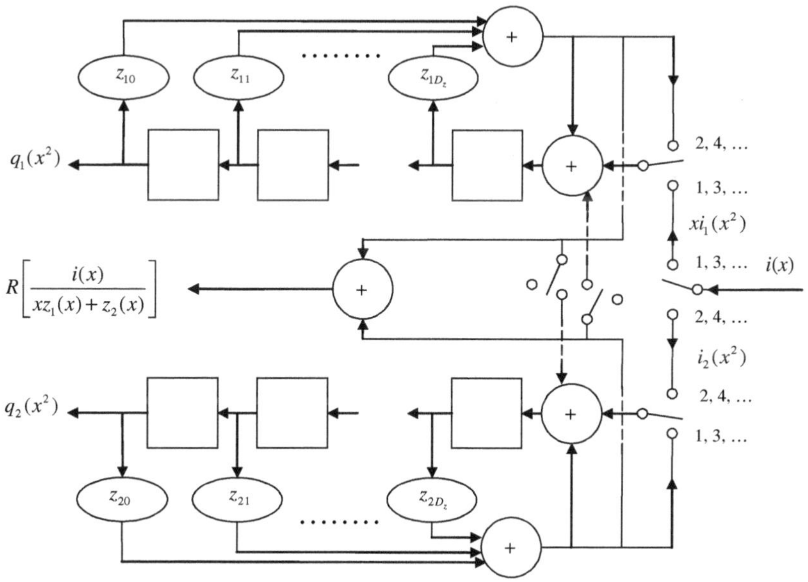

Fig. C.12 Circuit able to perform an interleaved division in controller arrangement

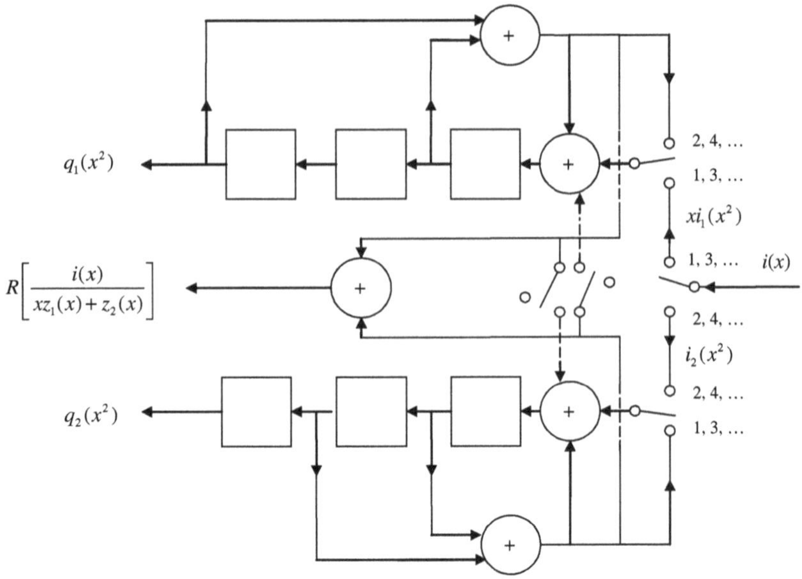

Fig. C.13 Circuit able to perform the interleaved division of an input sequence by $z_1(x) = x^3 + x^2 + 1$ and $z_2(x) = x^3 + x^2 + x$

Proof After (B.30), it is possible to write

$$\frac{i(x)}{xz_1(x) + z_2(x)} = xq_1(x^2) + q_2(x^2) + \frac{1}{xz_1(x) + z_2(x)} R\left[\frac{i(x)}{xz_1(x) + z_2(x)}\right]$$

where $q_1(x^2)$ and $q_2(x^2)$ are the two ordered quotients. The upper shift register calculates $q_1(x^2)$, which shifts along its memory elements. The lower shift register operates analogously regarding $q_2(x^2)$. They both contribute to the unique remainder, which can be taken from the central ex-OR sum. Since also the current remainder is unique, the sequence from the two feed-back wires must be summed to the input sequence in either odd or even time symbols. The switches acting on the cross feed-back wires assure that proper timing is satisfied in this operation, according to the input switch, whose purpose is to subdivide the input sequence as $i(x) = xi_1(x^2) + i_2(x^2)$. □

Example C.5 Taking the situation described in Example B.37, that is $z_1(x) = x^3 + x^2 + 1$ and $z_2(x) = x^3 + x^2 + x$, we obtain the circuit shown in Fig. C.13. Both such polynomials are characterized by $D_z = 3$. Remark that the wires associated to the highest-order power coefficients of the divisors are absent in the circuit, as usual.

References

Anderson JB, Mohan S (1991) Source and channel coding. Kluwer, Boston
Johannesson R, Zigangirov KS (1999) Fundamentals of convolutional coding. IEEE Press, New York
Lin S, Costello DJ (2004) Error control coding. Pearson/Prentice-Hall, Upper Saddle River
MacWilliams FJ, Sloane NJA (1977) The theory of error correcting codes. North Holland, New York
Oppenheim AV, Schafer RW (1974) Digital Signal Processing. Prentice-Hall, Englewood Cliffs
Peterson WW, Weldon EJ (1972) Error-correcting codes. MIT Press, Boston
Ryan WE, Lin S (2009) Channel codes: classical and modern. Cambridge University Press, Cambridge

Appendix D
Survey on the Main Performance of Error Correcting Codes

Purpose of Channel Coding

In digital transmissions or digital recording-reproduction, errors may occur, leading to a degradation of the information transferred from a source to a destination. Against this phenomenon, it is possible to insert additional binary symbols in the flux of those carrying information, so exploiting redundancy in order to reach a form of protection. This procedure has the consequence of reducing the time devoted to transfer each binary symbol, and hence of increasing the binary symbol transfer speed.

Definition D.1 A *binary channel* represents the model adopted for taking into account information flux and its degradation. The operation performed in order to protect information against such degradation during the transfer, by proper insertion in the flux of additional binary symbols, is called *channel coding*. Such operation is possible, paying proper costs, after having fixed precise rules, known by both the source and the destination, for additional symbol insertion and interpretation.

Although there are several types of binary channels (non-linear, with memory, affected by signal fading, time-variable, ...), we will consider, in this introductory treatment, just the one described below. Its intuitive scheme is depicted in Fig. D.1, where the information source is assumed to be on the right hand side and the two symbols, denoted 0, 1, before and after the channel, have the same probability, equal to 1/2. This scheme is called binary symmetric channel, and is characterized by two transition probabilities: $(1 - p)$ for a correct transition, p for a wrong transition. In a channel of this type the information degradation may be originated by different causes. A further restriction is in the specification of which type of noise influences the channel. Also the statistical properties of the symbol sequence are to be established.

Definition D.2 We will assume that an *uncorrelated binary symbol sequence*, made by equally probable symbols 0, 1, comes out from the information source and feeds a *linear memoryless channel*, affected by *Additive White Gaussian Noise* (*AWGN*).

This means that the flux of binary information symbols resembles that of a completely uncorrelated two-level antipodal signal (Proakis 1983), in which both

G. Cancellieri, *Polynomial Theory of Error Correcting Codes*,
Signals and Communication Technology, DOI 10.1007/978-3-319-01727-3

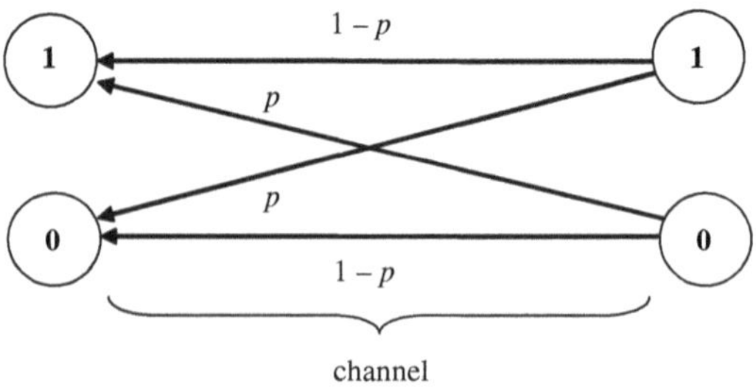

Fig. D.1 Intuitive scheme for the binary symmetric channel here considered

signal levels have the same probability 1/2 and are equally affected by noise. In this situation the channel introduces errors, whose probability can be estimated by means of a Gaussian random variable density distribution (Papoulis 1965).

Definition D.3 Under the assumption introduced in Definition D.2, the expression of the *uncoded binary symbol error probability p* is

$$p = \frac{1}{2} erfc\left(\sqrt{SNR}\right), \tag{D.1}$$

where *erfc*() represents the complementary error function (Abramowitz and Stegun 1968), and *SNR* is the *signal-to-noise power ratio.* Parameter *p* is sometimes also called *BER* (*Bit Error Ratio*), meaning the average number of errors on a large amount of transferred binary symbols.

Uncoded error probability means that it is calculated without the variation produced by channel coding. Since *erfc*() is a decreasing function, the *SNR* can be considered as a sort of cost to be faced in order to reach the desired quality of information transfer. But it is not the only cost. Considering also the introduction of channel coding, we have four main costs on the whole, which can be listed as follows:

i. signal-to-noise ratio *SNR*,
ii. bandwidth *B*,
iii. computational complexity of the encoder and of the decoder,
iv. time latency.

One fixed a certain value for *p*, depending on the particular application under examination, a reduction in the required *SNR*, maintaining acceptable the other costs from *ii* to *iv*, is the typical goal of channel coding.

Parameters of a Code

A coded channel, on average, takes *k* binary symbols from the information source, and gives back the same *k* binary symbols to the information destination, but it employs, on average during the transfer, *n* binary symbols, instead of *k*, with

$n > k$. So there are $(n - k)$ additional redundant binary symbols, still on average. In order to allow this operation, the bandwidth of the channel is to be enlarged by a factor of n/k, proportional to an equal increase in the binary symbol transfer speed. Since the signal power is maintained constant, an augment in the noise power, proportional to the enlarged bandwidth B (after assumption that the noise power spectrum is white, that is uniform), entails a reduction in the *SNR* by a factor of k/n. Sometimes the same concept is expressed adopting, as the independent variable, the ratio E_b/N_o, being E_b the average energy per binary symbol, and N_o the noise power spectral density. In a binary channel *SNR* and E_b/N_o represent the same parameter.

The scheme depicted in Fig. D.2 describes the process of information transfer from a source S to a destination D, without and with coding, the latter being provided by a pair of electronic devices, called respectively the encoder (inserted between the information source and the channel), and the decoder (inserted between the channel and the information destination).

Definition D.4 Under the assumptions regarding parameters k and n considered so far, the ratio k/n is called the *code rate*.

The binary symbol error probability, after introduction of the code, but without any benefit from the decoding procedure, then becomes

$$p' = \frac{1}{2} erfc\left(\sqrt{\frac{k}{n} SNR}\right), \tag{D.2}$$

where the new *SNR* has been considered. Since the code rate k/n is always smaller than one, keeping in mind that *erfc*() is a decreasing function, we have $p' > p$. Nevertheless the action of the decoder allows to obtain, in proper conditions, the expected benefit of reducing the error probability. This performance depends on the type of decision adopted on the received flux, in establishing the identity of any binary symbol.

Definition D.5 We can assume two types of decision in decoding: *hard-decision* decoding (in which the identities of the binary symbols are firstly established, and only then the decoder acts on them), or *soft-decision* decoding (in which the identities of the binary symbols are established after the action of the decoder).

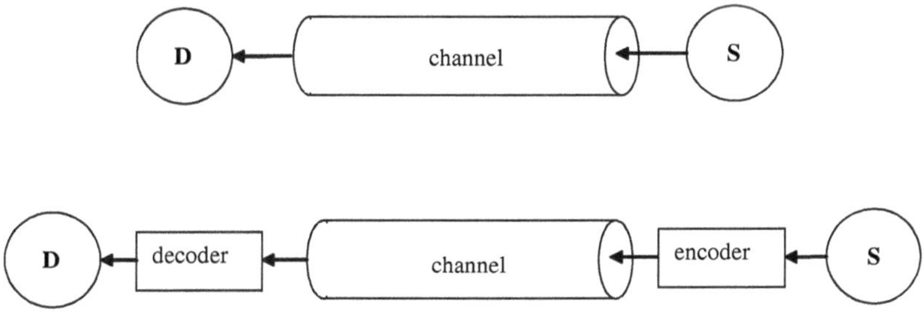

Fig. D.2 Information transfer without coding (*up*) and with coding (*down*)

Coming back to the introduction of parameters k and n in the description of a channel code, let us consider now that they are no longer average parameters, but represent the numbers of binary symbols in two symbol sequences, the one before encoding and the other after encoding. In this case the code assures a one-to-one correspondence between sequences of k binary information symbols at the encoder input and sequences of n binary symbols at the encoder output. Clearly the reverse correspondence is exploited by the decoder, in order to recover the original sequence of k binary information symbols.

Definition D.6 When the binary symbol flux along the coded channel is subdivided into adjacent blocks, called *code words*, the code is defined a *block code*. So code words are permitted binary symbol sequences, of length n, according to the particular block code adopted by the encoder (and by the decoder). Between consecutive code words there is no correlation.

The decoder knows the exact timing of such code words, which means that it is synchronized with the encoder (Peterson and Weldon 1972), and is compliant with the particular code adopted in forming them by the encoder. So, considering also the degradation introduced by the channel, it tries to find the most likely sequence of k binary symbols associated to each received code word.

Definition D.7 Considering two code words, the number of positions in which they differ is defined the *Hamming distance* between them. Taking two code words with minimum number of positions in which they differ, over the whole ensemble of code words, the Hamming distance d between them is called the *minimum distance of the code*. It represents a sort of protection offered by the code against information degradation.

Example D.1 Assuming a block code whose four code words are 00000, 01011, 10110, 11000, between the first and the last code words there is a Hamming distance 2, and $d = 2$ is also the minimum distance of such code.

Roughly speaking, the minimum distance d is a good feature introduced by the code, since it contributes to better separate the code words. It is preferable that this separation is uniformly distributed among any couple of code words. In this sense, it is important the following restriction.

Definition D.8 A *linear code* is a code characterized by the property that any ex-OR sum of code words gives rise to another code word. This definition regards block codes but can be extended also to more general codes, simply substituting the term code words with permitted symbol sequences.

Example D.2 The code introduced in Example D.1 is not linear, in fact ex-OR sum of the second and third code word yields 11101, which is not a code word.

Example D.3 Let us take the code whose four code words are: 00000, 01011, 10110, 11101. This is linear, and exhibits a minimum distance $d = 3$.

It is useful to add a further property, regarding the possibility of an ex-OR sum of a code word with itself. If one requires that, also in this case, it is possible to obtain, as the result, a code word, it is necessary to include, in the ensemble of all the code words of the considered code, also the code word made by all 0-symbols. When this is not, the code cannot be defined linear. A code, whose code words are obtained from those of a linear code by addition of an n-symbol sequence, not permitted by the code itself, is often called the translate of a true linear code (MacWilliams and Sloane 1977).

Theorem D.1 *In a linear code, the distribution of Hamming distances, taking one code word as a reference, from it to all the other code words, is invariant with respect to changes in the code word chosen.*

Proof Let us consider three code words, represented by underlined capital letters: a code word $\underline{A}$, the code word consisting of all 0-symbols (denoted $\underline{0}$), and one code word of the remaining $(2^k - 2)$, denoted $\underline{B}_j$. Assuming $\underline{0}$ as the reference point, the distribution of Hamming distances from it can be written as $d(\underline{A}, \underline{0}), d(\underline{B}_j, \underline{0}), j = 1, 2, \ldots$. Taking a new reference point in $\underline{A}$, the new distribution of Hamming distances will result in $d(\underline{0}, \underline{A}), d(\underline{B}_j, \underline{A}), j = 1, 2, \ldots$. Clearly it is $d(\underline{0}, \underline{A}) = d(\underline{A}, \underline{0})$. Furthermore, being the code linear, after Definition D.8 the ex-OR sum of the code words $\underline{A}$ and $\underline{B}_j$ yields a third code word, say $\underline{B}_i$, for which $d(\underline{B}_j, \underline{A}) = d(\underline{B}_i, \underline{0})$. Then the change in the reference point leads to an order permutation in the list of code word distances, but not to any variation in its structure. □

Theorem D.1 can be considered as a sort of invariance property in the choice of the point of view (MacWilliams and Sloane 1977): if an observer takes position in a code word and looks at all the other code words, he will see, in the space around him, the same constellation of code words (with the respective Hamming distances from the code word chosen), independently of the position assumed.

Definition D.9 A sequence of binary symbols exhibits a *weight* w which equals the number of 1-symbols in it.

Example D.4 The binary sequence 11001010101001 has weight $w = 7$.

Lemma D.1 *The distribution of Hamming distances in a linear code coincides with the weight distribution of its code words.*

Proof Taking the code word made by all 0-symbols as a reference point, we obtain a weight distribution that, owing to Theorem D.1, is unique for the linear code considered, and expresses its Hamming distance distribution. □

The word weight distribution of a linear code will be described by the coefficients $A(w)$. This distribution has great importance in predicting code potential performance, even before the choice of the more suitable decoding procedure to be applied on it.

Example D.5 The word weight distribution of the linear code introduced in Example D.3, is $A(0) = 1$, $A(3) = 2$, $A(4) = 1$.

We can observe that the overall number of code words is 2^k. They are permitted binary sequences of length n, chosen in the set comprehending all the 2^n possible binary n-tuples.

In principle, a non-linear code could exhibit whichever number of code words, also different from an integer power of 2 (like 2^k is). However such code words have still to be chosen in the set of all the 2^n possible n-tuples. In any case, a good code would realize the most uniform distribution possible of its code words in the n-dimension binary hyper-space, where the n-tuples are represented as points. Following this intent, the minimum distance d will be maximized if the distribution of the distances does not give privilege to any code word. Although there are good non-linear codes with 2^k code words, for which the invariance of the point of view is satisfied too (MacWilliams and Sloane 1977), in the following we will restrict our attention to linear codes only.

Code Performance with a Hard-Decision Decoding

In soft-decision decoding the minimum distance d is directly involved in the benefit gained by the decoding operation. In hard-decision decoding, it is necessary to introduce an intermediate parameter, still related to d.

Definition D.10 Considering a hard-decision decoding, integer t represents the *maximum number of correctable single errors* (or simply *error correction capability*) of the code under examination. A single error is that affecting one precise position in the code word.

Remark that, in a binary alphabet, if the decoder obtains, after a proper computation, the position of a single error, its correction will be immediate, by a simple exchange of its identity. Typically, when a non-binary alphabet is employed, the decoder firstly operates single error localizations and then it performs the true correction by proper substitutions. Either in a binary or in a non-binary alphabet, sometimes, the decoder can argue that the received code word is wrong, without having the possibility of determining the exact positions of the single errors. In this case we deal with a function called error detection and not with the function of error correction.

A code expresses a particular geometrical distribution of the code words in the hyper-space of all the possible binary sequences having length n (possible n-tuples). The same code can be decoded by means of either a hard-decision or a soft-decision procedure, although its geometrical structure often suggests that one of the two procedures is preferable, as a trade-off between benefits and costs (especially in decoding computational complexity).

Lemma D.2 *When d is odd, the following relationship holds, between parameter t and the minimum distance d of the considered code,*

$$t = \frac{d-1}{2}, (d \text{ odd}). \tag{D.3}$$

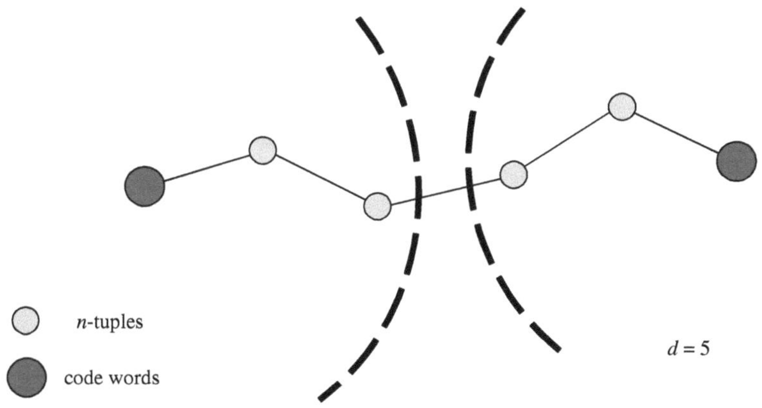

Fig. D.3 Decoding hyper-spheres centred in two code words at the minimum distance $d = 5$ each other

Proof In Fig. D.3, a sequence of tilted line segments represents the shortest link between a couple of code words, separated by the minimum distance d of the considered code. Any segment expresses a difference in a precise position between the two code words and then a single error in passing from the one to the other. Let us imagine two hyper-spheres, centred in the two code words. Each of them includes the n-tuples which the decoder will suppose, according to a maximum likelihood estimation, as deriving from the code word in its centre. The boundaries of such hyper-spheres are denoted as dashed curves. In the example described by the Figure, we assume $d = 5$. Clearly the radius of the two hyper-spheres, in this case, is 2. There are no n-tuples out of such hyper-spheres along the link. This is consequence of the fact that d is odd. In general, the radius of the hyper-sphere is t, and the link is made by d subsequent tilted segments. The number of n-tuples along the link is $(d - 1)$. Being d odd, $(d - 1)/2$ of them turn out to be included in one hyper-sphere and the remaining $(d - 1)/2$ in the other. □

Definition D.11 An *error pattern* (or briefly an *error*) is an n-tuple which, added to a certain code word, gives the received wrong n-tuple. The *weight of the error* is the number of non-null symbols (1-symbols in a binary alphabet) characterizing the error pattern, that is the number of single errors forming the error pattern.

After Definitions D.10 and D.11, we conclude that a code can correct all the errors whose weight is at most equal to t. Nevertheless, along certain directions in the hyper-space, it often can correct even more.

Sometimes, instead of performing a forward error correction (briefly called FEC), the coded channel can be organized in such a way as the decoder is intentionally limited to provide error detection only. Once ascertained that a certain code word is affected by errors, the information destination can request the information source to send the considered code word again (in this case the code words are identified by a progressive number, in a proper context). Alternatively a second code can be engaged for determining the correct code word, after error detection performed by a first code, still in the framework of a FEC procedure. If a code, characterized by minimum distance d, is devoted to only detect errors, clearly it will be able detect any pattern of up to $(d - 1)$ single errors (Klove 2007).

More frequently, depending also on the value of d, a code can correct single errors up to its correction capability t, but it could also detect a larger number of single errors. Taking into account this possibility, still by means of a hard-decision decoding procedure, the following statement regards the case of a minimum distance d which is an even number.

Lemma D.3 *When d is even, the relationship between parameter t and the minimum distance d of the code is*

$$t = \frac{d-2}{2}, \quad (d\,\text{even}), \tag{D.4}$$

but the code can also detect all the error patterns whose weight is d/2 .

Proof In Fig. D.4, two code words are separated by the minimum distance $d = 6$ characterizing the considered code. This situation is similar to that depicted in Fig. D.3, with the only difference consisting in the presence, here, of an n-tuple out of both the hyper-spheres. It cannot be associated, on the basis of a maximum likelihood estimation, to any of the two code words, because such n-tuple shows three single errors with respect to both. Nevertheless this triple error can be detected. An ambiguity condition of the type here described is due to the fact that d is even. In general, for d even, the radius of the two decoding hyper-spheres is $t = (d-2)/2$, but $d/2$ single errors, although not correctable, can be always detected. □

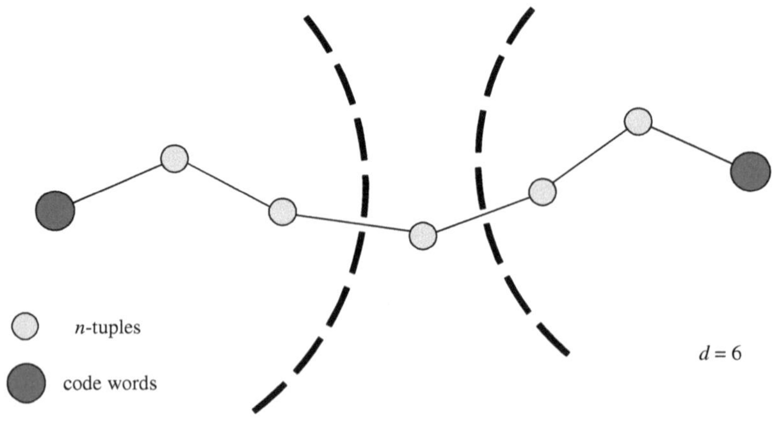

Fig. D.4 Decoding hyper-spheres centred in two code words at the minimum distance $d = 6$ each other

Definition D.12 The most exhaustive method for implementing a hard-decision decoding is that based on a table, called *standard array*. It is formed by 2^k columns and by 2^{n-k} rows. In the first row the code words are listed, with the all-0-symbol code word occupying the leftmost position. In the other rows all the possible n-tuples not permitted by the code are reported. In the leftmost column of the table the correctable (or at least detectable) errors are listed in increasing-weight order. In all the remaining positions of the table, we have the ex-OR sums between the code word at the top of the column and the error in the leftmost position of the row.

Example D.6 The code considered in Example D.3, for which $d = 3$ and hence $t = 1$, is characterized by the standard array table shown in Fig. D.5. The leftmost position in the first row, where the code words are listed, is occupied by 00000. In the leftmost column, under this code word, we have firstly the five errors with weight 1, and then two errors with weight 2. Such two errors have been chosen among the n-tuples not obtained by ex-OR sum in the preceding five rows. The decoding procedure operates as follows. Once received an n-tuple, the decoder searches for it in the table. If the weight of the error is within the error correction capability $t = 1$ of this code, the received n-tuple will belong to the upper part of the table, and the maximum likelihood estimate for the original code word will be that present at the top of the column. On the contrary, if the weight of the error is beyond its error correction capability $t = 1$, there will be more than one code word, which could be taken as the maximum likelihood estimate. This ambiguity cannot be resolved by the decoder, because it regards the lower part of the table. For instance, the 5-tuple 01111 certainly derives from the code word 01011, with a single error in its third position, whereas the 5-tuple 01110 may derive either from the code word 10110 or from the code word 01011, showing two single errors with respect to both.

The code discussed in Example D.6 thus can detect only some double errors, because its geometrical structure has some links (for example the one between code word 00000 and code word 11101) whose length is 4. On the contrary, in a code

2^{n-k} ↓ / 2^k →				
00000	**01011**	**10110**	**11101**	← code word
00001	01010	10111	11100	upper part
00010	01001	10100	11111	upper part
00100	01111	10010	11001	upper part
01000	00011	11110	10101	upper part
10000	11011	00110	01101	upper part
11000	10011	01110	00101	lower part
01100	00111	11010	10001	lower part

Fig. D.5 Standard array table for the code with $n = 5$ and $k = 2$ considered in Example D.6

whose minimum distance is $d = 4$, besides the possibility of correcting single errors (higher part of the standard array table), since all the links between couples of code words are at least 4 segments long, any 2-weight error can be detected too.

In a more general situation, we can exploit also the possibility of correcting some errors having weight more than t. Standard array hard-decision decoding allows this opportunity, in certain codes, owing to the presence of some particular rows in the lower part of the table, where ambiguity does not occur (Torrieri 1988). Nevertheless this procedure is often rather complicate, and able to give only modest improvements.

The main problem in the implementation of a hard-decision decoding procedure, based on the standard array table, is in its computational complexity, which grows as 2^n, becoming rapidly unfeasible. If one ignores the possibility of treating ambiguities, or correcting beyond the code correction capability t, and restricts the operation performed by the decoder to correct at most t single errors, the decoding procedure becomes in general computationally feasible. This can be done following different possible algorithms, more or less efficient depending on the particular code considered. Once adopted a choice of this type, we can evaluate the binary symbol error probability, after decoding, by means of rather simple approximate expression.

Lemma D.4 *To a code, with parameters n, k, and characterized by maximum number t of correctable single errors, let us apply a hard-decision decoding procedure, devoted to error correction only (not error detection) within its correction capability. The average binary symbol error probability, after decoding, which is so obtained, can be approximately expressed as*

$$p'' \cong \frac{d}{n} \sum_{j=t+1}^{d} \binom{n}{j} p'^j (1 - p')^{n-j}, \tag{D.5}$$

where p' is given by (D.2).

Proof It is possible to adopt a model based on a binomial random variable (Papoulis 1965). Since our code corrects up to t single errors in a code word, the first term in the expression giving the average word error probability will be that having order $(t + 1)$. Such average word error probability then results in

$$P_W = \sum_{j=t+1}^{n} \binom{n}{j} p'^j (1 - p')^{n-j}. \tag{D.6}$$

If p' is small (of the order of 10^{-2} or less), the first few terms are dominant. Since t is related to d by means of (D.3) or (D.4), it is reasonable to consider an approximate expression for P_W, restricting summation to the only terms up to $j = d$. So we are left with

$$P_W \cong \sum_{j=t+1}^{d} \binom{n}{j} p'^j (1-p')^{n-j}.$$

From this expression, in order to obtain the binary symbol error probability, we can assume that the wrong code word reached after the decoding procedure is at a distance d from the right original code word. So, on average, the number of single errors after wrong decoding will be d on a total amount of n positions. Therefore we will take $p'' \cong \frac{d}{n} P_W$, leading to (D.5). □

It is interesting to note that, in expression (D.5), we do not distinguish between information binary symbols and control binary symbols. In this respect, we have to stress that often there are not k precise positions, among the n total positions in a code word, devoted to information symbols, being the remaining $(n - k)$ positions reserved to control symbols.

Definition D.13 A linear code is called *systematic* (in wide sense) if it shows the information binary symbols in k precise positions of its code words.

This definition holds in strict sense if the k positions collecting information symbols are the first ones. The possibility of a direct look-up of the information symbols in the code words represents a practical advantage in particular concatenated coding schemes or in channel management under very severe conditions (Clark and Cain 1983).

Example D.7 The code described in Example D.3 and Example D.6, whose code words are 00000, 01011, 10110, 11101, can be assumed to be in strict-sense systematic form, assigning the first two positions to as many binary information symbols. A wide-sense systematic form can be obtained also with other choices, provided that the pair of information symbols are properly assigned to a same pair of fixed positions.

It is however important to stress that a possible reordering in the sequence of the n binary symbols of all the code words, after encoding, does not change the structure and performance of the considered code, but simply modifies how information symbol sequences are mapped on the code words.

Code Performance with a Soft-Decision Decoding

When the decoder exploits a soft-decision decoding, a quantization of the signals to be processed is needed. In the following, we will assume that the number of quantization levels adopted is so high that the performance obtained resembles the one reached when the signal samples belong to a continuous domain. For a binary alphabet, and p not lower than 10^{-10}, $2^4 = 16$ quantization levels are typically enough for achieving this result (Clark and Cain 1983).

Lemma D.5 *To a code, with parameters n, k, and characterized by minimum distance d, let us apply a soft-decision decoding procedure with quantization level number high enough. The average binary symbol error probability, after decoding, which is so obtained, can be approximately expressed as*

$$P'' \cong \sum_{w=d}^{2d-1} \frac{w}{n} A(w) \frac{1}{2} erfc\left(\sqrt{w \frac{k}{n} SNR}\right), \tag{D.7}$$

where $A(w)$ is the code word weight distribution.

Proof The code can be thought as an alphabet of 2^k symbols, each represented by a different sequence of n binary symbols. The decoder takes n subsequent signal samples, affected by noise, and processes the whole sequence at a time. The signal-to-noise ratio to be considered in this enlarged alphabet depends on all the possible couples in the ensemble of the 2^k symbols. Assuming the all-0-symbol code word as a reference signal, the signal-to-noise ratio calculated taking the signal separation between it and a code word whose weight is w, can be written as $SNR' = w\frac{k}{n}SNR$. In fact k/n accounts for the worsening due to bandwidth expansion (like in p' for a hard-decision decoding), whereas w is due to the separation introduced by the code. Then an event of error affecting a code word, whose weight is w, can be associated to the approximate probability

$$P_W(w) \cong A(w) \frac{1}{2} erfc\left(\sqrt{SNR'}\right) = A(w) \frac{1}{2} erfc\left(\sqrt{w \frac{k}{n} SNR}\right). \tag{D.8}$$

In this expression, multiplication by $A(w)$ takes into account that a number of distinct events of errors can occur, equal to that of the code words having weight w. Although there is a small but finite probability that some events of this type are not independent, we can restrict our attention to independent events only, and do not subtract possible joint probabilities. In this sense, the approximation present in a summation of probabilities like (D.8), varying w, yields an over-estimation for the resulting word error probability, sometimes called Union bound (Proakis 1983). In order to obtain the binary symbol error probability, after such a soft-decision decoding, we will assume that w errors, on a total amount of n positions, affect a not correctly decoded word, whose weight is w (still taking the all-0-symbol code word as the original). So we are left with

$$p'' \cong \sum_{w=d}^{2d-1} \frac{w}{n} P_W(w),$$

which leads to (D.7). Truncation of this summation after the term characterized by $w = 2d - 1$ is due to the smallness of the remaining terms (like for a hard-decision decoding), but also to the fact that many good codes, for which soft-decision decoding has acceptable computational complexity, show a strong change in the structure of their weight distribution $A(w)$, for $w \geq 2d$. This is the case, for instance, of convolutional codes. □

Remark that, also here, we have not taken in any consideration the distinction between information binary symbols and control binary symbols. In this sense, expressions (D.5) and (D.7) are comparable. It is finally important to stress that, when the worsening due to bandwidth expansion overcomes the benefit introduced by exponent $(t + 1)$ or by the minimum distance d, in such expressions, the code becomes useless.

Example D.8 Assuming the code already treated in Examples D.3, D.5, D.6, whose four code words are 00000, 01011, 10110, 11101, we will compare (D.5) and (D.7), restricting our attention to the first term only in the two summations. This approximation is acceptable for *SNR* values of the order of 8 dB or more, but this choice is adopted here only in order to make the comparison easier. Expression (D.5), in which $n = 5$, $k = 2$, $t + 1 = 2$ are to be considered, for a hard-decision decoding, yields

$$p'' \cong \frac{3}{5}\binom{5}{2}\left[\frac{1}{2}erfc\left(\sqrt{\frac{2}{5}SNR}\right)\right]^2 = 6\left[\frac{1}{2}erfc\left(\sqrt{\frac{2}{5}SNR}\right)\right]^2.$$

In the considered range of values for the *SNR*, we can take the following approximation (Abramowitz and Stegun 1968) of the complementary error function

$$erfc(x) \cong \frac{1}{x\sqrt{\pi}}\exp(-x^2). \tag{D.9}$$

Therefore comparison between p'' and the uncoded error probability p yields

$$p'' \cong 6\frac{1}{4SNR^2\pi}\exp\left(-\frac{4}{5}SNR\right), p \cong 2\frac{1}{2SNR\sqrt{\pi}}\exp(SNR),$$

where the exponential dependence on the *SNR* is dominant. So we can conclude that this coding-decoding scheme is useless. Instead, from (D.7), in which $w = d = 3$, $k = 2$, $n = 5$, $A(3) = 2$ are to be taken, for a soft-decision decoding, we obtain

$$p'' \cong \frac{3}{5}2\frac{1}{2}erfc\left(\sqrt{3\frac{2}{5}SNR}\right) = \frac{3}{5}erfc\left(\sqrt{3\frac{2}{5}SNR}\right).$$

From (D.9), taken again to approximate the complementary error function, we are left with

$$p'' \cong \frac{6}{5}\frac{1}{2SNR\sqrt{\pi}}\exp\left(-\frac{6}{5}SNR\right),$$

leading to an asymptotic modest reduction with respect to the uncoded error probability p.

It is important to stress that, in the above approximate expressions, the dominant behaviour is imposed by the argument in the exponential functions. The poor results obtained in Example D.8 are due to the strong bandwidth expansion $n/k = 5/2$, which, in the considered code, frustrates the benefit produced by the presence of a certain error correction capability or minimum distance.

Example D.9 Let us consider now a more powerful code, characterized by the following parameters $n = 23$, $k = 12$, $d = 7$ and hence $t = 3$, finally $A(7) = 253$. From (D.5) and (D.9), for a hard-decision decoding, with the only fundamental term, we obtain

$$p \cong \frac{7}{23}\binom{23}{4}\left[\frac{1}{2}erfc\left(\sqrt{\frac{12}{23}SNR}\right)\right]^4 \cong 2695\frac{1}{16SNR^4\pi^2}\exp\left(-\frac{48}{23}SNR\right).$$

From (D.7) and (D.9), instead, for a soft-decision decoding, still with the only fundamental term, the result is

$$p \cong \frac{7}{23}253\frac{1}{2}erfc\left(\sqrt{7\frac{12}{23}SNR}\right) \cong \frac{1771}{23}\frac{1}{2SNR\sqrt{\pi}}\exp\left(-\frac{84}{23}SNR\right).$$

Both these probabilities must be compared with the uncoded binary symbol error probability p. It is possible to recognize an improvement in both cases. This is due to the factor in the exponent, remarkably greater than 1, also having taken into account the code rate k/n. In this respect, it is possible to note how this factor is greater in the case of a soft-decision decoding, and becomes remarkable especially for high values of d.

The code introduced in Example D.9, called binary Golay code, is very important from the theoretical point of view, as we will discuss in the following. Nevertheless, its code word length n is too small for attaining significant results in practical applications.

Coding Gain

A typical behaviour of the curves reporting the binary symbol error probability versus the signal-to-noise ratio *SNR*, for the coded and uncoded channel, with reference to a certain code and to a proper decoding procedure, is depicted in Fig. D.6. The intersection between these two curves (called break-even point) may occur at *SNR* values from about 1 dB (if the multiplicative factors of various terms in p'' are small) to about 4 dB or even more (if they are large). Its position is strongly influenced also by the code rate. Typically lower code rates allow the obtainment of smaller *SNR* values at the break-even point. Only on the right hand side of the break-even point, p'' becomes lower than p, and code employment turns out to be efficient.

Fig. D.6 Typical curves of p and p'' versus *SNR*

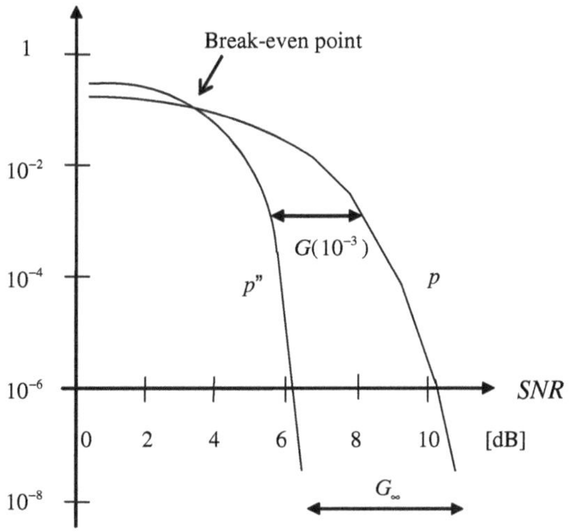

Definition D.14 Fixed a certain value for p, the reduction in the *SNR* allowed by the introduction of the code is called *coding gain*. It is function of p, therefore can be expressed as $G(p)$.

Instead of evaluating the reduction of p'' with respect to p, it is preferable to fix a certain value for the binary symbol error probability (imposed in turn by the application to which the channel is devoted), and measure the reduction in the *SNR*. In most cases the coding gain is an increasing function for p decreasing, but typically it reaches an asymptotic value (Clark and Cain 1983), which can be defined as follows.

Definition D.15 The *asymptotic coding gain* G_∞ is that reached when p tends to 0 (or when the *SNR* tends to infinity). Nevertheless often it is practically reached long before, when the negative slopes of the two curves become approximately coincident.

In the example shown in Fig. D.6, it is possible to find $G(10^{-3}) \cong 2$ dB and $G_\infty \cong 4$ dB, the latter being reached approximately for $p \leq 10^{-8}$.

Theorem D.2 *Given a code characterized by parameters n, k, d, the expressions for the asymptotic coding gain, with a hard-decision or a soft-decision decoding, are respectively*

$$G_\infty = \frac{k}{n}(t+1), \text{ (hard decision)}, \tag{D.10}$$

$$G_\infty = \frac{k}{n}d, \text{ (soft decision)}, \tag{D.11}$$

in the first of which parameter t is related to the minimum distanced by means of (D.3) or (D.4).

Proof First of all, asymptotically, just the dominant term in the summations expressing p'', respectively (D.5) and (D.7) in the two cases, is to be considered. Then, in order to evaluate G_∞, the external multiplicative constants in such simplified expressions for p'' are not influent. They influence only how rapidly the asymptotic behaviour is reached. In conclusion, just the ratio between the arguments in the one-term expression for p'' (coded channel) and for p (uncoded channel) is to be considered. So (D.11) comes out from the comparison between the argument in the dominant term of (D.7), taken for p'', and the argument of (D.1), taken for p. As regards (D.10), we can observe that an approximation like (D.9) is here certainly valid. So, substitution of (D.2) into (D.5), restricting our attention to the only dominant term, enables us to extract an exponential function of the type $\exp\left[-\frac{k}{n}(t+1)SNR\right]$ in the asymptotic evaluation of p''. The ratio between such exponent and that appearing in a similar approximation for p, that is the simple $\exp(-SNR)$, finally yields (D.10). □

Example D.10 The code introduced in Example D.9 is characterized by asymptotic coding gains given by

$$G_\infty = \frac{48}{23} \cong 2, \text{(hard decision)},$$
$$G_\infty = \frac{84}{23} \cong 3.6, \text{(soft decision)},$$

that is respectively about 3 dB and about 5.5 dB. Such values are approached rather slowly, because of the high external multiplicative constants in p''. The break-even point occurs at rather high *SNR* values, in both cases, owing to the contribution of several terms in the summation, besides the dominant term.

Sometimes an anomalous behaviour for the function $G(p)$, only in the case of a soft-decision decoding, may appear. It is characterized by a break-even point located at very small values of the *SNR*. The decreasing slope of the encoded channel error probability p'' is then extremely steep in its first part, leading to a phenomenon called water fall. A theoretical prediction of this behaviour is very difficult (Lin and Costello 2004), since it is due to many terms in expression (D.7) for p''. They are typically characterized by rather high values of the weight w. After this steep descent, there is an unavoidable reduction in the slope, due to the effect of the term related to the minimum distance d of the code, typically not extremely high, but associated to a very small value for coefficient $A(d)$. So we have a maximum in the $G(p)$ function, for medium values of p, whereas its asymptotic value G_∞ turns out to be smaller. If the application, to which the channel is devoted, requires a quality satisfied by such medium values of the error probability, the purpose of the code can be considered reached.

An example of this anomalous behaviour for the $G(p)$ function is depicted in Fig. D.7. The maximum in $G(p)$ is near $G(10^{-5})$ and is characterized by a value as high as about 6 dB. G_∞, on the other hand, is only of about 2 dB. The p'' curve experiences here three different slopes.

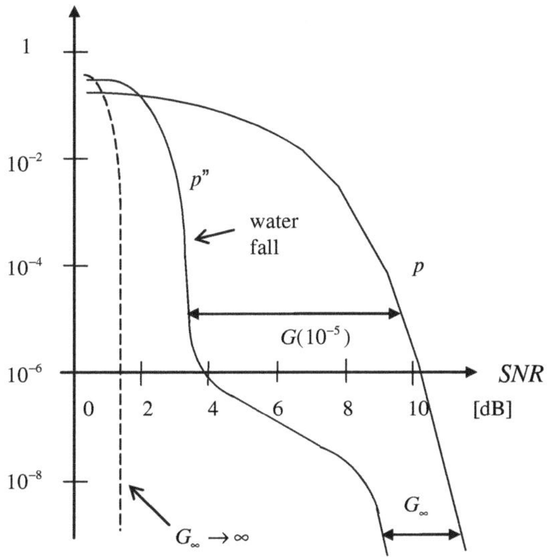

Fig. D.7 Two types of anomalous behaviour in the $G(p)$ function

Another possible anomalous behaviour can be described as follows. Some families of codes, interesting just from a theoretical point of view, can attain an infinite value for their G_∞. This result is reached at the expense of a very strong bandwidth expansion. In (D.10) or (D.11), for such families of codes, t or d increases more rapidly than n/k, so leading to this surprising result. An example of the behaviour characterizing a code so constructed, theoretically with an infinite G_∞, is described by a dashed curve in Fig. D.7. This $p''(SNR)$ function approaches a vertical unlimited descent.

Theoretical Bounds

The minimum distance of a code is constrained by the number of control symbols introduced. The maximum number of correctable single errors (when d is odd) and also the maximum number of those which can be detected (when d is even) depend on the geometric packing of the decoding hyper-spheres introduced in Fig. D.3 and Fig. D.4. Such geometric packing can be more or less optimized. Finally a coded channel able to yield an error probability, after decoding, small as one wishes, must satisfy a constraint in the couple of parameters signal-to-noise ratio SNR and bandwidth B. In the following we will present three fundamental results related to the above considerations.

Theorem D.3 (Singleton bound) *Binary and non-binary linear codes must satisfy the following bound* (Singleton 1964), *relating the minimum distance d and the number of control symbols* $(n-k)$ *in a code word,*

$$d \leq n - k + 1. \qquad \text{(D.12)}$$

Proof Let us consider the all-0-symbol code word, and a code word with just one non-null symbol out of the $(n - k)$ control symbols. The larger separation which the code can produce between these two code words is that characterized by all the control symbols coincident with as many non-null symbols. Such two code words exhibit a Hamming distance equal to $(n - k + 1)$. This observation can be generalized augmenting the number of non-null information symbols taken initially in the code word considered. Also in this case, the all-0-symbol code word and the new code word so constructed are characterized by a Hamming distance equal to $(n - k + 1)$. Linearity assures that the same property holds for any couple of code words in the code. The construction described so far allows to verify (D.12) with equality sign, and this is the maximum value for d which can be obtained with the adopted number of control symbols. □

Example D.11 In a binary alphabet, the only codes satisfying (D.12) with equality sign belong to two families: those characterized by $k = n - 1$, for which $d = 2$, called parity check codes, and those characterized by $k = 1$, for which $d = n$, called repetition codes. They will be described with more details in the following.

In a non-binary alphabet, characterized by Q symbols, there are more degrees of freedom. This makes it possible to enlarge the number of code words in the condition of being maximally separated, and also the ensemble of different codes satisfying Singleton bound with equality sign can be enriched with further elements.

Example D.12 Taking $Q = 3$, it is possible to show that, besides the two families of codes present also in the binary alphabet, a ternary code exists, characterized by $n = 4$, $k = 2$ and $d = 3$, able to satisfy (D.12) with equality sign too. This code represents a point of intersection between two important families of codes, end shows many symmetry properties, as will be better described in the remainder of this brief review, treating non-binary codes.

Although the goal of reaching the maximum possible value for d, given a certain number of control symbols, would appear an interesting challenge from the theoretical point of view, not always this fact is convenient in practice. Let us consider, for example, the case of $Q = 4$. Each of the four symbols of this quaternary alphabet can be represented by a pair of binary symbols. The four possible pairs are 00, 01, 10, 11. One could expect that, starting from two code words at Hamming distance d in the quaternary alphabet, the transformation in the correspondent binary code originates two code words, whose length is $2n$, at Hamming distance $2d$ between them. Unfortunately this is not, as one immediately can verify, taking, as usual, the all-0-symbol code word as a reference: the presence of the pairs 01 and 10 clearly makes a double minimum distance in the correspondent binary code impossible.

Theorem D.4 (Hamming bound with d odd) *Given a binary linear code characterized by parameters n, k and $d = 2t + 1$, the following bound holds*

$$2^{n-k} \geq 1 + \binom{n}{1} + \cdots + \binom{n}{t}, (d\ odd). \quad \text{(D.13)}$$

Proof Let us consider a code word as a point in the hyper-space. Around it, in the framework of a hard-decision decoding model, there is a decoding hyper-sphere, whose radius is t. Therefore the n-tuples included in it, besides the code word itself (occupying its centre) can be enumerated following combinatorial algebra. The right hand side of (D.13) is so justified. After Theorem D.1, the same situation regards any code word. If all the 2^n n-tuples of our hyper-space are equally subdivided into the 2^k hyper-spheres centred in as many code words, which means that there are no n-tuples in possible interstices, we have the following equality

$$2^n = 2^k\left[1 + \binom{n}{1} + \cdots \binom{n}{t}\right].$$

This expression coincides with (D.13) in the case of equality sign. In a situation not so optimized, that is with some n-tuples in the interstices, the left hand side of (D.13) will be greater. □

This theoretical result is sometimes called hyper-sphere packing bound (Gallager 1968). It can be interpreted also taking into account the standard array table, provided that t is odd. Equality sign in (D.13) means that the number of rows of such table is exactly equal to the number of correctable errors. In other terms, there are no rows with ambiguity situations and no rows where it is possible to correct an error with weight greater than t.

Example D.13 In a binary alphabet, the only codes satisfying (D.13) with equality sign are those belonging to two families of codes plus an isolated code. More precisely, we can list the family of repetition codes (already mentioned in Example D.12) with d odd, the family called Hamming codes, whose minimum distance is $d = 3$, and the binary Golay code (already treated in Examples D.9 and D.10).

Theorem D.5 (Hamming bound with d even) *Given a binary linear code characterized by parameters* n, k and $d = 2t + 2$, *the following bound holds*

$$2^{n-k} \geq 1 + \binom{n}{1} + \cdots + \binom{n}{t} + \binom{n-1}{t}, (d\,\text{even}). \qquad \text{(D.14)}$$

Proof Let us consider a code word and the decoding hyper-sphere around it, whose radius is t. Here we have also ambiguous n-tuples along the links connecting the centres of adjacent hyper-spheres. The n-tuples inside each hyper-sphere, including the code word itself (occupying its centre), can be enumerated as in (D.13). There is an additional term taking into account the n-tuples outside the hyper-spheres, which can be so calculated. In an optimized hyper-sphere packing (Games 1983), all the combinations of $d/2$ 1-symbols on a total amount of n positions are to be equally subdivided among $n/(d/2)$ links connecting the centre of the hyper-sphere with as many centres of adjacent hyper-spheres. This number is

$$\Delta = \binom{n}{d/2}\frac{d/2}{n} = \binom{n-1}{d/2-1} = \binom{n-1}{t},$$

where we have used also (D.4). Finally, after Theorem D.1, this discussion can be repeated for all the code words, leading to

$$2^n = 2^k\left[1+\binom{n}{1}+\cdots\binom{n}{t}+\Delta\right] = 2^k\left[1+\binom{n}{1}+\cdots\binom{n}{t}+\binom{n-1}{t}\right],$$

which coincides with (D.14) in the case of equality sign. In any situation different from that here described, some n-tuples will be out of the two ensembles described above (inside the hyper-spheres and in the central positions of the links), so the left hand side of (D.14) will be greater. □

Example D.14 In a binary alphabet, the only codes satisfying (D.14) with equality sign can be grouped as follows: the family of repetition codes with d even, the family of extended Hamming codes and the extended binary Golay code. The procedure of code extension is a particular form of direct code concatenation, useful when the minimum distance d of the original code is odd, in order to increase this parameter by one unit. It is noticeable the fact that all the three situations considered here represent as many extensions of the codes satisfying (D.13) with equality sign.

It could be of some interest to revisit now the organization of a standard array table, in order to stress a property always valid, although with different characteristics for d odd or d even. In order to do this, we can treat the smallest non-trivial extended Hamming code.

Example D.15 The extended Hamming code characterized by $n = 8$, is a code with $k = 4$ and $d = 4$. Its word weight distribution is $A(0) = 1$, $A(4) = 14$, $A(8) = 1$. There are different possibilities for arranging its 16 code words. A possible choice is as follows

$$\begin{array}{llll} 00000000, & 00010111, & 00101101, & 00111010, \\ 01001110, & 01011001, & 01100011, & 01110100, \\ 10001011, & 10011100, & 10100110, & 10110001, \\ 11000101, & 11010010, & 11101000, & 11111111. \end{array}$$

The left half of the standard array table of this code is shown in Table D.1. The remaining part of this table (on the right hand side) con be obtained, from the previous left half, forming code words and n-tuples, by exchanging 1-symbols with 0-symbols and vice versa. The table exhibits three horizontal layers: the first row, where the code words are listed, the subsequent eight rows (associated to single errors), where the n-tuples have minimum weights 1 or 3, and the last seven rows (associated to double errors), where all the n-tuples have minimum weights 2. This characteristic (absence of further rows, after those where the n-tuples have minimum distance $d/2$) is typical for a code satisfying (D.14) with equality sign.

In general, the layer of the table associated to single errors contains n-tuples whose minimum and maximum weights are 1 or $(d-1)$. If the code has an error correction capability $t > 1$, there will be further layers of this type, in which the two minimum

Table D.1 Left half of the standard array table for the extended Hamming code with $n = 8$

00000000	**00010111**	**00101101**	**00111010**	**01001110**	**01011001**	**01100011**	**01110100**	...
00000001	00010110	00101100	00111011	01001111	01011000	01100010	01110101	
00000010	00010101	00101111	00111000	01001100	01011011	01100001	01110110	
00000100	00010011	00101001	00111110	01001010	01011101	01100111	01110000	
00001000	00011111	00100101	00110010	01000110	01010001	01101011	01111100	
00010000	00000111	00111101	00101010	01011110	01001001	01110011	01100100	...
00100000	00110111	00001101	00011010	01101110	01111001	01000011	01010100	
01000000	01010111	01101101	01111010	00001110	00011001	00100011	00110100	
10000000	10010111	10101101	10111010	11001110	11011001	11100011	11110100	
00000011	00010100	00101110	00111001	01001101	01001001	01100000	01110111	
00000110	00010001	00101011	00111100	01001000	01011111	01100101	01110010	
00001100	00011011	00100001	00110110	01000010	01010101	01101111	01111000	
00011000	00001111	00110101	00100010	01010110	01000001	01111011	01101100	...
00110000	00100111	00011101	00001010	01111110	01101001	01010011	01000100	
01010000	01000111	01111101	01101010	00011110	00001001	00110011	00100100	
11000000	11010111	11101101	11111010	10001110	10011001	10100011	10110100	

n-tuple weights progressively converge. The behaviour described so far occurs for either d odd or d even. A different behaviour characterizes the $(t + 1)$th layer of the table. If d is odd, except when it is absent, that is when (D.13) is satisfied with equality sign, such layer shows n-tuples whose weight is higher than t. On the contrary, if d is even, such layer is characterized by ambiguity situations, showing n-tuples whose minimum weights are to $d/2 = t + 1$. This condition enables error detection of all the patterns of $d/2$ single errors.

A sort of degenerate situation regarding the constraint expressed by Theorem D.5 is that related to the family of parity check codes (already mentioned in Example D.11). Such codes, characterized by $d = 2$, do not correct but simply detect single errors. So their standard array table is made by just two rows, the upper one with the code words (whose weight is even), and the lower one with unpermitted n-tuples (whose weight is odd). In the evaluation of parameter Δ, employed in the proof of (D.14), it is necessary to consider only the first equality, which, assuming $d = 2$, gives $\Delta = 1$. So (D.14) can be considered satisfied with equality sign, although in degenerate way, as $2^1 = 1 + 1$.

Proper widened forms of Theorem D.4 and Theorem D.5, able to take into account also non-binary codes, will be discussed later. There are several other theoretical bounds, for instance due to Gilbert-Varshamov (Gilbert 1952; Varshamov 1957), Plotkin (Plotkin 1960), Griesmer (Griesmer 1960), but their treatment is beyond the purpose of the present short review. For a comparative analysis of all the bounds, see for instance (Sweeney 2002).

Shannon Limit

We will present here two important results of information theory (Shannon 1949) without demonstrating them.

Theorem D.6 (Shannon limit for hard-decision decoding) *In a binary symmetric channel affected by the error transition probability p, the maximum information per binary symbol transferred on average, with a residual error probability as small as one wishes, adopting a hard-decision decoding, is*

$$R_o = 1 - p\log_2\left(\tfrac{1}{p}\right) - (1-p)\log_2\left(\tfrac{1}{1-p}\right). \tag{D.15}$$

This quantity, expressed in bit/symbol, is certainly smaller than 1, and is also called the capacity of the considered channel (whose scheme is reported in Fig. D.1) with a hard-decision decoding. Any practical channel code will work with a code rate *k*/*n* lower than this, in the same condition (residual error probability p'' as small as one wishes).

In an AWGN channel, where p is related to the *SNR* through (D.1), parameter p' (the binary symbol error probability before error correction) is related to the *SNR* and to the code rate *k*/*n* by means of (D.2). It is then possible to find a relationship between R_o and the *SNR* (Proakis 1983). Substitution of R_o to *k*/*n* in (D.2) yields

$$p' = \frac{1}{2} erfc\left(\sqrt{R_0 SNR}\right). \tag{D.16}$$

On the other hand, application of the limit condition (D.15) to the coded channel gives us

$$R_o = 1 - p'\log_2\left(\frac{1}{p'}\right) - (1-p')\log_2\left(\frac{1}{1-p'}\right). \tag{D.17}$$

Substitution of (D.16) into (D.17) produces a transcendental relationship between the capacity R_o and the *SNR*. The result of a numerical computation is shown by the lower curve in Fig. D.8. expected, R_o increases with increasing the

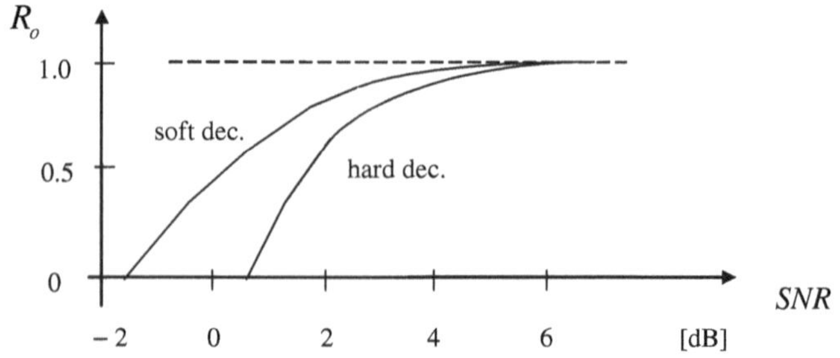

Fig. D.8 Relationship between the capacity R_o and the *SNR*

SNR, approaching its asymptotic value 1.

For soft-decision decoding, we have to consider more complicated expressions. Ignoring the problem of a possible finite number in the quantization levels, that is considering a continuous domain for the signals, the following statement holds.

Theorem D.7 (Shannon limit for soft-decision decoding) *In a binary channel, with one symbol associated to the signal level* $-a$*, and the other symbol associated to the signal level* $+a$*, the maximum information per binary symbol transferred on average, with a residual error probability as small as one wishes, adopting a soft-decision decoding, is*

$$\begin{aligned} R_o = & \frac{1}{2}\int_{-\infty}^{+\infty} p(y|-a)\log_2\left[\frac{2p(y|-a)}{p(y|-a)+p(y|+a)}\right]dy \\ & + \frac{1}{2}\int_{-\infty}^{+\infty} p(y|+a)\log_2\left[\frac{2p(y|+a)}{p(y|-a)+p(y|+a)}\right]dy, \end{aligned} \tag{D.18}$$

where $p(y|-a)$ *is the conditional probability density that the signal to be decided is* y *if that emitted by the source was* $-a$*, and* $p(y|+a)$ *is the conditional probability density that the signal to be decided is* y *if that emitted by the source was* $+a$*. Also the quantity* R_o given by (D.18), *expressed in bit/symbol, is certainly smaller than 1, and is called the capacity of the considered channel with a soft-decision decoding.*

In this case too, taking an AWGN channel, it is possible to represent the right hand side of expression (D.18) as a function of the product R_oSNR, which is inside the expressions of the above conditional probability densities (Proakis 1983). The result of a numerical computation is shown by the upper curve in Fig. D.8. The new behaviour is similar to the previous one, but interests values somewhat higher.

Since, by definition, R_o describes a condition characterized by a residual error probability as small as one wishes, such theoretical behaviours must be compared with the code rates k/n of practical codes able to produce very small values for p''. Keeping in mind the curves reported in Fig. D.7, Shannon limit would represent a limit condition characterized by infinite asymptotic coding gain G_∞, like that denoted with a dashed curve. The vertical unlimited descent of $p''(SNR)$ in Fig. D.7 occurs for *SNR* values, established by the theoretical limits which are expressed by the curves $SNR(R_o)$ in Fig. D.8. In modern coding, certain very good practical codes are able to closely approach these limits, but only with values of p'' rather high (of the order of 10^{-6} or higher).

In particular, for R_o tending to 0, there are analytical solutions in the *SNR*, respectively

$$\lim_{R_o \to 0} SNR = \frac{\pi}{2}\ln 2, \text{(hard decision)}, \tag{D.19}$$

which results in about 0.37 dB, and

$$\lim_{R_o \to 0} SNR = \ln 2, \text{(soft decision)}, \tag{D.20}$$

which results in about −1.6 dB.

Some Important Families of Binary Block Codes

A linear block code is conventionally identified by a tern of parameters included in square brackets, of the type [n, k, d], where, as already introduced, n expresses the code word length, k the number of binary information symbols and d the minimum distance. The specification of having a block characterized by a precise length is needed because the properties of such codes depend strongly on this constraint. Other types of linear codes, for instance convolutional codes, can have a variable length, without an appreciable effect on the code performance, provided that the length of the encoded binary sequence, here called frame, is long enough.

Definition D.16 The family of *binary repetition codes* is characterized by the tern [n, 1, n]. They exhibit just two code words: that with all 0-symbols, and that with all 1-symbols.

Hard-decision decoding, applied to such codes, gives a result worse than that characterizing the uncoded channel, leading to a negative coding gain. Soft-decision decoding yields a result coincident to that characterizing the uncoded channel, in fact the code rate $1/n$ exactly balances the minimum distance $d = n$, in expression (D.11). So the interest for such family of codes is only in the description of a limit condition.

Definition D.17 The family of *binary parity check codes* is characterized by the tern [n, $n - 1$, 2]. Their name is due to the fact that the unique control symbol is chosen in such a way as to force the overall number n of binary symbols in any code word to show an even number of 1-symbols.

Such family of codes is employed for single error detection, or sometimes in forward error correction with a soft-decision decoding. In the first application, the smaller the n, the better the performance. In fact it is possible to detect one single error over a total amount of n binary symbols, therefore a value of n not too high appears preferable (provided that the bandwidth expansion does not deteriorate too much p' with respect to p). In the second application, an asymptotic coding gain of the order of 3 dB can be approached, but very slowly, because of the high number $A(2)$ of code words having weight 2. Often a code of this type is employed in a proper code concatenation scheme.

Lemma D.6 *The complete word weight distribution of an* [n, $n - 1$, 2] *binary parity check code is given by the simple expression*

$$A(w) = \binom{n}{w}, \quad w = 0, 2, 4, \ldots. \tag{D.21}$$

Proof This binomial coefficient enumerates all the possible w-tuples of 1-symbols over a total number n of positions. Restriction to w even is due to the particular selection rule adopted in constructing the words. □

Example D.16 The [5,4,2] binary parity check code is characterized by the following word weight distribution $A(0) = 1$, $A(2) = 10$, $A(4) = 5$. The all-1-symbol code word is absent (this property always holds if n is odd). The [6,5,2] binary parity check code is characterized by the following symmetric word weight distribution $A(0) = 1$, $A(2) = 15$, $A(4) = 15$, $A(6) = 1$. Here the all-1-symbol code word is present (this property always holds if n is even).

Definition D.18 The family of *binary Hamming codes* is characterized by the tern $[2^r - 1, 2^r - r - 1, 3]$, where parameter r is an integer not lower then 2, coincident with the number of binary control symbols $r = n - k$ (Hamming 1950).

It is possible to observe that, with $r = 2$, the [3,1,3] Hamming code is also a repetition code. The number $A(3)$ of minimum-distance code words, for a Hamming code, can be exactly evaluated (MacWilliams and Sloane 1977) and results in $A(3) = \frac{1}{3}\binom{n}{2}$.

Example D.17 The Hamming codes associated to $r = 3, 4, 5$ are respectively expressed by the following terns: [7,4,3], [15,11,3], [31,26,3]. In Table D.2 a comparison regarding some performance parameters of the above three Hamming codes is shown. We can observe that the difference between $G_\infty^{(soft)}$ and $G_\infty^{(hard)}$ is always 1.8 dB, which represents the dB equivalent of the ratio $d/(t + 1) = 3/2$.

The asymptotic gain is approached very slowly, owing to the binomial coefficient in (D.5) for hard-decision decoding, and to the number of code words (binomial as well) having weight $d = 3$ in expression (D.7), for soft-decision decoding.

Definition D.19 A generalization of the family of binary Hamming codes is that of the so called *primitive binary BCH codes*, which can correct a maximum number of single errors t greater than 1 (Hocquenghem 1959; Bose and Ray-Chaudhuri 1960). They are characterized by values of n still equal to integer powers of 2 minus one. For them, we can adopt a tern of the type $[2^m - 1, 2^m - r - 1, 2t + 1]$, where the new integer m is related to the number r of control symbols and to the maximum number t of correctable single errors, through the following inequality

$$r \leq mt. \tag{D.22}$$

Table D.2 Comparison of performance parameters for three Hamming codes

Code	k/n	$G_\infty^{(hard)}$ (dB)	$G_\infty^{(soft)}$ (dB)	$A(3)$
[7,4,3]	4/7	0.4	2.2	7
[15,11,3]	11/15	1.7	3.5	35
[31,26,3]	26/31 ≅ 5/6	2.2	4.0	155

For a certain m, which identifies a set of primitive BCH codes, with respect to the equality condition (always satisfied by Hamming codes, that is for $t = 1$), as long as t increases, also r increases but more and more slowly.

Example D.18 The only BCH codes characterized by $m = 3$ are a [7,4,3] code, which is also a Hamming code, and a [7,1,7] code, which is also a repetition code.

Example D.19 The only BCH codes characterized by $m = 4$ are expressed by the terns: [15,11,3] (also a Hamming code), [15,7,5], [15,5,7] and [15,1,15], the last being also a repetition code.

Example D.20 The only BCH codes characterized by $m = 5$ are expressed by the terns: [31,26,3] (also a Hamming code), [31,21,5], [31,16,7], [31,11,11], [31,6,15], and the [31,1,31], the last being also a repetition code. In spite of the apparent regularity, we can observe that the differential increase in d grows, as long as the condition for having a repetition code is approached.

Example D.21 The first five BCH codes characterized by $m = 6$ are expressed by the terns: [63,57,3] (also a Hamming code), [63,51,5], [63,45,7], [63,39,9] and [63,36,11]. The last one shows a differential reduction in parameter k smaller than the previous ones. Then there are several further BCH codes with code rate lower and lower, up to the [63,1,63] repetition code.

Definition D.20 A solitary code, not belonging to a family, like the ones described so far, is the *binary Golay code*, described by the tern [23,12,7], already introduced in Example D.9 (Golay 1949). Its word weight distribution is symmetric and results in $A(0) = 1$, $A(7) = 253$, $A(8) = 506$, $A(11) = 1288$, $A(12) = 1288$, $A(15) = 506$, $A(16) = 253$, $A(23) = 1$.

Let us consider now some families of codes, or solitary codes, having even minimum distance.

Definition D.21 It is possible to add a parity check binary symbol to an odd-distance code, and obtain a new code with minimum distance increased by one unit. This operation is called *code extension*, as already noted, and originates many interesting families of binary codes.

Definition D.22 From the family of binary Hamming codes, we obtain the *extended Hamming codes*, characterized by the tern $[2^r, 2^r - r - 1, 4]$. In Example D.15 we have already discussed the [8,4,4] extended Hamming code.

Definition D.23 From the family of primitive BCH codes, we obtain the *extended primitive BCH codes*, characterized by the tern $[2^m, 2^m - r - 1, 2t + 2]$.

Definition D.24 From the binary Golay code [23,12,7], it is possible to derive the *extended binary Golay code* [24,12,8], whose word weight distribution turns out to be $A(0) = 1$, $A(8) = 759$, $A(12) = 2576$, $A(16) = 759$, $A(24) = 1$. Also this weight distribution is symmetric, and it is easy to recognize how, in the new coefficient $A(w)$, the old coefficients associated to the weights $(w - 1)$ and w are collected.

Example D.22 The extended Golay code [24,12,8] exhibits an asymptotic coding gain G_∞ = 8/2 = 4, that is about 6 dB, with a soft-decision decoding. The computational complexity of this decoding procedure is acceptable, owing to an efficient organization of the algorithm (Forney 1988).

Example D.23 The repetition codes with even length n are generally obtained independently, but they can be interpreted also as extensions of the correspondent repetition codes having length smaller by one unit, in fact $(n - 1)$ is odd.

Unfortunately, extension is efficient for increasing the minimum distance d just one time in a binary alphabet. This constraint is no longer valid in a non-binary alphabet.

Definition D.25 The family of *simplex codes*, in a binary alphabet, is characterized by the tern $[2^k - 1, k, 2^{k-1}]$. Their word weight distribution is very particular: $A(0) = 1, A(2^{k-1}) = 2^k - 1$. In other terms, all the non-null code words have the same weight, which equals the code minimum distance.

Example D.24 The [7,3,4] simplex code, for instance, exhibits the following eight code words

0000000, 0011101, 0100111, 0111010,
1001110, 1010011, 1101001, 1110100.

It is possible to recognize that there is a unique pattern of 1-symbols which shifts by one position each time according to a circular permutation, in the seven non-null code words. This is not the only way for constructing such a code.

Definition D.26 The family of pth-order Reed-Muller codes (Reed 1954; Muller 1954), in a binary alphabet, is characterized by the tern $[2^m, s(p), 2^{m-p}]$, where p is an integer and s is the following function of p

$$s = \sum_{j=0}^{p} \binom{m}{j}, \quad p = 1, 2, \ldots m - 2. \tag{D.23}$$

First-order Reed-Muller codes are then characterized by the tern $[2^m, 1 + m, 2^{m-1}]$. Their word weight distribution has the following symmetric behaviour (MacWilliams and Sloane 1977): $A(0) = 1, A(2^{m-1}) = 2^{m+1} - 2, A(2^m) = 1$.

Example D.25 With m = 3, we have only one first-order Reed-Muller code, characterized by the tern [8,4,4], which is also an extended Hamming code, already treated in Example D.15 (see Definition D.22 too). With m = 4, we have the first-order Reed-Muller code [16,5,8] and the second-order Reed-Muller code [16,11,4], which is also an extended Hamming code again.

Simplex codes and first-order Reed-Muller codes approach limit (D.20), assuming k or m tending to infinity (Viterbi 1964). This means an infinite bandwidth (or a code rate tending to 0). In practice, an unlimited bandwidth growth pays an unlimited asymptotic coding gain, of the type denoted with a dashed curve in

Fig. D.7. In this case, such behaviour occurs for the minimum possible value of the *SNR*.

There is another family of binary codes able to approach limit (D.20), under the same conditions. They are called orthogonal codes, and can be obtained from simplex codes after addition of one position in the code words, always occupied by a 0-symbol. Alternatively they can be constructed from first-order Reed-Muller codes, after eliminating one half of the code words. From both procedures we obtain the tern $[2^k, k, 2^{k-1}]$. Although they are worse than simplex and Reed-Muller codes, sometimes are preferred, owing to easier encoding and decoding procedures.

In modulated transmissions, a non-binary alphabet can be set up by means of proper sequences of antipodal elementary signals. In this application, first-order Reed-Muller codes are also known as bi-orthogonal sequences. Analogously simplex codes are known as trans-orthogonal sequences (Viterbi 1964). These denominations derive from the property of being a variant of orthogonal sequences, which clearly correspond to orthogonal codes.

Outline of Non-binary Block Codes

Many theoretical results hold in a non-binary alphabet, and can guide to find good codes reaching or approaching expected bounds. Nevertheless, when we come back to a binary alphabet, as already noted, it is possible to verify that binary codes always exist, able to give better performance. On the other hand, non-binary codes are preferred, for instance, for allowing the possibility of processing packets of binary symbols in parallel, any packet being related to a symbol in the considered non-binary alphabet. This solution can reduce time latency.

Before looking for a non-binary code in an alphabet with Q symbols, it is necessary to be sure that a Q-ary finite field exists. This condition is satisfied for Q prime, or integer power of a prime (Peterson and Weldon 1972). In a finite field two operations are defined: addition (correspondent to ex-OR sum in a binary finite field) and multiplication. For a binary finite field, see Appendix A.

In Fig. D.9, the tables describing such two operations for a ternary and for a quaternary field are reported. In the quaternary field (like in any finite field with number of elements equal to an integer power of 2), the property that any symbol coincides with its addition opposite makes negative sign useless. This property is responsible for the presence of all 0-symbols along the diagonal in the table reported down-left in Fig. D.9.

The relationships between d and t, given by (D.3) for d odd and by (D.4) for d even, are still valid in a Q-ary alphabet. In fact the number of non-null symbols does not influence the separations between neighbouring code words, but only their multiplicity. The following proposition represents a proper widening of Hamming bound, for either d odd or d even (Games 1983).

$Q = 3$, addition:

+	0	1	2
0	0	1	2
1	1	2	0
2	2	0	1

$Q = 3$, multiplication:

·	0	1	2
0	0	0	0
1	0	1	2
2	0	2	1

$Q = 4$, addition:

+	0	1	2	3
0	0	1	2	3
1	1	0	3	2
2	2	3	0	1
3	3	2	1	0

$Q = 4$, multiplication:

·	0	1	2	3
0	0	0	0	0
1	0	1	2	3
2	0	2	3	1
3	0	3	1	2

addition multiplication

Fig. D.9 Addition and multiplication in finite fields with $Q = 3$ and $Q = 4$

Theorem D.8 Given a Q-ary linear code characterized by parameters n, k and $d = 2t + 1$ (when d is odd), $d = 2t + 2$ (when d is even), the following bounds hold

$$Q^{n-k} \geq 1 + \binom{n}{1}(Q-1) + \cdots + \binom{n}{t}(Q-1)^t, \quad (d\,\text{odd}), \tag{D.24}$$

$$Q^{n-k} \geq 1 + \binom{n}{1}(Q-1) + \cdots + \binom{n}{t}(Q-1)^t + \binom{n-1}{t}(Q-1)^{t+1}, \, (d\,\text{even}).$$

Proof Expressions (D.13) and (D.14), which hold for $Q = 2$, are to be corrected taking into account that any sequence of j symbols must be accounted for, with $(Q-1)^j$ different combinations, j being the weight of the considered sequence. So, multiplying any terms by this factor, from (D.13) and (D.14), the two inequalities (D.24) are directly obtained. □

In the following we will adopt a tern for describing a Q-ary block code, characterized by parameters, n, k, d, so modified $[n, k, d]_Q$ with respect to the binary case, where subscript $Q = 2$ is omitted for simplicity.

Definition D.27 *Q-ary Hamming codes* are represented by the following tern $\left[\frac{Q^r-1}{Q-1}, \frac{Q^r-1}{Q-1} - r, 3\right]_Q$. They satisfy the bound expressed by the first of (D.24) with equality sign (Golay 1958).

Definition D.28 *Q-ary simplex codes* are represented by the tern $\left[\frac{Q^k-1}{Q-1}, k, Q^{k-1}\right]_Q$. Their word weight distribution is made by just one coefficient (Peterson and

Weldon 1972), besides $A(0) = 1$. It is $A\left(Q^{k-1}\right) = Q^k - 1$. The property of having d even, shown by binary simplex codes, in general, is no longer valid here.

Example D.26 The ternary Hamming code, characterized by $r = 2$, is represented by the tern $[4, 2, 3]_3$. An example for its 9 code words is as follows

$$0000, 0122, 0211, 1220, 2110, 2202, 1101, 2021, 1012.$$

It is possible to note that all the 8 non-null code words have $w = 3$. So, besides satisfaction of (D.24), d odd, with equality sign, this code can be classified also as a simplex code with $k = 2$. Furthermore it satisfies Singleton bound (D.12) with equality sign. Also a different choice can be adopted in constructing its code words, for instance assuming 0112 as the first code word in the above list. The two codes so obtained are equivalent, that is have the same tern and an equal word weight distribution.

Example D.27 The quaternary Hamming code, characterized by $r = 2$, is represented by the tern $[5, 3, 3]_4$. This code satisfies Hamming bound, expressed by the first of (D.24), as well as Singleton bound (D.12), both with equality sign, but it is not a simplex code. Extending this $[5, 3, 3]_4$ Hamming code, we obtain instead a $[6, 3, 4]_4$ code, which satisfies the second of (D.24), and still Singleton bound (D.12), both with equality sign.

In Fig. D.10 parameters k and $r = n - k$ are reported on two orthogonal axes, in order to map (with small circles) the codes satisfying Singleton bound (D.12) with equality sign, for $Q = 2, 3, 4$. As already stressed, just binary repetition codes (rc)

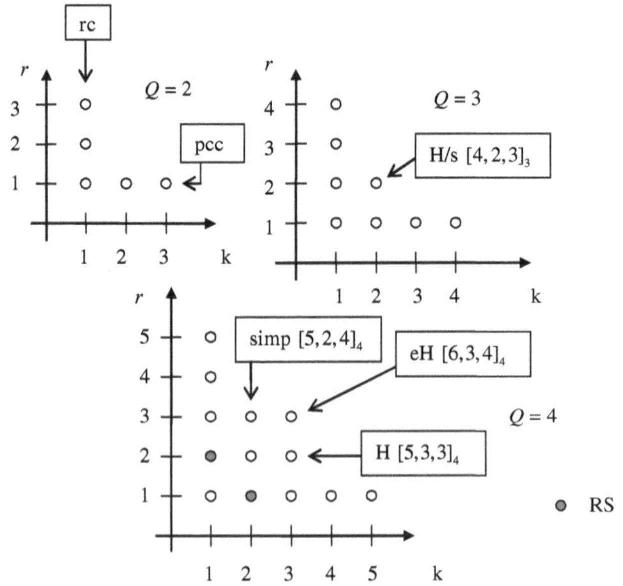

Fig. D.10 Map of MDS codes in a $k \times r$ plane, for $Q = 2, 3, 4$

and binary parity check codes (pcc) belong to such a class of codes. For $Q = 3$, the $[4, 2, 3]_3$ Hamming/simplex code, described in Example D.26, can be added. For $Q = 4$, we have three further codes, two of which already treated in Example D.27. The third is a $[5, 2, 4]_4$ simplex code.

Definition D.29 *Q-ary repetition codes* are represented by the tern $[n, 1, n]_Q$. They exhibit $(Q - 1)$ non-null code words of weight $w = d = n$. Such code words can be formed by all equal symbols, but also other choices are possible.

Example D.28 The ternary $[3, 1, 3]_3$ repetition code can be arranged in such a way as to form the following three code words: 000, 111, 222. Nevertheless also these further choices are possible: 000, 121, 212 or 000, 112, 221.

Definition D.30 *Q-ary parity check codes* are represented by the tern $[n, n - 1, 2]_Q$. The control symbol to be added is chosen taking into account also all the possible non-null multiplicative factors. The property of having just even-weight code words is no longer true, in general, for these codes in a non-binary alphabet. For this reason they do not satisfy the second of (D.24), with equality sign.

Example D.29 The quaternary $[3, 2, 2]_4$ parity check code, obtained after the choice of 011 as a code word, exhibits the following 16 code words

$$\begin{array}{cccccccc} 000, & 011, & 022, & 033, & 101, & 202, & 303, & 110, \\ 220, & 330, & 123, & 132, & 213, & 231, & 312, & 321, \end{array}$$

Therefore we have this word weight distribution

$$A(0) = 1, A(2) = 9, A(3) = 6,$$

showing how the minimum distance is $d = 2$, but there are also 3-weight code words.

Definition D.31 The codes able to satisfy Singleton bound with equality sign are called *Maximum Distance Separable* (*MDS codes*). This definition is due to the fact that they exhibit maximum possible minimum distance d, and, within the n positions of a code word, one can always distinguish between information and control symbols (information symbols can be separated). The number of code words having weight equal to the minimum distance can be exactly evaluated in such codes (MacWilliams and Sloane 1977), and is $\mathrm{A}(d) = (Q - 1)\binom{n}{d}$.

In performing a code extension, taking non-null symbols from a Q-ary alphabet, it is necessary to widen Definition D.21, valid only for code extension in a binary alphabet. The availability of a larger alphabet allows to perform more than one extension operation. In principle, subsequent extension operations can be as many as the non-null symbols, that is $(Q - 1)$. Nevertheless, only sometimes we obtain an increase in the minimum distance.

Definition D.32 A solitary code, able to satisfy (D.24), for d odd, with equality sign, is the $[11, 6, 5]_3$ *ternary Golay code* (Golay 1949). It is possible to extend it obtaining the $[12, 6, 6]_3$ extended ternary Golay code, able to satisfy (D.24), for

d even, with equality sign. None of these codes is an MDS code. A further possible extension does not increase the minimum distance.

The family of binary BCH codes can be generalized originating that of Q-ary BCH codes (Gorenstein and Zierler 1961). Among them, an interesting and widely employed set of codes, which instead are MDS, can be so introduced.

Definition D.33 Taking $n = Q - 1$, it is possible to form MDS codes whose tern is $[Q - 1, k, Q - k]_Q$. They are called *Reed-Solomon codes* (or briefly *RS*) (Reed and Solomon 1960).

Example D.30 In Fig. D.10, regarding MDS codes, let us consider the case of $Q = 4$. The codes characterized by $k = 2$ and $r = 1$ and by $k = 1$ and $r = 2$ (denoted by solid small circles), besides to be interpreted as a quaternary parity check and as a quaternary repetition code respectively, can be considered also RS codes.

Main Characteristics of Binary Convolutional Codes

Binary convolutional codes are characterized by very long code words, called frames. A frame is subdivided into subsequent periods, whose length is n_o. In a block code, between two subsequent code words, there is no correlation. On the contrary, in a convolutional code, between two subsequent periods a correlation is present. This correlation has a length, measured by a parameter m, which represents the number of memory elements in the shift register employed in the encoder circuit. About the length of a frame and its termination, we will give some further detail at the end of this brief introduction. Typically a soft-decision decoding is performed. Its computational complexity increases exponentially with m. On the other hand, the minimum distance d also increases with m, although more slowly.

Definition D.34 The *code rateof a convolutional code* results in the ratio between the number of information symbols k_o in any period and the period length n_o. This code rate is effectively reached only if the frame is very long with respect to m.

Typically k_o and n_o are small co-prime numbers. The class of 1/2-rate convolutional codes represents a separation element between low-rate convolutional codes, characterized by $k_o < n_o/2$, and high-rate convolutional codes, characterized by $k_o > n_o/2$. The performance of a convolutional code with respect to a block code, having comparable code rate, can be summarized as follows. Since the low-weight code frames, up to $w = 2d - 1$, are few, the break-even point, where the curve of the coded error probability intersects that of the uncoded error probability, occurs for rather small values of the *SNR*. The asymptotic coding gain is reached almost rapidly, but its value is typically smaller than the one characterizing block codes with analogous code rate.

The decoding algorithm more frequently employed is due to Viterbi (1967). It can be applied adopting either a hard or a soft decision. By far the latter solution is preferred, resulting to be a good trade-off between performance and computational

Table D.3 Performance of the best 1/2-rate convolutional codes

m	Best value for d	Asymptotic coding gain G_∞ (soft decision) (dB)
1	3	1.8
2	5	4.0
3	6	5.8
4	7	5.5
5	8	6.0
6	10	7.0
7	10	7.0

complexity. It operates a choice between two paths in any node, at every stage of a binary trellis.

In Table D.3, the performance of the best 1/2-rate convolutional codes are listed, reporting their minimum distance d and asymptotic coding gain with a soft-decision decoding. It is possible to note an increase in the minimum distance d with increasing parameter m. Nevertheless, for m greater than 7, Viterbi algorithm is characterized by a too high computational complexity, and other possible decoding procedures, called comprehensively sequential decoding algorithms, have to be adopted. Among them we can mention, for instance, that proposed by Fano (1963) or that due to Zigangirov-Jelinek (Zigangirov 1966; Jelinek 1969), sometimes called also stack algorithm. For a discussion about such alternative decoding algorithms, see for instance (Bossert 1999). They exhibit reduced computational complexity, but also typically poorer performance. Finally, the number of quantization levels in the practical implementation of soft-decision decoding may become more or less influent in the resulting performance (Clark and Cain 1983).

In Table D.4 the same parameters are shown for the best low-rate convolutional codes (rate $1/n_o$, with $n_o = 3$ and $n_o = 4$). It is possible to observe here a more rapid increase of the minimum distance d with increasing parameter m.

In order to design a high-rate convolutional code, we can follow two ways. The first exploits a code puncturation (that is elimination of some control symbols), starting from of a good 1/2-rate convolutional code. This solution maintains the same decoding computational complexity as that of the original 1/2-rate code. The second is based on the use of an encoder circuit with more than one shift register

Table D.4 Performance of the best low-rate convolutional codes

Code rate	m	Best value for d	Asymptotic coding gain G_∞ (soft decision) (dB)
1/3	1	5	2.2
	2	8	4.2
	3	10	5.2
	4	12	6.0
1/4	1	6	1.8
	2	8	3.0
	3	11	4.4

Table D.5 Performance of the best high-rate punctured convolutional codes

Code rate	m	Best value for d	Asymptotic coding gain G_∞ (soft decision) (dB)
2/3	1	3	3.0
	2	4	4.2
	3	5	5.2
	4	6	6.0
3/4	1	3	3.5
	2	4	4.8
	3	4	4.8
	4	5	5.7

(Johannesson and Zigangirov 1999). The latter possibility allows to obtain better performance, but often at the expense of an increased decoding computational complexity. Table D.5 shows the performance of the best punctured convolutional codes (Anderson and Mohan 1991), with code rate 2/3 and 3/4. With respect to the performance reported in Table D.3, we can observe a slower and slower increase of the minimum distance d with increasing parameter m, as long as the code rate grows.

After a certain number of periods, depending on the particular application in which the convolutional code is involved, the code frame must be interrupted. Nevertheless, differently from what occurs for block codes, the performance of a convolutional code is not influenced by the frame length, that is by the overall number of periods in the frame, provided that this number is high enough. A correct frame termination requires that the m memory elements of the shift register in the encoder circuit are all filled by 0-symbols. This fact implies that some additional control symbols have to be added at the end of the frame. So we can conclude that the true code rate approaches the nominal code rate k_o/n_o, provided that the number of periods in the frame is much higher than parameter m.

Error Control and Erasure Control

Sometimes the identity of a received symbol, taken by itself, is so uncertain that it can be more convenient to avoid its decision, and attempt an estimate exploiting the constraint offered by a code word. In this situation, the uncertain symbol is firstly maintained undecided, and its identity is reconstructed in decoding. Although so simply described, this choice of operation requires a change in the model of the channel. For instance, in the scheme of the binary symmetric channel depicted in Fig. D.1, it is necessary to add a third symbol in output and proper branches describing the transition probabilities for reaching it, from the 0-symbol and from the 1-symbol in input.

Definition D.35 In a code word, if some symbols, in precise positions, are undecided, the number of single events of this type are called *erasures*, and the positions where they occur are called *erasure positions.*

An improved version of hard-decision decoding is able to treat erasures. In this case, the output symbol alphabet, before decoding, is enlarged by one unit. Besides the Q input symbols, there is a further output symbol: the erasure. This can be done by proper arrangement of the thresholds separating the zones where the signals affected by noise are collected (Peterson and Weldon 1972).

Lemma D.7 *Given a linear code with minimum distance d, by means of a hard-decision decoding, considering also erasures, it is possible to correct a maximum number t of single errors together with a maximum number t_e of single erasures so constrained*

$$2t + t_e < d. \tag{D.25}$$

Proof Two code words differ, at least, in d positions. If t_e of them are erasure positions, it is possible to define a reduced minimum distance $d' = d - t_e$. If d' is odd, from (D.3), we are with $d = 2t + t_e + 1$. If d' is even, from (D.4), we are with $d = 2t + t_e + 2$. In both cases constraint (D.25) is satisfied. □

When both errors and erasures are to be corrected, properly partitioning d between t and t_e, we have a simultaneous control of errors and erasures. A particular situation is that characterized by the whole code correction capability devoted to correct erasures only.

Lemma D.8 *When the whole minimum distance d of the code is devoted to correct erasures only, the maximum number of single erasures t_e which can be corrected results in*

$$t_e = d - 1. \tag{D.26}$$

Proof From (D.25), assuming $t = 0$, we obtain $t_e < d$. Only when t_e exactly equals d, we can have an ambiguity situation at least between two code words. So (D.26) holds, independently of the fact that d is even or odd. □

Example D.31 Assuming a [7,3,4] binary simplex code, we can correct up to $t_e = 3$ single erasures and no errors. Taking the code introduced in Example D.24, which is characterized by the 1-symbol pattern 11101, after the 7-tuple 01e1ee0, where any e-symbol denotes an erasure position, it is possible to discover, by attempts, that the correct code word was 0111010. Nevertheless, adding a fourth erasure, it is possible to verify that sometimes also this situation is not ambiguous. In this sense, we can conclude that particular erasure patterns exceeding (D.26) can be corrected too.

Example D.32 Assuming a $[4, 2, 3]_3$ ternary Hamming code, of the type introduced in Example D.26, we can correct all the pattern of 2 single erasures and no errors. After the 4-tuple 1e2e, it is possible to discover, by attempts, that the correct code word was 1220. In this case there are no patterns of 3 single erasures which can be corrected.

As already noted, a code devoted to error detection only is able to detect up to $(d-1)$ single errors. This performance could be confused with that fixed by Lemma D.8 regarding erasure correction. On the contrary, they are two very different situations: error detection is performed in the framework of a true hard-decision decoding, when d is small with respect to the number of single errors, erasure correction represents, on the contrary, a first step towards a soft-decision decoding. In fact, instead of only Q symbols, the decoder is here arranged in such a way as to distinguish at least one further symbol, not coincident with any of the Q symbols in the alphabet adopted, but too affected by noise and then too uncertain. Clearly a situation like that described by (D.26) is somehow extreme. Nevertheless some particular channels (for instance the photon counting channel) are well represented by this model (McEliece 1981).

References

Abramowitz M, Stegun IA (1968) Handbook of mathematical functions. Dover, New York
Anderson JB, Mohan S (1991) Source and channel coding. Kluwer, Boston
Bose RC, Ray-Chaudhuri DK (1960) On a class of error correcting binary group codes. Inf Contr 3:68–79
Bossert M (1999) Channel coding for telecommunications. Wiley, Weinheim
Clark GC, Cain JB (1983) Error correcting coding for digital communications. Plenum, New York
Fano RM (1963) A heuristic discussion of probabilistic decoding. IEEE Trans Inf Th 9:64–74
Forney GD (1988) Coset codes—part II: binary lattices and related codes. IEEE Trans Inf Th 34:1152–1187
Gallager RG (1968) Information theory and reliable communications. Wiley, New York
Games RA (1983) The packing problem for projective geometries over GF(3) with dimensions greater than five. J Combinat Th A35:126–144
Gilbert EN (1952) A comparison of signalling alphabets. Bell Syst Tech J 31:504–522
Golay MJE (1949) Notes on digital coding. Proc IRE 37:657–657
Golay MJE (1958) Notes on the penny-weighing problem, lossless symbol coding with nonprimes, etc. IRE Trans Inf Th 8:103–109
Gorenstein DC, Zierler N (1961) A class of error correcting codes in p^m symbols. J SIAM 9:207–214
Griesmer JH (1960) A bound for error correcting codes. IBM J Res and Dev 4:532–542
Hamming RW (1950) Error detecting and error correcting codes. Bell Syst Tech J 27:147–160
Hocquenghem A (1959) Codes correcteurs d'erreurs (Error correcting codes). Chiffres 2:147–156
Jelinek F (1969) A fast sequential decoding algorithm using a stack. IBM J Res Dev 4:675–685
Johannesson R, Zigangirov KS (1999) Fundamentals of convolutional coding. IEEE Press, New York
Klove T (2007) Codes for error detection. World Scientific, Singapore
Lin S, Costello DJ (2004) Error control coding. Pearson/Prentice-Hall, Upper Saddle River
MacWilliams FJ, Sloane NJA (1977) The theory of error correcting codes. North Holland, New York
McEliece RJ (1981) Practical codes for photon communications. IEEE Trans Inf Th 27:393–398
Muller DE (1954) Application of Boolean algebra to switching circuit design and error detection. IRE Trans Electr Comp 3:6–12
Papoulis A (1965) Probability, random variables, and stochastic processes. McGraw Hill, New York
Peterson WW, Weldon EJ (1972) Error-correcting codes. MIT Press, Boston

Plotkin M (1960) Binary codes with specified minimum distance. IEEE Trans Inf Th 6:445–450
Proakis JC (1983) Digital Communications. McGraw Hill, Tokyo
Reed IS (1954) A class of multiple-error-correcting codes and decoding scheme. IRE Trans Inf Th 4:38–49
Reed IS, Solomon G (1960) Polynomial codes over certain finite fields. J SIAM 8:300–304
Shannon CE (1949) A mathematical theory of communication. Bell Syst Tech J 27:379–423, 623–653
Singleton RC (1964) Maximum distance q-nary codes. IEEE Trans Inf Th 10:116–118
Sweeney P (2002) Error control coding. Wiley, New York
Torrieri D. (1988) Information-bit, information-symbol, and decoded symbol error rates for linear block codes. IEEE Trans Comm 36:613–617
Varshamov RR (1957) Estimate of the number of signals in error correcting codes. Dokl Acad Nauk SSSR 117:739–741
Viterbi AJ (1964) Phase-coherent communications over the continuous Gaussian channel. In: Golomb SW (ed) Digital communications. Prentice Hall, Englewood Cliffs
Viterbi AJ (1967) Error bounds for convolutional codes and an asymptotically optimum decoding algorithm. IEEE Trans Inf Th 13:260–269
Zigangirov KS (1966) Some sequential decoding procedures. Probl Peredachi Inf 2:13–25

Index

G. Cancellieri, *Polynomial Theory of Error Correcting Codes*,
Signals and Communication Technology, DOI 10.1007/978-3-319-01727-3

MIX
Papier aus verantwortungsvollen Quellen
Paper from responsible sources
FSC® C105338

If you have any concerns about our products,
you can contact us on
ProductSafety@springernature.com

In case Publisher is established outside the EU,
the EU authorized representative is:
Springer Nature Customer Service Center GmbH
Europaplatz 3, 69115 Heidelberg, Germany

Printed by Libri Plureos GmbH
in Hamburg, Germany